INTRODUCTION TO CHEMICAL ENGINEERING THERMODYNAMICS

EIGHTH EDITION

J. M. Smith

Late Professor of Chemical Engineering
University of California, Davis

H. C. Van Ness

Late Professor of Chemical Engineering
Rensselaer Polytechnic Institute

M. M. Abbott

Late Professor of Chemical Engineering
Rensselaer Polytechnic Institute

M. T. Swihart

UB Distinguished Professor of Chemical and Biological Engineering
University at Buffalo, The State University of New York

Mc
Graw
Hill
Education

INTRODUCTION TO CHEMICAL ENGINEERING THERMODYNAMICS, EIGHTH EDITION

Published by McGraw-Hill Education, 2 Penn Plaza, New York, NY 10121. Copyright © 2018 by McGraw-Hill Education. All rights reserved. Printed in the United States of America. Previous editions © 2005, 2001, and 1996. No part of this publication may be reproduced or distributed in any form or by any means, or stored in a database or retrieval system, without the prior written consent of McGraw-Hill Education, including, but not limited to, in any network or other electronic storage or transmission, or broadcast for distance learning.

Some ancillaries, including electronic and print components, may not be available to customers outside the United States.

This book is printed on acid-free paper.

1 2 3 4 5 6 7 8 9 LCR 21 20 19 18 17

ISBN 978-1-259-69652-7
MHID 1-259-69652-9

Chief Product Officer, SVP Products & Markets: *G. Scott Virkler*
Vice President, General Manager, Products & Markets: *Marty Lange*
Vice President, Content Design & Delivery: *Kimberly Meriwether David*
Managing Director: *Thomas Timp*
Brand Manager: *Raghothaman Srinivasan/Thomas M. Scaife, Ph.D.*
Director, Product Development: *Rose Koos*
Product Developer: *Chelsea Haupt, Ph.D.*
Marketing Director: *Tamara L. Hodge*
Marketing Manager: *Shannon O'Donnell*
Director of Digital Content: *Chelsea Haupt, Ph.D.*
Digital Product Analyst: *Patrick Diller*

Digital Product Developer: *Joan Weber*
Director, Content Design & Delivery: *Linda Avenarius*
Program Manager: *Lora Neyens*
Content Project Managers: *Laura Bies, Rachael Hillebrand & Sandy Schnee*
Buyer: *Laura M. Fuller*
Design: *Egzon Shaqiri*
Content Licensing Specialists: *Melissa Homer & Melisa Seegmiller*
Cover Image: © *(Richard Megna) FUNDAMENTAL PHOTOGRAPHS, NYC*
Compositor: *SPi Global*
Printer: *LSC Communications*

All credits appearing on page or at the end of the book are considered to be an extension of the copyright page.

Library of Congress Cataloging-in-Publication Data

Names: Smith, J. M. (Joseph Mauk), 1916-2009, author. | Van Ness, H. C. (Hendrick C.), author. | Abbott, Michael M., author. | Swihart, Mark T. (Mark Thomas), author.
Title: Introduction to chemical engineering thermodynamics / J.M. Smith, Late Professor of Chemical Engineering, University of California, Davis; H.C. Van Ness, Late Professor of Chemical Engineering, Rensselaer Polytechnic Institute; M.M. Abbott, Late Professor of Chemical Engineering, Rensselaer Polytechnic Institute; M.T. Swihart, UB Distinguished Professor of Chemical and Biological Engineering, University at Buffalo, The State University of New York.
Description: Eighth edition. | Dubuque : McGraw-Hill Education, 2017.
Identifiers: LCCN 2016040832 | ISBN 9781259696527 (alk. paper)
Subjects: LCSH: Thermodynamics. | Chemical engineering.
Classification: LCC TP155.2.T45 S58 2017 | DDC 660/.2969—dc23
LC record available at https://lccn.loc.gov/2016040832

The Internet addresses listed in the text were accurate at the time of publication. The inclusion of a website does not indicate an endorsement by the authors or McGraw-Hill Education, and McGraw-Hill Education does not guarantee the accuracy of the information presented at these sites.

mheducation.com/highered

Contents

List of Symbols

A	Area
A	Molar or specific Helmholtz energy $\equiv U - TS$
A	Parameter, empirical equations, e.g., Eq. (4.4), Eq. (6.89), Eq. (13.29)
a	Acceleration
a	Molar area, adsorbed phase
a	Parameter, cubic equations of state
$\bar{a}_i$	Partial parameter, cubic equations of state
B	Second virial coefficient, density expansion
B	Parameter, empirical equations, e.g., Eq. (4.4), Eq. (6.89)
$\hat{B}$	Reduced second-virial coefficient, defined by Eq. (3.58)
B'	Second virial coefficient, pressure expansion
B^0, B^1	Functions, generalized second-virial-coefficient correlation
B_{ij}	Interaction second virial coefficient
b	Parameter, cubic equations of state
$\bar{b}_i$	Partial parameter, cubic equations of state
C	Third virial coefficient, density expansion
C	Parameter, empirical equations, e.g., Eq. (4.4), Eq. (6.90)
$\hat{C}$	Reduced third-virial coefficient, defined by Eq. (3.64)
C'	Third virial coefficient, pressure expansion
C^0, C^1	Functions, generalized third-virial-coefficient correlation
C_P	Molar or specific heat capacity, constant pressure
C_V	Molar or specific heat capacity, constant volume
C_P°	Standard-state heat capacity, constant pressure
ΔC_P°	Standard heat-capacity change of reaction
$\langle C_P \rangle_H$	Mean heat capacity, enthalpy calculations
$\langle C_P \rangle_S$	Mean heat capacity, entropy calculations
$\langle C_P^\circ \rangle_H$	Mean standard heat capacity, enthalpy calculations
$\langle C_P^\circ \rangle_S$	Mean standard heat capacity, entropy calculations
c	Speed of sound
D	Fourth virial coefficient, density expansion
D	Parameter, empirical equations, e.g., Eq. (4.4), Eq. (6.91)
D'	Fourth virial coefficient, pressure expansion
E_K	Kinetic energy
E_P	Gravitational potential energy
F	Degrees of freedom, phase rule
F	Force
$\mathcal{F}$	Faraday's constant

f_i	Fugacity, pure species i
f_i°	Standard-state fugacity
$\hat{f}_i$	Fugacity, species i in solution
G	Molar or specific Gibbs energy $\equiv H - TS$
G_i°	Standard-state Gibbs energy, species i
$\bar{G}_i$	Partial Gibbs energy, species i in solution
G^E	Excess Gibbs energy $\equiv G - G^{id}$
G^R	Residual Gibbs energy $\equiv G - G^{ig}$
ΔG	Gibbs-energy change of mixing
ΔG°	Standard Gibbs-energy change of reaction
ΔG_f°	Standard Gibbs-energy change of formation
g	Local acceleration of gravity
g_c	Dimensional constant $= 32.1740(\text{lb}_m)(\text{ft})(\text{lb}_f)^{-1}(\text{s})^{-2}$
H	Molar or specific enthalpy $\equiv U + PV$
$\mathcal{H}_i$	Henry's constant, species i in solution
H_i°	Standard-state enthalpy, pure species i
$\bar{H}_i$	Partial enthalpy, species i in solution
H^E	Excess enthalpy $\equiv H - H^{id}$
H^R	Residual enthalpy $\equiv H - H^{ig}$
$(H^R)^0, (H^R)^1$	Functions, generalized residual-enthalpy correlation
ΔH	Enthalpy change ("heat") of mixing; also, latent heat of phase transition
$\widetilde{\Delta H}$	Heat of solution
ΔH°	Standard enthalpy change of reaction
ΔH_0°	Standard heat of reaction at reference temperature T_0
ΔH_f°	Standard enthalpy change of formation
I	Represents an integral, defined, e.g., by Eq. (13.71)
K_j	Equilibrium constant, chemical reaction j
K_i	Vapor/liquid equilibrium ratio, species $i \equiv y_i / x_i$
k	Boltzmann's constant
k_{ij}	Empirical interaction parameter, Eq. (10.71)
$\mathcal{L}$	Molar fraction of system that is liquid
l	Length
l_{ij}	Equation-of-state interaction parameter, Eq. (15.31)
$\mathbf{M}$	Mach number
$\mathcal{M}$	Molar mass (molecular weight)
M	Molar or specific value, extensive thermodynamic property
$\bar{M}_i$	Partial property, species i in solution
M^E	Excess property $\equiv M - M^{id}$
M^R	Residual property $\equiv M - M^{ig}$
ΔM	Property change of mixing
ΔM°	Standard property change of reaction
ΔM_f°	Standard property change of formation
m	Mass
$\dot{m}$	Mass flow rate
N	Number of chemical species, phase rule
N_A	Avogadro's number

n	Number of moles
$\dot{n}$	Molar flow rate
$\tilde{n}$	Moles of solvent per mole of solute
n_i	Number of moles, species i
P	Absolute pressure
P°	Standard-state pressure
P_c	Critical pressure
P_r	Reduced pressure
P_r^0, P_r^1	Functions, generalized vapor-pressure correlation
P_0	Reference pressure
p_i	Partial pressure, species i
P_i^{sat}	Saturation vapor pressure, species i
Q	Heat
$\dot{Q}$	Rate of heat transfer
q	Volumetric flow rate
q	Parameter, cubic equations of state
q	Electric charge
$\bar{q}_i$	Partial parameter, cubic equations of state
R	Universal gas constant (Table A.2)
r	Compression ratio
r	Number of independent chemical reactions, phase rule
S	Molar or specific entropy
$\bar{S}_i$	Partial entropy, species i in solution
S^E	Excess entropy $\equiv S - S^{id}$
S^R	Residual entropy $\equiv S - S^{ig}$
$(S^R)^0, (S^R)^1$	Functions, generalized residual-entropy correlation
S_G	Entropy generation per unit amount of fluid
$\dot{S}_G$	Rate of entropy generation
ΔS	Entropy change of mixing
ΔS°	Standard entropy change of reaction
ΔS_f°	Standard entropy change of formation
T	Absolute temperature, kelvins or rankines
T_c	Critical temperature
T_n	Normal-boiling-point temperature
T_r	Reduced temperature
T_0	Reference temperature
T_σ	Absolute temperature of surroundings
T_i^{sat}	Saturation temperature, species i
t	Temperature, °C or (°F)
t	Time
U	Molar or specific internal energy
u	Velocity
V	Molar or specific volume
$\mathcal{V}$	Molar fraction of system that is vapor
$\bar{V}_i$	Partial volume, species i in solution
V_c	Critical volume

V_r	Reduced volume
V^E	Excess volume $\equiv V - V^{id}$
V^R	Residual volume $\equiv V - V^{ig}$
ΔV	Volume change of mixing; also, volume change of phase transition
W	Work
$\dot{W}$	Work rate (power)
W_{ideal}	Ideal work
$\dot{W}_{ideal}$	Ideal-work rate
W_{lost}	Lost work
$\dot{W}_{lost}$	Lost-work rate
W_s	Shaft work for flow process
$\dot{W}_s$	Shaft power for flow process
x_i	Mole fraction, species i, liquid phase or general
x^v	Quality
y_i	Mole fraction, species i, vapor phase
Z	Compressibility factor $\equiv PV/RT$
Z_c	Critical compressibility factor $\equiv P_c V_c/RT_c$
Z^0, Z^1	Functions, generalized compressibility-factor correlation
z	Adsorbed phase compressibility factor, defined by Eq. (15.38)
z	Elevation above a datum level
z_i	Overall mole fraction or mole fraction in a solid phase

Superscripts

E	Denotes excess thermodynamic property
av	Denotes phase transition from adsorbed phase to vapor
id	Denotes value for an ideal solution
ig	Denotes value for an ideal gas
l	Denotes liquid phase
lv	Denotes phase transition from liquid to vapor
R	Denotes residual thermodynamic property
s	Denotes solid phase
sl	Denotes phase transition from solid to liquid
t	Denotes a total value of an extensive thermodynamic property
v	Denotes vapor phase
∞	Denotes a value at infinite dilution

Greek letters

α	Function, cubic equations of state (Table 3.1)
α, β	As superscripts, identify phases
$\alpha\beta$	As superscript, denotes phase transition from phase α to phase β
β	Volume expansivity
β	Parameter, cubic equations of state
Γ_i	Integration constant
γ	Ratio of heat capacities C_P/C_V
γ_i	Activity coefficient, species i in solution
δ	Polytropic exponent

ε	Constant, cubic equations of state
ε	Reaction coordinate
η	Efficiency
κ	Isothermal compressibility
Π	Spreading pressure, adsorbed phase
Π	Osmotic pressure
π	Number of phases, phase rule
μ	Joule/Thomson coefficient
μ_i	Chemical potential, species i
ν_i	Stoichiometric number, species i
ρ	Molar or specific density $\equiv 1/V$
ρ_c	Critical density
ρ_r	Reduced density
σ	Constant, cubic equations of state
Φ_i	Ratio of fugacity coefficients, defined by Eq. (13.14)
ϕ_i	Fugacity coefficient, pure species i
$\hat{\phi}_i$	Fugacity coefficient, species i in solution
ϕ^0, ϕ^1	Functions, generalized fugacity-coefficient correlation
Ψ, Ω	Constants, cubic equations of state
ω	Acentric factor

Notes

cv	As a subscript, denotes a control volume
fs	As a subscript, denotes flowing streams
$^\circ$	As a superscript, denotes the standard state
$^-$	Overbar denotes a partial property
$^\cdot$	Overdot denotes a time rate
$^\wedge$	Circumflex denotes a property in solution
Δ	Difference operator

Preface

Thermodynamics, a key component of many fields of science and engineering, is based on laws of universal applicability. However, the most important applications of those laws, and the materials and processes of greatest concern, differ from one branch of science or engineering to another. Thus, we believe there is value in presenting this material from a chemical-engineering perspective, focusing on the application of thermodynamic principles to materials and processes most likely to be encountered by chemical engineers.

Although *introductory* in nature, the material of this text should not be considered simple. Indeed, there is no way to make it simple. A student new to the subject will find that a demanding task of discovery lies ahead. New concepts, words, and symbols appear at a bewildering rate, and a degree of memorization and mental organization is required. A far greater challenge is to develop the capacity to reason in the context of thermodynamics so that one can apply thermodynamic principles in the solution of practical problems. While maintaining the rigor characteristic of sound thermodynamic analysis, we have made every effort to avoid unnecessary mathematical complexity. Moreover, we aim to encourage understanding by writing in simple active-voice, present-tense prose. We can hardly supply the required motivation, but our objective, as it has been for all previous editions, is a treatment that may be understood by any student willing to put forth the required effort.

The text is structured to alternate between the development of thermodynamic principles and the correlation and use of thermodynamic properties as well as between theory and applications. The first two chapters of the book present basic definitions and a development of the first law of thermodynamics. Chapters 3 and 4 then treat the pressure/volume/temperature behavior of fluids and heat effects associated with temperature change, phase change, and chemical reaction, allowing early application of the first law to realistic problems. The second law is developed in Chap. 5, where its most basic applications are also introduced. A full treatment of the thermodynamic properties of pure fluids in Chap. 6 allows general application of the first and second laws, and provides for an expanded treatment of flow processes in Chap. 7. Chapters 8 and 9 deal with power production and refrigeration processes. The remainder of the book, concerned with fluid mixtures, treats topics in the unique domain of chemical-engineering thermodynamics. Chapter 10 introduces the framework of solution thermodynamics, which underlies the applications in the following chapters. Chapter 12 then describes the analysis of phase equilibria, in a mostly qualitative manner. Chapter 13 provides full treatment of vapor/liquid equilibrium. Chemical-reaction equilibrium is covered at length in Chap. 14. Chapter 15 deals with topics in phase equilibria, including liquid/liquid, solid/liquid, solid/vapor, gas adsorption, and osmotic equilibria. Chapter 16 treats the thermodynamic analysis of real processes, affording a review of much of the practical subject matter of thermodynamics.

The material of these 16 chapters is more than adequate for an academic-year undergraduate course, and discretion, conditioned by the content of other courses, is required in the choice of what is covered. The first 14 chapters include material considered necessary to any chemical engineer's education. Where only a single-semester course in chemical engineering thermodynamics is provided, these chapters may represent sufficient content.

The book is comprehensive enough to make it a useful reference both in graduate courses and for professional practice. However, length considerations have required a prudent selectivity. Thus, we do not include certain topics that are worthy of attention but are of a specialized nature. These include applications to polymers, electrolytes, and biomaterials.

We are indebted to many people—students, professors, reviewers—who have contributed in various ways to the quality of this eighth edition, directly and indirectly, through question and comment, praise and criticism, through seven previous editions spanning more than 65 years.

We would like to thank McGraw-Hill Education and all of the teams that contributed to the development and support of this project. In particular, we would like to thank the following editorial and production staff for their essential contributions to this eighth edition: Thomas Scaife, Chelsea Haupt, Nick McFadden, and Laura Bies. We would also like to thank Professor Bharat Bhatt for his much appreciated comments and advice during the accuracy check.

To all we extend our thanks.

J. M. Smith
H. C. Van Ness
M. M. Abbott
M. T. Swihart

A brief explanation of the authorship of the eighth edition

In December 2003, I received an unexpected e-mail from Hank Van Ness that began as follows: "I'm sure this message comes as a surprise you; so let me state immediately its purpose. We would like to invite you to discuss the possibility that you join us as the 4th author . . . of Introduction to Chemical Engineering Thermodynamics." I met with Hank and with Mike Abbott in summer 2004, and began working with them on the eighth edition in earnest almost immediately after the seventh edition was published in 2005. Unfortunately, the following years witnessed the deaths of Michael Abbott (2006), Hank Van Ness (2008), and Joe Smith (2009) in close succession. In the months preceding his death, Hank Van Ness worked diligently on revisions to this textbook. The reordering of content and overall structure of this eighth edition reflect his vision for the book.

I am sure Joe, Hank, and Michael would all be delighted to see this eighth edition in print and, for the first time, in a fully electronic version including Connect and SmartBook. I am both humbled and honored to have been entrusted with the task of revising this classic textbook, which by the time I was born had already been used by a generation of chemical engineering students. I hope that the changes we have made, from content revision and re-ordering to the addition of more structured chapter introductions and a concise synopsis at the end of each chapter, will improve the experience of using this text for the next generation of students, while maintaining the essential character of the text, which has made it the most-used chemical engineering textbook of all time. I look forward to receiving your feedback on the changes that have been made and those that you would like to see in the future, as well as what additional resources would be of most value in supporting your use of the text.

Mark T. Swihart, March 2016

McGraw-Hill Connect®
Learn Without Limits

Connect is a teaching and learning platform that is proven to deliver better results for students and instructors.

Connect empowers students by continually adapting to deliver precisely what they need, when they need it, and how they need it, so your class time is more engaging and effective.

73% of instructors who use **Connect** require it; instructor satisfaction **increases** by 28% when **Connect** is required.

Connect's Impact on Retention Rates, Pass Rates, and Average Exam Scores

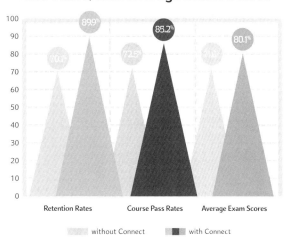

	Retention Rates	Course Pass Rates	Average Exam Scores
without Connect	70.1%	72.5%	71.5%
with Connect	89.9%	85.2%	80.1%

Using **Connect** improves retention rates by **19.8%**, passing rates by **12.7%**, and exam scores by **9.1%**.

Analytics———
Connect Insight®

Connect Insight is Connect's new one-of-a-kind visual analytics dashboard that provides at-a-glance information regarding student performance, which is immediately actionable. By presenting assignment, assessment, and topical performance results together with a time metric that is easily visible for aggregate or individual results, Connect Insight gives the user the ability to take a just-in-time approach to teaching and learning, which was never before available. Connect Insight presents data that helps instructors improve class performance in a way that is efficient and effective.

Impact on Final Course Grade Distribution

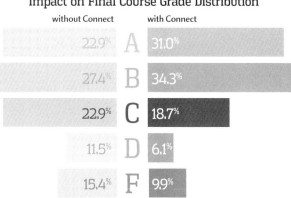

	without Connect	with Connect
A	22.9%	31.0%
B	27.4%	34.3%
C	22.9%	18.7%
D	11.5%	6.1%
F	15.4%	9.9%

Adaptive

©Getty Images/iStockphoto

THE **ADAPTIVE** **READING EXPERIENCE**
DESIGNED TO TRANSFORM
THE WAY STUDENTS READ

More students earn **A's** and **B's** when they use McGraw-Hill Education **Adaptive** products.

SmartBook®

Proven to help students improve grades and study more efficiently, SmartBook contains the same content within the print book, but actively tailors that content to the needs of the individual. SmartBook's adaptive technology provides precise, personalized instruction on what the student should do next, guiding the student to master and remember key concepts, targeting gaps in knowledge and offering customized feedback, and driving the student toward comprehension and retention of the subject matter. Available on tablets, SmartBook puts learning at the student's fingertips—anywhere, anytime.

Over **8 billion questions** have been answered, making McGraw-Hill Education products more intelligent, reliable, and precise.

www.mheducation.com

STUDENTS WANT SMARTBOOK®

95% of students reported **SmartBook** to be a more effective way of reading material

100% of students want to use the Practice Quiz feature available within **SmartBook** to help them study

100% of students reported having reliable access to off-campus wifi

90% of students say they would purchase **SmartBook** over print alone

95% reported that **SmartBook** would impact their study skills in a positive way

Mc Graw Hill Education

*Findings based on a 2015 focus group survey at Pellissippi State Community College administered by McGraw-Hill Education

Chapter 1

Introduction

By way of introduction, in this chapter we outline the origin of thermodynamics and its present scope. We also review a number of familiar, but basic, scientific concepts essential to the subject:

- Dimensions and units of measure
- Force and pressure
- Temperature
- Work and heat
- Mechanical energy and its conservation

1.1   THE SCOPE OF THERMODYNAMICS

The science of thermodynamics was developed in the 19th century as a result of the need to describe the basic operating principles of the newly invented steam engine and to provide a basis for relating the work produced to the heat supplied. Thus the name itself denotes power generated from heat. From the study of steam engines, there emerged two of the primary generalizations of science: *the First and Second Laws of Thermodynamics*. All of classical thermodynamics is implicit in these laws. Their statements are very simple, but their implications are profound.

The First Law simply says that *energy* is conserved, meaning that it is neither created nor destroyed. It provides no definition of energy that is both general and precise. No help comes from its common informal use where the word has imprecise meanings. However, in scientific and engineering contexts, energy is recognized as appearing in various forms, useful because each form has mathematical definition as a *function* of some recognizable and measurable characteristics of the real world. Thus kinetic energy is defined as a function of velocity, and gravitational potential energy as a function of elevation.

Conservation implies the transformation of one form of energy into another. Windmills have long operated to transform the kinetic energy of the wind into work that is used to raise water from land lying below sea level. The overall effect is to convert the kinetic energy of the wind into potential energy of water. Wind energy is now more widely converted to electrical energy. Similarly, the potential energy of water has long been transformed into work used to grind grain or saw lumber. Hydroelectric plants are now a significant source of electrical power.

The Second Law is more difficult to comprehend because it depends on *entropy*, a word and concept not in everyday use. Its consequences in daily life are significant with respect to environmental conservation and efficient use of energy. Formal treatment is postponed until we have laid a proper foundation.

The two laws of thermodynamics have no proof in a mathematical sense. However, they are universally observed to be obeyed. An enormous volume of experimental evidence demonstrates their validity. Thus, thermodynamics shares with mechanics and electromagnetism a basis in primitive laws.

These laws lead, through mathematical deduction, to a network of equations that find application in all branches of science and engineering. Included are calculation of heat and work requirements for physical, chemical, and biological processes, and the determination of equilibrium conditions for chemical reactions and for the transfer of chemical species between phases. Practical application of these equations almost always requires information on the properties of materials. Thus, the study and application of thermodynamics is inextricably linked with the tabulation, correlation, and prediction of properties of substances. Fig. 1.1 illustrates schematically how the two laws of thermodynamics are combined with information on material properties to yield useful analyses of, and predictions about, physical, chemical, and biological systems. It also notes the chapters of this text that treat each component.

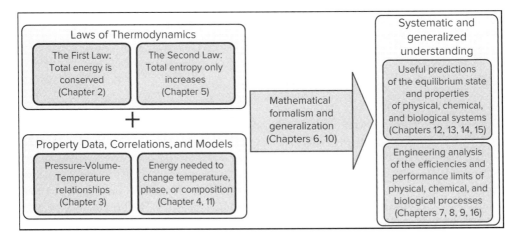

Figure 1.1: Schematic illustrating the combination of the laws of thermodynamics with data on material properties to produce useful predictions and analyses.

Examples of questions that can be answered on the basis of the laws of thermodynamics combined with property information include the following:

- How much energy is released when a liter of ethanol is burned (or metabolized)?

- What maximum flame temperature can be reached when ethanol is burned in air?

- What maximum fraction of the heat released in an ethanol flame can be converted to electrical energy or work?

- How do the answers to the preceding two questions change if the ethanol is burned with pure oxygen, rather than air?

- What is the maximum amount of electrical energy that can be produced when a liter of ethanol is reacted with O_2 to produce CO_2 and water in a fuel cell?

- In the distillation of an ethanol/water mixture, how are the vapor and liquid compositions related?

- When water and ethylene react at high pressure and temperature to produce ethanol, what are the compositions of the phases that result?

- How much ethylene is contained in a high-pressure gas cylinder for given temperature, pressure, and volume?

- When ethanol is added to a two-phase system containing toluene and water, how much ethanol goes into each phase?

- If a water/ethanol mixture is partially frozen, what are the compositions of the liquid and solid phases?

- What volume of solution results from mixing one liter of ethanol with one liter of water? (It *is not* exactly 2 liters!)

The application of thermodynamics to any real problem starts with the specification of a particular region of space or body of matter designated as the *system*. Everything outside the system is called the *surroundings*. The system and surroundings interact through transfer of material and energy across the system boundaries, but the system is the focus of attention. Many different thermodynamic systems are of interest. A pure vapor such as steam is the working medium of a power plant. A reacting mixture of fuel and air powers an internal-combustion engine. A vaporizing liquid provides refrigeration. Expanding gases in a nozzle propel a rocket. The metabolism of food provides the nourishment for life.

Once a system has been selected, we must describe its *state*. There are two possible points of view, the *macroscopic* and the *microscopic*. The former relates to quantities such as composition, density, temperature, and pressure. These *macroscopic coordinates* require no assumptions regarding the structure of matter. They are few in number, are suggested by our sense perceptions, and are measured with relative ease. A macroscopic description thus requires specification of a *few fundamental measurable properties*. The macroscopic point of view, as adopted in classical thermodynamics, reveals nothing of the microscopic (molecular) mechanisms of physical, chemical, or biological processes.

A microscopic description depends on the existence and behavior of molecules, is not directly related to our sense perceptions, and treats quantities that cannot routinely be directly measured. Nevertheless, it offers insight into material behavior and contributes to evaluation of thermodynamic properties. Bridging the length and time scales between the microscopic behavior of molecules and the macroscopic world is the subject of *statistical mechanics* or *statistical thermodynamics*, which applies the laws of quantum mechanics and classical mechanics to large ensembles of atoms, molecules, or other elementary objects to predict and interpret macroscopic behavior. Although we make occasional reference to the molecular basis for observed material properties, the subject of statistical thermodynamics is not treated in this book.[1]

1.2   INTERNATIONAL SYSTEM OF UNITS

Descriptions of thermodynamic states depend on the fundamental *dimensions* of science, of which length, time, mass, temperature, and amount of substance are of greatest interest here. These dimensions are *primitives*, recognized through our sensory perceptions, and are not definable in terms of anything simpler. Their use, however, requires the definition of arbitrary scales of measure, divided into specific *units* of size. Primary units have been set by international agreement, and are codified as the International System of Units (abbreviated SI, for Système International).[2] This is the primary system of units used throughout this book.

The *second*, symbol s, the SI unit of time, is the duration of 9,192,631,770 cycles of radiation associated with a specified transition of the cesium atom. The *meter*, symbol m, is the fundamental unit of length, defined as the distance light travels in a vacuum during 1/299,792,458 of a second. The *kilogram*, symbol kg, is the basic unit of mass, defined as the mass of a platinum/iridium cylinder kept at the International Bureau of Weights and Measures at Sèvres, France.[3] (The *gram*, symbol g, is 0.001 kg.) Temperature is a characteristic dimension of thermodynamics, and is measured on the Kelvin scale, as described in Sec. 1.4. The *mole*, symbol mol, is defined as the amount of a substance represented by as many elementary entities (e.g., molecules) as there are atoms in 0.012 kg of carbon-12.

The SI unit of force is the *newton*, symbol N, *derived* from Newton's second law, which expresses force F as the product of mass m and acceleration a: $F = ma$. Thus, a newton is the force that, when applied to a mass of 1 kg, produces an acceleration of 1 $m \cdot s^{-2}$, and is therefore a unit representing 1 $kg \cdot m \cdot s^{-2}$. This illustrates a key feature of the SI system, namely, that derived units always reduce to combinations of primary units. Pressure P (Sec. 1.5), defined as the normal force exerted by a fluid on a unit area of surface, is expressed in pascals, symbol Pa. With force in newtons and area in square meters, 1 Pa is equivalent to 1 $N \cdot m^{-2}$ or 1 $kg \cdot m^{-1} \cdot s^{-2}$. Essential to thermodynamics is the derived unit for energy, the joule, symbol J, defined as 1 N·m or 1 $kg \cdot m^2 \cdot s^{-2}$.

Multiples and decimal fractions of SI units are designated by prefixes, with symbol abbreviations, as listed in Table 1.1. Common examples of their use are the centimeter, 1 cm = 10^{-2} m, the kilopascal, 1 kPa = 10^3 Pa, and the kilojoule, 1 kJ = 10^3 J.

---

[1]Many introductory texts on statistical thermodynamics are available. The interested reader is referred to *Molecular Driving Forces: Statistical Thermodynamics in Chemistry & Biology*, by K. A. Dill and S. Bromberg, Garland Science, 2010, and many books referenced therein.

[2]In-depth information on the SI is provided by the National Institute of Standards and Technology (NIST) online at http://physics.nist.gov/cuu/Units/index.html.

[3]At the time of this writing, the International Committee on Weights and Measures has recommended changes that would eliminate the need for a standard reference kilogram and would base all units, including mass, on fundamental physical constants.

Table 1.1: Prefixes for SI Units

Multiple	Prefix	Symbol
10^{-15}	femto	f
10^{-12}	pico	p
10^{-9}	nano	n
10^{-6}	micro	μ
10^{-3}	milli	m
10^{-2}	centi	c
10^{2}	hecto	h
10^{3}	kilo	k
10^{6}	mega	M
10^{9}	giga	G
10^{12}	tera	T
10^{15}	peta	P

Two widely used units in engineering that are not part of SI, but are acceptable for use with it, are the bar, a pressure unit equal to 10^2 kPa, and the liter, a volume unit equal to 10^3 cm³. The bar closely approximates atmospheric pressure. Other acceptable units are the minute, symbol min; hour, symbol h; day, symbol d; and the metric ton, symbol t; equal to 10^3 kg.

Weight properly refers to the force of gravity on a body, expressed in newtons, and not to its mass, expressed in kilograms. Force and mass are, of course, directly related through Newton's law, with a body's weight defined as its mass times the local acceleration of gravity. The comparison of masses by a balance is called "weighing" because it also compares gravitational forces. A spring scale provides correct mass readings only when used in the gravitational field of its calibration.

Although the SI is well established throughout most of the world, use of the U.S. Customary system of units persists in daily commerce in the United States. Even in science and engineering, conversion to SI is incomplete, though globalization is a major incentive. U.S. Customary units are related to SI units by fixed conversion factors. Those units most likely to be useful are defined in Appendix A. Conversion factors are listed in Table A.1.

Example 1.1

An astronaut weighs 730 N in Houston, Texas, where the local acceleration of gravity is $g = 9.792$ m·s⁻². What are the astronaut's mass and weight on the moon, where $g = 1.67$ m·s⁻²?

Solution 1.1

By Newton's law, with acceleration equal to the acceleration of gravity, g,

$$m = \frac{F}{g} = \frac{730 \text{ N}}{9.792 \text{ m·s}^{-2}} = 74.55 \text{ N·m}^{-1}\text{·s}^2$$

Because $1 \text{ N} = 1 \text{ kg·m·s}^{-2}$,

$$m = 74.55 \text{ kg}$$

This *mass* of the astronaut is independent of location, but *weight* depends on the local acceleration of gravity. Thus on the moon the astronaut's weight is:

$$F(\text{moon}) = m \times g(\text{moon}) = 74.55 \text{ kg} \times 1.67 \text{ m·s}^{-2}$$

or

$$F(\text{moon}) = 124.5 \text{ kg·m·s}^{-2} = 124.5 \text{ N}$$

1.3 MEASURES OF AMOUNT OR SIZE

Three measures of amount or size of a homogeneous material are in common use:

- Mass, m • Number of moles, n • Total volume, V^t

These measures for a specific system are in direct proportion to one another. Mass may be divided by the *molar mass* $\mathcal{M}$ (formerly called molecular weight) to yield number of moles:

$$n = \frac{m}{\mathcal{M}} \qquad \text{or} \qquad m = \mathcal{M} n$$

Total volume, representing the size of a system, is a defined quantity given as the product of three lengths. It may be divided by the mass or number of moles of the system to yield *specific* or *molar* volume:

- Specific volume: $V \equiv \dfrac{V^t}{m} \qquad \text{or} \qquad V^t = mV$

- Molar volume: $V \equiv \dfrac{V^t}{n} \qquad \text{or} \qquad V^t = nV$

Specific or molar density is defined as the reciprocal of specific or molar volume: $\rho \equiv V^{-1}$.

These quantities (V and ρ) are independent of the size of a system, and are examples of *intensive* thermodynamic variables. For a given state of matter (solid, liquid, or gas) they are functions of temperature, pressure, and composition, additional quantities independent of system size. Throughout this text, the same symbols will generally be used for both molar and specific quantities. Most equations of thermodynamics apply to both, and when distinction is necessary, it can be made based on the context. The alternative of introducing separate notation for each leads to an even greater proliferation of variables than is already inherent in the study of chemical thermodynamics.

1.4 TEMPERATURE

The notion of temperature, based on sensory perception of heat and cold, needs no explanation. It is a matter of common experience. However, giving temperature a scientific role requires a scale that affixes numbers to the perception of hot and cold. This scale must also extend far beyond the range of temperatures of everyday experience and perception. Establishing such a scale and devising measuring instruments based on this scale has a long and intriguing history. A simple instrument is the common liquid-in-glass thermometer, wherein the liquid expands when heated. Thus a uniform tube, partially filled with mercury, alcohol, or some other fluid, and connected to a bulb containing a larger amount of fluid, indicates degree of hotness by the length of the fluid column.

The scale requires definition and the instrument requires calibration. The Celsius[4] scale was established early and remains in common use throughout most of the world. Its scale is defined by fixing zero as the *ice point* (freezing point of water saturated with air at standard atmospheric pressure) and 100 as the *steam point* (boiling point of pure water at standard atmospheric pressure). Thus a thermometer when immersed in an ice bath is marked zero and when immersed in boiling water is marked 100. Dividing the length between these marks into 100 equal spaces, called *degrees*, provides a scale, which may be extended with equal spaces below zero and above 100.

Scientific and industrial practice depends on the *International Temperature Scale of 1990* (ITS−90).[5] This is the Kelvin scale, based on assigned values of temperature for a number of reproducible *fixed points*, that is, states of pure substances like the ice and steam points, and on *standard instruments* calibrated at these temperatures. Interpolation between the fixed-point temperatures is provided by formulas that establish the relation between readings of the standard instruments and values on ITS-90. The platinum-resistance thermometer is an example of a standard instrument; it is used for temperatures from −259.35°C (the triple point of hydrogen) to 961.78°C (the freezing point of silver).

The Kelvin scale, which we indicate with the symbol T, provides SI temperatures. An *absolute* scale, it is based on the concept of a lower limit of temperature, called absolute zero. Its unit is the *kelvin*, symbol K. Celsius temperatures, with symbol t, are defined in relation to Kelvin temperatures:

$$t°C = T \text{ K} − 273.15$$

The unit of Celsius temperature is the degree Celsius, °C, which is equal in size to the kelvin.[6] However, temperatures on the Celsius scale are 273.15 degrees lower than on the Kelvin scale. Thus absolute zero on the Celsius scale occurs at −273.15°C. Kelvin temperatures

---

[4]Anders Celsius, Swedish astronomer (1701–1744). See: http://en.wikipedia.org/wiki/Anders_Celsius.

[5]The English-language text describing ITS-90 is given by H. Preston-Thomas, *Metrologia*, vol. 27, pp. 3–10, 1990. It is also available at http://www.its-90.com/its-90.html.

[6]Note that neither the word *degree* nor the *degree sign* is used for temperatures in kelvins, and that the word *kelvin* as a unit is not capitalized.

are used in thermodynamic calculations. Celsius temperatures can only be used in thermodynamic calculations involving *temperature differences*, which are of course the same in both degrees Celsius and kelvins.

1.5 PRESSURE

The primary standard for pressure measurement is the dead-weight gauge in which a known force is balanced by fluid pressure acting on a piston of known area: $P \equiv F/A$. The basic design is shown in Fig. 1.2. Objects of known mass ("weights") are placed on the pan until the pressure of the oil, which tends to make the piston rise, is just balanced by the force of gravity on the piston and all that it supports. With this force given by Newton's law, the pressure exerted by the oil is:

$$P = \frac{F}{A} = \frac{mg}{A}$$

where m is the mass of the piston, pan, and "weights"; g is the local acceleration of gravity; and A is the cross-sectional area of the piston. This formula yields *gauge pressures*, the difference between the pressure of interest and the pressure of the surrounding atmosphere. They are converted to *absolute pressures* by addition of the local barometric pressure. Gauges in common use, such as Bourdon gauges, are calibrated by comparison with dead-weight gauges. Absolute pressures are used in thermodynamic calculations.

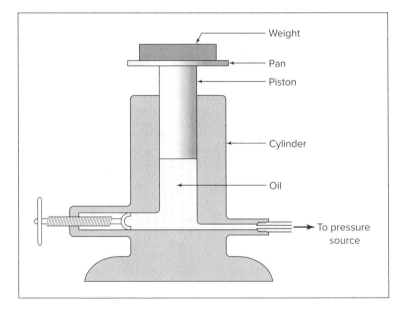

Figure 1.2:
Dead-weight gauge.

Because a vertical column of fluid under the influence of gravity exerts a pressure at its base in direct proportion to its height, pressure may be expressed as the equivalent height of a fluid column. This is the basis for the use of manometers for pressure measurement. Conversion of height to force per unit area follows from Newton's law applied to the force of gravity acting on the mass of fluid in the column. The mass is given by: $m = Ah\rho$, where A is the cross-sectional area of the column, h is its height, and ρ is the fluid density. Therefore,

$$P = \frac{F}{A} = \frac{mg}{A} = \frac{Ah\rho g}{A}$$

Thus,

$$P = h\rho g \qquad (1.1)$$

The pressure to which a fluid height corresponds is determined by the density of the fluid (which depends on its identity and temperature) and the local acceleration of gravity.

A unit of pressure in common use (but not an SI unit) is the *standard atmosphere*, representing the average pressure exerted by the earth's atmosphere at sea level, and defined as 101.325 kPa.

Example 1.2

A dead-weight gauge with a piston diameter of 1 cm is used for the accurate measurement of pressure. If a mass of 6.14 kg (including piston and pan) brings it into balance, and if $g = 9.82$ m·s^{-2}, what is the *gauge* pressure being measured? For a barometric pressure of 0.997 bar, what is the *absolute* pressure?

Solution 1.2

The force exerted by gravity on the piston, pan, and "weights" is:

$$F = mg = 6.14 \text{ kg} \times 9.82 \text{ m·s}^{-2} = 60.295 \text{ N}$$

$$\text{Gauge pressure} = \frac{F}{A} = \frac{60.295}{(\frac{1}{4})(\pi)(0.01)^2} = 7.677 \times 10^5 \text{ N·m}^{-2} = 767.7 \text{ kPa}$$

The absolute pressure is therefore:

$$P = 7.677 \times 10^5 + 0.997 \times 10^5 = 8.674 \times 10^5 \text{ N·m}^{-2}$$

or

$$P = 867.4 \text{ kPa}$$

Example 1.3

At 27°C the reading on a manometer filled with mercury is 60.5 cm. The local acceleration of gravity is 9.784 m·s^{-2}. To what pressure does this height of mercury correspond?

Solution 1.3

As discussed above, and summarized in Eq. (1.1): $P = h\rho g$. At 27°C the density of mercury is 13.53 g·cm^{-3}. Then,

$$P = 60.5 \text{ cm} \times 13.53 \text{ g·cm}^{-3} \times 9.784 \text{ m·s}^{-2} = 8009 \text{ g·m·s}^{-2}\text{·cm}^{-2}$$
$$= 8.009 \text{ kg·m·s}^{-2}\text{·cm}^{-2} = 8.009 \text{ N·cm}^{-2}$$
$$= 0.8009 \times 10^5 \text{ N·m}^{-2} = 0.8009 \text{ bar} = 80.09 \text{ kPa}$$

1.6   WORK

Work, W, is performed whenever a force acts through a distance. By its definition, the quantity of work is given by the equation:

$$dW = F \, dl \tag{1.2}$$

where F is the component of force acting along the line of the displacement dl. The SI unit of work is the newton·meter or joule, symbol J. When integrated, Eq. (1.2) yields the work of a finite process. By convention, work is regarded as positive when the displacement is in the same direction as the applied force and negative when they are in opposite directions.

Work is done when pressure acts on a surface and displaces a volume of fluid. An example is the movement of a piston in a cylinder so as to cause compression or expansion of a fluid contained in the cylinder. The force exerted by the piston on the fluid is equal to the product of the piston area and the pressure of the fluid. The displacement of the piston is equal to the total volume change of the fluid divided by the area of the piston. Equation (1.2) therefore becomes:

$$dW = -PA \, d\frac{V^t}{A} = -P \, dV^t \tag{1.3}$$

Integration yields:

$$W = -\int_{V_1^t}^{V_2^t} P \, dV^t \tag{1.4}$$

The minus signs in these equations are made necessary by the sign convention adopted for work. When the piston moves into the cylinder so as to compress the fluid, the applied force and its displacement are in the same direction; the work is therefore positive. The minus sign is required because the volume change is negative. For an expansion process, the applied force and its displacement are in opposite directions. The volume change in this case is positive, and the minus sign is again required to make the work negative.

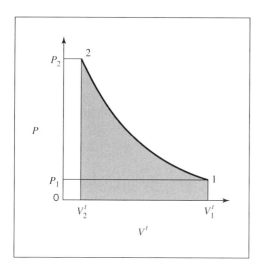

Figure 1.3: Diagram showing a P vs. V^t path.

Equation (1.4) expresses the work done by a finite compression or expansion process.[7] Figure 1.3 shows a *path* for compression of a gas from point 1, initial volume V_1^t at pressure P_1, to point 2, volume V_2^t at pressure P_2. This path relates the pressure at any point of the process to the volume. The work required is given by Eq. (1.4) and is proportional to the area under the curve of Fig. 1.3.

1.7 ENERGY

The general principle of conservation of energy was established about 1850. The germ of this principle as it applies to mechanics was implicit in the work of Galileo (1564–1642) and Isaac Newton (1642–1726). Indeed, it follows directly from Newton's second law of motion once work is defined as the product of force and displacement.

Kinetic Energy

When a body of mass m, acted upon by a force F, is displaced a distance dl during a differential interval of time dt, the work done is given by Eq. (1.2). In combination with Newton's second law this equation becomes:

$$dW = ma \, dl$$

By definition the acceleration is $a \equiv du/dt$, where u is the velocity of the body. Thus,

$$dW = m\frac{du}{dt}dl = m\frac{dl}{dt}du$$

Because the definition of velocity is $u \equiv dl/dt$, this expression for work reduces to:

$$dW = mu \, du$$

[7]However, as explained in Sec. 2.6, there are important limitations on its use.

Integration for a finite change in velocity from u_1 to u_2 gives:

$$W = m \int_{u_1}^{u_2} u \, du = m \left(\frac{u_2^2}{2} - \frac{u_1^2}{2} \right)$$

or

$$W = \frac{mu_2^2}{2} - \frac{mu_1^2}{2} = \Delta \left(\frac{mu^2}{2} \right) \tag{1.5}$$

Each of the quantities $\frac{1}{2}mu^2$ in Eq. (1.5) is a *kinetic energy*, a term introduced by Lord Kelvin[8] in 1856. Thus, by definition,

$$E_K \equiv \frac{1}{2}mu^2 \tag{1.6}$$

Equation (1.5) shows that the work done *on* a body in accelerating it from an initial velocity u_1 to a final velocity u_2 is equal to the change in kinetic energy of the body. Conversely, if a moving body is decelerated by the action of a resisting force, the work done *by* the body is equal to its change in kinetic energy. With mass in kilograms and velocity in meters/second, kinetic energy E_K is in joules, where $1 \, \text{J} = 1 \, \text{kg·m}^2\text{·s}^{-2} = 1 \, \text{N·m}$. In accord with Eq. (1.5), this is the unit of work.

Potential Energy

When a body of mass m is raised from an initial elevation z_1 to a final elevation z_2, an upward force at least equal to the weight of the body is exerted on it, and this force moves through the distance $z_2 - z_1$. Because the weight of the body is the force of gravity on it, the minimum force required is given by Newton's law:

$$F = ma = mg$$

where g is the local acceleration of gravity. The minimum work required to raise the body is the product of this force and the change in elevation:

$$W = F(z_2 - z_1) = mg(z_2 - z_1)$$

or

$$W = mz_2g - mz_1g = mg\Delta z \tag{1.7}$$

We see from Eq. (1.7) that work done *on* a body in raising it is equal to the change in the quantity mzg. Conversely, if a body is lowered against a resisting force equal to its weight, the work done *by* the body is equal to the change in the quantity mzg. Each of the quantities mzg in Eq. (1.7) is a *potential energy*.[9] Thus, by definition,

$$E_P = mzg \tag{1.8}$$

---

[8]Lord Kelvin, or William Thomson (1824–1907), was an English physicist who, along with the German physicist Rudolf Clausius (1822–1888), laid the foundations for the modern science of thermodynamics. See http://en.wikipedia.org/wiki/William_Thomson,_1st_Baron_Kelvin. See also http://en.wikipedia.org/wiki/Rudolf_Clausius.

[9]This term was proposed in 1853 by the Scottish engineer William Rankine (1820–1872). See http://en.wikipedia.org/wiki/William_John_Macquorn_Rankine.

With mass in kg, elevation in m, and the acceleration of gravity in m·s^{-2}, E_P is in joules, where $1\,J = 1\,kg\cdot m^2\cdot s^{-2} = 1\,N\cdot m$. In accord with Eq. (1.7), this is the unit of work.

Energy Conservation

The utility of the energy-conservation principle was alluded to in Sec. 1.1. The definitions of kinetic energy and gravitational potential energy of the preceding section provide for limited quantitative applications. Equation (1.5) shows that the work done on an accelerating body produces a change in its *kinetic energy:*

$$W = \Delta E_K = \Delta\left(\frac{mu^2}{2}\right)$$

Similarly, Eq. (1.7) shows that the work done on a body in elevating it produces a change in its *potential energy:*

$$W = E_P = \Delta(mzg)$$

One simple consequence of these definitions is that an elevated body, allowed to fall freely (i.e., without friction or other resistance), gains in kinetic energy what it loses in potential energy. Mathematically,

$$\Delta E_K + \Delta E_P = 0$$

or

$$\frac{mu_2^2}{2} - \frac{mu_1^2}{2} + mz_2g - mz_1g = 0$$

The validity of this equation has been confirmed by countless experiments. Thus the development of the concept of energy led logically to the principle of its conservation for all *purely mechanical processes*, that is, processes without friction or heat transfer.

Other forms of mechanical energy are recognized. Among the most obvious is potential energy of configuration. When a spring is compressed, work is done by an external force. Because the spring can later perform this work against a resisting force, it possesses potential energy of configuration. Energy of the same form exists in a stretched rubber band or in a bar of metal deformed in the elastic region.

The generality of the principle of conservation of energy in mechanics is increased if we look upon work itself as a form of energy. This is clearly permissible because both kinetic- and potential-energy changes are equal to the work done in producing them [Eqs. (1.5) and (1.7)]. However, work is energy *in transit* and is never regarded as residing in a body. When work is done and does not appear simultaneously as work elsewhere, it is converted into another form of energy.

With the body or assemblage on which attention is focused as the *system* and all else as the *surroundings*, work represents energy transferred from the surroundings to the system, or the reverse. It is only during this transfer that the form of energy known as work exists. In contrast, kinetic and potential energy reside with the system. Their values, however, are measured with reference to the surroundings; that is, kinetic energy depends on velocity with respect to the surroundings, and potential energy depends on elevation with respect to a datum level. *Changes* in kinetic and potential energy do not depend on these reference conditions, provided they are fixed.

Example 1.4

An elevator with a mass of 2500 kg rests at a level 10 m above the base of an elevator shaft. It is raised to 100 m above the base of the shaft, where the cable holding it breaks. The elevator falls freely to the base of the shaft and strikes a strong spring. The spring is designed to bring the elevator to rest and, by means of a catch arrangement, to hold the elevator at the position of maximum spring compression. Assuming the entire process to be frictionless, and taking $g = 9.8$ m·s^{-2}, calculate:

(a) The potential energy of the elevator in its initial position relative to its base.

(b) The work done in raising the elevator.

(c) The potential energy of the elevator in its highest position.

(d) The velocity and kinetic energy of the elevator just before it strikes the spring.

(e) The potential energy of the compressed spring.

(f) The energy of the system consisting of the elevator and spring (1) at the start of the process, (2) when the elevator reaches its maximum height, (3) just before the elevator strikes the spring, (4) after the elevator has come to rest.

Solution 1.4

Let subscript 1 denote the initial state; subscript 2, the state when the elevator is at its greatest elevation; and subscript 3, the state just before the elevator strikes the spring, as indicated in the figure.

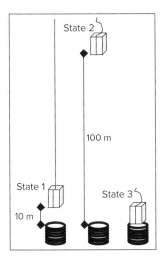

(*a*) Potential energy is defined by Eq. (1.8):

$$E_{P_1} = mz_1g = 2500 \text{ kg} \times 10 \text{ m} \times 9.8 \text{ m·s}^{-2}$$
$$= 245{,}000 \text{ kg·m}^2\text{·s}^{-2} = 245{,}000 \text{ J}$$

(*b*) Work is computed by Eq. (1.7). Units are as in the preceding calculation:

$$W = mg(z_2 - z_1) = (2500)(9.8)(100 - 10)$$
$$= 2,205,000 \text{ J}$$

(*c*) Again by Eq. (1.8),

$$E_{P_2} = mz_2g = (2500)(100)(9.8) = 2,450,000 \text{ J}$$

Note that $W = E_{P_2} - E_{P_1}$.

(*d*) The sum of the kinetic- and potential-energy changes during the process from state 2 to state 3 is zero; that is,

$$\Delta E_{K_{2 \to 3}} + \Delta E_{P_{2 \to 3}} = 0 \quad \text{or} \quad E_{K_3} - E_{K_2} + E_{P_3} - E_{P_2} = 0$$

However, E_{K_2} and E_{P_3} are zero; hence $E_{K_3} = E_{P_2} = 2,450,000$ J.

With $E_{K_3} = \frac{1}{2}mu_3^2$

$$u_3^2 = \frac{2E_{K_3}}{m} = \frac{2 \times 2,450,000 \text{ J}}{2500 \text{ kg}} = \frac{2 \times 2,450,000 \text{ kg·m}^2 \text{·s}^{-2}}{2500 \text{ kg}} = 1960 \text{ m}^2 \text{·s}^{-2}$$

and

$$u_3 = 44.272 \text{ m·s}^{-1}$$

(*e*) The changes in the potential energy of the spring and the kinetic energy of the elevator must sum to zero:

$$\Delta E_P(\text{spring}) + \Delta E_K(\text{elevator}) = 0$$

The initial potential energy of the spring and the final kinetic energy of the elevator are zero; therefore, the final potential energy of the spring equals the kinetic energy of the elevator just before it strikes the spring. Thus the final potential energy of the spring is 2,450,000 J.

(*f*) With the elevator and spring as the system, the initial energy is the potential energy of the elevator, or 245,000 J. The only energy change of the system occurs when work is done in raising the elevator. This amounts to 2,205,000 J, and the energy of the system when the elevator is at maximum height is 245,000 + 2,205,000 = 2,450,000 J. Subsequent changes occur entirely within the system, without interaction with the surroundings, and the total energy of the system remains constant at 2,450,000 J. It merely changes from potential energy of position (elevation) of the elevator to kinetic energy of the elevator to potential energy of configuration of the spring.

This example illustrates the conservation of mechanical energy. However, the entire process is assumed to occur without friction, and the results obtained are exact only for such an idealized process.

Example 1.5

A team from *Engineers Without Borders* constructs a system to supply water to a mountainside village located 1800 m above sea level from a spring in the valley below at 1500 m above sea level.

(a) When the pipe from the spring to the village is full of water, but no water is flowing, what is the pressure difference between the end of the pipe at the spring and the end of the pipe in the village?

(b) What is the change in gravitational potential energy of a liter of water when it is pumped from the spring to the village?

(c) What is the minimum amount of work required to pump a liter of water from the spring to the village?

Solution 1.5

(a) Take the density of water as 1000 kg·m^{-3} and the acceleration of gravity as 9.8 m·s^{-2}. By Eq. (1.1):

$$P = h\rho g = 300 \text{ m} \times 1000 \text{ kg·m}^{-3} \times 9.8 \text{ m·s}^{-2} = 29.4 \times 10^5 \text{ kg·m}^{-1}\text{·s}^{-2}$$

Whence $P = 29.4 \text{ bar}$ or   2940 kPa

(b) The mass of a liter of water is approximately 1 kg, and its potential-energy change is:

$$\Delta E_P = \Delta(mzg) = mg\Delta z = 1 \text{ kg} \times 9.8 \text{ m·s}^{-2} \times 300 \text{ m} = 2940 \text{ N·m} = 2940 \text{ J}$$

(c) The minimum amount of work required to lift each liter of water through an elevation change of 300 m equals the potential-energy change of the water. It is a minimum value because it takes no account of fluid friction that results from finite-velocity pipe flow.

1.8   HEAT

At the time when the principle of conservation of mechanical energy emerged, heat was considered an indestructible fluid called *caloric*. This concept was firmly entrenched, and it limited the application of energy conservation to frictionless mechanical processes. Such a limitation is now long gone. Heat, like work, is recognized as energy in transit. A simple example is the braking of an automobile. When its speed is reduced by the application of brakes, heat generated by friction is transferred to the surroundings in an amount equal to the change in kinetic energy of the vehicle.[10]

----

[10]Many modern electric or hybrid cars employ *regenerative braking*, a process through which some of the kinetic energy of the vehicle is converted to electrical energy and stored in a battery or capacitor for later use, rather than simply being transferred to the surroundings as heat.

We know from experience that a hot object brought into contact with a cold object becomes cooler, whereas the cold object becomes warmer. A reasonable view is that something is transferred from the hot object to the cold one, and we call that something heat Q.[11] Thus we say that heat always flows from a higher temperature to a lower one. This leads to the concept of temperature as the *driving force* for the transfer of energy as heat. When no temperature difference exists, no spontaneous heat transfer occurs, a condition of thermal equilibrium. In the thermodynamic sense, heat is never regarded as being stored within a body. Like work, it exists only as energy *in transit* from one body to another; in thermodynamics, from system to surroundings. When energy in the form of heat is added to a system, it is stored not as heat but as kinetic and potential energy of the atoms and molecules making up the system.

A kitchen refrigerator running on electrical energy must transfer this energy to the surroundings as heat. This may seem counterintuitive, as the interior of the refrigerator is maintained at temperatures below that of the surroundings, resulting in heat transfer *into* the refrigerator. But hidden from view (usually) is a heat exchanger that transfers heat to the surroundings in an amount equal to the sum of the electrical energy supplied to the refrigerator and the heat transfer into the refrigerator. Thus the net result is heating of the kitchen. A room air conditioner, operating in the same way, extracts heat from the room, but the heat exchanger is external, exhausting heat to the outside air, thus cooling the room.

In spite of the transient nature of heat, it is often viewed in relation to its effect on the system from which or to which it is transferred. Until about 1930 the definitions of units of heat were based on temperature changes of a unit mass of water. Thus the *calorie* was defined as that quantity of heat which, when transferred to one gram of water, raised its temperature one degree Celsius.[12] With heat now understood to be a form of energy, its SI unit is the joule. The SI unit of power is the watt, symbol W, defined as an energy rate of one joule per second. The tables of Appendix A provide relevant conversion factors.

1.9 SYNOPSIS

After studying this chapter, including the end-of-chapter problems, one should be able to:

- Describe qualitatively the scope and structure of thermodynamics
- Solve problems involving the pressure exerted by a column of fluid
- Solve problems involving conservation of mechanical energy
- Use SI units and convert from U.S. Customary to SI units
- Apply the concept of work as the transfer of energy accompanying the action of a force through a distance, and by extension to the action of pressure (force per area) acting through a volume (distance times area)

---

[11]An equally reasonable view would regard something called *cool* as being transferred from the cold object to the hot one.

[12]A unit reflecting the caloric theory of heat, but not in use with the SI system. The calorie used by nutritionists to measure the energy content of food is 1000 times larger.

1.10 PROBLEMS

1.1. *Electric current* is the fundamental SI electrical dimension, with the *ampere* (A) as its unit. Determine units for the following quantities as combinations of *fundamental* SI units.

    (*a*) Electric power
    (*b*) Electric charge
    (*c*) Electric potential difference
    (*d*) Electric resistance
    (*e*) Electric capacitance

1.2. Liquid/vapor saturation pressure P^{sat} is often represented as a function of temperature by the Antoine equation, which can be written in the form:

$$\log_{10} P^{sat}/(\text{torr}) = a - \frac{b}{t/°\text{C} + c}$$

Here, parameters a, b, and c are substance-specific constants. Suppose this equation is to be rewritten in the equivalent form:

$$\ln P^{sat}/\text{kPa} = A - \frac{B}{T/\text{K} + C}$$

Show how the parameters in the two equations are related.

1.3. Table B.2 in Appendix B provides parameters for computing the vapor pressure of many substances by the Antoine equation (see Prob. 1.2). For one of these substances, prepare two plots of P^{sat} versus T over the range of temperature for which the parameters are valid. One plot should present P^{sat} on a linear scale and the other should present P^{sat} on a log scale.

1.4. At what absolute temperature do the Celsius and Fahrenheit temperature scales give the same numerical value? What is the value?

1.5. The SI unit of *luminous intensity* is the candela (abbreviated cd), which is a primary unit. The derived SI unit of *luminous flux* is the lumen (abbreviated lm). These are based on the sensitivity of the human eye to light. Light sources are often evaluated based on their *luminous efficacy*, which is defined as the luminous flux divided by the power consumed and is measured in lm·W^{-1}. In a physical or online store, find manufacturer's specifications for representative incandescent, halogen, high-temperature-discharge, LED, and fluorescent lamps of similar luminous flux and compare their luminous efficacy.

1.6. Pressures up to 3000 bar are measured with a dead-weight gauge. The piston diameter is 4 mm. What is the approximate mass in kg of the weights required?

1.7. Pressures up to 3000(atm) are measured with a dead-weight gauge. The piston diameter is 0.17(in). What is the approximate mass in (lb_m) of the weights required?

1.8. The reading on a mercury manometer at 25°C (open to the atmosphere at one end) is 56.38 cm. The local acceleration of gravity is 9.832 m·s^{-2}. Atmospheric pressure is 101.78 kPa. What is the absolute pressure in kPa being measured? The density of mercury at 25°C is 13.534 g·cm^{-3}.

1.9. The reading on a mercury manometer at 70(°F) (open to the atmosphere at one end) is 25.62(in). The local acceleration of gravity is 32.243(ft)·(s)$^{-2}$. Atmospheric pressure is 29.86(in Hg). What is the absolute pressure in (psia) being measured? The density of mercury at 70(°F) is 13.543 g·cm^{-3}.

1.10. An absolute pressure gauge is submerged 50 m (1979 inches) below the surface of the ocean and reads $P = 6.064$ bar. This is $P = 2434$(inches of H_2O), according to the unit conversions built into a particular calculator. Explain the apparent discrepancy between the pressure measurement and the actual depth of submersion.

1.11. Liquids that boil at relatively low temperatures are often stored as liquids under their vapor pressures, which at ambient temperature can be quite large. Thus, *n*-butane stored as a liquid/vapor system is at a pressure of 2.581 bar for a temperature of 300 K. Large-scale storage (>50 m^3) of this kind is sometimes done in *spherical* tanks. Suggest two reasons why.

1.12. The first accurate measurements of the properties of high-pressure gases were made by E. H. Amagat in France between 1869 and 1893. Before developing the dead-weight gauge, he worked in a mineshaft and used a mercury manometer for measurements of pressure to more than 400 bar. Estimate the height of manometer required.

1.13. An instrument to measure the acceleration of gravity on Mars is constructed of a spring from which is suspended a mass of 0.40 kg. At a place on earth where the local acceleration of gravity is 9.81 m·s^{-2}, the spring extends 1.08 cm. When the instrument package is landed on Mars, it radios the information that the spring is extended 0.40 cm. What is the Martian acceleration of gravity?

1.14. The variation of fluid pressure with height is described by the differential equation:

$$\frac{dP}{dz} = -\rho g$$

Here, ρ is specific density and g is the local acceleration of gravity. For an *ideal gas*, $\rho = \mathcal{M}P/RT$, where $\mathcal{M}$ is molar mass and R is the universal gas constant. Modeling the atmosphere as an isothermal column of ideal gas at 10°C, estimate the ambient pressure in Denver, where $z = 1$(mile) relative to sea level. For air, take $\mathcal{M} = 29$ g·mol^{-1}; values of R are given in App. A.

1.15. A group of engineers has landed on the moon, and they wish to determine the mass of some rocks. They have a spring scale calibrated to read pounds *mass* at a location where the acceleration of gravity is 32.186(ft)(s)$^{-2}$. One of the moon rocks gives a reading of 18.76 on this scale. What is its mass? What is its weight on the moon? Take g(moon) $= 5.32$(ft)(s)$^{-2}$.

1.16. In medical contexts, *blood pressure* is often given simply as numbers without units.

    (*a*) In taking blood pressure, what physical quantity is actually being measured?
    (*b*) What are the units in which blood pressure is typically reported?
    (*c*) Is the reported blood pressure an absolute pressure or a gauge pressure?
    (*d*) Suppose an ambitious zookeeper measures the blood pressure of a standing adult male giraffe (18 feet tall) in its front leg, just above the hoof, and in its neck, just below the jaw. By about how much are the two readings expected to differ?
    (*e*) What happens to the blood pressure in a giraffe's neck when it stoops to drink?
    (*f*) What adaptations do giraffes have that allow them to accommodate pressure differences related to their height?

1.17. A 70 W outdoor security light burns, on average, 10 hours a day. A new bulb costs $5.00, and the lifetime is about 1000 hours. If electricity costs $0.10 per kW·h, what is the yearly price of "security," per light?

1.18. A gas is confined in a 1.25(ft) diameter cylinder by a piston, on which rests a weight. The mass of the piston and weight together is 250(lb_m). The local acceleration of gravity is 32.169(ft)(s)$^{-2}$, and atmospheric pressure is 30.12(in Hg).

    (*a*) What is the force in (lb_f) exerted on the gas by the atmosphere, the piston, and the weight, assuming no friction between the piston and cylinder?
    (*b*) What is the pressure of the gas in (psia)?
    (*c*) If the gas in the cylinder is heated, it expands, pushing the piston and weight upward. If the piston and weight are raised 1.7(ft), what is the work done by the gas in (ft)(lb_f)? What is the change in potential energy of the piston and weight?

1.19. A gas is confined in a 0.47 m diameter cylinder by a piston, on which rests a weight. The mass of the piston and weight together is 150 kg. The local acceleration of gravity is 9.813 m·s^{-2}, and atmospheric pressure is 101.57 kPa.

    (*a*) What is the force in newtons exerted on the gas by the atmosphere, the piston, and the weight, assuming no friction between the piston and cylinder?
    (*b*) What is the pressure of the gas in kPa?
    (*c*) If the gas in the cylinder is heated, it expands, pushing the piston and weight upward. If the piston and weight are raised 0.83 m, what is the work done by the gas in kJ? What is the change in potential energy of the piston and weight?

1.20. Verify that the SI unit of kinetic and potential energy is the joule.

1.21. An automobile having a mass of 1250 kg is traveling at 40 m·s^{-1}. What is its kinetic energy in kJ? How much work must be done to bring it to a stop?

1.22. The turbines in a hydroelectric plant are fed by water falling from a 50 m height. Assuming 91% efficiency for conversion of potential to electrical energy, and 8% loss of the resulting power in transmission, what is the mass flow rate of water required to power a 200 W light bulb?

1.23. A wind turbine with a rotor diameter of 77 m produces 1.5 MW of electrical power at a wind speed of 12 m·s⁻¹. What fraction of the kinetic energy of the air passing through the turbine is converted to electrical power? You may assume a density of 1.25 kg·m⁻³ for air at the operating conditions.

1.24. The annual average insolation (energy of sunlight per unit area) striking a fixed solar panel in Buffalo, New York, is 200 W·m⁻², while in Phoenix, Arizona, it is 270 W·m⁻². In each location, the solar panel converts 15% of the incident energy into electricity. Average annual electricity use in Buffalo is 6000 kW·h at an average cost of $0.15 kW·h, while in Phoenix it is 11,000 kW·h at a cost of $0.09 kW·h.

 (a) In each city, what area of solar panel is needed to meet the average electrical needs of a residence?
 (b) In each city, what is the current average annual cost of electricity?
 (c) If the solar panel has a lifetime of 20 years, what price per square meter of solar panel can be justified in each location? Assume that future increases in electricity prices offset the cost of borrowing funds for the initial purchase, so that you need not consider the time value of money in this analysis.

1.25. Following is a list of approximate conversion factors, useful for "back-of-the-envelope" estimates. None of them is exact, but most are accurate to within about ±10%. Use Table A.1 (App. A) to establish the exact conversions.

 • 1(atm) ≈ 1 bar
 • 1(Btu) ≈ 1 kJ
 • 1(hp) ≈ 0.75 kW
 • 1(inch) ≈ 2.5 cm
 • 1(lbm) ≈ 0.5 kg
 • 1(mile) ≈ 1.6 km
 • 1(quart) ≈ 1 liter
 • 1(yard) ≈ 1 m

Add your own items to the list. The idea is to keep the conversion factors simple and easy to remember.

1.26. Consider the following proposal for a decimal calendar. The fundamental unit is the decimal year (Yr), equal to the number of conventional (SI) seconds required for the earth to complete a circuit of the sun. Other units are defined in the following table. Develop, where possible, factors for converting decimal calendar units to conventional calendar units. Discuss pros and cons of the proposal.

Decimal Calendar Unit	Symbol	Definition
Second	Sc	10^{-6} Yr
Minute	Mn	10^{-5} Yr
Hour	Hr	10^{-4} Yr
Day	Dy	10^{-3} Yr
Week	Wk	10^{-2} Yr
Month	Mo	10^{-1} Yr
Year	Yr	

1.27. Energy costs vary greatly with energy source: coal @ \$35.00/ton, gasoline @ a pump price of \$2.75/gal, and electricity @ \$0.100/kW·h. Conventional practice is to put these on a common basis by expressing them in $\$\cdot GJ^{-1}$. For this purpose, assume gross heating values of 29 MJ·kg^{-1} for coal and 37 GJ·m^{-3} for gasoline.

(*a*) Rank order the three energy sources with respect to energy cost in $\$\cdot GJ^{-1}$.

(*b*) Explain the large disparity in the numerical results of part (*a*). Discuss the advantages and disadvantages of the three energy sources.

1.28. Chemical-plant equipment costs rarely vary in proportion to size. In the simplest case, cost C varies with size S according to the *allometric* equation

$$C = \alpha S^{\beta}$$

The size exponent β is typically between 0 and 1. For a wide variety of equipment types it is approximately 0.6.

(*a*) For $0 < \beta < 1$, show that cost per *unit size* decreases with increasing size. ("Economy of scale.")

(*b*) Consider the case of a spherical storage tank. The size is commonly measured by internal volume V_i^t. Show why one might expect that $\beta = 2/3$. On what parameters or properties would you expect quantity α to depend?

1.29. A laboratory reports the following vapor-pressure (P^{sat}) data for a particular organic chemical:

$t/°C$	P^{sat}/kPa
−18.5	3.18
−9.5	5.48
0.2	9.45
11.8	16.9
23.1	28.2
32.7	41.9
44.4	66.6
52.1	89.5
63.3	129.
75.5	187.

Correlate the data by fitting them to the Antoine equation:

$$\ln P^{\text{sat}}/\text{kPa} = A - \frac{B}{T/\text{K} + C}$$

That is, find numerical values of parameters A, B, and C by an appropriate regression procedure. Discuss the comparison of correlated with experimental values. What is the predicted normal boiling point (i.e. temperature at which the vapor pressure is 1(atm)) of this chemical?

Chapter 2

The First Law and Other Basic Concepts

In this chapter, we introduce and apply the first law of thermodynamics, one of the two fundamental laws upon which all of thermodynamics rests. Thus, in this chapter we:

- Introduce the concept of internal energy; i.e., energy stored within a substance
- Present the first law of thermodynamics, which reflects the observation that energy is neither created nor destroyed
- Develop the concepts of thermodynamic equilibrium, state functions, and the thermodynamic state of a system
- Develop the concept of reversible processes connecting equilibrium states
- Introduce *enthalpy*, another measure of energy stored within a substance, particularly useful in analyzing open systems
- Use heat capacities to relate changes in the internal energy and enthalpy of a substance to changes in its temperature
- Illustrate the construction of energy balances for open systems

2.1 JOULE'S EXPERIMENTS

The present-day concept of heat developed following crucial experiments carried out in the 1840s by James P. Joule.[1] In the most famous series of measurements, he placed known amounts of water, oil, or mercury in an insulated container and agitated the fluid with a rotating stirrer. The amounts of work done on the fluid by the stirrer and the resulting temperature changes of the fluid were accurately and precisely measured. Joule showed that for each fluid a fixed amount of work per unit mass was required for each degree of temperature rise caused by the stirring, and that the original temperature of the fluid was restored by the transfer of heat through simple

[1]http://en.wikipedia.org/wiki/James_Prescott_Joule. See also: Encyclopaedia Britannica, 1992, Vol. 28, p. 612.

contact with a cooler object. These experiments demonstrated the existence of a quantitative relationship between work and heat, and thereby showed that heat is a form of energy.

2.2 INTERNAL ENERGY

In experiments like those of Joule, energy added to a fluid as work is later transferred from the fluid as heat. Where does this energy reside after its addition to, and before its transfer from, the fluid? A rational answer to this question is that it is contained within the fluid in another form, which we call *internal energy*.

The internal energy of a substance does not include energy that it may possess as a result of its gross position or movement as a whole. Rather it refers to energy of the molecules comprising the substance. Because of their ceaseless motion, all molecules possess kinetic energy of translation (motion through space); except for monatomic substances, they also possess kinetic energy of rotation and of internal vibration. The addition of heat to a substance increases molecular motion, and thus causes an increase in the internal energy of the substance. Work done on the substance can have the same effect, as was shown by Joule. The internal energy of a substance also includes the potential energy associated with intermolecular forces. Molecules attract or repel one another, and potential energy is stored through these interactions, just as potential energy of configuration is stored in a compressed or stretched spring. On a submolecular scale, energy is associated with the interactions of electrons and nuclei of atoms, which includes the energy of chemical bonds that hold atoms together as molecules.

This energy is named *internal* to distinguish it from the kinetic and potential energy associated with a substance because of its macroscopic position, configuration, or motion, which can be thought of as *external* forms of energy.

Internal energy has no concise thermodynamic definition. It is a thermodynamic *primitive*. It cannot be directly measured; there are no internal-energy meters. As a result, absolute values are unknown. However, this is not a disadvantage in thermodynamic analysis because only *changes* in internal energy are required. In the context of classical thermodynamics, the details of how internal energy is stored are immaterial. This is the province of statistical thermodynamics, which treats the relationship of macroscopic properties such as internal energy to molecular motions and interactions.

2.3 THE FIRST LAW OF THERMODYNAMICS

Recognition of heat and internal energy as forms of energy makes possible the generalization of the principle of conservation of mechanical energy (Sec. 1.7) to include heat and internal energy in addition to work and external potential and kinetic energy. Indeed, the generalization can be extended to still other forms, such as surface energy, electrical energy, and magnetic energy. Overwhelming evidence of the validity of this generalization has raised its stature to that of a law of nature, known as the first law of thermodynamics. One formal statement is:

> **Although energy assumes many forms, the total quantity of energy is constant, and when energy disappears in one form it appears simultaneously in other forms.**

In application of this law to a given process, the sphere of influence of the process is divided into two parts, the *system* and its *surroundings*. The region in which the process occurs is set apart as the system; everything with which the system interacts is its surroundings. A system may be of any size; its boundaries may be real or imaginary, rigid or flexible. Frequently a system consists of a single substance; in other cases it may be complex. In any event, the equations of thermodynamics are written with reference to a well-defined system. This focuses attention on the particular process of interest and on the equipment and material directly involved in the process. However, the first law applies to the system *and* its surroundings; not to the system alone. For any process, the first law requires:

$$\Delta(\text{Energy of the system}) + \Delta(\text{Energy of surroundings}) = 0 \qquad (2.1)$$

where the difference operator "Δ" signifies finite changes in the quantities enclosed in parentheses. The system may change in its internal energy, in its potential or kinetic energy, and in the potential or kinetic energy of its finite parts.

In the context of thermodynamics, heat and work represent energy *in transit across the boundary* dividing the system from its surroundings, and are never *stored* or *contained* in the system. Potential, kinetic, and internal energy, on the other hand, reside with and are stored with matter. Heat and work represent *energy flows* to or from a system, while potential, kinetic, and internal energy represent *quantities of energy* associated with a system. In practice Eq. (2.1) assumes special forms suitable to specific applications. The development of these forms and their subsequent application are the subject of the remainder of this chapter.

2.4   ENERGY BALANCE FOR CLOSED SYSTEMS

If the boundary of a system does not permit the transfer of matter between the system and its surroundings, the system is said to be *closed*, and its mass is necessarily constant. The development of basic concepts in thermodynamics is facilitated by a careful examination of closed systems. For this reason they are treated in detail here. Far more important for industrial practice are processes in which matter crosses the system boundary as streams that enter and leave process equipment. Such systems are said to be *open*, and they are treated later in this chapter, once the necessary foundation material has been presented.

Because no streams enter or leave a closed system, no energy associated with matter is transported across the boundary that divides the system from its surroundings. All energy exchange between a closed system and its surroundings is in the form of heat or work, and the total energy change of the surroundings equals the net energy transferred to or from it as heat and work. The second term of Eq. (2.1) may therefore be replaced by variables representing heat and work, to yield

$$\Delta(\text{Energy of surroundings}) = \pm Q \pm W$$

Heat Q and work W always refer to the system, and the choice of sign for numerical values of these quantities depends on which direction of energy transfer with respect to the system is regarded as positive. We adopt the convention that makes the numerical values of both quantities positive for transfer *into* the system from the surroundings. The corresponding quantities taken with reference to the surroundings, Q_{surr} and W_{surr}, have the opposite sign, i.e., $Q_{surr} = -Q$ and $W_{surr} = -W$. With this understanding:

$$\Delta(\text{Energy of surroundings}) = Q_{surr} + W_{surr} = -Q - W$$

Equation (2.1) now becomes:[2]

$$\Delta(\text{Energy of the system}) = Q + W \qquad (2.2)$$

This equation states that the total energy change of a closed system equals the net energy transferred into it as heat and work.

Closed systems often undergo processes during which only the *internal* energy of the system changes. For such processes, Eq. (2.2) reduces to:

$$\boxed{\Delta U^t = Q + W} \qquad (2.3)$$

where U^t is the total internal energy of the system. Equation (2.3) applies to processes of *finite* change in the internal energy of the system. For *differential* changes in U^t:

$$\boxed{dU^t = dQ + dW} \qquad (2.4)$$

In Eqs. (2.3) and (2.4) the symbols Q, W, and U^t pertain to the entire system, which may be of any size, but must be clearly defined. All terms require expression in the same energy units. In the SI system the unit is the joule.

Total volume V^t and total internal energy U^t depend on the quantity of material in a system, and are called *extensive* properties. In contrast, temperature and pressure, the principal thermodynamic coordinates for pure homogeneous fluids, are independent of the quantity of material, and are known as *intensive* properties. For a homogeneous system, an alternative means of expression for the extensive properties, such as V^t and U^t, is:

$$V^t = mV \quad \text{or} \quad V^t = nV \qquad \text{and} \qquad U^t = mU \quad \text{or} \quad U^t = nU$$

where the plain symbols V and U represent the volume and internal energy of a unit amount of material, either a unit mass or a mole. These are *specific* or *molar* properties, respectively, and they are *intensive*, independent of the quantity of material actually present.

Although V^t and U^t for a homogeneous system of arbitrary size are extensive properties, specific and molar volume V and specific and molar internal energy U are intensive.

Note that the intensive coordinates T and P have no extensive counterparts.

For a closed system of n moles, Eqs. (2.3) and (2.4) may now be written:

$$\boxed{\Delta(nU) = n\,\Delta U = Q + W} \qquad (2.5)$$

$$\boxed{d(nU) = n\,dU = dQ + dW} \qquad (2.6)$$

---

[2]The sign convention used here is recommended by the International Union of Pure and Applied Chemistry. However, the original choice of sign for work and the one used in the first four editions of this text was the opposite, and the right side of Eq. (2.2) was then written $Q - W$.

In this form, these equations show explicitly the amount of substance comprising the system.

The equations of thermodynamics are often written for a representative unit amount of material, either a unit mass or a mole. Thus, for $n = 1$, Eqs. (2.5) and (2.6) become:

$$\Delta U = Q + W \qquad \text{and} \qquad dU = dQ + dW$$

The *basis* for Q and W is always implied by the mass or number of moles associated with the left side of the energy equation.

These equations do not provide a *definition* of internal energy. Indeed, they presume prior affirmation of the existence of internal energy, as expressed in the following axiom:

> *Axiom 1:* **There exists a form of energy, known as internal energy U, which is an intrinsic property of a system, functionally related to the measurable coordinates that characterize the system. For a closed system, not in motion, changes in this property are given by Eqs. (2.5) and (2.6).**

Equations (2.5) and (2.6) not only supply the means for calculation of *changes* in internal energy from experimental measurements, but they also enable us to derive further *property relations* that supply connections to readily measurable characteristics (e.g., temperature and pressure). Moreover, they have a dual purpose, because once internal-energy values are known, they provide for the calculation of heat and work quantities for practical processes. Having accepted the preceding axiom and associated definitions of a system and its surroundings, one may state the first law of thermodynamics concisely as a second axiom:

> *Axiom 2:* **(The First Law of Thermodynamics) The total energy of any system and its surroundings is conserved.**

These two axioms cannot be proven, nor can they be expressed in a simpler way. When changes in internal energy are computed in accord with Axiom 1, then Axiom 2 is universally observed to be true. The profound importance of these axioms is that they are the basis for formulation of energy balances applicable to a vast number of processes. Without exception, they predict the behavior of real systems.[3]

Example 2.1

The Niagara river, separating the United States from Canada, flows from Lake Erie to Lake Ontario. These lakes differ in elevation by about 100 m. Most of this drop occurs over Niagara Falls and in the rapids just above and below the falls, creating a natural opportunity for hydroelectric power generation. The Robert Moses hydroelectric power plant draws water from the river well above the falls and discharges it well below them. It has a peak capacity of 2,300,000 kW at a maximum water flow of 3,100,000 kg·s^{-1}. In the following, take 1 kg of water as the system.

---

[3]For a down-to-earth treatment designed to help the student over the very difficult early stages of an introduction to thermodynamics, see a short paperback by H. C. Van Ness, *Understanding Thermodynamics*; DoverPublications.com.

(a) What is the potential energy of the water flowing out of Lake Erie, relative to the surface of Lake Ontario?

(b) At peak capacity, what fraction of this potential energy is converted to electrical energy in the Robert Moses power plant?

(c) If the temperature of the water is unchanged in the overall process, how much heat flows to or from it?

Solution 2.1

(a) Gravitational potential energy is related to height by Eq. (1.8). With g equal to its standard value, this equation yields:

$$E_P = mzg = 1 \text{ kg} \times 100 \text{ m} \times 9.81 \text{ m·s}^{-2}$$
$$= 981 \text{ kg·m}^2 \cdot \text{s}^{-2} = 981 \text{ N·m} = 981 \text{ J}$$

(b) Recalling that $1 \text{ kW} = 1000 \text{ J·s}^{-1}$, we find the electrical energy generated per kg water is:

$$\frac{2.3 \times 10^6 \text{ kW}}{3.1 \times 10^6 \text{ kg·s}^{-1}} = 0.742 \text{ kW·s·kg}^{-1} = 742 \text{ J·kg}^{-1}$$

The fraction of the potential energy converted to electrical energy is $742/981 = 0.76$.

This conversion efficiency would be higher but for the dissipation of potential energy in the flow upstream and downstream of the power plant.

(c) If the water leaves the process at the same temperature at which it enters, then its internal energy is unchanged. Neglecting also any change in kinetic energy, we write the first law, in the form of Eq. (2.2), as

$$\Delta(\text{Energy of the system}) = \Delta E_p = Q + W$$

For each kilogram of water, $W = -742 \text{ J}$ and $\Delta E_P = -981 \text{ J}$. Then

$$Q = \Delta E_p - W = -981 + 742 = -239 \text{ J}$$

This is heat lost from the system.

Example 2.2

A typical industrial-scale wind turbine has a peak efficiency of about 0.44 for a wind speed of 9 m·s⁻¹. That is, it converts about 44% of the kinetic energy of the wind approaching it into usable electrical energy. The total air flow impinging on such a turbine with a rotor diameter of 43 m is about 15,000 kg·s⁻¹ for the given wind speed.

(*a*) How much electrical energy is produced when 1 kg of air passes through the turbine?

(*b*) What is the power output of the turbine?

(*c*) If there is no heat transferred to the air, and if its temperature remains unchanged, what is its change in speed upon passing through the turbine?

Solution 2.2

(*a*) The kinetic energy of the wind on the basis of 1 kg of air is:

$$E_{K_1} = \frac{1}{2}mu^2 = \frac{(1 \text{ kg})(9 \text{ m·s}^{-1})^2}{2} = 40.5 \text{ kg·m}^2\text{·s}^{-2} = 40.5 \text{ J}$$

Thus, the electrical energy produced per kilogram of air is $0.44 \times 40.5 = 17.8$ J.

(*b*) The power output is:

$$17.8 \text{ J·kg}^{-1} \times 15,000 \text{ kg·s}^{-1} = 267,000 \text{ J·s}^{-1} = 267 \text{ kW}$$

(*c*) If the temperature and pressure of the air are unchanged, then its internal energy is unchanged. Changes in gravitational potential energy can also be neglected. Thus, with no heat transfer, the first law becomes

$$\Delta(\text{Energy of the system}) = \Delta E_K = E_{K_2} - E_{K_1} = W = -17.8 \text{ J·kg}^{-1}$$

$$E_{K_2} = 40.5 - 17.8 = 22.7 \text{ J·kg}^{-1} = 22.7 \text{ N·m·kg}^{-1} = 22.7 \text{ m}^2\text{·s}^{-2}$$

$$E_{K_2} = \frac{u_2^2}{2} = 22.7 \text{ m}^2\text{·s}^{-2} \qquad \text{and} \qquad u_2 = 6.74 \text{ m·s}^{-1}$$

The decrease in air speed is: $9.00 - 6.74 = 2.26$ m·s^{-1}.

2.5   EQUILIBRIUM AND THE THERMODYNAMIC STATE

Equilibrium is a word denoting a static condition, the absence of change. In thermodynamics it means not only the absence of change but the absence of any *tendency* toward change on a macroscopic scale. Because any tendency toward change is caused by a driving force of one kind or another, the absence of such a tendency indicates also the absence of any driving force. Hence, for a system at equilibrium all forces are in exact balance.

Different kinds ôf driving forces tend to bring about different kinds of change. For example, imbalance of mechanical forces such as pressure on a piston tend to cause energy transfer as work; temperature differences tend to cause the flow of heat; differences in chemical potential[4] tend to cause substances to be transferred from one phase to another. At equilibrium all such forces are in balance.

---

[4]Chemical potential is a thermodynamic property treated in Chapter 10.

Whether a change actually occurs in a system *not* at equilibrium depends on resistance as well as on driving force. Systems subject to appreciable driving forces may change at a negligible rate if the resistance to change is very large. For example, a mixture of hydrogen and oxygen at ordinary conditions is not in chemical equilibrium, because of the large driving force for the formation of water. However, if chemical reaction is not initiated, this system may exist in long-term thermal and mechanical equilibrium, and purely physical processes may be analyzed without regard to possible chemical reaction.

Likewise, living organisms are inherently far from overall thermodynamic equilibrium. They are constantly undergoing dynamic changes governed by the *rates* of competing bio-chemical reactions, which are outside the scope of thermodynamic analysis. Nonetheless, many local equilibria within organisms are amenable to thermodynamic analysis. Examples include the denaturing (unfolding) of proteins and the binding of enzymes to their substrates.

The systems most commonly found in chemical technology are fluids, for which the primary characteristics (properties) are temperature T, pressure P, specific or molar volume V, and composition. Such systems are known as *PVT* systems. They exist at *internal* equilibrium when their properties are uniform throughout the system, and conform to the following axiom:

Axiom 3: The macroscopic properties of a homogeneous *PVT* system at internal equilibrium can be expressed as a function of its temperature, pressure, and composition.

This axiom prescribes an idealization, a model that excludes the influence of fields (e.g., electric, magnetic, and gravitational) as well as surface effects and other less common effects. It is entirely satisfactory in a multitude of practical applications.

Associated with the concept of internal equilibrium is the concept of *thermodynamic state* for which a *PVT* system has a set of identifiable and reproducible properties, including not only P, V, and T, but also internal energy. However, the notation of Eqs. (2.3) through (2.6) suggests that the internal energy terms on the left are different in kind from the quantities on the right. Those on the left reflect *changes* in the thermodynamic state of the system as reflected by its properties. For a homogeneous pure substance we know from experience that fixing two of these properties also fixes all the others, and thus determines its thermodynamic state. For example, nitrogen gas at a temperature of 300 K and a pressure of 10^5 Pa (1 bar) has a fixed specific volume or density and a fixed molar internal energy. Indeed, it has a complete set of intensive thermodynamic properties. If this gas is heated or cooled, compressed or expanded, and then returned to its initial temperature and pressure, its intensive properties are restored to their initial values. They do not depend on the past history of the substance nor on the means by which it reaches a given state. They depend only on present conditions, however reached. Such quantities are known as *state functions*. For a homogeneous pure substance, if two state functions are held at fixed values the *thermodynamic state* of the substance is fully determined.[5] This means that a state function, such as specific internal energy, is a property that always has a value; it may therefore be expressed mathematically as a function of such coordinates as temperature and pressure, or temperature and density, and its values may be identified with points on a graph.

---

[5]For systems of greater complexity, the number of state functions that must be specified in order to define the state of the system may be different from two. The method of determining this number is found in Sec. 3.1.

On the other hand, the terms on the right sides of Eqs. (2.3) through (2.6), representing heat and work quantities, are not properties; they account for the energy changes that occur in the surroundings. They depend on the nature of the process, and they may be associated with areas rather than points on a graph, as suggested by Fig. 1.3. Although time is not a thermodynamic coordinate, the passage of time is inevitable whenever heat is transferred or work is accomplished.

The differential of a state function represents an infinitesimal *change* in its value. Integration of such a differential results in a finite difference between two of its values, e.g.:

$$\int_{V_1}^{V_2} dV = V_2 - V_1 = \Delta V \qquad \text{and} \qquad \int_{U_1}^{U_2} dU = U_2 - U_1 = \Delta U$$

The differentials of heat and work are not *changes*, but are infinitesimal *amounts*. When integrated, these differentials give not finite changes, but finite amounts. Thus,

$$\int dQ = Q \qquad \text{and} \qquad \int dW = W$$

For a closed system undergoing the same change in state by several processes, experiment shows that the amounts of heat and work required differ for different processes, but that the sum $Q + W$ [Eqs. (2.3) and (2.5)] is the same for all processes.

This is the basis for the identification of internal energy as a state function. The same value of ΔU^t is given by Eq. (2.3) regardless of the process, provided only that the change in the system is between the same initial and final states.

_____

Example 2.3

A gas is confined in a cylinder by a piston. The initial pressure of the gas is 7 bar, and the volume is 0.10 m³. The piston is held in place by latches.

(a) The whole apparatus is placed in a total vacuum. What is the energy change of the apparatus if the restraining latches are removed so that the gas suddenly expands to double its initial volume, the piston striking other latches at the end of the process?

(b) The process described in (a) is repeated, but in air at 101.3 kPa, rather than in a vacuum. What is the energy change of the apparatus? Assume the rate of heat exchange between the apparatus and the surrounding air is slow compared with the rate at which the process occurs.

Solution 2.3

Because the question concerns the entire apparatus, the system is taken as the gas, piston, and cylinder.

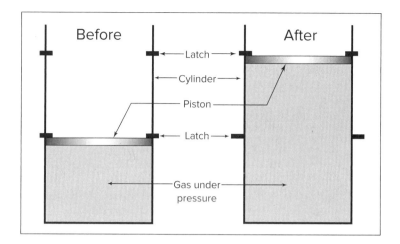

(*a*) No work is done during the process, because no force external to the system moves, and no heat is transferred through the vacuum surrounding the apparatus. Hence Q and W are zero, and the total energy of the system does not change. Without further information we can say nothing about the distribution of energy among the parts of the system. This may be different than the initial distribution.

(*b*) Here, work is done by the system in pushing back the atmosphere. It is evaluated as the product of the force of atmospheric pressure on the back side of the piston, $F = P_{atm}A$, and the displacement of the piston, $\Delta l = \Delta V^t/A$, where A is the area of the piston and ΔV^t is the volume change of the gas. This is work done by the system on the surroundings, and is a negative quantity; thus,

$$W = -F\,\Delta l = -P_{atm}\,\Delta V^t = -(101.3)(0.2 - 0.1)\ \text{kPa·m}^3 = -10.13\frac{\text{kN}}{\text{m}^2}\text{·m}^3$$

or

$$W = -10.13\ \text{kN·m} = -10.13\ \text{kJ}$$

Heat transfer between the system and surroundings is also possible in this case, but the problem is worked for the instant after the process has occurred and before appreciable heat transfer has had time to take place. Thus Q is assumed to be zero in Eq. (2.2), giving:

$$\Delta(\text{Energy of the system}) = Q + W = 0 - 10.13 = -10.13\ \text{kJ}$$

The total energy of the system has *decreased* by an amount equal to the work done on the surroundings.

Example 2.4

When a system is taken from state a to state b in the accompanying figure along path acb, 100 J of heat flows into the system and the system does 40 J of work.

    (a) How much heat flows into the system along path aeb if the work done by the system is 20 J?

    (b) The system returns from b to a along path bda. If the work done on the system is 30 J, does the system absorb or liberate heat? How much?

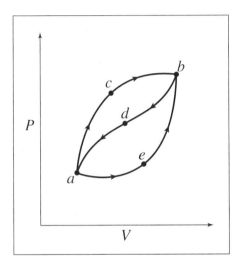

Solution 2.4

Assume that the system changes only in its internal energy and that Eq. (2.3) is applicable. For path acb, and thus for *any* path leading from a to b,

$$\Delta U_{ab}^t = Q_{acb} + W_{acb} = 100 - 40 = 60 \text{ J}$$

(a) For path aeb,

$$\Delta U_{ab}^t = 60 = Q_{aeb} + W_{aeb} = Q_{aeb} - 20 \qquad \text{whence} \qquad Q_{aeb} = 80 \text{ J}$$

(b) For path bda,

$$\Delta U_{ba}^t = -\Delta U_{ab}^t = -60 = Q_{bda} + W_{bda} = Q_{bda} + 30$$

and

$$Q_{bda} = -60 - 30 = -90 \text{ J}$$

Heat is therefore transferred from the system to the surroundings.

2.6 THE REVERSIBLE PROCESS

The development of thermodynamics is facilitated by the introduction of a special kind of closed-system process characterized as *reversible*:

A process is reversible when its direction can be reversed at any point by an infinitesimal change in external conditions.

The reversible process is ideal in that it produces a best possible result; it yields the minimum work input required or maximum work output attainable from a specified process. It represents a limit to the performance of an actual process that is never fully realized. Work is often calculated for hypothetical reversible processes, because the choice is between this calculation and no calculation at all. The reversible work as the limiting value may then be combined with an appropriate *efficiency* to yield a reasonable approximation to the work required for or produced by an actual process.[6]

The concept of the reversible processes also plays a key role in the derivation of thermodynamic relationships. In this context, we often compute changes in thermodynamic state functions along the path of a hypothetical reversible process. If the result is a relationship involving only state functions, then this relationship is valid for *any* process that results in the same *change of state*. Indeed, the primary use of the reversible process concept is for derivation of generally valid relationships among state functions.

Reversible Expansion of a Gas

The piston in Fig. 2.1 confines gas at a pressure just sufficient to balance the weight of the piston and all that it supports. In this equilibrium condition the system has no tendency to change. Imagine that mass m is slid from the piston to a shelf (at the same level). The piston assembly accelerates upward, reaching maximum velocity when the upward force on the piston just balances its weight. Momentum then carries it to a higher level, where it reverses direction. If the piston were held in the position of maximum elevation, its potential-energy increase would very nearly equal the expansion work done by the gas. However, when unconstrained, the piston assembly oscillates, with decreasing amplitude, ultimately coming to rest at a new equilibrium position Δl above the old.

Oscillations are damped out because the viscous nature of the gas gradually converts gross directed motion of the molecules into chaotic molecular motion. This *dissipative* process transforms some of the work done by the gas in raising the piston into internal energy of the gas. Once the process is initiated, no *infinitesimal* change in external conditions can reverse it; the process is *irreversible*.

The dissipative effects of the process have their origin in the sudden removal of a finite mass from the piston. The resulting imbalance of forces acting on the piston causes its acceleration and leads to its subsequent oscillation. The sudden removal of smaller mass increments reduces, but does not eliminate, this dissipative effect. Even the removal of an infinitesimal mass leads to piston oscillations of infinitesimal amplitude and a consequent dissipative effect. However, one may *imagine* a process wherein small mass increments are removed one after

---

[6]Quantitative analysis of the relationships between efficiency and irreversibility requires use of the second law of thermodynamics, and is treated in Chapter 5.

Figure 2.1: Expansion of a gas. The nature of reversible processes is illustrated by the expansion of gas in an idealized piston/cylinder arrangement. The apparatus shown is imagined to exist in an evacuated space. The gas trapped inside the cylinder is chosen as the system; all else is the surroundings. Expansion results when mass is removed from the piston. For simplicity, assume that the piston slides within the cylinder without friction and that the piston and cylinder neither absorb nor transmit heat. Moreover, because the density of the gas in the cylinder is low and its mass is small, we ignore the effects of gravity on the contents of the cylinder. This means that gravity-induced pressure gradients in the gas are very small relative to its total pressure and that changes in potential energy of the gas are negligible in comparison with the potential-energy changes of the piston assembly.

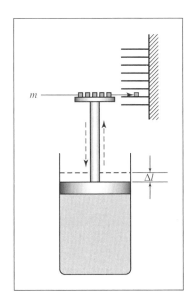

another at a rate such that the piston's rise is continuous, with minute oscillation only at the end of the process.

   The limiting case of removal of a succession of infinitesimal masses from the piston is approximated when the masses m in Fig. 2.1 are replaced by a pile of powder, blown in a very fine stream from the piston. During this process, the piston rises at a uniform but very slow rate, and the powder collects in storage at ever higher levels. The system is never more than differentially displaced from internal equilibrium or from equilibrium with its surroundings. If the removal of powder from the piston is stopped and the direction of transfer of powder is reversed, the process reverses direction and proceeds backwards along its original path. Both the system and its surroundings are ultimately restored to virtually their initial conditions. The original process approaches *reversibility*.

   Without the assumption of a frictionless piston, we cannot imagine a reversible process. If the piston sticks, a finite mass must be removed before the piston breaks free. Thus the equilibrium condition necessary to reversibility is not maintained. Moreover, friction between two sliding parts is a mechanism for the dissipation of mechanical energy into internal energy.

   This discussion has centered on a single closed-system process, the expansion of a gas in a cylinder. The opposite process, compression of a gas in a cylinder, is described in exactly the same way. There are, however, many processes that are driven by an imbalance of other than mechanical forces. For example, heat flow occurs when a temperature difference exists, electricity flows under the influence of an electromotive force, and chemical reactions occur in response to driving forces that arise from differences in the strengths and configurations of chemical bonds in molecules. The driving forces for chemical reactions and for transfer of substances between phases are complex functions of temperature, pressure, and composition, as will be described in detail in later chapters. In general, a process is reversible when the net force driving it is infinitesimal in size. Thus heat is transferred reversibly when it flows from a finite object at temperature T to another such object at temperature $T - dT$.

Reversible Chemical Reaction

The concept of a reversible chemical reaction is illustrated by the decomposition of solid calcium carbonate to form solid calcium oxide and carbon dioxide gas. At equilibrium, this system exerts a specific decomposition pressure of CO_2 for a given temperature. The chemical reaction is held in balance (in equilibrium) by the pressure of the CO_2. Any change of conditions, however slight, upsets the equilibrium and causes the reaction to proceed in one direction or the other.

If the mass m in Fig. 2.2 is minutely increased, the CO_2 pressure rises, and CO_2 combines with CaO to form $CaCO_3$, allowing the weight to fall. The heat given off by this reaction raises the temperature in the cylinder, and heat flows to the bath. Decreasing the weight sets off the opposite chain of events. The same results are obtained if the temperature of the bath is raised or lowered. Raising the bath temperature slightly causes heat transfer into the cylinder, and calcium carbonate decomposes. The CO_2 generated causes the pressure to rise, which in turn raises the piston and weight. This continues until the $CaCO_3$ is completely decomposed. A lowering of the bath temperature causes the system to return to its initial state. The imposition of differential changes causes only minute displacements of the system from equilibrium, and the resulting process is exceedingly slow and reversible.

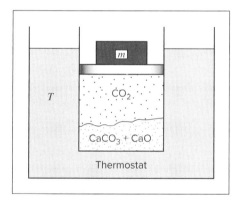

Figure 2.2: Reversibility of a chemical reaction. The cylinder is fitted with a frictionless piston and contains $CaCO_3$, CaO, and CO_2 in equilibrium. It is immersed in a constant-temperature bath, and thermal equilibrium assures equality of the system temperature with that of the bath. The temperature is adjusted to a value such that the decomposition pressure is just sufficient to balance the weight on the piston, a condition of mechanical equilibrium.

Some chemical reactions can be carried out in an electrolytic cell, and in this case they may be held in balance by an applied potential difference. For example, when a cell consisting of two electrodes, one of zinc and the other of platinum, is immersed in an aqueous solution of hydrochloric acid, the reaction that occurs is:

$$Zn + 2HCl \rightleftharpoons H_2 + ZnCl_2$$

The cell is held under fixed conditions of temperature and pressure, and the electrodes are connected externally to a potentiometer. If the electromotive force (emf) produced by the cell is exactly balanced by the potential difference of the potentiometer, the reaction is held in equilibrium. The reaction may be made to proceed in the forward direction by a slight decrease in the opposing potential difference, and it may be reversed by a corresponding increase in the potential difference above the emf of the cell.

Summary Remarks on Reversible Processes

A reversible process:

- Can be reversed at any point by an infinitesimal change in external conditions
- Is never more than minutely removed from equilibrium
- Traverses a succession of equilibrium states
- Is frictionless
- Is driven by forces whose imbalance is infinitesimal in magnitude
- Proceeds infinitely slowly
- When reversed, retraces its path, restoring the initial state of system and surroundings

Computing Work for Reversible Processes

Equation (1.3) gives the work of compression or expansion of a gas caused by the displacement of a piston in a cylinder:

$$dW = -P\,dV^t \tag{1.3}$$

The work done on the system is in fact given by this equation only when certain characteristics of the reversible process are realized. The first requirement is that the system be no more than infinitesimally displaced from a state of *internal* equilibrium, characterized by uniformity of temperature and pressure. The system then has an identifiable set of properties, including pressure P. The second requirement is that the system be no more than infinitesimally displaced from mechanical equilibrium with its surroundings. In this event, the internal pressure P is never more than minutely out of balance with the external force, and we may make the substitution $F = PA$ that transforms Eq. (1.2) into Eq. (1.3). Processes for which these requirements are met are said to be *mechanically reversible*, and Eq. (1.3) may be integrated:

$$W = -\int_{V_1^t}^{V_2^t} P\,dV^t \tag{1.4}$$

This equation gives the work for the mechanically reversible expansion or compression of a fluid in a piston/cylinder arrangement. Its evaluation clearly depends on the relation between P and V^t, i.e., on the "path" of the process, which must be specified. To find the work of an *irreversible* process for the same change in V^t, one needs an *efficiency*, which relates the actual work to the reversible work.

Example 2.5

A horizontal piston/cylinder arrangement is placed in a constant-temperature bath. The piston slides in the cylinder with negligible friction, and an external force holds it in place against an initial gas pressure of 14 bar. The initial gas volume is 0.03 m³. The external force on the piston is reduced gradually, and the gas expands isothermally as its volume doubles. If the volume of the gas is related to its pressure so that PV^t is constant, what is the work done by the gas in moving the external force?

Solution 2.5

The process as described is mechanically reversible, and Eq. (1.4) is applicable. If $PV^t = k$, a constant, then $P = k/V^t$. This specifies the path of the process, and leads to

$$W = -\int_{V_1^t}^{V_2^t} P\,dV^t = -k\int_{V_1^t}^{V_2^t} \frac{dV^t}{V^t} = -k\ln\frac{V_2^t}{V_1^t}$$

The value of k is given by:

$$k = PV^t = P_1 V_1^t = 14\times 10^5\,\text{Pa}\times 0.03\,\text{m}^3 = 42{,}000\,\text{J}$$

With $V_1^t = 0.03\ \text{m}^3$ and $V_2^t = 0.06\ \text{m}^3$,

$$W = -42{,}000\ln 2 = -29{,}112\,\text{J}$$

The final pressure is

$$P_2 = \frac{k}{V_2^t} = \frac{42{,}000}{0.06} = 700{,}000\,\text{Pa} \qquad \text{or} \qquad 7\,\text{bar}$$

Were the efficiency of such processes known to be about 80%, we could multiply the reversible work by this figure to get an estimate of the irreversible work, namely −23,290 J.

2.7 CLOSED-SYSTEM REVERSIBLE PROCESSES; ENTHALPY

We present here the analysis of closed-system mechanically reversible processes—not that such processes are common. Indeed they are of little interest for practical application. Their value lies in the simplicity they provide for the calculation of changes in state functions for a specific change of state. For a complex industrial process that brings about a particular change of state, the calculation of changes in state functions are not made for the path of the actual process. Rather, they are made for a simple closed-system reversible process that brings about the same change of state. This is possible because changes in state functions are independent of process. The closed-system mechanically reversible process is useful and important for this purpose, even though close approximations to such hypothetical processes are not often encountered in practice.

For 1 mole of a homogeneous fluid contained in a closed system, the energy balance of Eq. (2.6) is written:

$$dU = dQ + dW$$

The work for a mechanically reversible, closed-system process is given by Eq. (1.3), here written: $dW = -PdV$. Substitution into the preceding equation yields:

$$dU = dQ - PdV \tag{2.7}$$

This is the general energy balance for one mole or a unit mass of homogeneous fluid in a closed system undergoing a mechanically reversible process.

For a constant-volume change of state, the only possible mechanical work is that associated with stirring or mixing, which is excluded because it is inherently irreversible. Thus,

$$dU = dQ \quad (\text{const } V) \tag{2.8}$$

Integration yields:

$$\Delta U = Q \quad (\text{const } V) \tag{2.9}$$

The internal energy change for a mechanically reversible, constant-volume, closed-system process equals the amount of heat transferred into the system.

For a constant-pressure change of state:

$$dU + PdV = d(U + PV) = dQ$$

The group $U + PV$ naturally arises here and in many other applications. This suggests the definition, for convenience, of this combination as a new thermodynamic property. Thus, the mathematical (and only) **definition** of enthalpy[7] is:

$$\boxed{H \equiv U + PV} \tag{2.10}$$

where H, U, and V are molar or unit-mass values. The preceding energy balance becomes:

$$dH = dQ \quad (\text{const } P) \tag{2.11}$$

Integration yields:

$$\Delta H = Q \quad (\text{const } P) \tag{2.12}$$

The enthalpy change in a mechanically reversible, constant-pressure, closed-system process equals the amount of heat transferred into the system. Comparison of Eqs. (2.11) and (2.12) with Eqs. (2.8) and (2.9) shows that the enthalpy plays a role in constant-pressure processes analogous to the internal energy in constant-volume processes.

These equations suggest the usefulness of *enthalpy*, but its greatest use becomes fully apparent with its appearance in energy balances for *flow processes* as applied to heat exchangers, chemical and biochemical reactors, distillation columns, pumps, compressors, turbines, engines, etc., for calculation of heat and work.

The tabulation of Q and W for the infinite array of conceivable processes is impossible. The intensive *state functions*, however, such as molar or specific volume, internal energy, and enthalpy, are intrinsic properties of matter. Once determined for a particular substance, their values can be tabulated as functions of T and P for future use in the calculation of Q and W

---

[7]Originally and most properly pronounced en-**thal**′-py to distinguish it clearly from *entropy*, a property introduced in Chapter 5, and pronounced **en**′-tro-py. The word *enthalpy* was proposed by H. Kamerlingh Onnes, who won the 1913 Nobel Prize in physics (see: http://nobelprize.org/nobel_prizes/physics/laureates/1913/onnes-bio.html).

for any process involving that substance. The determination of numerical values for these state functions and their correlation and use are treated in later chapters.

All terms of Eq. (2.10) must be expressed in the same units. The product PV has units of energy per mole or per unit mass, as does U; therefore H also has units of energy per mole or per unit mass. In the SI system the basic unit of pressure is the pascal ($=1$ $N \cdot m^{-2}$), and that of molar volume is cubic meters per mol ($=1$ $m^3 \cdot mol^{-1}$). For the PV product we have 1 $N \cdot m \cdot mol^{-1} = 1$ $J \cdot mol^{-1}$.

Because U, P, and V are all state functions, H as defined by Eq. (2.10) is also a state function. Like U and V, H is an intensive property of matter. The differential form of Eq. (2.10) is:

$$dH = dU + d(PV) \qquad (2.13)$$

This equation applies for any differential change of state. Upon integration, it becomes an equation for a finite change of state:

$$\Delta H = \Delta U + \Delta(PV) \qquad (2.14)$$

Equations (2.10), (2.13), and (2.14) apply to a unit mass or mole of a substance.

Example 2.6

Calculate ΔU and ΔH for 1 kg of water when it is vaporized at the constant temperature of 100°C and the constant pressure of 101.33 kPa. The specific volumes of liquid and vapor water at these conditions are 0.00104 and 1.673 $m^3 \cdot kg^{-1}$, respectively. For this change, heat in the amount of 2256.9 kJ is added to the water.

Solution 2.6

We take the 1 kg of water as the system because it alone is of interest, and we imagine it contained in a cylinder by a frictionless piston that exerts a constant pressure of 101.33 kPa. As heat is added, the water evaporates, expanding from its initial to its final volume. Equation (2.12) as written for the 1 kg system is:

$$\Delta H = Q = 2256.9 \text{ kJ}$$

By Eq. (2.14),

$$\Delta U = \Delta H - \Delta(PV) = \Delta H - P \, \Delta V$$

For the final term:

$$P \, \Delta V = 101.33 \text{ kPa} \times (1.673 - 0.001) \text{ m}^3$$
$$= 169.4 \text{ kPa} \cdot \text{m}^3 = 169.4 \text{ kN} \cdot \text{m}^{-2} \cdot \text{m}^3 = 169.4 \text{ kJ}$$

Then

$$\Delta U = 2256.9 - 169.4 = 2087.5 \text{ kJ}$$

2.8   HEAT CAPACITY

Our recognition of heat as energy in transit was preceded historically by the idea that gases, liquids, and solids have *capacity* for heat. The smaller the temperature change caused in a substance by the transfer of a given quantity of heat, the greater its capacity. Indeed, a *heat capacity* might be defined as $C \equiv dQ/dT$. The difficulty with this is that it makes C, like Q, a process-dependent quantity rather than a state function. However, it does suggest the definition of two quantities that, although they retain this outmoded name, are in fact state functions, unambiguously related to other state functions. The discussion here is preliminary to more complete treatment in Chapter 4.

Heat Capacity at Constant Volume

The constant-volume heat capacity of a substance is **defined** as:

$$C_V \equiv \left(\frac{\partial U}{\partial T} \right)_V \tag{2.15}$$

Observe carefully the notation used here with the partial derivative. The parentheses and subscript V indicate that the derivative is taken with volume held constant; i.e., U is considered a function of T and V. This notation is widely used in this text and more generally in thermodynamics. It is needed because thermodynamic state functions, like U, can be written as functions of different sets of independent variables. Thus, we can write $U(T, V)$ and $U(T, P)$. Ordinarily in multivariable calculus, a set of independent variables is unambiguous, and a partial derivative with respect to one variable implies constancy of the others. Because thermodynamics reflects physical reality, one may deal with alternative sets of independent variables, introducing ambiguity unless the variables being held constant are explicitly specified.

The definition of Eq. (2.15) accommodates both the molar heat capacity and the specific heat capacity (usually called specific heat), depending on whether U is the molar or specific internal energy. Although this definition makes no reference to any process, it relates in an especially simple way to a constant-volume process in a closed system, for which Eq. (2.15) may be written:

$$dU = C_V \, dT \quad \text{(const } V) \tag{2.16}$$

Integration yields:

$$\Delta U = \int_{T_1}^{T_2} C_V \, dT \quad \text{(const } V) \tag{2.17}$$

This result with Eq. (2.9) for a mechanically reversible, constant-volume process (conditions that preclude stirring work) gives:

$$Q = \Delta U = \int_{T_1}^{T_2} C_V \, dT \quad \text{(const } V) \tag{2.18}$$

If the volume varies during the process but returns at the end of the process to its initial value, the process cannot rightly be called one of constant volume, even though $V_2 = V_1$ and $\Delta V = 0$. However, changes in state functions are fixed by the initial and final conditions, independent of path, and may therefore be calculated by equations for a truly constant-volume process regardless of the actual process. Equation (2.17) therefore has *general* validity, because U, C_V, T, and V are all state functions. On the other hand, Q and W depend on path. Thus, Eq. (2.18) is a valid expression for Q, and W is in general zero, only for a *constant-volume* process. This is the reason for emphasizing the distinction between state functions and path-dependent quantities such as Q and W. The principle that state functions are path- and process-independent is an essential concept in thermodynamics.

> **For the calculation of property changes, but not for Q and W, an actual process may be replaced by any other process that accomplishes the same change in state. The choice is made based on convenience, with simplicity a great advantage.**

Heat Capacity at Constant Pressure

The constant-pressure heat capacity is **defined** as:

$$C_P \equiv \left(\frac{\partial H}{\partial T}\right)_P \tag{2.19}$$

Again, the definition accommodates both molar and specific heat capacities, depending on whether H is the molar or specific enthalpy. This heat capacity relates in an especially simple way to a constant-pressure, closed-system process, for which Eq. (2.19) is equally well written:

$$dH = C_P \, dT \quad (\text{const } P) \tag{2.20}$$

whence

$$\Delta H = \int_{T_1}^{T_2} C_P \, dT \quad (\text{const } P) \tag{2.21}$$

For a mechanically reversible, constant-P process, this result may be combined with Eq. (2.12):

$$Q = \Delta H = \int_{T_1}^{T_2} C_P \, dT \quad (\text{const } P) \tag{2.22}$$

Because H, C_P, T, and P are state functions, Eq. (2.21) applies to any process for which $P_2 = P_1$ whether or not it is actually carried out at constant pressure. However, only for the mechanically reversible, constant-pressure process can the amount of heat transferred be calculated by Eq. (2.22) and work by Eq. (1.3), here written for 1 mole, $W = -P \, \Delta V$.

Example 2.7

Air at 1 bar and 298.15 K is compressed to 3 bar and 298.15 K by two different closed-system mechanically reversible processes:

(*a*) Cooling at constant pressure followed by heating at constant volume.

(*b*) Heating at constant volume followed by cooling at constant pressure.

Calculate the heat and work requirements and ΔU and ΔH of the air for each path. The following heat capacities for air may be assumed independent of temperature:

$$C_V = 20.785 \quad \text{and} \quad C_P = 29.100 \text{ J·mol}^{-1}\text{·K}^{-1}$$

Assume also that air remains a gas for which PV/T is a constant, regardless of the changes it undergoes. At 298.15 K and 1 bar the molar volume of air is 0.02479 m³· mol⁻¹.

Solution 2.7

In each case take the system as 1 mol of air contained in an imaginary piston/cylinder arrangement. Because the processes are mechanically reversible, the piston is imagined to move in the cylinder without friction. The final volume is:

$$V_2 = V_1\frac{P_1}{P_2} = 0.02479\left(\frac{1}{3}\right) = 0.008263 \text{ m}^3$$

The two paths are shown on the V vs. P diagram of Fig. 2.3(I) and on the T vs. P diagram of Fig. 2.3(II).

Figure 2.3: V vs. P and T vs. P diagrams for Ex. 2.7.

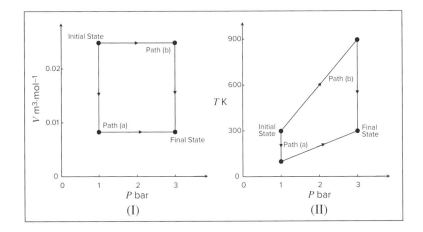

(a) During the first step of this path, air is cooled at the constant pressure of 1 bar until the final volume of 0.008263 m³ is reached. The temperature of the air at the end of this cooling step is:

$$T' = T_1\frac{V_2}{V_1} = 298.15\left(\frac{0.008263}{0.02479}\right) = 99.38 \text{ K}$$

Thus, for the first step,

$$Q = \Delta H = C_P \Delta T = (29.100)(99.38 - 298.15) = -5784 \text{ J}$$
$$W = -P \Delta V = -1 \times 10^5 \text{ Pa} \times (0.008263 - 0.02479) \text{ m}^3 = 1653 \text{ J}$$
$$\Delta U = \Delta H - \Delta(PV) = \Delta H - P \Delta V = -5784 + 1653 = -4131 \text{ J}$$

The second step is at constant V_2 with heating to the final state. Work $W = 0$, and for this step:

$$\Delta U = Q = C_V \Delta T = (20.785)(298.15 - 99.38) = 4131 \text{ J}$$
$$V \Delta P = 0.008263 \text{ m}^3 \times (2 \times 10^5) \text{ Pa} = 1653 \text{ J}$$
$$\Delta H = \Delta U + \Delta(PV) = \Delta U + V \Delta P = 4131 + 1653 = 5784 \text{ J}$$

For the overall process:

$$Q = -5784 + 4131 = -1653 \text{ J}$$
$$W = 1653 + 0 = 1653 \text{ J}$$
$$\Delta U = -4131 + 4131 = 0$$
$$\Delta H = -5784 + 5784 = 0$$

Notice that the first law, $\Delta U = Q + W$, applied to the overall process is satisfied.

(b) Two different steps of this path produce the same final state of the air. In the first step air is heated at a constant volume equal to V_1 until the final pressure of 3 bar is reached. The air temperature at the end of this step is:

$$T' = T_1 \frac{P_2}{P_1} = 298.15 \left(\frac{3}{1}\right) = 894.45 \text{ K}$$

For this first constant-volume step, $W = 0$, and

$$Q = \Delta U = C_V \Delta T = (20.785)(894.45 - 298.15) = 12,394 \text{ J}$$
$$V \Delta P = (0.02479)(2 \times 10^5) = 4958 \text{ J}$$
$$\Delta H = \Delta U + V \Delta P = 12,394 + 4958 = 17,352 \text{ J}$$

In the second step air is cooled at $P = 3$ bar to its final state:

$$Q = \Delta H = C_P \Delta T = (29.10)(298.15 - 894.45) = -17,352 \text{ J}$$
$$W = -P \Delta V = -(3 \times 10^5)(0.008263 - 0.02479) = 4958 \text{ J}$$
$$\Delta U = \Delta H - \Delta(PV) = \Delta H - P \Delta V = -17,352 + 4958 = -12,394 \text{ J}$$

For the two steps combined,

$$Q = 12,394 - 17,352 = -4958 \text{ J}$$
$$W = 0 + 4958 = 4958 \text{ J}$$
$$\Delta U = 12,394 - 12,394 = 0$$
$$\Delta H = 17,352 - 17,352 = 0$$

This example illustrates that changes in state functions (ΔU and ΔH) are independent of path for given initial and final states. On the other hand, Q and W depend on the path. Note also that the total changes in ΔU and ΔH are zero. This is because the input information provided makes U and H functions of temperature only, and $T_1 = T_2$. While the processes of this example are not of practical interest, state-function changes (ΔU and ΔH) for actual flow processes *are* calculated as illustrated in this example for processes that *are* of practical interest. This is possible because the state-function changes are the same for a reversible process, like the ones used here, as for a real process that connects the same states.

Example 2.8

Calculate the internal energy and enthalpy changes resulting if air changes from an initial state of 5°C and 10 bar, where its molar volume is 2.312×10^{-3} m^3·mol^{-1}, to a final state of 60°C and 1 bar. Assume also that air remains a gas for which PV/T is constant and that $C_V = 20.785$ and $C_P = 29.100$ J·mol^{-1}·K^{-1}.

Solution 2.8

Because property changes are independent of process, calculations may be based on any process that accomplishes the change. Here, we choose a two-step, mechanically reversible process wherein 1 mol of air is (a) cooled at constant volume to the final pressure, and (b) heated at constant pressure to the final temperature. Of course, other paths could be chosen, and would yield the same result.

$$T_1 = 5 + 273.15 = 278.15 \text{ K} \qquad T_2 = 60 + 273.15 = 333.15 \text{ K}$$

With $PV = kT$, the ratio T/P is constant for step (a). The intermediate temperature between the two steps is therefore:

$$T' = (278.15)(1/10) = 27.82 \text{ K}$$

and the temperature changes for the two steps are:

$$\Delta T_a = 27.82 - 278.15 = -250.33 \text{ K}$$
$$\Delta T_b = 333.15 - 27.82 = 305.33 \text{ K}$$

For step (a), by Eqs. (2.17) and (2.14),

$$\Delta U_a = C_V \Delta T_a = (20.785)(-250.33) = -5203.1 \text{ J}$$
$$\Delta H_a = \Delta U_a + V \Delta P_a$$
$$= -5203.1 \text{ J} + 2.312 \times 10^{-3} \text{ m}^3 \times \left(-9 \times 10^5\right) \text{ Pa} = -7283.9 \text{ J}$$

For step (b), the final volume of the air is:

$$V_2 = V_1 \frac{P_1 T_2}{P_2 T_1} = 2.312 \times 10^{-3} \left(\frac{10 \times 333.15}{1 \times 278.15}\right) = 2.769 \times 10^{-2} \text{ m}^3$$

By Eqs. (2.21) and (2.14),

$$\Delta H_b = C_P \Delta T_b = (29.100)(305.33) = 8885.1 \text{ J}$$
$$\Delta U_b = \Delta H_b - P \Delta V_b$$
$$= 8885.1 - (1 \times 10^5)(0.02769 - 0.00231) = 6347.1 \text{ J}$$

For the two steps together,

$$\Delta U = -5203.1 + 6347.1 = 1144.0 \text{ J}$$
$$\Delta H = -7283.9 + 8885.1 = 1601.2 \text{ J}$$

These values would be the same for *any* process that results in the same change of state.[8]

---

[8]You might be concerned that the path selected here goes through an intermediate state where, in reality, air would not be a gas, but would condense. Paths for thermodynamic calculations often proceed through such *hypothetical* states that cannot be physically realized but are nonetheless useful and appropriate for the calculation. Additional such states will be encountered repeatedly in later chapters.

2.9 MASS AND ENERGY BALANCES FOR OPEN SYSTEMS

Although the focus of the preceding sections has been on closed systems, the concepts presented find far more extensive application. The laws of mass and energy conservation apply to *all* processes, to open as well as to closed systems. Indeed, the open system includes the closed system as a special case. The remainder of this chapter is therefore devoted to the treatment of open systems and thus to the development of equations of wide practical application.

Measures of Flow

Open systems are characterized by flowing streams; there are four common measures of flow:

- Mass flow rate, $\dot{m}$ • Molar flow rate, $\dot{n}$ • Volumetric flow rate, q • Velocity, u

The measures of flow are interrelated:

$$\dot{m} = \mathcal{M}\dot{n} \qquad \text{and} \qquad q = uA$$

where $\mathcal{M}$ is molar mass and A is the cross-sectional area for flow. Importantly, mass and molar flow rates relate to velocity:

$$\dot{m} = uA\rho \qquad (2.23a) \qquad \dot{n} = uA\rho \qquad (2.23b)$$

The area for flow A is the cross-sectional area of a conduit, and ρ is specific or molar density. Although velocity is a *vector quantity*, its scalar magnitude u is used here as the average speed of a stream in the direction normal to A. Flow rates $\dot{m}$, $\dot{n}$, and q represent measures of quantity per unit of time. Velocity u is quite different in nature, as it does not suggest the magnitude of flow. Nevertheless, it is an important design parameter.

Example 2.9

In a major human artery with an internal diameter of 5 mm, the flow of blood, averaged over the cardiac cycle, is 5 cm^3·s^{-1}. The artery bifurcates (splits) into two identical blood vessels that are each 3 mm in diameter. What are the average velocity and the mass flow rate upstream and downstream of the bifurcation? The density of blood is 1.06 g·cm^{-3}.

Solution 2.9

The average velocity is given by the volumetric flow rate divided by the area for flow. Thus, upstream of the bifurcation, where the vessel diameter is 0.5 cm,

$$u_{up} = \frac{q}{A} = \frac{5 \text{ cm}^3 \cdot \text{s}^{-1}}{(\pi/4)(0.5^2 \text{ cm}^2)} = 25.5 \text{ cm} \cdot \text{s}^{-1}$$

Downstream of the bifurcation, the volumetric flow rate in each vessel is $2.5 \text{ cm}^3\cdot\text{s}^{-1}$, and the vessel diameter is 0.3 cm. Thus,

$$u_{\text{down}} = \frac{2.5 \text{ cm}^3\cdot\text{s}^{-1}}{(\pi/4)(0.3^2 \text{ cm}^2)} = 35.4 \text{ cm}\cdot\text{s}^{-1}$$

The mass flow rate in the upstream vessel is given by the volumetric flow rate times the density:

$$\dot{m}_{\text{up}} = 5 \text{ cm}^3\cdot\text{s}^{-1} \times 1.06 \text{ g}\cdot\text{cm}^{-3} = 5.30 \text{ g}\cdot\text{s}^{-1}$$

Similarly, for each downstream vessel:

$$\dot{m}_{\text{down}} = 2.5 \text{ cm}^3\cdot\text{s}^{-3} \times 1.06 \text{ g}\cdot\text{cm}^{-3} = 2.65 \text{ g}\cdot\text{s}^{-1}$$

which is of course half the upstream value.

Mass Balance for Open Systems

The region of space identified for analysis of open systems is called a *control volume;* it is separated from its surroundings by a *control surface*. The fluid within the control volume is the thermodynamic system for which mass and energy balances are written. Because mass is conserved, the rate of change of mass within the control volume, dm_{cv}/dt, equals the net rate of flow of mass into the control volume. The convention is that flow is positive when directed into the control volume and negative when directed out. The mass balance is expressed mathematically by:

$$\boxed{\frac{dm_{\text{cv}}}{dt} + \Delta\,(\dot{m})_{\text{fs}} = 0}\qquad(2.24)$$

For the control volume of Fig. 2.4, the second term is:

$$\Delta\,(\dot{m})_{\text{fs}} = \dot{m}_3 - \dot{m}_1 - \dot{m}_2$$

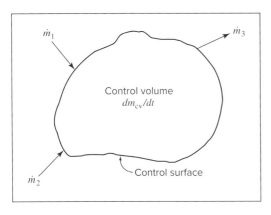

Figure 2.4: Schematic representation of a control volume. It is separated from its surroundings by an extensible control surface. Two streams with flow rates $\dot{m}_1$ and $\dot{m}_2$ are shown directed into the control volume, and one stream with flow rate $\dot{m}_3$ is directed out.

The difference operator Δ here signifies the difference between exit and entrance flows, and the subscript "fs" indicates that the term applies to all flowing streams. Note that this is a different usage of this operator compared with previous sections, where the difference was between an initial state and a final state. Both usages of the difference operator are common, and care must be taken to ensure that the correct sense is understood.

When the mass flow rate $\dot{m}$ is given by Eq. (2.23a), Eq. (2.24) becomes:

$$\frac{dm_{cv}}{dt} + \Delta(\rho uA)_{fs} = 0 \qquad (2.25)$$

In this form the mass-balance equation is often called the *continuity equation*.

Steady-state flow processes are those for which conditions within the control volume do not change with time. These are an important class of flow processes often encountered in practice. In a steady-state process, the control volume contains a constant mass of fluid, and the first or *accumulation* term of Eq. (2.24) is zero, reducing Eq. (2.25) to:

$$\Delta(\rho uA)_{fs} = 0$$

The term "steady state" does not necessarily imply that flow rates are *constant*, merely that the inflow of mass is exactly matched by the outflow of mass.

When there is a single entrance and a single exit stream, the mass flow rate $\dot{m}$ is the same for both streams; then,

$$\dot{m} = \text{const} = \rho_2 u_2 A_2 = \rho_1 u_1 A_1$$

Because specific volume is the reciprocal of density,

$$\boxed{\dot{m} = \frac{u_1 A_1}{V_1} = \frac{u_2 A_2}{V_2} = \frac{uA}{V}} \qquad (2.26)$$

This form of the continuity equation finds frequent use.

The General Energy Balance

Because energy, like mass, is conserved, the rate of change of energy within the control volume equals the net rate of energy transfer into the control volume. Streams flowing into and out of the control volume have associated with them energy in its internal, potential, and kinetic forms, and all may contribute to the energy change of the system. Each unit mass of a stream carries with it a total energy $U + \frac{1}{2}u^2 + zg$, where u is the average velocity of the stream, z is its elevation above a datum level, and g is the local acceleration of gravity. Thus, each stream transports energy at the rate $\left(U + \frac{1}{2}u^2 + zg\right)\dot{m}$. The net energy transported *into* the system by the flowing streams is therefore $-\Delta\left[\left(U + \frac{1}{2}u^2 + zg\right)\dot{m}\right]_{fs}$, where the effect of the minus sign with "Δ" is to make the term read *in – out*. The rate of energy accumulation within the control volume includes this quantity in addition to the heat transfer rate $\dot{Q}$ and work rate:

$$\frac{d(mU)_{cv}}{dt} = -\Delta\left[\left(U + \frac{1}{2}u^2 + zg\right)\dot{m}\right]_{fs} + \dot{Q} + \text{work rate}$$

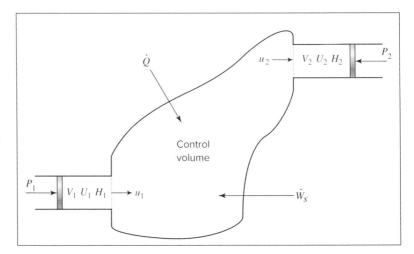

Figure 2.5:
Control volume
with one entrance
and one exit.

The work rate may include work of several forms. First, work is associated with moving the flowing streams through entrances and exits. The fluid at any entrance or exit has a set of average properties, P, V, U, H, etc. Imagine that a unit mass of fluid with these properties exists at an entrance or exit, as shown in Fig. 2.5. This unit mass of fluid is acted upon by additional fluid, here replaced by a piston that exerts the constant pressure P. The work done by this piston in moving the unit mass through the entrance is PV, and the work rate is $(PV)\dot{m}$. Because Δ denotes the difference between exit and entrance quantities, the net work done *on* the system when all entrance and exit sections are taken into account is $-\Delta[(PV)\dot{m}]_{\text{fs}}$.

Another form of work is the shaft work[9] indicated in Fig. 2.5 by rate $\dot{W}_s$. In addition, work may be associated with expansion or contraction of the entire control volume. These forms of work are all included in a rate term represented by $\dot{W}$. The preceding equation may now be written:

$$\frac{d(mU)_{\text{cv}}}{dt} = -\Delta\left[\left(U + \frac{1}{2}u^2 + zg\right)\dot{m}\right]_{\text{fs}} + \dot{Q} - \Delta[(PV)\dot{m}]_{\text{fs}} + \dot{W}$$

Combination of terms in accord with the definition of enthalpy, $H = U + PV$, leads to:

$$\frac{d(mU)_{\text{cv}}}{dt} = -\Delta\left[\left(H + \frac{1}{2}u^2 + zg\right)\dot{m}\right]_{\text{fs}} = \dot{Q} + \dot{W}$$

which is usually written:

$$\frac{d(mU)_{\text{cv}}}{dt} + \Delta\left[\left(H + \frac{1}{2}u^2 + zg\right)\dot{m}\right]_{\text{fs}} = \dot{Q} + \dot{W} \qquad (2.27)$$

The velocity u in the kinetic-energy terms is the bulk-mean velocity as defined by the equation $u = \dot{m}/(\rho A)$. Fluids flowing in pipes exhibit a velocity profile that rises from zero at

[9]Mechanical work added to or removed from the system without transfer of mass is called shaft work because it is often transferred by means of a rotating shaft, like that in a turbine or compressor. However, this term is used more broadly to include work transferred by other mechanical means as well.

the wall (the no-slip condition) to a maximum at the center of the pipe. The kinetic energy of a fluid in a pipe depends on its velocity profile. For the case of laminar flow, the profile is parabolic, and integration across the pipe shows that the kinetic-energy term should properly be u^2. In fully developed turbulent flow, the more common case in practice, the velocity across the major portion of the pipe is not far from uniform, and the expression $u^2/2$, as used in the energy equations, is more nearly correct.

Although Eq. (2.27) is an energy balance of reasonable generality, it has limitations. In particular, it reflects the tacit assumption that the center of mass of the control volume is stationary. Thus no terms for kinetic- and potential-energy changes of the fluid in the control volume are included. For virtually all applications of interest to chemical engineers, Eq. (2.27) is adequate. For many (but not all) applications, kinetic- and potential-energy changes in the flowing streams are also negligible, and Eq. (2.27) then simplifies to:

$$\frac{d(mU)_{cv}}{dt} + \Delta(H\dot{m})_{fs} = \dot{Q} + \dot{W} \tag{2.28}$$

Example 2.10

Show that Eq. (2.28) reduces to Eq. (2.3) for the case of a closed system.

Solution 2.10

The second term of Eq. (2.28) is omitted in the absence of flowing streams:

$$\frac{d(mU)_{cv}}{dt} = \dot{Q} + \dot{W}$$

Integration over time gives

$$\Delta(mU)_{cv} = \int_{t_1}^{t_2} \dot{Q}\, dt + \int_{t_1}^{t_2} \dot{W}\, dt$$

or

$$\Delta U^t = Q + W$$

The Q and W terms are defined by the integrals of the preceding equation.

Note here that Δ indicates a change over time, not from an inlet to an outlet. One must be aware of its context to discern its meaning.

Example 2.11

An insulated, electrically heated tank for hot water contains 190 kg of liquid water at 60°C. Imagine you are taking a shower using water from this tank when a power outage occurs. If water is withdrawn from the tank at a steady rate of $\dot{m} = 0.2$ kg·s^{-1}, how long will it take for the temperature of the water in the tank to drop from 60 to 35°C? Assume that cold water enters the tank at 10°C and that heat losses from the tank are negligible. Here, an excellent assumption for liquid water is that $C_V = C_P = C$, independent of T and P.

Solution 2.11

This is an example of the application of Eq. (2.28) to a transient process for which $\dot{Q} = \dot{W} = 0$. We assume perfect mixing of the contents of the tank; this implies that the properties of the water leaving the tank are those of the water in the tank. With the mass flow rate into the tank equal to the mass flow rate out, $\dot{m}_{cv}$ is constant; moreover, the differences between inlet and outlet kinetic and potential energies can be neglected. Equation (2.28) is therefore written:

$$m\frac{dU}{dt} + \dot{m}(H - H_1) = 0$$

where unsubscripted quantities refer to the contents of the tank (and therefore the water leaving the tank) and H_1 is the specific enthalpy of the water entering the tank. With $C_V = C_P = C$,

$$\frac{dU}{dt} = C\frac{dT}{dt} \qquad \text{and} \qquad H - H_1 = C(T - T_1)$$

The energy balance then becomes, on rearrangement,

$$dt = -\frac{m}{\dot{m}} \cdot \frac{dT}{T - T_1}$$

Integration from $t = 0$ (where $T = T_0$) to arbitrary time t yields:

$$t = -\frac{m}{\dot{m}}\ln\left(\frac{T - T_1}{T_0 - T_1}\right)$$

Substitution of numerical values into this equation gives, for the conditions of this problem,

$$t = -\frac{190}{0.2}\ln\left(\frac{35 - 10}{60 - 10}\right) = 658.5 \text{ s}$$

Thus, the water temperature in the tank will drop from 60 to 35°C after about 11 minutes.

Energy Balances for Steady-State Flow Processes

Flow processes for which the accumulation term of Eq. (2.27), $d(mU)_{cv}/dt$, is zero are said to occur at *steady state*. As discussed with respect to the mass balance, this means that the mass of the system within the control volume is constant; it also means that no changes occur with time in the properties of the fluid within the control volume nor at its entrances and exits. No expansion of the control volume is possible under these circumstances. The only work of the process is shaft work, and the general energy balance, Eq. (2.27), becomes:

$$\Delta\left[\left(H + \frac{1}{2}u^2 + zg\right)\dot{m}\right]_{fs} = \dot{Q} + \dot{W}_s \tag{2.29}$$

Although "steady state" does not necessarily imply "steady flow," the usual application of this equation is to steady-state, steady-flow processes, because such processes represent the industrial norm.[10]

A further specialization results when the control volume has one entrance and one exit. The same mass flow rate $\dot{m}$ then applies to both streams, and Eq. (2.29) reduces to:

$$\Delta\left(H + \frac{1}{2}u^2 + zg\right)\dot{m} = \dot{Q} + \dot{W}_s \tag{2.30}$$

where subscript "fs" has been omitted in this simple case and Δ denotes the change from entrance to exit. Division by $\dot{m}$ gives:

$$\Delta\left(H + \frac{1}{2}u^2 + zg\right) = \frac{\dot{Q}}{\dot{m}} + \frac{\dot{W}_s}{\dot{m}} = Q + W_s$$

or

$$\Delta H + \frac{\Delta u^2}{2} + g\Delta z = Q + W_s \tag{2.31}$$

This equation is the mathematical expression of the first law for a steady-state, steady-flow process between one entrance and one exit. All terms represent energy per unit mass of fluid. The energy unit is usually the joule.

In many applications, kinetic- and potential-energy terms are omitted because they are negligible compared with other terms.[11] For such cases, Eq. (2.31) reduces to:

$$\Delta H = Q + W_s \tag{2.32}$$

This expression of the first law for a steady-state, steady-flow process is analogous to Eq. (2.3) for a nonflow process. However, in Eq. (2.32), enthalpy rather than internal energy is the thermodynamic property of importance, and Δ refers to a change from inlet to outlet, rather than from before to after an event.

---

[10]An example of a steady-state process that is not steady flow is a water heater, in which variations in flow rate are exactly compensated by changes in the rate of heat transfer, so that temperatures throughout remain constant.

[11]Notable exceptions include applications to nozzles, metering devices, wind tunnels, and hydroelectric power stations.

A Flow Calorimeter for Enthalpy Measurements

The application of Eqs. (2.31) and (2.32) to the solution of practical problems requires enthalpy values. Because H is a state function, its values depend only on point conditions; once determined, they may be tabulated for subsequent use for the same sets of conditions. To this end, Eq. (2.32) may be applied to laboratory processes designed for enthalpy measurements.

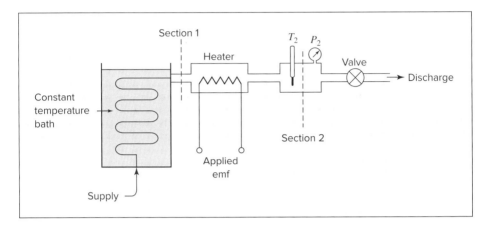

Figure 2.6: Flow calorimeter.

A simple flow calorimeter is illustrated schematically in Fig. 2.6. Its essential feature is an electric resistance heater immersed in a flowing fluid. The design provides for minimal velocity and elevation changes from section 1 to section 2, making kinetic- and potential-energy changes of the fluid negligible. With no shaft work entering the system, Eq. (2.32) reduces to $\Delta H = H_2 - H_1 = Q$. The rate of heat transfer to the fluid is determined from the resistance of the heater and the current passing through it. In practice a number of details require careful attention, but in principle the operation of the flow calorimeter is simple. Measurements of the heat transfer rate and flow rate allow calculation of the change ΔH between sections 1 and 2.

For example, enthalpies of both liquid and vapor H_2O are readily determined. The constant-temperature bath is filled with a mixture of crushed ice and water to maintain a temperature of 0°C. Liquid water is supplied to the apparatus, and the coil that carries it through the constant-temperature bath is long enough to bring it to an exit temperature of essentially 0°C. The temperature and pressure at section 2 are measured by suitable instruments. Values of the enthalpy of H_2O for various conditions at section 2 are given by:

$$H_2 = H_1 + Q$$

where Q is the heat added per unit mass of water flowing.

The pressure may vary from run to run, but in the range encountered here it has a negligible effect on the enthalpy of the entering water, and for practical purposes H_1 is a constant. Absolute values of enthalpy, like absolute values of internal energy, are unknown. An arbitrary value may therefore be assigned to H_1 as the *basis* for all other enthalpy values. Setting $H_1 = 0$ for liquid water at 0°C makes:

$$H_2 = H_1 + Q = 0 + Q = Q$$

Enthalpy values may be tabulated for the temperatures and pressures existing at section 2 for a large number of runs. In addition, specific-volume measurements made for these same conditions may be added to the table, along with corresponding values of the internal energy calculated by Eq. (2.10), $U = H - PV$. In this way, tables of thermodynamic properties are compiled over the entire useful range of conditions. The most widely used such tabulation is for H_2O and is known as the *steam tables*.[12]

The enthalpy may be taken as zero for some other state than liquid at 0°C. The choice is arbitrary. The equations of thermodynamics, such as Eqs. (2.31) and (2.32), apply to *changes* of state, for which the enthalpy *differences* are independent of the location of the zero point. However, once an arbitrary zero point is selected for the enthalpy, an arbitrary choice cannot be made for the internal energy, because internal energy is related to enthalpy by Eq. (2.10).

Example 2.12

For the flow calorimeter just discussed, the following data are taken with water as the test fluid:

$$\text{Flow rate} = 4.15 \text{ g·s}^{-1} \qquad t_1 = 0°C \qquad t_2 = 300°C \qquad P_2 = 3 \text{ bar}$$

$$\text{Rate of heat addition from resistance heater} = 12{,}740 \text{ W}$$

The water is completely vaporized in the process. Calculate the enthalpy of steam at 300°C and 3 bar based on $H = 0$ for liquid water at 0°C.

Solution 2.12

If Δz and Δu^2 are negligible and if W_s and H_1 are zero, then $H_2 = Q$, and

$$H_2 = \frac{12{,}740 \text{ J·s}^{-1}}{4.15 \text{ g·s}^{-1}} = 3070 \text{ J·g}^{-1} \qquad \text{or} \qquad 3070 \text{ kJ·kg}^{-1}$$

Example 2.13

Air at 1 bar and 25°C enters a compressor at low velocity, discharges at 3 bar, and enters a nozzle in which it expands to a final velocity of 600 m·s^{-1} at the initial conditions of pressure and temperature. If the work of compression is 240 kJ per kilogram of air, how much heat must be removed during compression?

---

[12]Steam tables adequate for many purposes are given in Appendix E. The Chemistry WebBook of NIST includes a fluid properties calculator with which one can generate tables for water and some 75 other substances: http://webbook.nist.gov/chemistry/fluid/

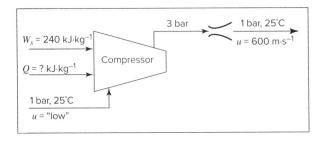

Solution 2.13

Because the air returns to its initial conditions of T and P, the overall process produces no change in enthalpy of the air. Moreover, the potential-energy change of the air is presumed negligible. Neglecting also the initial kinetic energy of the air, we write Eq. (2.31) as:

$$\Delta H + \frac{\Delta u^2}{2} + g\Delta z = 0 + \frac{u_2^2}{2} + 0 = Q + W_s$$

Then

$$Q = \frac{u_2^2}{2} - W_s$$

The kinetic-energy term is evaluated as follows:

$$\frac{1}{2}u_2^2 = \frac{1}{2}\left(600\,\frac{\text{m}}{\text{s}}\right)^2 = 180{,}000\,\frac{\text{m}^2}{\text{s}^2} = 180{,}000\,\frac{\text{m}^2}{\text{s}^2}\cdot\frac{\text{kg}}{\text{kg}}$$

$$= 180{,}000\ \text{N}\cdot\text{m}\cdot\text{kg}^{-1} = 180\ \text{kJ}\cdot\text{kg}^{-1}$$

Then

$$Q = 180 - 240 = -60\ \text{kJ}\cdot\text{kg}^{-1}$$

Heat in the amount of 60 kJ must be removed per kilogram of air compressed.

Example 2.14

Water at 90°C is pumped from a storage tank at a rate of 3 L·s^{-1}. The motor for the pump supplies work at a rate of 1.5 kJ·s^{-1}. The water goes through a heat exchanger, giving up heat at a rate of 670 kJ·s^{-1}, and is delivered to a second storage tank at an elevation 15 m above the first tank. What is the temperature of the water delivered to the second tank?

Solution 2.14

This is a steady-state, steady-flow process for which Eq. (2.31) applies. The initial and final velocities of water in the storage tanks are negligible, and the term $\Delta u^2/2$

may be omitted. All remaining terms are expressed in units of $kJ \cdot kg^{-1}$. At 90°C the density of water is $0.965 \ kg \cdot L^{-1}$ and the mass flow rate is:

$$\dot{m} = (3)(0.965) = 2.895 \ kg \cdot s^{-1}$$

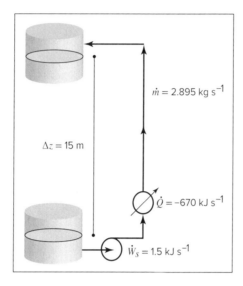

For the heat exchanger,

$$Q = -670/2.895 = -231.4 \ kJ \cdot kg^{-1}$$

For the shaft work of the pump,

$$W_s = 1.5/2.895 = 0.52 \ kJ \cdot kg^{-1}$$

If g is taken as the standard value of $9.8 \ m \cdot s^{-2}$, the potential-energy term is:

$$g\Delta z = (9.8)(15) = 147 \ m^2 \cdot s^{-2}$$
$$= 147 \ J \cdot kg^{-1} = 0.147 \ kJ \cdot kg^{-1}$$

Equation (2.31) now yields:

$$\Delta H = Q + W_s - g\Delta z = -231.4 + 0.52 - 0.15 = -231.03 \ kJ \cdot kg^{-1}$$

The steam-table value for the enthalpy of liquid water at 90°C is:

$$H_1 = 376.9 \ kJ \cdot kg^{-1}$$

Thus,

$$\Delta H = H_2 - H_1 = H_2 - 376.9 = -231.0$$

and

$$H_2 = 376.9 - 231.0 = 145.9 \ kJ \cdot kg^{-1}$$

The temperature of water having this enthalpy is found from the steam tables:

$$t = 34.83°C$$

In this example, W_s and $g\Delta z$ are small compared with Q, and for practical purposes could be neglected.

Example 2.15

A steam turbine operates adiabatically with a power output of 4000 kW. Steam enters the turbine at 2100 kPa and 475°C. The exhaust is saturated steam at 10 kPa that enters a condenser, where it is condensed and cooled to 30°C. What is the mass flow rate of the steam, and at what rate must cooling water be supplied to the condenser, if the water enters at 15°C and is heated to 25°C?

Solution 2.15

The enthalpies of entering and exiting steam from the turbine are found from the steam tables:

$$H_1 = 3411.3 \text{ kJ·kg}^{-1} \qquad \text{and} \qquad H_2 = 2584.8 \text{ kJ·kg}^{-1}$$

For a properly designed turbine, kinetic- and potential-energy changes are negligible, and for adiabatic operation $Q = 0$. Eq. (2.32) becomes simply $W_s = \Delta H$. Then $\dot{W}_s = \dot{m}(\Delta H)$, and

$$\dot{m}_{\text{steam}} = \frac{\dot{W}_s}{\Delta H} = \frac{-4000 \text{ kJ·s}^{-1}}{(2584.8 - 3411.3) \text{ kJ·kg}^{-1}} = 4.840 \text{ kg·s}^{-1}$$

For the condenser, the steam condensate leaving is subcooled water at 30°C, for which (from the steam tables) $H_3 = 125.7$ kJ·kg^{-1}. For the cooling water entering at 15°C and leaving at 25°C, the enthalpies are

$$H_{\text{in}} = 62.9 \text{ kJ·kg}^{-1} \qquad \text{and} \qquad H_{\text{out}} = 104.8 \text{ kJ·kg}^{-1}$$

Equation (2.29) here reduces to

$$\dot{m}_{\text{steam}}(H_3 - H_2) + \dot{m}_{\text{water}}(H_{\text{out}} - H_{\text{in}}) = 0$$
$$4.840(125.7 - 2584.8) + \dot{m}_{\text{water}}(104.8 - 62.9) = 0$$

Solution gives,

$$\dot{m}_{\text{water}} = 284.1 \text{ kg·s}^{-1}$$

2.10 SYNOPSIS

After studying this chapter, including the end-of-chapter problems, one should be able to:

- State and apply the first law of thermodynamics, making use of the appropriate sign conventions
- Explain and employ the concepts of internal energy, enthalpy, state function, equilibrium, and reversible process
- Explain the differences between state functions and path-dependent quantities such as heat and work
- Calculate changes in state variables for a real process by substituting a hypothetical reversible process connecting the same states
- Relate changes in the internal energy and enthalpy of a substance to changes in temperature, with calculations based on the appropriate heat capacity
- Construct and apply mass and energy balances for open systems

2.11 PROBLEMS

2.1. A nonconducting container filled with 25 kg of water at 20°C is fitted with a stirrer, which is made to turn by gravity acting on a weight of mass 35 kg. The weight falls slowly through a distance of 5 m in driving the stirrer. Assuming that all work done on the weight is transferred to the water and that the local acceleration of gravity is 9.8 m·s^{-2}, determine:

(*a*) The amount of work done on the water.
(*b*) The internal energy change of the water.
(*c*) The final temperature of the water, for which $C_P = 4.18$ kJ·kg^{-1}·°C^{-1}.
(*d*) The amount of heat that must be removed from the water to return it to its initial temperature.
(*e*) The total energy change of the universe because of (1) the process of lowering the weight, (2) the process of cooling the water back to its initial temperature, and (3) both processes together.

2.2. Rework Prob. 2.1 for an insulated container that changes in temperature along with the water and has a heat capacity equivalent to 5 kg of water. Work the problem with:

(*a*) The water and container as the system. (*b*) The water alone as the system.

2.3. An egg, initially at rest, is dropped onto a concrete surface and breaks. With the egg treated as the system,

(*a*) What is the sign of W?
(*b*) What is the sign of ΔE_P?

(c) What is ΔE_K?

(d) What is ΔU^t?

(e) What is the sign of Q?

In modeling this process, assume the passage of sufficient time for the broken egg to return to its initial temperature. What is the origin of the heat transfer of part (e)?

2.4. An electric motor under steady load draws 9.7 amperes at 110 volts, delivering 1.25(hp) of mechanical energy. What is the rate of heat transfer from the motor, in kW?

2.5. An electric hand mixer draws 1.5 amperes at 110 volts. It is used to mix 1 kg of cookie dough for 5 minutes. After mixing, the temperature of the cookie dough is found to have increased by 5°C. If the heat capacity of the dough is 4.2 kJ·kg^{-1}·K^{-1}, what fraction of the electrical energy used by the mixer is converted to internal energy of the dough? Discuss the fate of the remainder of the energy.

2.6. One mole of gas in a closed system undergoes a four-step thermodynamic cycle. Use the data given in the following table to determine numerical values for the missing quantities indicated by question marks.

Step	ΔU^t/J	Q/J	W/J
12	−200	?	−6000
23	?	−3800	?
34	?	−800	300
41	4700	?	?
12341	?	?	−1400

2.7. Comment on the feasibility of cooling your kitchen in the summer by opening the door to the electrically powered refrigerator.

2.8. A tank containing 20 kg of water at 20°C is fitted with a stirrer that delivers work to the water at the rate of 0.25 kW. How long does it take for the temperature of the water to rise to 30°C if no heat is lost from the water? For water, $C_P = 4.18$ kJ·kg^{-1}·°C^{-1}.

2.9. Heat in the amount of 7.5 kJ is added to a closed system while its internal energy decreases by 12 kJ. How much energy is transferred as work? For a process causing the same change of state but for which the work is zero, how much heat is transferred?

2.10. A steel casting weighing 2 kg has an initial temperature of 500°C; 40 kg of water initially at 25°C is contained in a perfectly insulated steel tank weighing 5 kg. The casting is immersed in the water and the system is allowed to come to equilibrium. What is its final temperature? Ignore the effects of expansion or contraction, and assume constant specific heats of 4.18 kJ·kg^{-1}·K^{-1} for water and 0.50 kJ·kg^{-1}·K^{-1} for steel.

2.11. An incompressible fluid (ρ = constant) is contained in an insulated cylinder fitted with a frictionless piston. Can energy as work be transferred to the fluid? What is the change in internal energy of the fluid when the pressure is increased from P_1 to P_2?

2.12. One kg of liquid water at 25°C, for which $C_P = 4.18$ kJ·kg^{-1}·°C^{-1}:

(a) Experiences a temperature increase of 1 K. What is ΔU^t, in kJ?

(b) Experiences a change in elevation Δz. The change in potential energy ΔE_P is the same as ΔU^t for part (a). What is Δz, in meters?

(c) Is accelerated from rest to final velocity u. The change in kinetic energy ΔE_K is the same as ΔU^t for part (a). What is u, in m·s^{-1}?

Compare and discuss the results of the three preceding parts.

2.13. An electric motor runs "hot" under load, owing to internal irreversibilities. It has been suggested that the associated energy loss be minimized by thermally insulating the motor casing. Comment critically on this suggestion.

2.14. A hydroturbine operates with a head of 50 m of water. Inlet and outlet conduits are 2 m in diameter. Estimate the mechanical power developed by the turbine for an outlet velocity of 5 m·s^{-1}.

2.15. A wind turbine with a rotor diameter of 40 m produces 90 kW of electrical power when the wind speed is 8 m·s^{-1}. The density of air impinging on the turbine is 1.2 kg·m^{-3}. What fraction of the kinetic energy of the wind impinging on the turbine is converted to electrical energy?

2.16. The battery in a laptop computer supplies 11.1 V and has a capacity of 56 W·h. In ordinary use, it is discharged after 4 hours. What is the average current drawn by the laptop, and what is the average rate of heat dissipation from it? You may assume that the temperature of the computer remains constant.

2.17. Suppose that the laptop of Prob. 2.16 is placed in an insulating briefcase with a fully charged battery, but it does not go into "sleep" mode, and the battery discharges as if the laptop were in use. If no heat leaves the briefcase, the heat capacity of the briefcase itself is negligible, and the laptop has a mass of 2.3 kg and an average specific heat of 0.8 kJ·kg^{-1}·°C^{-1}, estimate the temperature of the laptop after the battery has fully discharged.

2.18. In addition to heat and work flows, energy can be transferred as light, as in a photovoltaic device (solar cell). The energy content of light depends on both its wavelength (color) and its intensity. When sunlight impinges on a solar cell, some is reflected, some is absorbed and converted to electrical work, and some is absorbed and converted to heat. Consider an array of solar cells with an area of 3 m^2. The power of sunlight impinging upon it is 1 kW·m^{-2}. The array converts 17% of the incident power to electrical work, and it reflects 20% of the incident light. At steady state, what is the rate of heat removal from the solar cell array?

2.19. Liquid water at 180°C and 1002.7 kPa has an internal energy (on an arbitrary scale) of 762.0 kJ·kg^{-1} and a specific volume of 1.128 cm^3·g^{-1}.

(a) What is its enthalpy?
(b) The water is brought to the vapor state at 300°C and 1500 kPa, where its internal energy is 2784.4 kJ·kg^{-1} and its specific volume is 169.7 cm^3·g^{-1}. Calculate ΔU and ΔH for the process.

2.20. A solid body at initial temperature T_0 is immersed in a bath of water at initial temperature T_{w0}. Heat is transferred from the solid to the water at a rate $\dot{Q} = K \cdot (T_w - T)$, where K is a constant and T_w and T are *instantaneous* values of the temperatures of the water and solid. Develop an expression for T as a function of time τ. Check your result for the limiting cases, $\tau = 0$ and $\tau = \infty$. Ignore effects of expansion or contraction, and assume constant specific heats for both water and solid.

2.21. A list of common unit operations follows:

(a) Single-pipe heat exchanger
(b) Double-pipe heat exchanger
(c) Pump
(d) Gas compressor
(e) Gas turbine
(f) Throttle valve
(g) Nozzle

Develop a simplified form of the general steady-state energy balance appropriate for each operation. State carefully, and justify, any assumptions you make.

2.22. The Reynolds number Re is a dimensionless group that characterizes the intensity of a flow. For large Re, a flow is turbulent; for small Re, it is laminar. For pipe flow, Re $\equiv$ $u\rho D/\mu$, where D is pipe diameter and μ is dynamic viscosity.

(a) If D and μ are fixed, what is the effect of increasing mass flow rate $\dot{m}$ on Re?
(b) If $\dot{m}$ and μ are fixed, what is the effect of increasing D on Re?

2.23. An incompressible (ρ = constant) liquid flows steadily through a conduit of circular cross-section and increasing diameter. At location 1, the diameter is 2.5 cm and the velocity is 2 m·s^{-1}; at location 2, the diameter is 5 cm.

(a) What is the velocity at location 2?
(b) What is the kinetic-energy change (J·kg^{-1}) of the fluid between locations 1 and 2?

2.24. A stream of warm water is produced in a steady-flow mixing process by combining 1.0 kg·s^{-1} of cool water at 25°C with 0.8 kg·s^{-1} of hot water at 75°C. During mixing, heat is lost to the surroundings at the rate of 30 kJ·s^{-1}. What is the temperature of the warm water stream? Assume the specific heat of water is constant at 4.18 kJ·kg^{-1}·K^{-1}.

2.25. Gas is bled from a tank. Neglecting heat transfer between the gas and the tank, show that mass and energy balances produce the differential equation:

$$\frac{dU}{H' - U} = \frac{dm}{m}$$

Here, U and m refer to the gas remaining in the tank; H' is the specific enthalpy of the gas leaving the tank. Under what conditions can one assume $H' = H$?

2.26. Water at 28°C flows in a straight horizontal pipe in which there is no exchange of either heat or work with the surroundings. Its velocity is 14 m·s^{-1} in a pipe with an internal diameter of 2.5 cm until it flows into a section where the pipe diameter abruptly increases. What is the temperature change of the water if the downstream diameter is 3.8 cm? If it is 7.5 cm? What is the maximum temperature change for an enlargement in the pipe?

2.27. Fifty (50) kmol per hour of air is compressed from $P_1 = 1.2$ bar to $P_2 = 6.0$ bar in a steady-flow compressor. Delivered mechanical power is 98.8 kW. Temperatures and velocities are:

$$T_1 = 300 \text{ K} \qquad T_2 = 520 \text{ K}$$
$$u_1 = 10 \text{ m·s}^{-1} \qquad u_2 = 3.5 \text{ m·s}^{-1}$$

Estimate the rate of heat transfer from the compressor. Assume for air that $C_P = \frac{7}{2}R$ and that enthalpy is independent of pressure.

2.28. Nitrogen flows at steady state through a horizontal, insulated pipe with inside diameter of 1.5(in). A pressure drop results from flow through a partially opened valve. Just upstream from the valve the pressure is 100(psia), the temperature is 120(°F), and the average velocity is 20(ft)·s^{-1}. If the pressure just downstream from the valve is 20(psia), what is the temperature? Assume for air that PV/T is constant, $C_V = (5/2)R$, and $C_P = (7/2)R$. (Values for R, the ideal gas constant, are given in App. A.)

2.29. Air flows at steady state through a horizontal, insulated pipe with inside diameter of 4 cm. A pressure drop results from flow through a partially opened valve. Just upstream from the valve, the pressure is 7 bar, the temperature is 45°C, and the average velocity is 20 m·s^{-1}. If the pressure just downstream from the valve is 1.3 bar, what is the temperature? Assume for air that PV/T is constant, $C_V = (5/2)R$, and $C_P = (7/2)R$. (Values for R, the ideal gas constant, are given in App. A.)

2.30. Water flows through a horizontal coil heated from the outside by high-temperature flue gases. As it passes through the coil, the water changes state from liquid at 200 kPa and 80°C to vapor at 100 kPa and 125°C. Its entering velocity is 3 m·s^{-1} and its exit velocity is 200 m·s^{-1}. Determine the heat transferred through the coil per unit mass of water. Enthalpies of the inlet and outlet streams are:

Inlet: 334.9 kJ·kg^{-1}; Outlet: 2726.5 kJ·kg^{-1}

2.31. Steam flows at steady state through a converging, insulated nozzle, 25 cm long and with an inlet diameter of 5 cm. At the nozzle entrance (state 1), the temperature and pressure are 325°C and 700 kPa and the velocity is 30 m·s⁻¹. At the nozzle exit (state 2), the steam temperature and pressure are 240°C and 350 kPa. Property values are:

$$H_1 = 3112.5 \text{ kJ·kg}^{-1} \qquad V_1 = 388.61 \text{ cm}^3\text{·g}^{-1}$$

$$H_2 = 2945.7 \text{ kJ·kg}^{-1} \qquad V_2 = 667.75 \text{ cm}^3\text{·g}^{-1}$$

What is the velocity of the steam at the nozzle exit, and what is the exit diameter?

2.32. In the following take $C_V = 20.8$ and $C_P = 29.1$ J·mol⁻¹·°C⁻¹ for nitrogen gas:

(*a*) Three moles of nitrogen at 30°C, contained in a rigid vessel, is heated to 250°C. How much heat is required if the vessel has a negligible heat capacity? If the vessel weighs 100 kg and has a heat capacity of 0.5 kJ·kg⁻¹·°C⁻¹, how much heat is required?

(*b*) Four moles of nitrogen at 200°C is contained in a piston/cylinder arrangement. How much heat must be extracted from this system, which is kept at constant pressure, to cool it to 40°C if the heat capacity of the piston and cylinder is neglected?

2.33. In the following take $C_V = 5$ and $C_P = 7$ (Btu)(lb mole)⁻¹(°F)⁻¹ for nitrogen gas:

(*a*) Three pound moles of nitrogen at 70(°F), contained in a rigid vessel, is heated to 350(°F). How much heat is required if the vessel has a negligible heat capacity? If it weighs 200(lb$_m$) and has a heat capacity of 0.12(Btu)(lb$_m$)⁻¹(°F)⁻¹, how much heat is required?

(*b*) Four pound moles of nitrogen at 400(°F) is contained in a piston/cylinder arrangement. How much heat must be extracted from this system, which is kept at constant pressure, to cool it to 150(°F) if the heat capacity of the piston and cylinder is neglected?

2.34. Find an equation for the work of reversible, isothermal compression of 1 mol of gas in a piston/cylinder assembly if the molar volume of the gas is given by

$$V = \frac{RT}{P} + b$$

where b and R are positive constants.

2.35. Steam at 200(psia) and 600(°F) [state 1] enters a turbine through a 3-inch-diameter pipe with a velocity of 10(ft)·s⁻¹. The exhaust from the turbine is carried through a 10-inch-diameter pipe and is at 5(psia) and 200(°F) [state 2]. What is the power output of the turbine?

$$H_1 = 1322.6(\text{Btu})(\text{lb}_m)^{-1} \qquad V_1 = 3.058(\text{ft})^3(\text{lb}_m)^{-1}$$

$$H_2 = 1148.6(\text{Btu})(\text{lb}_m)^{-1} \qquad V_2 = 78.14(\text{ft})^3(\text{lb}_m)^{-1}$$

2.36. Steam at 1400 kPa and 350°C [state 1] enters a turbine through a pipe that is 8 cm in diameter, at a mass flow rate of 0.1 kg·s⁻¹. The exhaust from the turbine is carried through a 25-cm-diameter pipe and is at 50 kPa and 100°C [state 2]. What is the power output of the turbine?

$$H_1 = 3150.7 \text{ kJ·kg}^{-1} \qquad V_1 = 0.20024 \text{ m}^3\text{·kg}^{-1}$$

$$H_2 = 2682.6 \text{ kJ·kg}^{-1} \qquad V_2 = 3.4181 \text{ m}^3\text{·kg}^{-1}$$

2.37. Carbon dioxide gas enters a water-cooled compressor at conditions $P_1 = 1$ bar and $T_1 = 10°C$, and is discharged at conditions $P_2 = 36$ bar and $T_2 = 90°C$. The entering CO_2 flows through a 10-cm-diameter pipe with an average velocity of 10 m·s^{-1}, and is discharged through a 3-cm-diameter pipe. The power supplied to the compressor is 12.5 kJ·mol^{-1}. What is the heat-transfer rate from the compressor?

$$H_1 = 21.71 \text{ kJ·mol}^{-1} \qquad V_1 = 23.40 \text{ L·mol}^{-1}$$

$$H_2 = 23.78 \text{ kJ·mol}^{-1} \qquad V_2 = 0.7587 \text{ L·mol}^{-1}$$

2.38. Carbon dioxide gas enters a water-cooled compressor at conditions $P_1 = 15$(psia) and $T_1 = 50$(°F), and is discharged at conditions $P_2 = 520$(psia) and $T_2 = 200$(°F). The entering CO_2 flows through a 4-inch-diameter pipe with a velocity of 20(ft)·s^{-1}, and is discharged through a 1-inch-diameter pipe. The shaft work supplied to the compressor is 5360(Btu)(lb mole)$^{-1}$. What is the heat-transfer rate from the compressor in (Btu)·h^{-1}?

$$H_1 = 307(\text{Btu})(\text{lb}_\text{m})^{-1} \qquad V_1 = 9.25(\text{ft})^3(\text{lb}_\text{m})^{-1}$$

$$H_2 = 330(\text{Btu})(\text{lb}_\text{m})^{-1} \qquad V_2 = 0.28(\text{ft})^3(\text{lb}_\text{m})^{-1}$$

2.39. Show that W and Q for an *arbitrary* mechanically reversible nonflow process are given by:

$$W = \int V\, dp - \Delta(PV) \qquad Q = \Delta H - \int V\, dp$$

2.40. One kilogram of air is heated reversibly at constant pressure from an initial state of 300 K and 1 bar until its volume triples. Calculate W, Q, ΔU, and ΔH for the process. Assume for air that $PV/T = 83.14$ bar·cm^3·mol^{-1}·K^{-1} and $C_P = 29$ J·mol^{-1}·K^{-1}.

2.41. The conditions of a gas change in a steady-flow process from 20°C and 1000 kPa to 60°C and 100 kPa. Devise a reversible nonflow process (any number of steps) for accomplishing this change of state, and calculate ΔU and ΔH for the process on the basis of 1 mol of gas. Assume for the gas that PV/T is constant, $C_V = (5/2)R$, and $C_P = (7/2)R$.

2.42. A flow calorimeter like that shown in Figure 2.6 is used with a flow rate of 20 g·min^{-1} of the fluid being tested and a constant temperature of 0°C leaving the constant-temperature bath. The steady-state temperature at section two (T_2) is measured as a function of the power supplied to the heater (P), to obtain the data shown in the table below. What is the average specific heat of the substance tested over the temperature range from 0°C to 10°C? What is the average specific heat from 90°C to 100°C? What is the average specific heat over the entire range tested? Describe how you would use this data to derive an expression for the specific heat as a function of temperature.

$T_2/°C$	10	20	30	40	50	60	70	80	90	100
P/W	5.5	11.0	16.6	22.3	28.0	33.7	39.6	45.4	51.3	57.3

2.43. Like the flow calorimeter of Figure 2.6, a particular single-cup coffee maker uses an electric heating element to heat a steady flow of water from 22°C to 88°C. It heats 8 fluid ounces of water (with a mass of 237 g) in 60 s. Estimate the power requirement of the heater during this process. You may assume the specific heat of water is constant at 4.18 $J \cdot g^{-1} \cdot °C^{-1}$.

2.44. (a) An incompressible fluid (ρ = constant) flows through a pipe of constant cross-sectional area. If the flow is steady, show that velocity u and volumetric flow rate q are constant.
(b) A chemically reactive gas stream flows steadily through a pipe of constant cross-sectional area. Temperature and pressure vary with pipe length. Which of the following quantities are necessarily constant: $\dot{m}$, $\dot{n}$, q, u?

2.45. The *mechanical-energy balance* provides a basis for estimating pressure drop owing to friction in fluid flow. For steady flow of an incompressible fluid in a horizontal pipe of constant cross-sectional area, it may be written,

$$\frac{\Delta P}{\Delta L} + \frac{2}{D} f_F \rho u^2 = 0$$

where f_F is the *Fanning friction factor*. Churchill[13] gives the following expression for f_F for turbulent flow:

$$f_F = 0.3305 \left\{ \ln \left[0.27 \frac{\in}{D} + \left(\frac{7}{Re} \right)^{0.9} \right] \right\}^{-2}$$

Here, Re is the Reynolds number and $\in/D$ is the dimensionless pipe roughness. For pipe flow, $Re \equiv u\rho D/\mu$, where D is pipe diameter and μ is dynamic viscosity. The flow is turbulent for $Re > 3000$.

Consider the flow of liquid water at 25°C. For one of the sets of conditions given below, determine $\dot{m}$ (in $kg \cdot s^{-1}$) and $\Delta P/\Delta L$ (in $kPa \cdot m^{-1}$). Assume $\in/D = 0.0001$. For liquid water at 25°C, $\rho = 996$ $kg \cdot m^{-3}$, and $\mu = 9.0 \times 10^{-4}$ $kg \cdot m^{-1} \cdot s^{-1}$. Verify that the flow is turbulent.

(a) $D = 2$ cm, $u = 1$ $m \cdot s^{-1}$
(b) $D = 5$ cm, $u = 1$ $m \cdot s^{-1}$
(c) $D = 2$ cm, $u = 5$ $m \cdot s^{-1}$
(d) $D = 5$ cm, $u = 5$ $m \cdot s^{-1}$

2.46. Ethylene enters a turbine at 10 bar and 450 K, and exhausts at 1(atm) and 325 K. For $\dot{m} = 4.5$ $kg \cdot s^{-1}$, determine the cost C of the turbine. State any assumptions you make.

Data: $H_1 = 761.1$ $H_2 = 536.9$ $kJ \cdot kg^{-1}$ $C/\$ = (15,200) \left(|\dot{W}|/kW \right)^{0.573}$

2.47. The heating of a home to increase its temperature must be modeled as an open system because expansion of the household air at constant pressure results in leakage of air to the outdoors. Assuming that the molar properties of air leaving the home are the same as those of the air in the home, show that energy and mole balances yield the following differential equation:

$$\dot{Q} = -PV\frac{dn}{dt} + n\frac{dU}{dt}$$

Here, $\dot{Q}$ is the rate of heat transfer to the air in the home, and t is time. Quantities P, V, n, and U refer to the air in the home.

2.48. (*a*) Water flows through the nozzle of a garden hose. Find an expression for $\dot{m}$ in terms of line pressure P_1, ambient pressure P_2, inside hose diameter D_1, and nozzle outlet diameter D_2. Assume steady flow, and isothermal, adiabatic operation. For liquid water modeled as an incompressible fluid, $H_2 - H_1 = (P_2 - P_1)/\rho$ for constant temperature.

(*b*) In fact, the flow cannot be truly isothermal: we expect $T_2 > T_1$, owing to fluid friction. Hence, $H_2 - H_1 = C(T_2 - T_1) + (P_2 - P_1)/\rho$, where C is the specific heat of water. Directionally, how would inclusion of the temperature change affect the value of $\dot{m}$ as found in Part (*a*)?

Chapter 3

Volumetric Properties
of Pure Fluids

The equations of the preceding chapter provide the means for calculation of the heat and work quantities associated with various processes, but they are useless without knowledge of property values for internal energy or enthalpy. Such properties differ from one substance to another, and the laws of thermodynamics themselves do not provide any description or model of material behavior. Property values come from experiment, or from the correlated results of experiment, or from models grounded in and validated by experiment. Because there are no internal-energy or enthalpy meters, indirect measurement is the rule. For fluids, the most comprehensive procedure requires measurements of molar volume in relation to temperature and pressure. The resulting pressure/volume/temperature (*PVT*) data are most usefully correlated by *equations of state*, through which molar volume (or density), temperature, and pressure are functionally related.

In this chapter we:

- Present the phase rule, which relates the number of independent variables required to fix the thermodynamic state of a system to the number of chemical species and phases present

- Describe qualitatively the general nature of *PVT* behavior of pure substances

- Provide a detailed treatment of the ideal-gas state

- Treat equations of state, which are mathematical formulations of the *PVT* behavior of fluids

- Introduce generalized correlations that allow prediction of the *PVT* behavior of fluids for which experimental data are lacking

3.1 THE PHASE RULE

As indicated in Section 2.5, the state of a pure homogeneous fluid is fixed whenever two intensive thermodynamic properties are set at specific values. In contrast, when *two* phases of the same pure species are in equilibrium, the state of the system is fixed when only a single property is specified. For example, a system of steam and liquid water in equilibrium at

101.33 kPa can exist only at 100°C. It is impossible to change the temperature without also changing the pressure, if equilibrium between vapor and liquid phases is to be maintained. There is a single independent variable.

For a multiphase system at equilibrium, the number of independent variables that must be arbitrarily fixed to establish its *intensive* state is called the number of *degrees of freedom* of the system. This number is given by the phase rule of J. Willard Gibbs.[1] It is presented here without proof in the form applicable to nonreacting systems:[2]

$$\boxed{F = 2 - \pi + N} \tag{3.1}$$

where F is the number of degrees of freedom, π is the number of phases, and N is the number of chemical species present in the system.

The *intensive* state of a system at equilibrium is established when its temperature, pressure, and the compositions of all phases are fixed. These are the variables of the phase rule, but they are not all independent. The phase rule gives the number of variables from this set that must be specified to fix all remaining intensive variables, and thus the intensive state of the system.

A *phase* is a homogeneous region of matter. A gas or a mixture of gases, a liquid or a liquid solution, and a crystalline solid are examples of phases. An abrupt change in properties always occurs at the boundary between phases. Various phases can coexist, but they *must be in equilibrium* for the phase rule to apply. A phase need not be continuous; examples of discontinuous phases are a gas dispersed as bubbles in a liquid, a liquid dispersed as droplets in another liquid with which it is immiscible, and solid crystals dispersed in either a gas or a liquid. In each case a dispersed phase is distributed throughout a continuous phase.

As an example, the phase rule may be applied to an aqueous solution of ethanol in equilibrium with its vapor. Here $N = 2$, $\pi = 2$, and

$$F = 2 - \pi + N = 2 - 2 + 2 = 2$$

This is a system in vapor/liquid equilibrium, and it has two degrees of freedom. If the system exists at specified T and P (assuming this is possible), its liquid- and vapor-phase compositions are fixed by these conditions. A more common specification is of T and the liquid-phase composition, in which case P and the vapor-phase composition are fixed.

Intensive variables are independent of the size of the system and of the individual phases. Thus, the phase rule gives the same information for a large system as for a small one and for different relative amounts of the phases. Moreover, the phase rule applies only to individual-phase compositions, and not to the overall composition of a multiphase system. Note also that for a phase only $N - 1$ compositions are independent, because the mole or mass fractions of a phase must sum to unity.

The minimum number of degrees of freedom for any system is zero. When $F = 0$, the system is *invariant*; Eq. (3.1) becomes $\pi = 2 + N$. This value of π is the maximum number of phases that can coexist at equilibrium for a system containing N chemical species. When $N = 1$, this limit is reached for $\pi = 3$, characteristic of a triple point (Sec. 3.2). For example, the

---

[1]Josiah Willard Gibbs (1839–1903), American mathematical physicist, who deduced it in 1875. See http://en.wikipedia.org/wiki/Willard_Gibbs

[2]The theoretical justification of the phase rule for nonreacting systems is given in Sec. 12.2, and the phase rule for reacting systems is considered in Sec. 14.8.

triple point of H_2O, where liquid, vapor, and the common form of ice exist together in equilibrium, occurs at $0.01°C$ and 0.0061 bar. Any change from these conditions causes at least one phase to disappear.

Example 3.1

How many phase-rule variables must be specified to fix the thermodynamic state of each of the following systems?

(a) Liquid water in equilibrium with its vapor.

(b) Liquid water in equilibrium with a mixture of water vapor and nitrogen.

(c) A three-phase system of a saturated aqueous salt solution at its boiling point with excess salt crystals present.

Solution 3.1

(a) The system contains a single chemical species existing as two phases (one liquid and one vapor), and

$$F = 2 - \pi + N = 2 - 2 + 1 = 1$$

This result is in agreement with the fact that for a given pressure water has but one boiling point. Temperature or pressure, but not both, may be specified for a system comprised of water in equilibrium with its vapor.

(b) Two chemical species are present. Again there are two phases, and

$$F = 2 - \pi + N = 2 - 2 + 2 = 2$$

The addition of an inert gas to a system of water in equilibrium with its vapor changes the characteristics of the system. Now temperature and pressure may be independently varied, but once they are fixed the system described can exist in equilibrium only at a particular composition of the vapor phase. (If nitrogen is considered negligibly soluble in water, the liquid phase is pure water.)

(c) The three phases ($\pi = 3$) are crystalline salt, the saturated aqueous solution, and vapor generated at the boiling point. The two chemical species ($N = 2$) are water and salt. For this system,

$$F = 2 - 3 + 2 = 1$$

3.2 *PVT* BEHAVIOR OF PURE SUBSTANCES

Figure 3.1 displays the equilibrium conditions of P and T at which solid, liquid, and gas phases of a pure substance exist. Lines 1-2 and 2-*C* represent the conditions at which solid and liquid phases exist in equilibrium with a vapor phase. These *vapor pressure* versus temperature lines describe states of solid/vapor (line 1-2) and liquid/vapor (line 2-*C*) equilibrium. As indicated in Ex. 3.1(*a*), such systems have but a single degree of freedom. Similarly, solid/liquid equilibrium is represented by line 2-3. The three lines display conditions of *P* and *T* at which two phases may coexist, and they divide the diagram into single-phase regions. Line 1-2, the *sublimation curve*, separates the solid and gas regions; line 2-3, the *fusion curve*, separates the solid and liquid regions; line 2-*C*, the *vaporization curve*, separates the liquid and gas regions. Point *C* is known as the *critical point*; its coordinates P_c and T_c are the highest pressure and highest temperature at which a pure chemical species is observed to exist in vapor/liquid equilibrium.

The positive slope of the fusion line (2-3) represents the behavior of the vast majority of substances. Water, a very common substance, has some very uncommon properties, and exhibits a fusion line with negative slope.

The three lines meet at the *triple point*, where the three phases coexist in equilibrium. According to the phase rule the triple point is invariant ($F = 0$). If the system exists along any of the two-phase lines of Fig. 3.1, it is univariant ($F = 1$), whereas in the single-phase regions it is divariant ($F = 2$). Invariant, univariant, and divariant states appear as points, curves, and areas, respectively, on a *PT* diagram.

Changes of state can be represented by lines on the *PT* diagram: a constant-*T* change by a vertical line, and a constant-*P* change by a horizontal line. When such a line crosses a phase boundary, an abrupt change in properties of the fluid occurs at constant *T* and *P*; for example, vaporization for the transition from liquid to vapor.

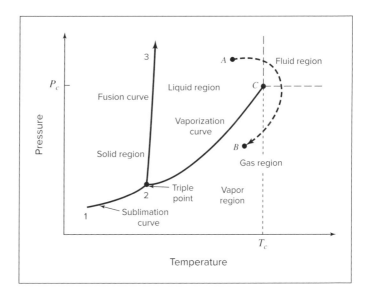

Figure 3.1: *PT* diagram for a pure substance.

Water in an open flask is obviously a liquid in contact with air. If the flask is sealed and the air is pumped out, water vaporizes to replace the air, and H_2O fills the flask. Though the pressure in the flask is much reduced, everything appears unchanged. The liquid water resides at the bottom of the flask because its density is much greater than that of water vapor (steam), and the two phases are in equilibrium at conditions represented by a point on curve 2-*C* of Fig. 3.1. Far from point *C*, the properties of liquid and vapor are very different. However, if the temperature is raised so that the equilibrium state progresses upward along curve 2-C, the properties of the two phases become more and more nearly alike; at point C they become identical, and the meniscus disappears. One consequence is that transitions from liquid to vapor may occur along paths that do not cross the vaporization curve 2-C, i.e., from A to B. The transition from liquid to gas is gradual and does not include the usual vaporization step.

The region existing at temperatures and pressures greater than T_c and P_c is marked off by dashed lines in Fig. 3.1; these do not represent phase boundaries, but rather are limits fixed by the meanings accorded the *words* liquid and gas. A phase is generally considered a liquid if vaporization results from pressure reduction at constant temperature. A phase is considered a gas if condensation results from temperature reduction at constant pressure. Since neither process can be initiated in the region beyond the dashed lines, it is called the *fluid region*.

The gas region is sometimes divided into two parts, as indicated by the dotted vertical line of Fig. 3.1. A gas to the left of this line, which can be condensed either by compression at constant temperature or by cooling at constant pressure, is called a vapor. A fluid existing at a temperature greater than T_c is said to be *supercritical*. An example is atmospheric air.

PV Diagram

Figure 3.1 does not provide any information about volume; it merely displays the boundaries between single-phase regions. On a *PV* diagram [Fig. 3.2(a)] these boundaries in turn become *regions* where two phases—solid/liquid, solid/vapor, and liquid/vapor—coexist in equilibrium. The curves that outline these two-phase regions represent single phases that are in equilibrium. Their relative amounts determine the molar (or specific) volumes within the two-phase regions. The triple point of Fig. 3.1 here becomes a triple *line*, where the three phases with different values of V coexist at a single temperature and pressure.

Figure 3.2(a), like Fig. 3.1, represents the behavior of the vast majority of substances wherein the transition from liquid to solid (freezing) is accompanied by a decrease in specific volume (increase in density), and the solid phase sinks in the liquid. Here again water displays unusual behavior in that freezing results in an increase in specific volume (decrease in density), and on Fig. 3.2(a) the lines labeled *solid* and *liquid* are interchanged for water. Ice therefore floats on liquid water. Were it not so, the conditions on the earth's surface would be vastly different.

Figure 3.2(b) is an expanded view of the liquid, liquid/vapor, and vapor regions of the *PV* diagram, with four isotherms (paths of constant *T*) superimposed. Isotherms on Fig. 3.1 are vertical lines, and at temperatures greater than T_c do not cross a phase boundary. On Fig. 3.2(b) the isotherm labeled $T > T_c$ is therefore smooth.

The lines labeled T_1 and T_2 are for subcritical temperatures and consist of three segments. The horizontal segment of each isotherm represents all possible mixtures of liquid and vapor in equilibrium, ranging from 100% liquid at the left end to 100% vapor at the right end. The locus of these end points is the dome-shaped curve labeled BCD, the left half of which (from B to C) represents single-phase liquids at their vaporization (boiling) temperatures and the right half (from C to D) single-phase vapors at their condensation temperatures. Liquids and vapors represented by BCD are said to be *saturated*, and coexisting phases are connected by the horizontal segment of the isotherm at the *saturation* pressure specific to the isotherm. Also called the vapor pressure, it is given by a point on Fig. 3.1 where an isotherm (vertical line) crosses the vaporization curve.

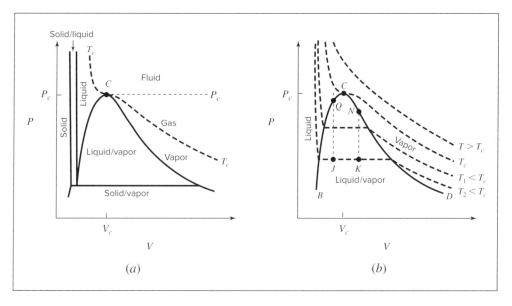

Figure 3.2: *PV* diagrams for a pure substance. (*a*) Showing solid, liquid, and gas regions. (*b*) Showing liquid, liquid/vapor, and vapor regions with isotherms.

The two-phase liquid/vapor region lies under dome *BCD*; the *subcooled-liquid* region lies to the left of the saturated-liquid curve *BC*, and the *superheated-vapor* region lies to the right of the saturated-vapor curve *CD*. Subcooled liquid exists at temperatures *below*, and superheated vapor, at temperatures *above* the boiling point for the given pressure. Isotherms in the subcooled-liquid region are very steep because liquid volumes change little with large changes in pressure.

The horizontal segments of the isotherms in the two-phase region become progressively shorter at higher temperatures, being ultimately reduced to a point at *C*. Thus, the critical isotherm, labeled T_c, exhibits a horizontal inflection at the critical point *C* at the top of the dome, where the liquid and vapor phases become indistinguishable.

Critical Behavior

Insight into the nature of the critical point is gained from a description of the changes that occur when a pure substance is heated in a sealed upright tube of constant volume. The dotted vertical lines of Fig. 3.2(*b*) indicate such processes. They may also be traced on the *PT* diagram of Fig. 3.3, where the solid line is the vaporization curve (Fig. 3.1), and the dashed lines are constant-volume paths in the single-phase regions. If the tube is filled with either liquid or vapor, the heating process produces changes that lie along the dashed lines of Fig. 3.3, for example, by the change from *E* to *F* (subcooled-liquid) and by the change from *G* to *H* (superheated-vapor). The corresponding vertical lines on Fig. 3.2(*b*) are not shown, but they lie to the left and right of *BCD* respectively.

If the tube is only partially filled with liquid (the remainder being vapor in equilibrium with the liquid), heating at first causes changes described by the vapor-pressure curve (solid

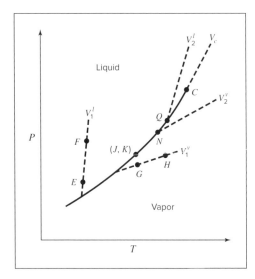

Figure 3.3: PT diagram for a pure fluid showing the vapor-pressure curve and constant-volume lines in the single-phase regions.

line of Fig. 3.3). For the process indicated by line JQ on Fig. 3.2(b), the meniscus is initially near the top of the tube (point J), and the liquid expands sufficiently upon heating to fill the tube (point Q). On Fig. 3.3 the process traces a path from (J, K) to Q, and with further heating departs from the vapor-pressure curve along the line of constant molar volume V_2^l.

The process indicated by line KN on Fig. 3.2(b) starts with a meniscus level closer to the bottom of the tube (point K), and heating vaporizes liquid, causing the meniscus to recede to the bottom (point N). On Fig. 3.3 the process traces a path from (J, K) to N. With further heating the path continues along the line of constant molar volume V_2^v.

For a unique filling of the tube, with a particular intermediate meniscus level, the heating process follows a vertical line on Fig. 3.2(b) that passes through the critical point C. Physically, heating does not produce much change in the level of the meniscus. As the critical point is approached, the meniscus becomes indistinct, then hazy, and finally disappears. On Fig. 3.3 the path first follows the vapor-pressure curve, proceeding from point (J, K) to the critical point C, where it enters the single-phase fluid region, and follows V_c, the line of constant molar volume equal to the critical volume of the fluid.[3]

PVT Surfaces

For a pure substance, existing as a single phase, the phase rule tells us that two state variables must be specified to determine the intensive state of the substance. Any two, from among P, V, and T, can be selected as the specified, or independent, variables, and the third can then be regarded as a function of those two. Thus, the relationship among P, V, and T for a pure substance can be represented as a surface in three dimensions. PT and PV diagrams like those illustrated in Figs. 3.1, 3.2, and 3.3 represent slices or projections of the three-dimensional PVT surface. Fig. 3.4 presents a view of the PVT surface for carbon dioxide over a region including liquid, vapor, and supercritical fluid states. Isotherms are superimposed on this

---

[3]A video illustrating this behavior is available in the online learning center at http://highered.mheducation.com/sites/1259696529.

surface. The vapor/liquid equilibrium curve is shown in white, with the vapor and liquid portions of it connected by the vertical segments of the isotherms. Note that for ease of visualization, the molar volume is given on a logarithmic scale, because the vapor volume at low pressure is several orders of magnitude larger than the liquid volume.

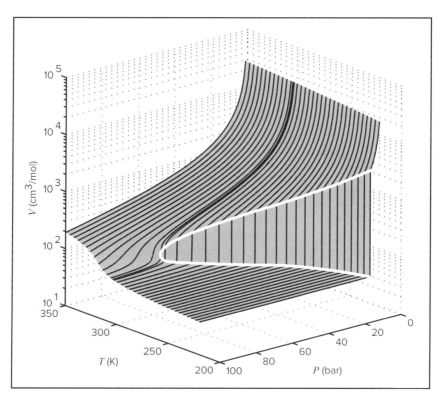

Figure 3.4: *PVT* surface for carbon dioxide, with isotherms shown in black and the vapor/liquid equilibrium curve in white.

Single-Phase Regions

For the regions of the diagram where a single phase exists, there is a unique relation connecting P, V, and T. Expressed analytically, as $f(P, V, T) = 0$, such a relation is known as a *PVT equation of state*. It relates pressure, molar or specific volume, and temperature for a pure homogeneous fluid at equilibrium. The simplest example, the equation for the ideal-gas state, $PV = RT$, has approximate validity for the low-pressure gas region and is discussed in detail in the following section.

An equation of state may be solved for any one of the three quantities P, V, or T, given values for other two. For example, if V is considered a function of T and P, then $V = V(T, P)$, and

$$dV = \left(\frac{\partial V}{\partial T}\right)_P dT + \left(\frac{\partial V}{\partial P}\right)_T dP \tag{3.2}$$

The partial derivatives in this equation have definite physical meanings and are related to two properties, commonly tabulated for liquids, and **defined** as follows:

- *Volume expansivity:*

$$\beta \equiv \frac{1}{V}\left(\frac{\partial V}{\partial T}\right)_P \tag{3.3}$$

- *Isothermal compressibility:*

$$\kappa \equiv -\frac{1}{V}\left(\frac{\partial V}{\partial P}\right)_T \tag{3.4}$$

Combining Eqs. (3.2) through (3.4) yields:

$$\frac{dV}{V} = \beta\, dT - \kappa\, dP \tag{3.5}$$

The isotherms for the liquid phase on the left side of Fig. 3.2(b) are very steep and closely spaced. Thus both $(\partial V/\partial T)_P$ and $(\partial V/\partial P)_T$ are small. Hence, both β and κ are small. This characteristic behavior of liquids (outside the critical region) suggests an idealization, commonly employed in fluid mechanics and known as the *incompressible fluid*, for which both β and κ are zero. No real fluid is truly incompressible, but the idealization is useful, because it provides a sufficiently realistic model of liquid behavior for many practical purposes. No equation of state exists for an incompressible fluid, because V is independent of T and P.

For liquids, β is almost always positive (liquid water between 0°C and 4°C is an exception), and κ is necessarily positive. At conditions not close to the critical point, β and κ are weak functions of temperature and pressure. Thus for small changes in T and P little error is introduced if they are assumed constant. Integration of Eq. (3.5) then yields:

$$\ln\frac{V_2}{V_1} = \beta(T_2 - T_1) - \kappa(P_2 - P_1) \tag{3.6}$$

This is a less restrictive approximation than the assumption of an incompressible fluid.

Example 3.2

For liquid acetone at 20°C and 1 bar,

$$\beta = 1.487 \times 10^{-3}\,°C^{-1} \quad \kappa = 62 \times 10^{-6}\,bar^{-1} \quad V = 1.287\,cm^3 \cdot g^{-1}$$

For acetone, find:

(a) The value of $(\partial P/\partial T)_V$ at 20°C and 1 bar.

(b) The pressure after heating at constant V from 20°C and 1 bar to 30°C.

(c) The volume change when T and P go from 20°C and 1 bar to 0°C and 10 bar.

Solution 3.2

(*a*) The derivative $(\partial P / \partial T)_V$ is determined by application of Eq. (3.5) to the case for which V is constant and $dV = 0$:

$$\beta\, dT - \kappa\, dP = 0 \quad (\text{const } V)$$

or

$$\left(\frac{\partial P}{\partial T}\right)_V = \frac{\beta}{\kappa} = \frac{1.487 \times 10^{-3}}{62 \times 10^{-6}} = 24 \text{ bar} \cdot {}^\circ C^{-1}$$

(*b*) If β and κ are assumed constant in the $10^\circ C$ temperature interval, then for constant volume Eq. (3.6) can be written:

$$P_2 = P_1 + \frac{\beta}{\kappa}(T_2 - T_1) = 1 \text{ bar} + 24 \text{ bar} \cdot {}^\circ C^{-1} \times 10^\circ C = 241 \text{ bar}$$

(*c*) Direct substitution into Eq. (3.6) gives:

$$\ln\frac{V_2}{V_1} = (1.487 \times 10^{-3})(-20) - (62 \times 10^{-6})(9) = -0.0303$$

$$\frac{V_2}{V_1} = 0.9702 \quad \text{and} \quad V_2 = (0.9702)(1.287) = 1.249 \text{ cm}^3 \cdot g^{-1}$$

Then,

$$\Delta V = V_2 - V_1 = 1.249 - 1.287 = -0.038 \text{ cm}^3 \cdot g^{-1}$$

The preceding example illustrates the fact that heating a liquid that completely fills a closed vessel can cause a substantial rise in pressure. On the other hand, liquid volume decreases very slowly with rising pressure. Thus, the very high pressure generated by heating a subcooled liquid at constant volume can be relieved by a very small volume increase, or a very small leak in the constant volume container.

3.3 IDEAL GAS AND IDEAL-GAS STATE

In the 19th century, scientists developed a rough experimental knowledge of the *PVT* behavior of gases at moderate conditions of temperature and pressure, leading to the equation $PV = RT$, wherein V is molar volume and R is a universal constant. This equation adequately describes *PVT* behavior of gases for many practical purposes near ambient conditions of T and P. However, more precise measurements show that for pressures appreciably above, and temperatures appreciably below, ambient conditions, deviations become pronounced. On the other hand, deviations become ever smaller as pressure decreases and temperature increases.

The equation $PV = RT$ is now understood to define an *ideal gas* and to represent a model of behavior more or less approximating the behavior of real gases. It is called the *ideal gas law*, but is in fact valid only for pressures approaching zero and temperatures approaching infinity. Thus, it is a *law* only at limiting conditions. As these limits are approached, the molecules making up a gas become more and more widely separated, and the volume of the molecules themselves becomes a smaller and smaller fraction of the total volume occupied by the gas. Furthermore, the forces of attraction between molecules become ever smaller because of the increasing distances between them. In the zero-pressure limit, molecules are separated by infinite distances. Their volumes become negligible compared with the total volume of the gas, and the intermolecular forces approach zero. The ideal gas concept extrapolates this behavior to all conditions of temperature and pressure.

The internal energy of a real gas depends on both pressure and temperature. Pressure dependence results from intermolecular forces. If such forces did not exist, no energy would be required to alter intermolecular distances, and no energy would be required to bring about pressure and volume changes in a gas at constant temperature. Thus, in the absence of intermolecular forces, internal energy would depend on temperature only.

These observations are the basis for the concept of a hypothetical state of matter designated the *ideal-gas state*. It is the state of a gas comprised of real molecules that have negligible molecular volume and no intermolecular forces at all temperatures and pressures. Although related to the ideal gas, it presents a different perspective. It is not the gas that is ideal, but the state, and this has practical advantages. Two equations are fundamental to this state, namely the "ideal-gas law" and an expression showing that internal energy depends on temperature alone:

- The equation of state:

$$\boxed{PV^{ig} = RT} \tag{3.7}$$

- Internal energy:

$$\boxed{U^{ig} = U(T)} \tag{3.8}$$

The superscript *ig* denotes properties for the ideal-gas state.

The property relations for this state are very simple, and at appropriate conditions of T and P they may serve as suitable approximations for direct application to the real-gas state. However, they have far greater importance as part of a general three-step procedure for calculation of property changes for real gases that includes a major step in the ideal-gas state. The three steps are as follows:

1. Evaluate property changes for the *mathematical* transformation of an initial real-gas state into the ideal-gas state at the same T and P.

2. Calculate property changes in the ideal-gas state for the T and P changes of the process.

3. Evaluate property changes for the *mathematical* transformation of the ideal-gas state back to the real-gas state at the final T and P.

This procedure calculates the primary property-value changes resulting from T and P changes by simple, but exact, equations for the ideal-gas state. The property-value changes for transitions between real and ideal-gas states are usually relatively minor corrections. These transition calculations are treated in Chapter 6. Here, we develop property-value calculations for the ideal-gas state.

Property Relations for the Ideal-Gas State

The definition of heat capacity at constant volume, Eq. (2.15), leads for the ideal-gas state to the conclusion that C_V^{ig} is a function of temperature only:

$$C_V^{ig} \equiv \left(\frac{\partial U^{ig}}{\partial T} \right)_V = \frac{dU^{ig}(T)}{dT} = C_V^{ig}(T) \tag{3.9}$$

The defining equation for enthalpy, Eq. (2.10), applied to the ideal-gas state, leads to the conclusion that H^{ig} is also a function only of temperature:

$$H^{ig} \equiv U^{ig} + PV^{ig} = U^{ig}(T) + RT = H^{ig}(T) \tag{3.10}$$

The heat capacity at constant pressure C_P^{ig}, defined by Eq. (2.19), like C_V^{ig}, is a function of temperature only:

$$C_P^{ig} \equiv \left(\frac{\partial H^{ig}}{\partial T} \right)_P = \frac{dH^{ig}(T)}{dT} = C_P^{ig}(T) \tag{3.11}$$

A useful relation between C_P^{ig} and C_V^{ig} for the ideal-gas state comes from differentiation of Eq. (3.10):

$$C_P^{ig} \equiv \frac{dH^{ig}}{dT} = \frac{dU^{ig}}{dT} + R = C_V^{ig} + R \tag{3.12}$$

This equation does not mean that C_P^{ig} and C_V^{ig} are themselves constant for the ideal-gas state, but only that they vary with temperature in such a way that their *difference* is equal to R. For any change in the ideal-gas state, Eqs. (3.9) and (3.11) lead to:

$dU^{ig} = C_V^{ig}dT$ (3.13a)	$\Delta U^{ig} = \displaystyle\int C_V^{ig}dT$ (3.13b)
$dH^{ig} = C_P^{ig}dT$ (3.14a)	$\Delta H^{ig} = \displaystyle\int C_P^{ig}dT$ (3.14b)

Because both U^{ig} and C_V^{ig} for the ideal-gas state are functions of temperature only, ΔU^{ig} for the ideal-gas state is *always* given by Eq. (3.13b), regardless of the kind of process causing the change. This is illustrated in Fig. 3.5, which shows a graph of internal energy as a function of V^{ig} at two different temperatures.

The dashed line connecting points a and b represents a constant-volume process for which the temperature increases from T_1 to T_2 and the internal energy changes by $\Delta U^{ig} = U_2^{ig} - U_1^{ig}$. This change in internal energy is given by Eq. (3.13b) as $\Delta U^{ig} = \int C_V^{ig}dT$. The dashed lines connecting points a and c and points a and d represent other processes not occurring at constant volume but which also lead from an initial temperature T_1 to a final temperature T_2. The graph shows that the change in U^{ig} for these processes is the same as for the constant-volume process, and it is therefore given by the same equation, namely, $\Delta U^{ig} = \int C_V^{ig}dT$. However, ΔU^{ig} is *not* equal to Q for these processes, because Q depends not only on T_1 and T_2 but also on the path of the process. An entirely analogous discussion applies to the enthalpy H^{ig} in the ideal-gas state.

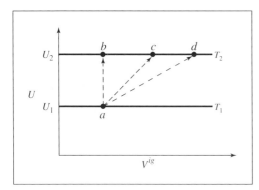

Figure 3.5: Internal energy changes for the ideal-gas state. Because U^{ig} is independent of V^{ig}, the plot of U^{ig} vs. V^{ig} at constant temperature is a horizontal line. For different temperatures, U^{ig} has different values, with a separate line for each temperature. Two such lines are shown, one for temperature T_1 and one for a higher temperature T_2.

Process Calculations for the Ideal-Gas State

Process calculations provide work and heat quantities. The work of a mechanically reversible closed-system process is given by Eq. (1.3), here written:

$$dW = -P\,dV^{ig} \tag{1.3}$$

For the ideal-gas state in any closed-system process, the first law as given by Eq. (2.6) written for a unit mass or a mole, may be combined with Eq. (3.13a) to give:

$$dQ + dW = C_V^{ig}\,dT$$

Substitution for dW by Eq. (1.3) and solution for dQ yields an equation valid for the ideal-gas state in any mechanically reversible closed-system process:

$$dQ = C_V^{ig}\,dT + P\,dV^{ig} \tag{3.15}$$

This equation contains the variables P, V^{ig}, and T, only two of which are independent. Working equations for dQ and dW depend on which pair of these variables is selected as independent; i.e., upon which variable is eliminated by Eq. (3.7). We consider two cases, eliminating first P, and second, V^{ig}. With $P = RT/V^{ig}$, Eqs. (3.15) and (1.3) become:

$$dQ = C_V^{ig}\,dT + RT\frac{dV^{ig}}{V^{ig}} \quad (3.16) \qquad dW = -RT\frac{dV^{ig}}{V^{ig}} \quad (3.17)$$

For $V^{ig} = RT/P$, $dV^{ig} = \dfrac{R}{P}(dT - T\dfrac{dP}{P})$. Substituting for dV^{ig} and for $C_V^{ig} = C_P^{ig} - R$ transforms Eqs. (3.15) and (1.3) into:

$$dQ = C_P^{ig}\,dT - RT\frac{dP}{P} \quad (3.18) \qquad dW = -R\,dT + RT\frac{dP}{P} \quad (3.19)$$

These equations apply to the ideal-gas state for various process calculations. The assumptions implicit in their derivation are that the system is closed and the process is mechanically reversible.

Isothermal Process

By Eqs. (3.13b) and (3.14b),

$$\Delta U^{ig} = \Delta H^{ig} = 0 \quad \text{(const } T)$$

By Eqs. (3.16) and (3.18),
$$Q = RT \ln \frac{V_2^{ig}}{V_1^{ig}} = RT \ln \frac{P_1}{P_2}$$

By Eqs. (3.17) and (3.19),
$$W = RT \ln \frac{V_1^{ig}}{V_2^{ig}} = RT \ln \frac{P_2}{P_1}$$

Because $Q = -W$, a result that also follows from Eq. (2.3), we can write in summary:

$$Q = -W = RT \ln \frac{V_2^{ig}}{V_1^{ig}} = RT \ln \frac{P_1}{P_2} \quad \text{(const } T) \tag{3.20}$$

Isobaric Process

By Eqs. (3.13b) and (3.19) with $dP = 0$,

$$\Delta U^{ig} = \int C_V^{ig} dT \quad \text{and} \quad W = -R(T_2 - T_1)$$

By Eqs. (3.14b) and (3.18),

$$Q = \Delta H^{ig} = \int C_P^{ig} dT \quad \text{(const } P) \tag{3.21}$$

Isochoric (Constant-V) Process

With $dV^{ig} = 0$, $W = 0$, and by Eqs. (3.13b) and (3.16),

$$Q = \Delta U^{ig} = \int C_V^{ig} dT \quad \text{(const } V^{ig}) \tag{3.22}$$

Adiabatic Process; Constant Heat Capacities

An adiabatic process is one for which there is no heat transfer between the system and its surroundings; i.e., $dQ = 0$. Each of Eqs. (3.16) and (3.18) may therefore be set equal to zero. Integration with C_V^{ig} and C_P^{ig} constant then yields simple relations among the variables T, P, and V^{ig}, valid for mechanically reversible adiabatic compression or expansion in the ideal-gas state with constant heat capacities. For example, Eq. (3.16) becomes:

$$\frac{dT}{T} = -\frac{R}{C_V^{ig}} \frac{dV^{ig}}{V^{ig}}$$

Integration with C_V^{ig} constant gives:

$$\frac{T_2}{T_1} = \left(\frac{V_1^{ig}}{V_2^{ig}}\right)^{R/C_V^{ig}}$$

Similarly, Eq. (3.18) leads to:

$$\frac{T_2}{T_1} = \left(\frac{P_2}{P_1}\right)^{R/C_P^{ig}}$$

These equations may also be expressed as:

$T(V^{ig})^{\gamma-1} = \text{const}$ (3.23a)	$TP^{(1-\gamma)/\gamma} = \text{const}$ (3.23b)	$P(V^{ig})^{\gamma} = \text{const}$ (3.23c)

where Eq. (3.23c) results by combining Eqs. (3.23a and 3.23b) and where by **definition,**[4]

$$\gamma \equiv \frac{C_P^{ig}}{C_V^{ig}} \qquad (3.24)$$

Equations (3.23) apply for the ideal-gas state with constant heat capacities and are restricted to mechanically reversible adiabatic expansion or compression.

The first law for an adiabatic process in a closed system combined with Eq. (3.13a) yields:

$$dW = dU = C_V^{ig}\, dT$$

For constant C_V^{ig},

$$W = \Delta U^{ig} = C_V^{ig}\, \Delta T \qquad (3.25)$$

Alternative forms of Eq. (3.25) result if C_V^{ig} is eliminated in favor of the heat-capacity ratio γ:

$$\gamma \equiv \frac{C_P^{ig}}{C_V^{ig}} = \frac{C_V^{ig} + R}{C_V^{ig}} = 1 + \frac{R}{C_V^{ig}} \qquad \text{or} \qquad C_V^{ig} = \frac{R}{\gamma - 1}$$

and

$$W = C_V^{ig}\,\Delta T = \frac{R\Delta T}{\gamma - 1}$$

---

[4]If C_P^{ig} and C_V^{ig} are constant, γ is necessarily constant. The assumption of constant γ is equivalent to the assumption that the heat capacities themselves are constant. This is the only way that the ratio C_P^{ig}/C_V^{ig} and the difference $C_P^{ig}-C_V^{ig} = R$ can *both* be constant. Except for the monatomic gases, both C_P^{ig} and C_V^{ig} actually increase with temperature, but the ratio γ is less sensitive to temperature than the heat capacities themselves.

Because $RT_1 = P_1 V_1^{ig}$ and $RT_2 = P_2 V_2^{ig}$, this expression may be written:

$$W = \frac{RT_2 - RT_1}{\gamma - 1} = \frac{P_2 V_2^{ig} - P_1 V_1^{ig}}{\gamma - 1} \tag{3.26}$$

Equations (3.25) and (3.26) are general for adiabatic compression and expansion processes in a closed system, whether reversible or not, because P, V^{ig}, and T are state functions, independent of path. However, T_2 and V_2^{ig} are usually unknown. Elimination of V_2^{ig} from Eq. (3.26) by Eq. (3.23c), valid only for mechanically reversible processes, leads to the expression:

$$W = \frac{P_1 V_1^{ig}}{\gamma - 1} \left[\left(\frac{P_2}{P_1} \right)^{(\gamma - 1)/\gamma} - 1 \right] = \frac{RT_1}{\gamma - 1} \left[\left(\frac{P_2}{P_1} \right)^{(\gamma - 1)/\gamma} - 1 \right] \tag{3.27}$$

The same result is obtained when the relation between P and V^{ig} given by Eq. (3.23c) is used for the integration, $W = -\int P\, dV^{ig}$.

Equation (3.27) is valid only for the ideal-gas state, for constant heat capacities, and for adiabatic, mechanically reversible, closed-system processes.

When applied to real gases, Eqs. (3.23) through (3.27) often yield satisfactory approximations, provided the deviations from ideality are relatively small. For monatomic gases, $\gamma = 1.67$; approximate values of γ are 1.4 for diatomic gases and 1.3 for simple polyatomic gases such as CO_2, SO_2, NH_3, and CH_4.

Irreversible Processes

All equations developed in this section have been *derived* for mechanically reversible, closed-system processes for the ideal-gas state. However, the equations for *property changes*— dU^{ig}, dH^{ig}, ΔU^{ig}, and ΔH^{ig}—are valid for the ideal-gas state *regardless of the process*. They apply equally to reversible and irreversible processes in both closed and open systems, because changes in properties depend only on initial and final states of the system. On the other hand, an equation for Q or W, unless it is equal to a property change, is subject to the restrictions of its derivation.

The work of an *irreversible* process is usually calculated by a two-step procedure. First, W is determined for a mechanically reversible process that accomplishes the same change of state as the actual irreversible process. Second, this result is multiplied or divided by an efficiency to give the actual work. If the process produces work, the absolute value for the reversible process is larger than the value for the actual irreversible process and must be multiplied by an efficiency. If the process requires work, the value for the reversible process is smaller than the value for the actual irreversible process and must be divided by an efficiency.

Applications of the concepts and equations of this section are illustrated in the examples that follow. In particular, the work of irreversible processes is treated in Ex. 3.5.

Example 3.3

Air is compressed from an initial state of 1 bar and 298.15 K to a final state of 3 bar and 298.15 K by three different mechanically reversible processes in a closed system:

(a) Heating at constant volume followed by cooling at constant pressure.

(b) Isothermal compression.

(c) Adiabatic compression followed by cooling at constant volume.

These processes are shown in the figure. We assume air to be in its ideal-gas state, and assume constant heat capacities, $C_V^{ig} = 20.785$ and $C_P^{ig} = 29.100$ J·mol^{-1}·K^{-1}. Calculate the work required, heat transferred, and the changes in internal energy and enthalpy of the air for each process.

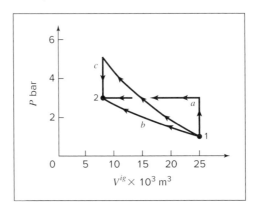

Solution 3.3

Choose the system as 1 mol of air. The initial and final states of the air are identical with those of Ex. 2.7. The molar volumes given there are

$$V_1^{ig} = 0.02479 \text{ m}^3 \qquad V_2^{ig} = 0.008263 \text{ m}^3$$

Because *T* is the same at the beginning and end of the process, in all cases,

$$\Delta U^{ig} = \Delta H^{ig} = 0$$

(a) The process here is exactly that of Ex. 2.7(b), for which:

$$Q = -4958 \text{ J} \quad \text{and} \quad W = 4958 \text{ J}$$

(b) Equation (3.20) for isothermal compression applies. The appropriate value of *R* here (from Table A.2 of App. A) is $R = 8.314$ J·mol^{-1}·K^{-1}.

$$Q = -W = RT \ln \frac{P_1}{P_2} = (8.314)(298.15) \ln \frac{1}{3} = -2723 \text{ J}$$

(c) The initial step of adiabatic compression takes the air to its final volume of 0.008263 m^3. By Eq. (3.23a), the temperature at this point is:

$$T' = T_1 \left(\frac{V_1^{ig}}{V_2^{ig}} \right)^{\gamma - 1} = (298.15) \left(\frac{0.02479}{0.008263} \right)^{0.4} = 462.69 \text{ K}$$

For this step, $Q = 0$, and by Eq. (3.25), the work of compression is:

$$W = C_V^{ig} \Delta T = C_V^{ig}(T' - T_1) = (20.785)(462.69 - 298.15) = 3420 \text{ J}$$

For the constant-volume step, no work is done; the heat transfer is:

$$Q = \Delta U^{ig} = C_V^{ig}(T_2 - T') = 20.785(298.15 - 462.69) = -3420 \text{ J}$$

Thus for process (c),

$$W = 3420 \text{ J} \quad \text{and} \quad Q = -3420 \text{ J}$$

Although the property changes ΔU^{ig} and ΔH^{ig} are zero for each process, Q and W are path-dependent, and here $Q = -W$. The figure shows each process on a PV^{ig} diagram. Because the work for each of these mechanically reversible processes is given by $W = -\int P d V^{ig}$, the work for each process is proportional to the total area below the paths on the PV^{ig} diagram from 1 to 2. The relative sizes of these areas correspond to the numerical values of W.

Example 3.4

A gas in its ideal-gas state undergoes the following sequence of mechanically reversible processes in a closed system:

(a) From an initial state of 70°C and 1 bar, it is compressed adiabatically to 150°C.

(b) It is then cooled from 150 to 70°C at constant pressure.

(c) Finally, it expands isothermally to its original state.

Calculate W, Q, ΔU^{ig}, and ΔH^{ig} for each of the three processes and for the entire cycle. Take $C_V^{ig} = 12.471$ and $C_P^{ig} = 20.785$ J·mol^{-1}·K^{-1}.

Solution 3.4

Take as a basis 1 mol of gas.

(a) For adiabatic compression, $Q = 0$, and

$$\Delta U^{ig} = W = C_V^{ig} \Delta T = (12.471)(150 - 70) = 998 \text{ J}$$
$$\Delta H^{ig} = C_P^{ig} \Delta T = (20.785)(150 - 70) = 1663 \text{ J}$$

Pressure P_2 is found from Eq. (3.23b):

$$P_2 = P_1 \left(\frac{T_2}{T_1} \right)^{\gamma/(\gamma - 1)} = (1) \left(\frac{150 + 273.15}{70 + 273.15} \right)^{2.5} = 1.689 \text{ bar}$$

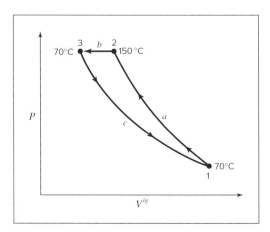

(*b*) For this constant-pressure process,

$$Q = \Delta H^{ig} = C_P^{ig} \, \Delta T = (20.785)(70 - 150) = -1663 \text{ J}$$

$$\Delta U = C_V^{ig} \, \Delta T = (12.471)(70 - 150) = -998 \text{ J}$$

$$W = \Delta U^{ig} - Q = -998 - (-1663) = 665 \text{ J}$$

(*c*) For this isothermal process, ΔU^{ig} and ΔH^{ig} are zero; Eq. (3.20) yields:

$$Q = -W = RT \ln \frac{P_3}{P_1} = RT \ln \frac{P_2}{P_1} = (8.314)(343.15) \ln \frac{1.689}{1} = 1495 \text{ J}$$

For the entire cycle,

$$Q = 0 - 1663 + 1495 = -168 \text{ J}$$
$$W = 998 + 665 - 1495 = 168 \text{ J}$$
$$\Delta U^{ig} = 998 - 998 + 0 = 0$$
$$\Delta H^{ig} = 1663 - 1663 + 0 = 0$$

The property changes ΔU^{ig} and ΔH^{ig} both are zero for the entire cycle because the initial and final states are identical. Note also that $Q = -W$ for the cycle. This follows from the first law with $\Delta U^{ig} = 0$.

Example 3.5

If the processes of Ex. 3.4 are carried out *irreversibly* but so as to accomplish exactly the same *changes of state*—the same changes in P, T, U^{ig}, and H^{ig}—then different values of Q and W result. Calculate Q and W if each step is carried out with a work efficiency of 80%.

Solution 3.5

If the same changes of state as in Ex. 3.4 are carried out by irreversible processes, the property changes for the steps are identical with those of Ex. 3.4. However, the values of Q and W change.

(*a*) For mechanically reversible, adiabatic compression, the work is $W_{rev} = 998$ J. If the process is 80% efficient compared with this, the actual work is larger, and $W = 998/0.80 = 1248$ J. This step cannot here be adiabatic. By the first law,

$$Q = \Delta U^{ig} - W = 998 - 1248 = -250 \text{ J}$$

(*b*) The work required for the mechanically reversible cooling process is 665 J. For the irreversible process, $W = 665/0.80 = 831$ J. From Ex. 3.4(*b*), $\Delta U^{ig} = -998$ J, and

$$Q = \Delta U^{ig} - W = -998 - 831 = -1829 \text{ J}$$

(*c*) As work is done *by* the system in this step, the irreversible work in absolute value is less than the reversible work of -1495 J, and the actual work done is:

$$W = (0.80)(-1495) = -1196 \text{ J}$$
$$Q = \Delta U^{ig} - W = 0 + 1196 = 1196 \text{ J}$$

For the entire cycle, ΔU^{ig} and ΔH^{ig} are zero, with

$$Q = -250 - 1829 + 1196 = -883 \text{ J}$$
$$W = 1248 + 831 - 1196 = 883 \text{ J}$$

A summary of these results and those for Ex. 3.4 is given in the following table; values are in joules.

	Mechanically reversible, Ex. 3.4				Irreversible, Ex. 3.5			
	ΔU^{ig}	ΔH^{ig}	Q	W	ΔU^{ig}	ΔH^{ig}	Q	W
(*a*)	998	1663	0	998	998	1663	-250	1248
(*b*)	-998	-1663	-1663	665	-998	-1663	-1829	831
(*c*)	0	0	1495	-1495	0	0	1196	-1196
Cycle	0	0	-168	168	0	0	-883	883

The cycle is one which requires work and produces an equal amount of heat. The striking feature of the comparison shown in the table is that the total work required when the cycle consists of three irreversible steps is more than five times the total work required when the steps are mechanically reversible, even though each irreversible step is assumed to be 80% efficient.

Example 3.6

Air flows at a steady rate through a horizontal pipe to a partly closed valve. The pipe leaving the valve is enough larger than the entrance pipe that the kinetic-energy change of the air as it flows through the valve is negligible. The valve and connecting pipes are well insulated. The conditions of the air upstream from the valve are 20°C and 6 bar, and the downstream pressure is 3 bar. If the air is in its ideal-gas state, what is the temperature of the air some distance downstream from the valve?

Solution 3.6

Flow through a partly closed valve is known as a *throttling process*. The system is insulated, making Q negligible; moreover, the potential-energy and kinetic-energy changes are negligible. No shaft work is accomplished, and $W_s = 0$. Hence, Eq. (2.31) reduces to:

$\Delta H^{ig} = H_2^{ig} - H_1^{ig} = 0$. Because H^{ig} is a function of temperature only, this requires that $T_2 = T_1$. The result that $\Delta H^{ig} = 0$ is general for a throttling process, because the assumptions of negligible heat transfer and potential- and kinetic-energy changes are usually valid. For a fluid in its ideal-gas state, no temperature change occurs. The throttling process is inherently irreversible, but this is immaterial to the calculation because Eq. (3.14b) is valid for the ideal-gas state whatever the process.[5]

Example 3.7

If in Ex. 3.6 the flow rate of air is 1 mol·s^{-1} and if both upstream and downstream pipes have an inner diameter of 5 cm, what is the kinetic-energy change of the air and what is its temperature change? For air, $C_P^{ig} = 29.100$ J·mol^{-1} and the molar mass is $\mathcal{M} = 29$ g·mol^{-1}.

Solution 3.7

By Eq. (2.23b),

$$u = \frac{\dot{n}}{A\rho} = \frac{\dot{n}V^{ig}}{A}$$

where

$$A = \frac{\pi}{4}D^2 = \left(\frac{\pi}{4}\right)(5 \times 10^{-2})^2 = 1.964 \times 10^{-3} \text{ m}^2$$

The appropriate value here of the gas constant for calculation of the upstream molar volume is $R = 83.14 \times 10^{-6}$ bar·m^3·mol^{-1}·K^{-1}. Then

$$V_1^{ig} = \frac{RT_1}{P_1} = \frac{(83.14 \times 10^{-6})(293.15 \text{ K})}{6 \text{ bar}} = 4.062 \times 10^{-3} \text{ m}^3\text{·mol}^{-1}$$

Then,

$$u_1 = \frac{(1 \text{ mol·s}^{-1})(4.062 \times 10^{-3} \text{ m}^3\text{·mol}^{-1})}{1.964 \times 10^{-3} \text{ m}^2} = 2.069 \text{ m·s}^{-1}$$

If the downstream temperature is little changed from the upstream temperature, then to a good approximation:

$$V_2^{ig} = 2V_1^{ig} \quad \text{and} \quad u_2 = 2u_1 = 4.138 \text{ m·s}^{-1}$$

[5]The throttling of real gases may result in a relatively small temperature increase or decrease, known as the Joule/Thomson effect. A more detailed discussion is found in Chapter 7.

The rate of change in kinetic energy is therefore:

$$\dot{m} \Delta \left(\frac{1}{2}u^2\right) = \dot{n}.\mathcal{M} \Delta \left(\frac{1}{2}u^2\right)$$

$$= (1 \times 29 \times 10^{-3} \text{ kg·s}^{-1}) \frac{(4.138^2 - 2.069^2)\text{m}^2\text{·s}^{-2}}{2}$$

$$= 0.186 \text{ kg·m}^2\text{·s}^{-3} = 0.186 \text{ J·s}^{-1}$$

In the absence of heat transfer and work, the energy balance, Eq. (2.30), becomes:

$$\Delta \left(H^{ig} + \frac{1}{2}u^2 \right)\dot{m} = \dot{m}\Delta H^{ig} + \dot{m} \Delta \left(\frac{1}{2}u^2\right) = 0$$

$$(\dot{m}/\mathcal{M})C_P^{ig}\Delta T + \dot{m} \Delta \left(\frac{1}{2}u^2\right) = \dot{n}C_P^{ig} \Delta T + \dot{m} \Delta \left(\frac{1}{2}u^2\right) = 0$$

Then

$$(1)(29.100)\Delta T = -\dot{m} \Delta \left(\frac{1}{2}u^2\right) = -0.186$$

and

$$\Delta T = -0.0064 \text{ K}$$

Clearly, the assumption of negligible temperature change across the valve is justified. Even for an upstream pressure of 10 bar and a downstream pressure of 1 bar and for the same flow rate, the temperature change is only -0.076 K. We conclude that, except for very unusual conditions, $\Delta H^{ig} = 0$ is a satisfactory energy balance.

3.4 VIRIAL EQUATIONS OF STATE

Volumetric data for fluids are useful for many purposes, from the metering of fluids to the sizing of tanks. Data for V as a function of T and P can of course be given as tables. However, expression of the functional relation $f(P, V, T) = 0$ by equations is much more compact and convenient. The virial equations of state for gases are uniquely suited to this purpose.

Isotherms for gases and vapors, lying to the right of the saturated-vapor curve CD in Fig. 3.2(b), are relatively simple curves for which V decreases as P increases. Here, the product PV for a given T varies much more slowly than either of its members, and hence is more easily represented analytically as a function of P. This suggests expressing PV for an isotherm by a power series in P:

$$PV = a + bP + cP^2 + \ldots$$

If we define, $b \equiv aB'$, $c \equiv aC'$, etc., then,

$$PV = a\left(1 + B'P + C'P^2 + D'P^3 + \ldots\right) \tag{3.28}$$

where B', C', etc., are constants for a given temperature and a given substance.

Ideal-Gas Temperature; Universal Gas Constant

Parameters B', C', etc., in Eq. (3.28) are species-dependent functions of temperature, but parameter a is found by experiment to be the same function of temperature for all chemical species. This is shown by isothermal measurements of V as a function of P for various gases. Extrapolations of the PV product to zero pressure, where Eq. (3.28) reduces to $PV = a$, show that a is the same function of T for all gases. Denoting this zero-pressure limit by an asterisk provides:

$$(PV)^* = a = f(T)$$

This property of gases serves as a basis for an absolute temperature scale. It is defined by arbitrary assignment of the functional relationship $f(T)$ and the assignment of a specific value to a single point on the scale. The simplest procedure, the one adopted internationally to define the Kelvin scale (Sec. 1.4):

- Makes $(PV)^*$ directly proportional to T, with R as the proportionality constant:

$$(PV)^* = a \equiv RT \tag{3.29}$$

- Assigns the value 273.16 K to the temperature of the triple point of water (denoted by subscript t):

$$(PV)_t^* = R \times 273.16 \text{ K} \tag{3.30}$$

Division of Eq. (3.29) by Eq. (3.30) gives:

$$T\text{ K} = 273.16 \frac{(PV)^*}{(PV)_t^*} \tag{3.31}$$

This equation provides the experimental basis for the *ideal-gas temperature scale* throughout the temperature range for which values of $(PV)^*$ are experimentally accessible. The Kelvin temperature scale is defined so as to be in as close agreement as possible with this scale.

The proportionality constant R in Eqs. (3.29) and (3.30) is the *universal gas constant*. Its numerical value is found from Eq. (3.30):

$$R = \frac{(PV)_t^*}{273.16 \text{ K}}$$

The accepted experimental value of $(PV)_t^*$ is 22,711.8 bar·cm^3·mol^{-1}, from which[6]

$$R = \frac{22{,}711.8 \text{ bar·cm}^3\text{·mol}^{-1}}{273.16 \text{ K}} = 83.1446 \text{ bar·cm}^3\text{·mol}^{-1}\text{·K}^{-1}$$

Through the use of conversion factors, R may be expressed in various units. Commonly used values are given in Table A.2 of App. A.

Two Forms of the Virial Equation

A useful auxiliary thermodynamic property is **defined** by the equation:

$$\boxed{Z \equiv \frac{PV}{RT} = \frac{V}{V^{ig}}} \tag{3.32}$$

---

[6]See http://physics.nist.gov/constants.

This dimensionless ratio is called the *compressibility factor*. It is a measure of the deviation of the real-gas molar volume from its ideal-gas value. For the ideal-gas state, $Z = 1$. At moderate temperatures its value is usually <1, though at elevated temperatures it may be >1. Figure 3.6 shows the compressibility factor of carbon dioxide as a function of T and P. This figure presents the same information as Fig. 3.4, except that it is plotted in terms of Z rather than V. It shows that at low pressure, Z approaches 1, and at moderate pressures, Z decreases roughly linearly with pressure.

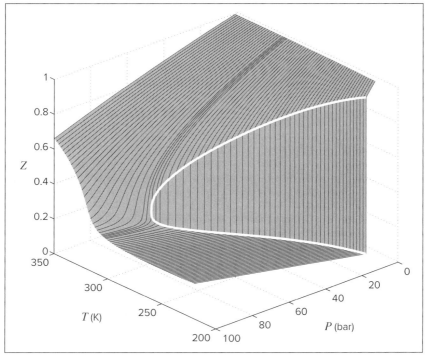

Figure 3.6: *PZT* surface for carbon dioxide, with isotherms shown in black and the vapor/liquid equilibrium curve in white.

With Z defined by Eq. (3.32) and with $a = RT$ [Eq. (3.29)], Eq. (3.28) becomes:

$$Z = 1 + B'P + C'P^2 + D'P^3 + \dots \tag{3.33}$$

An alternative expression for Z is also in common use:[7]

$$Z = 1 + \frac{B}{V} + \frac{C}{V^2} + \frac{D}{V^3} + \dots \tag{3.34}$$

Both of these equations are known as *virial expansions*, and the parameters B', C', D', etc., and B, C, D, etc., are called *virial coefficients*. Parameters B' and B are *second* virial coefficients; C' and C are *third* virial coefficients; and so on. For a given gas the virial coefficients are functions of temperature only.

---

[7]Proposed by H. Kamerlingh Onnes, "Expression of the Equation of State of Gases and Liquids by Means of Series," *Communications from the Physical Laboratory of the University of Leiden*, no. 71, 1901.

The two sets of coefficients in Eqs. (3.33) and (3.34) are related as follows:

$$B' = \frac{B}{RT} \quad (3.35a) \qquad C' = \frac{C - B^2}{(RT)^2} \quad (3.35b) \qquad D' = \frac{D - 3BC + 2B^3}{(RT)^3} \quad (3.35c)$$

To derive these relations, we set $Z = PV/RT$ in Eq. (3.34) and solve for P. This allows elimination of P on the right side of Eq. (3.33). The resulting equation reduces to a power series in $1/V$ which may be compared term by term with Eq. (3.34) to yield the given relations. They hold *exactly* only for the two virial expansions as infinite series, but they are acceptable approximations for the truncated forms used in practice.

Many other equations of state have been proposed for gases, but the virial equations are the only ones firmly based on statistical mechanics, which provides physical significance to the virial coefficients. Thus, for the expansion in $1/V$, the term B/V arises on account of interactions between pairs of molecules; the C/V^2 term, on account of three-body interactions; etc. Because, at gas-like densities, two-body interactions are many times more common than three-body interactions, and three-body interactions are many times more numerous than four-body interactions, the contributions to Z of the successively higher-ordered terms decrease rapidly.

3.5 APPLICATION OF THE VIRIAL EQUATIONS

The two forms of the virial expansion given by Eqs. (3.33) and (3.34) are infinite series. For engineering purposes their use is practical only where convergence is very rapid, that is, where two or three terms suffice for reasonably close approximations to the values of the series. This is realized for gases and vapors at low to moderate pressures.

Figure 3.7 shows a compressibility-factor graph for methane. All isotherms originate at $Z = 1$ and $P = 0$, and are nearly straight lines at low pressures. Thus the tangent to an isotherm at $P = 0$ is a good approximation of the isotherm from $P \to 0$ to some finite pressure. Differentiation of Eq. (3.33) for a given temperature gives:

$$\left(\frac{\partial Z}{\partial P} \right)_T = B' + 2C'P + 3D'P^2 + \dots$$

from which,

$$\left(\frac{\partial Z}{\partial P} \right)_{T;P=0} = B'$$

Thus the equation of the tangent line is $Z = 1 + B'P$, a result also given by truncating Eq. (3.33) to two terms.

A more common form of this equation results from substitution for B' by Eq. (3.35a):

$$Z = \frac{PV}{RT} = 1 + \frac{BP}{RT} \qquad (3.36)$$

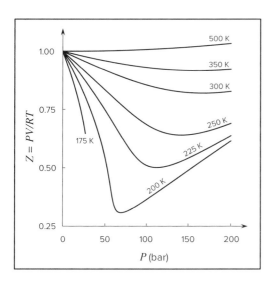

Figure 3.7: Compressibility-factor graph for methane. Shown are isotherms of the compressibility factor Z, as calculated from *PVT* data for methane by the defining equation $Z = PV/RT$. They are plotted vs. pressure for a number of constant temperatures, and they show graphically what the virial expansion in *P* represents analytically.

This equation expresses direct linearity between Z and P and is often applied to vapors at subcritical temperatures up to their saturation pressures. At higher temperatures it often provides a reasonable approximation for gases up to a pressure of several bars, with the pressure range increasing as the temperature increases.

Equation (3.34) as well may be truncated to two terms for application at low pressures:

$$Z = \frac{PV}{RT} = 1 + \frac{B}{V} \tag{3.37}$$

However, Eq. (3.36) is more convenient in application and is normally at least as accurate as Eq. (3.37). Thus when the virial equation is truncated to two terms, Eq. (3.36) is preferred.

The second virial coefficient B is substance dependent and a function of temperature. Experimental values are available for a number of gases.[8] Moreover, estimation of second virial coefficients is possible where no data are available, as discussed in Sec. 3.7.

For pressures above the range of applicability of Eq. (3.36) but below the critical pressure, the virial equation truncated to three terms often provides excellent results. In this case Eq. (3.34), the expansion in $1/V$, is far superior to Eq. (3.33). Thus when the virial equation is truncated to three terms, the appropriate form is:

$$Z = \frac{PV}{RT} = 1 + \frac{B}{V} + \frac{C}{V^2} \tag{3.38}$$

This equation is explicit in pressure but is cubic in volume. Analytic solution for V is possible, but solution by an iterative scheme, as illustrated in Ex. 3.8, is often more convenient.

Values of C, like those of B, depend on the gas and on temperature. Much less is known about third virial coefficients than about second virial coefficients, though data for a number of gases are found in the literature. Because virial coefficients beyond the third are rarely known and because the virial expansion with more than three terms becomes unwieldy, its use is uncommon.

[8]J. H. Dymond and E. B. Smith, *The Virial Coefficients of Pure Gases and Mixtures*, Clarendon Press, Oxford, 1980.

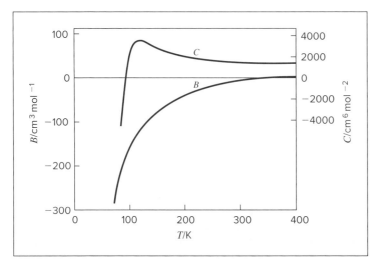

Figure 3.8: Virial coefficients B and C for nitrogen.

Figure 3.8 illustrates the effect of temperature on the virial coefficients B and C for nitrogen; although numerical values are different for other gases, the trends are similar. The curve of Fig. 3.8 suggests that B increases monotonically with T; however, at temperatures much higher than shown, B reaches a maximum and then slowly decreases. The temperature dependence of C is more difficult to establish experimentally, but its main features are clear: C is negative at low temperatures, passes through a maximum at a temperature near the critical temperature, and thereafter decreases slowly with increasing T.

Example 3.8

Reported values for the virial coefficients of isopropanol vapor at 200°C are:

$$B = -388 \text{ cm}^3\cdot\text{mol}^{-1} \quad C = -26{,}000 \text{ cm}^6\cdot\text{mol}^{-2}$$

Calculate V and Z for isopropanol vapor at 200°C and 10 bar:

(a) For the ideal-gas state; (b) By Eq. (3.36); (c) By Eq. (3.38).

Solution 3.8

The absolute temperature is $T = 473.15$ K, and the appropriate value of the gas constant is $R = 83.14 \text{ bar}\cdot\text{cm}^3\cdot\text{mol}^{-1}\cdot\text{K}^{-1}$.

(a) For the ideal-gas state, $Z = 1$, and

$$V^{ig} = \frac{RT}{P} = \frac{(83.14)(473.15)}{10} = 3934 \text{ cm}^3\cdot\text{mol}^{-1}$$

(b) From the second equality of Eq. (3.36), we have

$$V = \frac{RT}{P} + B = 3934 - 388 = 3546 \text{ cm}^3\cdot\text{mol}^{-1}$$

and

$$Z = \frac{PV}{RT} = \frac{V}{RT/P} = \frac{V}{V^{ig}} = \frac{3546}{3934} = 0.9014$$

(*c*) If solution by iteration is intended, Eq. (3.38) may be written:

$$V_{i+1} = \frac{RT}{P}\left(1 + \frac{B}{V_i} + \frac{C}{V_i^2}\right)$$

where *i* is the iteration number. Iteration is initiated with the ideal-gas state value V^{ig}. Solution yields:

$$V = 3488 \ \mathrm{cm^3 \cdot mol^{-1}}$$

from which $Z = 0.8866$. In comparison with this result, the ideal-gas-state value is 13% too high, and Eq. (3.36) gives a value 1.7% too high.

3.6 CUBIC EQUATIONS OF STATE

If an equation of state is to represent the *PVT* behavior of both liquids and vapors, it must encompass a wide range of temperatures, pressures, and molar volumes. Yet it must not be so complex as to present excessive numerical or analytical difficulties in application. Polynomial equations that are cubic in molar volume offer a compromise between generality and simplicity that is suitable to many purposes. Cubic equations are in fact the simplest equations capable of representing both liquid and vapor behavior.

The van der Waals Equation of State

The first practical cubic equation of state was proposed by J. D. van der Waals[9] in 1873:

$$P = \frac{RT}{V - b} - \frac{a}{V^2} \tag{3.39}$$

Here, *a* and *b* are positive constants, specific to a particular species; when they are zero, the equation for the ideal-gas state is recovered. The purpose of the term a/V^2 is to account for the attractive forces between molecules, which make the pressure lower than that which would be exerted in the ideal-gas state. The purpose of constant *b* is to account for the finite size of molecules, which makes the volume larger than in the ideal-gas state.

Given values of *a* and *b* for a particular fluid, one can calculate *P* as a function of *V* for various values of *T*. Figure 3.9 is a schematic *PV* diagram showing three such isotherms.

---

[9]Johannes Diderik van der Waals (1837–1923), Dutch physicist who won the 1910 Nobel Prize for physics. See: http://www.nobelprize.org/nobel_prizes/physics/laureates/1910/waals-bio.html.

Superimposed is the "dome" representing states of saturated liquid and saturated vapor.[10] For the isotherm $T_1 > T_c$, pressure decreases with increasing molar volume. The critical isotherm (labeled T_c) contains the horizontal inflection at C characteristic of the critical point. For the isotherm $T_2 < T_c$, the pressure decreases rapidly in the subcooled-liquid region with increasing V; after crossing the saturated-liquid line, it goes through a minimum, rises to a maximum, and then decreases, crossing the saturated-vapor line and continuing downward into the superheated-vapor region.

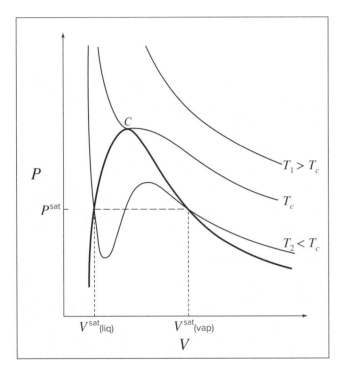

Figure 3.9: *PV* isotherms as given by a cubic equation of state for *T* above, at, and below the critical temperature. The superimposed darker curve shows the locus of saturated vapor and liquid volumes.

Experimental isotherms do not exhibit the smooth transition from saturated liquid to saturated vapor characteristic of equations of state; rather, they contain a horizontal segment within the two-phase region where saturated liquid and saturated vapor coexist in varying proportions at the saturation (or vapor) pressure. This behavior, shown by the dashed line in Fig. 3.9, cannot be represented by an equation of state, and we accept as inevitable the unrealistic behavior of equations of state in the two-phase region.

Actually, the *PV* behavior predicted in the two-phase region by proper cubic equations of state is not wholly fictitious. If pressure is decreased on a saturated liquid devoid of vapor nucleation sites in a carefully controlled experiment, vaporization does not occur, and liquid persists alone to pressures well below its vapor pressure. Similarly, raising the pressure on a saturated vapor in a suitable experiment does not cause condensation, and vapor persists

---

[10]Though far from obvious, the equation of state also provides the basis for calculation of the saturated liquid- and vapor-phase volumes that determine the location of the "dome." This is explained in Sec. 13.7.

alone to pressures well above the vapor pressure. These nonequilibrium or metastable states of superheated liquid and subcooled vapor are approximated by those portions of the *PV* isotherm which lie in the two-phase region adjacent to the states of saturated liquid and saturated vapor.[11]

Cubic equations of state have three volume roots, of which two may be complex. Physically meaningful values of *V* are always real, positive, and greater than constant *b*. For an isotherm at $T > T_c$, reference to Fig. 3.9 shows that solution for *V* at any value of *P* yields only one such root. For the critical isotherm ($T = T_c$), this is also true, except at the critical pressure, where there are three roots, all equal to V_c. For isotherms at $T < T_c$, the equation may exhibit one or three real roots, depending on the pressure. Although these roots are real and positive, they are not physically stable states for the portion of an isotherm lying between saturated liquid and saturated vapor (under the "dome"). Only for the *saturation* pressure P^{sat} are the roots, $V^{sat}(liq)$ and $V^{sat}(vap)$, stable states, lying at the ends of the horizontal portion of the true isotherm. For any pressure other than P^{sat}, there is only a single physically meaningful root, corresponding to either a liquid or a vapor molar volume.

A Generic Cubic Equation of State

A mid-20th-century development of cubic equations of state was initiated in 1949 by publication of the Redlich/Kwong (RK) equation:[12]

$$P = \frac{RT}{V - b} - \frac{a(T)}{V(V + b)} \qquad (3.40)$$

Subsequent enhancements have produced an important class of equations, represented by a *generic cubic equation of state*:

$$\boxed{P = \frac{RT}{V - b} - \frac{a(T)}{(V + \varepsilon b)(V + \sigma b)}} \qquad (3.41)$$

The assignment of appropriate parameters leads not only to the van der Waals (vdW) equation and the Redlich/Kwong (RK) equation, but also to the Soave/Redlich/Kwong (SRK)[13] and the Peng/Robinson (PR) equations.[14] For a given equation, ε and σ are pure numbers, the same for all substances, whereas parameters $a(T)$ and *b* are substance dependent. The temperature dependence of $a(T)$ is specific to each equation of state. The SRK equation is identical to the RK equation, except for the *T* dependence of $a(T)$. The PR equation takes different values for ε and σ, as indicated in Table 3.1.

[11]The heating of liquids in a microwave oven can lead to a dangerous condition of superheated liquid, which can "flash" explosively.

[12]Otto Redlich and J. N. S. Kwong, *Chem. Rev.*, vol. 44, pp. 233–244, 1949.

[13]G. Soave, *Chem. Eng. Sci.*, vol. 27, pp. 1197–1203, 1972.

[14]D.-Y. Peng and D. B. Robinson, *Ind. Eng. Chem. Fundam.*, vol. 15, pp. 59–64, 1976.

Determination of Equation-of-State Parameters

The parameters b and $a(T)$ of Eq. (3.41) can in principle be found from *PVT* data, but sufficient data are rarely available. They are in fact usually found from values for the critical constants T_c and P_c. Because the critical isotherm exhibits a horizontal inflection at the critical point, we may impose the mathematical conditions:

$$\left(\frac{\partial P}{\partial V}\right)_{T;cr} = 0 \quad (3.42) \qquad \left(\frac{\partial^2 P}{\partial V^2}\right)_{T;cr} = 0 \quad (3.43)$$

Subscript "cr" denotes the critical point. Differentiation of Eq. (3.41) yields expressions for both derivatives, which may be equated to zero for $P = P_c$, $T = T_c$, and $V = V_c$. The equation of state may itself be written for the critical conditions. These three equations contain five constants: P_c, V_c, T_c, $a(T_c)$, and b. Of the several ways to treat these equations, the most suitable is elimination of V_c to yield expressions relating $a(T_c)$ and b to P_c and T_c. The reason is that P_c and T_c are more widely available and more accurately known than V_c.

The algebra is intricate, but it leads eventually to the following expressions for parameters b and $a(T_c)$:

$$b = \Omega \frac{RT_c}{P_c} \tag{3.44}$$

and

$$a(T_c) = \Psi \frac{R^2 T_c^2}{P_c}$$

This result is extended to temperatures other than T_c by the introduction of a dimensionless function $\alpha(T_r; \omega)$ that becomes unity at the critical temperature:

$$a(T) = \Psi \frac{\alpha(T_r; \omega) R^2 T_c^2}{P_c} \tag{3.45}$$

In these equations Ω and Ψ are pure numbers, independent of substance but specific to a particular equation of state. Function $\alpha(T_r; \omega)$ is an empirical expression, wherein by definition $T_r \equiv T/T_c$, and ω is a parameter specific to a given chemical species, defined and discussed further below.

This analysis also shows that each equation of state implies a value of the critical compressibility factor Z_c that is the same for all substances. Different values are found for different equations. Unfortunately, Z_c values calculated from experimental values of T_c, P_c, and V_c differ from one species to another, and they agree in general with none of the fixed values predicted by common cubic equations of state. Experimental values are almost all smaller than any of the predicted values.

Roots of the Generic Cubic Equation of State

Equations of state are commonly transformed into expressions for the compressibility factor. An equation for Z equivalent to Eq. (3.41) is obtained by substituting $V = ZRT/P$. In addition, we **define** two dimensionless quantities that lead to simplification:

$$\beta \equiv \frac{bP}{RT} \quad (3.46) \qquad q \equiv \frac{a(T)}{bRT} \quad (3.47)$$

With these substitutions, Eq. (3.41) assumes the dimensionless form:

$$Z = 1 + \beta - q\beta \frac{Z - \beta}{(Z + \varepsilon\beta)(Z + \sigma\beta)} \tag{3.48}$$

Although the three roots of this equation may be found analytically, they are usually calculated by iterative procedures built into mathematical software packages. Convergence problems are most likely avoided when the equation is arranged to a form suited to the solution for a particular root.

Equation (3.48) is particularly adapted to solving for vapor and vapor-like roots. Iterative solution starts with the value $Z = 1$ substituted on the right side. The calculated value of Z is returned to the right side and the process continues to convergence. The final value of Z yields the volume root through $V = ZRT/P = ZV^{ig}$.

An alternative equation for Z is obtained when Eq. (3.48) is solved for the Z in the numerator of the final fraction, yielding:

$$Z = \beta + (Z + \varepsilon\beta)(Z + \sigma\beta)\left(\frac{1 + \beta - Z}{q\beta}\right) \tag{3.49}$$

This equation is particularly suited to solving for liquid and liquid-like roots. Iterative solution starts with the value $Z = \beta$ substituted on the right side. Once Z is known, the volume root is again $V = ZRT/P = ZV^{ig}$.

Experimental compressibility-factor data show that values of Z for different fluids exhibit similar behavior when correlated as a function of *reduced temperature* T_r and *reduced pressure* P_r, where by **definition** $T_r \equiv T/T_c$ and $P_r \equiv P/P_c$. It is therefore common to evaluate equation-of-state parameters through these dimensionless variables. Thus, Eq. (3.46) and Eq. (3.47) in combination with Eq. (3.44) and (3.45) yield:

$$\beta = \Omega \frac{P_r}{T_r} \quad (3.50) \qquad q = \frac{\Psi\alpha(T_r; \omega)}{\Omega T_r} \quad (3.51)$$

With parameters β and q evaluated by these equations, Z becomes a function of T_r and P_r and the equation of state is said to be *generalized* because of its general applicability to all gases and liquids. The numerical assignments for parameters ε, σ, Ω, and Ψ for the equations of interest are summarized in Table 3.1. Expressions are also given for $\alpha(T_r; \omega)$ for the SRK and PR equations.

Table 3.1: Parameter Assignments for Equations of State

Eqn. of State	$\alpha(T_r)$	σ	ε	Ω	Ψ	Z_c
vdW (1873)	1	0	0	1/8	27/64	3/8
RK (1949)	$T_r^{-1/2}$	1	0	0.08664	0.42748	1/3
SRK (1972)	$\alpha_{SRK}(T_r; \omega)^\dagger$	1	0	0.08664	0.42748	1/3
PR (1976)	$\alpha_{PR}(T_r; \omega)^\ddagger$	$1 + \sqrt{2}$	$1 - \sqrt{2}$	0.07780	0.45724	0.30740

$^\dagger \alpha_{SRK}(T_r; \omega) = \left[1 + (0.480 + 1.574\,\omega - 0.176\,\omega^2)(1 - T_r^{1/2})\right]^2$

$^\ddagger \alpha_{PR}(T_r; \omega) = \left[1 + (0.37464 + 1.54226\,\omega - 0.26992\,\omega^2)(1 - T_r^{1/2})\right]^2$

Example 3.9

Given that the vapor pressure of n-butane at 350 K is 9.4573 bar, find the molar volumes of (a) saturated-vapor and (b) saturated-liquid n-butane at these conditions as given by the Redlich/Kwong equation.

Solution 3.9

Values of T_c and P_c for n-butane from App. B yield:

$$T_r = \frac{350}{425.1} = 0.8233 \quad \text{and} \quad P_r = \frac{9.4573}{37.96} = 0.2491$$

Parameter q is given by Eq. (3.51) with Ω, Ψ, and $\alpha(T_r)$ for the RK equation from Table 3.1:

$$q = \frac{\Psi T_r^{-1/2}}{\Omega T_r} = \frac{\Psi}{\Omega}T_r^{-3/2} = \frac{0.42748}{0.08664}(0.8233)^{-3/2} = 6.6048$$

Parameter β is found from Eq. (3.50):

$$\beta = \Omega \frac{P_r}{T_r} = \frac{(0.08664)(0.2491)}{0.8233} = 0.026214$$

(a) For the saturated vapor, write the RK form of Eq. (3.48) that results upon substitution of appropriate values for ε and σ from Table 3.1:

$$Z = 1 + \beta - q\beta\frac{(Z - \beta)}{Z(Z + \beta)}$$

or

$$Z = 1 + 0.026214 - (6.6048)(0.026214)\frac{(Z - 0.026214)}{Z(Z + 0.026214)}$$

Solution yields $Z = 0.8305$, and

$$V^v = \frac{ZRT}{P} = \frac{(0.8305)(83.14)(350)}{9.4573} = 2555 \text{ cm}^3 \cdot \text{mol}^{-1}$$

An experimental value is 2482 cm^3·mol^{-1}.

(*b*) For the saturated liquid, apply Eq. (3.49) in its RK form:

$$Z = \beta + Z(Z + \beta)\left(\frac{1 + \beta - Z}{q\beta}\right)$$

or

$$Z = 0.026214 + Z(Z + 0.026214)\frac{(1.026214 - Z)}{(6.6048)(0.026214)}$$

Solution yields $Z = 0.04331$, and

$$V^l = \frac{ZRT}{P} = \frac{(0.04331)(83.14)(350)}{9.4573} = 133.3 \text{ cm}^3 \cdot \text{mol}^{-1}$$

An experimental value is 115.0 cm^3·mol^{-1}.

For comparison, values of V^v and V^l calculated for the conditions of Ex. 3.9 by all four of the cubic equations of state considered here are summarized as follows:

V^v/cm^3·mol^{-1}					V^l/cm^3·mol^{-1}				
Exp.	vdW	RK	SRK	PR	Exp.	vdW	RK	SRK	PR
2482	2667	2555	2520	2486	115.0	191.0	133.3	127.8	112.6

Corresponding States; Acentric Factor

The dimensionless thermodynamic coordinates T_r and P_r provide the basis for the simplest *corresponding-states correlations*:

All fluids, when compared at the same reduced temperature and reduced pressure, have approximately the same compressibility factor, and all deviate from ideal-gas behavior to about the same degree.

Two-parameter corresponding-states correlations of Z require the use of only two reducing parameters T_c and P_c. Although these correlations are very nearly exact for the *simple fluids* (argon, krypton, and xenon), systematic deviations are observed for more complex fluids. Appreciable improvement results from the introduction of a third corresponding-states parameter (in addition to T_c and P_c), characteristic of molecular structure. The most popular such parameter is the *acentric factor*, ω, introduced by K. S. Pitzer and coworkers.[15]

---

[15]Fully described in K. S. Pitzer, *Thermodynamics*, 3rd ed., App. 3, McGraw-Hill, New York, 1995.

The acentric factor for a pure chemical species is defined with reference to its vapor pressure. The logarithm of the vapor pressure of a pure fluid is approximately linear in the reciprocal of absolute temperature. This linearity can be expressed as

$$\frac{d\log P_r^{sat}}{d(1/T_r)} = S$$

where P_r^{sat} is reduced vapor pressure, T_r is reduced temperature, and S is the slope of a plot of $\log P_r^{sat}$ vs. $1/T_r$. Note that "log" here denotes the base 10 logarithm.

If two-parameter corresponding-states correlations were generally valid, the slope S would be the same for all pure fluids. This is observed not to be true; within a limited range, each fluid has its own characteristic value of S, which could in principle serve as a third corresponding-states parameter. However, Pitzer noted that all vapor-pressure data for the simple fluids (Ar, Kr, Xe) lie on the same line when plotted as $\log P_r^{sat}$ vs. $1/T_r$ and that the line passes through $\log P_r^{sat} = -1.0$ at $T_r = 0.7$. This is illustrated in Fig. 3.10. Data for other fluids define other lines whose locations can be fixed relative to the line for the simple fluids (SF) by the difference:

$$\log P_r^{sat}(SF) - \log P_r^{sat}$$

The acentric factor is **defined** as this difference evaluated at $T_r = 0.7$:

$$\boxed{\omega \equiv -1.0 - \log(P_r^{sat})_{T_r = 0.7}} \tag{3.52}$$

Therefore ω can be determined for any fluid from T_c, P_c, and a single vapor-pressure measurement made at $T_r = 0.7$. Values of ω and the critical constants T_c, P_c, and V_c for a number of substances are listed in App. B.

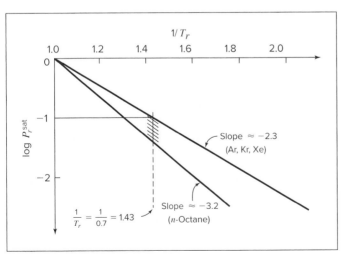

Figure 3.10: Approximate temperature dependence of the reduced vapor pressure.

The definition of ω makes its value zero for argon, krypton, and xenon, and experimental data yield compressibility factors for all three fluids that are correlated by the same curves when Z is plotted as a function of T_r and P_r. This is the basic premise of *three-parameter corresponding-states correlations*:

All fluids having the same value of acentric factor, when compared at the same reduced temperature and reduced pressure, have about the same compressibility factor, and all deviate from ideal-gas behavior to about the same degree.

Equations of state that express Z as a function of only T_r and P_r yield two-parameter corresponding-states correlations. The van der Waals and Redlich/Kwong equations are examples. The Soave/Redlich/Kwong (SRK) and Peng/Robinson equations, in which the acentric factor enters through function $\alpha(T_r; \omega)$ as an additional parameter, yield three-parameter corresponding-states correlations.

3.7 GENERALIZED CORRELATIONS FOR GASES

Generalized correlations find widespread use. Most popular are correlations of the kind developed by Pitzer and coworkers for the compressibility factor Z and for the second virial coefficient B.[16]

Pitzer Correlations for the Compressibility Factor

The correlation for Z is:

$$Z = Z^0 + \omega Z^1 \tag{3.53}$$

where Z^0 and Z^1 are functions of both T_r and P_r. When $\omega = 0$, as is the case for the simple fluids, the second term disappears, and Z^0 becomes identical with Z. Thus a generalized correlation for Z as a function of T_r and P_r based only on data for argon, krypton, and xenon provides the relationship $Z^0 = F^0(T_r, P_r)$. By itself, this represents a *two*-parameter corresponding-states correlation for Z.

Equation (3.53) is a simple linear relation between Z and ω for given values of T_r and P_r. Experimental data for Z for nonsimple fluids plotted vs. ω at constant T_r and P_r do indeed yield approximately straight lines, and their slopes provide values for Z^1 from which the generalized function $Z^1 = F^1(T_r, P_r)$ can be constructed.

Of the Pitzer-type correlations available, the one developed by Lee and Kesler[17] is the most widely used. It takes the form of tables that present values of Z^0 and Z^1 as functions of T_r and P_r. These are given in App. D as Tables D.1 through D.4. Use of these tables requires interpolation, as demonstrated at the beginning of App. E. The nature of the correlation is indicated by Fig. 3.11, a plot of Z^0 vs. P_r for six isotherms.

Fig. 3.12 shows Z^0 vs. P_r and T_r as a three-dimensional surface with isotherms and isobars superimposed. The saturation curve, where there is a discontinuity in Z, is not precisely defined in this plot, which is based on the data of Table D.1 and D.3. One should proceed with caution when applying the Lee/Kesler tables near the saturation curve. Although the tables contain values for liquid and vapor phases, the boundary between these will in general not be the same as the saturation curve for a given real substance. The tables should not be used to predict whether a substance is vapor or liquid at given conditions. Rather, one must know the

[16]See Pitzer, *op. cit.*

[17]B. I. Lee and M. G. Kesler, *AIChE J.*, vol. 21, pp. 510–527, 1975.

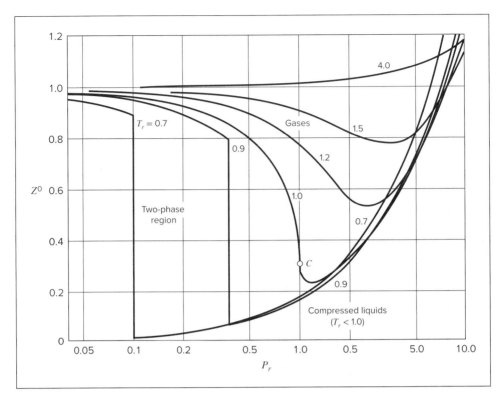

Figure 3.11: The Lee/Kesler correlation for $Z^0 = F^0(T_r, P_r)$.

phase of the substance, and then take care to interpolate or extrapolate only from points in the table that represent the appropriate phase.

The Lee/Kesler correlation provides reliable results for gases which are nonpolar or only slightly polar; for these, errors of no more than 2 or 3 percent are typical. When applied to highly polar gases or to gases that associate, larger errors can be expected.

The quantum gases (e.g., hydrogen, helium, and neon) do not conform to the same corresponding-states behavior as do normal fluids. Their treatment by the usual correlations is sometimes accommodated by use of temperature-dependent *effective* critical parameters.[18] For hydrogen, the quantum gas most commonly found in chemical processing, the recommended equations are:

$$T_c/\mathrm{K} = \frac{43.6}{1 + \dfrac{21.8}{2.016T}} \quad \text{(for } H_2\text{)} \tag{3.54}$$

$$P_c/\mathrm{bar} = \frac{20.5}{1 + \dfrac{44.2}{2.016T}} \quad \text{(for } H_2\text{)} \tag{3.55}$$

---

[18]J. M. Prausnitz, R. N. Lichtenthaler, and E. G. de Azevedo, *Molecular Thermodynamics of Fluid-Phase Equilibria*, 3d ed., pp. 172–173, Prentice Hall PTR, Upper Saddle River, NJ, 1999.

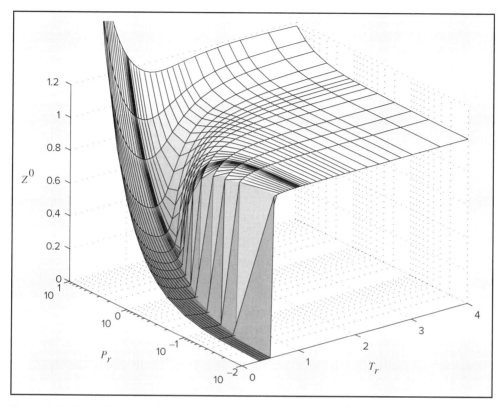

Figure 3.12: Three-dimensional plot of the Lee/Kesler correlation for $Z^0 = F^0(T_r, P_r)$ as given in Table D.1 and D.3.

$$V_c/\text{cm}^3 \cdot \text{mol}^{-1} = \frac{51.5}{1 - \dfrac{9.91}{2.016T}} \quad (\text{for H}_2) \tag{3.56}$$

where T is absolute temperature in kelvins. Use of these *effective* critical parameters for hydrogen requires the further specification that $\omega = 0$.

Pitzer Correlations for the Second Virial Coefficient

The tabular nature of the generalized compressibility-factor correlation is a disadvantage, but the complexity of the functions Z^0 and Z^1 precludes their accurate representation by simple equations. Nonetheless, we can give approximate analytical expression to these functions for a limited range of pressures. The basis for this is Eq. (3.36), the simplest form of the virial equation:

$$Z = 1 + \frac{BP}{RT} = 1 + \left(\frac{BP_c}{RT_c}\right)\frac{P_r}{T_r} = 1 + \hat{B}\frac{P_r}{T_r} \tag{3.57}$$

The reduced (and dimensionless) second virial coefficient and the Pitzer correlation for it are:

$$\hat{B} \equiv \frac{BP_c}{RT_c} \quad (3.58) \qquad \hat{B} = B^0 + \omega B^1 \quad (3.59)$$

Equations (3.57) and (3.59) together become:

$$Z = 1 + B^0 \frac{P_r}{T_r} + \omega B^1 \frac{P_r}{T_r}$$

Comparison of this equation with Eq. (3.53) provides the following identifications:

$$Z^0 = 1 + B^0 \frac{P_r}{T_r} \tag{3.60}$$

and

$$Z^1 = B^1 \frac{P_r}{T_r}$$

Second virial coefficients are functions of temperature only, and similarly B^0 and B^1 are functions of reduced temperature only. They are adequately represented by the Abbott equations:[19]

$$B^0 = 0.083 - \frac{0.422}{T_r^{1.6}} \quad (3.61) \qquad B^1 = 0.139 - \frac{0.172}{T_r^{4.2}} \quad (3.62)$$

The simplest form of the virial equation has validity only at low to moderate pressures where Z is linear in pressure. The generalized virial-coefficient correlation is therefore useful only where Z^0 and Z^1 are at least approximately linear functions of reduced pressure. Figure 3.13 compares the linear relation of Z^0 to P_r as given by Eqs. (3.60) and (3.61) with values of Z^0 from the Lee/Kesler compressibility-factor correlation, Tables D.1 and D.3 of App. D. The two correlations differ by less than 2% in the region above the dashed line of the figure. For reduced temperatures greater than $T_r \approx 3$, there appears to be no limitation on the pressure. For lower values of T_r the allowable pressure range decreases with decreasing temperature. A point is reached, however, at $T_r \approx 0.7$ where the pressure range is limited by the saturation pressure.[20] This is indicated by the left-most segment of the dashed line. The minor contributions of Z^1 to the correlations are here neglected. In view of the uncertainty associated with any generalized correlation, deviations of no more than 2% in Z^0 are not significant.

The relative simplicity of the generalized second-virial-coefficient correlation does much to recommend it. Moreover, temperatures and pressures of many chemical-processing

---

[19]These correlations first appeared in 1975 in the third edition of this book, attributed as a personal communication to M. M. Abbott, who developed them.

[20]Although the Lee/Kesler tables of Appendix D list values for superheated vapor and subcooled liquid, they do not provide values at saturation conditions.

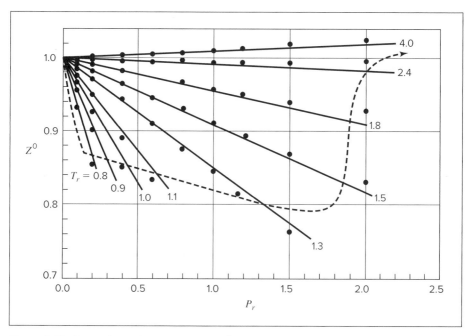

Figure 3.13: Comparison of correlations for Z^0. The virial-coefficient correlation is represented by the straight lines; the Lee/Kesler correlation, by the points. In the region above the dashed line the two correlations differ by less than 2%.

operations lie within the region appropriate to the compressibility-factor correlation. Like the parent correlation, it is most accurate for nonpolar species and least accurate for highly polar and associating molecules.

Correlations for the Third Virial Coefficient

Accurate data for third virial coefficients are far less common than for second virial coefficients. Nevertheless, generalized correlations for third virial coefficients do appear in the literature.

Equation (3.38) may be written in reduced form as:

$$Z = 1 + \hat{B}\frac{P_r}{T_r Z} + \hat{C}\left(\frac{P_r}{T_r Z}\right)^2 \qquad (3.63)$$

where the reduced second virial coefficient $\hat{B}$ is defined by Eq. (3.58). The reduced (and dimensionless) third virial coefficient and the Pitzer correlation for it are:

$$\hat{C} \equiv \frac{CP_c^2}{R^2 T_c^2} \quad (3.64) \qquad \hat{C} = C^0 + \omega C^1 \quad (3.65)$$

An expression for C^0 as a function of reduced temperature is given by Orbey and Vera:[21]

$$C^0 = 0.01407 + \frac{0.02432}{T_r} - \frac{0.00313}{T_r^{10.5}} \tag{3.66}$$

The expression for C^1 given by Orbey and Vera is replaced here by one that is algebraically simpler, but essentially equivalent numerically:

$$C^1 = -0.02676 + \frac{0.05539}{T_r^{2.7}} - \frac{0.00242}{T_r^{10.5}} \tag{3.67}$$

Equation (3.63) is cubic in Z, and it cannot be expressed in the form of Eq. (3.53). With T_r and P_r specified, Z may be found by iteration. An initial value of $Z = 1$ on the right side of Eq. (3.63) usually leads to rapid convergence.

The Ideal-Gas State as a Reasonable Approximation

The question often arises as to when the ideal-gas state may be a reasonable approximation to reality. Figure 3.14 can serve as a guide.

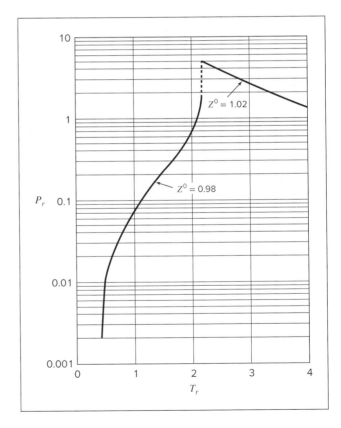

Figure 3.14: In the region lying below the curves, where Z^0 lies between 0.98 and 1.02, the ideal-gas state is a reasonable approximation.

[21]H. Orbey and J. H. Vera, *AIChE J.*, vol. 29, pp. 107–113, 1983.

Example 3.10

Determine the molar volume of n-butane at 510 K and 25 bar based on each of the following:

(a) The ideal-gas state.

(b) The generalized compressibility-factor correlation.

(c) Equation (3.57), with the generalized correlation for $\hat{B}$.

(d) Equation (3.63), with the generalized correlations for $\hat{B}$ and $\hat{C}$.

Solution 3.10

(a) For the ideal-gas state,

$$V = \frac{RT}{P} = \frac{(83.14)(510)}{25} = 1696.1 \text{ cm}^3 \cdot \text{mol}^{-1}$$

(b) With values of T_c and P_c given in Table B.1 of App. B,

$$T_r = \frac{510}{425.1} = 1.200 \qquad P_r = \frac{25}{37.96} = 0.659$$

Interpolation in Tables D.1 and D.2 then provides:

$$Z^0 = 0.865 \qquad Z^1 = 0.038$$

By Eq. (3.53) with $\omega = 0.200$,

$$Z = Z^0 + \omega Z^1 = 0.865 + (0.200)(0.038) = 0.873$$
$$V = \frac{ZRT}{P} = \frac{(0.873)(83.14)(510)}{25} = 1480.7 \text{ cm}^3 \cdot \text{mol}^{-1}$$

If Z^1, the secondary term, is neglected, $Z = Z^0 = 0.865$. This two-parameter corresponding-states correlation yields $V = 1467.1 \text{ cm}^3 \cdot \text{mol}^{-1}$, which is less than 1% lower than the value given by the three-parameter correlation.

(c) Values of B^0 and B^1 are given by Eqs. (3.61) and (3.62):

$$B^0 = -0.232 \qquad B^1 = 0.059$$

Equations (3.59) and (3.57) then yield:

$$\hat{B} = B^0 + \omega B^1 = -0.232 + (0.200)(0.059) = -0.220$$
$$Z = 1 + (-0.220)\frac{0.659}{1.200} = 0.879$$

from which $V = 1489.1 \text{ cm}^3 \cdot \text{mol}^{-1}$, a value less than 1% higher than that given by the compressibility-factor correlation.

(*d*) Values of C^0 and C^1 are given by Eqs. (3.66) and (3.67):

$$C^0 = 0.0339 \qquad C^1 = 0.0067$$

Equation (3.65) then yields:

$$\hat{C} = C^0 + \omega C^1 = 0.0339 + (0.200)(0.0067) = 0.0352$$

With this value of $\hat{C}$ and the value of $\hat{B}$ from part (*c*), Eq. (3.63) becomes,

$$Z = 1 + (-0.220)\left(\frac{0.659}{1.200Z}\right) + (0.0352)\left(\frac{0.659}{1.200Z}\right)^2$$

Solution for Z yields $Z = 0.876$ and $V = 1485.8 \text{ cm}^3 \cdot \text{mol}^{-1}$. The value of V differs from that of part (*c*) by about 0.2%. An experimental value for V is $1480.7 \text{ cm}^3 \cdot \text{mol}^{-1}$. Significantly, the results of parts (*b*), (*c*), and (*d*) are in excellent agreement. Mutual agreement at these conditions is suggested by Fig. 3.13.

Example 3.11

What pressure is generated when 500 mol of methane is stored in a volume of 0.06 m^3 at 50°C? Base calculations on each of the following:

(*a*) The ideal-gas state.

(*b*) The Redlich/Kwong equation.

(*c*) A generalized correlation.

Solution 3.11

The molar volume of the methane is $V = 0.06/500 = 0.0012 \text{ m}^3 \cdot \text{mol}^{-1}$.

(*a*) For the ideal-gas state, with $R = 8.314 \times 10^{-5} \text{ bar} \cdot \text{m}^3 \cdot \text{mol}^{-1} \cdot \text{K}^{-1}$:

$$P = \frac{RT}{V} = \frac{(8.314 \times 10^{-5})(323.15)}{0.00012} = 223.9 \text{ bar}$$

(*b*) The pressure as given by the Redlich/Kwong equation is:

$$P = \frac{RT}{V - b} - \frac{a(T)}{V(V + b)} \tag{3.40}$$

Values of b and $a(T)$ come from Eqs. (3.44) and (3.45), with Ω, Ψ, and $\alpha(T_r) = T_r^{-1/2}$ from Table 3.1. With values of T_c and P_c from Table B.1, we have:

$$T_r = \frac{323.15}{190.6} = 1.695$$

$$b = 0.08664 \frac{(8.314 \times 10^{-5})(190.6)}{45.99} = 2.985 \times 10^{-5} \text{ m}^3 \cdot \text{mol}^{-1}$$

$$a = 0.42748 \frac{(1.695)^{-0.5}(8.314 \times 10^{-5})^2(190.6)^2}{45.99} = 1.793 \times 10^{-6} \text{ bar} \cdot \text{m}^6 \cdot \text{mol}^{-2}$$

Substitution of numerical values into the Redlich/Kwong equation now yields:

$$P = \frac{(8.314 \times 10^{-5})(323.15)}{0.00012 - 2.985 \times 10^{-5}} - \frac{1.793 \times 10^{-6}}{0.00012(0.00012 + 2.985 \times 10^{-5})} = 198.3 \text{ bar}$$

(c) Because the pressure here is high, the generalized compressibility-factor correlation is the proper choice. In the absence of a known value for P_r, an iterative procedure is based on the following equation:

$$P = \frac{ZRT}{V} = \frac{Z(8.314 \times 10^{-5})(323.15)}{0.00012} = 223.9 \, Z$$

Because $P = P_c P_r = 45.99 \, P_r$, this equation becomes:

$$Z = \frac{45.99 \, P_r}{223.9} = 0.2054 \, P_r \qquad \text{or} \qquad P_r = \frac{Z}{0.2054}$$

One now assumes a starting value for Z, say $Z = 1$. This gives $P_r = 4.68$, and allows a new value of Z to be calculated by Eq. (3.53) from values interpolated in Tables D.3 and D.4 at the reduced temperature of $T_r = 1.695$. With this new value of Z, a new value of P_r is calculated, and the procedure continues until no significant change occurs from one step to the next. The final value of Z so found is 0.894 at $P_r = 4.35$. This may be confirmed by substitution into Eq. (3.53) of values for Z^0 and Z^1 from Tables D.3 and D.4 interpolated at $P_r = 4.35$ and $T_r = 1.695$. With $\omega = 0.012$,

$$Z = Z^0 + \omega Z^1 = 0.891 + (0.012)(0.268) = 0.894$$
$$P = \frac{ZRT}{V} = \frac{(0.894)(8.314 \times 10^{-5})(323.15)}{0.00012} = 200.2 \text{ bar}$$

Because the acentric factor is small, the two- and three-parameter compressibility-factor correlations are little different. The Redlich/Kwong equation and the generalized compressibility-factor correlation give answers within 2% of the experimental value of 196.5 bar.

Example 3.12

A mass of 500 g of gaseous ammonia is contained in a vessel of 30,000 cm³ volume and immersed in a constant-temperature bath at 65°C. Calculate the pressure of the gas by:

(a) The ideal-gas state;

(b) A generalized correlation.

SOLUTION 3.12

The molar volume of ammonia in the vessel is:

$$V = \frac{V^t}{n} = \frac{V^t}{m/\mathcal{M}} = \frac{30,000}{500/17.02} = 1021.2 \text{ cm}^3 \cdot \text{mol}^{-1}$$

(*a*) For the ideal-gas state,

$$P = \frac{RT}{V} = \frac{(83.14)(65 + 273.15)}{1021.2} = 27.53 \text{ bar}$$

(*b*) Because the reduced pressure is low ($P_r \approx 27.53/112.8 = 0.244$), the generalized virial-coefficient correlation should suffice. Values of B^0 and B^1 are given by Eqs. (3.61) and (3.62). With $T_r = 338.15/405.7 = 0.834$,

$$B^0 = -0.482 \qquad B^1 = -0.232$$

Substitution into Eq. (3.59) with $\omega = 0.253$ yields:

$$\hat{B} = -0.482 + (0.253)(-0.232) = -0.541$$

$$B = \frac{\hat{B}RT_c}{P_c} = \frac{-(0.541)(83.14)(405.7)}{112.8} = -161.8 \text{ cm}^3 \cdot \text{mol}^{-1}$$

By the second equality of Eq. (3.36):

$$P = \frac{RT}{V - B} = \frac{(83.14)(338.15)}{1021.2 + 161.8} = 23.76 \text{ bar}$$

An iterative solution is not necessary because B is independent of pressure. The calculated P corresponds to a reduced pressure of $P_r = 23.76/112.8 = 0.211$. Reference to Fig. 3.13 confirms the suitability of the generalized virial-coefficient correlation.

Experimental data indicate that the pressure is 23.82 bar at the given conditions. Thus the ideal-gas state yields an answer high by about 15%, whereas the virial-coefficient correlation gives an answer in substantial agreement with experiment, even though ammonia is a polar molecule.

3.8   GENERALIZED CORRELATIONS FOR LIQUIDS

Although the molar volumes of liquids can be calculated by means of generalized cubic equations of state, the results are often not of high accuracy. However, the Lee/Kesler correlation includes data for subcooled liquids, and Fig. 3.11 illustrates curves for both liquids and gases. Values for both phases are provided in Tables D.1 through D.4 of App. D. Recall, however, that this correlation is most suitable for nonpolar and slightly polar fluids.

In addition, generalized equations are available for the estimation of molar volumes of *saturated* liquids. The simplest equation, proposed by Rackett,[22] is an example:

$$V^{\text{sat}} = V_c Z_c^{(1-T_r)^{2/7}} \tag{3.68}$$

An alternative form of this equation is sometimes useful:

$$Z^{\text{sat}} = \frac{P_r}{T_r} Z_c^{[1+(1-T_r)^{2/7}]} \tag{3.69}$$

[22]H. G. Rackett, *J. Chem. Eng. Data*, vol. 15, pp. 514–517, 1970; see also C. F. Spencer and S. B. Adler, *ibid.*, vol. 23, pp. 82–89, 1978, for a review of available equations.

The only data required are the critical constants, given in Table B.1 of App. B. Results are usually accurate to 1 or 2%.

Lydersen, Greenkorn, and Hougen[23] developed a two-parameter corresponding-states correlation for estimation of liquid volumes. It provides a correlation of reduced density ρ_r as a function of reduced temperature and pressure. By definition,

$$\rho_r \equiv \frac{\rho}{\rho_c} = \frac{V_c}{V} \tag{3.70}$$

where ρ_c is the density at the critical point. The generalized correlation is shown by Fig. 3.15. This figure may be used directly with Eq. (3.70) for determination of liquid volumes if the value of the critical volume is known. A better procedure is to make use of a single known liquid volume (state 1) by the identity,

$$V_2 = V_1 \frac{\rho_{r_1}}{\rho_{r_2}} \tag{3.71}$$

where

$$V_2 = \text{required volume}$$
$$V_1 = \text{known volume}$$
$$\rho_{r_1}, \rho_{r_2} = \text{reduced densities read from Fig. 3.15}$$

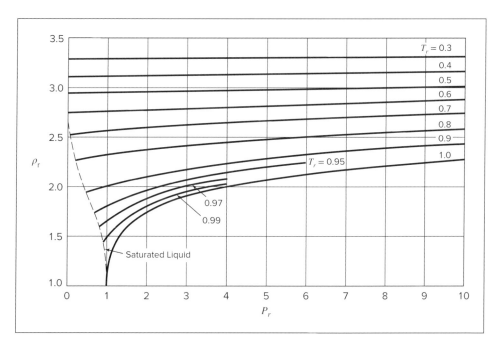

Figure 3.15: Generalized density correlation for liquids.

[23]A. L. Lydersen, R. A. Greenkorn, and O. A. Hougen, "Generalized Thermodynamic Properties of Pure Fluids," *Univ. Wisconsin, Eng. Expt. Sta. Rept. 4*, 1955.

This method gives good results and requires only experimental data that are usually available. Figure 3.15 makes clear the increasing effects of both temperature and pressure on liquid density as the critical point is approached.

Correlations for the molar densities as functions of temperature are given for many pure liquids by Daubert and coworkers.[24]

Example 3.13

For ammonia at 310 K, estimate the density of:

   (a) The saturated liquid;

   (b) The liquid at 100 bar.

Solution 3.13

(a) Apply the Rackett equation at the reduced temperature, $T_r = 310/405.7 = 0.7641$. With $V_c = 72.47$ and $Z_c = 0.242$ (from Table B.1),

$$V^{\text{sat}} = V_c Z_c^{(1-T_r)^{2/7}} = (72.47)(0.242)^{(0.2359)^{2/7}} = 28.33 \text{ cm}^3 \cdot \text{mol}^{-1}$$

For comparison, the experimental value is $29.14 \text{ cm}^3 \cdot \text{mol}^{-1}$, a 2.7% difference.

(b) The reduced conditions are:

$$T_r = 0.764 \qquad P_r = \frac{100}{112.8} = 0.887$$

Substituting the value, $\rho_r = 2.38$ (from Fig. 3.15), and V_c into Eq. (3.70) gives:

$$V = \frac{V_c}{\rho_r} = \frac{72.47}{2.38} = 30.45 \text{ cm}^3 \cdot \text{mol}^{-1}$$

In comparison with the experimental value of $28.6 \text{ cm}^3 \cdot \text{mol}^{-1}$, this result is higher by 6.5%.

If we start with the experimental value of $29.14 \text{ cm}^3 \cdot \text{mol}^{-1}$ for saturated liquid at 310 K, Eq. (3.71) may be used. For the saturated liquid at $T_r = 0.764$, $\rho_{r_1} = 2.34$ (from Fig. 3.15). Substitution of known values into Eq. (3.71) gives:

$$V_2 = V_1 \frac{\rho_{r_1}}{\rho_{r_2}} = (29.14)\left(\frac{2.34}{2.38}\right) = 28.65 \text{ cm}^3 \cdot \text{mol}^{-1}$$

This result is in essential agreement with the experimental value.

Direct application of the Lee/Kesler correlation with values of Z^0 and Z^1 interpolated from Tables D.1 and D.2 leads to a value of $33.87 \text{ cm}^3 \cdot \text{mol}^{-1}$, which is significantly in error, no doubt owing to the highly polar nature of ammonia.

[24]T. E. Daubert, R. P. Danner, H. M. Sibul, and C. C. Stebbins, *Physical and Thermodynamic Properties of Pure Chemicals: Data Compilation*, Taylor & Francis, Bristol, PA, extant 1995.

3.9 SYNOPSIS

After studying this chapter, including the end-of-chapter problems, one should be able to:

- State and apply the phase rule for nonreacting systems
- Interpret PT and PV diagrams for a pure substance, identifying the solid, liquid, gas, and fluid regions; the fusion (melting), sublimation, and vaporization curves; and the critical and triple points
- Draw isotherms on a PV diagram for temperatures above and below the critical temperature
- Define isothermal compressibility and volume expansivity and use them in calculations for liquids and solids
- Make use of the facts that for the ideal-gas state U^{ig} and H^{ig} depend only on T (not on P and V^{ig}), and that $C_P^{ig} = C_V^{ig} + R$
- Compute heat and work requirements and property changes for mechanically reversible isothermal, isobaric, isochoric, and adiabatic processes in the ideal-gas state
- Define and use the compressibility factor Z
- Intelligently select an appropriate equation of state or generalized correlation for application in a given situation, as indicated by the following chart:

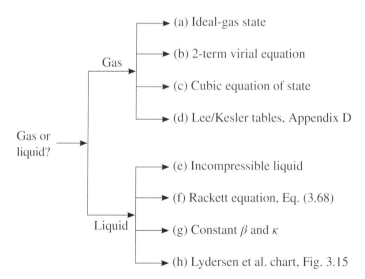

- Apply the two-term virial equation of state, written in terms of pressure or molar density
- Relate the second and third virial coefficients to the slope and curvature of a plot of the compressibility factor versus molar density
- Write the van der Waals and generic cubic equations of state, and explain how the equation-of-state parameters are related to critical properties

- Define and use T_r, P_r, and ω

- Explain the basis for the two- and three-parameter corresponding-states correlations

- Compute parameters for the Redlich/Kwong, Soave/Redlich/Kwong, and Peng/Robinson equations of state from critical properties

- Solve any of the cubic equations of state, where appropriate, for the vapor or vapor-like and/or liquid or liquid-like molar volumes at given T and P

- Apply the Lee/Kesler correlation with data from Appendix D

- Determine whether the Pitzer correlation for the second virial coefficient is applicable for given T and P, and use it if appropriate

- Estimate liquid-phase molar volumes by generalized correlations

3.10   PROBLEMS

3.1. How many phase rule variables must be specified to fix the thermodynamic state of each of the following systems?

    (*a*) A sealed flask containing a liquid ethanol-water mixture in equilibrium with its vapor.

    (*b*) A sealed flask containing a liquid ethanol-water mixture in equilibrium with its vapor and nitrogen.

    (*c*) A sealed flask containing ethanol, toluene, and water as two liquid phases plus vapor.

3.2. A renowned laboratory reports quadruple-point coordinates of 10.2 Mbar and 24.1°C for four-phase equilibrium of allotropic solid forms of the exotic chemical β-miasmone. Evaluate the claim.

3.3. A closed, nonreactive system contains species 1 and 2 in vapor/liquid equilibrium. Species 2 is a very light gas, essentially insoluble in the liquid phase. The vapor phase contains both species 1 and 2. Some additional moles of species 2 are added to the system, which is then restored to its initial T and P. As a result of the process, does the total number of moles of liquid increase, decrease, or remain unchanged?

3.4. A system comprised of chloroform, 1,4-dioxane, and ethanol exists as a two-phase vapor/liquid system at 50°C and 55 kPa. After the addition of some pure ethanol, the system can be returned to two-phase equilibrium at the initial T and P. In what respect has the system changed, and in what respect has it not changed?

3.5. For the system described in Prob. 3.4:

    (*a*) How many phase-rule variables in addition to T and P must be chosen so as to fix the compositions of both phases?

    (*b*) If the temperature and pressure are to remain the same, can the *overall* composition of the system be changed (by adding or removing material) without affecting the compositions of the liquid and vapor phases?

3.6. Express the volume expansivity and the isothermal compressibility as functions of density ρ and its partial derivatives. For water at 50°C and 1 bar, $\kappa = 44.18 \times 10^{-6}$ bar^{-1}. To what pressure must water be compressed at 50°C to change its density by 1%? Assume that κ is independent of P.

3.7. Generally, volume expansivity β and isothermal compressibility κ depend on T and P. Prove that:

$$\left(\frac{\partial \beta}{\partial P}\right)_T = -\left(\frac{\partial \kappa}{\partial T}\right)_P$$

3.8. The Tait equation for liquids is written for an isotherm as:

$$V = V_0\left(1 - \frac{AP}{B + P}\right)$$

where V is molar or specific volume, V_0 is the hypothetical molar or specific volume at zero pressure, and A and B are positive constants. Find an expression for the isothermal compressibility consistent with this equation.

3.9. For liquid water the isothermal compressibility is given by:

$$\kappa = \frac{c}{V(P + b)}$$

where c and b are functions of temperature only. If 1 kg of water is compressed isothermally and reversibly from 1 to 500 bar at 60°C, how much work is required? At 60°C, $b = 2700$ bar and $c = 0.125$ cm^3·g^{-1}.

3.10. Calculate the reversible work done in compressing 1(ft)^3 of mercury at a constant temperature of 32(°F) from 1(atm) to 3000(atm). The isothermal compressibility of mercury at 32(°F) is:

$$\kappa/(\text{atm})^{-1} = 3.9 \times 10^{-6} - 0.1 \times 10^{-9}\, P/(\text{atm})$$

3.11. Five kilograms of liquid carbon tetrachloride undergo a mechanically reversible, isobaric change of state at 1 bar during which the temperature changes from 0°C to 20°C. Determine ΔV^t, W, Q, ΔH^t, and ΔU^t. The properties for liquid carbon tetrachloride at 1 bar and 0°C may be assumed independent of temperature: $\beta = 1.2 \times 10^{-3}$ K^{-1}, $C_P = 0.84$ kJ·kg^{-1}·K^{-1}, and $\rho = 1590$ kg·m^{-3}.

3.12. Various species of hagfish, or slime eels, live on the ocean floor, where they burrow inside other fish, eating them from the inside out and secreting copious amounts of slime. Their skins are widely used to make eelskin wallets and accessories. Suppose a hagfish is caught in a trap at a depth of 200 m below the ocean surface, where the water temperature is 10°C, then brought to the surface where the temperature is 15°C. If the isothermal compressibility and volume expansivity are assumed constant and equal to the values for water,

$$(\beta = 10^{-4}\,\text{K}^{-1} \text{ and } \kappa = 4.8 \times 10^{-5}\,\text{bar}^{-1})$$

what is the fractional change in the volume of the hagfish when it is brought to the surface?

Table 3.2 provides the specific volume, isothermal compressibility, and volume expansivity of several liquids at 20°C and 1 bar[25] for use in Problems 3.13 to 3.15, where β and κ may be assumed constant.

Table 3.2: Volumetric Properties of Liquids at 20°C

Molecular Formula	Chemical Name	Specific Volume $V/L \cdot kg^{-1}$	Isothermal Compressibility $\kappa/10^{-5} \, bar^{-1}$	Volume Expansivity $\beta/10^{-3} \cdot °C^{-1}$
$C_2H_4O_2$	Acetic Acid	0.951	9.08	1.08
C_6H_7N	Aniline	0.976	4.53	0.81
CS_2	Carbon Disulfide	0.792	9.38	1.12
C_6H_5Cl	Chlorobenzene	0.904	7.45	0.94
C_6H_{12}	Cyclohexane	1.285	11.3	1.15
$C_4H_{10}O$	Diethyl Ether	1.401	18.65	1.65
C_2H_5OH	Ethanol	1.265	11.19	1.40
$C_4H_8O_2$	Ethyl Acetate	1.110	11.32	1.35
C_8H_{10}	m-Xylene	1.157	8.46	0.99
CH_3OH	Methanol	1.262	12.14	1.49
CCl_4	Tetrachloromethane	0.628	10.5	1.14
C_7H_8	Toluene	1.154	8.96	1.05
$CHCl_3$	Trichloromethane	0.672	9.96	1.21

3.13. For one of the substances in Table 3.2, compute the change in volume and work done when one kilogram of the substance is heated from 15°C to 25°C at a constant pressure of 1 bar.

3.14. For one of the substances in Table 3.2, compute the change in volume and work done when one kilogram of the substance is compressed from 1 bar to 100 bar at a constant temperature of 20°C.

3.15. For one of the substances in Table 3.2, compute the final pressure when the substance is heated from 15°C and 1 bar to 25°C at constant volume.

3.16. A substance for which κ is a constant undergoes an isothermal, mechanically reversible process from initial state (P_1, V_1) to final state (P_2, V_2), where V is molar volume.

(a) Starting with the definition of κ, show that the path of the process is described by:

$$V = A(T)\exp(-\kappa P)$$

(b) Determine an exact expression which gives the isothermal work done on 1 mol of this constant-κ substance.

[25]*CRC Handbook of Chemistry and Physics*, 90th Ed., pp. 6-140-6-141 and p. 15–25, CRC Press, Boca Raton, Florida, 2010.

3.17. One mole of an ideal gas with $C_P = (7/2)R$ and $C_V = (5/2)R$ expands from $P_1 = 8$ bar and $T_1 = 600$ K to $P_2 = 1$ bar by each of the following paths:

(a) Constant volume;
(b) Constant temperature;
(c) Adiabatically.

Assuming mechanical reversibility, calculate W, Q, ΔU, and ΔH for each process. Sketch each path on a single PV diagram.

3.18. One mole of an ideal gas with $C_P = (5/2)R$ and $C_V = (3/2)R$ expands from $P_1 = 6$ bar and $T_1 = 800$ K to $P_2 = 1$ bar by each of the following paths:

(a) Constant volume
(b) Constant temperature
(c) Adiabatically

Assuming mechanical reversibility, calculate W, Q, ΔU, and ΔH for each process. Sketch each path on a single PV diagram.

3.19. An ideal gas initially at 600 K and 10 bar undergoes a four-step mechanically reversible cycle in a closed system. In step 12, pressure decreases isothermally to 3 bar; in step 23, pressure decreases at constant volume to 2 bar; in step 34, volume decreases at constant pressure; and in step 41, the gas returns adiabatically to its initial state. Take $C_P = (7/2)R$ and $C_V = (5/2)R$.

(a) Sketch the cycle on a PV diagram.
(b) Determine (where unknown) both T and P for states 1, 2, 3, and 4.
(c) Calculate Q, W, ΔU, and ΔH for each step of the cycle.

3.20. An ideal gas initially at 300 K and 1 bar undergoes a three-step mechanically reversible cycle in a closed system. In step 12, pressure increases isothermally to 5 bar; in step 23, pressure increases at constant volume; and in step 31, the gas returns adiabatically to its initial state. Take $C_P = (7/2)R$ and $C_V = (5/2)R$.

(a) Sketch the cycle on a PV diagram.
(b) Determine (where unknown) V, T, and P for states 1, 2, and 3.
(c) Calculate Q, W, ΔU, and ΔH for each step of the cycle.

3.21. The state of an ideal gas with $C_P = (5/2)R$ is changed from $P = 1$ bar and $V_1^t = 12$ m^3 to $P_2 = 12$ bar and $V_2^t = 1$ m^3 by the following mechanically reversible processes:

(a) Isothermal compression.
(b) Adiabatic compression followed by cooling at constant pressure.
(c) Adiabatic compression followed by cooling at constant volume.
(d) Heating at constant volume followed by cooling at constant pressure.
(e) Cooling at constant pressure followed by heating at constant volume.

Calculate Q, W, ΔU^t, and ΔH^t for each of these processes, and sketch the paths of all processes on a single PV diagram.

3.22. The *environmental lapse rate* dT/dz characterizes the local variation of temperature with elevation in the earth's atmosphere. Atmospheric pressure varies with elevation according to the hydrostatic formula,

$$\frac{dP}{dz} = -\mathcal{M}\rho g$$

where $\mathcal{M}$ is molar mass, ρ is molar density, and g is the local acceleration of gravity. Asssume that the atmosphere is an ideal gas, with T related to P by the polytropic formula:

$$TP^{(1-\delta)/\delta} = \text{constant}$$

Develop an expression for the environmental lapse rate in relation to $\mathcal{M}$, g, R, and δ.

3.23. An evacuated tank is filled with gas from a constant-pressure line. Develop an expression relating the temperature of the gas in the tank to the temperature T' of the gas in the line. Assume the gas is ideal with constant heat capacities, and ignore heat transfer between the gas and the tank.

3.24. A tank of 0.1 m³ volume contains air at 25°C and 101.33 kPa. The tank is connected to a compressed-air line which supplies air at the constant conditions of 45°C and 1500 kPa. A valve in the line is cracked so that air flows slowly into the tank until the pressure equals the line pressure. If the process occurs slowly enough that the temperature in the tank remains at 25°C, how much heat is lost from the tank? Assume air to be an ideal gas for which $C_P = (7/2)R$ and $C_V = (5/2)R$.

3.25. Gas at constant T and P is contained in a supply line connected through a valve to a closed tank containing the same gas at a lower pressure. The valve is opened to allow flow of gas into the tank, and then is shut again.

(a) Develop a general equation relating n_1 and n_2, the moles (or mass) of gas in the tank at the beginning and end of the process, to the properties U_1 and U_2, the internal energy of the gas in the tank at the beginning and end of the process, and H', the enthalpy of the gas in the supply line, and to Q, the heat transferred to the material in the tank during the process.

(b) Reduce the general equation to its simplest form for the special case of an ideal gas with constant heat capacities.

(c) Further reduce the equation of (b) for the case of $n_1 = 0$.

(d) Further reduce the equation of (c) for the case in which, in addition, $Q = 0$.

(e) Treating nitrogen as an ideal gas for which $C_P = (7/2)R$, apply the appropriate equation to the case in which a steady supply of nitrogen at 25°C and 3 bar flows into an evacuated tank of 4 m³ volume, and calculate the moles of nitrogen that flow into the tank to equalize the pressures for two cases:

1. Assume that no heat flows from the gas to the tank or through the tank walls.
2. Assume that the tank weighs 400 kg, is perfectly insulated, has an initial temperature of 25°C, has a specific heat of 0.46 kJ·kg⁻¹·K⁻¹, and is heated by the gas so as always to be at the temperature of the gas in the tank.

3.26. Develop equations that may be solved to give the final temperature of the gas remaining in a tank after the tank has been bled from an initial pressure P_1 to a final pressure P_2. Known quantities are initial temperature, tank volume, heat capacity of the gas, total heat capacity of the containing tank, P_1, and P_2. Assume the tank to be always at the temperature of the gas remaining in the tank and the tank to be perfectly insulated.

3.27. A rigid, nonconducting tank with a volume of 4 m^3 is divided into two unequal parts by a thin membrane. One side of the membrane, representing 1/3 of the tank, contains nitrogen gas at 6 bar and 100°C, and the other side, representing 2/3 of the tank, is evacuated. The membrane ruptures and the gas fills the tank.

(a) What is the final temperature of the gas? How much work is done? Is the process reversible?

(b) Describe a reversible process by which the gas can be returned to its initial state. How much work is done?

Assume nitrogen is an ideal gas for which $C_P = (7/2)R$ and $C_V = (5/2)R$.

3.28. An ideal gas, initially at 30°C and 100 kPa, undergoes the following cyclic processes in a closed system:

(a) In mechanically reversible processes, it is first compressed adiabatically to 500 kPa, then cooled at a constant pressure of 500 kPa to 30°C, and finally expanded isothermally to its original state.

(b) The cycle traverses exactly the same changes of state, but each step is irreversible with an efficiency of 80% compared with the corresponding mechanically reversible process. *Note*: The initial step can no longer be adiabatic.

Calculate Q, W, ΔU, and ΔH for each step of the process and for the cycle. Take $C_P = (7/2)R$ and $C_V = (5/2)R$.

3.29. One cubic meter of an ideal gas at 600 K and 1000 kPa expands to five times its initial volume as follows:

(a) By a mechanically reversible, isothermal process.

(b) By a mechanically reversible, adiabatic process.

(c) By an adiabatic, irreversible process in which expansion is against a restraining pressure of 100 kPa.

For each case calculate the final temperature, pressure, and the work done by the gas. Take $C_P = 21$ J·mol^{-1}·K^{-1}.

3.30. One mole of air, initially at 150°C and 8 bar, undergoes the following mechanically reversible changes. It expands isothermally to a pressure such that when it is cooled at constant volume to 50°C its final pressure is 3 bar. Assuming air is an ideal gas for which $C_P = (7/2)R$ and $C_V = (5/2)R$, calculate W, Q, ΔU, and ΔH.

3.31. An ideal gas flows through a horizontal tube at steady state. No heat is added and no shaft work is done. The cross-sectional area of the tube changes with length, and this causes the velocity to change. Derive an equation relating the temperature to the velocity of the gas. If nitrogen at 150°C flows past one section of the tube at a velocity of 2.5 m·s^{-1}, what is its temperature at another section where its velocity is 50 m·s^{-1}? Assume $C_P = (7/2)R$.

3.32. One mole of an ideal gas, initially at 30°C and 1 bar, is changed to 130°C and 10 bar by three different mechanically reversible processes:

- The gas is first heated at constant volume until its temperature is 130°C; then it is compressed isothermally until its pressure is 10 bar.
- The gas is first heated at constant pressure until its temperature is 130°C; then it is compressed isothermally to 10 bar.
- The gas is first compressed isothermally to 10 bar; then it is heated at constant pressure to 130°C.

Calculate Q, W, ΔU, and ΔH in each case. Take $C_P = (7/2)R$ and $C_V = (5/2)R$. Alternatively, take $C_P = (5/2)R$ and $C_V = (3/2)R$.

3.33. One mole of an ideal gas, initially at 30°C and 1 bar, undergoes the following mechanically reversible changes. It is compressed isothermally to a point such that when it is heated at constant volume to 120°C its final pressure is 12 bar. Calculate Q, W, ΔU, and ΔH for the process.

3.34. One mole of an ideal gas in a closed system, initially at 25°C and 10 bar, is first expanded adiabatically, then heated isochorically to reach a final state of 25°C and 1 bar. Assuming these processes are mechanically reversible, compute T and P after the adiabatic expansion, and compute Q, W, ΔU, and ΔH for each step and for the overall process. Take $C_P = (7/2)R$ and $C_V = (5/2)R$.

3.35. A process consists of two steps: (1) One mole of air at $T = 800$ K and $P = 4$ bar is cooled at constant volume to $T = 350$ K. (2) The air is then heated at constant pressure until its temperature reaches 800 K. If this two-step process is replaced by a single isothermal expansion of the air from 800 K and 4 bar to some final pressure P, what is the value of P that makes the work of the two processes the same? Assume mechanical reversibility and treat air as an ideal gas with $C_P = (7/2)R$ and $C_V = (5/2)R$.

3.36. One cubic meter of argon is taken from 1 bar and 25°C to 10 bar and 300°C by each of the following two-step paths. For each path, compute Q, W, ΔU, and ΔH for each step and for the overall process. Assume mechanical reversibility and treat argon as an ideal gas with $C_P = (5/2)R$ and $C_V = (3/2)R$.

- (*a*) Isothermal compression followed by isobaric heating.
- (*b*) Adiabatic compression followed by isobaric heating or cooling.
- (*c*) Adiabatic compression followed by isochoric heating or cooling.
- (*d*) Adiabatic compression followed by isothermal compression or expansion.

3.37. A scheme for finding the internal volume V_B^t of a gas cylinder consists of the following steps. The cylinder is filled with a gas to a low pressure P_1, and connected through a small line and valve to an evacuated reference tank of known volume V_A^t. The valve is opened, and gas flows through the line into the reference tank. After the system returns to its initial temperature, a sensitive pressure transducer provides a value for the pressure change ΔP in the cylinder. Determine the cylinder volume V_B^t from the following data:

- $V_A^t = 256 \text{ cm}^3$

- $\Delta P/P_1 = -0.0639$

3.38. A closed, nonconducting, horizontal cylinder is fitted with a nonconducting, frictionless, floating piston that divides the cylinder into Sections A and B. The two sections contain equal masses of air, initially at the same conditions, $T_1 = 300$ K and $P_1 = 1(\text{atm})$. An electrical heating element in Section A is activated, and the air temperatures slowly increase: T_A in Section A because of heat transfer, and T_B in Section B because of adiabatic compression by the slowly moving piston. Treat air as an ideal gas with $C_P = \frac{7}{2}R$, and let n_A be the number of moles of air in Section A. For the process as described, evaluate one of the following sets of quantities:

(a) T_A, T_B, and Q/n_A, if $P(\text{final}) = 1.25(\text{atm})$
(b) T_B, Q/n_A, and P (final), if $T_A = 425$ K
(c) T_A, Q/n_A, and P (final), if $T_B = 325$ K
(d) T_A, T_B, and P (final), if $Q/n_A = 3$ kJ·mol^{-1}

3.39. One mole of an ideal gas with constant heat capacities undergoes an arbitrary mechanically reversible process. Show that:

$$\Delta U = \frac{1}{\gamma - 1}\Delta(PV)$$

3.40. Derive an equation for the work of mechanically reversible, isothermal compression of 1 mol of a gas from an initial pressure P_1 to a final pressure P_2 when the equation of state is the virial expansion [Eq. (3.33)] truncated to:

$$Z = 1 + B'P$$

How does the result compare with the corresponding equation for an ideal gas?

3.41. A certain gas is described by the equation of state:

$$PV = RT + \left(b - \frac{\theta}{RT}\right)P$$

Here, b is a constant and θ is a function of T only. For this gas, determine expressions for the isothermal compressibility κ and the thermal pressure coefficient $(\partial P/\partial T)_V$. These expressions should contain only T, P, θ, $d\theta/dT$, and constants.

3.42. For methyl chloride at 100°C the second and third virial coefficients are:

$$B = -242.5 \text{ cm}^3 \cdot \text{mol}^{-1} \quad C = 25{,}200 \text{ cm}^6 \cdot \text{mol}^{-2}$$

Calculate the work of mechanically reversible, isothermal compression of 1 mol of methyl chloride from 1 bar to 55 bar at 100°C. Base calculations on the following forms of the virial equation:

(a) $$Z = 1 + \frac{B}{V} + \frac{C}{V^2}$$

(b) $$Z = 1 + B'P + C'P^2$$

where $$B' = \frac{B}{RT} \quad \text{and} \quad C' = \frac{C - B^2}{(RT)^2}$$

Why don't both equations give exactly the same result?

3.43. Any equation of state valid for gases in the zero-pressure limit implies a full set of virial coefficients. Show that the second and third virial coefficients implied by the generic cubic equation of state, Eq. (3.41), are:

$$B = b - \frac{a(T)}{RT} \quad C = b^2 + \frac{(\varepsilon + \sigma)ba(T)}{RT}$$

Specialize the result for B to the Redlich/Kwong equation of state, express it in reduced form, and compare it numerically with the generalized correlation for B for simple fluids, Eq. (3.61). Discuss what you find.

3.44. Calculate Z and V for ethylene at 25°C and 12 bar by the following equations:

(a) The truncated virial equation [Eq. (3.38)] with the following experimental values of virial coefficients:

$$B = -140 \text{ cm}^3 \cdot \text{mol}^{-1} \quad C = 7200 \text{ cm}^6 \cdot \text{mol}^{-2}$$

(b) The truncated virial equation [Eq. (3.36)], with a value of B from the generalized Pitzer correlation [Eqs. (3.58)–(3.62)]

(c) The Redlich/Kwong equation

(d) The Soave/Redlich/Kwong equation

(e) The Peng/Robinson equation

3.45. Calculate Z and V for ethane at 50°C and 15 bar by the following equations:

(a) The truncated virial equation [Eq. (3.38)] with the following experimental values of virial coefficients:

$$B = -156.7 \text{ cm}^3 \cdot \text{mol}^{-1} \quad C = 9650 \text{ cm}^6 \cdot \text{mol}^{-2}$$

(b) The truncated virial equation [Eq. (3.36)], with a value of B from the generalized Pitzer correlation [Eqs. (3.58)–(3.62)]

(c) The Redlich/Kwong equation

(d) The Soave/Redlich/Kwong equation

(e) The Peng/Robinson equation

3.46. Calculate Z and V for sulfur hexafluoride at 75°C and 15 bar by the following equations:

(a) The truncated virial equation [Eq. (3.38)] with the following experimental values of virial coefficients:

$$B = -194 \text{ cm}^3 \cdot \text{mol}^{-1} \quad C = 15,300 \text{ cm}^6 \cdot \text{mol}^{-2}$$

(b) The truncated virial equation [Eq. (3.36)], with a value of B from the generalized Pitzer correlation [Eqs. (3.58)–(3.62)]

(c) The Redlich/Kwong equation

(d) The Soave/Redlich/Kwong equation

(e) The Peng/Robinson equation

For sulfur hexafluoride, $T_c = 318.7$ K, $P_c = 37.6$ bar, $V_c = 198$ cm$^3 \cdot$mol^{-1}, and $\omega = 0.286$.

3.47. Calculate Z and V for ammonia at 320 K and 15 bar by the following equations:

(a) The truncated virial equation [Eq. (3.38)] with the following values of virial coefficients:

$$B = -208 \text{ cm}^3 \cdot \text{mol}^{-1} \quad C = 4378 \text{ cm}^6 \cdot \text{mol}^{-2}$$

(b) The truncated virial equation [Eq. (3.36)], with a value of B from the generalized Pitzer correlation [Eqs. (3.58)–(3.62)]

(c) The Redlich/Kwong equation

(d) The Soave/Redlich/Kwong equation

(e) The Peng/Robinson equation

3.48. Calculate Z and V for boron trichloride at 300 K and 1.5 bar by the following equations:

(a) The truncated virial equation [Eq. (3.38)] with the following values of virial coefficients:

$$B = -724 \text{ cm}^3 \cdot \text{mol}^{-1} \quad C = -93,866 \text{ cm}^6 \cdot \text{mol}^{-2}$$

(b) The truncated virial equation [Eq. (3.36)], with a value of B from the generalized Pitzer correlation [Eqs. (3.58)–(3.62)]

(c) The Redlich/Kwong equation

(d) The Soave/Redlich/Kwong equation

(e) The Peng/Robinson equation

For BCl$_3$, $T_c = 452$ K, $P_c = 38.7$ bar, and $\omega = 0.086$.

3.49. Calculate Z and V for nitrogen trifluoride at 300 K and 95 bar by the following equations:

(a) The truncated virial equation [Eq. (3.38)] with the following values of virial coefficients:

$$B = -83.5 \text{ cm}^3 \cdot \text{mol}^{-1} \quad C = -5592 \text{ cm}^6 \cdot \text{mol}^{-2}$$

(b) The truncated virial equation [Eq. (3.36)], with a value of B from the generalized Pitzer correlation [Eqs. (3.58)–(3.62)]

(c) The Redlich/Kwong equation

(d) The Soave/Redlich/Kwong equation

(e) The Peng/Robinson equation

For NF_3, $T_c = 234$ K, $P_c = 44.6$ bar, and $\omega = 0.126$.

3.50. Determine Z and V for steam at 250°C and 1800 kPa by the following:

(a) The truncated virial equation [Eq. (3.38)] with the following experimental values of virial coefficients:

$$B = -152.5 \text{ cm}^3 \cdot \text{mol}^{-1} \quad C = -5800 \text{ cm}^6 \cdot \text{mol}^{-2}$$

(b) The truncated virial equation [Eq. (3.36)], with a value of B from the generalized Pitzer correlation [Eqs. (3.58)–(3.62)].

(c) The steam tables (App. E).

3.51. With respect to the virial expansions, Eqs. (3.33) and (3.34), show that:

$$B' = \left(\frac{\partial Z}{\partial P}\right)_{T,P=0} \quad \text{and} \quad B = \left(\frac{\partial Z}{\partial \rho}\right)_{T,\rho=0}$$

where $\rho \equiv 1/V$.

3.52. Equation (3.34) when truncated to *four* terms accurately represents the volumetric data for methane gas at 0°C with:

$$B = -53.4 \text{ cm}^3 \cdot \text{mol}^{-1} \quad C = 2620 \text{ cm}^6 \cdot \text{mol}^{-2} \quad D = 5000 \text{ cm}^9 \cdot \text{mol}^{-3}$$

(a) Use these data to prepare a plot of Z vs. P for methane at 0°C from 0 to 200 bar.

(b) To what pressures do Eqs. (3.36) and (3.37) provide good approximations?

3.53. Calculate the molar volume of saturated liquid and the molar volume of saturated vapor by the Redlich/Kwong equation for one of the following and compare results with values found by suitable generalized correlations.

(a) Propane at 40°C where $P^{\text{sat}} = 13.71$ bar

(b) Propane at 50°C where $P^{\text{sat}} = 17.16$ bar

(c) Propane at 60°C where $P^{\text{sat}} = 21.22$ bar

(d) Propane at 70°C where $P^{\text{sat}} = 25.94$ bar

(e) n-Butane at 100°C where $P^{\text{sat}} = 15.41$ bar
(f) n-Butane at 110°C where $P^{\text{sat}} = 18.66$ bar
(g) n-Butane at 120°C where $P^{\text{sat}} = 22.38$ bar
(h) n-Butane at 130°C where $P^{\text{sat}} = 26.59$ bar
(i) Isobutane at 90°C where $P^{\text{sat}} = 16.54$ bar
(j) Isobutane at 100°C where $P^{\text{sat}} = 20.03$ bar
(k) Isobutane at 110°C where $P^{\text{sat}} = 24.01$ bar
(l) Isobutane at 120°C where $P^{\text{sat}} = 28.53$ bar
(m) Chlorine at 60°C where $P^{\text{sat}} = 18.21$ bar
(n) Chlorine at 70°C where $P^{\text{sat}} = 22.49$ bar
(o) Chlorine at 80°C where $P^{\text{sat}} = 27.43$ bar
(p) Chlorine at 90°C where $P^{\text{sat}} = 33.08$ bar
(q) Sulfur dioxide at 80°C where $P^{\text{sat}} = 18.66$ bar
(r) Sulfur dioxide at 90°C where $P^{\text{sat}} = 23.31$ bar
(s) Sulfur dioxide at 100°C where $P^{\text{sat}} = 28.74$ bar
(t) Sulfur dioxide at 110°C where $P^{\text{sat}} = 35.01$ bar
(u) Boron trichloride at 400 K where $P^{\text{sat}} = 17.19$ bar
      For BCl_3, $T_c = 452$ K, $P_c = 38.7$ bar, and $\omega = 0.086$.
(v) Boron trichloride at 420 K where $P^{\text{sat}} = 23.97$ bar
(w) Boron trichloride at 440 K where $P^{\text{sat}} = 32.64$ bar
(x) Trimethylgallium at 430 K where $P^{\text{sat}} = 13.09$ bar
      For $Ga(CH_3)_3$, $T_c = 510$ K, $P_c = 40.4$ bar, and $\omega = 0.205$.
(y) Trimethylgallium at 450 K where $P^{\text{sat}} = 18.27$ bar
(z) Trimethylgallium at 470 K where $P^{\text{sat}} = 24.55$ bar

3.54. Use the Soave/Redlich/Kwong equation to calculate the molar volumes of saturated liquid and saturated vapor for the substance and conditions given by one of the parts of Prob. 3.53 and compare results with values found by suitable generalized correlations.

3.55. Use the Peng/Robinson equation to calculate the molar volumes of saturated liquid and saturated vapor for the substance and conditions given by one of the parts of Prob. 3.53 and compare results with values found by suitable generalized correlations.

3.56. Estimate the following:

(a) The volume occupied by 18 kg of ethylene at 55°C and 35 bar.
(b) The mass of ethylene contained in a 0.25 m³ cylinder at 50°C and 115 bar.

3.57. The vapor-phase molar volume of a particular compound is reported as 23,000 cm³·mol⁻¹ at 300 K and 1 bar. No other data are available. Without assuming ideal-gas behavior, determine a reasonable estimate of the molar volume of the vapor at 300 K and 5 bar.

3.58. To a good approximation, what is the molar volume of ethanol vapor at 480°C and 6000 kPa? How does this result compare with the ideal-gas value?

3.59. A 0.35 m³ vessel is used to store liquid propane at its vapor pressure. Safety considerations dictate that at a temperature of 320 K the liquid must occupy no more than 80% of the total volume of the vessel. For these conditions, determine the mass of vapor and the mass of liquid in the vessel. At 320 K the vapor pressure of propane is 16.0 bar.

3.60. A 30 m³ tank contains 14 m³ of liquid *n*-butane in equilibrium with its vapor at 25°C. Estimate the mass of *n*-butane vapor in the tank. The vapor pressure of *n*-butane at the given temperature is 2.43 bar.

3.61. Estimate:

   (*a*) The mass of ethane contained in a 0.15 m³ vessel at 60°C and 14,000 kPa.
   (*b*) The temperature at which 40 kg of ethane stored in a 0.15 m³ vessel exerts a pressure 20,000 kPa.

3.62. A size D compressed gas cylinder has an internal volume of 2.40 liters. Estimate the pressure in a size D cylinder if it contains 454 g of one of the following semiconductor process gases at 20°C:

   (*a*) Phosphine, PH_3, for which $T_c = 324.8$ K, $P_c = 65.4$ bar, and $\omega = 0.045$
   (*b*) Boron trifluoride, BF_3, for which $T_c = 260.9$ K, $P_c = 49.9$ bar, and $\omega = 0.434$
   (*c*) Silane, SiH_4, for which $T_c = 269.7$ K, $P_c = 48.4$ bar, and $\omega = 0.094$
   (*d*) Germane, GeH_4, for which $T_c = 312.2$ K, $P_c = 49.5$ bar, and $\omega = 0.151$
   (*e*) Arsine, AsH_3, for which $T_c = 373$ K, $P_c = 65.5$ bar, and $\omega = 0.011$
   (*f*) Nitrogen trifluoride, NF_3, for which $T_c = 234$ K, $P_c = 44.6$ bar, and $\omega = 0.120$

3.63. For one of the substances in Prob. 3.62, estimate the mass of the substance contained in the size D cylinder at 20°C and 25 bar.

3.64. Recreational scuba diving using air is limited to depths of 40 m. Technical divers use different gas mixes at different depths, allowing them to go much deeper. Assuming a lung volume of 6 liters, estimate the mass of air in the lungs of:

   (*a*) A person at atmospheric conditions.
   (*b*) A recreational diver breathing air at a depth of 40 m below the ocean surface.
   (*c*) A near-world-record technical diver at a depth of 300 m below the ocean surface, breathing 10 mol% oxygen, 20 mol% nitrogen, 70 mol% helium.

3.65. To what pressure does one fill a 0.15 m³ vessel at 25°C in order to store 40 kg of ethylene in it?

3.66. If 15 kg of H_2O in a 0.4 m³ container is heated to 400°C, what pressure is developed?

3.67. A 0.35 m³ vessel holds ethane vapor at 25°C and 2200 kPa. If it is heated to 220°C, what pressure is developed?

3.68. What is the pressure in a 0.5 m³ vessel when it is charged with 10 kg of carbon dioxide at 30°C?

3.69. A rigid vessel, filled to one-half its volume with liquid nitrogen at its normal boiling point, is allowed to warm to 25°C. What pressure is developed? The molar volume of liquid nitrogen at its normal boiling point is 34.7 cm³·mol⁻¹.

3.70. The specific volume of isobutane liquid at 300 K and 4 bar is 1.824 cm³·g⁻¹. Estimate the specific volume at 415 K and 75 bar.

3.71. The density of liquid *n*-pentane is 0.630 g·cm⁻³ at 18°C and 1 bar. Estimate its density at 140°C and 120 bar.

3.72. Estimate the density of liquid ethanol at 180°C and 200 bar.

3.73. Estimate the volume change of vaporization for ammonia at 20°C. At this temperature the vapor pressure of ammonia is 857 kPa.

3.74. *PVT* data may be taken by the following procedure: A mass m of a substance of molar mass $\mathcal{M}$ is introduced into a thermostated vessel of known total volume V^t. The system is allowed to equilibrate, and the temperature T and pressure P are measured.

   (a) Approximately what percentage errors are allowable in the measured variables (m, $\mathcal{M}$, V^t, T, and P) if the maximum allowable error in the calculated compressibility factor Z is $\pm 1\%$?
   (b) Approximately what percentage errors are allowable in the measured variables if the maximum allowable error in calculated values of the second virial coefficient B is $\pm 1\%$? Assume that $Z \simeq 0.9$ and that values of B are calculated by Eq. (3.37).

3.75. For a gas described by the Redlich/Kwong equation and for a temperature greater than T_c, develop expressions for the two limiting slopes,

$$\lim_{P \to 0} \left(\frac{\partial Z}{\partial P} \right)_T \qquad \lim_{P \to \infty} \left(\frac{\partial Z}{\partial P} \right)_T$$

Note that in the limit as $P \to 0$, $V \to \infty$, and that in the limit as $P \to \infty$, $V \to b$.

3.76. If 140(ft)³ of methane gas at 60(°F) and 1(atm) is equivalent to 1(gal) of gasoline as fuel for an automobile engine, what would be the volume of the tank required to hold methane at 3000(psia) and 60(°F) in an amount equivalent to 10(gal) of gasoline?

3.77. Determine a good estimate for the compressibility factor Z of saturated hydrogen vapor at 25 K and 3.213 bar. For comparison, an experimental value is $Z = 0.7757$.

3.78. The *Boyle temperature* is the temperature for which:

$$\lim_{P \to 0} \left(\frac{\partial Z}{\partial P} \right)_T = 0$$

   (a) Show that the second virial coefficient B is zero at the Boyle temperature.
   (b) Use the generalized correlation for B, Eqs. (3.58)–(3.62), to estimate the *reduced* Boyle temperature for simple fluids.

3.79. Natural gas (assume pure methane) is delivered to a city via pipeline at a volumetric rate of 150 million standard cubic feet per day. Average delivery conditions are 50(°F) and 300(psia). Determine:

(a) The volumetric delivery rate in *actual* cubic feet per day.
(b) The molar delivery rate in kmol per hour.
(c) The gas velocity at delivery conditions in $m \cdot s^{-1}$.

The pipe is 24(in) schedule-40 steel with an inside diameter of 22.624(in). Standard conditions are 60(°F) and 1(atm).

3.80. Some corresponding-states correlations use the critical compressibility factor Z_c, rather than the acentric factor ω, as a third parameter. The two types of correlation (one based on T_c, P_c, and Z_c, the other on T_c, P_c, and ω) would be equivalent were there a one-to-one correspondence between Z_c and ω. The data of App. B allow a test of this correspondence. Prepare a plot of Z_c vs. ω to see how well Z_c correlates with ω. Develop a linear correlation ($Z_c = a + b\omega$) for nonpolar substances.

3.81. Figure 3.3 suggests that the isochores (paths of constant volume) are approximately straight lines on a *P-T* diagram. Show that the following models imply linear isochores.

(a) Constant-β, κ equation for liquids
(b) Ideal-gas equation
(c) Van der Waals equation

3.82. An ideal gas, initially at 25°C and 1 bar, undergoes the following cyclic processes in a closed system:

(a) In mechanically reversible processes, it is first compressed adiabatically to 5 bar, then cooled at a constant pressure of 5 bar to 25°C, and finally expanded isothermally to its original pressure.
(b) The cycle is irreversible, and each step has an efficiency of 80% compared with the corresponding mechanically reversible process. The cycle still consists of an adiabatic compression step, an isobaric cooling step, and an isothermal expansion.

Calculate Q, W, ΔU, and ΔH for each step of the process and for the cycle. Take $C_P = (7/2)R$ and $C_V = (5/2)R$.

3.83. Show that the density-series second virial coefficients can be derived from isothermal volumetric data via the expression:

$$B = \lim_{\rho \to 0}(Z - 1)/\rho \qquad \rho(\text{molar density}) \equiv 1/V$$

3.84. Use the equation of the preceding problem and data from Table E.2 of App. E to obtain a value of B for water at one of the following temperatures:

(a) 300°C
(b) 350°C
(c) 400°C

3.85. Derive the values of Ω, Ψ, and Zc given in Table 3.1 for:

    (*a*) The Redlich/Kwong equation of state.
    (*b*) The Soave/Redlich/Kwong equation of state.
    (*c*) The Peng/Robinson equation of state.

3.86. Suppose Z vs. P_r data are available at constant T_r. Show that the reduced density-series second virial coefficient can be derived from such data via the expression:

$$\hat{B} = \lim_{P_r \to 0} (Z - 1)ZT_r/P_r$$

Suggestion: Base the development on the full virial expansion in density, Eq. (3.34)

3.87. Use the result of the preceding problem and data from Table D.1 of App. D to obtain a value of $\hat{B}$ for simple fluids at $T_r = 1$. Compare the result with the value implied by Eq. (3.61).

3.88. The following conversation was overheard in the corridors of a large engineering firm.

New engineer: "Hi, boss. Why the big smile?"

Old-timer: "I finally won a wager with Harry Carey, from Research. He bet me that I couldn't come up with a quick but accurate estimate for the molar volume of argon at 30°C and 300 bar. Nothing to it; I used the ideal-gas equation, and got about 83 cm^3·mol^{-1}. Harry shook his head, but paid up. What do you think about that?"

New engineer (consulting his thermo text): "I think you must be living right."

Argon at the stated conditions is **not** an ideal gas. Demonstrate numerically why the old-timer won his wager.

3.89. Five mol of calcium carbide are combined with 10 mol of water in a closed, rigid, high-pressure vessel of 1800 cm^3 internal empty volume. Acetylene gas is produced by the reaction:

$$CaC_2(s) + 2H_2O(l) \to C_2H_2(g) + Ca(OH)_2(s)$$

The vessel contains packing with a porosity of 40% to prevent explosive decomposition of the acetylene. Initial conditions are 25°C and 1 bar, and the reaction goes to completion. The reaction is exothermic, but owing to heat transfer, the final temperature is only 125°C. Determine the final pressure in the vessel.
Note: At 125°C, the molar volume of Ca(OH)$_2$ is 33.0 cm^3·mol^{-1}. Ignore the effects of any gases (e.g., air) initially present in the vessel.

3.90. Storage is required for 35,000 kg of propane, received as a gas at 10°C and 1(atm). Two proposals have been made:

    (*a*) Store it as a gas at 10°C and 1(atm).
    (*b*) Store it as a liquid in equilibrium with its vapor at 10°C and 6.294(atm). For this mode of storage, 90% of the tank volume is occupied by liquid.

Compare the two proposals, discussing pros and cons of each. Be quantitative where possible.

3.91. The definition of compressibility factor Z, Eq. (3.32), may be written in the more intuitive form:

$$Z \equiv \frac{V}{V(\text{ideal gas})}$$

where both volumes are at the same T and P. Recall that an ideal gas is a model substance comprising particles with no intermolecular forces. Use the intuitive definition of Z to argue that:

(a) Intermolecular attractions promote values of $Z < 1$.
(b) Intermolecular repulsions promote values of $Z > 1$.
(c) A balance of attractions and repulsions implies that $Z = 1$. (Note that an ideal gas is a special case for which there are *no* attractions or repulsions.)

3.92. Write the general form of an equation of state as:

$$Z = 1 + Z_{\text{rep}}(\rho) - Z_{\text{attr}}(T, \rho)$$

where $Z_{\text{rep}}(\rho)$ represents contributions from repulsions, and $Z_{\text{attr}}(T, \rho)$ represents contributions from attractions. What are the repulsive and attractive contributions to the van der Waals equation of state?

3.93. Given below are four proposed modifications of the van der Waals equation of state. Are any of these modifications *reasonable*? Explain carefully; statements such as, "It isn't cubic in volume" do not qualify.

(a) $P = \dfrac{RT}{V - b} - \dfrac{a}{V}$

(b) $P = \dfrac{RT}{(V - b)^2} - \dfrac{a}{V}$

(c) $P = \dfrac{RT}{V(V - b)} - \dfrac{a}{V^2}$

(d) $P = \dfrac{RT}{V} - \dfrac{a}{V^2}$

3.94. With reference to Prob. 2.47, assume air to be an ideal gas, and develop an expression giving the household air temperature as a function of time.

3.95. A garden hose with the water valve shut and the nozzle closed sits in the sun, full of liquid water. Initially, the water is at $10°C$ and 6 bar. After some time the temperature of the water rises to $40°C$. Owing to the increase in temperature and pressure and the elasticity of the hose, the internal diameter of the hose increases by 0.35%. Estimate the final pressure of the water in the hose.
Data: $\beta(\text{ave}) = 250 \times 10^{-6} \text{ K}^{-1}$; $\kappa(\text{ave}) = 45 \times 10^{-6} \text{ bar}^{-1}$

Chapter 4

Heat Effects

Heat effects refer to physical and chemical phenomena that are associated with heat transfer to or from a system or that result in temperature changes within a system, or both. The simplest example of a heat effect is the heating or cooling of a fluid by the purely physical direct transfer of heat to or from the fluid. The temperature changes that occur are known as *sensible* heat effects, because they may be detected by our sense perception of temperature. Phase changes, physical processes occurring for a pure substance at constant temperature and pressure, are accompanied by *latent* heats. Chemical reactions are characterized by *heats of reaction*, which for combustion reactions evolve heat. Every chemical or biochemical process is associated with one or more heat effects. The metabolism of the human body, for example, generates heat which is either transferred to its surroundings or used to maintain or increase body temperature.

Chemical manufacturing processes may include a number of heat effects. Ethylene glycol (a coolant and antifreeze) is made by catalytic partial oxidation of ethylene to form ethylene oxide, followed by a hydration reaction:

$$C_2H_4 + \tfrac{1}{2}O_2 \rightarrow C_2H_4O$$
$$C_2H_4O + H_2O \rightarrow C_2H_4(OH)_2$$

The oxidation reaction is carried out near 250°C, and the reactants must be heated to this temperature, a sensible heat effect. The oxidation reaction tends to raise the temperature, and the heat of reaction is removed from the reactor to keep the temperature near 250°C. The ethylene oxide is hydrated to glycol by absorption in water. Heat is evolved because of the phase change and dissolution, and also because of the hydration reaction. Finally, the glycol is purified by distillation, a process of vaporization and condensation, resulting in the separation of glycol from solution. Virtually all of the important heat effects are included in this process. Most of these are treated in the present chapter, although heat effects related to mixing processes must be delayed until Chap. 11, after thermodynamics of solutions have been introduced in Chap. 10. The following important heat effects are considered in this chapter:

- Sensible heat effects, characterized by temperature changes

- Heat capacities as a function of temperature and their use through defined functions

- Heats of phase transition, i.e., latent heats of pure substances

- Heats of reaction, combustion, and formation

- Heats of reaction as a function of temperature

- The calculation of heat effects for industrial reactions

4.1 SENSIBLE HEAT EFFECTS

Heat transfer to or from a system in which there are no phase transitions, no chemical reactions, and no changes in composition causes a sensible heat effect, i.e., the temperature of the system is caused to change. The need here is for a relation between the quantity of heat transferred and the resulting temperature change.

When the system is a homogeneous substance of constant composition, the phase rule indicates that fixing the values of two intensive properties establishes its state. The molar or specific internal energy of a substance may therefore be expressed as a *function of two other state variables*. The key thermodynamic variable is temperature. With molar or specific volume chosen arbitrarily, we have $U = U(T, V)$. Then

$$dU = \left(\frac{\partial U}{\partial T}\right)_V dT + \left(\frac{\partial U}{\partial V}\right)_T dV$$

With the definition of C_V provided by Eq. (2.15) this becomes:

$$dU = C_V dT + \left(\frac{\partial U}{\partial V}\right)_T dV$$

The final term is zero in two circumstances:

- For any closed-system constant-volume process.

- Whenever the internal energy is independent of volume, as for the ideal-gas state and the incompressible liquid.

In either case, $dU = C_V dT$

and $$\Delta U = \int_{T_1}^{T_2} C_V dT \qquad (4.1)$$

Although real liquids are to some degree compressible, far below their critical temperature they can often be treated as incompressible fluids. The ideal-gas state is also of interest, because actual gases at low pressures approach ideality. The only possible mechanically reversible constant-volume process is simple heating (stirring work is inherently irreversible), for which $Q = \Delta U$, and Eq. (2.18) written for a unit mass or a mole becomes:

$$Q = \Delta U = \int_{T_1}^{T_2} C_V dT$$

The enthalpy may be treated similarly, with molar or specific enthalpy expressed most conveniently as a function of temperature and pressure. Then $H = H(T, P)$, and

$$dH = \left(\frac{\partial H}{\partial T}\right)_P dT + \left(\frac{\partial H}{\partial P}\right)_T dP$$

With the definition of C_P provided by Eq. (2.19),

$$dH = C_P dT + \left(\frac{\partial H}{\partial P}\right)_T dP$$

Again, the final term is zero for two situations:

- For any constant-pressure process.

- When the enthalpy is independent of pressure, regardless of process. This is exactly true for the ideal-gas state and approximately true for real gases at low pressure and high temperature.

In either case,
$$dH = C_P dT$$

and
$$\Delta H = \int_{T_1}^{T_2} C_P dT \tag{4.2}$$

Moreover, $Q = \Delta H$ for mechanically reversible, constant-pressure, closed-system processes [Eq. (2.22)] and for the transfer of heat in steady-flow processes where ΔE_P and ΔE_K are negligible and $W_s = 0$ [Eq. (2.32)]. In either case,

$$Q = \Delta H = \int_{T_1}^{T_2} C_P dT \tag{4.3}$$

This equation finds frequent application for flow processes designed for simple heating and cooling of gases, liquids, and solids.

Temperature Dependence of the Heat Capacity

Evaluation of the integral in Eq. (4.3) requires knowledge of the temperature dependence of the heat capacity. This is usually given by an empirical equation; the two simplest expressions of practical value are:

$$\frac{C_P}{R} = \alpha + \beta T + \gamma T^2 \qquad \text{and} \qquad \frac{C_P}{R} = a + bT + cT^{-2}$$

where α, β, and γ and a, b, and c are constants characteristic of the particular substance. With the exception of the last term, these equations are of the same form. We therefore combine them to provide a single expression:

$$\frac{C_P}{R} = A + BT + CT^2 + DT^{-2} \tag{4.4}$$

where either C or D is usually zero, depending on the substance considered.[1] Because the ratio C_P/R is dimensionless, the units of C_P are governed by the choice of R. The parameters are independent of temperature, but, at least in principle, depend on the value of the constant pressure. However, for liquids and solids the effect of pressure is usually very small. Values of the constants for selected solids and liquids are given in Tables C.2 and C.3 of App. C. The heat capacities of solids and liquids are usually found by direct measurement. Correlations for the heat capacities of many solids and liquids are given by Perry and Green and in the DIPPR Project 801 collection.[2]

Heat Capacity in the Ideal-Gas State

We noted in Section 3.3 that as $P \to 0$ a gas approaches the ideal-gas state, wherein molecular volumes and intermolecular forces are negligible. If these conditions are imagined to persist with increasing pressure, a hypothetical ideal-gas state continues to exist at finite pressures. The gas still has properties reflective of its internal molecular configuration, just as does a real gas, but without the influence of intermolecular interactions. Accordingly, ideal-gas-state heat capacities, designated by C_P^{ig} and C_V^{ig}, are functions of temperature, but independent of pressure, providing for ease of correlation. Fig. 4.1 illustrates the temperature dependence of C_P^{ig} for several representative substances.

Statistical mechanics provides a basic equation for the temperature dependence of the ideal-gas-state internal energy:

$$U^{ig} = \frac{3}{2} RT + f(T)$$

Equation (3.10), for the ideal-gas state, $H^{ig} = U^{ig} + RT$, becomes:

$$H^{ig} = \frac{5}{2} RT + f(T)$$

In view of Eq. (2.19),

$$C_P^{ig} \equiv \left(\frac{\partial H^{ig}}{\partial T} \right)_P = \frac{5}{2} R + \left(\frac{\partial f(T)}{\partial T} \right)_P$$

The first term on the right represents translational kinetic energy of the molecule, whereas the second combines all rotational and vibrational kinetic energies associated with the molecule. Because the molecules of a monatomic gas have no energies of rotation or vibration, $f(T)$ in the preceding equation is zero. Thus, in Fig. 4.1 the value of C_P^{ig}/R for argon is constant at a value of 5/2. For diatomic and polyatomic gases, $f(T)$ contributes importantly at all temperatures of practical importance. Diatomic molecules have a contribution equal to RT from their two rotational modes of motion. Thus, in Fig. 4.1, C_P^{ig}/R for N_2 is about $7/2\,R$ at moderate temperature, and it increases at higher temperatures as intramolecular vibration begins to contribute. Nonlinear polyatomic molecules have a contribution of $3/2\,R$ from their three rotational modes

---

[1]The NIST Chemistry Webbook, http://webbook.nist.gov/ uses the Shomate equation for heat capacities, which also includes a T^3 term as well as all four terms of Eq. (4.4).

[2]R. H. Perry and D. Green, *Perry's Chemical Engineers' Handbook*, 8th ed., Sec. 2, McGraw-Hill, New York, 2008; Design Institute for Physical Properties, Project 801, http://www.aiche.org/dippr/projects/801.

of motion, and in addition usually have low-frequency vibrational modes that make an additional contribution at moderate temperature. The contribution becomes larger the more complex the molecule and increases monotonically with temperature, as is evident from the curves in Fig. 4.1 for H_2O and CO_2. The trend with molecular size and complexity is illustrated by the values of C_P^{ig}/R at 298 K in Table C.1 of App. C.

Figure 4.1: Ideal-gas-state heat capacities of argon, nitrogen, water, and carbon dioxide. Ideal-gas-state heat capacities increase smoothly with increasing temperature toward an upper limit, which is reached when all translational, rotational, and vibrational modes of molecular motion are fully excited.

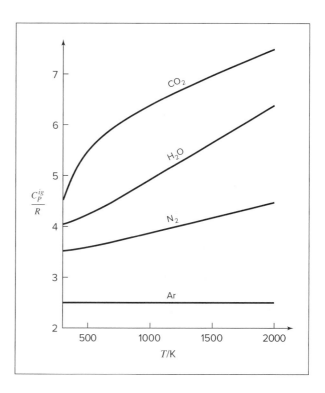

The temperature dependence of C_P^{ig} or C_V^{ig} is determined by experiment, most often from spectroscopic data and knowledge of molecular structure through calculations based on statistical mechanics.[3] Increasingly, quantum chemistry calculations, rather than spectroscopy experiments, are used to provide the molecular structure, and they often permit the calculation of heat capacities with precision comparable to experimental measurement. Where experimental data are not available, and quantum chemistry calculations are not warranted, methods of estimation are employed, as described by Prausnitz, Poling, and O'Connell.[4]

---

[3]D. A. McQuarrie, *Statistical Mechanics*, pp. 136–137, HarperCollins, New York, 1973.

[4]B. E. Poling, J. M. Prausnitz, and J. P. O'Connell, *The Properties of Gases and Liquids*, 5th ed., chap. 3, McGraw-Hill, New York, 2001.

Temperature dependence is expressed analytically by equations such as Eq. (4.4), here written:

$$\frac{C_P^{ig}}{R} = A + BT + CT^2 + DT^{-2} \qquad (4.5)$$

Values of the constants are given in Table C.1 of App. C for a number of common organic and inorganic gases. More accurate but more complex equations are found in the literature.[5] As a result of Eq. (3.12), the two ideal-gas-state heat capacities are related:

$$\frac{C_V^{ig}}{R} = \frac{C_P^{ig}}{R} - 1 \qquad (4.6)$$

The temperature dependence of C_V^{ig}/R follows from the temperature dependence of C_P^{ig}/R.

Although ideal-gas-state heat capacities are exactly correct for real gases only at zero pressure, the departure of real gases from the ideal-gas state is seldom significant at pressures below several bars, and here C_P^{ig} and C_V^{ig} are usually good approximations to their true heat capacities. Reference to Fig. 3.14 indicates a vast range of conditions at $P_r < 0.1$ for which assumption of the ideal-gas state is usually a suitable approximation. For most substances P_c exceeds 30 bar, which means that ideal-gas state behavior is often closely approximated up to a pressure of at least 3 bar.

Example 4.1

The parameters listed in Table C.1 of Appendix C require use of Kelvin temperatures in Eq. (4.5). Equations of the same form may also be developed for use with temperatures in °C, but the parameter values are different. The molar heat capacity of methane in the ideal-gas state is given as a function of temperature in kelvins by:

$$\frac{C_P^{ig}}{R} = 1.702 + 9.081 \times 10^{-3}\,T - 2.164 \times 10^{-6}\,T^2$$

where the parameter values are from Table C.1. Develop an equation for C_P^{ig}/R for use with temperatures in °C.

Solution 4.1

The relation between the two temperature scales is: $T\,\mathrm{K} = t\,°\mathrm{C} + 273.15$.
Therefore, as a function of t,

$$\frac{C_P^{ig}}{R} = 1.702 + 9.081 \times 10^{-3}(t + 273.15) - 2.164 \times 10^{-6}(t + 273.15)^2$$

or $\quad \dfrac{C_P^{ig}}{R} = 4.021 + 7.899 \times 10^{-3}\,t - 2.164 \times 10^{-6}\,t^2$

---

[5]See F. A. Aly and L. L. Lee, *Fluid Phase Equilibria*, vol. 6, pp. 169–179, 1981, and its bibliography; see also Design Institute for Physical Properties, Project 801, http://www.aiche.org/dippr/projects/801, and the Shomate equation employed by the NIST Chemistry Webbook, http://webbook.nist.gov.

Gas mixtures of constant composition behave exactly as do pure gases. In the ideal-gas state, molecules in mixtures have no influence on one another, and each gas exists independent of the others. The ideal-gas-state heat capacity of a mixture is therefore the mole-fraction-weighted sum of the heat capacities of the individual gases. Thus, for gases A, B, and C, the molar heat capacity of a mixture in the ideal-gas state is:

$$C_{P_{\text{mixture}}}^{ig} = y_A C_{P_A}^{ig} + y_B C_{P_B}^{ig} + y_C C_{P_C}^{ig} \tag{4.7}$$

where $C_{P_A}^{ig}$, $C_{P_B}^{ig}$, and $C_{P_C}^{ig}$ are the molar heat capacities of pure A, B, and C in the ideal-gas state, and y_A, y_B, and y_C are mole fractions. Because the heat-capacity polynomial, Eq. (4.5), is linear in the coefficients, the coefficients A, B, C, and D for a gas mixture are similarly given by mole-fraction weighted sums of the coefficients for the pure species.

Evaluation of the Sensible-Heat Integral

Evaluation of the integral $\int C_P dT$ is accomplished by substitution for C_P as a function of T by Eq. (4.4), followed by formal integration. For temperature limits of T_0 and T the result is:

$$\int_{T_0}^{T} \frac{C_P}{R} dT = A(T - T_0) + \frac{B}{2}(T^2 - T_0^2) + \frac{C}{3}(T^3 - T_0^3) + D\left(\frac{T - T_0}{TT_0}\right) \tag{4.8}$$

Given T_0 and T, the calculation of Q or ΔH is straightforward. Less direct is the calculation of T, given T_0 and Q or ΔH. Here, an iteration scheme may be useful. Factoring $(T - T_0)$ from each term on the right side of Eq. (4.8) gives:

$$\int_{T_0}^{T} \frac{C_P}{R} dT = \left[A + \frac{B}{2}(T + T_0) + \frac{C}{3}(T^2 + T_0^2 + TT_0) + \frac{D}{TT_0}\right](T - T_0)$$

We identify the quantity in square brackets as $\langle C_P \rangle_H / R$, where $\langle C_P \rangle_H$ is defined as a *mean heat capacity* for the temperature range from T_0 to T:

$$\frac{\langle C_P \rangle_H}{R} = A + \frac{B}{2}(T + T_0) + \frac{C}{3}(T^2 + T_0^2 + TT_0) + \frac{D}{TT_0} \tag{4.9}$$

Equation (4.2) may therefore be written:

$$\Delta H = \langle C_P \rangle_H (T - T_0) \tag{4.10}$$

The angular brackets enclosing C_P identify it as a mean value; subscript H denotes a mean value specific to enthalpy calculations and distinguishes this mean heat capacity from a similar quantity introduced in the next chapter.

Solution of Eq. (4.10) for T gives:

$$T = \frac{\Delta H}{\langle C_P \rangle_H} + T_0 \tag{4.11}$$

With a starting value for T, one can first evaluate $\langle C_P \rangle_H$ by Eq. (4.9). Substitution into Eq. (4.11) provides a new value of T from which to reevaluate $\langle C_P \rangle_H$. Iteration continues to convergence on a final value of T. Of course, such iteration is readily automated with built-in functions in a spreadsheet or a numerical analysis software package.

Example 4.2

Calculate the heat required to raise the temperature of 1 mol of methane from 260 to 600°C in a steady-flow process at a pressure sufficiently low that the ideal-gas state is a suitable approximation for methane.

Solution 4.2

Equations (4.3) and Eq. (4.8) together provide the required result. Parameters for C_P^{ig}/R are from Table C.1; $T_0 = 533.15$ K and $T = 873.15$ K.

Then

$$Q = \Delta H = R \int_{533.15}^{873.15} \frac{C_P^{ig}}{R} dT$$

$$Q = (8.314)\left[1.702(T - T_0) + \frac{9.081 \times 10^{-3}}{2}(T^2 - T_0^2)\right.$$

$$\left. - \frac{2.164 \times 10^{-6}}{3}(T^3 - T_0^3)\right] = 19{,}778 \text{ J}$$

Use of Defined Functions

The integral $\int (C_P/R)dT$ appears often in thermodynamic calculations. As a matter of convenience, we therefore define the right side of Eq. (4.8) as the function, ICPH(T_0, T; A, B, C, D), and presume the availability of a computer routine for its evaluation.[6] Equation (4.8) then becomes:

$$\int_{T_0}^{T} \frac{C_P}{R} dT \equiv \text{ICPH}(T_0, T; A, B, C, D)$$

The function name is ICPH (I indicates an integral), and the quantities in parentheses are the variables T_0 and T, followed by parameters A, B, C, and D. When these quantities are assigned numerical values, the notation represents a value for the integral. Thus, for the evaluation of Q in Ex. 4.2:

$$Q = 8.314 \times \text{ICPH}(533.15, 873.15; 1.702, 9.081 \times 10^{-3}, -2.164 \times 10^{-6}, 0.0) = 19{,}778 \text{ J}$$

Also useful is a defined function for the dimensionless mean value $\langle C_P \rangle_H/R$ given by Eq. (4.9). The function name is MCPH (M indicates a mean). The right side defines the function, MCPH(T_0, T; A, B, C, D). With this definition, Eq. (4.9) becomes:

$$\frac{\langle C_P \rangle_H}{R} = \text{MCPH}(T_0, T; A, B, C, D)$$

---

[6]Examples of these defined functions implemented in Microsoft Excel, Matlab, Maple, Mathematica, and Mathcad are provided in the online learning center at http://highered.mheducation.com:80/sites/1259696529.

A specific numerical value of this function is:

MCPH(533.15, 873.15; 1.702, 9.081×10^{-3}, −2.164 × 10^{-6}, 0.0) = 6.9965

representing $\langle C_P \rangle_H / R$ for methane in the calculation of Ex. 4.2. By Eq. (4.10),

$$\Delta H = (8.314)(6.9965)(873.15 - 533.15) = 19{,}778 \text{ J}$$

Example 4.3

What is the final temperature when heat in the amount of 400×10^6 J is added to 11×10^3 mol of ammonia initially at 530 K in a steady-flow process at 1 bar?

Solution 4.3

If ΔH is the enthalpy change for 1 mol, $Q = n \, \Delta H$, and

$$\Delta H = \frac{Q}{n} = \frac{400 \times 10^6}{11{,}000} = 36{,}360 \text{ J·mol}^{-1}$$

Then for any value of T, with parameters from Table C.1 and $R = 8.314 \text{ J·mol}^{-1}\text{·K}^{-1}$:

$$\frac{\langle C_P \rangle_H}{R} = \text{MCPH}(530, \, T; \, 3.578, \, 3.020 \times 10^{-3}, \, 0.0, \, -0.186 \times 10^5)$$

This equation and Eq. (4.11) together may be solved for T, yielding $T = 1234$ K.

A trial procedure is an alternative approach to solution of this problem. One sets up an equation for Q by combining Eqs. (4.3) and (4.8), with T as an unknown on the right. With Q known, one merely substitutes a rational succession of values for T until the value of Q is reproduced. Microsoft Excel's Goal Seek function is an example of an automated version of this procedure.

4.2 LATENT HEATS OF PURE SUBSTANCES

When a pure substance is liquefied from the solid state or vaporized from the liquid or solid at constant pressure, no change in temperature occurs; however, these processes require the transfer of finite amounts of heat to the substance. These heat effects are called *latent heats*: of fusion, of vaporization, and of sublimation. Similarly, there are heats of transition accompanying the change of a substance from one allotropic solid state to another; for example, the heat absorbed when rhombic crystalline sulfur changes to the monoclinic structure at 95°C and 1 bar is 11.3 J·g^{-1}.

The characteristic feature of all these processes is the coexistence of two phases. According to the phase rule, the intensive state of a two-phase system consisting of a single species is fixed by specification of just one intensive property. Thus the latent heat accompanying a phase change is a function of temperature only, and is related to other system properties by an exact thermodynamic equation:

$$\Delta H = T \Delta V \frac{dP^{\text{sat}}}{dT} \tag{4.12}$$

where for a pure species at temperature T,

ΔH = latent heat = enthalpy change accompanying the phase change
ΔV = volume change accompanying the phase change
P^{sat} = saturation pressure, i.e., the pressure at which the phase change occurs, which is
   a function only of T

The derivation of this equation, known as the **Clapeyron equation,** is given in Section 6.5.

When Eq. (4.12) is applied to the vaporization of a pure liquid, dP^{sat}/dT is the slope of the vapor pressure-versus-temperature curve at the temperature of interest, ΔV is the difference between molar volumes of saturated vapor and saturated liquid, and ΔH is the latent heat of vaporization. Thus, ΔH may be calculated from vapor-pressure and volumetric data, yielding an energy value with units of pressure $\times$ volume.

Latent heats are also measured calorimetrically. Experimental values are reported at selected temperatures for many substances.[7] Empirical correlations for the latent heats of many compounds as a function of temperature are given by Perry and Green and in the DIPPR Project 801 collection.[8] When required data are not available, approximate methods can provide estimates of the heat effect accompanying a phase change. Because heats of vaporization are by far the most important in practice, they have received the most attention. Predictions are most often made by group-contribution methods.[9] Alternative empirical methods serve one of two purposes:

- Prediction of the heat of vaporization at the normal boiling point, i.e., at a pressure of 1 standard atmosphere, defined as 101,325 Pa.

- Estimation of the heat of vaporization at any temperature from the known value at a single temperature.

Rough estimates of latent heats of vaporization for pure liquids at their normal boiling points (indicated by subscript n) are given by *Trouton's rule*:

$$\frac{\Delta H_n}{RT_n} \sim 10$$

where T_n is the absolute temperature of the normal boiling point. The units of ΔH_n, R, and T_n are chosen so that $\Delta H_n/RT_n$ is dimensionless. Dating from 1884, this empirical rule provides a simple check on whether values calculated by other methods are reasonable. Representative experimental values for this ratio are Ar, 8.0; N_2, 8.7; O_2, 9.1; HCl, 10.4; C_6H_6, 10.5; H_2S, 10.6; and H_2O, 13.1. The high value for water reflects the existence of intermolecular hydrogen bonds that rupture during vaporization.

[7]V. Majer and V. Svoboda, IUPAC Chemical Data Series No. 32, Blackwell, Oxford, 1985; R. H. Perry and D. Green, *Perry's Chemical Engineers' Handbook*, 8th ed., Sec. 2, McGraw-Hill, New York, 2008.

[8]R. H. Perry and D. Green, *Perry's Chemical Engineers' Handbook*, 8th ed., Sec. 2, McGraw-Hill, New York, 2008; Design Institute for Physical Properties, Project 801, http://www.aiche.org/dippr/projects/801.

[9]See, for example, M. Klüppel, S. Schulz, and P. Ulbig, *Fluid Phase Equilibria*, vol. 102, pp. 1–15, 1994.

Also for the normal boiling point, but not quite so simple, is the equation proposed by Riedel:[10]

$$\frac{\Delta H_n}{RT_n} = \frac{1.092(\ln P_c - 1.013)}{0.930 - T_{r_n}} \tag{4.13}$$

where P_c is the critical pressure in bars and T_{r_n} is the reduced temperature at T_n. Equation (4.13) is surprisingly accurate for an empirical expression; errors rarely exceed 5 percent. Applied to water it gives:

$$\frac{\Delta H_n}{RT_n} = \frac{1.092(\ln 220.55 - 1.013)}{0.930 - 0.577} = 13.56$$

from which $\Delta H_n = (13.56)(8.314)(373.15) = 42,065 \text{ J·mol}^{-1}$

This corresponds to 2334 J·g^{-1}; the steam-table value of 2257 J·g^{-1} is lower by 3.4 percent. Estimates of the latent heat of vaporization of a pure liquid at any temperature from the known value at a single temperature are given by the method of Watson.[11] The basis may be a known experimental value or a value estimated by Eq. (4.13):

$$\frac{\Delta H_2}{\Delta H_1} = \left(\frac{1 - T_{r_2}}{1 - T_{r_1}}\right)^{0.38} \tag{4.14}$$

This empirical equation is simple and fairly accurate; its use is illustrated in the following example.

Example 4.4

Given that the latent heat of vaporization of water at 100°C is 2257 J·g^{-1}, estimate the latent heat at 300°C.

Solution 4.4

Let ΔH_1 = latent heat at 100°C = 2257 J·g^{-1}
   ΔH_2 = latent heat at 300°C
   $T_{r_1} = 373.15/647.1 = 0.577$
   $T_{r_2} = 573.15/647.1 = 0.886$

Then by Eq. (4.14),

$$\Delta H_2 = (2257)\left(\frac{1 - 0.886}{1 - 0.577}\right)^{0.38} = (2257)(0.270)^{0.38} = 1371 \text{ J·g}^{-1}$$

The value given in the steam tables is 1406 J·g^{-1}.

---

[10]L. Riedel, *Chem. Ing. Tech.*, vol. 26, pp. 679–683, 1954.

[11]K. M. Watson, *Ind. Eng. Chem.*, vol. 35, pp. 398–406, 1943.

4.3   STANDARD HEAT OF REACTION

Heat effects of chemical processes are fully as important as those for physical processes. Chemical reactions are accompanied by the transfer of heat, by temperature changes during reaction, or by both. The ultimate cause lies in the difference between the molecular configurations of products and reactants. For an *adiabatic* combustion reaction, reactants and products possess the same energy, requiring an elevated temperature for the products. For the corresponding *isothermal* reaction, heat is necessarily transferred to the surroundings. Between these two extremes an infinite combination of effects is possible. Each reaction carried out in a particular way is accompanied by particular heat effects. Their complete tabulation is impossible. Our object is therefore to devise methods of calculation of the heat effects for reactions carried out in diverse ways from data for reactions carried out in an arbitrarily defined *standard* way, thus leading to *standard heats of reaction*. This reduces the required data to a minimum.

Heats of reaction are based on experimental measurements. Most easily measured are *heats of combustion*, because of the nature of such reactions. A simple procedure is provided by a flow calorimeter. Fuel is mixed with air at a temperature T, and the mixture flows into a combustion chamber where reaction occurs. The combustion products enter a water-jacketed section in which they are cooled to temperature T. Because no shaft work is produced and the calorimeter is designed to eliminate potential- and kinetic-energy changes, the overall energy balance, Eq. (2.32), reduces to

$$\Delta H = Q$$

Thus the enthalpy change caused by the combustion reaction is equal in magnitude to the heat flowing from the reaction products to the water, and may be calculated from the temperature rise and flow rate of the water. The enthalpy change of reaction ΔH is called the *heat of reaction*. If the reactants and products are in their *standard states*, then the heat effect is the **standard** *heat of reaction*.

The definition of a *standard state* is straightforward. For a given temperature,

> **A standard state is *defined* as the state of a substance at specified pressure, composition, and physical condition as, e.g., gas, liquid, or solid.**

The standard states in use throughout the world have been established by general agreement. They are based on a *standard-state pressure* of 1 bar (10^5 Pa). With respect to composition, the standard states used in this chapter are states of *pure* species. For liquids and solids it is the actual state of the pure species at the standard-state pressure. Nothing could be simpler. However, for gases there is a small complication, as the chosen physical state is the ideal-gas state, for which we have already established heat capacities. In summary, the standard states used in this chapter are:

- *Gases*: The pure substance in the ideal-gas state at 1 bar.
- *Liquids and solids*: The real pure liquid or solid at 1 bar.

> **One must understand that standard states apply *at any temperature*. There is no specification of temperature for any standard state. *Reference* temperatures, also in use with heats of reaction, are entirely independent of standard states.**

With respect to the chemical reaction, $aA + bB \rightarrow lL + mM$, the *standard* heat of reaction at temperature T is defined as the enthalpy change when a moles of A and b moles of B in their *standard states at temperature T* react to form l moles of L and m moles of M in their *standard states at the same temperature T*. The mechanism of this change is immaterial to the calculation of the enthalpy change. One may view the process shown in Fig. 4.2 as occurring in a "box of tricks." If the properties of reactants and products in their standard states are not significantly different from the properties of actual reactants and products, the standard heat of reaction is a reasonable approximation to the actual heat of reaction. If this is not the case, then additional steps must be incorporated into the calculation scheme to account for any differences. The most common difference is caused by pressures higher than appropriate to the ideal-gas state (as for the ammonia-synthesis reaction). In this case, enthalpy changes for transformation from real-gas states to ideal-gas states and the reverse are required. These are readily made, as shown in Chapter 6.

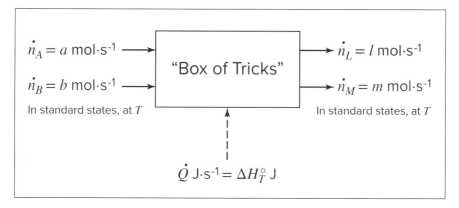

Figure 4.2: Schematic representation of the standard heat of reaction at temperature T.

Property values in the standard state are denoted by the degree symbol. For example, C_P° is the standard-state heat capacity. Because the standard state for gases is the ideal-gas state, C_P° is identical with C_P^{ig}, and the data of Table C.1 apply to the standard state for gases.

> **All conditions for a standard state are fixed except temperature, which is always the temperature of the system. Standard-state properties are therefore functions of temperature only.**

The standard state chosen for gases is hypothetical or fictitious because at 1 bar actual gases deviate from the ideal-gas state. However, they seldom deviate much, and for most purposes enthalpies for the real-gas state at 1 bar and the ideal-gas state are negligibly different.

When a heat of reaction is given for a particular reaction, it applies for the stoichiometric coefficients as written. If each stoichiometric coefficient is doubled, the heat of reaction is doubled. For example, two versions of the ammonia synthesis reaction are written:

$$\tfrac{1}{2}N_2 + \tfrac{3}{2}H_2 \rightarrow NH_3 \qquad \Delta H^\circ_{298} = -46{,}110 \text{ J}$$
$$N_2 + 3H_2 \rightarrow 2NH_3 \qquad \Delta H^\circ_{298} = -92{,}220 \text{ J}$$

The symbol ΔH°_{298} indicates that the heat of reaction is the *standard* value for a temperature of 298.15 K (25°C) and for the reaction as written.

4.4   STANDARD HEAT OF FORMATION

Tabulation of data for just one temperature and for just the *standard* heats of reaction for all of the vast number of possible reactions is impractical. Fortunately, the standard heat of any reaction at temperature T can be calculated if the *standard heats of formation* at the same temperature are known for the compounds taking part in the reaction. A *formation* reaction is **defined** as a reaction that forms a single compound *from its constituent elements*. For example, the reaction $C + \frac{1}{2}O_2 + 2H_2 \rightarrow CH_3OH$ is a formation reaction for methanol. The reaction $H_2O + SO_3 \rightarrow H_2SO_4$ is *not* a formation reaction, because it forms sulfuric acid not from the elements but from other compounds. Formation reactions are understood to produce 1 mol of product; the heat of formation is therefore based on 1 mol *of the compound formed*.

Heats of reaction at any temperature can be calculated from heat-capacity data if the value for one temperature is known; the tabulation of data can therefore be reduced to the compilation of *standard heats of formation at a single temperature*. The usual choice for this **reference** temperature is 298.15 K or 25°C. The standard heat of formation of a compound at this temperature is represented by the symbol $\Delta H^{\circ}_{f_{298}}$. The degree symbol denotes the standard-state value, subscript f identifies a heat of formation, and the 298 is the rounded absolute temperature in kelvins. Tables of these values for common substances may be found in standard handbooks, but the most extensive compilations available are in specialized reference works.[12] An abridged list of values is given in Table C.4 of App. C, and values for many additional compounds are given in publicly available online databases.[13]

When chemical reactions are combined by addition, the standard heats of reaction may also be added to give the standard heat of the resulting reaction. This is possible because enthalpy is a state function, and its changes for given initial and final states are independent of path. In particular, formation reactions and standard heats of formation may be combined to produce any desired reaction (not itself a formation reaction) and its accompanying standard heat of reaction. Reactions written for this purpose often include an indication of the physical state of each reactant and product—that is, the letter g, l, or s is placed in parentheses after the chemical formula to show whether it is a gas, a liquid, or a solid. This might seem unnecessary because a pure chemical species at a particular temperature and 1 bar can usually exist only in one physical state. However, fictitious states (e.g., the ideal-gas state) are often employed for convenience in such calculations.

---

[12]For example, see *TRC Thermodynamic Tables—Hydrocarbons* and *TRC Thermodynamic Tables—Non-Hydrocarbons*, serial publications of the Thermodynamics Research Center, Texas A & M Univ. System, College Station, Texas; "The NBS Tables of Chemical Thermodynamic Properties," *J. Physical and Chemical Reference Data*, vol. 11, supp. 2, 1982; and the DIPPR Project 801 Database, http://www.aiche.org/dippr/projects/801. Where data are unavailable, estimates based only on molecular structure may be found by the methods of S. W. Benson, *Thermochemical Kinetics*, 2nd ed., John Wiley & Sons, New York, 1976. An improved version of this method is implemented online at http://webbook.nist.gov/chemistry/grp-add/.

[13]Values for more than 7000 compounds are available at http://webbook.nist.gov/.

The water-gas-shift reaction, $CO_2(g) + H_2(g) \rightarrow CO(g) + H_2O(g)$ at 25°C, is commonly encountered in the chemical industry. Although it takes place only at temperatures well above 25°C, the data are for 25°C, and the initial step in any calculation of heat effects for this reaction is evaluation of the standard heat of reaction at 25°C. The pertinent formation reactions and the corresponding heats of formation from Table C.4 are:

$$CO_2(g): \quad C(s) + O_2(g) \rightarrow CO_2(g) \qquad \Delta H^\circ_{f_{298}} = -393{,}509 \text{ J}$$

$$H_2(g): \quad \text{Because hydrogen is an element} \quad \Delta H^\circ_{f_{298}} = 0$$

$$CO(g): \quad C(s) + \tfrac{1}{2}O_2(g) \rightarrow CO(g) \qquad \Delta H^\circ_{f_{298}} = -110{,}525 \text{ J}$$

$$H_2O(g): \quad H_2(g) + \tfrac{1}{2}O_2(g) \rightarrow H_2O(g) \qquad \Delta H^\circ_{f_{298}} = -241{,}818 \text{ J}$$

Because the reaction is actually carried out entirely in the gas phase at high temperature, convenience dictates that the standard states of all products and reactants at 25°C be taken as the ideal-gas state at 1 bar, even though water does not actually exist as a gas at these conditions.[14]

Writing the formation reactions so that their sum yields the desired reaction requires that the formation reaction for CO_2 be written in reverse; the heat of reaction is then of opposite sign to its standard heat of formation:

$$CO_2(g) \rightarrow C(s) + O_2(g) \qquad \Delta H^\circ_{298} = 393{,}509 \text{ J}$$

$$C(s) + \tfrac{1}{2}O_2(g) \rightarrow CO(g) \qquad \Delta H^\circ_{298} = -110{,}525 \text{ J}$$

$$\underline{H_2(g) + \tfrac{1}{2}O_2(g) \rightarrow H_2O(g) \qquad \Delta H^\circ_{298} = -241{,}818 \text{ J}}$$

$$CO_2(g) + H_2(g) \rightarrow CO(g) + H_2O(g) \qquad \Delta H^\circ_{298} = 41{,}166 \text{ J}$$

The meaning of this result is that the enthalpy of 1 mol of CO plus 1 mol of H_2O is greater than the enthalpy of 1 mol of CO_2 plus 1 mol of H_2 by 41,166 J when each product and reactant is taken as the pure gas at 25°C in its ideal-gas state at 1 bar.

In this example the standard heat of formation of H_2O is available for its hypothetical ideal-gas state at 25°C. One might expect the value of the heat of formation of water to be listed for its actual state as a liquid at 1 bar and 25°C. As a matter of fact, values for both states are given in Table C.4 because they are both often used. This is true for many compounds that normally exist as liquids at 25°C and 1 bar. Cases do arise, however, in which a value is given only for the standard state as a liquid or for the ideal-gas state when what is needed is the other value. Suppose this were the case for the preceding example, with only the standard heat of formation of liquid H_2O available. We would then include an equation for the physical change that transforms water from its standard state as a liquid into its ideal-gas state. The enthalpy change for this physical process is the difference between the heats of formation of water in its two standard states:

$$-241{,}818 - (-285{,}830) = 44{,}012 \text{ J}$$

---

[14]One might wonder about the origin of data for such hypothetical states, as it would appear difficult to make measurements for states that cannot exist. For the case of water vapor in the ideal gas state at 25°C and 1 bar, obtaining the enthalpy is straightforward. While water cannot exist as a gas at these conditions, it is a gas at 25°C at sufficiently low pressure. In the ideal gas state, enthalpy is independent of pressure, so the enthalpy measured in the limit of low pressure is exactly the enthalpy in the desired hypothetical state.

This is approximately the latent heat of vaporization of water at 25°C. The sequence of steps is now:

$$CO_2(g) \rightarrow C(s) + O_2(g) \qquad \Delta H^\circ_{298} = 393,509 \text{ J}$$

$$C(s) + \tfrac{1}{2}O_2(g) \rightarrow CO(g) \qquad \Delta H^\circ_{298} = -110,525 \text{ J}$$

$$H_2(g) + \tfrac{1}{2}O_2(g) \rightarrow H_2O(l) \qquad \Delta H^\circ_{298} = -285,830 \text{ J}$$

$$H_2O(l) \rightarrow H_2O(g) \qquad \Delta H^\circ_{298} = 44,012 \text{ J}$$

$$\overline{CO_2(g) + H_2(g) \rightarrow CO(g) + H_2O(g) \quad \Delta H^\circ_{298} = 41,166 \text{ J}}$$

This result is of course in agreement with the previous answer.

Example 4.5

Calculate the standard heat of reaction at 25°C for the following reaction:

$$4HCl(g) + O_2(g) \rightarrow 2H_2O(g) + 2Cl_2(g)$$

Solution 4.5

Standard heats of formation at 298.15 K from Table C.4 are:

$$HCl(g):-92,307 \text{ J} \qquad\qquad H_2O(g):-241,818 \text{ J}$$

The following combination gives the desired result:

$$4HCl(g) \rightarrow 2H_2(g) + 2Cl_2(g) \qquad \Delta H^\circ_{298} = (4)(92,307)$$

$$2H_2(g) + O_2(g) \rightarrow 2H_2O(g) \qquad \Delta H^\circ_{298} = (2)(-241,818)$$

$$\overline{4HCl(g) + O_2(g) \rightarrow 2\,H_2O(g) + 2Cl_2(g) \quad \Delta H^\circ_{298} = -114,408 \text{ J}}$$

4.5   STANDARD HEAT OF COMBUSTION

Only a few *formation* reactions can actually be carried out at the conditions of interest, and therefore data for these reactions are determined indirectly. One kind of reaction that readily lends itself to experiment is the combustion reaction, and many standard heats of formation come from standard heats of combustion, measured calorimetrically. A combustion reaction is **defined** as a reaction of an element or compound with oxygen to form specified combustion products. For organic compounds consisting only of carbon, hydrogen, and oxygen, the products are carbon dioxide and water, but the state of the water may be either vapor or liquid. The value for liquid water product is called the *higher heat of combustion*, while that with water vapor as product is the *lower heat of combustion*. Data are always based on *1 mol of the substance burned*.

A reaction such as the formation of *n*-butane:

$$4C(s) +5H_2(g) \rightarrow C_4H_{10}(g)$$

is not feasible in practice. However, this equation results from combination of the following combustion reactions:

$$4C(s) + 4O_2(g) \rightarrow 4CO_2(g) \qquad \Delta H^\circ_{298} = (4)(-393{,}509)$$
$$5H_2(g) + 2\tfrac{1}{2}O_2(g) \rightarrow 5H_2O(l) \qquad \Delta H^\circ_{298} = (5)(-285{,}830)$$
$$\underline{4CO_2(g) + 5H_2O(l) \rightarrow C_4H_{10}(g) + 6\tfrac{1}{2}O_2(g) \qquad \Delta H^\circ_{298} = 2{,}877{,}396}$$
$$4C(s) + 5H_2(g) \rightarrow C_4H_{10}(g) \qquad \Delta H^\circ_{298} = -125{,}790 \text{ J}$$

This result is the standard heat of formation of *n*-butane listed in Table C.4 of App. C.

4.6 TEMPERATURE DEPENDENCE OF ΔH°

In the foregoing sections, standard heats of reaction were discussed for an arbitrary reference temperature of 298.15 K. In this section we treat the calculation of standard heats of reaction at other temperatures from knowledge of the value at the reference temperature.

A general chemical reaction may be written:

$$|\nu_1|A_1 + |\nu_2|A_2 + \ldots \rightarrow |\nu_3|A_3 + |\nu_4|A_4 + \ldots$$

where ν_i is a stoichiometric coefficient and A_i stands for a chemical formula. The species on the left are reactants; those on the right, products. The sign convention for ν_i is as follows:

$$\text{positive } (+) \text{ for products} \quad \text{and} \quad \text{negative } (-) \text{ for reactants}$$

For example, when the ammonia synthesis reaction is written:

$$N_2 + 3H_2 \rightarrow 2NH_3$$

then
$$\nu_{N_2} = -1 \qquad \nu_{H_2} = -3 \qquad \nu_{NH_3} = 2$$

This sign convention allows the definition of a standard heat of reaction to be expressed mathematically by the simple equation:

$$\Delta H^\circ \equiv \sum_i \nu_i H_i^\circ \tag{4.15}$$

where H_i° is the enthalpy of species i in its standard state and the summation is over all products and reactants. The standard-state enthalpy of a chemical *compound* is equal to its heat of formation plus the standard-state enthalpies of its constituent elements. If the standard-state enthalpies of all *elements* are arbitrarily set equal to zero as the basis of calculation, then the standard-state enthalpy of each compound is simply its heat of formation. In this event, $H_i^\circ = \Delta H_{f_i}^\circ$ and Eq. (4.15) becomes:

$$\Delta H^\circ = \sum_i \nu_i \Delta H_{fi}^\circ \tag{4.16}$$

where the summation is over all products and reactants. This formalizes the procedure described in the preceding section for calculation of standard heats of other reactions from standard heats of formation. Applied to the reaction,

$$4HCl(g) + O_2(g) \rightarrow 2H_2O(g) + 2Cl_2(g)$$

Eq. (4.16) is written:

$$\Delta H^\circ = 2\Delta H^\circ_{f_{\mathrm{H_2O}}} - 4\Delta H^\circ_{f_{\mathrm{HCl}}}$$

With data from Table C.4 of App. C. for 298.15 K, this becomes:

$$\Delta H^\circ_{298} = (2)(-241{,}818) - (4)(-92{,}307) = -114{,}408 \text{ J}$$

in agreement with the result of Ex. 4.5. Note that for pure elemental gases that normally exist as dimers (e.g., O_2, N_2, H_2), it is the dimer form that is arbitrarily assigned a standard-state enthalpy of zero.

For standard reactions, products and reactants are always at the standard-state pressure of 1 bar. Standard-state enthalpies are therefore functions of temperature only, and by Eq. (2.20),

$$dH^\circ_i = C^\circ_{P_i} \, dT$$

where subscript i identifies a particular product or reactant. Multiplying by ν_i and summing over all products and reactants gives:

$$\sum_i \nu_i \, dH^\circ_i = \sum_i \nu_i C^\circ_{P_i} \, dT$$

Because ν_i is a constant, it may be placed inside the differential:

$$\sum_i d(\nu_i H^\circ_i) = \sum_i \nu_i C^\circ_{P_i} \, dT \quad \text{or} \quad d\sum_i \nu_i H^\circ_i = \sum_i \nu_i C^\circ_{P_i} \, dT$$

The term $\sum_i \nu_i H^\circ_i$ is the standard heat of reaction, defined by Eq. (4.15) as ΔH°. The standard heat-capacity change of reaction is defined similarly:

$$\Delta C^\circ_P \equiv \sum_i \nu_i C^\circ_{P_i} \tag{4.17}$$

From these definitions,

$$\boxed{d\Delta H^\circ = \Delta C^\circ_P \, dT} \tag{4.18}$$

This is the fundamental equation relating heats of reaction to temperature.

Integration of Eq. (4.18) yields:

$$\Delta H^\circ - \Delta H^\circ_0 = \int_{T_0}^{T} \Delta C^\circ_P \, dT$$

where ΔH° and ΔH°_0 are the standard heats of reaction at temperature T and at reference temperature T_0 respectively. This equation is more conveniently expressed as:

$$\Delta H^\circ = \Delta H^\circ_0 + R \int_{T_0}^{T} \frac{\Delta C^\circ_P}{R} \, dT \tag{4.19}$$

The reference temperature T_0 must be a temperature for which the heat of reaction is known or can be calculated as described in the two preceding sections, most often 298.15 K. What Eq. (4.19) provides is the means for calculation of a heat of reaction at temperature T from a known value at temperature T_0.

If the temperature dependence of the heat capacity of each product and reactant is given by Eq. (4.5), then the integral is given by the analog of Eq. (4.8):

$$\int_{T_0}^{T} \frac{\Delta C^\circ_P}{R} \, dT = \Delta A (T - T_0) + \frac{\Delta B}{2}(T^2 - T_0^2) + \frac{\Delta C}{3}(T^3 - T_0^3) + \Delta D \left(\frac{T - T_0}{T T_0} \right) \tag{4.20}$$

where by definition,
$$\Delta A \equiv \sum_i \nu_i A_i$$

with analogous definitions for ΔB, ΔC, and ΔD.

An alternative formulation results when a mean heat capacity change of reaction is defined in analogy to Eq. (4.9):

$$\frac{\langle \Delta C_P^\circ \rangle_H}{R} = \Delta A + \frac{\Delta B}{2}(T + T_0) + \frac{\Delta C}{3}(T^2 + T_0^2 + TT_0) + \frac{\Delta D}{TT_0} \qquad (4.21)$$

Equation (4.19) then becomes:

$$\Delta H^\circ = \Delta H_0^\circ + \langle \Delta C_P^\circ \rangle_H (T - T_0) \qquad (4.22)$$

The integral of Eq. (4.20) is of the same form as that of Eq. (4.8), and in analogous fashion may be set equal to a defined function:[15]

$$\int_{T_0}^{T} \frac{\Delta C_P^\circ}{R} dT = \text{IDCPH (T0, T; DA, DB, DC, DD)}$$

where "D" denotes "Δ." The analogy requires simple replacement of C_P by ΔC_P° and of A, B, etc. by ΔA, ΔB, etc. The same computer routine serves for evaluation of either integral. The only difference is in the function name.

Just as function MCPH is defined to represent $\langle C_P \rangle_H / R$, so function MDCPH by analogy is defined to represent $\langle \Delta C_P^\circ \rangle_H / R$; thus,

$$\frac{\langle \Delta C_P^\circ \rangle_H}{R} = \text{MDCPH (T0, T; DA, DB, DC, DD)}$$

The calculation represented by both Eqs. (4.19) and (4.22) is represented schematically in Fig. 4.3.

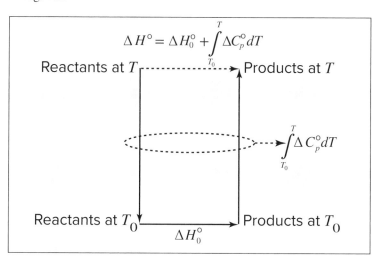

Figure 4.3: Path representing the procedure for calculating a standard heat of reaction at temperature T from the value at reference temperature T_0.

---

[15]Again, templates for these defined functions are available in the online learning center at http://highered.mheducation .com:80/sites/1259696529.

Example 4.6

Calculate the standard heat of the following methanol-synthesis reaction at 800°C:

$$CO(g) + 2H_2(g) \rightarrow CH_3OH(g)$$

Solution 4.6

Apply Eq. (4.16) to this reaction for reference temperature $T_0 = 298.15$ K and with heat-of-formation data from Table C.4:

$$\Delta H_0^{\circ} = \Delta H_{298}^{\circ} = -200,660 - (-110,525) = -90,135 \text{ J}$$

Evaluation of the parameters in Eq. (4.20) is based on data taken from Table C.1:

i	ν_i	A	$10^3 B$	$10^6 C$	$10^{-5} D$
CH_3OH	1	2.211	12.216	−3.450	0.000
CO	−1	3.376	0.557	0.000	−0.031
H_2	−2	3.249	0.422	0.000	0.083

From its definition,

$$\Delta A = (1)(2.211) + (-1)(3.376) + (-2)(3.249) = -7.663$$

Similarly,

$$\Delta B = 10.815 \times 10^{-3} \quad \Delta C = -3.450 \times 10^{-6} \quad \Delta D = -0.135 \times 10^5$$

The value of the integral of Eq. (4.20) for $T = 1{,}073.15$ K is represented by:

$$\text{IDCPH}(298.15, 1073.15; -7.663, 10.815 \times 10^{-3}, -3.450 \times 10^{-6}, -0.135 \times 10^5)$$

The value of this integral is -1615.5 K, and by Eq. (4.19),

$$\Delta H^{\circ} = -90{,}135 + 8.314(-1615.5) = -103{,}566 \text{ J}$$

4.7   HEAT EFFECTS OF INDUSTRIAL REACTIONS

The preceding sections have dealt with the *standard* heat of reaction. Industrial reactions are rarely carried out under standard-state conditions. Furthermore, in actual reactions the reactants may not be present in stoichiometric proportions, the reaction may not go to completion, and the final temperature may differ from the initial temperature. Moreover, inert species may be present, and several reactions may occur simultaneously. Nevertheless, calculations of the heat effects of actual reactions are based on the principles already considered and are illustrated by the following examples, **wherein the ideal-gas state is assumed for all gases.**

Example 4.7

What is the maximum temperature that can be reached by the combustion of methane with 20% excess air? Both the methane and the air enter the burner at 25°C.

Solution 4.7

The reaction is $CH_4 + 2O_2 \rightarrow CO_2 + 2H_2O(g)$ for which,

$$\Delta H^\circ_{298} = -393{,}509 + (2)(-241{,}818) - (-74{,}520) = -802{,}625 \text{ J}$$

Because the maximum attainable temperature (called the *theoretical flame temperature*) is sought, assume that the combustion reaction goes to completion adiabatically ($Q = 0$). If the kinetic- and potential-energy changes are negligible and if $W_s = 0$, the overall energy balance for the process reduces to $\Delta H = 0$. For purposes of calculation of the final temperature, any convenient path between the initial and final states may be used. The path chosen is indicated in the diagram.

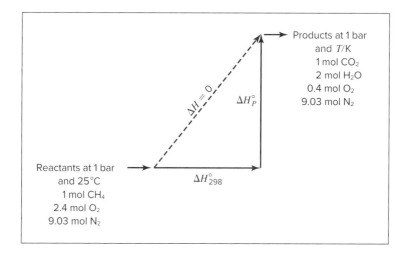

When one mole of methane burned is the basis for all calculations, the following quantities of oxygen and nitrogen are supplied by the entering air:

$$\text{Moles } O_2 \text{ required} = 2.0$$
$$\text{Moles excess } O_2 = (0.2)(2.0) = 0.4$$
$$\text{Moles } N_2 \text{ entering} = (2.4)(79/21) = 9.03$$

The mole numbers n_i of the gases in the product stream leaving the burner are 1 mol CO_2, 2 mol $H_2O(g)$, 0.4 mol O_2, and 9.03 mol N_2. Because the enthalpy change must be independent of path,

$$\Delta H^\circ_{298} + \Delta H^\circ_P = \Delta H = 0 \tag{A}$$

where all enthalpies are on the basis of 1 mol CH_4 burned. The enthalpy change of the products as they are heated from 298.15 K to T is:

$$\Delta H^\circ_P = \langle C^\circ_P \rangle_H (T - 298.15) \tag{B}$$

where we define $\langle C^\circ_P \rangle_H$ as the mean heat capacity for the *total* product stream:

$$\langle C^\circ_P \rangle_H \equiv \sum_i n_i \langle C^\circ_{P_i} \rangle_H$$

The simplest procedure here is to sum the mean-heat-capacity equations for the products, each multiplied by its appropriate mole number. Because $C = 0$ for each product gas (Table C.1), Eq. (4.9) yields:

$$\langle C_P^\circ \rangle_H = \sum_i n_i \langle C_{P_i}^\circ \rangle_H = R\left[\sum_i n_i A_i + \frac{\sum_i n_i B_i}{2}(T - T_0) + \frac{\sum_i n_i D_i}{T T_0}\right]$$

Data from Table C.1 are combined as follows:

$$A = \sum_i n_i A_i = (1)(5.457) + (2)(3.470) + (0.4)(3.639) + (9.03)(3.280) = 43.471$$

Similarly, $B = \sum_i n_i B_i = 9.502 \times 10^{-3}$ and $D = \sum_i n_i D_i = -0.645 \times 10^5$.

For the product stream $\langle C_P^\circ \rangle_H / R$ is therefore represented by:

$$\text{MCPH}(298.15, T; 43.471, 9.502 \times 10^{-3}, 0.0, -0.645 \times 10^5)$$

Equations (A) and (B) are combined and solved for T:

$$T = 298.15 - \frac{\Delta H_{298}^\circ}{\langle C_P^\circ \rangle_H}$$

Because the mean heat capacities depend on T, one first evaluates $\langle C_P^\circ \rangle_H$ for an assumed value of $T > 298.15$, then substitutes the result in the preceding equation. This yields a new value of T for which $\langle C_P^\circ \rangle_H$ is reevaluated. The procedure continues to convergence on the final value,

$$T = 2066 \text{ K} \qquad \text{or} \qquad 1793°\text{C}$$

Again, solution can be easily automated with the Goal Seek or Solver function in a spreadsheet or similar solve routines in other software packages.

Example 4.8

One method for the manufacture of "synthesis gas" (a mixture of CO and H_2) is the catalytic reforming of CH_4 with steam at high temperature and atmospheric pressure:

$$CH_4(g) + H_2O(g) \quad \rightarrow \quad CO(g) + 3H_2(g)$$

The only other reaction considered here is the water-gas-shift reaction:

$$CO(g) + H_2O(g) \quad \rightarrow \quad CO_2(g) + H_2(g)$$

Reactants are supplied in the ratio 2 mol steam to 1 mol CH_4, and heat is added to the reactor to bring the products to a temperature of 1300 K. The CH_4 is completely converted, and the product stream contains 17.4 mol-% CO. Assuming the reactants to be preheated to 600 K, calculate the heat requirement for the reactor.

Solution 4.8

The standard heats of reaction at 25°C for the two reactions are calculated from the data of Table C.4:

$$CH_4(g) + H_2O(g) \rightarrow CO(g) + 3H_2(g) \qquad \Delta H^{\circ}_{298} = 205{,}813 \text{ J}$$

$$CO(g) + H_2O(g) \rightarrow CO_2(g) + H_2(g) \qquad \Delta H^{\circ}_{298} = -41{,}166 \text{ J}$$

These two reactions may be added to give a third reaction:

$$CH_4(g) + 2H_2O(g) \rightarrow CO_2(g) + 4H_2(g) \qquad \Delta H^{\circ}_{298} = 164{,}647 \text{ J}$$

Any pair of the three reactions constitutes an independent set. The third reaction is not independent; it is obtained by combination of the other two. The reactions most convenient to work with here are the first and third:

$$CH_4(g) + H_2O(g) \rightarrow CO(g) + 3H_2(g) \qquad \Delta H^{\circ}_{298} = 205{,}813 \text{ J} \qquad (A)$$

$$CH_4(g) + 2H_2O(g) \rightarrow CO_2(g) + 4H_2(g) \qquad H^{\circ}_{298} = 164{,}647 \text{ J} \qquad (B)$$

First one must determine the fraction of CH_4 converted by each of these reactions. As a basis for calculations, let 1 mol CH_4 and 2 mol steam be fed to the reactor. If x mol CH_4 reacts by Eq. (A), then $1 - x$ mol reacts by Eq. (B). On this basis the products of the reaction are:

$$
\begin{array}{ll}
\text{CO:} & x \\
\text{H}_2: & 3x + 4(1 - x) = 4 - x \\
\text{CO}_2: & 1 - x \\
\text{H}_2\text{O:} & 2 - x - 2(1 - x) = x \\
\hline
\text{Total:} & 5 \text{ mol products}
\end{array}
$$

The mole fraction of CO in the product stream is $x/5 = 0.174$; whence $x = 0.870$. Thus, on the basis chosen, 0.870 mol CH_4 reacts by Eq. (A) and 0.130 mol reacts by Eq. (B). Furthermore, the amounts of the species in the product stream are:

$$
\begin{array}{l}
\text{Moles CO} = x = 0.87 \\
\text{Moles H}_2 = 4 - x = 3.13 \\
\text{Moles CO}_2 = 1 - x = 0.13 \\
\text{Moles H}_2\text{O} = x = 0.87
\end{array}
$$

We now devise a path, for purposes of calculation, to proceed from reactants at 600 K to products at 1300 K. Because data are available for the standard heats of reaction at 25°C, the most convenient path is the one which includes the reactions at 25°C (298.15 K). This is shown schematically in the accompanying diagram. The dashed line represents the actual path for which the enthalpy change is ΔH. Because this enthalpy change is independent of path,

$$\Delta H = \Delta H^{\circ}_R + \Delta H^{\circ}_{298} + \Delta H^{\circ}_P$$

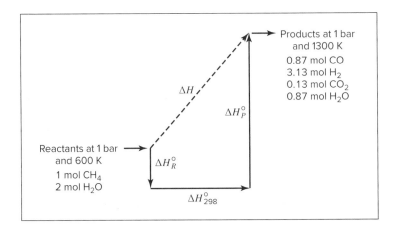

For the calculation of ΔH_{298}°, reactions (A) and (B) must both be taken into account. Because 0.87 mol CH_4 reacts by (A) and 0.13 mol reacts by (B),

$$\Delta H_{298}^{\circ} = (0.87)(205{,}813) + (0.13)(164{,}647) = 200{,}460 \text{ J}$$

The enthalpy change of the reactants cooled from 600 K to 298.15 K is:

$$\Delta H_R^{\circ} = \left(\sum_i n_i \langle C_{P_i}^{\circ} \rangle_H \right)(298.15 - 600)$$

where subscript i denotes reactants. The values of $\langle C_{P_i}^{\circ} \rangle_H / R$ are:

CH_4: MCPH(298.15, 600; 1.702, 9.081 × 10^{-3}, −2.164 × 10^{-6}, 0.0) = 5.3272
H_2O: MCPH(298.15, 600; 3.470, 1.450 × 10^{-3}, 0.0, 0.121 × 10^5) = 4.1888

and
$$\Delta H_R^{\circ} = (8.314)[(1)(5.3272) + (2)(4.1888)](298.15 - 600) = -34{,}390 \text{ J}$$

The enthalpy change of the products as they are heated from 298.15 to 1300 K is calculated similarly:

$$\Delta H_P^{\circ} = \left(\sum_i n_i \langle C_{P_i}^{\circ} \rangle_H \right)(1300 - 298.15)$$

where subscript i here denotes products. The $\langle C_{P_i}^{\circ} \rangle_H / R$ values are:

CO :   MCPH(298.15, 1300; 3.376, 0.557 × 10^{-3}, 0.0, −0.031 × 10^5) = 3.8131
H_2 :   MCPH(298.15, 1300; 3.249, 0.422 × 10^{-3}, 0.0, −0.083 × 10^5) = 3.6076
CO_2 :   MCPH(298.15, 1300; 5.457, 1.045 × 10^{-3}, 0.0, −1.157 × 10^5) = 5.9935
H_2O :   MCPH(298.15, 1300; 3.470, 1.450 × 10^{-3}, 0.0, 0.121 × 10^5) = 4.6599

Whence,

$$\Delta H_P^{\circ} = (8.314)[(0.87)(3.8131) + (3.13)(3.6076)$$
$$+ (0.13)(5.9935) + (0.87)(4.6599)] \times (1300 - 298.15)$$
$$= 161{,}940 \text{ J}$$

Therefore,

$$\Delta H = -34{,}390 + 200{,}460 + 161{,}940 = 328{,}010 \text{ J}$$

The process is one of steady flow for which W_s, Δz, and $\Delta u^2/2$ are presumed negligible. Thus,

$$Q = \Delta H = 328{,}010 \text{ J}$$

This result is on the basis of 1 mol CH_4 fed to the reactor.

Example 4.9

Solar-grade silicon can be manufactured by thermal decomposition of silane at moderate pressure in a fluidized-bed reactor, in which the overall reaction is:

$$SiH_4(g) \rightarrow Si(s) + 2H_2(g)$$

When pure silane is preheated to 300°C, and heat is added to the reactor to promote a reasonable reaction rate, 80% of the silane is converted to silicon and the products leave the reactor at 750°C. How much heat must be added to the reactor for each kilogram of silicon produced?

Solution 4.9

For a continuous-flow process with no shaft work and negligible changes in kinetic and potential energy, the energy balance is simply $Q = \Delta H$, and the heat added is the enthalpy change from reactant at 300°C to products at 750°C. A convenient path for calculation of the enthalpy change is to (1) cool the reactant to 298.15 K, (2) carry out the reaction at 298.15 K, and (3) heat the products to 750°C.

On the basis of 1 mol SiH_4, the products consist of 0.2 mol SiH_4, 0.8 mol Si, and 1.6 mol H_2. Thus, for the three steps we have:

$$\Delta H_1 = \int_{573.15K}^{298.15K} C_P^\circ(SiH_4)\, dT$$

$$\Delta H_2 = 0.8 \times \Delta H_{298}^\circ$$

$$\Delta H_3 = \int_{298.15K}^{1023.15K} [0.2 \times C_P^\circ(SiH_4) + 0.8 \times C_P^\circ(Si) + 1.6 \times C_P^\circ(H_2)]\, dT$$

Data needed for this example are not included in App. C, but are readily obtained from the NIST Chemistry Webbook (http://webbook.nist.gov). The reaction here is the reverse of the formation reaction for silane, and its standard heat of reaction at 298.15 K is $\Delta H_{298}^\circ = -34{,}310$ J. Thus, the reaction is mildly exothermic.

Heat capacity in the NIST Chemistry Workbook is expressed by the Shomate equation, a polynomial of different form from that used in this text. It includes a T^3 term, and is written in terms of $T/1000$, with T in K:

$$C_P^\circ = A + B\left(\frac{T}{1000}\right) + C\left(\frac{T}{1000}\right)^2 + D\left(\frac{T}{1000}\right)^3 + E\left(\frac{T}{1000}\right)^{-2}$$

Formal integration of this equation gives the enthalpy change:

$$\Delta H = \int_{T_0}^{T} C_P^\circ \, dT$$

$$\Delta H = 1000\left[A\left(\frac{T}{1000}\right) + \frac{B}{2}\left(\frac{T}{1000}\right)^2 + \frac{C}{3}\left(\frac{T}{1000}\right)^3 + \frac{D}{4}\left(\frac{T}{1000}\right)^4 - E\left(\frac{T}{1000}\right)^{-1}\right]_{T_0}^{T}$$

The first three rows in the accompaning table give parameters, on a molar basis, for SiH$_4$, crystalline silicon, and hydrogen. The final entry is for the collective products, represented for example by:

$$A(\text{products}) = (0.2)(6.060) + (0.8)(22.817) + (1.6)(33.066) = 72.3712$$

with corresponding equations for B, C, D, and E.

Species	A	B	C	D	E
SiH$_4$(g)	6.060	139.96	−77.88	16.241	0.1355
Si(s)	22.817	3.8995	−0.08289	0.04211	−0.3541
H$_2$(g)	33.066	−11.363	11.433	−2.773	−0.1586
Products	72.3712	12.9308	2.6505	−1.1549	−0.5099

For these parameters, and with T in Kelvins, the equation for ΔH yields values in joules. For the three steps making up the solution to this problem, the following results are obtained:

1. Substitution of the parameters for 1 mol of SiH$_4$ into the equation for ΔH leads upon evaluation to: $\Delta H_1 = -14,860$ J

2. Here, $\Delta H_2 = (0.8)(-34,310) = -27,450$ J

3. Substitution of the parameters for the total product stream into the equation for ΔH leads upon evaluation to: $\Delta H_3 = 58,060$ J

For the three steps the sum is:

$$\Delta H = -14,860 - 27,450 + 58,060 = 15,750 \text{ J}$$

This enthalpy change equals the heat input per mole of SiH$_4$ fed to the reactor. A kilogram of silicon, with a molar mass of 28.09, is 35.60 mol. Producing a kilogram of silicon therefore requires a feed of 35.60/0.8 or 44.50 mol of SiH$_4$. The heat requirement per kilogram of silicon produced is therefore $(15,750)(44.5) = 700,900$ J.

Example 4.10

A boiler is fired with a high-grade fuel oil (consisting only of hydrocarbons) having a standard heat of combustion of $-43,515$ J·g^{-1} at 25°C with $CO_2(g)$ and $H_2O(l)$ as products. The temperature of the fuel and air entering the combustion chamber is 25°C. The air is assumed dry. The flue gases leave at 300°C, and their average analysis (on a dry basis) is 11.2% CO_2, 0.4% CO, 6.2% O_2, and 82.2% N_2. Calculate the fraction of the heat of combustion of the oil that is transferred as heat to the boiler.

Solution 4.10

Take as a basis 100 mol dry flue gases, consisting of:

CO_2	11.2 mol
CO	0.4 mol
O_2	6.2 mol
N_2	82.2 mol
Total	100.0 mol

This analysis, on a dry basis, does not take into account the H_2O vapor present in the flue gases. The amount of H_2O formed by the combustion reaction is found from an oxygen balance. The O_2 supplied in the air represents 21 mol-% of the air stream. The remaining 79% is N_2, which goes through the combustion process unchanged. Thus the 82.2 mol N_2 appearing in 100 mol dry flue gases is supplied with the air, and the O_2 accompanying this N_2 is:

$$\text{Moles } O_2 \text{ entering in air} = (82.2)(21/79) = 21.85$$

and

$$\text{Total moles } O_2 \text{ in the dry flue gases } = 11.2 + 0.4/2 + 6.2 = 17.60$$

The difference between these figures is the moles of O_2 that react to form H_2O. Therefore on the basis of 100 mol dry flue gases,

$$\text{Moles } H_2O \text{ formed} = (21.85 - 17.60)(2) = 8.50$$

$$\text{Moles } H_2 \text{ in the fuel} = \text{moles of water formed} = 8.50$$

The amount of C in the fuel is given by a carbon balance:

$$\text{Moles C in flue gases} = \text{moles C in fuel} = 11.2 + 0.4 = 11.60$$

These amounts of C and H_2 together give:

$$\text{Mass of fuel burned} = (8.50)(2) + (11.6)(12) = 156.2 \text{ g}$$

If this amount of fuel is burned completely to $CO_2(g)$ and $H_2O(l)$ at 25°C, the heat of combustion is:

$$\Delta H^\circ_{298} = (-43,515)(156.2) = -6,797,040 \text{ J}$$

However, the reaction actually occurring does not represent complete combustion, and the H_2O is formed as vapor rather than as liquid. The 156.2 g of fuel, consisting of 11.6 mol of C and 8.5 mol of H_2, is represented by the empirical formula $C_{11.6}H_{17}$. Omit the 6.2 mol O_2 and 82.2 mol N_2 which enter and leave the reactor unchanged, and write the reaction:

$$C_{11.6}H_{17}(l) + 15.65\,O_2(g) \rightarrow 11.2\,CO_2(g) + 0.4\,CO(g) + 8.5\,H_2O(g)$$

This result is obtained by addition of the following reactions, for each of which the standard heat of reaction at 25°C is known:

$$C_{11.6}H_{17}(l) + 15.85\,O_2(g) \rightarrow 11.6\,CO_2(g) + 8.5\,H_2O(l)$$
$$8.5\,H_2O(l) \rightarrow 8.5\,H_2O(g)$$
$$0.4\,CO_2(g) \rightarrow 0.4\,CO(g) + 0.2\,O_2(g)$$

The sum of these reactions yields the actual reaction, and the sum of the ΔH°_{298} values gives the standard heat of the reaction occurring at 25°C:

$$\Delta H^\circ_{298} = -6{,}797{,}040 + (44{,}012)\,(8.5) + (282{,}984)\,(0.4) = -6{,}309{,}740\ \text{J}$$

The actual process leading from reactants at 25°C to products at 300°C is represented by the dashed line in the accompanying diagram. For purposes of calculating ΔH for this process, we may use any convenient path. The one drawn with solid lines is a logical one: ΔH°_{298} has already been calculated and ΔH°_P is easily evaluated.

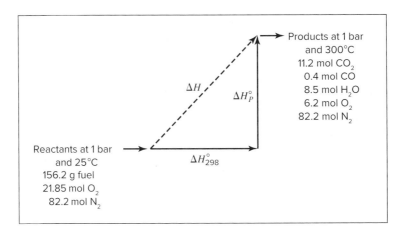

The enthalpy change caused by heating the products of reaction from 25 to 300°C is:

$$\Delta H^\circ_P = \left(\sum_i n_i \langle C^\circ_{P_i} \rangle_H \right)(573.15 - 298.15)$$

where subscript i denotes products. The $\langle C^\circ_{P_i} \rangle_H / R$ values are:

CO_2: MCPH(298.15, 573.15; 5.457, 1.045 × 10^{-3}, 0.0, -1.157 × 10^5) = 5.2352

CO: MCPH(298.15, 573.15; 3.376, 0.557 × 10^{-3}, 0.0, -0.031 × 10^5) = 3.6005

H_2O: MCPH(298.15, 573.15; 3.470, 1.450 × 10^{-3}, 0.0, 0.121 × 10^5) = 4.1725

O_2: MCPH(298.15, 573.15; 3.639, 0.506 × 10^{-3}, 0.0, -0.227 × 10^5) = 3.7267

N_2: MCPH(298.15, 573.15; 3.280, 0.593 × 10^{-3}, 0.0, 0.040 × 10^5) = 3.5618

Therefore,

$$\Delta H_P^\circ = (8.314)[(11.2)(5.2352) + (0.4)(3.6005) + (8.5)(4.1725) + (6.2)(3.7267) + \\ (82.2)(3.5618)](573.15 - 298.15) = 940{,}660 \text{ J}$$

and

$$\Delta H = \Delta H_{298}^\circ + \Delta H_P^\circ = -6{,}309{,}740 + 940{,}660 = -5{,}369{,}080 \text{ J}$$

Because the process is one of steady flow for which the shaft work and kinetic- and potential-energy terms in the energy balance [Eq. (2.32)] are zero or negligible, $\Delta H = Q$. Thus, $Q = -5369$ kJ, and this amount of heat is transferred to the boiler for every 100 mol dry flue gases formed. This represents

$$\frac{5{,}369{,}080}{6{,}797{,}040}(100) = 79.0\%$$

of the higher heat of combustion of the fuel.

In the foregoing examples of reactions that occur at approximately 1 bar, we have tacitly assumed that the heat effects of reaction are the same whether gases are mixed or pure, an acceptable procedure for low pressures. For reactions at elevated pressures, this may not be the case, and it may be necessary to account for the effects of pressure and of mixing on the heat of reaction. However, these effects are usually small. For reactions occurring in the liquid phase, the effects of mixing are generally more important. They are treated in detail in Chapter 11.

For biological reactions, occurring in aqueous solution, the effects of mixing are particularly important. The enthalpies and other properties of biomolecules in solution usually depend not only on temperature and pressure, but also on the pH, ionic strength, and concentrations of specific ions in solution. Table C.5 in App. C provides enthalpies of formation of a variety of molecules in dilute aqueous solution at zero ionic strength. These can be used to estimate heat effects of enzymatic or biological reactions involving such species. However, corrections for effects of pH, ionic strength, and finite concentration may be significant.[16] Heat capacities are often unknown for such species, but in dilute aqueous solution the overall specific heat is usually well approximated by the specific heat of water. Moreover, the temperature range of interest for biological reactions is quite narrow. The following example illustrates estimation of heat effects for a biological reaction.

---

[16]For analysis of these effects, see Robert A. Alberty, *Thermodynamics of Biochemical Reactions*, John Wiley & Sons, Hoboken, NJ, 2003.

Example 4.11

A dilute solution of glucose enters a continuous fermentation process, where yeast cells convert it to ethanol and carbon dioxide. The aqueous stream entering the reactor is at 25°C and contains 5 wt% glucose. Assuming this glucose is fully converted to ethanol and carbon dioxide, and that the product stream leaves the reactor at 35°C, estimate the amount of heat added or removed per kg of ethanol produced. Assume that the carbon dioxide remains dissolved in the product stream.

Solution 4.11

For this constant pressure process with no shaft work, the heat effect is simply equal to the enthalpy change from the feed stream to the product stream. The fermentation reaction is:

$$C_6H_{12}O_6(aq) \rightarrow 2\,C_2H_5OH(aq) + 2\,CO_2(aq)$$

The standard enthalpy of reaction at 298 K obtained using the heats of formation in dilute aqueous solution from Table C.5 is:

$$\Delta H^\circ_{298} = (2)(-288.3) + (2)(-413.8) - (-1262.2) = -142.0 \text{ kJ·mol}^{-1}$$

One kg of ethanol is $1/(0.046069 \text{ kg·mol}^{-1}) = 21.71$ mol ethanol. Each mole of glucose produces two moles of ethanol, so 10.85 mol of reaction must occur to produce 1 kg of ethanol. The standard enthalpy of reaction per kg ethanol is then $(10.85)(-142.0) = -1541$ kJ·kg^{-1}. The mass of glucose required to produce 1 kg ethanol is 10.85 mol $\times 0.18016$ kg·mol$^{-1} = 1.955$ kg glucose. If the feed stream is 5 wt% glucose, then the total mass of solution fed to the reactor per kg ethanol produced is $1.955/0.05 = 39.11$ kg. Assuming that the product stream has the specific heat of water, about 4.184 kJ·kg^{-1}·K^{-1}, then the enthalpy change per kg ethanol for heating the product stream from 25°C to 35°C is:

$$4.184 \text{ kJ·kg}^{-1}\cdot\text{K}^{-1} \times 10 \text{ K} \times 39.11 \text{ kg} = 1636 \text{ kJ}.$$

Adding this to the heat of reaction per kg ethanol gives the total enthalpy change from feed to product, which is also the total heat effect:

$$Q = \Delta H = -1541 + 1636 = 95 \text{ kJ·(kg ethanol)}^{-1}$$

This estimate leads to the conclusion that a small amount of heat must be added to the reactor because the reaction exothermicity is not quite sufficient to heat the feed stream to the product temperature. In an actual process, the glucose would not be fully converted to ethanol. Some fraction of the glucose must be directed to other products of cellular metabolism. This means that somewhat more than 1.955 kg glucose will be needed per kg of ethanol produced. The heat release from other reactions may be somewhat higher or lower than that for the production of ethanol, which would change the estimate. If some of the CO_2 leaves the reactor as a gas, then the heat requirement will be slightly higher because the enthalpy of $CO_2(g)$ is higher than that of aqueous CO_2.

4.8 SYNOPSIS

After studying this chapter, including the end-of-chapter problems, one should be able to:

- Define sensible heat effects, latent heat, heat of reaction, heat of formation, and heat of combustion

- Formulate a heat-capacity integral, decide whether to use C_P or C_V in it, and evaluate it with the heat capacity expressed as a polynomial in temperature

- Use a heat-capacity integral in an energy balance to determine the energy input required to achieve a given temperature change or to determine the temperature change that will result from a given energy input

- Look up or estimate latent heats of phase change and apply them in energy balances

- Apply the Clayperon equation

- Compute a standard heat of reaction at arbitrary temperature from heats of formation and heat capacities

- Compute standard heats of reaction from standard heats of combustion

- Compute heat requirements for a process with specified chemical reactions and specified inlet and outlet temperatures

4.9 PROBLEMS

4.1. For steady flow in a heat exchanger at approximately atmospheric pressure, what is the heat transferred:

(a) When 10 mol of SO_2 is heated from 200 to 1100°C?
(b) When 12 mol of propane is heated from 250 to 1200°C?
(c) When 20 kg of methane is heated from 100 to 800°C?
(d) When 10 mol of *n*-butane is heated from 150 to 1150°C?
(e) When 1000 kg of air is heated from 25 to 1000°C?
(f) When 20 mol of ammonia is heated from 100 to 800°C?
(g) When 10 mol of water is heated from 150 to 300°C?
(h) When 5 mol of chlorine is heated from 200 to 500°C?
(i) When 10 kg of ethylbenzene is heated from 300 to 700°C?

4.2. For steady flow through a heat exchanger at approximately atmospheric pressure, what is the final temperature,

(a) When heat in the amount of 800 kJ is added to 10 mol of ethylene initially at 200°C?
(b) When heat in the amount of 2500 kJ is added to 15 mol of 1-butene initially at 260°C?
(c) When heat in the amount of 10^6(Btu) is added to 40(lb mol) of ethylene initially at 500(°F)?

4.3. For a steady-flow heat exchanger with a feed temperature of 100°C, compute the outlet stream temperature when heat in the amount of 12 kJ·mol^{-1} is added to the following substances.

(*a*) methane, (*b*) ethane, (*c*) propane, (*d*) *n*-butane, (*e*) *n*-hexane, (*f*) *n*-octane, (*g*) propylene, (*h*) 1-pentene, (*i*) 1-heptene, (*j*) 1-octene, (*k*) acetylene, (*l*) benzene, (*m*) ethanol, (*n*) styrene, (*o*) formaldehyde, (*p*) ammonia, (*q*) carbon monoxide, (*r*) carbon dioxide, (*s*) sulfur dioxide, (*t*) water, (*u*) nitrogen, (*v*) hydrogen cyanide

4.4. If $250(\text{ft})^3(\text{s})^{-1}$ of air at $122(°\text{F})$ and approximately atmospheric pressure is preheated for a combustion process to $932(°\text{F})$, what rate of heat transfer is required?

4.5. How much heat is required when 10,000 kg of $CaCO_3$ is heated at atmospheric pressure from 50°C to 880°C?

4.6. If the heat capacity of a substance is correctly represented by an equation of the form,

$$C_P = A + BT + CT^2$$

show that the error resulting when $\langle C_P \rangle_H$ is assumed equal to C_P evaluated at the arithmetic mean of the initial and final temperatures is $C(T_2 - T_1)^2/12$.

4.7. If the heat capacity of a substance is correctly represented by an equation of the form,

$$C_P = A + BT + DT^{-2}$$

show that the error resulting when $\langle C_P \rangle_H$ is assumed equal to C_P evaluated at the arithmetic mean of the initial and final temperatures is:

$$\frac{D}{T_1 T_2} \left(\frac{T_2 - T_1}{T_2 + T_1} \right)^2$$

4.8. Calculate the heat capacity of a gas sample from the following information: The sample comes to equilibrium in a flask at 25°C and 121.3 kPa. A stopcock is opened briefly, allowing the pressure to drop to 101.3 kPa. With the stopcock closed, the flask warms, returning to 25°C, and the pressure is measured as 104.0 kPa. Determine C_P in J·mol^{-1}·K^{-1} assuming the gas to be ideal and the expansion of the gas remaining in the flask to be reversible and adiabatic.

4.9. A process stream is heated as a gas from 25°C to 250°C at constant P. A quick estimate of the energy requirement is obtained from Eq. (4.3), with C_P taken as constant and equal to its value at 25°C. Is the estimate of Q likely to be low or high? Why?

4.10. (*a*) For one of the compounds listed in Table B.2 of App. B, evaluate the latent heat of vaporization ΔH_n by Eq. (4.13). How does this result compare with the value listed in Table B.2?
     (*b*) Handbook values for the latent heats of vaporization at 25°C of four compounds are given in the table. For one of these, calculate ΔH_n using Eq. (4.14), and compare the result with the value given in Table B.2.

Latent heats of vaporization at 25°C in $J \cdot g^{-1}$

n-Pentane 366.3	Benzene	433.3
n-Hexane 366.1	Cyclohexane	392.5

4.11. Table 9.1 lists the thermodynamic properties of saturated liquid and vapor tetrafluoroethane. Making use of the vapor pressures as a function of temperature and of the saturated-liquid and saturated-vapor volumes, calculate the latent heat of vaporization by Eq. (4.12) at one of the following temperatures and compare the result with the latent heat of vaporization calculated from the enthalpy values given in the table.

(a) $-16°C$, (b) $0°C$, (c) $12°C$, (d) $26°C$, (e) $40°C$.

4.12. Handbook values for the latent heats of vaporization in $J \cdot g^{-1}$ are given in the table for three pure liquids at $0°C$.

	ΔH^{lv} at $0°C$
Chloroform	270.9
Methanol	1189.5
Tetrachloromethane	217.8

For one of these substances, calculate:

(a) The value of the latent heat at T_n by Eq. (4.14), given the value at $0°C$.
(b) The value of the latent heat at T_n by Eq. (4.13).

By what percentages do these results differ from the value listed in Table B.2 of App. B?

4.13. Table B.2 of App. B provides parameters for an equation that gives P^{sat} as a function of T for a number of pure compounds. For one of them, determine the heat of vaporization at its normal boiling point by application of Eq. (4.12), the Clapeyron equation. Evaluate dP^{sat}/dT from the given vapor-pressure equation, and use generalized correlations from Chapter 3 to estimate ΔV. Compare the computed value with the value of ΔH_n listed in Table B.2. Note that normal boiling points are listed in the last column of Table B.2.

4.14. A method for determination of the second virial coefficient of a pure gas is based on the Clapeyron equation and measurements of the latent heat of vaporization ΔH^{lv}, the molar volume of saturated liquid V^l, and the vapor pressure P^{sat}. Determine B in $cm^3 \cdot mol^{-1}$ for methyl ethyl ketone at $75°C$ from the following data at this temperature:

$$\Delta H^{lv} = 31{,}600 \ J \cdot mol^{-1} \qquad V^l = 96.49 \ cm^3 \cdot mol^{-1}$$

$$\ln P^{sat}/kPa = 48.158 - 5623/T - 4.705 \ln T \qquad [T = K]$$

4.15. One hundred kmol per hour of subcooled liquid at 300 K and 3 bar is superheated to 500 K in a steady-flow heat exchanger. Estimate the exchanger duty (in kW) for one of the following:

(a) Methanol, for which $T^{sat} = 368.0$ K at 3 bar.
(b) Benzene, for which $T^{sat} = 392.3$ K at 3 bar.
(c) Toluene, for which $T^{sat} = 426.9$ K at 3 bar.

4.16. For each of the following substances, compute the final temperature when heat in the amount of 60 kJ·mol^{-1} is added to the subcooled liquid at 25°C at atmospheric pressure.

(a) methanol
(b) ethanol
(c) benzene
(d) toluene
(e) water

4.17. Saturated-liquid benzene at pressure $P_1 = 10$ bar $(T_1^{sat} = 451.7K)$ is throttled in a steady-flow process to a pressure $P_2 = 1.2$ bar $(T_2^{sat} = 358.7K)$, where it is a liquid/vapor mixture. Estimate the molar fraction of the exit stream that is vapor. For liquid benzene, $C_P = 162$ J·mol^{-1}·K^{-1}. Ignore the effect of pressure on the enthalpy of liquid benzene.

4.18. Estimate $\Delta H_{f_{298}}^{\circ}$ for one of the following compounds as a *liquid* at 25°C.

(a) Acetylene, (b) 1,3-Butadiene, (c) Ethylbenzene, (d) n-Hexane, (e) Styrene.

4.19. Reversible compression of 1 mol of an ideal gas in a piston/cylinder device results in a pressure increase from 1 bar to P_2 and a temperature increase from 400 K to 950 K. The path followed by the gas during compression is given by $PV^{1.55} = $ const, and the molar heat capacity of the gas is given by:

$$C_P/R = 3.85 + 0.57 \times 10^{-3} T \qquad [T = K]$$

Determine the heat transferred during the process and the final pressure.

4.20. Hydrocarbon fuels can be produced from methanol by reactions such as the following, which yields 1-hexene:

$$6CH_3OH(g) \rightarrow C_6H_{12}(g) + 6H_2O(g)$$

Compare the standard heat of combustion at 25°C of $6CH_3OH(g)$ with the standard heat of combustion at 25°C of $C_6H_{12}(g)$ for reaction products $CO_2(g)$ and $H_2O(g)$.

4.21. Calculate the theoretical flame temperature when ethylene at 25°C is burned with:

(a) The theoretical amount of air at 25°C.
(b) 25% excess air at 25°C.
(c) 50% excess air at 25°C.
(d) 100% excess air at 25°C.
(e) 50% excess air preheated to 500°C.
(f) The theoretical amount of pure oxygen.

4.22. What is the standard heat of combustion of each of the following gases at 25°C if the combustion products are $H_2O(l)$ and $CO_2(g)$? Compute both the molar and specific heat of combustion in each case.

(*a*) methane, (*b*) ethane, (*c*) ethylene, (*d*) propane, (*e*) propylene, (*f*) *n*-butane, (*g*) 1-butene, (*h*) ethylene oxide, (*i*) acetaldehyde, (*j*) methanol, (*k*) ethanol.

4.23. Determine the standard heat of each of the following reactions at 25°C:

(*a*) $N_2(g) + 3H_2(g) \rightarrow 2NH_3(g)$
(*b*) $4NH_3(g) + 5O_2(g) \rightarrow 4NO(g) + 6H_2O(g)$
(*c*) $3NO_2(g) + H_2O(l) \rightarrow 2HNO_3(l) + NO(g)$
(*d*) $CaC_2(s) + H_2O(l) \rightarrow C_2H_2(g) + CaO(s)$
(*e*) $2Na(s) + 2H_2O(g) \rightarrow 2NaOH(s) + H_2(g)$
(*f*) $6NO_2(g) + 8NH_3(g) \rightarrow 7N_2(g) + 12H_2O(g)$
(*g*) $C_2H_4(g) + \frac{1}{2}O_2(g) \rightarrow \langle(CH_2)_2\rangle O(g)$
(*h*) $C_2H_2(g) + H_2O(g) \rightarrow \langle(CH_2)_2\rangle O(g)$
(*i*) $CH_4(g) + 2H_2O(g) \rightarrow CO_2(g) + 4H_2(g)$
(*j*) $CO_2(g) + 3H_2(g) \rightarrow CH_3OH(g) + H_2O(g)$
(*k*) $CH_3OH(g) + \frac{1}{2}O_2(g) \rightarrow HCHO(g) + H_2O(g)$
(*l*) $2H_2S(g) + 3O_2(g) \rightarrow 2H_2O(g) + 2SO_2(g)$
(*m*) $H_2S(g) + 2H_2O(g) \rightarrow 3H_2(g) + SO_2(g)$
(*n*) $N_2(g) + O_2(g) \rightarrow 2NO(g)$
(*o*) $CaCO_3(s) \rightarrow CaO(s) + CO_2(g)$
(*p*) $SO_3(g) + H_2O(l) \rightarrow H_2SO_4(l)$
(*q*) $C_2H_4(g) + H_2O(l) \rightarrow C_2H_5OH(l)$
(*r*) $CH_3CHO(g) + H_2(g) \rightarrow C_2H_5OH(g)$
(*s*) $C_2H_5OH(l) + O_2(g) \rightarrow CH_3COOH(l) + H_2O(l)$
(*t*) $C_2H_5CH:CH_2(g) \rightarrow CH_2:CHCH:CH_2(g) + H_2(g)$
(*u*) $C_4H_{10}(g) \rightarrow CH_2:CHCH:CH_2(g) + 2H_2(g)$
(*v*) $C_2H_5CH:CH_2(g) + \frac{1}{2}O_2(g) \rightarrow CH_2:CHCH:CH_2(g) + H_2O(g)$
(*w*) $4NH_3(g) + 6NO(g) \rightarrow 6H_2O(g) + 5N_2(g)$
(*x*) $N_2(g) + C_2H_2(g) \rightarrow 2HCN(g)$
(*y*) $C_6H_5C_2H_5(g) \rightarrow C_6H_5CH:CH_2(g) + H_2(g)$
(*z*) $C(s) + H_2O(l) \rightarrow H_2(g) + CO(g)$

4.24. Determine the standard heat for one of the reactions of Prob. 4.23: part (*a*) at 600°C, part (*b*) at 50°C, part (*f*) at 650°C, part (*i*) at 700°C, part (*j*) at 590(°F), part (*l*) at 770(°F), part (*m*) at 850 K, part (*n*) at 1300 K, part (*o*) at 800°C, part (*r*) at 450°C, part (*t*) at 860(°F), part (*u*) at 750 K, part (*v*) at 900 K, part (*w*) at 400°C, part (*x*) at 375°C, part (*y*) at 1490(°F).

4.25. Develop a general equation for the standard heat of reaction as a function of temperature for one of the reactions given in parts (*a*), (*b*), (*e*), (*f*), (*g*), (*h*), (*j*), (*k*), (*l*), (*m*), (*n*), (*o*), (*r*), (*t*), (*u*), (*v*), (*w*), (*x*), (*y*), and (*z*) of Prob. 4.23.

4.26. Compute the standard heat of reaction for each of the following reactions taking place at 298.15 K in dilute aqueous solution at zero ionic strength.

(*a*) D-Glucose + $ATP^{2-} \rightarrow$ D-Glucose 6-phosphate$^-$ + ADP$^-$
(*b*) D-Glucose 6-phosphate$^- \rightarrow$ D-Fructose 6-phosphate$^-$

(c) D-Fructose 6-phosphate$^-$ + ATP^{2-} → D-Fructose 1,6-biphosphate^{2-} + ADP$^-$
(d) D-Glucose + 2 ADP$^-$ + 2H$_2$PO$_4^-$ + 2 NAD$^+$ → 2 Pyruvate$^-$+ 2 ATP^{2-} +
      2 NADH + 4H$^+$ + 2H$_2$O
(e) D-Glucose + 2 ADP$^-$ + 2H$_2$PO$_4^-$ → 2 Lactate$^-$ + 2 ATP^{2-}+ 2H$^+$ + 2H$_2$O
(f) D-Glucose + 2 ADP$^-$ + 2H$_2$PO$_4^-$ → 2CO$_2$ + 2 Ethanol + 2 ATP^{2-} + 2 H$_2$O
(g) 2 NADH + O$_2$ + 2H$^+$ → 2 NAD$^+$ + 2H$_2$O
(h) ADP$^-$ + H$_2$PO$_4^-$ →ATP^{2-} + H$_2$O
(i) 2 NADH + 2 ADP$^-$+ 2H$_2$PO$_4^-$ + O$_2$ + 2H$^+$ → 2 NAD$^+$ + 2 ATP^{2-} + 4H$_2$O
(j) D-Fructose + 2 ADP$^-$ + 2H$_2$PO$_4^-$ → 2CO$_2$ + 2 Ethanol +2 ATP^{2-} + 2H$_2$O
(k) D-Galactose + 2 ADP$^-$ + 2H$_2$PO$_4^-$ → 2CO$_2$ + 2
(l) Ethanol + 2 ATP^{2-} + 2H$_2$O
      NH$_4^+$ + L-aspartate$^-$ + ATP^{2-} → L-asparagine + ADP$^-$ + H$_2$PO$_4^-$

4.27. The first step in the metabolism of ethanol is dehydrogenation by reaction with nicotinamide-adenine dinucleotide (NAD):

$$C_2H_5OH + NAD^+ → C_2H_4O + NADH$$

What is the heat effect of this reaction upon metabolizing 10 g of ethanol from a typical cocktail? What is the total heat effect for complete metabolism of the 10 g of ethanol to CO$_2$ and water? How, if at all, is the perception of warmth that accompanies ethanol consumption related to these heat effects? For computing heat effects, you may neglect the temperature, pH, and ionic strength dependence of the enthalpy of reaction (i.e. apply the enthalpies of formation from Table C.5 of App. C at physiological conditions).

4.28. Natural gas (assume pure methane) is delivered to a city via pipeline at a volumetric rate of 150 million standard cubic feet per day. If the selling price of the gas is $5.00 per GJ of higher heating value, what is the expected revenue in dollars per day? Standard conditions are 60(°F) and 1(atm).

4.29. Natural gases are rarely pure methane; they usually also contain other light hydrocarbons and nitrogen. Determine an expression for the standard higher heat of combustion as a function of composition for a natural gas containing methane, ethane, propane, and nitrogen. Assume liquid water as a product of combustion. Which of the following natural gases has the highest heat of combustion?

(a) $y_{CH_4} = 0.95, y_{C_2H_6} = 0.02, y_{C_3H_8} = 0.02, y_{N_2} = 0.01$.
(b) $y_{CH_4} = 0.90, y_{C_2H_6} = 0.05, y_{C_3H_8} = 0.03, y_{N_2} = 0.02$.
(c) $y_{CH_4} = 0.85, y_{C_2H_6} = 0.07, y_{C_3H_8} = 0.03, y_{N_2} = 0.05$.

4.30. If the heat of combustion of urea, (NH$_2$)$_2$CO(s), at 25°C is 631,660 J·mol^{-1} when the products are CO$_2$(g), H$_2$O(l), and N$_2$(g), what is $\Delta H^\circ_{f_{298}}$ for urea?

4.31. The *higher heating value* (HHV) of a fuel is its standard heat of combustion at 25°C with liquid water as a product; the *lower heating value* (LHV) is for water vapor as product.

(a) Explain the origins of these terms.
(b) Determine the HHV and the LHV for natural gas, modeled as pure methane.

(c) Determine the HHV and the LHV for a home-heating oil, modeled as pure liquid *n*-decane. For *n*-decane as a liquid

$$\Delta H^{\circ}_{f_{298}} = -249{,}700 \text{ J·mol}^{-1}.$$

4.32. A light fuel oil with an average chemical composition of $C_{10}H_{18}$ is burned with oxygen in a bomb calorimeter. The heat evolved is measured as 43,960 J·g^{-1} for the reaction at 25°C. Calculate the standard heat of combustion of the fuel oil at 25°C with $H_2O(g)$ and $CO_2(g)$ as products. Note that the reaction in the bomb occurs at constant volume, produces liquid water as a product, and goes to completion.

4.33. Methane gas is burned completely with 30% excess air at approximately atmospheric pressure. Both the methane and the air enter the furnace at 30°C saturated with water vapor, and the flue gases leave the furnace at 1500°C. The flue gases then pass through a heat exchanger from which they emerge at 50°C. Per mole of methane, how much heat is lost from the furnace, and how much heat is transferred in the heat exchanger?

4.34. Ammonia gas enters the reactor of a nitric acid plant mixed with 30% more dry air than is required for the complete conversion of the ammonia to nitric oxide and water vapor. If the gases enter the reactor at 75°C, if conversion is 80%, if no side reactions occur, and if the reactor operates adiabatically, what is the temperature of the gases leaving the reactor? Assume ideal gases.

4.35. Ethylene gas and steam at 320°C and atmospheric pressure are fed to a reaction process as an equimolar mixture. The process produces ethanol by the reaction:

$$C_2H_4(g) + H_2O(g) \rightarrow C_2H_5OH(l)$$

The liquid ethanol exits the process at 25°C. What is the heat transfer associated with this overall process per mole of ethanol produced?

4.36. A gas mixture of methane and steam at atmospheric pressure and 500°C is fed to a reactor, where the following reactions occur:

$$CH_4 + H_2O \rightarrow CO + 3\,H_2 \quad \text{and} \quad CO + H_2O \rightarrow CO_2 + H_2$$

The product stream leaves the reactor at 850°C. Its composition (mole fractions) is:

$$y_{CO_2} = 0.0275 \quad y_{CO} = 0.1725 \quad y_{H_2O} = 0.1725 \quad y_{H_2} = 0.6275$$

Determine the quantity of heat added to the reactor per mole of product gas.

4.37. A fuel consisting of 75 mol-% methane and 25 mol-% ethane enters a furnace with 80% excess air at 30°C. If 8×10^5 kJ·kmol^{-1} fuel is transferred as heat to boiler tubes, at what temperature does the flue gas leave the furnace? Assume complete combustion of the fuel.

4.38. The gas stream from a sulfur burner consists of 15 mol-% SO_2, 20 mol-% O_2, and 65 mol-% N_2. The gas stream at atmospheric pressure and 400°C enters a catalytic converter where 86% of the SO_2 is further oxidized to SO_3. On the basis of 1 mol of gas entering, how much heat must be added to or removed from the converter so that the product gases leave at 500°C?

4.39. Hydrogen is produced by the reaction: $CO(g) + H_2O(g) \rightarrow CO_2(g) + H_2(g)$. The feed stream to the reactor is an equimolar mixture of carbon monoxide and steam, and it enters the reactor at 125°C and atmospheric pressure. If 60% of the H_2O is converted to H_2 and if the product stream leaves the reactor at 425°C, how much heat must be transferred to or from the reactor?

4.40. A direct-fired dryer burns a fuel oil with a lower heating value of 19,000(Btu) $(lb_m)^{-1}$. [Products of combustion are $CO_2(g)$ and $H_2O(g)$.] The composition of the oil is 85% carbon, 12% hydrogen, 2% nitrogen, and 1% water by weight. The flue gases leave the dryer at 400(°F), and a partial analysis shows that they contain 3 mol-% CO_2 and 11.8 mol-% CO on a dry basis. The fuel, air, and material being dried enter the dryer at 77(°F). If the entering air is saturated with water and if 30% of the net heating value of the oil is allowed for heat losses (including the sensible heat carried out with the dried product), how much water is evaporated in the dryer per (lb_m) of oil burned?

4.41. An equimolar mixture of nitrogen and acetylene enters a steady-flow reactor at 25°C and atmospheric pressure. The only reaction occurring is: $N_2(g) + C_2H_2(g) \rightarrow 2HCN$ (g). The product gases leave the reactor at 600°C and contain 24.2 mol-% HCN. How much heat is supplied to the reactor per mole of product gas?

4.42. Chlorine is produced by the reaction: $4HCl(g) + O_2(g) \rightarrow 2H_2O(g) + 2Cl_2(g)$. The feed stream to the reactor consists of 60 mol-% HCl, 36 mol-% O_2, and 4 mol-% N_2, and it enters the reactor at 550°C. If the conversion of HCl is 75% and if the process is isothermal, how much heat must be transferred to or from the reactor per mole of the entering gas mixture?

4.43. A gas consisting only of CO and N_2 is made by passing a mixture of flue gas and air through a bed of incandescent coke (assume pure carbon). The two reactions that occur both go to completion:

$$CO_2 + C \rightarrow 2CO \quad \text{and} \quad 2C + O_2 \rightarrow 2CO$$

The flue gas composition is 12.8 mol-% CO, 3.7 mol-% CO_2, 5.4 mol-% O_2, and 78.1 mol-% N_2. The flue gas/air mixture is so proportioned that the heats of the two reactions cancel, and the temperature of the coke bed is therefore constant. If this temperature is 875°C, if the feed stream is preheated to 875°C, and if the process is adiabatic, what ratio of moles of flue gas to moles of air is required, and what is the composition of the gas produced?

4.44. A fuel gas consisting of 94 mol-% methane and 6 mol-% nitrogen is burned with 35% excess air in a continuous water heater. Both fuel gas and air enter dry at 77(°F). Water is heated at a rate of 75(lb$_m$)(s)$^{-1}$ from 77(°F) to 203(°F). The flue gases leave the heater at 410(°F). Of the entering methane, 70% burns to carbon dioxide and 30% burns to carbon monoxide. What volumetric flow rate of fuel gas is required if there are no heat losses to the surroundings?

4.45. A process for the production of 1,3-butadiene results from the catalytic dehydrogenation at atmospheric pressure of 1-butene according to the reaction:

$$C_4H_8(g) \rightarrow C_4H_6(g) + H_2(g)$$

To suppress side reactions, the 1-butene feed stream is diluted with steam in the ratio of 10 moles of steam per mole of 1-butene. The reaction is carried out *isothermally* at 525°C, and at this temperature 33% of the 1-butene is converted to 1,3-butadiene. How much heat is transferred to or from the reactor per mole of entering 1-butene?

4.46. (*a*) An air-cooled condenser transfers heat at a rate of 12(Btu)·s^{-1} to ambient air at 70(°F). If the air temperature is raised 20(°F), what is the required volumetric flow rate of the air?

(*b*) Rework part (*a*) for a heat-transfer rate of 12 kJ·s^{-1}, ambient air at 24°C, and a temperature rise of 13°C.

4.47. (*a*) An air-conditioning unit cools 50(ft)^3·s^{-1} of air at 94(°F) to 68(°F). What is the required heat-transfer rate in (Btu)·s^{-1}?

(*b*) Rework part (*a*) for a flow rate of 1.5 m^3·s^{-1}, a temperature change from 35°C to 25°C, and units of kJ·s^{-1}.

4.48. A propane-fired water heater delivers 80% of the standard heat of combustion of the propane [at 25°C with CO$_2$(g) and H$_2$O(g) as products] to the water. If the price of propane is $2.20 per gallon as measured at 25°C, what is the heating cost in $ per million (Btu)? In $ per MJ?

4.49. Determine the heat transfer (J·mol^{-1}) when one of the gases identified below is heated in a steady-flow process from 25°C to 500°C at atmospheric pressure.

(*a*) Acetylene; (*b*) Ammonia; (*c*) *n*-Butane; (*d*) Carbon dioxide; (*e*) Carbon monoxide; (*f*) Ethane; (*g*) Hydrogen; (*h*) Hydrogen chloride; (*i*) Methane; (*j*) Nitric oxide; (*k*) Nitrogen; (*l*) Nitrogen dioxide; (*m*) Nitrous oxide; (*n*) Oxygen; (*o*) Propylene

4.50. Determine the final temperature for one of the gases of the preceding problem if heat in the amount of 30,000 J·mol^{-1} is transferred to the gas, initially at 25°C, in a steady-flow process at atmospheric pressure.

4.51. Quantitative thermal analysis has been suggested as a technique for monitoring the composition of a binary gas stream. To illustrate the principle, do one of the following problems.

  (*a*) A methane/ethane gas mixture is heated from 25°C to 250°C at 1(atm) in a steady-flow process. If $Q = 11{,}500$ J·mol^{-1}, what is the composition of the mixture?
  (*b*) A benzene/cyclohexane gas mixture is heated from 100°C to 400°C at 1(atm) in a steady-flow process. If $Q = 54{,}000$ J·mol^{-1}, what is the composition of the mixture?
  (*c*) A toluene/ethylbenzene gas mixture is heated from 150°C to 250°C at 1(atm) in a steady-flow process. If $Q = 17{,}500$ J·mol^{-1}, what is the composition of the mixture?

4.52. Saturated steam at 1(atm) and 100°C is continuously generated from liquid water at 1(atm) and 25°C by thermal contact with hot air in a counterflow heat exchanger. The air flows steadily at 1(atm). Determine values for $\dot{m}$ (steam)/$\dot{n}$(air) for two cases:

  (*a*) Air enters the exchanger at 1000°C.
  (*b*) Air enters the exchanger at 500°C.

For both cases, assume a minimum approach ΔT for heat exchange of 10°C.

4.53. Saturated water vapor, i.e., *steam*, is commonly used as a heat source in heat-exchanger applications. Why *saturated* vapor? Why saturated *water* vapor? In a plant of any reasonable size, several varieties of saturated steam are commonly available; for example, saturated steam at 4.5, 9, 17, and 33 bar. But the higher the pressure the lower the useful energy content (why?), and the greater the unit cost. Why then is higher-pressure steam used?

4.54. The oxidation of glucose provides the principal source of energy for animal cells. Assume the reactants are glucose [$C_6H_{12}O_6(s)$] and oxygen [$O_2(g)$]. The products are $CO_2(g)$ and $H_2O(l)$.

  (*a*) Write a balanced equation for glucose oxidation, and determine the standard heat of reaction at 298 K.
  (*b*) During a day an average person consumes about 150 kJ of energy per kg of body mass. Assuming glucose the sole source of energy, estimate the mass (grams) of glucose required daily to sustain a person of 57 kg.
  (*c*) For a population of 275 million persons, what mass of CO_2 (a greenhouse gas) is produced daily by mere respiration. *Data*: For glucose, $\Delta H^\circ_{f_{298}} = -1274.4$ kJ·mol^{-1}. Ignore the effect of temperature on the heat of reaction.

4.55. A natural-gas fuel contains 85 mol-% methane, 10 mol-% ethane, and 5 mol-% nitrogen.

  (*a*) What is the standard heat of combustion (kJ·mol^{-1}) of the fuel at 25°C with $H_2O(g)$ as a product?
  (*b*) The fuel is supplied to a furnace with 50% excess air, both entering at 25°C. The products leave at 600°C. If combustion is complete and if no side reactions occur, how much heat (kJ per mol of fuel) is transferred in the furnace?

Chapter 5

The Second Law of Thermodynamics

Thermodynamics treats the principles of energy transformations, and the laws of thermodynamics establish the bounds within which these transformations are observed to occur. The first law states the principle of energy conservation, leading to energy balances in which work and heat are included as simple additive terms. Yet work and heat are otherwise quite different. Work is directly useful in ways that heat is not, e.g., for the elevation of a weight or for the acceleration of a mass. Evidently, work is a form of energy intrinsically more valuable than an equal quantity of heat. This difference is reflected in a second fundamental law which, together with the first law, makes up the foundation upon which the edifice of thermodynamics is built. The purpose of this chapter is to:

- Introduce the concept of entropy, an essential thermodynamic property

- Present the second law of thermodynamics, which reflects the observation that limits exist to what can be accomplished even by reversible processes

- Apply the second law to some familiar processes

- Relate changes in entropy to T and P for substances in the ideal-gas state

- Present entropy balances for open systems

- Demonstrate the calculation of ideal work and lost work for flow processes

- Relate entropy to the microscopic world of molecules

5.1 AXIOMATIC STATEMENTS OF THE SECOND LAW

The two axioms presented in Chapter 2 in relation to the first law have counterparts with respect to the second law. They are:

> ***Axiom 4:*** **There exists a property called entropy[1] S, which for systems at internal equilibrium is an intrinsic property, functionally related to the**

[1]Pronounced **en'**-tro-py to distinguish it clearly from en-**thal'**-py.

measurable state variables that characterize the system. Differential changes in this property are given by the equation:

$$dS^t = dQ_{\text{rev}}/T \tag{5.1}$$

where S^t is the *system* (rather than the molar) entropy.

Axiom 5: (The Second Law of Thermodynamics) The entropy change of any system and its surroundings, considered together, and resulting from any real process, is positive, approaching zero when the process approaches reversibility. Mathematically,

$$\Delta S_{\text{total}} \geq 0 \tag{5.2}$$

The second law affirms that every process proceeds in such a direction that the *total* entropy change associated with it is positive, the limiting value of zero being attained only by a reversible process. No process is possible for which the *total* entropy decreases. The practical utility of the second law is illustrated by application to two very common processes. The first shows its consistency with our everyday experience that heat flows from hot to cold. The second shows how it establishes limits to the conversion of heat to work by any device.

Application of the Second Law to Simple Heat Transfer

First, consider direct heat transfer between two *heat reservoirs*, bodies imagined capable of absorbing or rejecting unlimited quantities of heat without temperature change.[2] The equation for the entropy change of a heat reservoir follows from Eq. (5.1). Because T is constant, integration gives:

$$\Delta S = \frac{Q}{T}$$

A quantity of heat Q is transferred to or from a reservoir at temperature T. From the reservoir's point of view the transfer is reversible, because its effect on the reservoir is the same regardless of source or sink of the heat.

Let the temperatures of the reservoirs be T_H and T_C with $T_H > T_C$. Heat quantity Q, transferred from one reservoir to the other, is the same for both reservoirs. However, Q_H and Q_C have opposite signs: positive for the heat added to one reservoir and negative for the heat extracted from the other. Therefore $Q_H = -Q_C$, and the entropy changes of the reservoirs at T_H and at T_C are:

$$\Delta S^t_H = \frac{Q_H}{T_H} = \frac{-Q_C}{T_H} \qquad \text{and} \qquad \Delta S^t_C = \frac{Q_C}{T_C}$$

These two entropy changes are added to give:

$$\Delta S_{\text{total}} = \Delta S^t_H + \Delta S^t_C = \frac{-Q_C}{T_H} + \frac{Q_C}{T_C} = Q_C\left(\frac{T_H - T_C}{T_H T_C}\right)$$

[2]The firebox of a furnace is in effect a hot reservoir, and the surrounding atmosphere, a cold reservoir.

Because the heat-transfer process is irreversible, Eq. (5.2) requires a positive value for ΔS_{total}, and therefore

$$Q_C(T_H - T_C) > 0$$

With the temperature difference positive, Q_C must also be positive, which means that heat flows *into* the reservoir at T_C, i.e., from the higher to the lower temperature. This result conforms to universal experience that heat flows from higher to lower temperature. A formal statement conveys this result:

> **No process is possible which consists *solely* of the transfer of heat from one temperature level to a higher one.**

Note also that ΔS_{total} becomes smaller as the temperature difference decreases. When T_H is only infinitesimally higher than T_C, the heat transfer is reversible, and ΔS_{total} approaches zero.

Application of the Second Law to Heat Engines

Heat can be used far more usefully than by simple transfer from one temperature level to a lower one. Indeed, useful work is produced by countless engines that employ the flow of heat as their energy source. The most common examples are the internal-combustion engine and the steam power plant. Collectively, these are *heat engines.* They all rely on a high-temperature source of heat, and all discard heat to the environment.

The second law imposes restrictions on how much of their heat intake can be converted into work, and our object now is to establish quantitatively this relationship. We imagine that the engine receives heat from a higher-temperature heat reservoir at T_H and discards heat to a lower-temperature reservoir T_C. The engine is taken as the system and the two heat reservoirs comprise the surroundings. The work and heat quantities in relation to both the engine and the heat reservoirs are shown in Fig. 5.1(a).

With respect to the engine, the first law as given by Eq. (2.3) becomes:

$$\Delta U = Q + W = Q_H + Q_C + W$$

Because the engine inevitably operates in cycles, its properties over a cycle do not change. Therefore $\Delta U = 0$, and $W = -Q_H - Q_C$.

The entropy change of the surroundings equals the sum of the entropy changes of the reservoirs. Because the entropy change of the engine over a cycle is zero, the total entropy change is that of the heat reservoirs. Therefore

$$\Delta S_{total} = -\frac{Q_H}{T_H} - \frac{Q_C}{T_C}$$

Note that Q_C with respect to the *engine* is a negative number, whereas Q_H is positive. Combining this equation with the equation for W to eliminate Q_H yields:

$$W = T_H \Delta S_{total} + Q_C\left(\frac{T_H - T_C}{T_C}\right)$$

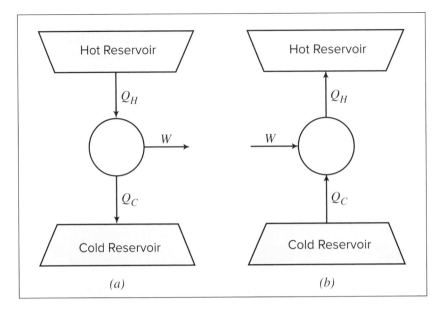

Figure 5.1: Schematic diagrams: (*a*) Carnot engine; (*b*) Carnot heat pump or refrigerator.

This result gives the work output of a heat engine within two limits. If the engine is totally ineffective, $W = 0$; the equation then reduces to the result obtained for simple heat transfer between the two heat reservoirs, i.e.:

$$\Delta S_{\text{total}} = -Q_C\left(\frac{T_H - T_C}{T_H T_C}\right)$$

The difference in sign here simply reflects the fact that Q_C is with respect to the engine, whereas previously it was with respect to the lower-temperature reservoir.

If the process is reversible in all respects, then $\Delta S_{\text{total}} = 0$, and the equation reduces to:

$$W = Q_C\left(\frac{T_H}{T_C} - 1\right) \tag{5.3}$$

A heat engine operating as described in a completely reversible manner is very special and is called a *Carnot engine*. The characteristics of such an ideal engine were first described by N. L. S. Carnot[3] in 1824.

Note again that Q_C is a negative number, as it represents heat transferred *from* the engine. This makes W negative, in accord with the fact that work is not added to, but is *produced by*, the engine. Clearly, for any finite value of W, Q_C is also finite. This means that a portion of the heat transferred from the higher-temperature reservoir must inevitably be exhausted to the lower-temperature reservoir. This observation can be given formal statement:

It is impossible to construct an engine that, operating in a cycle, produces *no* effect (in system and surroundings) other than the extraction of heat from a reservoir and the performance of an equivalent amount of work.

[3]Nicolas Leonard Sadi Carnot (1796–1832), French engineer. See http://en.wikipedia.org/wiki/Nicolas_Leonard_Sadi_Carnot.

The second law does not prohibit the continuous production of work from heat, but it does place a limit on how much of the heat taken into a cyclic process can be converted into work.

Combining this equation with $W = -Q_H - Q_C$ to eliminate first W and then Q_C leads to *Carnot's equations:*

$$\frac{-Q_C}{T_C} = \frac{Q_H}{T_H} \tag{5.4}$$

$$\frac{W}{Q_H} = \frac{T_C}{T_H} - 1 \tag{5.5}$$

Note that in application to a Carnot engine Q_H, representing heat transferred to the engine, is a positive number, making the work produced (W) negative. In Eq. (5.4) the smallest possible value of Q_C is zero; the corresponding value of T_C is absolute zero on the Kelvin scale, which corresponds to $-273.15°C$.

The *thermal* efficiency of a heat engine is **defined** as the ratio of the work produced to the heat supplied to the engine. With respect to the engine, the work W is negative. Thus,

$$\eta \equiv \frac{-W}{Q_H} \tag{5.6}$$

In view of Eq. (5.5) the thermal efficiency of a Carnot engine is:

$$\eta_{\text{Carnot}} = 1 - \frac{T_C}{T_H} \tag{5.7}$$

Although a Carnot engine operates reversibly in all respects, and cannot be improved, its efficiency approaches unity only when T_H approaches infinity or T_C approaches zero. Neither condition exists on earth; all terrestrial heat engines therefore operate with thermal efficiencies less than unity. The cold reservoirs available on earth are the atmosphere, lakes, rivers, and oceans, for which $T_C \simeq 300$ K. Hot reservoirs are objects such as furnaces where the temperature is maintained by combustion of fossil fuels or by fission of radioactive elements, and for which $T_H \simeq 600$ K. With these values, $\eta = 1 - 300/600 = 0.5$, an approximate realistic limit for the thermal efficiency of a Carnot engine. Actual heat engines are irreversible, and η rarely exceeds 0.35.

Example 5.1

A central power plant, rated at 800,000 kW, generates steam at 585 K and discards heat to a river at 295 K. If the thermal efficiency of the plant is 70% of the maximum possible value, how much heat is discarded to the river at rated power?

Solution 5.1

The maximum possible thermal efficiency is given by Eq. (5.7). With T_H as the steam-generation temperature and T_C as the river temperature:

$$\eta_{\text{Carnot}} = 1 - \frac{295}{585} = 0.4957 \qquad \text{and} \qquad \eta = (0.7)(0.4957) = 0.3470$$

where η is the actual thermal efficiency. Combining Eq. (5.6) with the first law, written $W = -Q_H - Q_C$, to eliminate Q_H, yields:

$$Q_C = \left(\frac{1-\eta}{\eta}\right)W = \left(\frac{1-0.347}{0.347}\right)(-800,000) = -1,505,475 \text{ kW}$$

This rate of heat transfer to a modest river would cause a temperature rise of several °C.

5.2   HEAT ENGINES AND HEAT PUMPS

The following steps make up the cycle of any Carnot engine:

- **Step 1:** A system at an initial temperature of a cold reservoir at T_C undergoes a *reversible adiabatic* process that causes its temperature to rise to that of a hot reservoir at T_H.

- **Step 2:** The system maintains contact with the hot reservoir at T_H and undergoes a *reversible isothermal* process during which heat Q_H is absorbed from the hot reservoir.

- **Step 3:** The system undergoes a *reversible adiabatic* process in the opposite direction of Step 1 that brings its temperature back to that of the cold reservoir at T_C.

- **Step 4:** The system maintains contact with the reservoir at T_C, and undergoes a *reversible isothermal* process in the opposite direction of Step 2 that returns it to its initial state with rejection of heat Q_C to the cold reservoir.

This set of processes can in principle be performed on any kind of system, but only a few, yet to be described, are of practical interest.

A Carnot engine operates between two heat reservoirs in such a way that all heat absorbed is transferred at the constant temperature of the hot reservoir and all heat rejected is transferred at the constant temperature of the cold reservoir. Any *reversible* engine operating between two heat reservoirs is a Carnot engine; an engine operating on a different cycle must necessarily transfer heat across finite temperature differences and therefore cannot be reversible. Two important conclusions inevitably follow from the nature of the Carnot engine:

- **Its efficiency depends only on the temperature levels and not upon the working substance of the engine.**

- **For two given heat reservoirs no engine can have a thermal efficiency higher than that of a Carnot engine.**

We provide further treatment of practical heat engines in Chapter 8.

Because a Carnot engine is reversible, it may be operated in reverse; the Carnot cycle is then traversed in the opposite direction, and it becomes a reversible *heat pump* operating between the same temperature levels and with the same quantities Q_H, Q_C, and W as for the engine but reversed in direction, as illustrated in Fig. 5.1(b). Here, work is required, and it is used to "pump" heat from the lower-temperature heat reservoir to the higher-temperature heat reservoir. Refrigerators are heat pumps with the "cold box" as the lower-temperature reservoir and some portion of the environment as the higher-temperature reservoir. The usual measure of quality of a heat pump is the *coefficient of performance,* **defined** as the heat extracted at the lower temperature divided by the work required, both of which are positive quantities with respect to the heat pump:

$$\omega \equiv \frac{Q_C}{W} \qquad (5.8)$$

For a Carnot heat pump this coefficient can be obtained by combining Eq. (5.4) and Eq. (5.5) to eliminate Q_H:

$$\omega_{\text{Carnot}} \equiv \frac{T_C}{T_H - T_C} \qquad (5.9)$$

For a refrigerator at 4°C and heat transfer to an environment at 24°C, Eq. (5.9) yields:

$$\omega_{\text{Carnot}} = \frac{4 + 273.15}{24 - 4} = 13.86$$

Any actual refrigerator would operate irreversibly with a lower value of ω. The practical aspects of refrigeration are treated in Chapter 9.

5.3 CARNOT ENGINE WITH IDEAL-GAS-STATE WORKING FLUID

The cycle traversed by a working fluid in its ideal-gas state in a Carnot engine is shown on a *PV* diagram in Fig. 5.2. It consists of four *reversible* processes corresponding to Steps 1 through 4 of the general Carnot cycle described in the preceding section:

- $a \rightarrow b$ Adiabatic compression with temperature rising from T_C to T_H.
- $b \rightarrow c$ Isothermal expansion to arbitrary point c with absorption of heat Q_H.
- $c \rightarrow d$ Adiabatic expansion with temperature decreasing to T_C.
- $d \rightarrow a$ Isothermal compression to the initial state with rejection of heat Q_C.

In this analysis the ideal-gas-state working fluid is regarded as the system. For the isothermal steps $b \rightarrow c$ and $d \rightarrow a$, Eq. (3.20) yields:

$$Q_H = RT_H \ln \frac{V_c^{ig}}{V_b^{ig}} \qquad \text{and} \qquad Q_C = RT_C \ln \frac{V_a^{ig}}{V_d^{ig}}$$

Dividing the first equation by the second gives:

$$\frac{Q_H}{Q_C} = \frac{T_H \ln\left(V_c^{ig}/V_b^{ig}\right)}{T_C \ln\left(V_a^{ig}/V_d^{ig}\right)}$$

For an adiabatic process Eq. (3.16) with $dQ = 0$ becomes,

$$-\frac{C_V^{ig}}{R}\frac{dT}{T} = \frac{dV^{ig}}{V^{ig}}$$

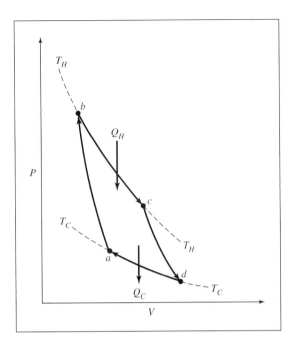

Figure 5.2: *PV* diagram showing a Carnot cycle for a working fluid in the ideal-gas state.

For steps $a \to b$ and $c \to d$, integration gives:

$$\int_{T_C}^{T_H} \frac{C_V^{ig}}{R} \frac{dT}{T} = \ln \frac{V_a^{ig}}{V_b^{ig}} \qquad \text{and} \qquad \int_{T_C}^{T_H} \frac{C_V^{ig}}{R} \frac{dT}{T} = \ln \frac{V_d^{ig}}{V_c^{ig}}$$

Because the left sides of these two equations are the same, the adiabatic steps are related by:

$$\ln \frac{V_d^{ig}}{V_c^{ig}} = \ln \frac{V_a^{ig}}{V_b^{ig}} \qquad \text{or} \qquad \ln \frac{V_c^{ig}}{V_b^{ig}} = -\ln \frac{V_a^{ig}}{V_d^{ig}}$$

Combining the second expression with the equation relating the two isothermal steps gives:

$$\frac{Q_H}{Q_C} = -\frac{T_H}{T_C} \qquad \text{or} \qquad \frac{-Q_C}{T_C} = \frac{Q_H}{T_H}$$

This last equation is identical with Eq. (5.4).

5.4 ENTROPY

Points A and B on the PV^t diagram of Fig. 5.3 represent two equilibrium states of a particular fluid, and paths ACB and ADB represent two arbitrary *reversible* processes connecting these points. Integration of Eq. (5.1) for each path gives:

$$\Delta S^t = \int_{ACB} \frac{dQ_{\text{rev}}}{T} \qquad \text{and} \qquad \Delta S^t = \int_{ADB} \frac{dQ_{\text{rev}}}{T}$$

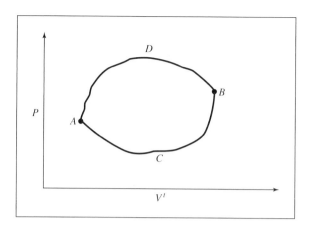

Figure 5.3: Two reversible paths joining equilibrium states *A* and *B*.

Because ΔS^t is a property change, it is independent of path and is given by $S_B^t - S_A^t$.

If the fluid is changed from state *A* to state *B* by an *irreversible* process, the entropy change is again $\Delta S^t = S_B^t - S_A^t$, but experiment shows that this result is *not* given by $\int dQ/T$ evaluated for the irreversible process itself, because the calculation of entropy changes by this integral must, in general, be along reversible paths.

The entropy change of a *heat reservoir*, however, is always given by Q/T, where Q is the quantity of heat transferred to or from the reservoir at temperature *T*, whether the transfer is reversible or irreversible. As noted earlier, the effect of heat transfer on a heat reservoir is the same regardless of the temperature of the source or sink of the heat.

If a process is reversible and adiabatic, $dQ_{\text{rev}} = 0$; then by Eq. (5.1), $dS^t = 0$. Thus the entropy of a system is constant during a reversible adiabatic process, and the process is said to be *isentropic*.

The characteristics of entropy may be summarized as follows:

• Entropy relates to the second law in much the same way that internal energy relates to the the first law. Equation (5.1) is the ultimate source of all equations that connect entropy to measurable quantities. It does not represent a definition of entropy; there is none in the context of classical thermodynamics. What it provides is the means for calculating *changes* in this property.

• The change in entropy of any system undergoing a finite *reversible* process is given by the integral form of Eq. (5.1). When a system undergoes an *irreversible* process between two equilibrium states, the entropy change of the system ΔS^t is evaluated by application of Eq. (5.1) to *an arbitrarily chosen reversible process* that accomplishes the same change of state as the actual process. Integration is *not* carried out for the irreversible path. Because entropy is a state function, the entropy changes of the irreversible and reversible processes are identical.

• In the special case of a *mechanically reversible* process (Sec. 2.8), the entropy change of the system is correctly evaluated from $\int dQ/T$ applied to the actual process, even though the *heat transfer* between system and surroundings represents an *external* irreversibility. The reason is that it is immaterial, as far as the system is concerned, whether the temperature difference causing the heat transfer is infinitesimal (making the process externally reversible) or finite. The entropy change of a system *resulting only from the transfer of heat* can always be calculated by $\int dQ/T$, whether the heat transfer is accomplished reversibly or irreversibly. However, when a process is irreversible on account of finite differences in other driving forces, such as pressure, the entropy change is not caused solely by the heat transfer, and for its calculation one must devise a mechanically reversible means of accomplishing the same change of state.

5.5 ENTROPY CHANGES FOR THE IDEAL-GAS STATE

For one mole or a unit mass of fluid undergoing a mechanically reversible process in a closed system, the first law, Eq. (2.7), becomes:

$$dU = dQ_{rev} - P\,dV$$

Differentiation of the defining equation for enthalpy, $H = U + PV$, yields:

$$dH = dU + P\,dV + V\,dP$$

Eliminating dU gives:

$$dH = dQ_{rev} - P\,dV + P\,dV + V\,dP$$

or

$$dQ_{rev} = dH - V\,dP$$

For the ideal-gas state, $dH^{ig} = C_P^{ig}\,dT$ and $V^{ig} = RT/P$. With these substitutions and division by T,

$$\frac{dQ_{rev}}{T} = C_P^{ig}\frac{dT}{T} - R\frac{dP}{P}$$

As a result of Eq. (5.1), this becomes:

$$dS^{ig} = C_P^{ig}\frac{dT}{T} - R\frac{dP}{P} \qquad \text{or} \qquad \frac{dS^{ig}}{R} = \frac{C_P^{ig}}{R}\frac{dT}{T} - d\ln P$$

where S^{ig} is the molar entropy for the ideal-gas state. Integration from an initial state at conditions T_0 and P_0 to a final state at conditions T and P gives:

$$\frac{\Delta S^{ig}}{R} = \int_{T_0}^{T}\frac{C_P^{ig}}{R}\frac{dT}{T} - \ln\frac{P}{P_0} \qquad\qquad (5.10)$$

Although *derived* for a mechanically reversible process, this equation relates properties only, independent of the process causing the change of state, and is therefore a *general* equation for the calculation of entropy changes in the ideal-gas state.

Example 5.2

For the ideal-gas state and constant heat capacities, Eq. (3.23b) for a reversible adiabatic (and therefore isentropic) process can be written:

$$\frac{T_2}{T_1} = \left(\frac{P_2}{P_1}\right)^{(\gamma-1)/\gamma}$$

Show that this same equation results from application of Eq. (5.10) with $\Delta S^{ig} = 0$.

Solution 5.2

Because C_P^{ig} is constant, Eq. (5.10) becomes:

$$0 = \frac{C_P^{ig}}{R} \ln\frac{T_2}{T_1} - \ln\frac{P_2}{P_1} = \ln\frac{T_2}{T_1} - \frac{R}{C_P^{ig}} \ln\frac{P_2}{P_1}$$

By Eq. (3.12) for the ideal-gas state, with $\gamma = C_P^{ig}/C_V^{ig}$:

$$C_P^{ig} = C_V^{ig} + R \qquad \text{or} \qquad \frac{R}{C_P^{ig}} = \frac{\gamma-1}{\gamma}$$

Whence,

$$\ln\frac{T_2}{T_1} = \frac{\gamma-1}{\gamma} \ln\frac{P_2}{P_1}$$

Exponentiating both sides of this equation leads to the given equation.

Equation (4.5) for the temperature dependence of the molar heat capacity C_P^{ig} allows integration of the first term on the right of Eq. (5.10). The result can be written as:

$$\int_{T_0}^{T} \frac{C_P^{ig}\,dT}{R\,T} = A \ln\frac{T}{T_0} + \left[B + \left(C + \frac{D}{T_0^2 T^2}\right)\left(\frac{T+T_0}{2}\right)\right](T - T_0) \qquad (5.11)$$

As with the integral $\int(C_P/R)dT$ of Eq. (4.8), this integral is often evaluated; for computational purposes we define the right side of Eq. (5.11) as the function ICPS(T_0, T; A, B, C, D) and presume the existence of a computer routine for its evaluation.[4] Equation (5.11) then becomes:

$$\int_{T_0}^{T} \frac{C_P^{ig}\,dT}{R\,T} = \text{ICPS}(T_0, T; \text{A, B, C, D})$$

Also useful is a *mean heat capacity*, here **defined** as:

$$\langle C_P^{ig}\rangle_S = \frac{\displaystyle\int_{T_0}^{T} C_P^{ig}\,dT/T}{\ln(T/T_0)} \qquad (5.12)$$

In accord with this equation, division of Eq. (5.11) by $\ln(T/T_0)$ yields:

$$\frac{\langle C_P^{ig}\rangle_S}{R} = A + \left[B + \left(C + \frac{D}{T_0^2 T^2}\right)\left(\frac{T+T_0}{2}\right)\right]\frac{T-T_0}{\ln(T/T_0)} \qquad (5.13)$$

---

[4]Examples of these defined functions implemented in Microsoft Excel, Matlab, Maple, Mathematica, and Mathcad are provided in the online learning center at http://highered.mheducation.com:80/sites/1259696529.

The right side of this equation is defined as another function, MCPS (T_0, T; A, B, C, D). Equation (5.13) is then written:

$$\frac{\left\langle C_P^{ig} \right\rangle_S}{R} = \text{MCPS}(T_0,\ T;\ \text{A, B, C, D})$$

The subscript S denotes a mean value specific to entropy calculations. Comparison of this mean value with the mean value specific to enthalpy calculations, as defined by Eq. (4.9), shows the two means to be quite different. This is inevitable because they are defined for the purpose of evaluating entirely different integrals.

Solving for the integral in Eq. (5.12) gives:

$$\int_{T_0}^{T} C_P^{ig} \frac{dT}{T} = \left\langle C_P^{ig} \right\rangle_S \ln \frac{T}{T_0}$$

and Eq. (5.10) becomes:

$$\boxed{\frac{\Delta S^{ig}}{R} = \frac{\left\langle C_P^{ig} \right\rangle_S}{R} \ln \frac{T}{T_0} - \ln \frac{P}{P_0}} \qquad (5.14)$$

This form of the equation for entropy changes for the ideal-gas state is often useful in iterative calculations where the final temperature is unknown.

___

Example 5.3

Methane gas at 550 K and 5 bar undergoes a reversible adiabatic expansion to 1 bar. Assuming ideal-gas-state methane at these conditions, find its final temperature.

Solution 5.3

For this process $\Delta S^{ig} = 0$, and Eq. (5.14) becomes:

$$\frac{\left\langle C_P^{ig} \right\rangle_S}{R} \ln \frac{T_2}{T_1} = \ln \frac{P_2}{P_1} = \ln \frac{1}{5} = -1.6094$$

Because $\left\langle C_P^{ig} \right\rangle_S$ depends on T_2, we rearrange this equation for iterative solution:

$$\ln \frac{T_2}{T_1} = \frac{-1.6094}{\left\langle C_P^{ig} \right\rangle_S / R} \qquad \text{or} \qquad T_2 = T_1 \exp \left(\frac{-1.6094}{\left\langle C_P^{ig} \right\rangle_S / R} \right)$$

With constants from Table C.1 of App. C, $\left\langle C_P^{ig} \right\rangle_S / R$ is evaluated by Eq. (5.13) written in its functional form:

$$\frac{\left\langle C_P^{ig} \right\rangle_S}{R} = \text{MCPS}\left(550,\ T_2;\ 1.702,\ 9.081 \times 10^{-3},\ -2.164 \times 10^{-6},\ 0.0\right)$$

For an initial value of $T_2 < 550$, compute a value of $\left\langle C_P^{ig} \right\rangle_S /R$ for substitution into the equation for T_2. This new value of T_2 allows recalculation of $\left\langle C_P^{ig} \right\rangle_S /R$, and the process continues to convergence on a final value of $T_2 = 411.34$ K.

As with Ex. 4.3 a trial procedure is an alternative approach, with Microsoft Excel's Goal Seek providing a prototypical automated version.

Example 5.4

A 40 kg steel casting ($C_P = 0.5$ kJ·kg^{-1}·K^{-1}) at a temperature of 450°C is quenched in 150 kg of oil ($C_P = 2.5$ kJ·kg^{-1}·K^{-1}) at 25°C. If there are no heat losses, what is the change in entropy of (a) the casting, (b) the oil, and (c) both considered together?

Solution 5.4

The final temperature t of the oil and the steel casting is found by an energy balance. Because the change in energy of the oil and steel together must be zero,

$$(40)(0.5)(t-450) + (150)(2.5)(t-25) = 0$$

Solution yields $t = 46.52°C$.

(*a*) Change in entropy of the casting:

$$\Delta S^t = m \int \frac{C_P dT}{T} = mC_P \ln \frac{T_2}{T_1}$$

$$= (40)(0.5) \ln \frac{273.15 + 46.52}{273.15 + 450} = -16.33 \text{ kJ·K}^{-1}$$

(*b*) Change in entropy of the oil:

$$\Delta S^t = (150)(2.5) \ln \frac{273.15 + 46.52}{273.15 + 25} = 26.13 \text{ kJ·K}^{-1}$$

(*c*) Total entropy change:

$$\Delta S_{total} = -16.33 + 26.13 = 9.80 \text{ kJ·K}^{-1}$$

Note that although the total entropy change is positive, the entropy of the casting has decreased.

5.6 ENTROPY BALANCE FOR OPEN SYSTEMS

Just as an energy balance is written for processes in which fluid enters, exits, or flows through a control volume (Sec. 2.9), so too may an entropy balance be written. There is, however, an important difference: *Entropy is not conserved.* The second law states that the *total* entropy change associated with any process must be positive, with a limiting value of zero for a reversible process. This requirement is taken into account by writing the entropy balance for both the system and its surroundings, considered together, and by including an *entropy-generation* term to account for the irreversibilities of the process. This term is the sum of three others: one for the difference in entropy between exit and entrance streams, one for entropy change within the control volume, and one for entropy change in the surroundings. If the process is reversible, these three terms sum to zero, making $\Delta S_{total} = 0$. If the process is irreversible, they sum to a positive quantity, the entropy-generation term.

The statement of balance, expressed as rates, is therefore:

$$
\left\{ \begin{array}{c} \text{Time rate of} \\ \text{change of} \\ \text{entropy} \\ \text{in control} \\ \text{volume} \end{array} \right\} + \left\{ \begin{array}{c} \text{Net rate of} \\ \text{change in} \\ \text{entropy of} \\ \text{flowing streams} \end{array} \right\} + \left\{ \begin{array}{c} \text{Time rate of} \\ \text{change of} \\ \text{entropy in} \\ \text{surroundings} \end{array} \right\} = \left\{ \begin{array}{c} \text{Total rate} \\ \text{of entropy} \\ \text{generation} \end{array} \right\}
$$

The equivalent *equation of entropy balance* is

$$
\frac{d(mS)_{cv}}{dt} + \Delta(S\dot{m})_{fs} + \frac{dS^t_{surr}}{dt} = \dot{S}_G \geq 0 \tag{5.15}
$$

where $\dot{S}_G$ is the rate of entropy generation. This equation is the general *rate* form of the entropy balance, applicable at any instant. Each term can vary with time. The first term is the time rate of change of the total entropy of the fluid contained within the control volume. The second term is the net rate of gain in entropy of the flowing streams, i.e., the difference between the total entropy transported out by exit streams and the total entropy transported in by entrance streams. The third term is the time rate of change of the entropy of the surroundings, resulting from heat transfer between system and surroundings.

Let rate of heat transfer $\dot{Q}_j$ with respect to a particular part of the control surface be associated with $T_{\sigma,j}$ where subscript σ,j denotes a temperature in the surroundings. The rate of entropy change in the surroundings as a result of this transfer is then $-\dot{Q}_j/T_{\sigma,j}$. The minus sign converts $\dot{Q}_j$, defined with respect to the system, to a heat transfer rate with respect to the surroundings. The third term in Eq. (5.15) is therefore the sum of all such quantities:

$$
\frac{dS^t_{surr}}{dt} = -\sum_j \frac{\dot{Q}_j}{T_{\sigma,j}}
$$

Equation (5.15) is now written:

$$
\frac{d(mS)_{cv}}{dt} + \Delta(S\dot{m})_{fs} - \sum_j \frac{\dot{Q}_j}{T_{\sigma,j}} = \dot{S}_G \geq 0 \tag{5.16}
$$

The final term, representing the *rate of entropy generation* $\dot{S}_G$, reflects the second-law requirement that it be positive for irreversible processes. There are two sources of irreversibility: (*a*) those *within* the control volume, i.e., *internal* irreversibilities, and (*b*) those resulting from heat transfer across finite temperature differences between system and surroundings, i.e., *external* thermal irreversibilities. The limiting case for which $\dot{S}_G = 0$ applies when the process is *completely reversible,* implying:

- **The process is internally reversible within the control volume**

- **Heat transfer between the control volume and its surroundings is reversible**

The second item means either that heat reservoirs are included in the surroundings with temperatures equal to those of the control surface or that work-producing Carnot engines are interposed in the surroundings between the control-surface temperatures and the heat-reservoir temperatures.

For a *steady-state* flow process the mass and entropy of the fluid in the control volume are constant, and $d(mS)_{cv}/dt$ is zero. Equation (5.16) then becomes:

$$\Delta(\dot{S}m)_{\text{fs}} - \sum_j \frac{\dot{Q}_j}{T_{\sigma,j}} = \dot{S}_G \geq 0 \tag{5.17}$$

If in addition there is but one entrance and one exit, $\dot{m}$ is the same for both streams, and dividing through by $\dot{m}$ yields:

$$\Delta S - \sum_j \frac{Q_j}{T_{\sigma,j}} = S_G \geq 0 \tag{5.18}$$

Each term in Eq. (5.18) is based on a unit mass of fluid flowing through the control volume.

Example 5.5

In a steady-state flow process carried out at atmospheric pressure, 1 mol·s^{-1} of air at 600 K is continuously mixed with 2 mol·s^{-1} of air at 450 K. The product stream is at 400 K and 1 atm. A schematic representation of the process is shown in Fig. 5.4. Determine the rate of heat transfer and the rate of entropy generation for the process. Assume the ideal-gas state for air with $C_P^{ig} = (7/2)R$, that the surroundings are at 300 K, and that kinetic- and potential-energy changes of the streams are negligible.

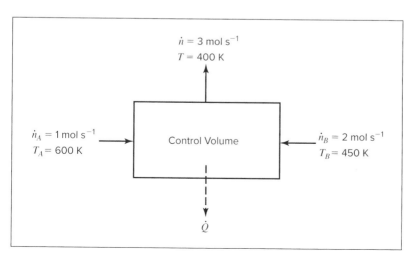

Figure 5.4: Process described in Ex. 5.5.

Solution 5.5

We start by applying an energy balance to determine the rate of heat transfer, which we will need to know in order to compute the rate of entropy generation. Writing the energy balance, Eq. (2.29), with $\dot{m}$ replaced by $\dot{n}$, and then replacing $\dot{n}$ with $\dot{n}_A + \dot{n}_B$,

$$\dot{Q} = \dot{n}H^{ig} - \dot{n}_A H_A^{ig} - \dot{n}_B H_B^{ig} = \dot{n}_A\left(H^{ig} - H_A^{ig}\right) + \dot{n}_B\left(H^{ig} - H_B^{ig}\right)$$

$$\dot{Q} = \dot{n}_A C_P^{ig}(T - T_A) + \dot{n}_B C_P^{ig}(T - T_B) = C_P^{ig}[\dot{n}_A(T - T_A) + \dot{n}_B(T - T_B)]$$
$$= (7/2)(8.314)[(1)(400 - 600) + (2)(400 - 450)] = -8729.7 \text{ J·s}^{-1}$$

The steady-state entropy balance, Eq. (5.17), with $\dot{m}$ again replaced by $\dot{n}$, can similarly be written as

$$\dot{S}_G = \dot{n}S^{ig} - \dot{n}_A S_A^{ig} - \dot{n}_B S_B^{ig} - \frac{\dot{Q}}{T_\sigma} = \dot{n}_A\left(S^{ig} - S_A^{ig}\right) + \dot{n}_B\left(S^{ig} - S_B^{ig}\right) - \frac{\dot{Q}}{T_\sigma}$$

$$= \dot{n}_A C_P^{ig}\ln\frac{T}{T_A} + \dot{n}_B C_P^{ig}\ln\frac{T}{T_B} - \frac{\dot{Q}}{T_\sigma} = C_P^{ig}\left(\dot{n}_A\ln\frac{T}{T_A} + \dot{n}_B\ln\frac{T}{T_B}\right) - \frac{\dot{Q}}{T_\sigma}$$

$$= (7/2)(8.314)\left[(1)\ln\frac{400}{600} + (2)\ln\frac{400}{450}\right] + \frac{8729.7}{300} = 10.446 \text{ J·K}^{-1}\text{·s}^{-1}$$

The rate of entropy generation is positive, as it must be for any real process.

Example 5.6

An inventor claims to have devised a process which takes in only saturated steam at 100°C and which by a complicated series of steps makes heat continuously available at a temperature level of 200°C, with 2000 kJ of energy available at 200°C for every kilogram of steam taken into the process. Show whether or not this process is possible. To give this process the most favorable conditions, assume cooling water available in unlimited quantity at a temperature of 0°C.

Solution 5.6

For any process to be even theoretically possible, it must meet the requirements of the first and second laws of thermodynamics. The detailed mechanism need not be known to determine whether this is true; only the overall result is required. If the claims of the inventor satisfy the laws of thermodynamics, fulfilling the claims is theoretically possible. The determination of a mechanism is then a matter of ingenuity. Otherwise, the process is impossible, and no mechanism for carrying it out can be devised.

In the present instance, a continuous process takes in saturated steam at 100°C, and makes heat Q continuously available at a temperature level of 200°C. Because cooling water is available at 0°C, the maximum possible use is made of the steam if it is condensed and cooled (without freezing) to this exit temperature and discharged at atmospheric pressure. This is a limiting case (most favorable to the inventor), and requires a heat-exchange surface of infinite area.

It is not possible in this process for heat to be liberated *only* at the 200°C temperature level, because as we have shown no process is possible that does nothing except transfer heat from one temperature level to a higher one. We must therefore suppose that some heat Q_σ is transferred to the cooling water at 0°C. Moreover, the process must satisfy the first law; thus by Eq. (2.32):

$$\Delta H = Q + Q_\sigma + W_s$$

where ΔH is the enthalpy change of the steam as it flows through the system. Because no shaft work accompanies the process, $W_s = 0$. The surroundings consist of cooling water, which acts as a heat reservoir at the constant temperature of $T_\sigma = 273.15$ K, and a heat reservoir at $T = 473.15$ K to which heat in the amount of 2000 kJ is transferred for each kilogram of steam entering the system. The diagram of Fig. 5.5 indicates overall results of the process based on the inventor's claim.

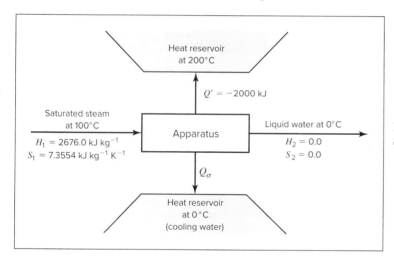

Figure 5.5: Process described in Ex. 5.6.

The values of H and S shown on Fig. 5.5 for liquid water at 0°C and for saturated steam at 100°C are taken from the steam tables (App. E). Note that the values for liquid water at 0°C are for *saturated* liquid ($P^{sat} = 0.61$ kPa), but the effect of an increase in pressure to atmospheric pressure is insignificant. On the basis of 1 kg of entering steam, the first law becomes:

$$\Delta H = H_2 - H_1 = 0.0 - 2676.0 = -2000 + Q_\sigma \qquad \text{and} \qquad Q_\sigma = -676.0 \text{ kJ}$$

The negative value for Q_σ indicates heat transfer is from the system to the cooling water.

We now examine this result in the light of the second law to determine whether the entropy generation is greater than or less than zero for the process. Equation (5.18) is here written:

$$\Delta S - \frac{Q_\sigma}{T_\sigma} - \frac{Q}{473.15} = S_G$$

For 1 kg of steam,

$$\Delta S = S_2 - S_1 = 0.0000 - 7.3554 = -7.3554 \text{ kJ·K}^{-1}$$

The entropy generation is then:

$$S_G = -7.3554 - \frac{-676.0}{273.15} - \frac{-2000}{473.15}$$

$$= -7.3554 + 4.2270 + 2.4748 = -0.6536 \text{ kJ·K}^{-1}$$

This negative result means that the process as described is impossible; the second law in the form of Eq. (5.18) requires $S_G \geq 0$.

The negative result of the preceding example does not mean that all processes of this general nature are impossible; only that the inventor claimed too much. Indeed, the maximum amount of heat which can be transferred to the hot reservoir at 200°C is readily calculated. The energy balance is:

$$Q + Q_\sigma = \Delta H \tag{A}$$

The maximum heat transfer to the hot reservoir occurs when the process is completely reversible, in which case $S_G = 0$, and Eq. (5.18) becomes

$$\frac{Q}{T} + \frac{Q_\sigma}{T_\sigma} = \Delta S \tag{B}$$

Combination of Eqs. (A) and (B) and solution for Q yields:

$$Q = \frac{T}{T - T_\sigma}(\Delta H - T_\sigma \Delta S)$$

With respect to the preceding example,

$$Q = \frac{473.15}{200}\left[-2676.0 - (273.15)(-7.3554)\right] = -1577.7 \text{ kJ·kg}^{-1}$$

This value is *smaller* in magnitude than the -2000 kJ·kg^{-1} claimed.

5.7   CALCULATION OF IDEAL WORK

In any steady-state flow process requiring work, an absolute minimum amount of work is required to produce the desired change of state in the fluid flowing through the control volume. In a process producing work, an absolute maximum amount of work can be realized from a given change of state of the fluid flowing through the control volume. In either case, the limiting value results when the change of state associated with the process is accomplished *completely reversibly*. For such a process, the entropy generation is zero, and Eq. (5.17), written for a uniform surroundings temperature T_σ, becomes:

$$\Delta(S\dot{m})_{\text{fs}} - \frac{\dot{Q}}{T_\sigma} = 0 \qquad \text{or} \qquad \dot{Q} = T_\sigma \Delta(S\dot{m})_{\text{fs}}$$

Substituting this expression for $\dot{Q}$ in the energy balance, Eq. (2.29):

$$\Delta\left[\left(H + \tfrac{1}{2}u^2 + zg\right)\dot{m}\right]_{fs} = T_\sigma\Delta(S\dot{m})_{fs} + \dot{W}_s(rev)$$

The shaft work, $\dot{W}_s(rev)$, is here the work of a completely reversible process. If $\dot{W}_s(rev)$ is given the name *ideal work*, $\dot{W}_{ideal}$, the preceding equation may be rewritten:

$$\dot{W}_{ideal} = \Delta\left[\left(H + \tfrac{1}{2}u^2 + zg\right)\dot{m}\right]_{fs} - T_\sigma\Delta(S\dot{m})_{fs} \tag{5.19}$$

In most applications to chemical processes, the kinetic- and potential-energy terms are negligible compared with the others; in this event Eq. (5.19) reduces to:

$$\boxed{\dot{W}_{ideal} = \Delta(H\dot{m})_{fs} - T_\sigma\Delta(S\dot{m})_{fs}} \tag{5.20}$$

For the special case of a single stream flowing through the control volume, $\dot{m}$ may be factored. The resulting equation may then be divided by $\dot{m}$ to express it on the basis of a unit amount of fluid flowing through the control volume. Thus,

$$\boxed{\dot{W}_{ideal} = \dot{m}(\Delta H - T_\sigma\Delta S) \quad (5.21) \quad \Big| \quad W_{ideal} = \Delta H - T_\sigma\Delta S \quad (5.22)}$$

A completely reversible process is hypothetical, devised here solely for determination of the ideal work associated with a given change of state.

> **The sole connection between an actual process and an imagined hypothetical reversible process employed for determining ideal work is that they both apply to the same *changes of state*.**

Our objective is to compare the actual work of a process $\dot{W}_s$ (or W_s) as given by an energy balance to the ideal work as given by Eqs. (5.19) through (5.22) for a hypothetical reversible process that produces the same property changes. No description of the hypothetical process is required, as it may always be imagined (see Ex. 5.7).

When $\dot{W}_{ideal}$ (or W_{ideal}) is positive, it is the *minimum work required* to bring about a given change in the properties of the flowing streams, and is smaller than $\dot{W}_s$. In this case a thermodynamic efficiency η_t is defined as the ratio of the ideal work to the actual work:

$$\eta_t(\text{work required}) = \frac{\dot{W}_{ideal}}{\dot{W}_s} \tag{5.23}$$

When $\dot{W}_{ideal}$ (or W_{ideal}) is negative, $|\dot{W}_{ideal}|$ is the *maximum work obtainable* from a given change in the properties of the flowing streams, and is larger than $|\dot{W}_s|$. In this case, the thermodynamic efficiency is defined as the ratio of the actual work to the ideal work:

$$\eta_t(\text{work produced}) = \frac{\dot{W}_s}{\dot{W}_{ideal}} \tag{5.24}$$

Example 5.7

What is the maximum work that can be obtained in a steady-state flow process from 1 mol of nitrogen in its ideal-gas state at 800 K and 50 bar? Take the temperature and pressure of the surroundings as 300 K and 1.0133 bar.

Solution 5.7

The maximum possible work is obtained from any completely reversible process that reduces the nitrogen to the temperature and pressure of the surroundings, i.e., to 300 K and 1.0133 bar. Required here is the calculation of W_{ideal} by Eq. (5.22), in which ΔS and ΔH are the molar entropy and enthalpy changes of the nitrogen as its state changes from 800 K and 50 bar to 300 K and 1.0133 bar. For the ideal-gas state, enthalpy is independent of pressure, and its change is given by:

$$\Delta H^{ig} = \int_{T_1}^{T_2} C_P^{ig} dT$$

The value of this integral is found from Eq. (4.8), and is represented by:

$$8.314 \times \text{ICPH}(800, 300; 3.280, 0.593 \times 10^{-3}, 0.0, 0.040 \times 10^{-5}) = -15{,}060 \text{ J·mol}^{-1}$$

where the parameters for nitrogen come from Table C.1 of App. C.
Similarly, the entropy change is found from Eq. (5.10), here written:

$$\Delta S^{ig} = \int_{T_1}^{T_2} C_P^{ig} \frac{dT}{T} - R \ln \frac{P_2}{P_1}$$

The value of the integral, found from Eq. (5.11), is represented by:

$$8.314 \times \text{ICPS}(800, 300; 3.280, 0.593 \times 10^{-3}, 0.0, 0.040 \times 10^{-5}) = -29.373 \text{ J·mol}^{-1}\cdot\text{K}^{-1}$$

Whence, $\Delta S^{ig} = -29.373 - 8.314 \ln \dfrac{1.0133}{50} = 3.042 \text{ J·mol}^{-1}\cdot\text{K}^{-1}$

With these values of ΔH^{ig} and ΔS^{ig}, Eq. (5.22) becomes:

$$W_{\text{ideal}} = -15{,}060 - (300)(3.042) = -15{,}973 \text{ J·mol}^{-1}$$

One can easily devise a specific reversible process to bring about the given change of state of the preceding example. Suppose the nitrogen is continuously changed to its final state at 1.0133 bar and $T_2 = T_\sigma = 300$ K by the following two-step steady-flow process:

- **Step 1:** Reversible, adiabatic expansion (as in a turbine) from initial state P_1, T_1, H_1 to 1.0133 bar. Let T' denote the discharge temperature.

- **Step 2:** Cooling (or heating, if T' is less than T_2) to the final temperature T_2 at a constant pressure of 1.0133 bar.

For step 1, with $Q = 0$, the energy balance is $W_s = \Delta H = (H' - H_1)$, where H' is the enthalpy at the intermediate state of T' and 1.0133 bar.

For maximum work production, step 2 must also be reversible, with heat exchanged reversibly with the surroundings at T_σ. This requirement is satisfied when Carnot engines take heat from the nitrogen, produce work W_{Carnot}, and reject heat to the surroundings at T_σ. Because the temperature of the nitrogen decreases from T' to T_2, Eq. (5.5) for the work of a Carnot engine is written in differential form:

$$dW_{Carnot} = \left(\frac{T_\sigma}{T} - 1\right)(-dQ) = \left(1 - \frac{T_\sigma}{T}\right)dQ$$

Here dQ refers to the nitrogen, which is taken as the system, rather than to the engine. Integration yields:

$$W_{Carnot} = Q - T_\sigma \int_{T'}^{T_2} \frac{dQ}{T}$$

The first term on the right is the heat transferred with respect to the nitrogen, given by $Q = H_2 - H'$. The integral is the change in entropy of the nitrogen as it is cooled by the Carnot engines. Because step 1 occurs at constant entropy, the integral also represents ΔS for both steps. Hence,

$$W_{Carnot} = (H_2 - H') - T_\sigma \Delta S$$

The sum of W_s and W_{Carnot} gives the ideal work; thus,

$$W_{ideal} = (H' - H_1) + (H_2 - H') - T_\sigma \Delta S = (H_2 - H_1) - T_\sigma \Delta S$$

or

$$W_{ideal} = \Delta H - T_\sigma \Delta S$$

which is the same as Eq. (5.22).

This derivation makes clear the difference between W_s, the reversible adiabatic shaft work of the turbine, and W_{ideal}. The ideal work includes not only W_s, but also all work obtainable from Carnot engines for the reversible exchange of heat with the surroundings at T_σ. In practice, work produced by a turbine can be as much as 80% of the reversible adiabatic work, but usually no mechanism is present for extraction of W_{Carnot}.[5]

Example 5.8

Rework Ex. 5.6, making use of the equation for ideal work.

Solution 5.8

The procedure here is to calculate the maximum possible work W_{ideal} which can be obtained from 1 kg of steam in a flow process as it undergoes a change in state from saturated steam at 100°C to liquid water at 0°C. The problem then reduces to the question of whether this amount of work is sufficient to operate a Carnot

---

[5]A *cogeneration plant* produces power both from a gas turbine and from a steam turbine operating on steam generated by heat from the gas-turbine exhaust.

refrigerator rejecting 2000 kJ as heat at 200°C and taking heat from the unlimited supply of cooling water at 0°C. From Ex. 5.6, we have

$$\Delta H = -2676.0 \text{ kJ·kg}^{-1} \quad \text{and} \quad \Delta S = -7.3554 \text{ kJ·K}^{-1}\text{·kg}^{-1}$$

With negligible kinetic- and potential-energy terms, Eq. (5.22) yields:

$$W_{\text{ideal}} = \Delta H - T_\sigma \Delta S = -2676.0 - (273.15)(-7.3554) = -666.9 \text{ kJ·kg}^{-1}$$

If this amount of work, numerically the maximum obtainable from the steam, is used to drive the Carnot refrigerator operating between the temperatures of 0°C and 200°C, the heat rejected is found from Eq. (5.5), solved for Q_H:

$$Q_H = W_{\text{ideal}} \left(\frac{T}{T_\sigma - T} \right) = 666.9 \left(\frac{200 + 273.15}{200 - 0} \right) = 1577.7 \text{ kJ}$$

As calculated in Ex. 5.6, this is the maximum possible heat release at 200°C; it is less than the claimed value of 2000 kJ. As in Ex. 5.6, we conclude that the process as described is not possible.

5.8 LOST WORK

Work that is wasted as the result of irreversibilities in a process is called *lost work*, W_{lost}, and is defined as the difference between the actual work of a change of state and the ideal work for the same change of state. Thus by definition,

$$W_{\text{lost}} \equiv W_s - W_{\text{ideal}} \tag{5.25}$$

In terms of rates,

$$\dot{W}_{\text{lost}} \equiv \dot{W}_s - \dot{W}_{\text{ideal}} \tag{5.26}$$

The actual work rate comes from the energy balance, Eq. (2.29), and the ideal work rate is obtained using Eq. (5.19):

$$\dot{W}_s = \Delta \left[\left(H + \frac{1}{2}u^2 + zg \right)\dot{m} \right]_{\text{fs}} - \dot{Q}$$

$$\dot{W}_{\text{ideal}} = \Delta \left[\left(H + \frac{1}{2}u^2 + zg \right)\dot{m} \right]_{\text{fs}} - T_\sigma \Delta(S\dot{m})_{\text{fs}}$$

By difference, as given by Eq. (5.26),

$$\boxed{\dot{W}_{\text{lost}} = T_\sigma \Delta(S\dot{m})_{\text{fs}} - \dot{Q}} \tag{5.27}$$

For the case of a single surroundings temperature T_σ, the steady-state entropy balance, Eq. (5.17), becomes:

$$\dot{S}_G = \Delta(S\dot{m})_{\text{fs}} - \frac{\dot{Q}}{T_\sigma} \tag{5.28}$$

Multiplication by T_σ gives: $T_\sigma \dot{S}_G = T_\sigma \Delta (S\dot{m})_{\text{fs}} - \dot{Q}$

The right sides of this equation and Eq. (5.27) are identical; therefore,

$$\dot{W}_{\text{lost}} = T_\sigma \dot{S}_G \quad , \tag{5.29}$$

As a consequence of the second law, $\dot{S}_G \geq 0$; it follows that $\dot{W}_{\text{lost}} \geq 0$. When a process is completely reversible, the equality holds, and $\dot{W}_{\text{lost}} = 0$. For irreversible processes the inequality holds, and $\dot{W}_{\text{lost}}$, i.e., the energy that becomes unavailable for work, is positive.

The engineering significance of this result is clear: The greater the irreversibility of a process, the greater the rate of entropy production and the greater the amount of energy made unavailable for work. Thus every irreversibility carries with it a price. Minimizing entropy production is essential for conservation of the earth's resources.

For the special case of a single stream flowing through the control volume, $\dot{m}$ factors and becomes a multiplier of the entropy difference in Eqs. (5.27) and (5.28). Division by $\dot{m}$ converts all terms to the basis of a unit amount of fluid flowing through the control volume. Thus,

$\dot{W}_{\text{lost}} = \dot{m} T_\sigma \Delta S - \dot{Q}$ (5.30)	$W_{\text{lost}} = T_\sigma \Delta S - Q$ (5.31)
$\dot{S}_G = \dot{m} \Delta S - \dfrac{Q}{T_\sigma}$ (5.32)	$S_G = \Delta S - \dfrac{Q}{T_\sigma}$ (5.33)

Recall here that $\dot{Q}$ and Q represent heat exchange with the surroundings, but with the sign referred to the system. Equations (5.31) and (5.33) combine for a unit amount of fluid to give:

$$W_{\text{lost}} = T_\sigma S_G \tag{5.34}$$

Again, because $S_G \geq 0$, it follows that $W_{\text{lost}} \geq 0$.

Example 5.9

The two basic types of steady-flow heat exchanger are characterized by their flow patterns: *cocurrent* and *countercurrent*. Temperature profiles for the two types are indicated in Fig. 5.6. In cocurrent flow, heat is transferred from a hot stream, flowing from left to right, to a cold stream flowing in the same direction, as indicated by arrows. In countercurrent flow, the cold stream, again flowing from left to right, receives heat from the hot stream flowing in the opposite direction.

The lines relate the temperatures of the hot and cold streams, T_H and T_C respectively, to $\dot{Q}_C$, the accumulated rate of heat addition to the cold stream as it progresses through the exchanger from the left end to an arbitrary downstream location. The following specifications apply to both cases:

$$T_{H_1} = 400 \text{ K} \qquad T_{H_2} = 350 \text{ K} \qquad T_{C_1} = 300 \text{ K} \qquad \dot{n}_H = 1 \text{ mol·s}^{-1}$$

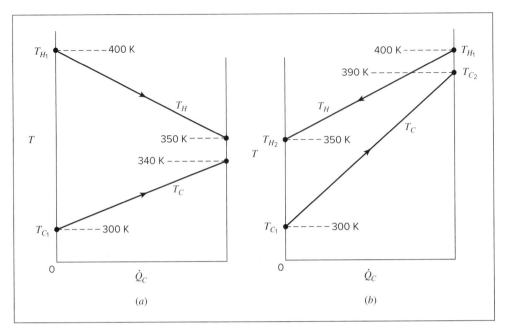

Figure 5.6: Heat exchangers. (*a*) Case I, cocurrent. (*b*) Case II, countercurrent.

The minimum temperature difference between the flowing streams is 10 K. Assume the ideal-gas state for both streams with $C_P = (7/2)R$. Find the lost work for both cases. Take $T_\sigma = 300$ K.

Solution 5.9

With the assumption of negligible kinetic- and potential-energy changes, with $\dot{W}_s = 0$, and with $\dot{Q} = 0$ (there is no heat exchange with the surroundings) the energy balance [Eq. (2.29)] can be written:

$$\dot{n}_H(\Delta H^{ig})_H + \dot{n}_C(\Delta H^{ig})_C = 0$$

With constant molar heat capacity, this becomes:

$$\dot{n}_H C_P^{ig}\left(T_{H_2} - T_{H_1}\right) + \dot{n}_C C_P^{ig}\left(T_{C_2} - T_{C_1}\right) = 0 \tag{A}$$

The total rate of entropy change for the flowing streams is:

$$\Delta\left(S^{ig}\dot{n}\right)_{fs} = \dot{n}_H(\Delta S^{ig})_H + \dot{n}_C(\Delta S^{ig})_C$$

By Eq. (5.10), with the assumption of negligible pressure change in the flowing streams,

$$\Delta\left(S^{ig}\dot{n}\right)_{fs} = \dot{n}_H C_P^{ig}\left(\ln\frac{T_{H_2}}{T_{H_1}} + \frac{\dot{n}_C}{\dot{n}_H}\ln\frac{T_{C_2}}{T_{C_1}}\right) \tag{B}$$

Finally, by Eq. (5.27), with $\dot{Q} = 0$,

$$\dot{W}_{\text{lost}} = T_\sigma \Delta \left(S^{ig}\dot{n} \right)_{\text{fs}} \tag{C}$$

These equations apply to both cases.

- **Case I:** Cocurrent flow. By Eqs. (*A*), (*B*), and (*C*), respectively:

$$\frac{\dot{n}_C}{\dot{n}_H} = \frac{400 - 350}{340 - 300} = 1.25$$

$$\Delta (S^{ig}\dot{n})_{\text{fs}} = (1)(7/2)(8.314)\left(\ln \frac{350}{400} + 1.25 \ln \frac{340}{300} \right) = 0.667 \ \text{J·K}^{-1}\text{·s}^{-1}$$

$$\dot{W}_{\text{lost}} = (300)(0.667) = 200.1 \ \text{J·s}^{-1}$$

- **Case II:** Countercurrent flow. By Eqs. (*A*), (*B*), and (*C*), respectively:

$$\frac{\dot{n}_C}{\dot{n}_H} = \frac{400 - 350}{390 - 300} = 0.5556$$

$$\Delta (S^{ig}\dot{n})_{\text{fs}} = (1)(7/2)(8.314)\left(\ln \frac{350}{400} + 0.5556 \ln \frac{390}{300} \right) = 0.356 \ \text{J·K}^{-1}\text{·s}^{-1}$$

$$\dot{W}_{\text{lost}} = (300)(0.356) = 106.7 \ \text{J·s}^{-1}$$

Although the total rate of heat transfer is the same for both exchangers, the temperature rise of the cold stream in countercurrent flow is more than twice that for cocurrent flow. Thus, its entropy increase per unit mass is larger than for the cocurrent case. However, its flow rate is less than half that of the cold stream in cocurrent flow, so that the total entropy increase of the cold stream is less for countercurrent flow. From a thermodynamic point of view, the countercurrent case is much more efficient. Because $\Delta(S^{ig}\dot{n})_{\text{fs}} = \dot{S}_G$, both the rate of entropy generation and the lost work for the countercurrent case are only about half the value for the cocurrent case. Greater efficiency in the countercurrent case would be anticipated based on Fig. 5.6, which shows that heat is transferred across a smaller temperature difference (less irreversibly) in the countercurrent case.

5.9 THE THIRD LAW OF THERMODYNAMICS

Measurements of heat capacities at very low temperatures provide data for the calculation from Eq. (5.1) of entropy changes down to 0 K. When these calculations are made for different crystalline forms of the same chemical species, the entropy at 0 K appears to be the same for all forms. When the form is noncrystalline, e.g., amorphous or glassy, calculations show that the entropy of the disordered form is greater than that of the crystalline form. Such calculations, which are summarized elsewhere,[6] lead to the postulate that *the absolute entropy is zero for all perfect crystalline substances at absolute zero temperature.* While this essential idea was advanced by Nernst and

---

[6]K. S. Pitzer, *Thermodynamics*, 3d ed., chap. 6, McGraw-Hill, New York, 1995.

Planck at the beginning of the twentieth century, more recent studies at very low temperatures have increased confidence in this postulate, which is now accepted as the third law of thermodynamics.

If the entropy is zero at $T = 0$ K, then Eq. (5.1) lends itself to the calculation of absolute entropies. With $T = 0$ as the lower limit of integration, the absolute entropy of a gas at temperature T based on calorimetric data is:

$$S = \int_0^{T_f} \frac{(C_P)_s}{T} dT + \frac{\Delta H_f}{T_f} + \int_{T_f}^{T_v} \frac{(C_P)_l}{T} dT + \frac{\Delta H_v}{T_v} + \int_{T_v}^{T} \frac{(C_P)_g}{T} dT \qquad (5.35)$$

This equation[7] is based on the supposition that no solid-state transitions take place and thus no heats of transition need appear. The only constant-temperature heat effects are those of melting at T_f and vaporization at T_v. When a solid-phase transition occurs, a term $\Delta H_t / T_t$ is added.

Note that although the third law implies that absolute values of entropy are obtainable, for most thermodynamic analyses, only relative values are needed. As a result, reference states other than the perfect crystal at 0 K are commonly used. For example, in the steam tables of App. E, saturated liquid water at 273.16 K is taken as the reference state and assigned zero entropy. However, the absolute or "third law" entropy of saturated liquid water at 273.16 K is 3.515 kJ·kg^{-1}·K^{-1}.

5.10   ENTROPY FROM THE MICROSCOPIC VIEWPOINT

Because molecules in the ideal-gas state do not interact, their internal energy resides with individual molecules. This is not true of the entropy, which is inherently a collective property of a large number of molecules or other entities. The microscopic interpretation of entropy is suggested by the following example.

Suppose an insulated container, partitioned into two equal volumes, contains Avogadro's number N_A of molecules in one section and no molecules in the other. When the partition is withdrawn, the molecules quickly distribute themselves uniformly throughout the total volume. The process is an adiabatic expansion that accomplishes no work. Therefore,

$$\Delta U = C_V^{ig} \Delta T = 0$$

and the temperature does not change. However, the pressure of the gas decreases by half, and the entropy change, as given by Eq. (5.10), is:

$$\Delta S^{ig} = -R \ln \frac{P_2}{P_1} = R \ln 2$$

Because this is the total entropy change, the process is clearly irreversible.

At the instant when the partition is removed, the molecules occupy only half the space available to them. In this momentary initial state, the molecules are not randomly distributed over the total volume to which they have access but are crowded into just half the total volume. In this sense they are more ordered than they are in the final state of uniform distribution throughout the entire volume. Thus, the final state can be regarded as a more random, or more disordered, state than the initial state. Generalizing from this example, and from many other

---

[7]Evaluation of the first term on the right is not a problem for crystalline substances, because C_P/T remains finite as $T \to 0$.

similar observations, one is led to the notion that increasing disorder (or decreasing structure) at the molecular level corresponds to increasing entropy.

The means for expressing disorder in a quantitative way was developed by L. Boltzmann and J. W. Gibbs through a quantity Ω, defined as the *number of different ways* that microscopic particles can be distributed among the "states" accessible to them. It is given by the general formula:

$$\Omega = \frac{N!}{(N_1!)(N_2!)(N_3!)\ldots} \tag{5.36}$$

where N is the total number of particles, and N_1, N_2, N_3, etc., represent the numbers of particles in "states" 1, 2, 3, etc. The term "state" denotes a condition of the microscopic particles, and the quotation marks distinguish this idea of state from the usual thermodynamic meaning as applied to a macroscopic system.

With respect to our example there are but two "states" for each molecule, representing location of the given molecule in one half or the other of the container. The total number of particles is N_A molecules, and initially they are all in a single "state." Thus

$$\Omega_1 = \frac{N_A!}{(N_A!)(0!)} = 1$$

This result confirms that initially the molecules can be distributed between the two accessible "states" in just one way. They are all in a given "state," all in just one half of the container. For an assumed final condition of uniform distribution of the molecules between the two halves of the container, $N_1 = N_2 = N_A/2$, and

$$\Omega_2 = \frac{N_A!}{\left[(N_A/2)!\right]^2}$$

This expression gives a very large number for Ω_2, indicating that the molecules can be distributed uniformly between the two "states" in many different ways. Many other values of Ω_2 are possible, each one of which is associated with a particular *nonuniform* distribution of the molecules between the two halves of the container. The ratio of a particular Ω_2 to the sum of all possible values is the probability of that particular distribution.

The connection established by Boltzmann between entropy S and Ω is given by the equation:

$$S = k \ln\Omega \tag{5.37}$$

Boltzmann's constant k equals R/N_A. The entropy difference between states 1 and 2 is:

$$S_2 - S_1 = k \ln\frac{\Omega_2}{\Omega_1}$$

Substituting values for Ω_1 and Ω_2 from our example into this expression gives:

$$S_2 - S_1 = k \ln \frac{N_A!}{\left[(N_A/2)!\right]^2} = k[\ln N_A! - 2 \ln (N_A/2)!]$$

Because N_A is very large, we take advantage of Stirling's formula for the logarithms of factorials of large numbers:

$$\ln X! = X \ln X - X$$

Whence,

$$S_2 - S_1 = k \left[N_A \ln N_A - N_A - 2 \left(\frac{N_A}{2} \ln \frac{N_A}{2} - \frac{N_A}{2} \right) \right]$$

$$= kN_A \ln \frac{N_A}{N_A/2} = kN_A \ln 2 = R \ln 2$$

This value for the entropy change of the expansion process is the same as that given by Eq. (5.10), the classical thermodynamic formula for the ideal-gas state.

Equations (5.36) and (5.37) provide a basis for relating macroscopic thermodynamic properties to the microscopic configurations of molecules. Here we have applied them to a very simple, and somewhat contrived, situation involving molecules in the ideal gas state, in order to provide a simple illustration of this connection between molecular configurations and macroscopic properties. The field of science and engineering devoted to studying and exploiting this connection is called *statistical thermodynamics* or *statistical mechanics.* The methods of statistical thermodynamics are well developed, and are now routinely applied, in combination with computational simulation of the behavior of molecules, to make useful predictions of thermodynamic properties of real substances without recourse to experiment.[8]

5.11   SYNOPSIS

After studying this chapter, including the end-of-chapter problems, one should be able to:

- Comprehend the existence of entropy as a state function, related to observable properties of a system, for which changes are computed from:

$$dS^t = dQ_{rev}/T \tag{5.1}$$

- State the second law of thermodynamics in words and as an inequality in terms of entropy.

- Define and distinguish between thermal efficiency and thermodynamic efficiency.

- Compute the thermal efficiency of a reversible heat engine.

- Compute entropy changes for the ideal-gas state with heat capacity expressed as a polynomial in temperature.

- Construct and apply entropy balances for open systems.

- Determine whether a specified process violates the second law.

- Compute ideal work, lost work, and thermodynamic efficiencies of processes.

[8]Many introductory texts on statistical thermodynamics are available. The interested reader is referred to *Molecular Driving Forces: Statistical Thermodynamics in Chemistry & Biology,* by K.A. Dill and S. Bromberg, Garland Science, 2010, and many books referenced therein.

5.12 PROBLEMS

5.1. Prove that it is impossible for two lines representing reversible, adiabatic processes on a *PV* diagram to intersect. (*Hint:* Assume that they do intersect, and complete the cycle with a line representing a reversible, isothermal process. Show that performance of this cycle violates the second law.)

5.2. A Carnot engine receives 250 kJ·s^{-1} of heat from a heat-source reservoir at 525°C and rejects heat to a heat-sink reservoir at 50°C. What are the power developed and the heat rejected?

5.3. The following heat engines produce power of 95,000 kW. Determine in each case the rates at which heat is absorbed from the hot reservoir and discarded to the cold reservoir.

(*a*) A Carnot engine operates between heat reservoirs at 750 K and 300 K.
(*b*) A practical engine operates between the same heat reservoirs but with a thermal efficiency $\eta = 0.35$.

5.4. A particular power plant operates with a heat-source reservoir at 350°C and a heat-sink reservoir at 30°C. It has a thermal efficiency equal to 55% of the Carnot-engine thermal efficiency for the same temperatures.

(*a*) What is the thermal efficiency of the plant?
(*b*) To what temperature must the heat-source reservoir be raised to increase the thermal efficiency of the plant to 35%? Again η is 55% of the Carnot-engine value.

5.5. An egg, initially at rest, is dropped onto a concrete surface; it breaks. Prove that the process is irreversible. In modeling this process treat the egg as the system, and assume the passage of sufficient time for the egg to return to its initial temperature.

5.6. Which is the more effective way to increase the thermal efficiency of a Carnot engine: to increase T_H with T_C constant, or to decrease T_C with T_H constant? For a real engine, which would be the more *practical* way?

5.7. Large quantities of liquefied natural gas (LNG) are shipped by ocean tanker. At the unloading port, provision is made for vaporization of the LNG so that it may be delivered to pipelines as gas. The LNG arrives in the tanker at atmospheric pressure and 113.7 K, and represents a possible heat sink for use as the cold reservoir of a heat engine. For unloading of LNG as a vapor at the rate of 9000 m^3·s^{-1}, as measured at 25°C and 1.0133 bar, and assuming the availability of an adequate heat source at 30°C, what is the maximum possible power obtainable and what is the rate of heat transfer from the heat source? Assume that LNG at 25°C and 1.0133 bar is an ideal gas with the molar mass of 17. Also assume that the LNG vaporizes only, absorbing only its latent heat of 512 kJ·kg^{-1} at 113.7 K.

5.8. With respect to 1 kg of liquid water:

(a) Initially at 0°C, it is heated to 100°C by contact with a heat reservoir at 100°C. What is the entropy change of the water? Of the heat reservoir? What is ΔS_{total}?

(b) Initially at 0°C, it is first heated to 50°C by contact with a heat reservoir at 50°C and then to 100°C by contact with a reservoir at 100°C. What is ΔS_{total}?

(c) Explain how the water might be heated from 0°C to 100°C so that $\Delta S_{\text{total}} = 0$.

5.9. A rigid vessel of 0.06 m³ volume contains an ideal gas, $C_V = (5/2)R$, at 500 K and 1 bar.

(a) If heat in the amount of 15,000 J is transferred to the gas, determine its entropy change.

(b) If the vessel is fitted with a stirrer that is rotated by a shaft so that work in the amount of 15,000 J is done on the gas, what is the entropy change of the gas if the process is adiabatic? What is ΔS_{total}? What is the irreversible feature of the process?

5.10. An ideal gas, $C_P = (7/2)R$, is heated in a steady-flow heat exchanger from 70°C to 190°C by another stream of the same ideal gas which enters at 320°C. The flow rates of the two streams are the same, and heat losses from the exchanger are negligible.

(a) Calculate the molar entropy changes of the two gas streams for both parallel and countercurrent flow in the exchanger.

(b) What is ΔS_{total} in each case?

(c) Repeat parts (a) and (b) for countercurrent flow if the heating stream enters at 200°C.

5.11. For an ideal gas with constant heat capacities, show that:

(a) For a temperature change from T_1 to T_2, ΔS of the gas is greater when the change occurs at constant pressure than when it occurs at constant volume.

(b) For a pressure change from P_1 to P_2, the sign of ΔS for an isothermal change is opposite that for a constant-volume change.

5.12. For an ideal gas prove that:

$$\frac{\Delta S}{R} = \int_{T_0}^{T} \frac{C_V^{ig}}{R} \frac{dT}{T} + \ln \frac{V}{V_0}$$

5.13. A Carnot engine operates between two *finite* heat reservoirs of total heat capacity C_H^t and C_C^t.

(a) Develop an expression relating T_C to T_H at any time.

(b) Determine an expression for the work obtained as a function of C_H^t, C_C^t, T_H, and the initial temperatures T_{H_0} and T_{C_0}.

(c) What is the *maximum* work obtainable? This corresponds to infinite time, when the reservoirs attain the same temperature.

In approaching this problem, use the differential form of Carnot's equation,

$$\frac{dQ_H}{dQ_C} = -\frac{T_H}{T_C}$$

and a differential energy balance for the engine,

$$dW - dQ_C - dQ_H = 0$$

Here, Q_C and Q_H refer to the *reservoirs*.

5.14. A Carnot engine operates between an infinite hot reservoir and a *finite* cold reservoir of total heat capacity C_C^t.

(a) Determine an expression for the work obtained as a function of C_C^t, T_H (= constant), T_C, and the initial cold-reservoir temperature T_{C_0}.

(b) What is the *maximum* work obtainable? This corresponds to infinite time, when T_C becomes equal to T_H.

The approach to this problem is the same as for Prob. 5.13.

5.15. A heat engine operating in outer space may be assumed equivalent to a Carnot engine operating between reservoirs at temperatures T_H and T_C. The only way heat can be discarded from the engine is by radiation, the rate of which is given (approximately) by:

$$|\dot{Q}_C| = kAT_C^4$$

where k is a constant and A is the area of the radiator. Prove that, for fixed power output $|\dot{W}|$ and for fixed temperature T_H, the radiator area A is a minimum when the temperature ratio T_C/T_H is 0.75.

5.16. Imagine that a stream of fluid in steady-state flow serves as a heat source for an infinite set of Carnot engines, each of which absorbs a differential amount of heat from the fluid, causing its temperature to decrease by a differential amount, and each of which rejects a differential amount of heat to a heat reservoir at temperature T_σ. As a result of the operation of the Carnot engines, the temperature of the fluid decreases from T_1 to T_2. Equation (5.8) applies here in differential form, wherein η is defined as:

$$\eta \equiv dW/dQ$$

where Q is heat transfer with respect to the flowing fluid. Show that the total work of the Carnot engines is given by:

$$W = Q - T_\sigma \Delta S$$

where ΔS and Q both refer to the fluid. In a particular case, the fluid is an ideal gas, with $C_P = (7/2)R$, and the operating temperatures are $T_1 = 600$ K and $T_2 = 400$ K. If $T_\sigma = 300$ K, what is the value of W in J·mol^{-1}? How much heat is discarded to the heat reservoir at T_σ? What is the entropy change of the heat reservoir? What is ΔS_{total}?

5.17. A Carnot engine operates between temperature levels of 600 K and 300 K. It drives a Carnot refrigerator, which provides cooling at 250 K and discards heat at 300 K. Determine a numerical value for the ratio of heat extracted by the refrigerator ("cooling load") to the heat delivered to the engine ("heating load").

5.18. An ideal gas with constant heat capacity undergoes a change of state from conditions T_1, P_1 to conditions T_2, P_2. Determine ΔH (J·mol^{-1}) and ΔS (J·mol^{-1}·K^{-1}) for one of the following cases.

(a) $T_1 = 300$ K, $P_1 = 1.2$ bar, $T_2 = 450$ K, $P_2 = 6$ bar, $C_P/R = 7/2$.
(b) $T_1 = 300$ K, $P_1 = 1.2$ bar, $T_2 = 500$ K, $P_2 = 6$ bar, $C_P/R = 7/2$.
(c) $T_1 = 450$ K, $P_1 = 10$ bar, $T_2 = 300$ K, $P_2 = 2$ bar, $C_P/R = 5/2$.
(d) $T_1 = 400$ K, $P_1 = 6$ bar, $T_2 = 300$ K, $P_2 = 1.2$ bar, $C_P/R = 9/2$.
(e) $T_1 = 500$ K, $P_1 = 6$ bar, $T_2 = 300$ K, $P_2 = 1.2$ bar, $C_P/R = 4$.

5.19. An ideal gas, $C_P = (7/2)R$ and $C_V = (5/2)R$, undergoes a cycle consisting of the following mechanically reversible steps:

- An adiabatic compression from P_1, V_1, T_1 to P_2, V_2, T_2.
- An isobaric expansion from P_2, V_2, T_2 to $P_3 = P_2$, V_3, T_3.
- An adiabatic expansion from P_3, V_3, T_3 to P_4, V_4, T_4.
- A constant-volume process from P_4, V_4, T_4 to P_1, $V_1 = V_4$, T_1.

Sketch this cycle on a PV diagram and determine its thermal efficiency if $T_1 = 200°C$, $T_2 = 1000°C$, and $T_3 = 1700°C$.

5.20. The infinite heat reservoir is an abstraction, often approximated in engineering applications by large bodies of air or water. Apply the closed-system form of the energy balance [Eq. (2.3)] to such a reservoir, treating it as a constant-volume system. How is it that heat transfer to or from the reservoir can be nonzero, yet the temperature of the reservoir remains constant?

5.21. One mole of an ideal gas, $C_P = (7/2)R$ and $C_V = (5/2)R$, is compressed adiabatically in a piston/cylinder device from 2 bar and 25°C to 7 bar. The process is irreversible and requires 35% more work than a reversible, adiabatic compression from the same initial state to the same final pressure. What is the entropy change of the gas?

5.22. A mass m of liquid water at temperature T_1 is mixed adiabatically and isobarically with an equal mass of liquid water at temperature T_2. Assuming constant C_P, show

$$\Delta S^t = \Delta S_{\text{total}} = S_G = 2mC_P \ln \frac{(T_1 + T_2)/2}{(T_1 T_2)^{1/2}}$$

and prove that this is positive. What would be the result if the masses of the water were *different*, say, m_1 and m_2?

5.23. Reversible adiabatic processes are *isentropic*. Are isentropic processes necessarily reversible and adiabatic? If so, explain why; if not, give an illustrative example.

5.24. Prove that the mean heat capacities $\langle C_P \rangle_H$ and $\langle C_P \rangle_S$ are inherently *positive*, whether $T > T_0$ or $T < T_0$. Explain why they are well defined for $T = T_0$.

5.25. A reversible cycle executed by 1 mol of an ideal gas for which $C_P = (5/2)R$ and $C_V = (3/2)R$ consists of the following:

- Starting at $T_1 = 700$ K and $P_1 = 1.5$ bar, the gas is cooled at constant pressure to $T_2 = 350$ K.
- From 350 K and 1.5 bar, the gas is compressed isothermally to pressure P_2.
- The gas returns to its initial state along a path for which $PT = $ constant.

What is the thermal efficiency of the cycle?

5.26. One mole of an ideal gas is compressed isothermally but irreversibly at 130°C from 2.5 bar to 6.5 bar in a piston/cylinder device. The work required is 30% greater than the work of reversible, isothermal compression. The heat transferred from the gas during compression flows to a heat reservoir at 25°C. Calculate the entropy changes of the gas, the heat reservoir, and ΔS_{total}.

5.27. For a steady-flow process at approximately atmospheric pressure, what is the entropy change of the gas:

(a) When 10 mol of SO_2 is heated from 200 to 1100°C?
(b) When 12 mol of propane is heated from 250 to 1200°C?
(c) When 20 kg of methane is heated from 100 to 800°C?
(d) When 10 mol of *n*-butane is heated from 150 to 1150°C?
(e) When 1000 kg of air is heated from 25 to 1000°C?
(f) When 20 mol of ammonia is heated from 100 to 800°C?
(g) When 10 mol of water is heated from 150 to 300°C?
(h) When 5 mol of chlorine is heated from 200 to 500°C?
(i) When 10 kg of ethylbenzene is heated from 300 to 700°C?

5.28. What is the entropy change of the gas, heated in a steady-flow process at approximately atmospheric pressure,

(a) When 800 kJ is added to 10 mol of ethylene initially at 200°C?
(b) When 2500 kJ is added to 15 mol of 1-butene initially at 260°C?
(c) When 10^6(Btu) is added to 40(lb mol) of ethylene initially at 500(°F)?

5.29. A device with no moving parts provides a steady stream of chilled air at −25°C and 1 bar. The feed to the device is compressed air at 25°C and 5 bar. In addition to the stream of chilled air, a second stream of warm air flows from the device at 75°C and 1 bar. Assuming adiabatic operation, what is the ratio of chilled air to warm air that the device produces? Assume that air is an ideal gas for which $C_P = (7/2)R$.

5.30. An inventor has devised a complicated nonflow process in which 1 mol of air is the working fluid. The net effects of the process are claimed to be:

- A change in state of the air from 250°C and 3 bar to 80°C and 1 bar.
- The production of 1800 J of work.
- The transfer of an undisclosed amount of heat to a heat reservoir at 30°C.

Determine whether the claimed performance of the process is consistent with the second law. Assume that air is an ideal gas for which $C_P = (7/2)R$.

5.31. Consider the heating of a house by a furnace, which serves as a heat-source reservoir at a high temperature T_F. The house acts as a heat-sink reservoir at temperature T, and heat $|Q|$ must be added to the house during a particular time interval to maintain this temperature. Heat $|Q|$ can of course be transferred directly from the furnace to the house, as is the usual practice. However, a third heat reservoir is readily available, namely, the surroundings at temperature T_σ, which can serve as another heat source, thus reducing the amount of heat required from the furnace. Given that $T_F = 810$ K, $T = 295$ K, $T_\sigma = 265$ K, and $|Q| = 1000$ kJ, determine the minimum amount of heat $|Q_F|$ which must be extracted from the heat-source reservoir (furnace) at T_F. No other sources of energy are available.

5.32. Consider the air conditioning of a house through use of solar energy. At a particular location, experiment has shown that solar radiation allows a large tank of pressurized water to be maintained at 175°C. During a particular time interval, heat in the amount of 1500 kJ must be extracted from the house to maintain its temperature at 24°C when the surroundings temperature is 33°C. Treating the tank of water, the house, and the surroundings as heat reservoirs, determine the minimum amount of heat that must be extracted from the tank of water by any device built to accomplish the required cooling of the house. No other sources of energy are available.

5.33. A refrigeration system cools a brine from 25°C to −15°C at a rate of 20 kg·s⁻¹. Heat is discarded to the atmosphere at a temperature of 30°C. What is the power requirement if the thermodynamic efficiency of the system is 0.27? The specific heat of the brine is 3.5 kJ·kg⁻¹·°C⁻¹.

5.34. An electric motor under steady load draws 9.7 amperes at 110 volts; it delivers 1.25(hp) of mechanical energy. The temperature of the surroundings is 300 K. What is the total rate of entropy generation in W·K⁻¹?

5.35. A 25-ohm resistor at steady state draws a current of 10 amperes. Its temperature is 310 K; the temperature of the surroundings is 300 K. What is the total rate of entropy generation $\dot{S}_G$? What is its origin?

5.36. Show how the general rate form of the entropy balance, Eq. (5.16), reduces to Eq. (5.2) for the case of a closed system.

5.37. A list of common unit operations follows:

>   (a) Single-pipe heat exchanger;
>   (b) Double-pipe heat exchanger;
>   (c) Pump;
>   (d) Gas compressor;
>   (e) Gas turbine (expander);
>   (f) Throttle valve;
>   (g) Nozzle.

Develop a simplified form of the general steady-state entropy balance appropriate to each operation. State carefully, and justify, any assumptions you make.

5.38. Ten kmol per hour of air is throttled from upstream conditions of 25°C and 10 bar to a downstream pressure of 1.2 bar. Assume air to be an ideal gas with $C_P = (7/2)R$.

>   (a) What is the downstream temperature?
>   (b) What is the entropy change of the air in $J \cdot mol^{-1} \cdot K^{-1}$?
>   (c) What is the rate of entropy generation in $W \cdot K^{-1}$?
>   (d) If the surroundings are at 20°C, what is the lost work?

5.39. A steady-flow adiabatic turbine (expander) accepts gas at conditions T_1, P_1, and discharges at conditions T_2, P_2. Assuming ideal gases, determine (per mole of gas) W, W_{ideal}, W_{lost}, and S_G for one of the following cases. Take $T_\sigma = 300$ K.

>   (a) $T_1 = 500$ K, $P_1 = 6$ bar, $T_2 = 371$ K, $P_2 = 1.2$ bar, $C_P/R = 7/2$.
>   (b) $T_1 = 450$ K, $P_1 = 5$ bar, $T_2 = 376$ K, $P_2 = 2$ bar, $C_P/R = 4$.
>   (c) $T_1 = 525$ K, $P_1 = 10$ bar, $T_2 = 458$ K, $P_2 = 3$ bar, $C_P/R = 11/2$.
>   (d) $T_1 = 475$ K, $P_1 = 7$ bar, $T_2 = 372$ K, $P_2 = 1.5$ bar, $C_P/R = 9/2$.
>   (e) $T_1 = 550$ K, $P_1 = 4$ bar, $T_2 = 403$ K, $P_2 = 1.2$ bar, $C_P/R = 5/2$.

5.40. Consider the direct heat transfer from a heat reservoir at T_1 to another heat reservoir at temperature T_2, where $T_1 > T_2 > T_\sigma$. It is not obvious why the lost work of this process should depend on T_σ, the temperature of the surroundings, because the surroundings are not involved in the actual heat-transfer process. Through appropriate use of the Carnot-engine formula, show for the transfer of an amount of heat equal to $|Q|$ that

$$W_{lost} = T_\sigma |Q| \frac{T_1 - T_2}{T_1 T_2} = T_\sigma S_G$$

5.41. An ideal gas at 2500 kPa is throttled adiabatically to 150 kPa at the rate of 20 mol·s⁻¹. Determine $\dot{S}_G$ and $\dot{W}_{lost}$ if $T_\sigma = 300$ K.

5.42. An inventor claims to have devised a cyclic engine which exchanges heat with reservoirs at 25°C and 250°C, and which produces 0.45 kJ of work for each kJ of heat extracted from the hot reservoir. Is the claim believable?

5.43. Heat in the amount of 150 kJ is transferred directly from a hot reservoir at $T_H = 550$ K to two cooler reservoirs at $T_1 = 350$ K and $T_2 = 250$ K. The surroundings temperature is $T_\sigma = 300$ K. If the heat transferred to the reservoir at T_1 is half that transferred to the reservoir at T_2, calculate:

(a) The entropy generation in kJ·K^{-1}.
(b) The lost work.

How could the process be made reversible?

5.44. A nuclear power plant generates 750 MW; the reactor temperature is 315°C and a river with water temperature of 20°C is available.

(a) What is the maximum possible thermal efficiency of the plant, and what is the minimum rate at which heat must be discarded to the river?
(b) If the actual thermal efficiency of the plant is 60% of the maximum, at what rate must heat be discarded to the river, and what is the temperature rise of the river if it has a flow rate of 165 m^3·s^{-1} ?

5.45. A single gas stream enters a process at conditions T_1, P_1, and leaves at pressure P_2. The process is adiabatic. Prove that the outlet temperature T_2 for the actual (irreversible) adiabatic process is greater than that for a *reversible* adiabatic process. Assume the gas is ideal with constant heat capacities.

5.46. A Hilsch vortex tube operates with no moving mechanical parts and splits a gas stream into two streams: one warmer and the other cooler than the entering stream. One such tube is reported to operate with air entering at 5 bar and 20°C, and air streams leaving at 27°C and –22°C, both at 1(atm). The mass flow rate of warm air leaving is six times that of the cool air. Are these results possible? Assume air to be an ideal gas at the conditions given.

5.47. (a) Air at 70(°F) and 1(atm) is cooled at the rate of 100,000(ft)^3(hr)$^{-1}$ to 20(°F) by refrigeration. For a surroundings temperature of 70(°F), what is the minimum power requirement in (hp)?
(b) Air at 25°C and 1(atm) is cooled at the rate of 3000 m^3·hr^{-1} to –8°C by refrigeration. For a surroundings temperature of 25°C, what is the minimum power requirement in kW?

5.48. A flue gas is cooled from 2000 to 300(°F), and the heat is used to generate saturated steam at 212(°F) in a boiler. The flue gas has a heat capacity given by:

$$\frac{C_P}{R} = 3.83 + 0.000306\, T/(R)$$

Water enters the boiler at 212(°F) and is vaporized at this temperature; its latent heat of vaporization is 970.3(Btu)(lb$_m$)$^{-1}$.

(a) With reference to a surroundings temperature of 70(°F), what is the lost work of this process in (Btu)(lb mole)$^{-1}$ of flue gas?

(b) With reference to a surroundings temperature of 70(°F), what is the maximum work, in (Btu)(lb mole)$^{-1}$ of flue gas, that can be accomplished by the saturated steam at 212(°F) if it condenses only, and does not subcool?

(c) How does the answer to part (b) compare with the maximum work theoretically obtainable from the flue gas itself as it is cooled from 2000 to 300(°F)?

5.49. A flue gas is cooled from 1100 to 150°C, and the heat is used to generate saturated steam at 100°C in a boiler. The flue gas has a heat capacity given by:

$$\frac{C_P}{R} = 3.83 + 0.000551 \ T/K$$

Water enters the boiler at 100°C and is vaporized at this temperature; its latent heat of vaporization is 2256.9 kJ·kg^{-1}.

(a) With reference to a surroundings temperature of 25°C, what is the lost work of this process in kJ·mol^{-1} of flue gas?

(b) With reference to a surroundings temperature of 25°C, what is the maximum work, in kJ·mol^{-1} of flue gas, that can be accomplished by the saturated steam at 100°C if it condenses only, and does not subcool?

(c) How does the answer to part (b) compare with the maximum work theoretically obtainable from the flue gas itself as it is cooled from 1100 to 150°C?

5.50. Ethylene vapor is cooled at atmospheric pressure from 830 to 35°C by direct heat transfer to the surroundings at a temperature of 25°C. With respect to this surroundings temperature, what is the lost work of the process in kJ·mol^{-1}? Show that the same result is obtained as the work which can be derived from reversible heat engines operating with the ethylene vapor as heat source and the surroundings as sink. The heat capacity of ethylene is given in Table C.1 of App. C.

Chapter 6

Thermodynamic Properties of Fluids

Application of thermodynamics to practical problems requires numerical values of thermodynamic properties. A very simple example is calculation of the work required for a steady-state gas compressor. If designed to operate adiabatically with the purpose of raising the pressure of a gas from P_1 to P_2, this work can be determined by an energy balance [Eq. (2.32)], wherein the small kinetic- and potential-energy changes of the gas are neglected:

$$W_s = \Delta H = H_2 - H_1$$

The shaft work is simply ΔH, the difference between inlet and outlet values for the enthalpy of the gas. The necessary enthalpy values must come from experimental data or by estimation. It is our purpose in this chapter to:

- Develop from the first and second laws the fundamental property relations which underlie the mathematical structure of applied thermodynamics for systems of constant composition

- Derive equations that allow calculation of enthalpy and entropy values from PVT and heat-capacity data

- Illustrate and discuss the types of diagrams and tables used to present property values for convenient use

- Develop generalized correlations that provide estimates of property values in the absence of complete experimental information

6.1 FUNDAMENTAL PROPERTY RELATIONS

Equation (2.6), the first law for a closed system of n moles of a substance, may be written for the special case of a reversible process:

$$d(nU) = dQ_{\text{rev}} + dW_{\text{rev}}$$

210

Equations (1.3) and (5.1) as applied to this process are:

$$dW_{rev} = -P\,d(nV) \qquad dQ_{rev} = T\,d(nS)$$

These three equations combine to give:

$$d(nU) = T\,d(nS) - P\,d(nV) \qquad (6.1)$$

where U, S, and V are molar values of the internal energy, entropy, and volume. All of the *primitive* thermodynamic properties—P, V, T, U, and S—are included in this equation. It is a **fundamental property relation** connecting these properties for closed PVT systems. All other equations relating properties of such systems derive from it.

Additional thermodynamic properties, beyond those appearing in Eq. (6.1), are **defined** as a matter of convenience, in relation to the primary properties. Thus, the enthalpy, defined and applied in Chap. 2, is joined here by two others. The three, all with recognized names and useful applications, are:

$$\textbf{Enthalpy } H \equiv U + PV \qquad (6.2)$$

$$\textbf{Helmholtz energy } A \equiv U - TS \qquad (6.3)$$

$$\textbf{Gibbs energy } G \equiv U + PV - TS = H - TS \qquad (6.4)$$

The Helmholtz energy and Gibbs energy,[1] with their origins here, find application in phase- and chemical-equilibrium calculations and in statistical thermodynamics.

Multiplication of Eq. (6.2) by n, followed by differentiation, yields the general expression

$$d(nH) = d(nU) + P\,d(nV) + nV\,dP$$

Substitution for $d(nU)$ by Eq. (6.1) reduces this result to

$$d(nH) = T\,d(nS) + nV\,dP \qquad (6.5)$$

The differentials of nA and nG are obtained similarly:

$$d(nA) = -nS\,dT - P\,d(nV) \qquad (6.6)$$

$$d(nG) = -nS\,dT + nV\,dP \qquad (6.7)$$

Equations (6.1) and (6.5) through (6.7) are equivalent fundamental property relations. They are *derived* for a reversible process. However, they contain only *properties* of the system, which depend only on the state of the system, and not the path by which it reached that state. These equations are therefore not restricted in *application* to reversible processes. However, the restrictions placed on the *nature of the system* cannot be relaxed.

Application is to *any* process in a closed *PVT* system resulting in a differential change from one *equilibrium* state to another.

The system may consist of a single phase (a homogeneous system), or it may comprise several phases (a heterogeneous system); it may be chemically inert, or it may undergo

---

[1]These have traditionally been called Helmholtz *free* energy and the Gibbs *free* energy. The word *free* originally had the connotation of energy available to perform useful work, under appropriate conditions. However, in current usage, the word *free* adds nothing, and is best omitted. Since 1988, the IUPAC-recommended terminology omits the word *free*.

chemical reaction. The choice of which equation to use in a particular application is dictated by convenience. However, the Gibbs energy G is special, because of its unique functional relation to T and P, the variables of primary interest by virtue of the ease of their measurement and control.

An immediate application of these equations is to one mole (or to a unit mass) of a homogeneous fluid of constant composition. For this case, $n = 1$, and they simplify to:

$dU = T\,dS - P\,dV$	(6.8)	$dH = T\,dS + V\,dP$	(6.9)
$dA = -S\,dT - P\,dV$	(6.10)	$dG = -S\,dT + V\,dP$	(6.11)

Implicit in each of these equations is a functional relationship that expresses a molar (or unit mass) property as a function of a natural or special pair of independent variables:

$$U = U(S, V) \quad H = H(S, P) \quad A = A(T, V) \quad G = G(T, P)$$

These variables are said to be *canonical*,[2] and a thermodynamic property known as a function of its canonical variables has a unique characteristic:

All other thermodynamic properties may be evaluated from it by simple mathematical operations.

Equations (6.8) through (6.11) lead to another set of property relations because they are *exact* differential expressions. In general, if $F = F(x, y)$, then the total differential of F is defined as:

$$dF \equiv \left(\frac{\partial F}{\partial x}\right)_y dx + \left(\frac{\partial F}{\partial y}\right)_x dy$$

or
$$dF = M\,dx + N\,dy \tag{6.12}$$

where
$$M \equiv \left(\frac{\partial F}{\partial x}\right)_y \qquad\qquad N \equiv \left(\frac{\partial F}{\partial y}\right)_x$$

Then
$$\left(\frac{\partial M}{\partial y}\right)_x = \frac{\partial^2 F}{\partial y\,\partial x} \qquad\qquad \left(\frac{\partial N}{\partial x}\right)_y = \frac{\partial^2 F}{\partial x\,\partial y}$$

The order of differentiation in mixed second derivatives is immaterial, and these equations combine to give:

$$\left(\frac{\partial M}{\partial y}\right)_x = \left(\frac{\partial N}{\partial x}\right)_y \tag{6.13}$$

Because F is a function of x and y, the right side of Eq. (6.12) is an *exact differential expression*, and Eq. (6.13) correctly relates the partial derivatives. This equation serves in fact as the criterion of exactness. For example, $y\,dx + x\,dy$ is a simple differential expression which is exact because the derivatives of Eq. (6.13) are equal, i.e., $1 = 1$. The function that produces this expression is clearly $F(x, y) = xy$. On the other hand, $y\,dx - x\,dy$ is an equally simple differential expression, but the derivatives of Eq. (6.13) are not equal, i.e., $1 \neq -1$, and it is not exact. There is *no* function of x and y whose differential yields the original expression.

---

[2]*Canonical* here means that the variables conform to a general rule that is both simple and clear.

The thermodynamic properties U, H, A, and G are *known* to be functions of the canonical variables on the right sides of Eqs. (6.8) through (6.11). For each of these exact differential expressions, we may write the relationship of Eq. (6.13), producing the *Maxwell relations*.[3]

$\left(\dfrac{\partial T}{\partial V}\right)_S = -\left(\dfrac{\partial P}{\partial S}\right)_V$	(6.14)	$\left(\dfrac{\partial T}{\partial P}\right)_S = \left(\dfrac{\partial V}{\partial S}\right)_P$	(6.15)
$\left(\dfrac{\partial S}{\partial V}\right)_T = \left(\dfrac{\partial P}{\partial T}\right)_V$	(6.16)	$-\left(\dfrac{\partial S}{\partial P}\right)_T = \left(\dfrac{\partial V}{\partial T}\right)_P$	(6.17)

Expression of U, H, A, and G as functions of their canonical variables does not preclude the validity of other functional relationships for application to particular systems. Indeed, Axiom 3 of Sec. 2.5, applied to *homogeneous PVT systems of constant composition*, asserts their dependence on T and P. The restrictions exclude heterogeneous and reacting systems, except for G, for which T and P are the canonical variables. A simple example is a system comprised of a pure liquid in equilibrium with its vapor. Its molar internal energy depends on the relative amounts of liquid and vapor present, and this is in no way reflected by T and P. However, the canonical variables S and V also depend on the relative amounts of the phases, giving $U = U(S, V)$ its greater generality. On the other hand, T and P are the canonical variables for the Gibbs energy, and $G = G(T, P)$ is general. Thus G is fixed for given T and P, regardless of the relative amounts of the phases, and provides the fundamental basis for the working equations of phase equilibria.

Equations (6.8) through (6.11) lead not only to the Maxwell relations but also to many other equations relating thermodynamic properties. The remainder of this section develops those most useful for evaluation of thermodynamic properties from experimental data.

Enthalpy and Entropy as Functions of T and P

In engineering practice, enthalpy and entropy are often the thermodynamic properties of interest, and T and P are the most common measurable properties of a substance or system. Thus, their mathematical connections, expressing the variation of H and S with changes in T and P, are needed. This information is contained in the derivatives $(\partial H/\partial T)_P$, $(\partial S/\partial T)_P$, $(\partial H/\partial P)_T$, and $(\partial S/\partial P)_T$, with which we can write:

$$dH = \left(\frac{\partial H}{\partial T}\right)_P dT + \left(\frac{\partial H}{\partial P}\right)_T dP \qquad dS = \left(\frac{\partial S}{\partial T}\right)_P dT + \left(\frac{\partial S}{\partial P}\right)_T dP$$

Our goal here is to express these four partial derivatives in terms of measurable properties.

The definition of heat capacity at constant pressure is:

$$\left(\frac{\partial H}{\partial T}\right)_P = C_P \tag{2.19}$$

Another expression for this quantity is obtained by applying Eq. (6.9) to changes with respect to T at constant P:

$$\left(\frac{\partial H}{\partial T}\right)_P = T\left(\frac{\partial S}{\partial T}\right)_P$$

[3]After James Clerk Maxwell (1831–1879). See http://en.wikipedia.org/wiki/James_Clerk_Maxwell.

Combination of this equation with Eq. (2.19) gives:

$$\left(\frac{\partial S}{\partial T}\right)_P = \frac{C_P}{T} \tag{6.18}$$

The pressure derivative of entropy results directly from Eq. (6.17):

$$\left(\frac{\partial S}{\partial P}\right)_T = -\left(\frac{\partial V}{\partial T}\right)_P \tag{6.19}$$

The corresponding derivative for enthalpy is found by applying Eq. (6.9) to changes with respect to P at constant T:

$$\left(\frac{\partial H}{\partial P}\right)_T = T\left(\frac{\partial S}{\partial P}\right)_T + V$$

As a result of Eq. (6.19) this becomes:

$$\left(\frac{\partial H}{\partial P}\right)_T = V - T\left(\frac{\partial V}{\partial T}\right)_P \tag{6.20}$$

With expressions for the four partial derivatives given by Eqs. (2.19) and (6.18) through (6.20), we can write the required functional relations as:

$$dH = C_P\, dT + \left[V - T\left(\frac{\partial V}{\partial T}\right)_P\right] dP \tag{6.21}$$

$$dS = C_P\frac{dT}{T} - \left(\frac{\partial V}{\partial T}\right)_P dP \tag{6.22}$$

These are general equations relating enthalpy and entropy to temperature and pressure for *homogeneous fluids of constant composition*. Eqs. (6.19) and (6.20) illustrate the utility of the Maxwell relations, particularly Eqs. (6.16) and (6.17), which relate changes in entropy that are not experimentally accessible to *PVT* data that are experimentally measurable.

The Ideal-Gas State

The coefficients of dT and dP in Eqs. (6.21) and (6.22) are evaluated from heat-capacity and *PVT* data. The ideal-gas state (denoted by superscript *ig*) provides an example of *PVT* behavior:

$$PV^{ig} = RT \qquad\qquad \left(\frac{\partial V^{ig}}{\partial T}\right)_P = \frac{R}{P}$$

Substituting these equations into Eqs. (6.21) and (6.22) reduces them to:

$$dH^{ig} = C_P^{ig} \, dT \qquad (6.23) \qquad dS^{ig} = C_P^{ig} \frac{dT}{T} - R\frac{dP}{P} \qquad (6.24)$$

These are restatements of equations for the ideal-gas state presented in Secs. 3.3 and 5.5.

Alternative Forms for Liquids

Alternative forms of Eqs. (6.19) and (6.20) result when $(\partial V/\partial T)_P$ is replaced by βV [Eq. (3.3)]:

$$\left(\frac{\partial S}{\partial P}\right)_T = -\beta V \qquad (6.25) \qquad \left(\frac{\partial H}{\partial P}\right)_T = (1 - \beta T)V \qquad (6.26)$$

These equations incorporating β, although general, are usually applied only to liquids. However, for liquids at conditions far from the critical point, both the volume and β are small. Thus at most conditions pressure has little effect on the properties of liquids. The important idealization of an *incompressible fluid* (Sec. 3.2) is considered in Ex. 6.2.
  Replacing $(\partial V/\partial T)_P$ in Eqs. (6.21) and (6.22) with βV yields:

$$dH = C_P \, dT + (1 - \beta T)V \, dP \qquad (6.27) \qquad dS = C_P \frac{dT}{T} - \beta V \, dP \qquad (6.28)$$

Because β and V are weak functions of pressure for liquids, they are usually assumed constant at appropriate average values for integration of the final terms.

Internal Energy as a Function of P

Internal energy is related to enthalpy by Eq. (6.2) as $U = H - PV$. Differentiation yields:

$$\left(\frac{\partial U}{\partial P}\right)_T = \left(\frac{\partial H}{\partial P}\right)_T - P\left(\frac{\partial V}{\partial P}\right)_T - V$$

Then by Eq. (6.20),

$$\left(\frac{\partial U}{\partial P}\right)_T = -T\left(\frac{\partial V}{\partial T}\right)_P - P\left(\frac{\partial V}{\partial P}\right)_T$$

An alternative form results if the derivatives on the right are replaced by βV [Eq. (3.3)] and $-\kappa V$ [Eq. (3.4)]:

$$\left(\frac{\partial U}{\partial P}\right)_T = (-\beta T + \kappa P)V \qquad (6.29)$$

Example 6.1

Determine the enthalpy and entropy changes of liquid water for a change of state from 1 bar and 25°C to 1000 bar and 50°C. Data for water are given in the following table.

t °C	P/bar	C_P/J·mol^{-1}·K^{-1}	V/cm^3·mol^{-1}	β/K^{-1}
25	1	75.305	18.071	256×10^{-6}
25	1000	. . . . . .	18.012	366×10^{-6}
50	1	75.314	18.234	458×10^{-6}
50	1000	. . . . . .	18.174	568×10^{-6}

Solution 6.1

For application to the change of state described, Eqs. (6.27) and (6.28) require integration. Enthalpy and entropy are state functions, and the path of integration is arbitrary; the path most suited to the given data is shown in Fig. 6.1. Because the data indicate that C_P is a weak function of T and that both V and β change relatively slowly with P, integration with arithmetic means is satisfactory. The integrated forms of Eqs. (6.27) and (6.28) that result are:

$$\Delta H = \langle C_P \rangle (T_2 - T_1) + (1 - \langle \beta \rangle T_2) \langle V \rangle (P_2 - P_1)$$

$$\Delta S = \langle C_P \rangle \ln \frac{T_2}{T_1} - \langle \beta \rangle \langle V \rangle (P_2 - P_1)$$

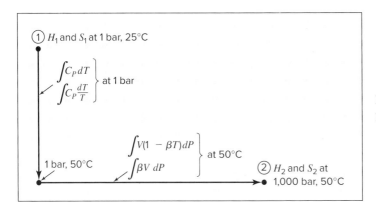

Figure 6.1: Calculation path for Ex. 6.1.

For $P = 1$ bar,

$$\langle C_P \rangle = \frac{75.305 + 75.314}{2} = 75.310 \text{ J·mol}^{-1}·\text{K}^{-1}$$

For $t = 50$°C,

$$\langle V \rangle = \frac{18.234 + 18.174}{2} = 18.204 \text{ cm}^3·\text{mol}^{-1}$$

and

$$\langle \beta \rangle = \frac{458 + 568}{2} \times 10^{-6} = 513 \times 10^{-6} \ K^{-1}$$

Substitution of these numerical values into the equation for ΔH gives:

$$\Delta H = 75.310(323.15 - 298.15) \ J \cdot mol^{-1}$$
$$+ \frac{\left[1 - (513 \times 10^{-6})(323.15)\right](18.204)(1000 - 1) \ bar \cdot cm^3 \cdot mol^{-1}}{10 \ bar \cdot cm^3 \cdot J^{-1}}$$
$$= 1883 + 1517 = 3400 \ J \cdot mol^{-1}$$

Similarly for ΔS,

$$\Delta S = 75.310 \ \ln \frac{323.15}{298.15} \ J \cdot mol^{-1} \cdot K^{-1}$$
$$- \frac{(513 \times 10^{-6})(18.204)(1000 - 1) \ bar \cdot cm^3 \cdot mol^{-1} \cdot K^{-1}}{10 \ bar \cdot cm^3 \cdot J^{-1}}$$
$$= 6.06 - 0.93 = 5.13 \ J \cdot mol^{-1} \cdot K^{-1}$$

Note that the effect of a pressure change of almost 1000 bar on the enthalpy and entropy of liquid water is less than that of a temperature change of only 25°C.

Internal Energy and Entropy as Functions of T and V

In some circumstances, temperature and volume may be more convenient independent variables than temperature and pressure. The most useful property relations are then for internal energy and entropy. Required here are the derivatives $(\partial U/\partial T)_V$, $(\partial U/\partial V)_T$, $(\partial S/\partial T)_V$, and $(\partial S/\partial V)_T$, with which we can write:

$$dU = \left(\frac{\partial U}{\partial T}\right)_V dT + \left(\frac{\partial U}{\partial V}\right)_T dV \qquad dS = \left(\frac{\partial S}{\partial T}\right)_V dT + \left(\frac{\partial S}{\partial V}\right)_T dV$$

The partial derivatives of U follow directly from Eq. (6.8):

$$\left(\frac{\partial U}{\partial T}\right)_V = T\left(\frac{\partial S}{\partial T}\right)_V \qquad \left(\frac{\partial U}{\partial V}\right)_T = T\left(\frac{\partial S}{\partial V}\right)_T - P$$

Combining the first of these with Eq. (2.15) and the second with Eq. (6.16) gives:

$$\left(\frac{\partial S}{\partial T}\right)_V = \frac{C_V}{T} \qquad (6.30) \qquad \left(\frac{\partial U}{\partial V}\right)_T = T\left(\frac{\partial P}{\partial T}\right)_V - P \qquad (6.31)$$

With expressions for the four partial derivatives given by Eqs. (2.15), (6.31), (6.30), and (6.16) we can write the required functional relations as:

$$dU = C_V \, dT + \left[T\left(\frac{\partial P}{\partial T}\right)_V - P \right] dV \tag{6.32}$$

$$dS = C_V \frac{dT}{T} + \left(\frac{\partial P}{\partial T}\right)_V dV \tag{6.33}$$

These are general equations relating the internal energy and entropy of homogeneous fluids of constant composition to temperature and volume.

Equation (3.5) applied to a change of state at constant volume becomes:

$$\left(\frac{\partial P}{\partial T}\right)_V = \frac{\beta}{\kappa} \tag{6.34}$$

Alternative forms of Eqs. (6.32) and (6.33) are therefore:

$$dU = C_V \, dT + \left(\frac{\beta}{\kappa}T - P\right) dV \quad (6.35) \qquad\qquad dS = \frac{C_V}{T} dT + \frac{\beta}{\kappa} dV \qquad (6.36)$$

Example 6.2

Develop the property relations appropriate to the *incompressible fluid,* a model fluid for which both β and κ are zero (Sec. 3.2). This is an idealization employed in fluid mechanics.

Solution 6.2

Equations (6.27) and (6.28) written for an incompressible fluid become:

$$dH = C_P \, dT + V \, dP \tag{A}$$

$$dS = C_P \frac{dT}{T}$$

The enthalpy of an incompressible fluid is therefore a function of both temperature and pressure, whereas the entropy is a function of temperature only, independent of P. With $\kappa = \beta = 0$, Eq. (6.29) shows that the internal energy is also a function of temperature only, and is therefore given by the equation, $dU = C_V \, dT$. Equation (6.13), the criterion of exactness, applied to Eq. (A), yields:

$$\left(\frac{\partial C_P}{\partial P}\right)_T = \left(\frac{\partial V}{\partial T}\right)_P$$

However, the definition of β, given by Eq. (3.3), shows that the derivative on the right equals βV, which is zero for an incompressible fluid. This means that C_P is a function of temperature only, independent of P.

The relation of C_P to C_V for an incompressible fluid is of interest. For a given change of state, Eqs. (6.28) and (6.36) must give the same value for dS; they are therefore equated. The resulting expression, after rearrangement, is:

$$(C_P - C_V)dT = \beta TV\, dP + \frac{\beta T}{\kappa}dV$$

Upon restriction to constant V, this reduces to:

$$C_P - C_V = \beta TV \left(\frac{\partial P}{\partial T}\right)_V$$

Elimination of the derivative by Eq. (6.34) yields:

$$C_P - C_V = \beta TV \left(\frac{\beta}{\kappa}\right) \qquad (B)$$

Because $\beta = 0$, the right side of this equation is zero, provided that the indeterminate ratio β/κ is finite. This ratio is indeed finite for real fluids, and a contrary presumption for the *model* fluid would be irrational. Thus the definition of the incompressible fluid presumes this ratio is finite, and we conclude for such a fluid that the heat capacities at constant V and at constant P are identical:

$$C_P = C_V = C$$

The Gibbs Energy as a Generating Function

The fundamental property relation for $G = G(T, P)$,

$$dG = V\, dP - S\, dT \qquad (6.11)$$

has an alternative form. It follows from the mathematical identity:

$$d\left(\frac{G}{RT}\right) \equiv \frac{1}{RT}dG - \frac{G}{RT^2}dT$$

Substitution for dG by Eq. (6.11) and for G by Eq. (6.4) gives, after algebraic reduction:

$$\boxed{d\left(\frac{G}{RT}\right) = \frac{V}{RT}dP - \frac{H}{RT^2}dT} \qquad (6.37)$$

The advantage of this equation is that all terms are dimensionless; moreover, in contrast to Eq. (6.11), the enthalpy rather than the entropy appears on the right side.

Equations such as Eqs. (6.11) and (6.37) are most readily applied in restricted form. Thus, from Eq. (6.37)

$$\frac{V}{RT} = \left[\frac{\partial(G/RT)}{\partial P}\right]_T \quad (6.38) \qquad \frac{H}{RT} = -T\left[\frac{\partial(G/RT)}{\partial T}\right]_P \quad (6.39)$$

Given G/RT as a function of T and P, V/RT and H/RT follow by simple differentiation. The remaining properties follow from defining equations. In particular,

$$\frac{S}{R} = \frac{H}{RT} - \frac{G}{RT} \qquad \frac{U}{RT} = \frac{H}{RT} - \frac{PV}{RT}$$

The Gibbs energy, *G* or *G/RT*, when given as a function of its canonical variables *T* and *P*, serves as a *generating function* for the other thermodynamic properties through simple mathematics, and implicitly represents *complete* property information.

Just as Eq. (6.11) leads to expressions for all thermodynamic properties, so Eq. (6.10), the fundamental property relation for the Helmholtz energy $A = A(T, V)$, leads to equations for all thermodynamic properties from knowledge of A as a function of T and V. This is particularly useful in connecting thermodynamic properties to statistical mechanics because closed systems at fixed volume and temperature are often most amenable to treatment by both theoretical methods of statistical mechanics and computational methods of molecular simulation based on statistical mechanics.

6.2   RESIDUAL PROPERTIES

Unfortunately, no experimental method for the measurement of numerical values of G or G/RT is known, and the equations which relate other properties to the Gibbs energy are of little direct practical use. However, the concept of the Gibbs energy as a generating function for other thermodynamic properties carries over to a closely related property for which numerical values *are* readily obtained. By **definition,** the *residual* Gibbs energy is: $G^R \equiv G - G^{ig}$, where G and G^{ig} are the actual and the ideal-gas-state values of the Gibbs energy at the same temperature and pressure. Other residual properties are defined in an analogous way. The residual volume, for example, is:

$$V^R \equiv V - V^{ig} = V - \frac{RT}{P}$$

Because $V = ZRT/P$, the residual volume and the compressibility factor are related:

$$V^R = \frac{RT}{P}(Z - 1) \tag{6.40}$$

The generic residual property[4] is defined by:

$$\boxed{M^R \equiv M - M^{ig}} \tag{6.41}$$

---

[4]Sometimes referred to as a departure function. *Generic* here denotes a class of properties with the same characteristics.

where M and M^{ig} are actual and ideal-gas-state properties at the same T and P. They represent molar values for any extensive thermodynamic property, e.g., V, U, H, S, or G.

The underlying purpose of this definition is more easily understood when it is written as:

$$M = M^{ig} + M^R$$

From a practical perspective this equation divides property calculations into two parts: first, simple calculations for properties in the ideal-gas state; second, calculations for the residual properties, which have the nature of corrections to the ideal-gas-state values. Properties for the ideal-gas state reflect real molecular configurations but presume the absence of intermolecular interactions. Residual properties account for for the effect of such interactions. Our purpose here is to develop equations for the calculation of residual properties from PVT data or from their representation by equations of state.

Equation (6.37), written for the ideal-gas state, becomes:

$$d\left(\frac{G^{ig}}{RT}\right) = \frac{V^{ig}}{RT}dP - \frac{H^{ig}}{RT^2}dT$$

Subtracting this equation from Eq. (6.37) itself gives:

$$d\left(\frac{G^R}{RT}\right) = \frac{V^R}{RT}dP - \frac{H^R}{RT^2}dT \tag{6.42}$$

This *fundamental residual-property relation* applies to fluids of constant composition. Useful restricted forms are:

$$\frac{V^R}{RT} = \left[\frac{\partial(G^R/RT)}{\partial P}\right]_T \tag{6.43} \qquad \frac{H^R}{RT} = -T\left[\frac{\partial(G^R/RT)}{\partial T}\right]_P \tag{6.44}$$

Equation (6.43) provides a direct link between the residual Gibbs energy and experiment.

Written

$$d\left(\frac{G^R}{RT}\right) = \frac{V^R}{RT}dP \quad (\text{const } T)$$

it may be integrated from zero pressure to arbitrary pressure P, yielding:

$$\frac{G^R}{RT} = \left(\frac{G^R}{RT}\right)_{P=0} + \int_0^P \frac{V^R}{RT}dP \quad (\text{const } T)$$

For convenience, define:

$$\left(\frac{G^R}{RT}\right)_{P=0} \equiv J$$

With this definition and elimination of V^R by Eq. (6.40),

$$\frac{G^R}{RT} = J + \int_0^P (Z-1)\frac{dP}{P} \quad (\text{const } T) \tag{6.45}$$

As explained in the Addendum to this chapter, J is a *constant*, independent of T, and the derivative of this equation in accord with Eq. (6.44) gives:

$$\boxed{\frac{H^R}{RT} = -T \int_0^P \left(\frac{\partial Z}{\partial T} \right)_P \frac{dP}{P} \quad \text{(const } T)} \tag{6.46}$$

The defining equation for the Gibbs energy, $G = H - TS$, may also be written for the ideal-gas state, $G^{ig} = H^{ig} - TS^{ig}$; by difference, $G^R = H^R - TS^R$, and

$$\frac{S^R}{R} = \frac{H^R}{RT} - \frac{G^R}{RT} \tag{6.47}$$

Combining this equation with Eqs. (6.45) and (6.46) gives:

$$\frac{S^R}{R} = -T \int_0^P \left(\frac{\partial Z}{\partial T} \right)_P \frac{dP}{P} - J - \int_0^P (Z - 1) \frac{dP}{P} \quad \text{(const } T)$$

In application, entropy always appears in *differences*. In accord with Eq. (6.41), we write: $S = S^{ig} + S^R$ for two different states. Then by difference:

$$\Delta S \equiv S_2 - S_1 = \left(S_2^{ig} - S_1^{ig} \right) + \left(S_2^R - S_1^R \right)$$

Because J is constant, it cancels from the final term, and its value is of no consequence. Constant J is therefore arbitrarily set equal to zero, and the working equation for S^R becomes:

$$\boxed{\frac{S^R}{R} = -T \int_0^P \left(\frac{\partial Z}{\partial T} \right)_P \frac{dP}{P} - \int_0^P (Z - 1) \frac{dP}{P} \quad \text{(const } T)} \tag{6.48}$$

and Eq. (6.45) is written:

$$\boxed{\frac{G^R}{RT} = \int_0^P (Z - 1) \frac{dP}{P} \quad \text{(const } T)} \tag{6.49}$$

Values of the compressibility factor $Z = PV/RT$ and of $(\partial Z/\partial T)_P$ may be calculated from experimental PVT data, with the two integrals in Eqs. (6.46), (6.48), and (6.49) evaluated by numerical or graphical methods. Alternatively, the two integrals may be expressed analytically with Z given as a function of T and P by a volume-explicit equation of state. This direct connection with experiment allows evaluation of the residual properties H^R and S^R for use in the calculation of enthalpy and entropy values.

Enthalpy and Entropy from Residual Properties

General expressions for H^{ig} and S^{ig} are found by integration of Eqs. (6.23) and (6.24) from an ideal-gas state at reference conditions T_0 and P_0 to an ideal-gas state at T and P:[5]

$$H^{ig} = H_0^{ig} + \int_{T_0}^T C_P^{ig} \, dT \qquad\qquad S^{ig} = S_0^{ig} + \int_{T_0}^T C_P^{ig} \frac{dT}{T} - R \, \ln \frac{P}{P_0}$$

[5]Thermodynamic properties for organic compounds in the ideal-gas state are given by M. Frenkel, G. J. Kabo, K. N. Marsh, G. N. Roganov, and R. C. Wilhoit, *Thermodynamics of Organic Compounds in the Gas State*, Thermodynamics Research Center, Texas A & M Univ. System, College Station, Texas, 1994. For many compounds, these data are also available via the NIST Chemistry Webbook, http://webbook.nist.gov.

Because $H = H^{ig} + H^R$ and $S = S^{ig} + S^R$:

$$H = H_0^{ig} + \int_{T_0}^{T} C_P^{ig}\, dT + H^R \tag{6.50}$$

$$S = S_0^{ig} + \int_{T_0}^{T} C_P^{ig}\, \frac{dT}{T} - R \ln \frac{P}{P_0} + S^R \tag{6.51}$$

Recall (Secs. 4.1 and 5.5) that for purposes of computation the integrals in Eqs. (6.50) and (6.51) are represented by:

$$\int_{T_0}^{T} C_P^{ig}\, dT = R \times \text{ICPH}(T_0, T; A, B, C, D)$$

$$\int_{T_0}^{T} C_P^{ig}\, \frac{dT}{T} = R \times \text{ICPS}(T_0, T; A, B, C, D)$$

Equations (6.50) and (6.51) have alternative forms when the integrals are replaced by equivalent terms that include the mean heat capacities introduced in Secs. 4.1 and 5.5:

$$H = H_0^{ig} + \langle C_P^{ig} \rangle_H (T - T_0) + H^R \tag{6.52}$$

$$S = S_0^{ig} + \langle C_P^{ig} \rangle_S \ln \frac{T}{T_0} - R \ln \frac{P}{P_0} + S^R \tag{6.53}$$

In Eqs. (6.50) through (6.53), H^R and S^R are given by Eqs. (6.46) and (6.48). Again, for computational purposes, the mean heat capacities are represented by:

$$\langle C_P^{ig} \rangle_H = R \times \text{MCPH}(T_0, T; A, B, C, D)$$

$$\langle C_P^{ig} \rangle_S = R \times \text{MCPS}(T_0, T; A, B, C, D)$$

Applications of thermodynamics require only *differences* in enthalpy and entropy, and these do not change when the scale of values is shifted by a constant amount. The reference-state conditions T_0 and P_0 are therefore selected for convenience, and values are assigned to H_0^{ig} and S_0^{ig} arbitrarily. The only information needed for application of Eqs. (6.52) and (6.53) are ideal-gas-state heat capacities and *PVT* data. Once V, H, and S are known at given conditions of T and P, the other thermodynamic properties follow from defining equations.

The great practical value of the ideal-gas state is now evident. It provides the base for calculation of real-gas properties.

Residual properties have validity for both gases and liquids. However, the advantage of Eqs. (6.50) and (6.51) in application to gases is that H^R and S^R, the terms which contain all the complex calculations, are *residuals* that are usually small. They act as corrections to the major terms, H^{ig} and S^{ig}. For liquids, this advantage is largely lost because H^R and S^R must include the large enthalpy and entropy changes of vaporization. Property changes of liquids are usually calculated by integrated forms of Eqs. (6.27) and (6.28), as illustrated in Ex. 6.1.

Example 6.3

Calculate the enthalpy and entropy of saturated isobutane vapor at 360 K from the following information:

1. Table 6.1 gives compressibility-factor data (values of Z) for isobutane vapor.

2. The vapor pressure of isobutane at 360 K is 15.41 bar.

3. Set $H_0^{ig} = 18,115.0 \; \text{J·mol}^{-1}$ and $S_0^{ig} = 295.976 \; \text{J·mol}^{-1}·\text{K}^{-1}$ for the reference state at 300 K and 1 bar. [These values are in accord with the bases adopted by R. D. Goodwin and W. M. Haynes, Nat. Bur. Stand. (U.S.), Tech. Note 1051, 1982.]

4. The ideal-gas-state heat capacity of isobutane vapor at temperatures of interest is:

$$C_P^{ig}/R = 1.7765 + 33.037 \times 10^{-3}\, T \quad (T\,\text{K})$$

Table 6.1: Compressibility Factor Z for Isobutane

P bar	340 K	350 K	360 K	370 K	380 K
0.10	0.99700	0.99719	0.99737	0.99753	0.99767
0.50	0.98745	0.98830	0.98907	0.98977	0.99040
2.00	0.95895	0.96206	0.96483	0.96730	0.96953
4.00	0.92422	0.93069	0.93635	0.94132	0.94574
6.00	0.88742	0.89816	0.90734	0.91529	0.92223
8.00	0.84575	0.86218	0.87586	0.88745	0.89743
10.0	0.79659	0.82117	0.84077	0.85695	0.87061
12.0	. . . . . .	0.77310	0.80103	0.82315	0.84134
14.0	. . . . . .	. . . . . .	0.75506	0.78531	0.80923
15.41	. . . . . .	. . . . . .	0.71727		

Solution 6.3

Calculation of H^R and S^R at 360 K and 15.41 bar by application of Eqs. (6.46) and (6.48) requires evaluation of two integrals:

$$\int_0^P \left(\frac{\partial Z}{\partial T} \right)_P \frac{dP}{P} \qquad \int_0^P (Z-1)\frac{dP}{P}$$

Graphical integration requires simple plots of $(\partial Z/\partial T)_P/P$ and $(Z-1)/P$ vs. P. Values of $(Z-1)/P$ are found from the compressibility-factor data at 360 K. The quantity $(\partial Z/\partial T)_P/P$ requires evaluation of the partial derivative $(\partial Z/\partial T)_P$, given by the slope of a plot of Z vs. T at constant pressure. For this purpose, separate plots are made of Z vs. T for each pressure at which compressibility-factor data are given, and a slope is determined at 360 K for each curve (for example, by construction of a tangent line at 360 K). Data for the required plots are shown in Table 6.2.

Table 6.2: Values of the Integrands Required in Ex. 6.3
Values in parentheses are by extrapolation.

P bar	$[(\partial Z/\partial T)_P/P] \times 10^4$ K^{-1}·bar^{-1}	$[-(Z-1)/P] \times 10^2$ bar^{-1}
0.00	(1.780)	(2.590)
0.10	1.700	2.470
0.50	1.514	2.186
2.00	1.293	1.759
4.00	1.290	1.591
6.00	1.395	1.544
8.00	1.560	1.552
10.0	1.777	1.592
12.0	2.073	1.658
14.0	2.432	1.750
15.41	(2.720)	(1.835)

The values of the two integrals, as determined from the plots, are:

$$\int_0^P \left(\frac{\partial Z}{\partial T}\right)_P \frac{dP}{P} = 26.37 \times 10^{-4} \text{ K}^{-1} \qquad \int_0^P (Z-1)\frac{dP}{P} = -0.2596$$

By Eq. (6.46),

$$\frac{H^R}{RT} = -(360)(26.37 \times 10^{-4}) = -0.9493$$

By Eq. (6.48),

$$\frac{S^R}{R} = -0.9493 - (-0.2596) = -0.6897$$

For $R = 8.314$ J·mol^{-1}·K^{-1},

$$H^R = (-0.9493)(8.314)(360) = -2841.3 \text{ J·mol} -1$$
$$S^R = (-0.6897)(8.314) = -5.734 \text{ J·mol} -1 \cdot \text{K} -1$$

Values of the integrals in Eqs. (6.50) and (6.51), with parameters from the given equation for C_P^{ig}/R, are:

$$8.314 \times \text{ICPH}(300, 360; 1.7765, 33.037 \times 10^{-3}, 0.0, 0.0) = 6324.8 \text{ J·mol}^{-1}$$
$$8.314 \times \text{ICPS}(300, 360; 1.7765, 33.037 \times 10^{-3}, 0.0, 0.0) = 19.174 \text{ J·mol}^{-1}\cdot\text{K}^{-1}$$

Substitution of numerical values into Eqs. (6.50) and (6.51) yields:

$$H = 18,115.0 + 6324.8 - 2841.3 = -21,598.5 \text{ J·mol}^{-1}$$
$$S = 295.976 + 19.174 - 8.314 \ln 15.41 - 5.734 = 286.676 \text{ J·mol}^{-1}\cdot\text{K}^{-1}$$

Although calculations are here carried out for just one state, enthalpies and entropies can be evaluated for any number of states, given adequate data. After having completed a set of calculations, one is not irrevocably committed to the particular values of H_0^{ig} and S_0^{ig} initially assigned. The scale of values for either the enthalpy or the entropy can be shifted by addition of a constant to all values. In this way one can give arbitrary values to H and S for some particular state so as to make the scales convenient for some particular purpose.

The calculation of thermodynamic properties is an exacting task, seldom required of an engineer. However, engineers do make practical use of thermodynamic properties, and an understanding of methods by which they are calculated should suggest that some uncertainty is associated with every property value. Inaccuracy derives partly from experimental error in the data, which are also frequently incomplete and must be extended by interpolation and extrapolation. Moreover, even with reliable PVT data, a loss of accuracy occurs in the differentiation process required in the calculation of derived properties. Thus data of high accuracy are required to produce enthalpy and entropy values suitable for engineering calculations.

6.3   RESIDUAL PROPERTIES FROM THE VIRIAL EQUATIONS OF STATE

The numerical or graphical evaluation of integrals, as in Eqs. (6.46) and (6.48), is often tedious and imprecise. An attractive alternative is analytical evaluation through equations of state. The procedure depends on whether the equation of state is *volume explicit*, i.e., expresses V (or Z) as a function of P at constant T, or *pressure explicit*, i.e., expresses P (or Z) as a function of V (or ρ) at constant T.[6] Equations (6.46) and (6.48) are directly applicable only for a volume-explicit equation, such as the two-term virial equation in P [Eq. (3.36)]. For pressure-explicit equations, such as the virial expansions in reciprocal volume [Eq. (3.38)], Eqs. (6.46), (6.48), and (6.49) require reformulation.

The two-term virial equation of state, Eq. (3.36), is volume explicit, $Z - 1 = BP/RT$. Differentiation yields $(\partial Z/\partial T)_P$. We therefore have the required expressions for substitution into Eqs. (6.46) and (6.48). Direct integration gives H^R/RT and S^R/R. An alternative procedure is to evaluate G^R/RT by Eq. (6.49):

$$\frac{G^R}{RT} = \frac{BP}{RT} \tag{6.54}$$

From this result H^R/RT is found from Eq. (6.44), and S^R/R is given by Eq. (6.47). Either way, we find:

$$\boxed{\frac{H^R}{RT} = \frac{P}{R}\left(\frac{B}{T} - \frac{dB}{dT}\right) \quad (6.55) \qquad \frac{S^R}{R} = -\frac{P}{R}\frac{dB}{dT} \quad (6.56)}$$

Evaluation of residual enthalpies and residual entropies by Eqs. (6.55) and (6.56) is straightforward for given values of T and P, provided one has sufficient information to evaluate B and dB/dT. The range of applicability of these equations is the same as for Eq. (3.36), as discussed in Sec. 3.5.

---

[6]The ideal-gas equation of state is both pressure and volume explicit.

Equations (6.46), (6.48), and (6.49) are incompatible with pressure-explicit equations of state and must be transformed such that P is no longer the variable of integration. In carrying out this transformation, the molar density ρ is a more convenient variable of integration than V, because ρ goes to zero, rather than to infinity, as P goes to zero. Thus, the equation $PV = ZRT$ is written in alternative form as

$$P = Z\rho RT \tag{6.57}$$

Differentiation at constant T gives:

$$dP = RT(Zd\rho + \rho dZ) \quad (\text{const } T)$$

Dividing this equation by Eq. (6.57) gives:

$$\frac{dP}{P} = \frac{d\rho}{\rho} + \frac{dZ}{Z} \quad (\text{const } T)$$

Upon substitution for dP/P, Eq. (6.49) becomes:

$$\boxed{\frac{G^R}{RT} = \int_0^\rho (Z - 1)\frac{d\rho}{\rho} + Z - 1 - \ln Z} \tag{6.58}$$

where the integral is evaluated at constant T. Note also that $\rho \to 0$ when $P \to 0$.

Solving Eq. (6.42) for its final term and substituting for V^R by Eq. (6.40) yields:

$$\frac{H^R}{RT^2}dT = (Z - 1)\frac{dP}{P} - d\left(\frac{G^R}{RT}\right)$$

Applying this for changes with respect to T at constant ρ gives:

$$\frac{H^R}{RT^2} = \frac{Z - 1}{P}\left(\frac{\partial P}{\partial T}\right)_\rho - \left[\frac{\partial(G^R/RT)}{\partial T}\right]_\rho$$

Differentiation of Eq. (6.57) provides the first derivative on the right, and differentiation of Eq. (6.58) provides the second. Substitution leads to:

$$\boxed{\frac{H^R}{RT} = -T\int_0^\rho \left(\frac{\partial Z}{\partial T}\right)_\rho \frac{d\rho}{\rho} + Z - 1} \tag{6.59}$$

The residual entropy is found from Eq. (6.47) in combination with Eqs. (6.58) and (6.59):

$$\boxed{\frac{S^R}{R} = \ln Z - T\int_0^\rho \left(\frac{\partial Z}{\partial T}\right)_\rho \frac{d\rho}{\rho} - \int_0^\rho (Z - 1)\frac{d\rho}{\rho}} \tag{6.60}$$

We now apply this to the pressure-explicit three-term virial equation:

$$Z - 1 = B\rho + C\rho^2 \tag{3.38}$$

Substitution into Eqs. (6.58) through (6.60) leads to:

$$\frac{G^R}{RT} = 2B\rho + \frac{3}{2}C\rho^2 - \ln Z \tag{6.61}$$

$$\frac{H^R}{RT} = T\left[\left(\frac{B}{T} - \frac{dB}{dT}\right)\rho + \left(\frac{C}{T} - \frac{1}{2}\frac{dC}{dT}\right)\rho^2\right] \tag{6.62}$$

$$\frac{S^R}{R} = \ln Z - T\left[\left(\frac{B}{T} - \frac{dB}{dT}\right)\rho + \frac{1}{2}\left(\frac{C}{T} + \frac{dC}{dT}\right)\rho^2\right] \tag{6.63}$$

Application of these equations, useful for gases up to moderate pressures, requires data for both the second and third virial coefficients.

6.4 GENERALIZED PROPERTY CORRELATIONS FOR GASES

Of the two kinds of data needed for evaluation of thermodynamic properties, heat capacities and *PVT* data, the latter are most frequently missing. Fortunately, the generalized methods developed in Sec. 3.7 for the compressibility factor are also applicable to residual properties.

Equations (6.46) and (6.48) are put into generalized form by substitution of the relations:

$$P = P_c P_r \quad T = T_c T_r$$
$$dP = P_c dP_r \quad dT = T_c dT_r$$

The resulting equations are:

$$\frac{H^R}{RT_c} = -T_r^2 \int_0^{P_r} \left(\frac{\partial Z}{\partial T_r}\right)_{P_r} \frac{dP_r}{P_r} \tag{6.64}$$

$$\frac{S^R}{R} = -T_r \int_0^{P_r} \left(\frac{\partial Z}{\partial T_r}\right)_{P_r} \frac{dP_r}{P_r} - \int_0^{P_r} (Z - 1)\frac{dP_r}{P_r} \tag{6.65}$$

The terms on the right sides of these equations depend only on the upper limit P_r of the integrals and on the reduced temperature at which they are evaluated. Thus, values of H^R/RT_c and S^R/R may be determined once and for all at any reduced temperature and pressure from generalized compressibility-factor data.

The correlation for Z is based on Eq. (3.53):

$$Z = Z^0 + \omega Z^1$$

Differentiation yields:

$$\left(\frac{\partial Z}{\partial T_r}\right)_{P_r} = \left(\frac{\partial Z^0}{\partial T_r}\right)_{P_r} + \omega\left(\frac{\partial Z^1}{\partial T_r}\right)_{P_r}$$

Substitution for Z and $(\partial Z/\partial T)_{P_r}$ in Eqs. (6.64) and (6.65) gives:

$$\frac{H^R}{RT_c} = -T_r^2 \int_0^{P_r} \left(\frac{\partial Z^0}{\partial T_r}\right)_{P_r} \frac{dP_r}{P_r} - \omega T_r^2 \int_0^{P_r} \left(\frac{\partial Z^1}{\partial T_r}\right)_{P_r} \frac{dP_r}{P_r}$$

$$\frac{S^R}{R} = -\int_0^{P_r} \left[T_r\left(\frac{\partial Z^0}{\partial T_r}\right)_{P_r} + Z^0 - 1\right]\frac{dP_r}{P_r} - \omega \int_0^{P_r} \left[T_r\left(\frac{\partial Z^1}{\partial T_r}\right)_{P_r} + Z^1\right]\frac{dP_r}{P_r}$$

The first integrals on the right sides of these two equations may be evaluated numerically or graphically for various values of T_r and P_r from the data for Z^0 given in Tables D.1 and D.3 of App. D, and the integrals which follow ω in each equation may be similarly evaluated from the data for Z^1 given in Tables D.2 and D.4.

If the first terms on the right sides of the preceding equations (including the minus signs) are represented by $(H^R)^0/RT_c$ and $(S^R)^0/R$ and if the terms which follow ω, together with the preceding minus signs, are represented by $(H^R)^1/RT_c$ and $(S^R)^1/R$, then:

$$\frac{H^R}{RT_c} = \frac{(H^R)^0}{RT_c} + \omega\frac{(H^R)^1}{RT_c} \quad (6.66) \qquad \frac{S^R}{R} = \frac{(S^R)^0}{R} + \omega\frac{(S^R)^1}{R} \quad (6.67)$$

Calculated values of the quantities $(H^R)^0/RT_c$, $(H^R)^1/RT_c$, $(S^R)^0/R$, and $(S^R)^1/R$ as determined by Lee and Kesler are given as functions of T_r and P_r in Tables D.5 through D.12. These values, together with Eqs. (6.66) and (6.67), allow estimation of residual enthalpies and entropies on the basis of the three-parameter corresponding-states principle as developed by Lee and Kesler (Sec. 3.7).

Tables D.5 and D.7 for $(H^R)^0/RT_c$ and Tables D.9 and D.11 for $(S^R)^0/R$, used alone, provide two-parameter corresponding-states correlations that quickly yield rough estimates of the residual properties. The nature of these correlations is indicated by Fig. 6.2, which shows a plot of $(H^R)^0/RT_c$ vs. P_r for six isotherms.

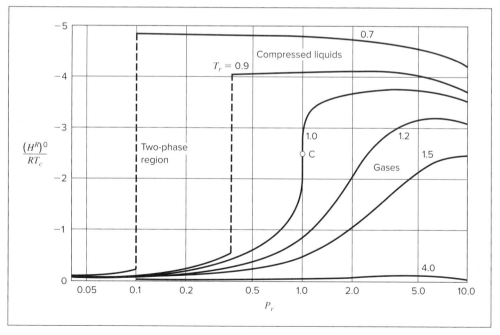

Figure 6.2: The Lee/Kesler correlation for $(H^R)^0/RT_c$ as a function of T_r and P_r.

As with the generalized compressibility-factor correlation, the complexity of the functions $(H^R)^0/RT_c$, $(H^R)^1/RT_c$, $(S^R)^0/R$, and $(S^R)^1/R$ precludes their general representation by simple equations. However, the generalized second-virial-coefficient correlation forms the basis for analytical correlations of the residual properties at low pressures. Recall Eqs. (3.58) and (3.59):

$$\hat{B} = \frac{BP_c}{RT_c} = B^0 + \omega B^1$$

Quantities $\hat{B}$, B^0, and B^1 are functions of T_r only. Hence,

$$\frac{d\hat{B}}{dT_r} = \frac{dB^0}{dT_r} + \omega\frac{dB^1}{dT_r}$$

Equations (6.55) and (6.56) may be written:

$$\frac{H^R}{RT_c} = P_r\left(\hat{B} - T_r\frac{d\hat{B}}{dT_r}\right) \qquad \frac{S^R}{R} = -P_r\frac{d\hat{B}}{dT_r}$$

Combining each of these equations with the previous two equations yields:

$$\frac{H^R}{RT_c} = P_r\left[B^0 - T_r\frac{dB^0}{dT_r} + \omega\left(B^1 - T_r\frac{dB^1}{dT_r}\right)\right] \tag{6.68}$$

$$\frac{S^R}{R} = -P_r\left(\frac{dB^0}{dT_r} + \omega\frac{dB^1}{dT_r}\right) \tag{6.69}$$

The dependence of B^0 and B^1 on reduced temperature is given by Eqs. (3.61) and (3.62). Differentiation of these equations provides expressions for dB^0/dT_r and dB^1/dT_r. Thus the equations required for application of Eqs. (6.68) and (6.69) are:

$B^0 = 0.083 - \dfrac{0.422}{T_r^{1.6}}$ (3.61)		$B^1 = 0.139 - \dfrac{0.172}{T_r^{4.2}}$ (3.62)	
$\dfrac{dB^0}{dT_r} = \dfrac{0.675}{T_r^{2.6}}$	(6.70)	$\dfrac{dB^1}{dT_r} = \dfrac{0.722}{T_r^{5.2}}$	(6.71)

Figure 3.10, drawn specifically for the compressibility-factor correlation, is also used as a guide to the reliability of the correlations of residual properties based on generalized second virial coefficients. However, all residual-property correlations are less accurate than the compressibility-factor correlations on which they are based and are, of course, least reliable for strongly polar and associating molecules.

The generalized correlations for H^R and S^R, together with ideal-gas heat capacities, allow calculation of enthalpy and entropy values of gases at any temperature and pressure by Eqs. (6.50) and (6.51). For a change from state 1 to state 2, write Eq. (6.50) for both states:

$$H_2 = H_0^{ig} + \int_{T_0}^{T_2} C_P^{ig}\,dT + H_2^R \qquad H_1 = H_0^{ig} + \int_{T_0}^{T_1} C_P^{ig}\,dT + H_1^R$$

The enthalpy change for the process, $\Delta H = H_2 - H_1$, is the difference between these two equations:

$$\Delta H = \int_{T_1}^{T_2} C_P^{ig}\, dT + H_2^R - H_1^R \tag{6.72}$$

Similarly, by Eq. (6.51),

$$\Delta S = \int_{T_1}^{T_2} C_P^{ig}\, \frac{dT}{T} - R\ln\frac{P_2}{P_1} + S_2^R - S_1^R \tag{6.73}$$

Written in alternative form, these equations become:

$$\Delta H = \langle C_P^{ig}\rangle_H (T_2 - T_1) + H_2^R - H_1^R \tag{6.74}$$

$$\Delta S = \langle C_P^{ig}\rangle_S \ln\frac{T_2}{T_1} - R\ln\frac{P_2}{P_1} + S_2^R - S_1^R \tag{6.75}$$

Just as we have given names to functions used in evaluation of the integrals in Eqs. (6.72) and (6.73) and the mean heat capacities in Eqs. (6.74) and (6.75), we also name functions useful for evaluation of H^R and S^R. Equations (6.68), (3.61), (6.70), (3.62), and (6.71) together provide a function for the evaluation of H^R/RT_c, named HRB(T_r, P_r, OMEGA):[7]

$$\frac{H^R}{RT_c} = \mathrm{HRB}\,(T_r, P_r, \mathrm{OMEGA})$$

A numerical value of H^R is therefore represented by:

$$RT_c \times \mathrm{HRB}\,(T_r, P_r, \mathrm{OMEGA})$$

Similarly, Eqs. (6.69) through (6.71) provide a function for the evaluation of S^R/R, named SRB(T_r, P_r, OMEGA):

$$\frac{S^R}{R} = \mathrm{SRB}\,(T_r, P_r, \mathrm{OMEGA})$$

A numerical value of S^R is therefore represented by:

$$R \times \mathrm{SRB}\,(T_r, P_r, \mathrm{OMEGA})$$

The terms on the right sides of Eqs. (6.72) through (6.75) are readily associated with steps in a *computational path* leading from an initial to a final state of a system. Thus, in Fig. 6.3, the actual path from state 1 to state 2 (dashed line) is replaced by a three-step computational path:

- **Step 1 $\rightarrow$ 1^{ig}:** A hypothetical process that transforms a real gas into an ideal gas at T_1 and P_1. The enthalpy and entropy changes for this process are:

$$H_1^{ig} - H_1 = -H_1^R \qquad\qquad S_1^{ig} - S_1 = -S_1^R$$

- **Step 1^{ig} $\rightarrow$ 2^{ig}:** Changes in the ideal-gas state from (T_1, P_1) to (T_2, P_2). For this process,

$$\Delta H^{ig} = H_2^{ig} - H_1^{ig} = \int_{T_1}^{T_2} C_P^{ig}\, dT \tag{6.76}$$

[7]Sample programs and spreadsheets for evaluation of these functions are available in the online learning center at http://highered.mheducation.com:80/sites/1259696529.

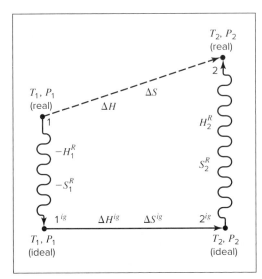

Figure 6.3: Computational path for property changes ΔH and ΔS.

$$\Delta S^{ig} = S_2^{ig} - S_1^{ig} = \int_{T_1}^{T_2} C_P^{ig} \frac{dT}{T} - R \ln \frac{P_2}{P_1} \tag{6.77}$$

- **Step** $2^{ig} \to 2$: Another hypothetical process that transforms the ideal gas back into a real gas at T_2 and P_2. Here,

$$H_2 - H_2^{ig} = H_2^R \qquad\qquad S_2 - S_2^{ig} = S_2^R$$

Equations (6.72) and (6.73) result from addition of the enthalpy and entropy changes for the three steps.

Example 6.4

Supercritical CO_2 is increasingly used as an environmentally friendly solvent for cleaning applications, ranging from dry cleaning clothing to degreasing machine parts to photoresist stripping. A key advantage of CO_2 is the ease with which it is separated from "dirt" and detergents. When its temperature and pressure are reduced below the critical temperature and vapor pressure respectively, it vaporizes, leaving dissolved substances behind. For a change in state of CO_2 from 70°C and 150 bar to 20°C and 15 bar, estimate the changes in its molar enthalpy and entropy.

Solution 6.4

We follow the three-step computational path of Fig. 6.3. Step 1 transforms the real fluid at 70°C and 150 bar into its ideal-gas state at the same conditions. Step 2 changes conditions in the ideal-gas state from the initial to the final conditions of T and P. Step 3 transforms the fluid from its ideal-gas state to the real-gas final state at 20°C and 15 bar.

The residual-property values required for calculating the changes of Steps 1 and 3 depend on the reduced conditions of the initial and final states. With critical properties from Table B.1 of App. B:

$$T_{r_1} = 1.128 \qquad P_{r_1} = 2.032 \qquad T_{r_2} = 0.964 \qquad P_{r_2} = 0.203$$

A check of Fig. 3.10 indicates that the Lee/Kesler tables are required for the initial state, whereas the second-virial-coefficient correlation should be suitable for the final state.

Thus, for Step 1, interpolation in Lee/Kesler tables D.7, D.8, D.11, and D.12 provides the values:

$$\frac{(H^R)^0}{RT_c} = -2.709, \quad \frac{(H^R)^1}{RT_c} = -0.921, \quad \frac{(S^R)^0}{R} = -1.846, \quad \frac{(S^R)^1}{R} = -0.938$$

Then:

$$\Delta H_1 = -H^R(343.15 \text{ K}, 150 \text{ bar})$$
$$= -(8.314)(304.2)[-2.709 + (0.224)(-0.921)] = 7372 \text{ J·mol}^{-1}$$
$$\Delta S_1 = -S^R(343.15 \text{ K}, 150 \text{ bar})$$
$$= -(8.314)[-1.846 + (0.224)(-0.938)] = 17.09 \text{ J·mol}^{-1}\cdot\text{K}^{-1}$$

For Step 2, the enthalpy and entropy changes are calculated by the usual heat-capacity integrals, with polynomial coefficients from Table C.1. The ideal-gas-state entropy change caused by the pressure change must also be included.

$$\Delta H_2 = 8.314 \times \text{ICPH}(343.15, 293.15; 5.547, 1.045 \times 10^{-3}, 0.0, -1.157 \times 10^5)$$
$$= -1978 \text{ J·mol}^{-1}$$
$$\Delta S_2 = 8.314 \times \text{ICPS}(343.15, 293.15; 5.547, 1.045 \times 10^{-3}, 0.0, -1.157 \times 10^5)$$
$$-(8.314) \ln (15/150)$$
$$= -6.067 + 19.144 = 13.08 \text{ J·mol}^{-1}\cdot\text{K}^{-1}$$

Finally, for Step 3,

$$\Delta H_3 = H^R(293.15 \text{ K}, 15 \text{ bar})$$
$$= 8.314 \times 304.2 \times \text{HRB}(0.964, 0.203, 0.224) = -660 \text{ J·mol}^{-1}$$
$$\Delta S_3 = S^R(293.15 \text{ K}, 15 \text{ bar})$$
$$= 8.314 \times \text{SRB}(0.964, 0.203, 0.224) = -1.59 \text{ J·mol}^{-1}\cdot\text{K}^{-1}$$

Sums over the three steps yield overall changes, $\Delta H = 4734 \text{ J·mol}^{-1}$ and $\Delta S = 28.6 \text{ J·mol}^{-1}\cdot\text{K}^{-1}$. The largest contribution here comes from the residual properties of the initial state, because the reduced pressure is high, and the super-critical fluid is far from its ideal-gas state. Despite the substantial reduction in temperature, the enthalpy actually increases in the overall process.

For comparison, the properties given in the NIST fluid-properties database, accessed through the NIST Chemistry Webbook, are:

$$H_1 = 16{,}776 \text{ J·mol}^{-1} \qquad S_1 = 67.66 \text{ J·mol}^{-1}\cdot\text{K}^{-1}$$
$$H_2 = 21{,}437 \text{ J·mol}^{-1} \qquad S_1 = 95.86 \text{ J·mol}^{-1}\cdot\text{K}^{-1}$$

From these values, considered accurate, overall changes are $\Delta H = 4661$ J·mol^{-1} and $\Delta S = 28.2$ J·mol 1·K^{-1}. Even though the changes in residual properties make up a substantial part of the total, the prediction from generalized correlations agrees with the NIST data to within 2 percent.

Extension to Gas Mixtures

Although no fundamental basis exists for extension of generalized correlations to mixtures, reasonable and useful approximate results for mixtures can often be obtained with *pseudocritical parameters* resulting from simple linear mixing rules according to the definitions:

$$\omega \equiv \sum_i y_i \omega_i \quad (6.78) \qquad T_{pc} \equiv \sum_i y_i T_{c_i} \quad (6.79) \qquad P_{pc} \equiv \sum_i y_i P_{c_i} \quad (6.80)$$

The values so obtained are the mixture ω and pseudocritical temperature and pressure, T_{pc} and P_{pc}, which replace T_c and P_c to define *pseudoreduced parameters*:

$$T_{pr} = \frac{T}{T_{pc}} \quad (6.81) \qquad P_{pr} = \frac{P}{P_{pc}} \quad (6.82)$$

These replace T_r and P_r for reading entries from the tables of App. D, and lead to values of Z by Eq. (3.57), H^R/RT_{pc} by Eq. (6.66), and S^R/R by Eq. (6.67).

Example 6.5

Estimate V, H^R, and S^R for an equimolar mixture of carbon dioxide(1) and propane(2) at 450 K and 140 bar by the Lee/Kesler correlations.

Solution 6.5

The pseudocritical parameters are found by Eqs. (6.78) through (6.80) with critical constants from Table B.1 of App. B:

$$\omega = y_1\omega_1 + y_2\omega_2 = (0.5)(0.224) + (0.5)(0.152) = 0.188$$
$$T_{pc} = y_1 T_{c_1} + y_2 T_{c_2} = (0.5)(304.2) + (0.5)(369.8) = 337.0 \text{ K}$$
$$P_{pc} = y_1 P_{c_1} + y_2 P_{c_2} = (0.5)(73.83) + (0.5)(42.48) = 58.15 \text{ bar}$$

Then,

$$T_{pr} = \frac{450}{337.0} = 1.335 \quad P_{pr} = \frac{140}{58.15} = 2.41$$

Values of Z^0 and Z^1 from Tables D.3 and D.4 at these reduced conditions are:

$$Z^0 = 0.697 \quad \text{and} \quad Z^1 = 0.205$$

By Eq. (3.57),

$$Z = Z^0 + \omega Z^1 = 0.697 + (0.188)(0.205) = 0.736$$

Whence,

$$V = \frac{ZRT}{P} = \frac{(0.736)(83.14)(450)}{140} = 196.7 \text{ cm}^3 \cdot \text{mol}^{-1}$$

Similarly, from Tables D.7 and D.8, with substitution into Eq. (6.66):

$$\left(\frac{H^R}{RT_{pc}}\right)^0 = -1.730 \quad \left(\frac{H^R}{RT_{pc}}\right)^1 = -0.169$$

$$\frac{H^R}{RT_{pc}} = -1.730 + (0.188)(-0.169) = -1.762$$

and

$$H^R = (8.314)(337.0)(-1.762) = -4937 \text{ J} \cdot \text{mol}^{-1}$$

From Tables D.11 and D.12 and substitution into Eq. (6.67),

$$\frac{S^R}{R} = -0.967 + (0.188)(-0.330) = -1.029$$

and

$$S^R = (8.314)(-1.029) = -8.56 \text{ J} \cdot \text{mol}^{-1} \cdot \text{K}^{-1}$$

6.5 TWO-PHASE SYSTEMS

The curves shown on the *PT* diagram of Fig. 3.1 represent phase boundaries for a pure sub-stance. A phase transition at constant temperature and pressure occurs whenever one of these curves is crossed, and as a result the molar or specific values of the extensive thermodynamic properties change abruptly. Thus the molar or specific volume of a saturated liquid is very different from the molar or specific volume of saturated vapor at the same *T* and *P*. This is true as well for internal energy, enthalpy, and entropy. The exception is the molar or specific Gibbs energy, which for a pure species does not change during a phase transition such as melting, vaporization, or sublimation.

Consider a pure liquid in equilibrium with its vapor and existing in a piston/cylinder arrangement that undergoes a differential evaporation at temperature *T* and corresponding vapor pressure P^{sat}. Equation (6.7) applied to the process reduces to $d(nG) = 0$. Because the number of moles n is constant, $dG = 0$, and this requires the molar (or specific) Gibbs energy of the vapor to be identical with that of the liquid. More generally, for two phases α and β of a pure species coexisting at equilibrium,

$$G^\alpha = G^\beta \tag{6.83}$$

where G^α and G^β are the molar or specific Gibbs energies of the individual phases.

The Clapeyron equation, introduced as Eq. (4.12), follows from this equality. If the temperature of a two-phase system is changed, then the pressure must also change in accord with the relation between vapor pressure and temperature if the two phases continue to coexist in equilibrium. Because Eq. (6.83) applies throughout this change,

$$dG^\alpha = dG^\beta$$

Substituting expressions for dG^α and dG^β as given by Eq. (6.11) yields:

$$V^\alpha dP^{\text{sat}} - S^\alpha dT = V^\beta dP^{\text{sat}} - S^\beta dT$$

which upon rearrangement becomes:

$$\frac{dP^{\text{sat}}}{dT} = \frac{S^\beta - S^\alpha}{V^\beta - V^\alpha} = \frac{\Delta S^{\alpha\beta}}{\Delta V^{\alpha\beta}}$$

The entropy change $\Delta S^{\alpha\beta}$ and the volume change $\Delta V^{\alpha\beta}$ are changes that occur when a unit amount of a pure chemical species is transferred from phase α to phase β at the equilibrium T and P. Integration of Eq. (6.9) for this change yields the latent heat of phase transition:

$$\Delta H^{\alpha\beta} = T\Delta S^{\alpha\beta} \tag{6.84}$$

Thus, $\Delta S^{\alpha\beta} = \Delta H^{\alpha\beta}/T$, and substitution in the preceding equation gives:

$$\frac{dP^{\text{sat}}}{dT} = \frac{\Delta H^{\alpha\beta}}{T\Delta V^{\alpha\beta}} \tag{6.85}$$

which is the Clapeyron equation.

For the particularly important case of phase transition from liquid l to vapor v, Eq. (6.85) is written:

$$\frac{dP^{\text{sat}}}{dT} = \frac{\Delta H^{lv}}{T\Delta V^{lv}} \tag{6.86}$$

But

$$\Delta V^{lv} = V^v - V^l = \frac{RT}{P^{\text{sat}}}(Z^v - Z^l) = \frac{RT}{P^{\text{sat}}}\Delta Z^{lv}$$

Combination of the last two equations gives, on rearrangement:

$$\frac{d \ln P^{\text{sat}}}{dT} = \frac{\Delta H^{lv}}{RT^2 \Delta Z^{lv}} \tag{6.87}$$

or

$$\frac{d \ln P^{\text{sat}}}{d(1/T)} = -\frac{\Delta H^{lv}}{R\Delta Z^{lv}} \tag{6.88}$$

Equations (6.86) through (6.88) are equivalent, exact forms of the Clapeyron equation for pure-species vaporization.

Example 6.6

The Clapeyron equation for vaporization at low pressure is often simplified through reasonable approximations, namely, that the vapor phase is an ideal gas and that the molar volume of the liquid is negligible compared with the molar volume of the vapor. How do these assumptions alter the Clapeyron equation?

Solution 6.6

For the assumptions made,

$$\Delta Z^{lv} = Z^v - Z^l = \frac{P^{\text{sat}} V^v}{RT} - \frac{P^{\text{sat}} V^l}{RT} = 1 - 0 = 1$$

Then by Eq. (6.87),

$$\Delta H^{lv} = -R \frac{d \ln P^{\text{sat}}}{d(1/T)}$$

Known as the Clausius/Clapeyron equation, it relates the latent heat of vaporization directly to the vapor-pressure curve. Specifically, it indicates a direct proportionality of ΔH^{lv} to the slope of a plot of $\ln P^{\text{sat}}$ vs. $1/T$. Such plots of experimental data produce lines for many substances that are very nearly straight. The Clausius/Clapeyron equation implies, in such cases, that ΔH^{lv} is constant, virtually independent of T. This is not in accord with experiment; indeed, ΔH^{lv} decreases monotonically with increasing temperature from the triple point to the critical point, where it becomes zero. The assumptions upon which the Clausius/Clapeyron equation are based have approximate validity only at low pressures.

Temperature Dependence of the Vapor Pressure of Liquids

The Clapeyron equation [Eq. (6.85)] is an exact thermodynamic relation, providing a vital connection between the properties of different phases. When applied to the calculation of latent heats of vaporization, its use presupposes knowledge of the vapor pressure-vs.-temperature relation. Because thermodynamics imposes no model of material behavior, either in general or for particular species, such relations are empirical. As noted in Ex. 6.6, a plot of $\ln P^{\text{sat}}$ vs. $1/T$ generally yields a line that is nearly straight:

$$\ln P^{\text{sat}} = A - \frac{B}{T} \tag{6.89}$$

where A and B are constants for a given species. This equation gives a rough approximation of the vapor-pressure relation for the entire temperature range from the triple point to the critical point. Moreover, it provides an excellent basis for interpolation between reasonably spaced values of T.

The Antoine equation, which is more satisfactory for general use, is:

$$\ln P^{\text{sat}} = A - \frac{B}{T + C} \tag{6.90}$$

A principal advantage of this equation is that values of the constants A, B, and C are readily available for a large number of species.[8] Each set of constants is valid for a specified temperature range, and should not be used much outside that range. The Antoine equation is sometimes written in terms of the base 10 logarithm, and the numerical values of the constants A, B, and C depend on the units selected for T and P. Thus, one must exercise care when using coefficients from different sources to ensure that the form of the equation and units are clear and consistent. Values of Antoine constants for selected substances are given in Table B.2 of App. B.

The accurate representation of vapor-pressure data over a wide temperature range requires an equation of greater complexity. The Wagner equation is one of the best available; it expresses the reduced vapor pressure as a function of reduced temperature:

$$\ln P_r^{sat} = \frac{A\tau + B\tau^{1.5} + C\tau^3 + D\tau^6}{1 - \tau} \tag{6.91}$$

where

$$\tau \equiv 1 - T_r$$

and A, B, C, and D are constants. Values of the constants either for this equation or for the Antoine equation are given by Prausnitz, Poling, and O'Connell[9] for many species.

Corresponding-States Correlations for Vapor Pressure

A number of corresponding-states correlations are available for the vapor pressure of non-polar, non-associating liquids. One of the simplest is that of Lee and Kesler.[10] It is a Pitzer-type correlation, of the form:

$$\ln P_r^{sat}(T_r) = \ln P_r^0(T_r) + \omega \ln P_r^1(T_r) \tag{6.92}$$

where

$$\ln P_r^0(T_r) = 5.92714 - \frac{6.09648}{T_r} - 1.28862 \ln T_r + 0.169347 T_r^6 \tag{6.93}$$

$$\ln P_r^1(T_r) = 15.2518 - \frac{15.6875}{T_r} - 13.4721 \ln T_r + 0.43577 T_r^6 \tag{6.94}$$

Lee and Kesler recommend that the value of ω used with Eq. (6.92) be found *from the correlation* by requiring that it reproduce the normal boiling point. In other words, ω for a particular substance is determined from Eq. (6.92) solved for ω:

$$\omega = \frac{\ln P_{r_n}^{sat} - \ln P_r^0(T_{r_n})}{\ln P_r^1(T_{r_n})} \tag{6.95}$$

where T_{r_n} is the reduced normal boiling point temperature and $P_{r_n}^{sat}$ is the reduced vapor pressure corresponding to 1 standard atmosphere (1.01325 bar).

---

[8]S. Ohe, *Computer Aided Data Book of Vapor Pressure*, Data Book Publishing Co., Tokyo, 1976; T. Boublik, V. Fried, and E. Hala, *The Vapor Pressures of Pure Substances*, Elsevier, Amsterdam, 1984; NIST Chemistry Webbook, http://webbook.nist.gov.

[9]J. M. Prausnitz, B. E. Poling, and J. P. O'Connell, *The Properties of Gases and Liquids*, 5th ed., App. A, McGraw-Hill, 2001.

[10]B. I. Lee and M. G. Kesler, *AIChE J.*, vol. 21, pp. 510–527, 1975.

Example 6.7

Determine the vapor pressure (in kPa) for liquid *n*-hexane at 0, 30, 60, and 90°C:

(a) With constants from App. B.2.   (b) From the Lee/Kesler correlation for P_r^{sat}.

Solution 6.7

(a) With constants from App. B.2, the Antoine equation for *n*-hexane is:

$$\ln P^{sat}/kPa = 13.8193 - \frac{2696.04}{t/°C + 224.317}$$

Substitution of temperatures yields the values of P^{sat} under the heading "Antoine" in the table below. These are presumed equivalent to good experimental values.

(b) We first determine ω from the Lee/Kesler correlation. At the normal boiling point of *n*-hexane (Table B.1),

$$T_{r_n} = \frac{341.9}{507.6} = 0.6736 \qquad \text{and} \qquad P_{r_n}^{sat} = \frac{1.01325}{30.25} = 0.03350$$

Application of Eq. (6.94) then gives the value of ω for use with the Lee/Kesler correlation: $\omega = 0.298$. With this value, the correlation produces the P^{sat} values shown in the table. The average difference from the Antoine values is about 1.5%.

$t\,°C$	P^{sat} kPa (Antoine)	P^{sat} kPa (Lee/Kesler)	$t\,°C$	P^{sat} kPa (Antoine)	P^{sat} kPa (Lee/Kesler)
0	6.052	5.835	30	24.98	24.49
60	76.46	76.12	90	189.0	190.0

Example 6.8

Estimate *V*, *U*, *H*, and *S* for 1-butene vapor at 200°C and 70 bar if *H* and *S* are set equal to zero for saturated liquid at 0°C. Assume that the only data available are:

$$T_c = 420.0 \text{ K} \qquad P_c = 40.43 \text{ bar} \qquad \omega = 0.191$$

$$T_n = 266.9 \text{ K} \quad \text{(normal boiling point)}$$

$$C_P^{ig}/R = 1.967 + 31.630 \times 10^{-3}\,T - 9.837 \times 10^{-6}\,T^2 \quad (T \text{ K})$$

Solution 6.8

The volume of 1-butene vapor at 200°C and 70 bar is calculated directly from the equation $V = ZRT/P$, where Z is given by Eq. (3.53) with values of Z^0 and Z^1 interpolated in Tables D.3 and D.4. For the reduced conditions,

$$T_r = \frac{200 + 273.15}{420.00} = 1.127 \qquad P_r = \frac{70}{40.43} = 1.731$$

the compressibility factor and molar volume are:

$$Z = Z^0 + \omega Z^1 = 0.485 + (0.191)(0.142) = 0.512$$

$$V = \frac{ZRT}{P} = \frac{(0.512)(83.14)(473.15)}{70} = 287.8 \text{ cm}^3 \cdot \text{mol}^{-1}$$

For H and S, we use a computational path like that of Fig. 6.3, leading from an initial state of saturated liquid 1-butene at 0°C, where H and S are zero, to the final state of interest. In this case, an initial vaporization step is required, leading to the four-step path shown by Fig. 6.4. The steps are:

(a) Vaporization at T_1 and $P_1 = P^{\text{sat}}$.
(b) Transition to the ideal-gas state at (T_1, P_1).
(c) Change to (T_2, P_2) in the ideal-gas state.
(d) Transition to the actual final state at (T_2, P_2).

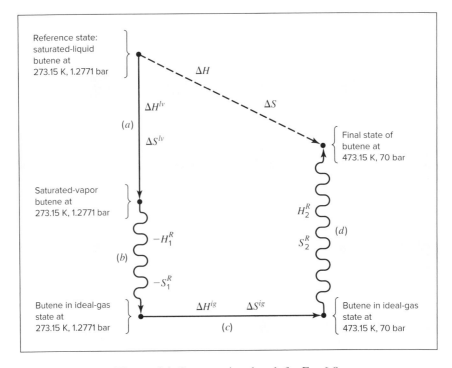

Figure 6.4: Computational path for Ex. 6.8.

- **Step (a):** Vaporization of saturated liquid 1-butene at 0°C. The vapor pressure must be estimated, as it is not given. One method is based on the equation:

$$\ln P^{\text{sat}} = A - \frac{B}{T}$$

The vapor-pressure curve contains both the normal boiling point, for which $P^{\text{sat}} = 1.0133$ bar at 266.9 K, and the critical point, for which $P^{\text{sat}} = 40.43$ bar at 420.0 K. For these two points,

$$\ln 1.0133 = A - \frac{B}{266.9} \qquad \ln 40.43 = A - \frac{B}{420.0}$$

Solution gives,

$$A = 10.1260 \quad B = 2699.11$$

For 0°C (273.15 K), $P^{\text{sat}} = 1.2771$ bar, a result used in steps (b) and (c). Here, the latent heat of vaporization is required. Equation (4.13) provides an estimate at the normal boiling point, where $T_{r_n} = 266.9/420.0 = 0.636$:

$$\frac{\Delta H_n^{lv}}{RT_n} = \frac{1.092\,(\ln P_c - 1.013)}{0.930 - T_{r_n}} = \frac{1.092\,(\ln 40.43 - 1.013)}{0.930 - 0.636} = 9.979$$

and

$$\Delta H_n^{lv} = (9.979)(8.314)(266.9) = 22{,}137 \text{ J·mol}^{-1}$$

The latent heat at $T_r = 273.15/420.0 = 0.650$, is given by Eq. (4.14):

$$\frac{\Delta H^{lv}}{\Delta H_n^{lv}} = \left(\frac{1 - T_r}{1 - T_{r_n}}\right)^{0.38}$$

or

$$\Delta H^{lv} = (22{,}137)(0.350/0.364)^{0.38} = 21{,}810 \text{ J·mol}^{-1}$$

Moreover, by Eq. (6.84),

$$\Delta S^{lv} = \Delta H^{lv}/T = 21{,}810/273.15 = 79.84 \text{ J·mol}^{-1}\text{·K}^{-1}$$

- **Step (b):** Transformation of saturated-vapor 1-butene to its ideal-gas state at the initial conditions (T_1, P_1). Because the pressure is relatively low, the values of H_1^R and S_1^R are estimated by Eqs. (6.68) and (6.69) for the reduced conditions, $T_r = 0.650$ and $P_r = 1.2771/40.43 = 0.0316$. The computational procedure is represented by:

$$\text{HRB}(0.650, 0.0316, 0.191) = -0.0985$$
$$\text{SRB}(0.650, 0.0316, 0.191) = -0.1063$$

and

$$H_1^R = (-0.0985)(8.314)(420.0) = -344 \text{ J·mol}^{-1}$$
$$S_1^R = (-0.1063)(8.314) = -0.88 \text{ J·mol}^{-1}\text{·K}^{-1}$$

As indicated in Fig. 6.4, the property changes for this step are $-H_1^R$ and $-S_1^R$, because the change is from the real to the ideal-gas state.

• **Step (c):** Change in ideal-gas state from (273.15 K, 1.2771 bar) to (473.15 K, 70 bar). Here, ΔH^{ig} and ΔS^{ig} are given by Eqs. (6.76) and (6.77), for which (Secs. 4.1 and 5.5):

$$8.314 \times \text{ICPH}(273.15, 473.15;\ 1.967,\ 31.630 \times 10^{-3},\ -9.837 \times 10^{-6},\ 0.0)$$
$$= 20{,}564\ \text{J·mol}^{-1}$$

$$8.314 \times \text{ICPS}(273.15, 473.15;\ 1.967,\ 31.630 \times 10^{-3},\ -9.837 \times 10^{-6},\ 0.0)$$
$$= 55.474\ \text{J·mol}^{-1}\text{·K}^{-1}$$

Thus, Eqs. (6.76) and (6.77) yield:

$$\Delta H^{ig} = 20{,}564\ \text{J·mol}^{-1}$$

$$\Delta S^{ig} = 55.474 - 8.314\ \ln\frac{70}{1.2771} = 22.18\ \text{J·mol}^{-1}\text{·K}^{-1}$$

• **Step (d):** Transformation of 1-butene from the ideal-gas state to the real-gas state at T_2 and P_2. The final reduced conditions are:

$$T_r = 1.127 \quad P_r = 1.731$$

At the higher pressure of this step, H_2^R and S_2^R are found by Eqs. (6.66) and (6.67), together with the Lee/Kesler correlation. With interpolated values from Tables D.7, D.8, D.11, and D.12, these equations give:

$$\frac{H_2^R}{RT_c} = -2.294 + (0.191)(-0.713) = -2.430$$

$$\frac{S_2^R}{R} = -1.566 + (0.191)(-0.726) = -1.705$$

$$H_2^R = (-2.430)(8.314)(420.0) = -8485\ \text{J·mol}^{-1}$$

$$S_2^R = (-1.705)(8.314) = -14.18\ \text{J·mol}^{-1}\text{·K}^{-1}$$

The sums of the enthalpy and entropy changes for the four steps give the total changes for the process leading from the initial reference state (where H and S are set equal to zero) to the final state:

$$H = \Delta H = 21{,}810 - (-344) + 20{,}564 - 8485 = 34{,}233\ \text{J·mol}^{-1}$$
$$S = \Delta S = 79.84 - (-0.88) + 22.18 - 14.18 = 88.72\ \text{J·mol}^{-1}\text{·K}^{-1}$$

The internal energy is:

$$U = H - PV = 34{,}233 - \frac{(70)(287.8)\ \text{bar·cm}^3\text{·mol}^{-1}}{10\ \text{bar·cm}^3\text{·J}^{-1}} = 32{,}218\ \text{J·mol}^{-1}$$

The residual properties at the final conditions make important contributions to the final values.

Two-Phase Liquid / Vapor Systems

When a system consists of saturated-liquid and saturated-vapor phases coexisting in equilibrium, the total value of any extensive property of the two-phase system is the sum of the total properties of the phases. Written for the volume, this relation is:

$$nV = n^l V^l + n^v V^v$$

where V is the molar volume for a system containing a total number of moles $n = n^l + n^v$. Division by n gives:

$$V = x^l V^l + x^v V^v$$

where x^l and x^v represent the mass fractions of the total system that are liquid and vapor. With $x^l = 1 - x^v$,

$$V = (1 - x^v)V^l + x^v V^v$$

In this equation the properties V, V^l, and V^v may be either molar or unit-mass values. The mass or molar fraction of the system that is vapor x^v is often called the *quality*, particularly when the fluid in question is water. Analogous equations can be written for the other extensive thermodynamic properties. All of these relations are represented by the generic equation

$$M = (1 - x^v)M^l + x^v M^v \tag{6.96a}$$

where M represents V, U, H, S, etc. An alternative form is sometimes useful:

$$M = M^l + x^v \Delta M^{lv} \tag{6.96b}$$

6.6 THERMODYNAMIC DIAGRAMS

A thermodynamic diagram is a graph showing a set of properties for a particular substance, e.g., T, P, V, H, and S. The most common diagrams are: TS, PH (usually $\ln P$ vs. H), and HS (called a *Mollier* diagram). The designations refer to the variables chosen for the coordinates. Other diagrams are possible, but are seldom used.

Figures 6.5 through 6.7 show the general features of these diagrams. Though based on data for water, their general character is similar for all substances. The two-phase states, represented by lines on the PT diagram of Fig. 3.1, lie over *areas* in these diagrams, and the triple point of Fig. 3.1 becomes a *line*. Lines of constant quality in a liquid/vapor region provide two-phase property values directly. The critical point is identified by the letter C, and the solid curve passing through it represents states of saturated liquid (to the left of C) and of saturated vapor (to the right of C). The Mollier diagram (Fig. 6.7) does not usually include volume data. In the vapor or gas region, lines for constant temperature and constant *superheat* appear. Superheat is a term denoting the difference between the actual temperature and the saturation temperature at the same pressure. Thermodynamic diagrams included in this book are the PH diagrams for methane and tetrafluoroethane, and the Mollier diagram for steam in App. F.

Some process paths are easily traced on particular thermodynamic diagrams. For example, the boiler of a steam power plant has liquid water as feed at a temperature below its boiling point, and superheated steam as product. Thus, water is heated at constant P to its saturation temperature (line 1–2 in Figs. 6.5 and 6.6), vaporized at constant T and P (line 2–3), and superheated at constant P (line 3–4). On a PH diagram (Fig. 6.5) the whole process is represented by a horizontal line

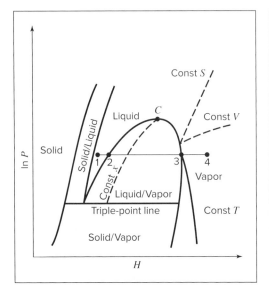

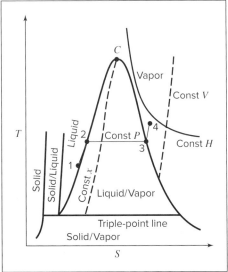

Figure 6.5: Simplified *PH* diagram representing the general features of such charts.

Figure 6.6: Simplified *TS* diagram representing the general features of such charts.

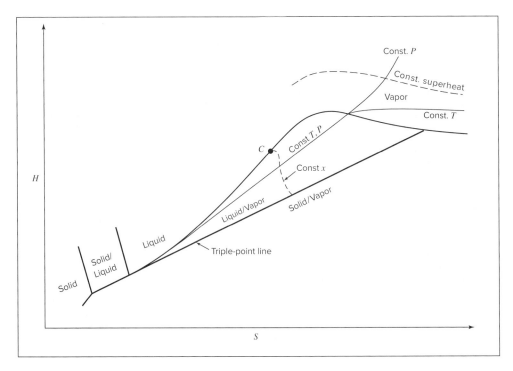

Figure 6.7: Simplified Mollier diagram illustrating the general features of such charts.

corresponding to the boiler pressure. The same process is shown on the *TS* diagram of Fig. 6.6. The compressibility of a liquid is small for temperatures well below T_c, and liquid-phase properties change very slowly with pressure. The constant-*P* lines on this diagram for the liquid region therefore lie very close together, and line 1–2 nearly coincides with the saturated-liquid curve. The isentropic path of the fluid in a reversible adiabatic turbine or compressor is represented on both the *TS* and *HS* (Mollier) diagrams by a vertical line from the initial to the final pressure.

6.7 TABLES OF THERMODYNAMIC PROPERTIES

In many instances thermodynamic properties are tabulated. This has the advantage that data can be presented more precisely than in diagrams, but the need for interpolation is introduced. Thermodynamic tables for saturated steam from its normal freezing point to the critical point and for superheated steam over a substantial pressure range appear in App. E. Values are given at intervals close enough that linear interpolation is satisfactory.[11] The first table shows the equilibrium properties of saturated liquid and saturated vapor at even increments of temperature. The enthalpy and entropy are arbitrarily assigned values of zero for the saturated-liquid state at the triple point. The second table is for the gas region and gives properties of superheated steam at temperatures higher than the saturation temperature for a given pressure. Volume (*V*), internal energy (*U*), enthalpy (*H*), and entropy (*S*) are tabulated as functions of pressure at various temperatures. The steam tables are the most thorough compilation of properties for any single material. However, tables are available for a number of other substances.[12] Electronic versions of such tables generally eliminate the need for manual interpolation.

Example 6.9

Superheated steam originally at P_1 and T_1 expands through a nozzle to an exhaust pressure P_2. Assuming the process is reversible and adiabatic, determine the downstream state of the steam and ΔH for $P_1 = 1000$ kPa, $t_1 = 250°C$, and $P_2 = 200$ kPa.

Solution 6.9

Because the process is both reversible and adiabatic, there is no change in the entropy of the steam. For the initial temperature of 250°C at 1000 kPa, no entries appear in the tables for superheated steam. Interpolation between values for 240°C and 260°C yields, at 1000 kPa,

$$H_1 = 2942.9 \text{ kJ·kg}^{-1} \qquad S_1 = 6.9252 \text{ kJ·kg}^{-1}·\text{K}^{-1}$$

---

[11]Procedures for linear interpolation are shown at the beginning of App. E.

[12]Data for many common chemicals are given by R. H. Perry and D. Green, *Perry's Chemical Engineers' Handbook*, 8th ed., Sec. 2, McGraw-Hill, New York, 2008. See also N. B. Vargaftik, *Handbook of Physical Properties of Liquids and Gases*, 2d ed., Hemisphere Publishing Corp., Washington, DC, 1975. Data for refrigerants appear in the *ASHRAE Handbook: Fundamentals*, American Society of Heating, Refrigerating, and Air-Conditioning Engineers, Inc., Atlanta, 2005. Electronic version of NIST Reference Database is available as REFPROP, ver. 9.1 which cover 121 pure fluids, 5 pseudo-pure fluids and mixtures of 20 components. Data for many common gases, refrigerants, and light hydrocarbons are available from the NIST Chemistry Webbook at http://webbook.nist.gov/chemistry/fluid/.

For the final state at 200 kPa,

$$S_2 = S_1 = 6.9252 \text{ kJ·kg}^{-1}\text{·K}^{-1}$$

Because the entropy of saturated vapor at 200 Pa is greater than S_2, the final state must lie in the two-phase liquid/vapor region. Thus t_2 is the saturation temperature at 200 kPa, given in the superheat tables as $t_2 = 120.23°C$. Equation (6.96a) applied to the entropy becomes:

$$S_2 = (1 - x_2^v)S_2^l + x_2^v S_2^v$$

Numerically,

$$6.9252 = 1.5301(1 - x_2^v) + 7.1268 x_2^v$$

where the values 1.5301 and 7.1268 are entropies of saturated liquid and saturated vapor at 200 kPa. Solving,

$$x_2^v = 0.9640$$

The mixture is 96.40% vapor and 3.60% liquid. Its enthalpy is obtained by further application of Eq. (6.96a):

$$H_2 = (0.0360)(504.7) + (0.9640)(2706.3) = 2627.0 \text{ kJ·kg}^{-1}$$

Finally,

$$\Delta H = H_2 - H_1 = 2627.0 - 2942.9 = -315.9 \text{ kJ·kg}^{-1}$$

For a nozzle, under the stated assumptions the steady-flow energy balance, Eq. (2.31), becomes

$$\Delta H + \frac{1}{2}\Delta u^2 = 0$$

Thus the decrease in enthalpy is exactly compensated by an increase in kinetic energy of the fluid. In other words, the velocity of a fluid increases as it flows through a nozzle, which is its usual purpose. Nozzles are treated further in Sec. 7.1.

Example 6.10

A 1.5 m³ tank contains 500 kg of liquid water in equilibrium with pure water vapor, which fills the remainder of the tank. The temperature and pressure are 100°C and 101.33 kPa. From a water line at a constant temperature of 70°C and a constant pressure somewhat above 101.33 kPa, 750 kg of liquid is bled into the tank. If the temperature and pressure in the tank are not to change as a result of the process, how much energy as heat must be transferred to the tank?

Solution 6.10

Choose the tank as the control volume. There is no work, and kinetic- and potential-energy changes are assumed negligible. Equation (2.28) therefore is written

$$\frac{d(mU)_{\text{tank}}}{dt} - H'\dot{m}' = \dot{Q}$$

where the prime denotes the state of the inlet stream. The mass balance, $\dot{m}' = dm_{\text{tank}}/dt$, may be combined with the energy balance to yield

$$\frac{d(mU)_{\text{tank}}}{dt} - H'\frac{dm_{\text{tank}}}{dt} = \dot{Q}$$

Multiplication by dt and integration over time (with H' constant) give

$$Q = \Delta(mU)_{\text{tank}} - H'\,\Delta m_{\text{tank}}$$

The definition of enthalpy may be applied to the entire contents of the tank to give

$$\Delta(mU)_{\text{tank}} = \Delta(mH)_{\text{tank}} - \Delta(PmV)_{\text{tank}}$$

Because total tank volume mV and P are constant, $\Delta(PmV)_{\text{tank}} = 0$. Then, with $\Delta(mH)_{\text{tank}} = (m_2H_2)_{\text{tank}} - (m_1H_1)_{\text{tank}}$, the two preceding equations combine to yield

$$Q = (m_2H_2)_{\text{tank}} - (m_1H_1)_{\text{tank}} - H'\,\Delta m_{\text{tank}} \qquad (A)$$

where Δm_{tank} is the 750 kg of water bled into the tank, and subscripts 1 and 2 refer to conditions in the tank at the beginning and end of the process. At the end of the process the tank still contains saturated liquid and saturated vapor in equilibrium at 100°C and 101.33 kPa. Hence, m_1H_1 and m_2H_2 each consist of two terms, one for the liquid phase and one for the vapor phase.

The numerical solution makes use of the following enthalpies taken from the steam tables:

$$H' \quad = 293.0 \text{ kJ·kg}^{-1}; \text{ saturated liquid at } 70°C$$
$$H^l_{\text{tank}} = 419.1 \text{ kJ·kg}^{-1}; \text{ saturated liquid at } 100°C$$
$$H^v_{\text{tank}} = 2676.0 \text{ kJ·kg}^{-1}; \text{ saturated vapor at } 100°C$$

The volume of vapor in the tank initially is 1.5 m³ minus the volume occupied by the 500 kg of liquid water. Thus,

$$m^v_1 = \frac{1.5 - (500)(0.001044)}{1.673} = 0.772 \text{ kg}$$

where 0.001044 and 1.673 m³·kg⁻¹ are the specific volumes of saturated liquid and saturated vapor at 100°C from the steam tables. Then,

$$(m_1H_1)_{\text{tank}} = m^l_1H^l_1 + m^v_1H^v_1 = 500(419.1) + 0.722(2676.0) = 211{,}616 \text{ kJ}$$

At the end of the process, the masses of liquid and vapor are determined by a mass balance and by the fact that the tank volume is still 1.5 m³:

$$m_2 = 500 + 0.722 + 750 = m^v_2 + m^l_2$$
$$1.5 = 1.673\,m^v_2 + 0.001044\,m^l_2$$

Solution gives:

$$m_2^l = 1250.65 \text{ kg} \quad \text{and} \quad m_2^v = 0.116 \text{ kg}$$

Then, with $H_2^l = H_1^l$ and $H_2^v = H_1^v$,

$$(m_2 H_2)_{\text{tank}} = (1250.65)(419.1) + (0.116)(2676.0) = 524{,}458 \text{ kJ}$$

Substitution of appropriate values into Eq. (A) gives:

$$Q = 524{,}458 - 211{,}616 - (750)(293.0) = 93{,}092 \text{ kJ}$$

Before conducting the rigorous analysis demonstrated here, one might make a reasonable estimate of the heat requirement by assuming that it is equal to the enthalpy change for heating 750 kg of water from 70°C to 100°C. This approach would give 94,575 kJ, which is slightly higher than the rigorous result, because it neglects the heat effect of condensing water vapor to accommodate the added liquid.

6.8 SYNOPSIS

After thorough study of this chapter, including multiple readings and working through example and end-of-chapter problems, one should be able to:

- Write and apply the fundamental property relations for internal energy, enthalpy, Gibbs energy, and Helmholtz energy in both a general form applicable to any closed PVT system [Eqs. (6.1), (6.5), (6.6), and (6.7)], and in a form applicable to one mole of a homogeneous substance [Eqs. (6.8)–(6.11)]

- Write the Maxwell relations [Eqs. (6.14)–(6.17)], and apply them to replace unmeasurable partial derivatives involving entropy with partial derivatives that can be determined from PVT data

- Recognize that knowledge of any thermodynamic energy measure as a function of its canonical variables [$U(S, V)$, $H(S, P)$, $A(T, V)$ or $G(T, P)$] implicitly provides complete property information

- Obtain any thermodynamic property from G/RT as a function of T and P

- Write thermodynamic functions in terms of alternative, noncanonical variables, including $H(T, P)$, $S(T, P)$, $U(T, V)$, and $S(T, V)$, and apply specialized forms of these for liquids, using the isothermal compressibility and volume expansivity

- Define and apply residual properties and relationships among them (e.g., the fundamental residual property relations)

- Estimate residual properties by:

    The volume-explicit two-term virial equation of state, Eqs. (6.54)–(6.56)
    The three-term pressure-explicit virial equation of state, Eqs. (6.61)–(6.63)
    The full Lee/Kesler correlations, Eqs. (6.66) and (6.67) and App. D
    The two-term virial equation with coefficients from the Abbott equations, Eq. (6.68)–(6.71)

- Explain the origins of the Clapeyron equation and apply it to estimate the change in phase transition pressure with temperature from latent heat data and vice versa

- Recognize equality of Gibbs energy, temperature, and pressure as a criterion for phase equilibrium of a pure substance

- Read common thermodynamic diagrams and trace the paths of processes on them

- Apply the Antoine equation and similar equations to determine vapor pressure at a given temperature and enthalpy of vaporization, via the Clapeyron equation

- Construct multi-step computational paths that allow one to compute property changes for arbitrary changes of state of a pure substance, making use of data or correlations for residual properties, heat capacities, and latent heats

6.9 ADDENDUM. RESIDUAL PROPERTIES IN THE ZERO-PRESSURE LIMIT

The constant J, omitted from Eqs. (6.46), (6.48), and (6.49), is the value of G^R/RT in the limit as $P \to 0$. The following treatment of residual properties in this limit provides background. Because a gas becomes ideal as $P \to 0$ (in the sense that $Z \to 1$), one might suppose that in this limit all residual properties are zero. This is not in general true, as is easily demonstrated for the residual volume.

Written for V^R in the limit of zero pressure, Eq. (6.41) becomes

$$\lim_{P \to 0} V^R = \lim_{P \to 0} V - \lim_{P \to 0} V^{ig}$$

Both terms on the right side of this equation are infinite, and their difference is indeterminate. Experimental insight is provided by Eq. (6.40):

$$\lim_{P \to 0} V^R = RT \lim_{P \to 0} \left(\frac{Z - 1}{P} \right) = RT \lim_{P \to 0} \left(\frac{\partial Z}{\partial p} \right)_T$$

The center expression arises directly from Eq. (6.40), and the rightmost expression is obtained by application of L'Hôpital's rule. Thus, V^R/RT in the limit as $P \to 0$ at a given T is proportional to the slope of the Z-versus-P isotherm at $P = 0$. Figure 3.7 shows clearly that these values are finite, and not, in general, zero.

For the internal energy, $U^R \equiv U - U^{ig}$. Because U^{ig} is a function of T only, a plot of U^{ig} vs. P for a given T is a horizontal line extending to $P = 0$. For a real gas with intermolecular forces, an isothermal expansion to $P \to 0$ results in a finite increase in U because the molecules move apart against the forces of intermolecular attraction. Expansion to $P = 0$ ($V = \infty$) reduces these forces to zero, exactly as in an ideal gas, and therefore at all temperatures,

$$\lim_{P \to 0} U = U^{ig} \quad \text{and} \quad \lim_{P \to 0} U^R = 0$$

From the definition of enthalpy,

$$\lim_{P \to 0} H^R = \lim_{P \to 0} U^R + \lim_{P \to 0} (P V^R)$$

Because both terms on the right are zero, $\lim\limits_{P\to 0} H^R = 0$ for all temperatures.

For the Gibbs energy, by Eq. (6.37),

$$d\left(\frac{G}{RT}\right) = \frac{V}{RT}dP \quad \text{(const } T)$$

For the ideal-gas state, $V = V^{ig} = RT/P$, and this becomes

$$d\left(\frac{G^{ig}}{RT}\right) = \frac{dP}{P} \quad \text{(const } T)$$

Integration from $P = 0$ to pressure P yields

$$\frac{G^{ig}}{RT} = \left(\frac{G^{ig}}{RT}\right)_{P=0} + \int_0^P \frac{dP}{P} = \left(\frac{G^{ig}}{RT}\right)_{P=0} + \ln P + \infty \quad \text{(const } T)$$

For finite values of G^{ig}/RT at $P > 0$, we must have $\lim\limits_{P\to 0}(G^{ig}/RT) = -\infty$. We cannot reasonably expect a different result for $\lim\limits_{P\to 0}(G/RT)$, and we conclude that

$$\lim_{P\to 0}\frac{G^R}{RT} = \lim_{P\to 0}\frac{G}{RT} - \lim_{P\to 0}\frac{G^{ig}}{RT} = \infty - \infty$$

Thus G^R/RT (and of course G^R) is, like V^R, indeterminate in the limit as $P \to 0$. In this case, however, no experimental means exists for finding the limiting value. However, we have no reason to presume it is zero, and therefore we regard it like $\lim\limits_{P\to 0} V^R$ as finite, and not in general zero.

Equation (6.44) provides an opportunity for further analysis. We write it for the limiting case of $P = 0$:

$$\left(\frac{H^R}{RT^2}\right)_{P=0} = -\left[\frac{\partial(G^R/RT)}{\partial T}\right]_{P=0}$$

As already shown, $H^R(P = 0) = 0$, and therefore the preceding derivative is zero. As a result,

$$\left(\frac{G^R}{RT}\right)_{P=0} = J$$

where J is a constant of integration, independent of T, justifying the derivation of Eq. (6.46).

6.10 PROBLEMS

6.1. Starting with Eq. (6.9), show that isobars in the vapor region of a Mollier (HS) diagram must have positive slope and positive curvature.

6.2. (*a*) Making use of the fact that Eq. (6.21) is an exact differential expression, show that:

$$(\partial C_P/\partial P)_T = -T\left(\partial^2 V/\partial T^2\right)_P$$

What is the result of application of this equation to an ideal gas?

(b) Heat capacities C_V and C_P are defined as temperature derivatives respectively of U and H. Because these properties are related, one expects the heat capacities also to be related. Show that the general expression connecting C_P to C_V is:

$$C_P = C_V + T \left(\frac{\partial P}{\partial T}\right)_V \left(\frac{\partial V}{\partial T}\right)_P$$

Show that Eq. (B) of Ex. 6.2 is another form of this expression.

6.3. If U is considered a function of T and P, the "natural" heat capacity is neither C_V nor C_P, but rather the derivative $(\partial U/\partial T)_P$. Develop the following connections between $(\partial U/\partial T)_P$, C_P, and C_V:

$$\left(\frac{\partial U}{\partial T}\right)_P = C_P - P\left(\frac{\partial V}{\partial T}\right)_P = C_P - \beta PV$$

$$= C_V + \left[T\left(\frac{\partial P}{\partial T}\right)_V - P\right]\left(\frac{\partial V}{\partial T}\right)_P = C_V + \frac{\beta}{\kappa}(\beta T - \kappa P)V$$

To what do these equations reduce for an ideal gas? For an incompressible liquid?

6.4. The PVT behavior of a certain gas is described by the equation of state:

$$P(V - b) = RT$$

where b is a constant. If in addition C_V is constant, show that:

(a) U is a function of T only.
(b) $\gamma = $ const.
(c) For a mechanically reversible process, $P(V - b)^\gamma = $ const.

6.5. A pure fluid is described by the *canonical equation of state:* $G = \Gamma(T) + RT \ln P$, where $\Gamma(T)$ is a substance-specific function of temperature. Determine for such a fluid expressions for V, S, H, U, C_P, and C_V. These results are consistent with those for an important model of gas-phase behavior. What is the model?

6.6. A pure fluid, described by the *canonical equation of state:* $G = F(T) + KP$, where $F(T)$ is a substance-specific function of temperature and K is a substance-specific constant. Determine for such a fluid expressions for V, S, H, U, C_P, and C_V. These results are consistent with those for an important model of liquid-phase behavior. What is the model?

6.7. Estimate the change in enthalpy and entropy when liquid ammonia at 270 K is compressed from its saturation pressure of 381 kPa to 1200 kPa. For saturated liquid ammonia at 270 K, $V^l = 1.551 \times 10^{-3} \ \text{m}^3\cdot\text{kg}^{-1}$, and $\beta = 2.095 \times 10^{-3} \ \text{K}^{-1}$.

6.8. Liquid isobutane is throttled through a valve from an initial state of 360 K and 4000 kPa to a final pressure of 2000 kPa. Estimate the temperature change and the entropy change of the isobutane. The specific heat of liquid isobutane at 360 K is 2.78 J·g^{-1}·°C^{-1}. Estimates of V and β may be found from Eq. (3.68).

6.9. One kilogram of water ($V_1 = 1003$ cm^3·kg^{-1}) in a piston/cylinder device at 25°C and 1 bar is compressed in a mechanically reversible, isothermal process to 1500 bar. Determine Q, W, ΔU, ΔH, and ΔS given that $\beta = 250 \times 10^{-6}$ K^{-1} and $\kappa = 45 \times 10^{-6}$ bar^{-1}. A satisfactory assumption is that V is constant at its arithmetic average value.

6.10. Liquid water at 25°C and 1 bar fills a rigid vessel. If heat is added to the water until its temperature reaches 50°C, what pressure is developed? The average value of β between 25 and 50°C is 36.2×10^{-5} K^{-1}. The value of κ at 1 bar and 50°C is 4.42×10^{-5} bar^{-1}, and may be assumed to be independent of P. The specific volume of liquid water at 25°C is 1.0030 cm^3·g^{-1}.

6.11. Determine expressions for G^R, H^R, and S^R implied by the three-term virial equation in volume, Eq. (3.38).

6.12. Determine expressions for G^R, H^R, and S^R implied by the van der Waals equation of state, Eq. (3.39).

6.13. Determine expressions for G^R, H^R, and S^R implied by the Dieterici equation:

$$P = \frac{RT}{V - b} \exp\left(-\frac{a}{VRT}\right)$$

Here, parameters a and b are functions of composition only.

6.14. Estimate the entropy change of vaporization of benzene at 50°C. The vapor pressure of benzene is given by the equation:

$$\ln P^{\text{sat}}/\text{kPa} = 13.8858 - \frac{2788.51}{t/°C + 220.79}$$

(*a*) Use Eq. (6.86) with an estimated value of ΔV^{lv}.
(*b*) Use the Clausius/Clapeyron equation of Ex. 6.6.

6.15. Let P_1^{sat} and P_2^{sat} be values of the saturation vapor pressure of a pure liquid at absolute temperatures T_1 and T_2. Justify the following interpolation formula for estimation of the vapor pressure P^{sat} at intermediate temperature T:

$$\ln P^{\text{sat}} = \ln P_1^{\text{sat}} + \frac{T_2(T - T_1)}{T(T_2 - T_1)} \ln \frac{P_2^{\text{sat}}}{P_1^{\text{sat}}}$$

6.16. Assuming the validity of Eq. (6.89), derive *Edmister's formula* for estimation of the acentric factor:

$$\omega = \frac{3}{7}\left(\frac{\theta}{1 - \theta}\right) \log P_c - 1$$

where $\theta \equiv T_n/T_c$, T_n is the normal boiling point, and P_c is in (atm).

6.17. Very pure liquid water can be subcooled at atmospheric pressure to temperatures well below 0°C. Assume that 1 kg has been cooled as a liquid to −6°C. A small ice crystal (of negligible mass) is added to "seed" the subcooled liquid. If the subsequent change occurs adiabatically at atmospheric pressure, what fraction of the system freezes, and what is the final temperature? What is ΔS_{total} for the process, and what is its irreversible feature? The latent heat of fusion of water at 0°C is 333.4 J·g^{-1}, and the specific heat of subcooled liquid water is 4.226 J·g^{-1}·°C^{-1}.

6.18. The state of 1(lb$_m$) of steam is changed from saturated vapor at 20 (psia) to superheated vapor at 50 (psia) and 1000(°F). What are the enthalpy and entropy changes of the steam? What would the enthalpy and entropy changes be if steam were an ideal gas?

6.19. A two-phase system of liquid water and water vapor in equilibrium at 8000 kPa consists of equal volumes of liquid and vapor. If the total volume $V^t = 0.15$ m^3, what is the total enthalpy H^t and what is the total entropy S^t?

6.20. A vessel contains 1 kg of H_2O as liquid and vapor in equilibrium at 1000 kPa. If the vapor occupies 70% of the volume of the vessel, determine H and S for the 1 kg of H_2O.

6.21. A pressure vessel contains liquid water and water vapor in equilibrium at 350(°F). The total mass of liquid and vapor is 3(lb$_m$). If the volume of vapor is 50 times the volume of liquid, what is the total enthalpy of the contents of the vessel?

6.22. Wet steam at 230°C has a density of 0.025 g·cm^{-3}. Determine x, H, and S.

6.23. A vessel of 0.15 m^3 volume containing saturated-vapor steam at 150°C is cooled to 30°C. Determine the final volume and mass of *liquid* water in the vessel.

6.24. Wet steam at 1100 kPa expands at constant enthalpy (as in a throttling process) to 101.33 kPa, where its temperature is 105°C. What is the quality of the steam in its initial state?

6.25. Steam at 2100 kPa and 260°C expands at constant enthalpy (as in a throttling process) to 125 kPa. What is the temperature of the steam in its final state, and what is its entropy change? What would be the final temperature and entropy change for an ideal gas?

6.26. Steam at 300 (psia) and 500(°F) expands at constant enthalpy (as in a throttling process) to 20 (psia). What is the temperature of the steam in its final state, and what is its entropy change? What would be the final temperature and entropy change for an ideal gas?

6.27. Superheated steam at 500 kPa and 300°C expands isentropically to 50 kPa. What is its final enthalpy?

6.28. What is the mole fraction of water vapor in air that is saturated with water at 25°C and 101.33 kPa? At 50°C and 101.33 kPa?

6.29. A rigid vessel contains 0.014 m³ of saturated-vapor steam in equilibrium with 0.021 m³ of saturated-liquid water at 100°C. Heat is transferred to the vessel until one phase just disappears, and a single phase remains. Which phase (liquid or vapor) remains, and what are its temperature and pressure? How much heat is transferred in the process?

6.30. A vessel of 0.25 m³ capacity is filled with saturated steam at 1500 kPa. If the vessel is cooled until 25% of the steam has condensed, how much heat is transferred, and what is the final pressure?

6.31. A vessel of 2 m³ capacity contains 0.02 m³ of liquid water and 1.98 m³ of water vapor at 101.33 kPa. How much heat must be added to the contents of the vessel so that the liquid water is just evaporated?

6.32. A rigid vessel of 0.4 m³ volume is filled with steam at 800 kPa and 350°C. How much heat must be transferred from the steam to bring its temperature to 200°C?

6.33. One kilogram of steam is contained in a piston/cylinder device at 800 kPa and 200°C.

(*a*) If it undergoes a mechanically reversible, isothermal expansion to 150 kPa, how much heat does it absorb?
(*b*) If it undergoes a reversible, adiabatic expansion to 150 kPa, what is its final temperature and how much work is done?

6.34. Steam at 2000 kPa containing 6% moisture is heated at constant pressure to 575°C. How much heat is required per kilogram?

6.35. Steam at 2700 kPa and with a quality of 0.90 undergoes a reversible, adiabatic expansion in a nonflow process to 400 kPa. It is then heated at constant volume until it is saturated vapor. Determine Q and W for the process.

6.36. Four kilograms of steam in a piston/cylinder device at 400 kPa and 175°C undergoes a mechanically reversible, isothermal compression to a final pressure such that the steam is just saturated. Determine Q and W for the process.

6.37. Steam undergoes a change from an initial state of 450°C and 3000 kPa to a final state of 140°C and 235 kPa. Determine ΔH and ΔS:

(*a*) From steam-table data.
(*b*) By equations for an ideal gas.
(*c*) By appropriate generalized correlations.

6.38. A piston/cylinder device operating in a cycle with steam as the working fluid executes the following steps:

- Steam at 550 kPa and 200°C is heated at constant volume to a pressure of 800 kPa.
- It then expands, reversibly and adiabatically, to the initial temperature of 200°C.
- Finally, the steam is compressed in a mechanically reversible, isothermal process to the initial pressure of 550 kPa.

What is the thermal efficiency of the cycle?

6.39. A piston/cylinder device operating in a cycle with steam as the working fluid executes the following steps:

- Saturated-vapor steam at 300 (psia) is heated at constant pressure to 900(°F).
- It then expands, reversibly and adiabatically, to the initial temperature of 417.35(°F).
- Finally, the steam is compressed in a mechanically reversible, isothermal process to the initial state.

What is the thermal efficiency of the cycle?

6.40. Steam entering a turbine at 4000 kPa and 400°C expands reversibly and adiabatically.

(*a*) For what discharge pressure is the exit stream a saturated vapor?
(*b*) For what discharge pressure is the exit stream a wet vapor with quality of 0.95?

6.41. A steam turbine, operating reversibly and adiabatically, takes in superheated steam at 2000 kPa and discharges at 50 kPa.

(*a*) What is the minimum superheat required so that the exhaust contains no moisture?
(*b*) What is the power output of the turbine if it operates under these conditions and the steam rate is 5 kg·s^{-1}?

6.42. An operating test of a steam turbine produces the following results. With steam supplied to the turbine at 1350 kPa and 375°C, the exhaust from the turbine at 10 kPa is saturated vapor. Assuming adiabatic operation and negligible changes in kinetic and potential energies, determine the turbine efficiency, i.e., the ratio of actual work of the turbine to the work of a turbine operating isentropically from the same initial conditions to the same exhaust pressure.

6.43. A steam turbine operates adiabatically with a steam rate of 25 kg·s^{-1}. The steam is supplied at 1300 kPa and 400°C and discharges at 40 kPa and 100°C. Determine the power output of the turbine and the efficiency of its operation in comparison with a turbine that operates *reversibly* and adiabatically from the same initial conditions to the same final pressure.

6.44. From steam-table data, estimate values for the residual properties V^R, H^R, and S^R for steam at 225°C and 1600 kPa, and compare with values found by a suitable generalized correlation.

6.45. From data in the steam tables:

(*a*) Determine values for G^l and G^v for saturated liquid and vapor at 1000 kPa. Should these be the same?
(*b*) Determine values for $\Delta H^{lv}/T$ and ΔS^{lv} at 1000 kPa. Should these be the same?
(*c*) Find values for V^R, H^R, and S^R for saturated vapor at 1000 kPa.
(*d*) Estimate a value for dP^{sat}/dT at 1000 kPa and apply the Clapeyron equation to evaluate ΔS^{lv} at 1000 kPa. Does this result agree with the steam-table value?

Apply appropriate generalized correlations for evaluation of V^R, H^R, and S^R for saturated vapor at 1000 kPa. Do these results agree with the values found in (*c*)?

6.46. From data in the steam tables:

(a) Determine values for G^l and G^v for saturated liquid and vapor at 150(psia). Should these be the same?

(b) Determine values for $\Delta H^{lv}/T$ and ΔS^{lv} at 150(psia). Should these be the same?

(c) Find values for V^R, H^R, and S^R for saturated vapor at 150(psia).

(d) Estimate a value for dP^{sat}/dT at 150(psia) and apply the Clapeyron equation to evaluate ΔS^{lv} at 150(psia). Does this result agree with the steam-table value?

Apply appropriate generalized correlations for evaluation of V^R, H^R, and S^R for saturated vapor at 150(psia). Do these results agree with the values found in (c)?

6.47. Propane gas at 1 bar and 35°C is compressed to a final state of 135 bar and 195°C. Estimate the molar volume of the propane in the final state and the enthalpy and entropy changes for the process. In its initial state, propane may be assumed an ideal gas.

6.48. Propane at 70°C and 101.33 kPa is compressed isothermally to 1500 kPa. Estimate ΔH and ΔS for the process by suitable generalized correlations.

6.49. A stream of propane gas is partially liquefied by throttling from 200 bar and 370 K to 1 bar. What fraction of the gas is liquefied in this process? The vapor pressure of propane is given by Eq. (6.91) with parameters: $A = -6.72219$, $B = 1.33236$, $C = -2.13868$, $D = -1.38551$.

6.50. Estimate the molar volume, enthalpy, and entropy for 1,3-butadiene as a saturated vapor and as a saturated liquid at 380 K. The enthalpy and entropy are set equal to zero for the ideal-gas state at 101.33 kPa and 0°C. The vapor pressure of 1,3-butadiene at 380 K is 1919.4 kPa.

6.51. Estimate the molar volume, enthalpy, and entropy for *n*-butane as a saturated vapor and as a saturated liquid at 370 K. The enthalpy and entropy are set equal to zero for the ideal-gas state at 101.33 kPa and 273.15 K. The vapor pressure of *n*-butane at 370 K is 1435 kPa.

6.52. The *total* steam demand of a plant over the period of an hour is 6000 kg, but instantaneous demand fluctuates from 4000 to 10,000 kg·h^{-1}. Steady boiler operation at 6000 kg·h^{-1} is accommodated by inclusion of an *accumulator*, essentially a tank containing mostly saturated liquid water that "floats on the line" between the boiler and the plant. The boiler produces saturated steam at 1000 kPa, and the plant operates with steam at 700 kPa. A control valve regulates the steam pressure upstream from the accumulator and a second control valve regulates the pressure downstream from the accumulator. When steam demand is less than boiler output, steam flows into and is largely condensed by liquid residing in the accumulator, in the process increasing the pressure to values greater than 700 kPa. When steam demand is greater than boiler output, water in the accumulator vaporizes and steam flows out, thus reducing the pressure to values less than 1000 kPa. What accumulator volume is required for this service if no more that 95% of its volume should be occupied by liquid?

6.53. Propylene gas at 127°C and 38 bar is throttled in a steady-state flow process to 1 bar, where it may be assumed to be an ideal gas. Estimate the final temperature of the propylene and its entropy change.

6.54. Propane gas at 22 bar and 423 K is throttled in a steady-state flow process to 1 bar. Estimate the entropy change of the propane caused by this process. In its final state, propane may be assumed to be an ideal gas.

6.55. Propane gas at 100°C is compressed isothermally from an initial pressure of 1 bar to a final pressure of 10 bar. Estimate ΔH and ΔS.

6.56. Hydrogen sulfide gas is compressed from an initial state of 400 K and 5 bar to a final state of 600 K and 25 bar. Estimate ΔH and ΔS.

6.57. Carbon dioxide expands at constant enthalpy (as in a throttling process) from 1600 kPa and 45°C to 101.33 kPa. Estimate ΔS for the process.

6.58. A stream of ethylene gas at 250°C and 3800 kPa expands isentropically in a turbine to 120 kPa. Determine the temperature of the expanded gas and the work produced if the properties of ethylene are calculated by:

(*a*) Equations for an ideal gas;
(*b*) Appropriate generalized correlations.

6.59. A stream of ethane gas at 220°C and 30 bar expands isentropically in a turbine to 2.6 bar. Determine the temperature of the expanded gas and the work produced if the properties of ethane are calculated by:

(*a*) Equations for an ideal gas;
(*b*) Appropriate generalized correlations.

6.60. Estimate the final temperature and the work required when 1 mol of *n*-butane is compressed isentropically in a steady-flow process from 1 bar and 50°C to 7.8 bar.

6.61. Determine the maximum amount of work obtainable in a flow process from 1 kg of steam at 3000 kPa and 450°C for surrounding conditions of 300 K and 101.33 kPa.

6.62. Liquid water at 325 K and 8000 kPa flows into a boiler at a rate of 10 kg·s^{-1} and is vaporized, producing saturated vapor at 8000 kPa. What is the maximum fraction of the heat added to the water in the boiler that can be converted into work in a process whose product is water at the initial conditions, if $T_\sigma = 300$ K? What happens to the rest of the heat? What is the rate of entropy change in the surroundings as a result of the work-producing process? In the system? Total?

6.63. Suppose the heat added to the water in the boiler in the preceding problem comes from a furnace at a temperature of 600°C. What is the total rate of entropy generation as a result of the heating process? What is $\dot{W}_{lost}$?

6.64. An ice plant produces 0.5 kg·s^{-1} of flake ice at 0°C from water at 20°C (T_σ) in a continuous process. If the latent heat of fusion of water is 333.4 kJ·kg^{-1} and if the thermodynamic efficiency of the process is 32%, what is the power requirement of the plant?

6.65. An inventor has developed a complicated process for making heat continuously available at an elevated temperature. Saturated steam at 100°C is the only source of energy. Assuming that there is plenty of cooling water available at 0°C, what is the maximum temperature level at which heat in the amount of 2000 kJ can be made available for each kilogram of steam flowing through the process?

6.66. Two boilers, both operating at 200(psia), discharge equal amounts of steam into the same steam main. Steam from the first boiler is superheated at 420(°F) and steam from the second is wet with a quality of 96%. Assuming adiabatic mixing and negligible changes in potential and kinetic energies, what is the equilibrium condition after mixing and what is S_G for each (lb$_m$) of discharge steam?

6.67. A rigid tank of 80(ft)^3 capacity contains 4180(lb$_m$) of saturated liquid water at 430(°F). This amount of liquid almost completely fills the tank, the small remaining volume being occupied by saturated-vapor steam. Because a bit more vapor space in the tank is wanted, a valve at the top of the tank is opened, and saturated-vapor steam is vented to the atmosphere until the temperature in the tank falls to 420(°F). Assuming no heat transfer to the contents of the tank, determine the mass of steam vented.

6.68. A tank of 50 m^3 capacity contains steam at 4500 kPa and 400°C. Steam is vented from the tank through a relief valve to the atmosphere until the pressure in the tank falls to 3500 kPa. If the venting process is adiabatic, estimate the final temperature of the steam in the tank and the mass of steam vented.

6.69. A tank of 4 m^3 capacity contains 1500 kg of liquid water at 250°C in equilibrium with its vapor, which fills the rest of the tank. A quantity of 1000 kg of water at 50°C is pumped into the tank. How much heat must be added during this process if the temperature in the tank is not to change?

6.70. Liquid nitrogen is stored in 0.5 m^3 metal tanks that are thoroughly insulated. Consider the process of filling an evacuated tank, initially at 295 K. It is attached to a line containing liquid nitrogen at its normal boiling point of 77.3 K and at a pressure of several bars. At this condition, its enthalpy is –120.8 kJ·kg^{-1}. When a valve in the line is opened, the nitrogen flowing into the tank at first evaporates in the process of cooling the tank. If the tank has a mass of 30 kg and the metal has a specific heat capacity of 0.43 kJ·kg^{-1}·K^{-1}, what mass of nitrogen must flow into the tank just to cool it to a temperature such that *liquid* nitrogen begins to accumulate in the tank? Assume that the nitrogen and the tank are always at the same temperature. The properties of saturated nitrogen vapor at several temperatures are given as follows:

T/K	P /bar	V^v /m³·kg⁻¹	H^v /kJ·kg⁻¹
80	1.396	0.1640	78.9
85	2.287	0.1017	82.3
90	3.600	0.06628	85.0
95	5.398	0.04487	86.8
100	7.775	0.03126	87.7
105	10.83	0.02223	87.4
110	14.67	0.01598	85.6

6.71. A well-insulated tank of 50 m³ volume initially contains 16,000 kg of water distributed between liquid and vapor phases at 25°C. Saturated steam at 1500 kPa is admitted to the tank until the pressure reaches 800 kPa. What mass of steam is added?

6.72. An insulated evacuated tank of 1.75 m³ volume is attached to a line containing steam at 400 kPa and 240°C. Steam flows into the tank until the pressure in the tank reaches 400 kPa. Assuming no heat flow from the steam to the tank, prepare graphs showing the mass of steam in the tank and its temperature as a function of pressure in the tank.

6.73. A 2 m³ tank initially contains a mixture of saturated-vapor steam and saturated-liquid water at 3000 kPa. Of the total mass, 10% is vapor. Saturated-liquid water is bled from the tank through a valve until the total mass in the tank is 40% of the initial total mass. If during the process the temperature of the contents of the tank is kept constant, how much heat is transferred?

6.74. A stream of water at 85°C, flowing at the rate of 5 kg·s⁻¹ is formed by mixing water at 24°C with saturated steam at 400 kPa. Assuming adiabatic operation, at what rates are the steam and water fed to the mixer?

6.75. In a desuperheater, liquid water at 3100 kPa and 50°C is sprayed into a stream of superheated steam at 3000 kPa and 375°C in an amount such that a single stream of saturated-vapor steam at 2900 kPa flows from the desuperheater at the rate of 15 kg·s⁻¹. Assuming adiabatic operation, what is the mass flow rate of the water? What is $\dot{S}_G$ for the process? What is the irreversible feature of the process?

6.76. Superheated steam at 700 kPa and 280°C flowing at the rate of 50 kg·s⁻¹ is mixed with liquid water at 40°C to produce steam at 700 kPa and 200°C. Assuming adiabatic operation, at what rate is water supplied to the mixer? What is $\dot{S}_G$ for the process? What is the irreversible feature of the process?

6.77. A stream of air at 12 bar and 900 K is mixed with another stream of air at 2 bar and 400 K with 2.5 times the mass flow rate. If this process were accomplished reversibly and adiabatically, what would be the temperature and pressure of the resulting air stream? Assume air to be an ideal gas for which $C_P = (7/2)R$.

6.78. Hot nitrogen gas at 750(°F) and atmospheric pressure flows into a waste-heat boiler at the rate of 40(lb$_m$)s^{-1}, and transfers heat to water boiling at 1(atm). The water feed to the boiler is saturated liquid at 1(atm), and it leaves the boiler as superheated steam at 1(atm) and 300(°F). If the nitrogen is cooled to 325(°F) and if heat is lost to the surroundings at a rate of 60(Btu) for each (lb$_m$) of steam generated, what is the steam-generation rate? If the surroundings are at 70(°F), what is $\dot{S}_G$ for the process? Assume nitrogen to be an ideal gas for which $C_P = (7/2)R$.

6.79. Hot nitrogen gas at 400°C and atmospheric pressure flows into a waste-heat boiler at the rate of 20 kg·s^{-1}, and transfers heat to water boiling at 101.33 kPa. The water feed to the boiler is saturated liquid at 101.33 kPa, and it leaves the boiler as superheated steam at 101.33 kPa and 150°C. If the nitrogen is cooled to 170°C and if heat is lost to the surroundings at a rate of 80 kJ for each kilogram of steam generated, what is the steam-generation rate? If the surroundings are at 25°C, what is $\dot{S}_G$ for the process? Assume nitrogen to be an ideal gas for which $C_P = (7/2)R$.

6.80. Show that isobars and isochores have positive slopes in the single-phase regions of a *TS* diagram. Suppose that $C_P = a + bT$, where a and b are positive constants. Show that the curvature of an isobar is also positive. For specified T and S, which is steeper: an isobar or an isochore? Why? Note that $C_P > C_V$.

6.81. Starting with Eq. (6.9), show that isotherms in the vapor region of a Mollier (*HS*) diagram have slopes and curvatures given by:

$$\left(\frac{\partial H}{\partial S}\right)_T = \frac{1}{\beta}(\beta T - 1) \qquad \left(\frac{\partial^2 H}{\partial S^2}\right)_T = -\frac{1}{\beta^3 V}\left(\frac{\partial \beta}{\partial P}\right)_T$$

Here, β is volume expansivity. If the vapor is described by the two-term virial equation in P, Eq. (3.36), what can be said about the *signs* of these derivatives? Assume that, for normal temperatures, B is negative and dB/dT is positive.

6.82. The temperature dependence of the second virial coefficient B is shown for nitrogen in Fig. 3.8. Qualitatively, the shape of $B(T)$ is the same for all gases; quantitatively, the temperature for which $B = 0$ corresponds to a reduced temperature of about $T_r = 2.7$ for many gases. Use these observations to show by Eqs. (6.54) through (6.56) that the residual properties G^R, H^R, and S^R are *negative* for most gases at modest pressures and normal temperatures. What can you say about the signs of V^R and C_P^R?

6.83. An equimolar mixture of methane and propane is discharged from a compressor at 5500 kPa and 90°C at the rate of 1.4 kg·s^{-1}. If the velocity in the discharge line is not to exceed 30 m·s^{-1}, what is the minimum diameter of the discharge line?

6.84. Estimate V^R, H^R, and S^R for one of the following by appropriate generalized correlations:

    (*a*) 1,3-Butadiene at 500 K and 20 bar.
    (*b*) Carbon dioxide at 400 K and 200 bar.
    (*c*) Carbon disulfide at 450 K and 60 bar.

(d) n-Decane at 600 K and 20 bar.
(e) Ethylbenzene at 620 K and 20 bar.
(f) Methane at 250 K and 90 bar.
(g) Oxygen at 150 K and 20 bar.
(h) n-Pentane at 500 K and 10 bar.
(i) Sulfur dioxide at 450 K and 35 bar.
(j) Tetrafluoroethane at 400 K and 15 bar.

6.85. Estimate Z, H^R, and S^R for one of the following *equimolar* mixtures by the Lee/Kesler correlations:

(a) Benzene/cyclohexane at 650 K and 60 bar.
(b) Carbon dioxide/carbon monoxide at 300 K and 100 bar.
(c) Carbon dioxide/n-octane at 600 K and 100 bar.
(d) Ethane/ethylene at 350 K and 75 bar.
(e) Hydrogen sulfide/methane at 400 K and 150 bar.
(f) Methane/nitrogen at 200 K and 75 bar.
(g) Methane/n-pentane at 450 K and 80 bar.
(h) Nitrogen/oxygen at 250 K and 100 bar.

6.86. For the reversible isothermal compression of a liquid for which β and κ may be assumed independent of pressure, show that:

(a) $W = P_1 V_1 - P_2 V_2 - \dfrac{V_2 - V_1}{\kappa}$

(b) $\Delta S = \dfrac{\beta}{\kappa}(V_2 - V_1)$

(c) $\Delta H = \dfrac{1 - \beta T}{\kappa}(V_2 - V_1)$

Do not assume that V is constant at an average value, but use Eq. (3.6) for its P dependence (with V_2 replaced by V). Apply these equations to the conditions stated in Prob. 6.9. What do the results suggest with respect to use of an average value for V?

6.87. In general for an arbitrary thermodynamic property of a pure substance, $M = M(T, P)$; whence

$$dM = \left(\frac{\partial M}{\partial T}\right)_P dT + \left(\frac{\partial M}{\partial P}\right)_T dP$$

For what *two* distinct conditions is the following equation true?

$$\Delta M = \int_{T_1}^{T_2} \left(\frac{\partial M}{\partial T}\right)_P dT$$

6.88. The enthalpy of a pure ideal gas depends on temperature only. Hence, H^{ig} is often said to be "independent of pressure," and one writes $(\partial H^{ig}/\partial P)_T = 0$. Determine expressions for $(\partial H^{ig}/\partial P)_V$ and $(\partial H^{ig}/\partial P)_S$. Why are these quantities not zero?

6.89. Prove that

$$dS = \frac{C_V}{T}\left(\frac{\partial T}{\partial P}\right)_V dP + \frac{C_P}{T}\left(\frac{\partial T}{\partial V}\right)_P dV$$

For an ideal gas with constant heat capacities, use this result to derive Eq. (3.23c).

6.90. The derivative $(\partial U/\partial V)_T$ is sometimes called the *internal pressure* and the product $T(\partial P/\partial T)_V$ the *thermal pressure*. Find equations for their evaluation for:

(*a*) An ideal gas;
(*b*) A van der Waals fluid;
(*c*) A Redlich/Kwong fluid.

6.91. (*a*) A pure substance is described by an expression for $G(T, P)$. Show how to determine Z, U, and C_V, in relation to G, T, and P and/or derivatives of G with respect to T and P.

(*b*) A pure substance is described by an expression for $A(T, V)$. Show how to determine Z, H, and C_P, in relation to A, T, and V and/or derivatives of A with respect to T and V.

6.92. Use steam tables to estimate a value of the acentric factor ω for water. Compare the result with the value given in Table B.1.

6.93. The critical coordinates for tetrafluoroethane (refrigerant HFC-134a) are given in Table B.1, and Table 9.1 shows saturation properties for the same refrigerant. From these data determine the acentric factor ω for HFC-134a, and compare it with the value given in Table B.1.

6.94. As noted in Ex. 6.6, ΔH^{lv} is not independent of T; in fact, it becomes zero at the critical point. Nor may saturated vapors *in general* be considered ideal gases. Why is it then that Eq. (6.89) provides a reasonable approximation to vapor-pressure behavior over the entire liquid range?

6.95. Rationalize the following approximate expressions for solid/liquid saturation pressures:

(*a*) $P_{sl}^{sat} = A + BT$;

(*b*) $P_{sl}^{sat} = A + B\ln T$

6.96. As suggested by Fig. 3.1, the slope of the sublimation curve at the triple point is generally greater than that of the vaporization curve at the same state. Rationalize this observation. Note that triple-point pressures are usually low; hence assume for this exercise that

$$\Delta Z^{sv} \approx \Delta Z^{lv} \approx 1.$$

6.97. Show that the Clapeyron equation for liquid/vapor equilibrium may be written in the reduced form:

$$\frac{d\ln P_r^{\text{sat}}}{dT_r} = \frac{\widehat{\Delta H}^{lv}}{T_r^2 \Delta Z^{lv}} \quad \text{where} \quad \widehat{\Delta H}^{lv} \equiv \frac{\Delta H^{lv}}{RT_c}$$

6.98. Use the result of the preceding problem to estimate the heat of vaporization at the normal boiling point for one of the substances listed below. Compare the result with the value given in Table B.2 of App. B.
Ground rules: Represent P_r^{sat} with Eqs. (6.92), (6.93), and (6.94), with ω given by Eq. (6.95). Use Eqs. (3.57), (3.58), (3.59), (3.61), and (3.62) for Z^v, and Eq. (3.69) for Z^l. Critical properties and and normal boiling points are given in Table B.1.

(*a*) Benzene;
(*b*) *iso*-Butane;
(*c*) Carbon tetrachloride;
(*d*) Cyclohexane;
(*e*) *n*-Decane;
(*f*) *n*-Hexane;
(*g*) *n*-Octane;
(*h*) Toluene;
(*i*) *o*-Xylene

6.99. Riedel proposed a third corresponding-states parameter α_c, related to the vapor-pressure curve by:

$$\alpha_c \equiv \left[\frac{d\ln P^{\text{sat}}}{d\ln T}\right]_{T=T_c}$$

For simple fluids, experiment shows that $\alpha_c \approx 5.8$; for non-simple fluids, α_c increases with increasing molecular complexity. How well does the Lee/Kesler correlation for P_r^{sat} accommodate these observations?

6.100. Triple-point coordinates for carbon dioxide are $T_t = 216.55$ K and $P_t = 5.170$ bar. Hence, CO_2 has no normal boiling point. (Why?) Nevertheless, one can define a *hypothetical* normal boiling point by extrapolation of the vapor-pressure curve.

(*a*) Use the Lee/Kesler correlation for P_r^{sat} in conjunction with the triple-point coordinates to estimate ω for CO_2. Compare it with the value in Table B.1.
(*b*) Use the Lee/Kesler correlation to estimate the hypothetical normal boiling point for CO_2. Comment on the likely reasonableness of this result.

Chapter 7

Applications of Thermodynamics to Flow Processes

The thermodynamics of flow is based on the mass, energy, and entropy balances developed in Chaps. 2 and 5. The application of these balances to specific processes is the subject of this chapter. The discipline underlying the study of flow is fluid mechanics,[1] which encompasses not only the balances of thermodynamics but also a momentum balance that arises from the laws of classical mechanics (Newton's laws). This makes fluid mechanics a broader field of study. The distinction between *thermodynamics problems* and *fluid-mechanics problems* depends on whether this momentum balance is required for solution. Those problems whose solutions depend only on mass conservation and on the laws of thermodynamics are commonly set apart from the study of fluid mechanics and are treated in courses on thermodynamics. Fluid mechanics then deals with the broad spectrum of problems which *require* application of the momentum balance. This division, though arbitrary, is traditional and convenient.

For example, if the states and thermodynamic properties of a gas entering and leaving a pipeline are known, then application of the first law establishes the quantity of the energy exchanged with the surroundings. The mechanism of the process, the details of flow, and the state path actually followed by the gas between entrance and exit are not pertinent to this calculation. On the other hand, if one has only incomplete knowledge of initial or final state of the gas, then more detailed information about the process is needed. Perhaps the exit pressure of the gas is not specified. In this case, one must apply the momentum balance of fluid mechanics, and this requires an empirical or theoretical expression for the shear stress at the pipe wall.

Flow is caused by pressure gradients within a fluid. Moreover, temperature, velocity, and even concentration gradients may exist within a flowing fluid. This contrasts with the uniform conditions that prevail at equilibrium in closed systems. The distribution of conditions in flow systems requires that properties be attributed to point masses of fluid. Thus we assume that intensive properties, such as density, specific enthalpy, specific entropy, etc.,

---

[1] Noel de Nevers, *Fluid Mechanics for Chemical Engineers*, 3rd ed., McGraw-Hill, New York, 2005. Fluid mechanics is treated as an integral part of transport processes by R. B. Bird, W. E. Stewart, and E. N. Lightfoot in *Transport Phenomena*, 2nd ed., John Wiley, New York, 2001; by J. L. Plawsky in *Transport Phenomena Fundamentals*, 2nd ed., CRC Press, 2009; and by D. Welty, C. E. Wicks, G. L. Rorrer, and R. E. Wilson, in *Fundamentals of Momentum, Heat, and Mass Transfer*, 5th ed., John Wiley, New York, 2007.

at a point are determined solely by the temperature, pressure, and composition at the point, uninfluenced by gradients that may exist at the point. Moreover, we assume that the fluid exhibits the same set of intensive properties at the point as it would if it existed at equilibrium at the same temperature, pressure, and composition. The implication is that an equation of state applies locally and instantaneously at any point in a fluid system, and that one may invoke a concept of *local state*, independent of the concept of equilibrium. Experience shows that this leads to results in accord with observation for a broad range of practical processes.

Thermodynamic analysis of flow processes is most often applied to processes involving gases or supercritical fluids. In these *compressible flow* processes the fluid properties change as a result of changes in pressure, and thermodynamic analysis provides relationships between these changes. Thus, the intent of this brief chapter is to:

- Develop the thermodynamic equations applicable to one-dimensional steady-state flow of compressible fluids in conduits.
- Apply these equations to flow (both subsonic and supersonic) in pipes and nozzles
- Treat throttling processes, i.e., flow through restrictions
- Calculate the work produced by turbines and expanders
- Examine compression processes as produced by compressors, pumps, blowers, fans, and vacuum pumps

The equations of balance for open systems presented in Chaps. 2 and 5 are summarized here in Table 7.1 for easy reference. In addition, Eqs. (7.1) and (7.2), restricted forms of the mass balance, are included. These equations are the basis for thermodynamic analysis of *processes* in this and the next two chapters. When combined with thermodynamic *property* statements, they allow calculation of system states and process energy requirements.

7.1 DUCT FLOW OF COMPRESSIBLE FLUIDS

Such problems as sizing of pipes and shaping of nozzles require application of the momentum balance of fluid mechanics, and therefore are not part of thermodynamics. However, thermodynamics does provide equations that interrelate the changes in pressure, velocity, cross-sectional area, enthalpy, entropy, and specific volume of a flowing stream. We consider here the adiabatic, steady-state, one-dimensional flow of a compressible fluid in the absence of shaft work and of changes in potential energy. The pertinent thermodynamic equations are first derived, and are then applied to flow in pipes and nozzles.

The appropriate energy balance is Eq. (2.31). With Q, W_s and Δz all set equal to zero,

$$\Delta H + \frac{\Delta u^2}{2} = 0$$

In differential form, $$dH = -u\,du \qquad (7.3)$$

The continuity equation, Eq. (2.26), is also applicable. Because $\dot{m}$ is constant, its differential form is:

$$d(uA/V) = 0$$

or $$\frac{dV}{V} - \frac{du}{u} - \frac{dA}{A} = 0 \qquad (7.4)$$

Table 7.1: Equations of Balance

General Equations of Balance	Balance Equations for Steady-Flow Processes	Balance Equations for Single-Stream Steady-Flow Processes
$\dfrac{dm_{cv}}{dt} + \Delta(\dot{m})_{fs} = 0 \qquad (2.25)$	$\Delta(\dot{m})_{fs} = 0 \qquad (7.1)$	$\dot{m}_1 = \dot{m}_2 = \dot{m} \qquad (7.2)$
$\dfrac{d(mU)_{cv}}{dt} + \Delta\left[\left(H + \tfrac{1}{2}u^2 + zg\right)\dot{m}\right]_{fs} = \dot{Q} + \dot{W} \qquad (2.27)$	$\Delta\left[\left(H + \tfrac{1}{2}u^2 + zg\right)\dot{m}\right]_{fs} = \dot{Q} + \dot{W} \qquad (2.29)$	$\Delta H + \dfrac{\Delta u^2}{2} + g\Delta z = Q + W_s \qquad (2.31)$
$\dfrac{d(mS)_{cv}}{dt} + \Delta(S\dot{m})_{fs} - \sum_j \dfrac{\dot{Q}_j}{T_{\sigma,j}} = \dot{S}_G \geq 0 \qquad (5.16)$	$\Delta(S\dot{m})_{fs} - \sum_j \dfrac{\dot{Q}_j}{T_{\sigma,j}} = \dot{S}_G \geq 0 \qquad (5.17)$	$\Delta S - \sum_j \dfrac{Q_j}{T_{\sigma,j}} = S_G \geq 0 \qquad (5.18)$

The fundamental property relation appropriate to this application is:

$$dH = T\,dS + V\,dP \tag{6.9}$$

In addition, the specific volume of the fluid may be considered a function of its entropy and pressure: $V = V(S, P)$. Then,

$$dV = \left(\frac{\partial V}{\partial S}\right)_P dS + \left(\frac{\partial V}{\partial P}\right)_S dP$$

This equation is put into more convenient form through the mathematical identity:

$$\left(\frac{\partial V}{\partial S}\right)_P = \left(\frac{\partial V}{\partial T}\right)_P \left(\frac{\partial T}{\partial S}\right)_P$$

Substituting for the two partial derivatives on the right by Eqs. (3.3) and (6.18) gives:

$$\left(\frac{\partial V}{\partial S}\right)_P = \frac{\beta V T}{C_P}$$

where β is the volume expansivity. The equation derived in physics for the speed of sound c in a fluid is:

$$c^2 = -V^2 \left(\frac{\partial P}{\partial V}\right)_S \qquad \text{or} \qquad \left(\frac{\partial V}{\partial P}\right)_S = -\frac{V^2}{c^2}$$

Substituting for the two partial derivatives in the equation for dV now yields:

$$\frac{dV}{V} = \frac{\beta T}{C_P} dS - \frac{V}{c^2} dP \tag{7.5}$$

Equations (7.3), (7.4), (6.9), and (7.5) relate the six differentials dH, du, dV, dA, dS, and dP. With but four equations, we treat dS and dA as independent, and we develop equations that express the remaining differentials as functions of these two. First, Eqs. (7.3) and (6.9) are combined:

$$T\,dS + V\,dP = -u\,du \tag{7.6}$$

Eliminating dV and du from Eq. (7.4) by Eqs. (7.5) and (7.6) gives upon rearrangement:

$$(1 - \mathbf{M}^2) V\,dP + \left(1 + \frac{\beta u^2}{C_P}\right) T\,dS - \frac{u^2}{A}\,dA = 0 \tag{7.7}$$

where $\mathbf{M}$ is the Mach number, defined as the ratio of the speed of the fluid in the duct to the speed of sound in the fluid, u/c. Equation (7.7) relates dP to dS and dA.

Equations (7.6) and (7.7) are combined to eliminate $V\,dP$:

$$u\,du - \left(\frac{\dfrac{\beta u^2}{C_P} + \mathbf{M}^2}{1 - \mathbf{M}^2}\right) T\,dS + \left(\frac{1}{1 - \mathbf{M}^2}\right)\frac{u^2}{A}\,dA = 0 \tag{7.8}$$

This equation relates du to dS and dA. Combined with Eq. (7.3) it relates dH to dS and dA, and combined with Eq. (7.4) it relates dV to these same independent variables.

The differentials Eq. (7.4) in the preceding equations represent changes in the fluid as it traverses a differential length of its path. If this length is dx, then each of the equations of flow may be divided through by dx. Equations (7.7) and (7.8) then become:

$$V(1 - \mathbf{M}^2)\frac{dP}{dx} + T\left(1 + \frac{\beta u^2}{C_P}\right)\frac{dS}{dx} - \frac{u^2}{A}\frac{dA}{dx} = 0 \tag{7.9}$$

$$u\frac{du}{dx} - T\left(\frac{\frac{\beta u^2}{C_P} + \mathbf{M}^2}{1 - \mathbf{M}^2}\right)\frac{dS}{dx} + \left(\frac{1}{1 - \mathbf{M}^2}\right)\frac{u^2}{A}\frac{dA}{dx} = 0 \tag{7.10}$$

According to the second law, the irreversibilities due to fluid friction in adiabatic flow cause an entropy increase in the fluid in the direction of flow. In the limit as the flow approaches reversibility, this increase approaches zero. In general, then,

$$\frac{dS}{dx} \geq 0$$

Pipe Flow

For the case of steady-state adiabatic flow of compressible fluids in a horizontal pipe of constant cross-sectional area, $dA/dx = 0$, and Eqs. (7.9) and (7.10) reduce to:

$$\frac{dP}{dx} = -\frac{T}{V}\left(\frac{1 + \frac{\beta u^2}{C_P}}{1 - \mathbf{M}^2}\right)\frac{dS}{dx} \qquad\qquad u\frac{du}{dx} = T\left(\frac{\frac{\beta u^2}{C_P} + \mathbf{M}^2}{1 - \mathbf{M}^2}\right)\frac{dS}{dx}$$

For subsonic flow, $\mathbf{M}^2 < 1$. All terms on the right sides of these equations are then positive, and

$$\frac{dP}{dx} < 0 \qquad \text{and} \qquad \frac{du}{dx} > 0$$

Thus the pressure decreases and the velocity increases in the direction of flow. However, the velocity cannot increase indefinitely. If the velocity were to exceed the sonic value, then the above inequalities would reverse. Such a transition is not possible in a pipe of constant cross-sectional area. For subsonic flow, the maximum fluid velocity obtainable in a pipe of constant cross section is the speed of sound, and this value is reached at the *exit* of the pipe. At this point dS/dx reaches its limiting value of zero. Given a discharge pressure low enough for the flow to become sonic, lengthening the pipe does not alter this result; the sonic velocity is still obtained at the outlet of the lengthened pipe.

The equations for pipe flow indicate that when flow is supersonic the pressure increases and the velocity decreases in the direction of flow. However, such a flow regime is unstable, and when a supersonic stream enters a pipe of constant cross section, a compression shock occurs, the result of which is an abrupt and finite increase in pressure and decrease in velocity to a subsonic value.

Example 7.1

For the steady-state, adiabatic, irreversible flow of an *incompressible* liquid in a horizontal pipe of constant cross-sectional area, show that:

(a) The velocity is constant.

(b) The temperature increases in the direction of flow.

(c) The pressure decreases in the direction of flow.

Solution 7.1

(a) The control volume here is simply a finite length of horizontal pipe, with entrance and exit sections identified as 1 and 2. By the continuity equation, Eq. (2.26),

$$\frac{u_2 A_2}{V_2} = \frac{u_1 A_1}{V_1}$$

However, $A_2 = A_1$ (constant cross-sectional area) and $V_2 = V_1$ (incompressible fluid). Hence, $u_2 = u_1$.

(b) The entropy balance of Eq. (5.18) here becomes simply $S_G = S_2 - S_1$. For an incompressible liquid with heat capacity C (see Ex. 6.2),

$$S_G = S_2 - S_1 = \int_{T_1}^{T_2} C \frac{dT}{T}$$

But S_G is positive (flow is irreversible) and hence, by the last equation, $T_2 > T_1$, and temperature increases in the direction of flow.

(c) As shown in (a), $u_2 = u_1$, and therefore the energy balance, Eq. (2.31), reduces for the stated conditions to $H_2 - H_1 = 0$. Combining this with the integrated form of Eq. (A) of Ex. 6.2 applied to an incompressible liquid yields:

$$H_2 - H_1 = \int_{T_1}^{T_2} C \, dT + V(P_2 - P_1) = 0$$

and

$$V(P_2 - P_1) = -\int_{T_1}^{T_2} C \, dT$$

As shown in (b), $T_2 > T_1$; thus by the last equation, $P_2 < P_1$, and pressure decreases in the direction of flow.

Repeating this example for the case of *reversible* adiabatic flow is instructive. In this case $u_2 = u_1$ as before, but $S_G = 0$. The entropy balance then shows that $T_2 = T_1$, in which case the energy balance yields $P_2 = P_1$. We conclude that the temperature increase of (b) and the pressure decrease of (c) *originate* from flow irreversibilities, specifically from the irreversibilities associated with fluid friction.

Nozzles

The limitations observed for flow of compressible fluids in pipes do not extend to properly designed nozzles, which bring about the interchange of internal and kinetic energy of a fluid as a result of changing cross-sectional area available for flow. The design of effective nozzles is a problem in fluid mechanics, but the flow through a well-designed nozzle is susceptible to thermodynamic analysis. In a properly designed nozzle the area changes with length in such a way as to make the flow nearly frictionless. In the limit of reversible flow, the rate of entropy increase approaches zero, and $dS/dx = 0$. In this event Eqs. (7.9) and (7.10) become:

$$\frac{dP}{dx} = \frac{u^2}{VA}\left(\frac{1}{1-\mathbf{M}^2}\right)\frac{dA}{dx} \qquad\qquad \frac{du}{dx} = -\frac{u}{A}\left(\frac{1}{1-\mathbf{M}^2}\right)\frac{dA}{dx}$$

The characteristics of flow depend on whether the flow is subsonic ($\mathbf{M} < 1$) or supersonic ($\mathbf{M} > 1$). The various cases are summarized in Table 7.2.

Table 7.2: Characteristics of Flow for a Nozzle

	Subsonic: $\mathbf{M} < 1$		Supersonic: $\mathbf{M} > 1$	
	Converging	Diverging	Converging	Diverging
$\dfrac{dA}{dx}$	−	+	−	+
$\dfrac{dP}{dx}$	−	+	+	−
$\dfrac{du}{dx}$	+	−	−	+

Thus, for subsonic flow in a converging nozzle, the velocity increases and the pressure decreases as the cross-sectional area diminishes. The maximum obtainable fluid velocity is the speed of sound, reached at the exit. Because of this, a converging subsonic nozzle can be used to deliver a constant flow rate into a region of variable pressure. Suppose a compressible fluid enters a converging nozzle at pressure P_1 and discharges from the nozzle into a chamber of variable pressure P_2. As this discharge pressure decreases below P_1, the flow rate and velocity increase. Ultimately, the pressure ratio P_2/P_1 reaches a critical value at which the velocity at the nozzle exit is sonic. Further reduction in P_2 has no effect on the conditions in the nozzle. The flow remains constant, and the velocity at the nozzle exit is sonic, regardless of the value of P_2/P_1, provided it is always less than the critical value. For steam, the critical value of this ratio is about 0.55 at moderate temperatures and pressures.

Supersonic velocities are readily attained in the diverging section of a properly designed converging/diverging nozzle (Fig. 7.1). With sonic velocity reached at the throat, a further increase in velocity and decrease in pressure requires an increase in cross-sectional area, a diverging section to accommodate increasing volume of flow. The transition occurs at the throat, where $dA/dx = 0$. The relationships between velocity, area, and pressure in a converging/diverging nozzle are illustrated numerically in Ex. 7.2.

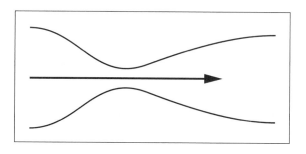

Figure 7.1:
Converging/diverging
nozzle.

The speed of sound is attained at the throat of a converging/diverging nozzle only when the pressure at the throat is low enough that the critical value of P_2/P_1 is reached. If insufficient pressure drop is available in the nozzle for the velocity to become sonic, the diverging section of the nozzle acts as a diffuser. That is, after the throat is reached, the pressure rises and the velocity decreases; this is the conventional behavior for subsonic flow in diverging sections. Of course, even when P_2/P_1 is low enough to achieve choked flow, the velocity cannot continue to increase indefinitely. Ultimately, it will return to subsonic velocity across a shock wave. As P_2/P_1 decreases, the location of this shock wave will move down the nozzle, away from the throat, until the shock is outside the nozzle and the flow exiting the nozzle is supersonic.

The relation of velocity to pressure in an isentropic nozzle can be expressed analytically for the ideal-gas state and constant heat capacities. Combination of Eqs. (6.9) and (7.3) for isentropic flow gives:

$$u\,du = -V\,dP$$

Integration, with nozzle entrance and exit conditions denoted by 1 and 2, yields:

$$u_2^2 - u_1^2 = -2\int_{P_1}^{P_2} V\,dP = \frac{2\gamma P_1 V_1}{\gamma - 1}\left[1 - \left(\frac{P_2}{P_1}\right)^{(\gamma-1)/\gamma}\right] \tag{7.11}$$

where the final term is obtained upon elimination of V by Eq. (3.23c), $PV^\gamma = \text{const.}$

Equation (7.11) may be solved for the pressure ratio P_2/P_1 for which u_2 reaches the speed of sound, i.e., where

$$u_2^2 = c^2 = -V^2\left(\frac{\partial P}{\partial V}\right)_S$$

The derivative is found by differentiation with respect to V of $PV^\gamma = \text{const.}$:

$$\left(\frac{\partial P}{\partial V}\right)_S = -\frac{\gamma P}{V}$$

Substitution then yields:

$$u_2^2 = \gamma P_2 V_2$$

With this value for u_2^2 in Eq. (7.11) and with $u_1 = 0$, solution for the pressure ratio at the throat gives:

$$\frac{P_2}{P_1} = \left(\frac{2}{\gamma + 1}\right)^{\gamma/(\gamma-1)} \tag{7.12}$$

Example 7.2

A high-velocity nozzle is designed to operate with steam at 700 kPa and 300°C. At the nozzle inlet the velocity is 30 m·s^{-1}. Calculate values of the ratio A/A_1 (where A_1 is the cross-sectional area of the nozzle inlet) for the sections where the pressure is 600, 500, 400, 300, and 200 kPa. Assume that the nozzle operates isentropically.

Solution 7.2

The required area ratios are determined by conservation of mass [Eq. (2.26)], and the velocity u is found from the integrated form of Eq. (7.3), a steady-state energy balance that includes enthalpy and kinetic energy terms:

$$\frac{A}{A_1} = \frac{u_1 V}{V_1 u} \qquad \text{and} \qquad u^2 = u_1^2 - 2(H - H_1)$$

For velocity units of m·s^{-1}, u^2 has the units, m^2·s^{-2}. Units of J·kg^{-1} for H are consistent with these, because 1 J = 1 kg·m^2·s^{-2}, and 1 J·kg^{-1} = 1 m^2·s^{-2}.

Initial values for entropy, enthalpy, and specific volume are obtained from the steam tables:

$$S_1 = 7.2997 \text{ kJ·kg}^{-1}\text{·K}^{-1} \qquad H_1 = 3059.8 \text{ kJ·kg}^{-1} \qquad V_1 = 371.39 \text{ cm}^3\text{·g}^{-1}$$

Thus,

$$\frac{A}{A_1} = \left(\frac{30}{371.39}\right)\frac{V}{u} \tag{A}$$

and

$$u^2 = 900 - 2\left(H - 3059.8 \times 10^3\right) \tag{B}$$

Because the expansion is isentropic, $S = S_1$; steam-table values at 600 kPa are:

$$S = 7.2997 \text{ kJ·kg}^{-1}\text{·K}^{-1} \qquad H = 3020.4 \text{ kJ·kg}^{-1} \qquad V = 418.25 \text{ cm}^3\text{·g}^{-1}$$

From Eq. (B), $u = 282.3 \text{ m·s}^{-1}$

By Eq. (A), $$\frac{A}{A_1} = \left(\frac{30}{371.39}\right)\left(\frac{418.25}{282.3}\right) = 0.120$$

Area ratios for other pressures are evaluated the same way, and the results are summarized in the following table.

P/kPa	V/cm^3·g^{-1}	u/m·s^{-1}	A/A_1	P/kPa	V/cm^3·g^{-1}	u/m·s^{-1}	A/A_1
700	371.39	30	1.0	400	571.23	523.0	0.088
600	418.25	282.3	0.120	300	711.93	633.0	0.091
500	481.26	411.2	0.095	200	970.04	752.2	0.104

The pressure at the throat of the nozzle is about 380 kPa. At lower pressures, the nozzle clearly diverges.

Example 7.3

Consider again the nozzle of Ex. 7.2, assuming now that steam exists in its ideal-gas state and constant heat capacity. Calculate:

(a) The critical pressure ratio and the velocity at the throat.

(b) The discharge pressure for a Mach number of 2.0 at the nozzle exhaust.

Solution 7.3

(a) The ratio of specific heats for steam is about 1.3. Substituting in Eq. (7.12),

$$\frac{P_2}{P_1} = \left(\frac{2}{1.3+1}\right)^{1.3/(1.3-1)} = 0.55$$

The velocity at the throat, equal to the speed of sound, is found from Eq. (7.11), which contains the product P_1V_1. For steam in its ideal-gas state:

$$P_1V_1 = \frac{RT_1}{\mathcal{M}} = \frac{(8.314)(573.15)}{0.01802} = 264{,}511 \text{ m}^2 \cdot \text{s}^{-2}$$

In this equation $R/\mathcal{M}$ has the units:

$$\frac{\text{J}}{\text{kg} \cdot \text{K}} = \frac{\text{N} \cdot \text{m}}{\text{kg} \cdot \text{K}} = \frac{\text{kg} \cdot \text{m} \cdot \text{s}^{-2} \text{m}}{\text{kg} \cdot \text{K}} = \frac{\text{m}^2 \cdot \text{s}^{-2}}{\text{K}}$$

Thus $RT/\mathcal{M}$, and hence P_1V_1, is in $\text{m}^2 \cdot \text{s}^{-2}$, the units of velocity squared. Substitution in Eq. (7.11) gives:

$$u^2_{\text{throat}} = (30)^2 + \frac{(2)(1.3)(264{,}511)}{1.3-1}[1 - (0.55)^{(1.3-1)/1.3}] = 296{,}322$$

$$u_{\text{throat}} = 544.35 \text{ m} \cdot \text{s}^{-1}$$

This result is in good agreement with the value obtained in Ex. 7.2, because the behavior of steam at these conditions closely approximates the ideal-gas state.

(b) For a Mach number of 2.0 (based on the velocity of sound at the nozzle throat) the discharge velocity is:

$$2u_{\text{throat}} = (2)(544.35) = 1088.7 \text{ m} \cdot \text{s}^{-1}$$

Substitution of this value in Eq. (7.11) allows calculation of the pressure ratio:

$$(1088.7)^2 - (30)^2 = \frac{(2)(1.3)(264{,}511)}{1.3-1}\left[1 - \left(\frac{P_2}{P_1}\right)^{(1.3-1)/1.3}\right]$$

$$\left(\frac{P_2}{P_1}\right)^{(1.3-1)/1.3} = 0.4834 \quad \text{and} \quad P_2 = (0.0428)(700) = 30.0 \text{ kPa}$$

Throttling Process

When a fluid flows through a restriction, such as an orifice, a partly closed valve, or a porous plug, without any appreciable change in kinetic or potential energy, the primary result of the process is a pressure drop in the fluid. Such a *throttling process* produces no shaft work, and in the absence of heat transfer, Eq. (2.31) reduces to

$$\Delta H = 0 \qquad \text{or} \qquad H_2 = H_1$$

The process therefore occurs at constant enthalpy.

Because enthalpy in the ideal-gas state depends on temperature only, a throttling process does not change the temperature in this state. For most real gases at moderate conditions of temperature and pressure, a reduction in pressure at constant enthalpy results in a decrease in temperature. For example, if steam at 1000 kPa and 300°C is throttled to 101.325 kPa (atmospheric pressure),

$$H_2 = H_1 = 3052.1 \text{ kJ·kg}^{-1}$$

Interpolation in the steam tables at this enthalpy and at a pressure of 101.325 kPa indicates a downstream temperature of 288.8°C. The temperature has decreased, but the effect is small.

Throttling of *wet* steam to sufficiently low pressure causes the liquid to evaporate and the vapor to become superheated. Thus if wet steam at 1000 kPa ($t^{\text{sat}} = 179.88°C$) with a quality of 0.96 is throttled to 101.325 kPa,

$$H_2 = H_1 = (0.04)(762.6) + (0.96)(2776.2) = 2695.7 \text{ kJ·kg}^{-1}$$

At 101.325 kPa steam with this enthalpy has a temperature of 109.8°C; it is therefore superheated ($t^{\text{sat}} = 100°C$). The considerable temperature drop here results from evaporation of liquid.

If a saturated liquid is throttled to a lower pressure, some of the liquid vaporizes or *flashes*, producing a mixture of saturated liquid and saturated vapor at the lower pressure. Thus if saturated liquid water at 1000 kPa ($t^{\text{sat}} = 179.88°C$) is flashed to 101.325 kPa ($t^{\text{sat}} = 100°C$),

$$H_2 = H_1 = 762.6 \text{ kJ·kg}^{-1}$$

At 101.325 kPa the quality of the resulting steam is found from Eq. (6.96a) with $M = H$:

$$762.6 = (1 - x)(419.1) + x(2676.0)$$
$$= 419.1 + x(2676.0 - 419.1)$$

Hence $x = 0.152$

Thus 15.2% of the original liquid vaporizes in the process. Again, the large temperature drop results from evaporation of liquid. Throttling processes find frequent application in refrigeration (Chap. 9).

Example 7.4

Propane gas at 20 bar and 400 K is throttled in a steady-state flow process to 1 bar. Estimate the final temperature of the propane and its entropy change. Properties of propane can be found from suitable generalized correlations.

Solution 7.4

To begin, we write the overall enthalpy change as the sum of three components: (1) removal of residual enthalpy at state 1, (2) sensible heat to take the substance from the ideal-gas state at the initial temperature to the ideal-gas state at the final temperature, and (3) adding the residual enthalpy at state 2. These must sum to zero for this constant-enthalpy process:

$$\Delta H = -H_1^R + \langle C_P^{ig} \rangle_H (T_2 - T_1) + H_2^R = 0$$

If propane in its final state at 1 bar is assumed to be in its ideal-gas state, $H_2^R = 0$, and the preceding equation, solved for T_2, becomes

$$T_2 = \frac{H_1^R}{\langle C_P^{ig} \rangle_H} + T_1 \tag{A}$$

For propane, $T_c = 369.8$ K $P_c = 42.48$ bar $\omega = 0.152$

Thus for the initial state,

$$T_{r_1} = \frac{400}{369.8} = 1.082 \qquad P_{r_1} = \frac{20}{42.48} = 0.471$$

At these conditions the generalized correlation based on second virial coefficients is satisfactory (Fig. 3.13), and calculation of H_1^R by Eqs. (6.68), (3.61), (6.70), (3.62), and (6.71) is represented by:

$$\frac{H_1^R}{RT_c} = \text{HRB}(1.082, 0.471, 0.152) = -0.452$$

and

$$H_1^R = (8.314)(369.8)(-0.452) = -1390 \text{ J·mol}^{-1}$$

The only remaining quantity in Eq. (A) to be evaluated is $\langle C_P^{ig} \rangle_H$. Data for propane from Table C.1 of App. C provide the heat-capacity equation:

$$\frac{C_P^{ig}}{R} = 1.213 + 28.785 \times 10^{-3} T - 8.824 \times 10^{-6} T^2$$

For an initial calculation, assume that $\langle C_P^{ig} \rangle_H$ equals the value of C_P^{ig} at the initial temperature of 400 K, i.e., $\langle C_P^{ig} \rangle_H = 94.07$ J·mol^{-1}·K^{-1}.

From Eq. (A), $T_2 = \dfrac{-1390}{94.07} + 400 = 385.2$ K

Clearly, the temperature change is small, and $\langle C_P^{ig} \rangle_H$ is reevaluated to an excellent approximation as C_P^{ig} at the arithmetic-mean temperature,

$$T_{am} = \frac{400 + 385.2}{2} = 392.6 \text{ K}$$

This gives: $\langle C_P^{ig} \rangle_H = 92.73$ J·mol^{-1}·K^{-1}

and recalculation of T_2 by Eq. (A) yields the final value: $T_2 = 385.0$ K.
    The entropy change of the propane is given by Eq. (6.75), which here becomes:

$$\Delta S = \langle C_P^{ig} \rangle_S \ln \frac{T_2}{T_1} - R \ln \frac{P_2}{P_1} - S_1^R$$

Because the temperature change is so small, to an excellent approximation,

$$\langle C_P^{ig} \rangle_S = \langle C_P^{ig} \rangle_H = 92.73 \text{ J·mol}^{-1} \text{·K}^{-1}$$

Calculation of S_1^R by Eqs. (6.69) through (6.71) is represented by:

$$\frac{S_1^R}{R} = \text{SRB}(1.082, 0.471, 0.152) = -0.2934$$

Then $S_1^R = (8.314)(-0.2934) = -2.439$ J·mol^{-1}·K^{-1}

and $\Delta S = 92.73 \ln \dfrac{385.0}{400} - 8.314 \ln \dfrac{1}{20} + 2.439 = 23.80$ J·mol^{-1}·K^{-1}

The positive value reflects the irreversibility of throttling processes.

Example 7.5

Throttling a real gas from conditions of moderate temperature and pressure usually results in a temperature decrease. Under what conditions would an *increase* in temperature be expected?

Solution 7.5

The sign of the temperature change is determined by the sign of the derivative $(\partial T / \partial P)_H$, called the *Joule/Thomson coefficient* μ:

$$\mu \equiv \left(\frac{\partial T}{\partial P} \right)_H$$

When μ is positive, throttling results in a temperature decrease; when negative, in a temperature increase.
    Because $H = f(T, P)$, the following equation relates the Joule/Thomson coefficient to other thermodynamic properties:[2]

$$\left(\frac{\partial T}{\partial P} \right)_H = -\left(\frac{\partial T}{\partial H} \right)_P \left(\frac{\partial H}{\partial P} \right)_T = -\left(\frac{\partial H}{\partial T} \right)_P^{-1} \left(\frac{\partial H}{\partial P} \right)_T$$

[2]Recall the general equation from differential calculus:

$$\left(\frac{\partial x}{\partial y} \right)_z = -\left(\frac{\partial x}{\partial z} \right)_y \left(\frac{\partial z}{\partial y} \right)_x$$

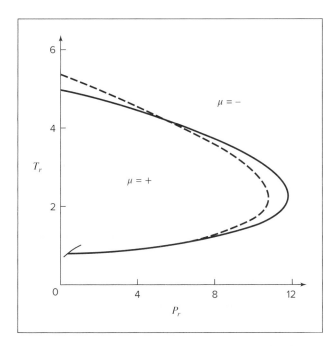

Figure 7.2: Inversion curves for reduced coordinates. Each line represents a locus of points for which $\mu = 0$. The solid curve is from a data correlation; the dashed curve, from the Redlich/Kwong equation. A temperature increase results from throttling in the region where μ is negative.

and by Eq. (2.19),
$$\mu = -\frac{1}{C_P}\left(\frac{\partial H}{\partial P}\right)_T \tag{A}$$

Because C_P is necessarily positive, the sign of μ is determined by the sign of $(\partial H/\partial P)_T$, which in turn is related to PVT behavior:

$$\left(\frac{\partial H}{\partial P}\right)_T = V - T\left(\frac{\partial V}{\partial T}\right)_P \tag{6.20}$$

Substituting $V = ZRT/P$ allows this equation to be rewritten in terms of Z as:

$$\left(\frac{\partial H}{\partial P}\right)_T = -\frac{RT^2}{P}\left(\frac{\partial Z}{\partial T}\right)_P$$

where Z is the compressibility factor. Substitution into Eq. (A) gives:

$$\mu = \frac{RT^2}{C_P P}\left(\frac{\partial Z}{\partial T}\right)_P$$

Thus, $(\partial Z/\partial T)_P$ and μ have the same sign. When $(\partial Z/\partial T)_P$ is zero, as for the ideal-gas state, then $\mu = 0$, and no temperature change accompanies throttling.

The condition $(\partial Z/\partial T)_P = 0$ may be satisfied locally for *real* gases. Such points define the Joule/Thomson *inversion curve*, which separates the region of positive μ from that of negative μ. Figure 7.2 shows *reduced* inversion curves giving the relation between T_r and P_r for which $\mu = 0$. The solid line correlates data for Ar, CH_4, N_2, CO, C_2H_4, C_3H_8, CO_2, and NH_3.[3] The dashed line is calculated from the condition $(\partial Z/\partial T_r)_{P_r} = 0$ applied to the Redlich/Kwong equation of state.

---

[3] D. G. Miller, *Ind. Eng. Chem. Fundam.*, vol. 9, pp. 585–589, 1970.

7.2 TURBINES (EXPANDERS)

The expansion of a gas in a nozzle to produce a high-velocity stream is a process that converts internal energy into kinetic energy, which in turn is converted into shaft work when the stream impinges on blades attached to a rotating shaft. Thus a turbine (or expander) consists of alternate sets of nozzles and rotating blades through which vapor or gas flows in a steady-state expansion process. The overall result is the conversion of the internal energy of a high-pressure stream into shaft work. When steam provides the motive force as in most power plants, the device is called a turbine; when it is a high-pressure gas, such as ammonia or ethylene in a chemical plant, the device is usually called an expander. The process is shown in Fig. 7.3.

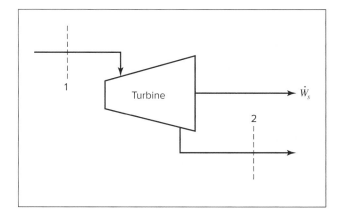

Figure 7.3: Steady-state flow through a turbine or expander.

Equations (2.30) and (2.31) are appropriate energy balances across an expander. However, the potential-energy term can be omitted because there is little change in elevation. Moreover, in any properly designed turbine, heat transfer is negligible and the inlet and exit pipes are sized to make fluid velocities roughly equal. For those conditions, Eqs. (2.30) and (2.31) reduce to:

$$\dot{W}_s = \dot{m}\, \Delta H = \dot{m}(H_2 - H_1) \quad (7.13) \qquad W_s = \Delta H = H_2 - H_1 \quad (7.14)$$

Usually, the inlet conditions T_1 and P_1 and the discharge pressure P_2 are fixed. Thus in Eq. (7.14) only H_1 is known; both H_2 and W_s are unknown, and the energy balance equation alone does not allow their calculation. However, if the fluid in the turbine expands *reversibly and adiabatically*, the process is isentropic, and $S_2 = S_1$. This second equation fixes the final state of the fluid and determines H_2. For this special case, W_s is given by Eq. (7.14), written:

$$W_s(\text{isentropic}) = (\Delta H)_S \qquad (7.15)$$

The shaft work $|W_s|$(isentropic) is the *maximum* that can be obtained from an adiabatic turbine with given inlet conditions and given discharge pressure. Actual turbines produce less work, because the actual expansion process is irreversible; we define a *turbine efficiency* as:

$$\eta \equiv \frac{W_s}{W_s(\text{isentropic})}$$

where W_s is the actual shaft work. By Eqs. (7.14) and (7.15),

$$\eta = \frac{\Delta H}{(\Delta H)_S} \qquad (7.16)$$

Values of η often fall in the range from 0.7 to 0.8. The *HS* diagram of Fig. 7.4 illustrates an actual expansion in a turbine and a reversible expansion for the same intake conditions and the same discharge pressure. The reversible path is the dashed vertical (constant-entropy) line from point 1 at intake pressure P_1 to point 2' at discharge pressure P_2. The solid line, representing the actual irreversible path, starts at point 1 and terminates at point 2 on the isobar for P_2. Because the process is adiabatic, irreversibilities cause an increase in entropy of the fluid, and the path is directed toward increasing entropy. The more irreversible the process, the further point 2 lies to the right on the P_2 isobar, and the lower the efficiency η of the process.

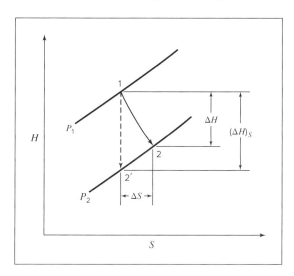

Figure 7.4: Adiabatic expansion process in a turbine or expander.

Example 7.6

A steam turbine with rated capacity of 56,400 kW (56,400 kJ·s^{-1}) operates with steam at inlet conditions of 8600 kPa and 500°C, and discharges into a condenser at a pressure of 10 kPa. Assuming a turbine efficiency of 0.75, determine the state of the steam at discharge and the mass rate of flow of the steam.

Solution 7.6

At the inlet conditions of 8600 kPa and 500°C, the steam tables provide:

$$H_1 = 3391.6 \text{ kJ·kg}^{-1} \qquad S_1 = 6.6858 \text{ kJ·kg}^{-1}\cdot\text{K}^{-1}$$

If the expansion to 10 kPa is isentropic, then, $S_2' = S_1 = 6.6858$ kJ·kg^{-1}·K^{-1}. Steam with this entropy at 10 kPa is wet. Applying the "lever rule" [Eq. (6.96b), with $M = S$ and $x^v = x_2'$], the quality is obtained as follows:

$$S_2' = S_2^l + x_2'(S_2^v - S_2^l)$$

Then, $6.6858 = 0.6493 + x_2'(8.1511 - 0.6493)$ $x_2' = 0.8047$

This is the quality (fraction vapor) of the discharge stream at point 2'. The enthalpy H_2' is also given by Eq. (6.96b), written:

$$H_2' = H_2^l + x_2'(H_2^v - H_2^l)$$

Thus, $H_2' = 191.8 + (0.8047)(2584.8 - 191.8) = 2117.4 \text{ kJ·kg}^{-1}$
           $(\Delta H)_S = H_2' - H_1 = 2117.4 - 3391.6 = -1274.2 \text{ kJ·kg}^{-1}$

and by Eq. (7.16),

$$\Delta H = \eta(\Delta H)_S = (0.75)(-1274.2) = -955.6 \text{ kJ·kg}^{-1}$$

Whence, $H_2 = H_1 + \Delta H = 3391.6 - 955.6 = 2436.0 \text{ kJ·kg}^{-1}$

Thus the steam in its actual final state is also wet, with its quality given by:

$$2436.0 = 191.8 + x_2(2584.8 - 191.8)   x_2 = 0.9378$$

Then $S_2 = 0.6493 + (0.9378)(8.1511 - 0.6493) = 7.6846 \text{ kJ·kg}^{-1}\text{·K}^{-1}$

This value may be compared with the initial value of $S_1 = 6.6858$.
   The steam rate $\dot{m}$ is given by Eq. (7.13). For a work rate of 56,400 kJ·s^{-1},

$$\dot{W}_s = -56,400 = \dot{m}(2436.0 - 3391.6)   \dot{m} = 59.02 \text{ kg·s}^{-1}$$

Example 7.6 was solved with data from the steam tables. When a comparable set of tables is not available for the working fluid, the generalized correlations of Sec. 6.4 may be used in conjunction with Eqs. (6.74) and (6.75), as illustrated in the following example.

Example 7.7

A stream of ethylene gas at 300°C and 45 bar is expanded adiabatically in a turbine to 2 bar. Calculate the isentropic work produced. Find the properties of ethylene by:

(a) Equations for an ideal gas.

(b) Appropriate generalized correlations.

Solution 7.7

The enthalpy and entropy changes for the process are:

$$\Delta H = \langle C_P^{ig} \rangle_H (T_2 - T_1) + H_2^R - H_1^R \tag{6.74}$$

$$\Delta S = \langle C_P^{ig} \rangle_S \ln \frac{T_2}{T_1} - R \ln \frac{P_2}{P_1} + S_2^R - S_1^R \tag{6.75}$$

Given values are $P_1 = 45$ bar, $P_2 = 2$ bar, and $T_1 = 300 + 273.15 = 573.15$ K.

(*a*) If ethylene is assumed to be in its ideal-gas state, then all residual properties are zero, and the preceding equations reduce to:

$$\Delta H = \langle C_P^{ig}\rangle_H (T_2 - T_1) \qquad \Delta S = \langle C_P^{ig}\rangle_S \ln\frac{T_2}{T_1} - R\ln\frac{P_2}{P_1}$$

For an isentropic process, $\Delta S = 0$, and the second equation becomes:

$$\frac{\langle C_P^{ig}\rangle_S}{R}\ln\frac{T_2}{T_1} = \ln\frac{P_2}{P_1} = \ln\frac{2}{45} = -3.1135$$

or

$$\ln T_2 = \frac{-3.1135}{\langle C_P^{ig}\rangle_S/R} + \ln 573.15$$

Then,

$$T_2 = \exp\left(\frac{-3.1135}{\langle C_P^{ig}\rangle_S/R} + 6.3511\right) \qquad\qquad (A)$$

Equation (5.13) provides an expression for $\langle C_P^{ig}\rangle_S/R$, which for computational purposes is represented by:

$$\frac{\langle C_P^{ig}\rangle_S}{R} = \text{MCPS}(573.15, T2; 1.424, 14.394\times 10^{-3}, -4.392\times 10^{-6}, 0.0)$$

where the constants for ethylene come from Table C.1 of App. C. Temperature T_2 is found by iteration. An initial value is assumed for evaluation of $\langle C_P^{ig}\rangle_S/R$. Equation (A) then provides a new value of T_2 from which to recompute $\langle C_P^{ig}\rangle_S/R$, and the procedure continues to convergence on the final value: $T_2 = 370.8$ K. The value of $\langle C_P^{ig}\rangle_H/R$, given by Eq. (4.9), is for computational purposes represented by:

$$\frac{\langle C_P^{ig}\rangle_H}{R} = \text{MCPH}(573.15, 370.8; 1.424, 14.394\times 10^{-3}, -4.392\times 10^{-6}, 0.0) = 7.224$$

Then

$$W_s(\text{isentropic}) = (\Delta H)_S = \langle C_P^{ig}\rangle_H (T_2 - T_1)$$

$$W_s(\text{isentropic}) = (7.224)(8.314)(370.8 - 573.15) = -12{,}153 \text{ J·mol}^{-1}$$

(*b*) For ethylene,

$$T_c = 282.3 \text{ K} \qquad P_c = 50.4 \text{ bar} \qquad \omega = 0.087$$

At the initial state,

$$T_{r_1} = \frac{573.15}{282.3} = 2.030 \qquad P_{r_1} = \frac{45}{50.4} = 0.893$$

According to Fig. 3.13, the generalized correlations based on second virial coefficients should be satisfactory. The computational procedures of Eqs. (6.68), (6.69), (3.61), (3.62), (6.70), and (6.71) are represented by:

$$\frac{H_1^R}{RT_c} = \text{HRB}(2.030, 0.893, 0.087) = -0.234$$

$$\frac{S_1^R}{R} = \mathrm{SRB}(2.030, 0.893, 0.087) = -0.097$$

Then, $$H_1^R = (-0.234)(8.314)(282.3) = -549 \text{ J·mol}^{-1}$$

$$S_1^R = (-0.097)(8.314) = -0.806 \text{ J·mol}^{-1}\text{·K}^{-1}$$

For an initial estimate of S_2^R, assume that $T_2 = 370.8$ K, the value determined in part (*a*). Then,

$$T_{r_2} = \frac{370.8}{282.3} = 1.314 \qquad P_{r_2} = \frac{2}{50.4} = 0.040$$

and $$\frac{S_2^R}{R} = \mathrm{SRB}(1.314, 0.040, 0.087) = -0.0139$$

Then $$S_2^R = (-0.0139)(8.314) = -0.116 \text{ J·mol}^{-1}\text{·K}^{-1}$$

If the expansion process is isentropic, Eq. (6.75) becomes:

$$0 = \langle C_P^{ig} \rangle_S \ln \frac{T_2}{573.15} - 8.314 \ln \frac{2}{45} - 0.116 + 0.806$$

Whence, $$\ln \frac{T_2}{573.15} = \frac{-26.576}{\langle C_P^{ig} \rangle_S}$$

or $$T_2 = \exp \left(\frac{-26.576}{\langle C_P^{ig} \rangle_S} + 6.3511 \right)$$

An iteration process exactly like that of part (*a*) yields the results

$$T_2 = 365.8 \text{ K} \qquad \text{and} \qquad T_{r_2} = 1.296$$

With this value of T_{r_2} and with $P_{r_2} = 0.040$,

$$\frac{S_2^R}{R} = \mathrm{SRB}(1.296, 0.040, 0.087) = -0.0144$$

and $$S_2^R = (-0.0144)(8.314) = -0.120 \text{ J·mol}^{-1}\text{·K}^{-1}$$

This result is so little changed from the initial estimate that recalculation of T_2 is unnecessary, and H_2^R is evaluated at the reduced conditions just established:

$$\frac{H_2^R}{RT_c} = \mathrm{HRB}(1.296, 0.040, 0.087) = -0.0262$$

$$H_2^R = (-0.0262)(8.314)(282.3) = -61.0 \text{ J·mol}^{-1}$$

By Eq. (6.74), $(\Delta H)_S = \langle C_P^{ig} \rangle_H (365.8 - 573.15) - 61.0 + 549$

Evaluation of $\langle C_P^{ig} \rangle_H$ as in part (a) with $T_2 = 365.8$ K gives:

$$\langle C_P^{ig} \rangle_H = 59.843 \text{ J·mol}^{-1}\text{·K}^{-1}$$

and

$$(\Delta H)_S = -11{,}920 \text{ J·mol}^{-1}$$

Then $\qquad W_s(\text{isentropic}) = (\Delta H)_S = -11{,}920 \text{ J·mol}^{-1}$

This differs from the ideal-gas state value by less than 2%.

7.3 COMPRESSION PROCESSES

Just as expansion processes result in pressure reductions in a flowing fluid, so compression processes bring about pressure increases. Compressors, pumps, fans, blowers, and vacuum pumps are all devices designed for this purpose. They are vital for the transport of fluids, for fluidization of particulate solids, for bringing fluids to the proper pressure for reaction or processing, etc. We are concerned here not with the design of such devices, but with specification of energy requirements for steady-state compression causing an increase in fluid pressure.

Compressors

The compression of gases may be accomplished in equipment with rotating blades (like a turbine operating in reverse) or in cylinders with reciprocating pistons. Rotary equipment is used for high-volume flow where the discharge pressure is not too high. For high pressures, reciprocating compressors are often required. The energy equations are independent of the type of equipment; indeed, they are the same as for turbines or expanders because here, too, potential and kinetic-energy changes are presumed negligible. Thus, Eqs. (7.13) through (7.15) apply to adiabatic compression, a process represented by Fig. 7.5.

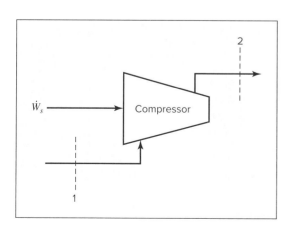

Figure 7.5: Steady-state compression process.

In a compression process, the isentropic work, as given by Eq. (7.15), is the *minimum* shaft work required for compression of a gas from a given initial state to a given discharge pressure. Thus we define a compressor efficiency as:

$$\eta \equiv \frac{W_s(\text{isentropic})}{W_s}$$

In view of Eqs. (7.14) and (7.15), this is also given by:

$$\eta \equiv \frac{(\Delta H)_S}{\Delta H} \tag{7.17}$$

Compressor efficiencies are usually in the range of 0.7 to 0.8.

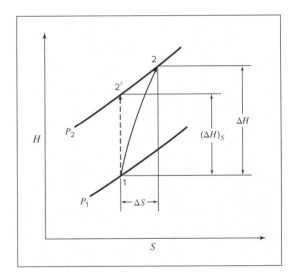

Figure 7.6: Adiabatic compression process.

The compression process is shown on an *HS* diagram in Fig. 7.6. The vertical dashed line rising from point 1 to point 2′ represents the reversible adiabatic (constant-entropy) compression process from P_1 to P_2. The actual irreversible compression process follows the solid line from point 1 upward and to the right in the direction of increasing entropy, terminating at point 2. The more irreversible the process, the further this point lies to the right on the P_2 isobar, and the lower the efficiency η of the process.

Example 7.8

Saturated-vapor steam at 100 kPa ($t^{\text{sat}} = 99.63°C$) is compressed adiabatically to 300 kPa. If the compressor efficiency is 0.75, what is the work required and what are the properties of the discharge stream?

Solution 7.8

For saturated steam at 100 kPa,

$$S_1 = 7.3598 \text{ kJ·kg}^{-1}\text{·K}^{-1} \qquad\qquad H_1 = 2675.4 \text{ kJ·kg}^{-1}$$

For isentropic compression to 300 kPa, $S_2' = S_1 = 7.3598$ kJ·kg^{-1}·K^{-1}. Interpolation in the tables for superheated steam at 300 kPa shows that steam with this entropy has the enthalpy: $H_2' = 2888.8$ kJ·kg^{-1}.

Thus, $\qquad\qquad (\Delta H)_S = 2888.8 - 2675.4 = 213.4$ kJ·kg^{-1}

By Eq. (7.17), $\qquad (\Delta H) = \dfrac{(\Delta H)_S}{\eta} = \dfrac{213.4}{0.75} = 284.5$ kJ·kg^{-1}

and $\qquad\qquad H_2 = H_1 + \Delta H = 2675.4 + 284.5 = 2959.9$ kJ·kg^{-1}

For superheated steam with this enthalpy, interpolation yields:

$$T_2 = 246.1°\text{C} \qquad S_2 = 7.5019 \text{ kJ·kg}^{-1}\text{·K}^{-1}$$

Moreover, by Eq. (7.14), the work required is:

$$W_s = \Delta H = 284.5 \text{ kJ·kg}^{-1}$$

The direct application of Eqs. (7.13) through (7.15) presumes the availability of tables of data or an equivalent thermodynamic diagram for the fluid being compressed. Where such information is not available, the generalized correlations of Sec. 6.4 may be used in conjunction with Eqs. (6.74) and (6.75), exactly as illustrated in Ex. 7.7 for an expansion process.

The assumption of the ideal-gas state leads to relatively simple equations. By Eq. (5.14):

$$\Delta S = \langle C_P \rangle_S \ln \frac{T_2}{T_1} - R \ln \frac{P_2}{P_1}$$

where for simplicity the superscript *ig* has been omitted from the mean heat capacity. If the compression is isentropic, $\Delta S = 0$, and this equation becomes:

$$T_2' = T_1 \left(\frac{P_2}{P_1} \right)^{R/\langle C_P' \rangle_S} \tag{7.18}$$

with T_2' the temperature that results when compression from T_1 and P_1 to P_2 is *isentropic* and where $\langle C_P' \rangle_S$ is the mean heat capacity for the temperature range from T_1 to T_2'.

Applied to isentropic compression, Eq. (4.10) here becomes:

$$(\Delta H)_S = \langle C_P' \rangle_H (T_2' - T_1)$$

In accord with Eq. (7.15), $\qquad W_s(\text{isentropic}) = \langle C_P' \rangle_H (T_2' - T_1) \tag{7.19}$

This result may be combined with the compressor efficiency to give:

$$W_S = \frac{W_S(\text{isentropic})}{\eta} \tag{7.20}$$

The *actual* discharge temperature T_2 resulting from compression is also found from Eq. (4.10), rewritten as:

$$\Delta H = \langle C_P \rangle_H (T_2 - T_1)$$

Whence, $$T_2 = T_1 + \frac{\Delta H}{\langle C_P \rangle_H} \qquad (7.21)$$

where by Eq. (7.14) $\Delta H = W_s$. Here $\langle C_P \rangle_H$ is the mean heat capacity for the temperature range from T_1 to T_2.

For the special case of the ideal-gas state and constant heat capacities,

$$\langle C_P' \rangle_H = \langle C_P \rangle_H = \langle C_P' \rangle_S = C_P$$

Equations (7.18) and (7.19) therefore become:

$$T_2' = T_1 \left(\frac{P_2}{P_1} \right)^{R/C_P} \qquad \text{and} \qquad W_s\,(\text{isentropic}) = C_P(T_2' - T_1)$$

Combining these equations gives:[4]

$$W_s\,(\text{isentropic}) = C_P T_1 \left[\left(\frac{P_2}{P_1} \right)^{R/C_P} - 1 \right] \qquad (7.22)$$

For monatomic gases, such as argon and helium, $R/C_P = 2/5 = 0.4$. For such diatomic gases as oxygen, nitrogen, and air at moderate temperatures, $R/C_P \approx 2/7 = 0.2857$. For gases of greater molecular complexity the ideal-gas heat capacity depends more strongly on temperature, and Eq. (7.22) is less likely to be suitable. One can easily show that the assumption of constant heat capacities also leads to the result:

$$T_2 = T_1 + \frac{T_2' - T_1}{\eta} \qquad (7.23)$$

Example 7.9

If methane (assumed to be in its ideal-gas state) is compressed adiabatically from 20°C and 140 kPa to 560 kPa, estimate the work requirement and the discharge temperature of the methane. The compressor efficiency is 0.75.

---

[4]Because $R = C_P - C_V$ for an ideal gas: $\dfrac{R}{C_P} = \dfrac{C_P - C_V}{C_P} = \dfrac{\gamma - 1}{\gamma}$. An alternative form of Eq. (7.22) is therefore:

$W_S(\text{isentropic}) = \dfrac{\gamma R T_1}{\gamma - 1} \left[\left(\dfrac{P_2}{P_1} \right)^{(\gamma - 1)/\gamma} - 1 \right]$. Although this is the form commonly encountered, Eq. (7.22) is simpler

and more easily applied.

Solution 7.9

Application of Eq. (7.18) requires evaluation of the exponent $R/\langle C_P'\rangle_S$. This is provided by Eq. (5.13), which for the present computation is represented by:

$$\frac{\langle C_P'\rangle_S}{R} = \text{MCPS}(293.15, T_2; 1.702, 9.081 \times 10^{-3}, -2.164 \times 10^{-6}, 0.0)$$

where the constants for methane are from Table C.1 of App. C. Choose a value for T_2' somewhat higher than the initial temperature $T_1 = 293.15$ K. The exponent in Eq. (7.18) is the reciprocal of $\langle C_P'\rangle_S/R$. With $P_2/P_1 = 560/140 = 4.0$ and $T_1 = 293.15$ K, find a new value of T_2'. The procedure is repeated until no further significant change occurs in the value of T_2'. This process produces the values:

$$\frac{\langle C_P'\rangle_S}{R} = 4.5574 \qquad \text{and} \qquad T_2' = 397.37 \text{ K}$$

For the same T_1 and T_2', evaluate $\langle C_P'\rangle_H/R$ by Eq. (4.9):

$$\frac{\langle C_P'\rangle_H}{R} = \text{MCPH}(293.15, 397.37; 1.702, 9.081 \times 10^{-3}, -2.164 \times 10^{-6}, 0.0) = 4.5774$$

Whence, $\qquad \langle C_P'\rangle_H = (4.5774)(8.314) = 38.506 \text{ J}\cdot\text{mol}^{-1}\cdot\text{K}^{-1}$

Then by Eq. (7.19),

$$W_s(\text{isentropic}) = (38.056)(397.37 - 293.15) = 3966.2 \text{ J}\cdot\text{mol}^{-1}$$

The actual work is found from Eq. (7.20):

$$W_s = \frac{3966.2}{0.75} = 5288.3 \text{ J}\cdot\text{mol}^{-1}$$

Application of Eq. (7.21) for the calculation of T_2 gives:

$$T_2 = 293.15 + \frac{5288.3}{\langle C_P\rangle_H}$$

Because $\langle C_P\rangle_H$ depends on T_2, we again iterate. With T_2' as a starting value, this leads to the results:

$$T_2 = 428.65 \text{ K} \qquad \text{or} \qquad t_2 = 155.5°\text{C}$$

and $\qquad\qquad\qquad \langle C_P\rangle_H = 39.027 \text{ J}\cdot\text{mol}^{-1}\cdot\text{K}^{-1}$

Pumps

Liquids are usually moved by pumps, which are generally rotating equipment. The same equations apply to adiabatic pumps as to adiabatic compressors. Thus, Eqs. (7.13) through (7.15) and Eq. (7.17) are valid. However, application of Eq. (7.14) for the calculation of $W_s = \Delta H$ requires values of the enthalpy of compressed (subcooled) liquids, and these are seldom available. The fundamental property relation, Eq. (6.9), provides an alternative. For an isentropic process,

$$dH = V\,dP \qquad (\text{const } S)$$

Combining this with Eq. (7.15) yields:

$$W_s(\text{isentropic}) = (\Delta H)_S = \int_{P_1}^{P_2} V\,dP$$

The usual assumption for liquids (at conditions well removed from the critical point) is that V is independent of P. Integration then gives:

$$W_s(\text{isentropic}) = (\Delta H)_S = V(P_2 - P_1) \qquad (7.24)$$

Also useful are the following equations from Chap. 6:

$$dH = C_P\,dT + V(1 - \beta T)dP \quad (6.27) \qquad dS = C_P\frac{dT}{T} - \beta V\,dP \qquad (6.28)$$

where the volume expansivity β is defined by Eq. (3.3). Because temperature changes in the pumped fluid are very small and because the properties of liquids are insensitive to pressure (again at conditions not close to the critical point), these equations are usually integrated on the assumption that C_P, V, and β are constant, usually at initial values. Thus, to a good approximation

$$\Delta H = C_P\Delta T + V(1 - \beta T)\,\Delta P \quad (7.25) \qquad \Delta S = C_P \ln\frac{T_2}{T_1} - \beta V\,\Delta P \qquad (7.26)$$

Example 7.10

Water at 45°C and 10 kPa enters an adiabatic pump and is discharged at a pressure of 8600 kPa. Assume the pump efficiency to be 0.75. Calculate the work of the pump, the temperature change of the water, and the entropy change of the water.

Solution 7.10

The following are properties for saturated liquid water at 45°C (318.15 K):

$$V = 1010\ \text{cm}^3\cdot\text{kg}^{-1} \qquad \beta = 425 \times 10^{-6}\ \text{K}^{-1} \qquad C_P = 4.178\ \text{kJ}\cdot\text{kg}^{-1}\cdot\text{K}^{-1}$$

By Eq. (7.24),

$$W_s(\text{isentropic}) = (\Delta H)_S = (1010)(8600 - 10) = 8.676 \times 10^6 \ \text{kPa·cm}^3\text{·kg}^{-1}$$

Because $1 \ \text{kJ} = 10^6 \ \text{kPa·cm}^3$,

$$W_s(\text{isentropic}) = (\Delta H)_S = 8.676 \ \text{kJ·kg}^{-1}$$

By Eq. (7.17), $\qquad \Delta H = \dfrac{(\Delta H)_S}{\eta} = \dfrac{8.676}{0.75} = 11.57 \ \text{kJ·kg}^{-1}$

and $\qquad\qquad\qquad W_s = \Delta H = 11.57 \ \text{kJ·kg}^{-1}$

The temperature change of the water during pumping, from Eq. (7.25):

$$11.57 = 4.178 \ \Delta T + 1010\left[1 - (425 \times 10^{-6})(318.15)\right]\frac{8590}{10^6}$$

Solution for ΔT gives:

$$\Delta T = 0.97 \ \text{K} \qquad \text{or} \qquad 0.97°\text{C}$$

The entropy change of the water is given by Eq. (7.26):

$$\Delta S = 4.178 \ \ln\frac{319.12}{318.15} - (425 \times 10^{-6})(1010)\frac{8590}{10^6} = 0.0090 \ \text{kJ·kg}^{-1}\text{·K}^{-1}$$

7.4 SYNOPSIS

After thorough study of this chapter, including working through example and end-of-chapter problems, one should be able to:

- Apply relationships between thermodynamic quantities to flow processes such as flow through pipes and nozzles

- Understand the concept of choked flow and the mechanism by which converging/diverging nozzles produce supersonic flows

- Analyze throttling processes, and define and apply the Joule/Thomson coefficient

- Compute the work produced by a turbine (expander) of given efficiency expanding a fluid from a known initial state to a known final pressure

- Define and apply isentropic efficiencies for both processes that produce work and processes that require work input

- Compute the work required to compress a gas from a given initial state to a final pressure, using a compressor with known efficiency

- Determine changes in all thermodynamic state variables for compression and expansion processes
- Compute work requirements for pumping liquids

7.5 PROBLEMS

7.1. Air expands adiabatically through a nozzle from a negligible initial velocity to a final velocity of 325 m·s^{-1}. What is the temperature drop of the air, if air is assumed to be an ideal gas for which $C_P = (7/2)R$?

7.2. In Ex. 7.5 an expression is found for the Joule/Thomson coefficient, $\mu = (\partial T/\partial P)_H$, that relates it to a heat capacity and equation-of-state information. Develop similar expressions for the derivatives:
(*a*) $(\partial T/\partial P)_S$ (*b*) $(\partial T/\partial V)_U$

What can you say about the *signs* of these derivatives? For what types of processes might these derivatives be important characterizing quantities?

7.3. The thermodynamic sound speed c is defined in Sec. 7.1. Prove that:

$$c = \sqrt{\frac{VC_P}{\mathcal{M}C_V\kappa}}$$

where V is molar volume and $\mathcal{M}$ is molar mass. To what does this general result reduce for: (*a*) an ideal gas? (*b*) an incompressible liquid? What do these results suggest qualitatively about the speed of sound in liquids relative to gases?

7.4. Steam enters a nozzle at 800 kPa and 280°C at negligible velocity and discharges at a pressure of 525 kPa. Assuming isentropic expansion of the steam in the nozzle, what is the exit velocity and what is the cross-sectional area at the nozzle exit for a flow rate of 0.75 kg·s^{-1}?

7.5. Steam enters a converging nozzle at 800 kPa and 280°C with negligible velocity. If expansion is isentropic, what is the minimum pressure that can be reached in such a nozzle, and what is the cross-sectional area at the nozzle throat at this pressure for a flow rate of 0.75 kg·s^{-1}?

7.6. A gas enters a converging nozzle at pressure P_1 with negligible velocity, expands isentropically in the nozzle, and discharges into a chamber at pressure P_2. Sketch graphs showing the velocity at the throat and the mass flow rate as functions of the pressure ratio P_2/P_1.

7.7. For isentropic expansion in a converging/diverging nozzle with negligible entrance velocity, sketch graphs of mass flow rate $\dot{m}$, velocity u, and area ratio A/A_1 versus the pressure ratio P/P_1. Here, A is the cross-sectional area of the nozzle at the point in the nozzle where the pressure is P, and subscript 1 denotes the nozzle entrance.

7.8. An ideal gas with constant heat capacities enters a converging/diverging nozzle with negligible velocity. If it expands isentropically within the nozzle, show that the throat velocity is given by:

$$u_{throat}^2 = \frac{\gamma R T_1}{\mathcal{M}}\left(\frac{2}{\gamma + 1}\right)$$

where T_1 is the temperature of the gas entering the nozzle, $\mathcal{M}$ is the molar mass, and R is the molar gas constant.

7.9. Steam expands isentropically in a converging/diverging nozzle from inlet conditions of 1400 kPa, 325°C, and negligible velocity to a discharge pressure of 140 kPa. At the throat, the cross-sectional area is 6 cm². Determine the mass flow rate of the steam and the state of the steam at the exit of the nozzle.

7.10. Steam expands adiabatically in a nozzle from inlet conditions of 130(psia), 420(°F), and a velocity of 230(ft)(s)$^{-1}$ to a discharge pressure of 35(psia) where its velocity is 2000(ft)(s)$^{-1}$. What is the state of the steam at the nozzle exit, and what is $\dot{S}_G$ for the process?

7.11. Air discharges from an adiabatic nozzle at 15°C with a velocity of 580 m·s^{-1}. What is the temperature at the entrance of the nozzle if the entrance velocity is negligible? Assume air to be an ideal gas for which $C_P = (7/2)R$.

7.12. Cool water at 15°C is throttled from 5(atm) to 1(atm), as in a kitchen faucet. What is the temperature change of the water? What is the lost work per kilogram of water for this everyday household happening? At 15°C and 1(atm), the volume expansivity β for liquid water is about 1.5×10^{-4} K^{-1}. The surroundings temperature T_σ is 20°C. State carefully any assumptions you make. The steam tables are a source of data.

7.13. For a pressure-explicit equation of state, prove that the Joule/Thomson inversion curve is the locus of states for which:

$$T\left(\frac{\partial Z}{\partial T}\right)_\rho = \rho\left(\frac{\partial Z}{\partial \rho}\right)_T$$

Apply this equation to (*a*) the van der Waals equation; (*b*) the Redlich/Kwong equation. Discuss the results.

7.14. Two nonconducting tanks of negligible heat capacity and of equal volume initially contain equal quantities of the same ideal gas at the same T and P. Tank A discharges to the atmosphere through a small turbine in which the gas expands isentropically; tank B discharges to the atmosphere through a porous plug. Both devices operate until discharge ceases.

(*a*) When discharge ceases, is the temperature in tank A less than, equal to, or greater than the temperature in tank B?

(*b*) When the pressures in both tanks have fallen to half the initial pressure, is the temperature of the gas discharging from the turbine less than, equal to, or greater than the temperature of the gas discharging from the porous plug?

(c) During the discharge process, is the temperature of the gas leaving the turbine less than, equal to, or greater than the temperature of the gas leaving tank A at the same instant?

(d) During the discharge process, is the temperature of the gas leaving the porous plug less than, equal to, or greater than the temperature of the gas leaving tank B at the same instant?

(e) When discharge ceases, is the mass of gas remaining in tank A less than, equal to, or greater than the mass of gas remaining in tank B?

7.15. A steam turbine operates adiabatically at a power level of 3500 kW. Steam enters the turbine at 2400 kPa and 500°C and exhausts from the turbine as saturated vapor at 20 kPa. What is the steam rate through the turbine, and what is the turbine efficiency?

7.16. A turbine operates adiabatically with superheated steam entering at T_1 and P_1 with a mass flow rate $\dot{m}$. The exhaust pressure is P_2 and the turbine efficiency is η. For one of the following sets of operating conditions, determine the power output of the turbine and the enthalpy and entropy of the exhaust steam.

(a) $T_1 = 450°C$, $P_1 = 8000$ kPa, $\dot{m} = 80$ kg·s^{-1}, $P_2 = 30$ kPa, $\eta = 0.80$.
(b) $T_1 = 550°C$, $P_1 = 9000$ kPa, $\dot{m} = 90$ kg·s^{-1}, $P_2 = 20$ kPa, $\eta = 0.77$.
(c) $T_1 = 600°C$, $P_1 = 8600$ kPa, $\dot{m} = 70$ kg·s^{-1}, $P_2 = 10$ kPa, $\eta = 0.82$.
(d) $T_1 = 400°C$, $P_1 = 7000$ kPa, $\dot{m} = 65$ kg·s^{-1}, $P_2 = 50$ kPa, $\eta = 0.75$.
(e) $T_1 = 200°C$, $P_1 = 1400$ kPa, $\dot{m} = 50$ kg·s^{-1}, $P_2 = 200$ kPa, $\eta = 0.75$.
(f) $T_1 = 900(°F)$, $P_1 = 1100$ (psia), $\dot{m} = 150(\text{lb}_m)(\text{s})^{-1}$, $P_2 = 2(\text{psia})$, $\eta = 0.80$.
(g) $T_1 = 800(°F)$, $P_1 = 1000$ (psia), $\dot{m} = 100(\text{lb}_m)(\text{s})^{-1}$, $P_2 = 4(\text{psia})$, $\eta = 0.75$.

7.17. Nitrogen gas initially at 8.5 bar expands isentropically to 1 bar and 150°C. Assuming nitrogen to be an ideal gas, calculate the *initial* temperature and the work produced per mole of nitrogen.

7.18. Combustion products from a burner enter a gas turbine at 10 bar and 950°C and discharge at 1.5 bar. The turbine operates adiabatically with an efficiency of 77%. Assuming the combustion products to be an ideal-gas mixture with a heat capacity of 32 J·mol^{-1}·K^{-1}, what is the work output of the turbine per mole of gas, and what is the temperature of the gases discharging from the turbine?

7.19. Isobutane expands adiabatically in a turbine from 5000 kPa and 250°C to 500 kPa at a rate of 0.7 kg mol·s^{-1}. If the turbine efficiency is 0.80, what is the power output of the turbine and what is the temperature of the isobutane leaving the turbine?

7.20. The steam rate to a turbine for variable output is controlled by a throttle valve in the inlet line. Steam is supplied to the throttle valve at 1700 kPa and 225°C. During a test run, the pressure at the turbine inlet is 1000 kPa, the exhaust steam at 10 kPa has a quality of 0.95, the steam flow rate is 0.5 kg·s^{-1}, and the power output of the turbine is 180 kW.

(a) What are the heat losses from the turbine?

(b) What would be the power output if the steam supplied to the throttle valve were expanded isentropically to the final pressure?

7.21. Carbon dioxide gas enters an adiabatic expander at 8 bar and 400°C and discharges at 1 bar. If the turbine efficiency is 0.75, what is the discharge temperature and what is the work output per mole of CO_2? Assume CO_2 to be an ideal gas at these conditions.

7.22. Tests on an adiabatic gas turbine (expander) yield values for inlet conditions (T_1, P_1) and outlet conditions (T_2, P_2). Assuming ideal gases with constant heat capacities, determine the turbine efficiency for one of the following:

    (a) $T_1 = 500$ K, $P_1 = 6$ bar, $T_2 = 371$ K, $P_2 = 1.2$ bar, $C_P/R = 7/2$.
    (b) $T_1 = 450$ K, $P_1 = 5$ bar, $T_2 = 376$ K, $P_2 = 2$ bar, $C_P/R = 4$.
    (c) $T_1 = 525$ K, $P_1 = 10$ bar, $T_2 = 458$ K, $P_2 = 3$ bar, $C_P/R = 11/2$.
    (d) $T_1 = 475$ K, $P_1 = 7$ bar, $T_2 = 372$ K, $P_2 = 1.5$ bar, $C_P/R = 9/2$.
    (e) $T_1 = 550$ K, $P_1 = 4$ bar, $T_2 = 403$ K, $P_2 = 1.2$ bar, $C_P/R = 5/2$.

7.23. The efficiency of a particular series of adiabatic gas turbines (expanders) correlates with power output according to the empirical expression: $\eta = 0.065 + 0.080 \ln |\dot{W}|$. Here, $|\dot{W}|$ is the absolute value of the *actual* power output in kW. Nitrogen gas is to be expanded from inlet conditions of 550 K and 6 bar to an outlet pressure of 1.2 bar. For a molar flow rate of 175 mol·s^{-1}, what is the delivered power in kW? What is the efficiency of the turbine? What is the rate of entropy generation $\dot{S}_G$? Assume nitrogen to be an ideal gas with $C_P = (7/2)R$.

7.24. A turbine operates adiabatically with superheated steam entering at 45 bar and 400°C. If the exhaust steam must be "dry," what is the minimum allowable exhaust pressure for a turbine efficiency, $\eta = 0.75$? Suppose the efficiency were 0.80. Would the minimum exhaust pressure be lower or higher? Why?

7.25. Turbines can be used to recover energy from high-pressure liquid streams. However, they are not used when the high-pressure stream is a *saturated* liquid. Why? Illustrate by determining the downstream state for isentropic expansion of saturated liquid water at 5 bar to a final pressure of 1 bar.

7.26. Liquid water enters an adiabatic hydroturbine at 5(atm) and 15°C, and exhausts at 1(atm). Estimate the power output of the turbine in J·kg^{-1} of water if its efficiency is $\eta = 0.55$. What is the outlet temperature of the water? Assume water to be an incompressible liquid.

7.27. An expander operates adiabatically with nitrogen entering at T_1 and P_1 with a molar flow rate $\dot{n}$. The exhaust pressure is P_2, and the expander efficiency is η. Estimate the power output of the expander and the temperature of the exhaust stream for one of the following sets of operating conditions.

    (a) $T_1 = 480$°C, $P_1 = 6$ bar, $\dot{n} = 200$ mol·s^{-1}, $P_2 = 1$ bar, $\eta = 0.80$.
    (b) $T_1 = 400$°C, $P_1 = 5$ bar, $\dot{n} = 150$ mol·s^{-1}, $P_2 = 1$ bar, $\eta = 0.75$.
    (c) $T_1 = 500$°C, $P_1 = 7$ bar, $\dot{n} = 175$ mol·s^{-1}, $P_2 = 1$ bar, $\eta = 0.78$.
    (d) $T_1 = 450$°C, $P_1 = 8$ bar, $\dot{n} = 100$ mol·s^{-1}, $P_2 = 2$ bar, $\eta = 0.85$.
    (e) $T_1 = 900$(°F), $P_1 = 95$(psia), $\dot{n} = 0.5$(lb mol)(s)$^{-1}$, $P_2 = 15$(psia), $\eta = 0.80$.

7.28. What is the ideal-work rate for the expansion process of Ex. 7.6? What is the thermo-dynamic efficiency of the process? What is the rate of entropy generation $\dot{S}_G$? What is $\dot{W}_{lost}$? Take $T_\sigma = 300$ K.

7.29. Exhaust gas at 400°C and 1 bar from internal-combustion engines flows at the rate of 125 mol·s^{-1} into a waste-heat boiler where saturated steam is generated at a pressure of 1200 kPa. Water enters the boiler at 20°C (T_σ), and the exhaust gases are cooled to within 10°C of the steam temperature. The heat capacity of the exhaust gases is $C_P/R = 3.34 + 1.12 \times 10^{-3} \, T/K$. The steam flows into an adiabatic turbine and exhausts at a pressure of 25 kPa. If the turbine efficiency η is 72%,

(a) What is $\dot{W}_S$, the power output of the turbine?
(b) What is the thermodynamic efficiency of the boiler/turbine combination?
(c) Determine $\dot{S}_G$ for the boiler and for the turbine.
(d) Express $\dot{W}_{lost}$ (boiler) and $\dot{W}_{lost}$ (turbine) as fractions of $|\dot{W}_{ideal}|$, the ideal work of the process.

7.30. A small adiabatic air compressor is used to pump air into a 20-m^3 insulated tank. The tank initially contains air at 25°C and 101.33 kPa, exactly the conditions at which air enters the compressor. The pumping process continues until the pressure in the tank reaches 1000 kPa. If the process is adiabatic and if compression is isentropic, what is the shaft work of the compressor? Assume air to be an ideal gas for which $C_P = (7/2)R$ and $C_V = (5/2)R$.

7.31. Saturated steam at 125 kPa is compressed adiabatically in a centrifugal compressor to 700 kPa at the rate of 2.5 kg·s^{-1}. The compressor efficiency is 78%. What is the power requirement of the compressor and what are the enthalpy and entropy of the steam in its final state?

7.32. A compressor operates adiabatically with air entering at T_1 and P_1 with a molar flow rate $\dot{n}$. The discharge pressure is P_2 and the compressor efficiency is η. Estimate the power requirement of the compressor and the temperature of the discharge stream for one of the following sets of operating conditions.

(a) $T_1 = 25$°C, $P_1 = 101.33$ kPa, $\dot{n} = 100$ mol·s^{-1}, $P_2 = 375$ kPa, $\eta = 0.75$.
(b) $T_1 = 80$°C, $P_1 = 375$ kPa, $\dot{n} = 100$ mol·s^{-1}, $P_2 = 1000$ kPa, $\eta = 0.70$.
(c) $T_1 = 30$°C, $P_1 = 100$ kPa, $\dot{n} = 150$ mol·s^{-1}, $P_2 = 500$ kPa, $\eta = 0.80$.
(d) $T_1 = 100$°C, $P_1 = 500$ kPa, $\dot{n} = 50$ mol·s^{-1}, $P_2 = 1300$ kPa, $\eta = 0.75$.
(e) $T_1 = 80$(°F), $P_1 = 14.7$ (psia), $\dot{n} = 0.5$ (lb mol)(s)$^{-1}$, $P_2 = 55$ (psia), $\eta = 0.75$.
(f) $T_1 = 150$(°F), $P_1 = 55$ (psia), $\dot{n} = 0.5$ (lb mol)(s)$^{-1}$, $P_2 = 135$ (psia), $\eta = 0.70$.

7.33. Ammonia gas is compressed from 21°C and 200 kPa to 1000 kPa in an adiabatic compressor with an efficiency of 0.82. Estimate the final temperature, the work required, and the entropy change of the ammonia.

7.34. Propylene is compressed adiabatically from 11.5 bar and 30°C to 18 bar at a rate of 1 kg mol·s^{-1}. If the compressor efficiency is 0.8, what is the power requirement of the compressor, and what is the discharge temperature of the propylene?

7.35. Methane is compressed adiabatically in a pipeline pumping station from 3500 kPa and 35°C to 5500 kPa at a rate of 1.5 kg mol·s^{-1}. If the compressor efficiency is 0.78, what is the power requirement of the compressor and what is the discharge temperature of the methane?

7.36. What is the ideal work for the compression process of Ex. 7.9? What is the thermodynamic efficiency of the process? What are S_G and W_{lost}? Take $T_\sigma = 293.15$ K.

7.37. A *fan* is (in effect) a gas compressor which moves large volumes of air at low pressure across small (1 to 15 kPa) pressure differences. The usual design equation is:

$$\dot{W} = \dot{n}\frac{RT_1}{\eta P_1}\Delta P$$

where subscript 1 denotes inlet conditions and η is the efficiency with respect to isentropic operation. Develop this equation. Show also how it follows from the usual equation for compression of an ideal gas with constant heat capacities.

7.38. For an adiabatic gas compressor, the efficiency with respect to isentropic operation η is a measure of internal irreversibilities; so is the dimensionless rate of entropy generation $S_G/R \equiv \dot{S}_G/(\dot{n}R)$. Assuming that the gas is ideal with constant heat capacities, show that η and S_G/R are related through the expression:

$$\frac{S_G}{R} = \frac{C_P}{R}\ln\left(\frac{\eta + \pi - 1}{\eta\pi}\right)$$

where

$$\pi \equiv (P_2/P_1)^{R/C_P}$$

7.39. Air at 1(atm) and 35°C is compressed in a staged reciprocating compressor (with intercooling) to a final pressure of 50(atm). For each stage, the inlet gas temperature is 35°C and the maximum allowable outlet temperature is 200°C. Mechanical power is the same for all stages, and isentropic efficiency is 65% for each stage. The volumetric flow rate of air is 0.5 m^3·s^{-1} at the inlet to the first stage.

 (*a*) How many stages are required?
 (*b*) What is the mechanical-power requirement per stage?
 (*c*) What is the heat duty for each intercooler?
 (*d*) Water is the coolant for the intercoolers. It enters at 25°C and leaves at 45°C. What is the cooling-water rate per intercooler?

Assume air is an ideal gas with $C_P = (7/2)R$.

7.40. Demonstrate that the power requirement for compressing a gas is smaller the more complex the gas. Assume fixed values of $\dot{n}$, η, T_1, P_1, and P_2, and that the gas is ideal with constant heat capacities.

7.41. Tests on an adiabatic gas compressor yield values for inlet conditions (T_1, P_1) and outlet conditions (T_2, P_2). Assuming ideal gases with constant heat capacities, determine the compressor efficiency for one of the following:

(a) $T_1 = 300$ K, $P_1 = 2$ bar, $T_2 = 464$ K, $P_2 = 6$ bar, $C_P/R = 7/2$.
(b) $T_1 = 290$ K, $P_1 = 1.5$ bar, $T_2 = 547$ K, $P_2 = 5$ bar, $C_P/R = 5/2$.
(c) $T_1 = 295$ K, $P_1 = 1.2$ bar, $T_2 = 455$ K, $P_2 = 6$ bar, $C_P/R = 9/2$.
(d) $T_1 = 300$ K, $P_1 = 1.1$ bar, $T_2 = 505$ K, $P_2 = 8$ bar, $C_P/R = 11/2$.
(e) $T_1 = 305$ K, $P_1 = 1.5$ bar, $T_2 = 496$ K, $P_2 = 7$ bar, $C_P/R = 4$.

7.42. Air is compressed in a steady-flow compressor, entering at 1.2 bar and 300 K and leaving at 5 bar and 500 K. Operation is *non*adiabatic, with heat transfer to the surroundings at 295 K. For the same change in state of the air, is the mechanical-power requirement per mole of air greater or less for nonadiabatic than for adiabatic operation? Why?

7.43. A boiler house produces a large excess of low-pressure [50(psig), 5(°F)-superheat] steam. An upgrade is proposed that would first run the low-pressure steam through an adiabatic steady-flow compressor, producing medium-pressure [150(psig)] steam. A young engineer expresses concern that compression could result in the formation of liquid water, damaging the compressor. Is there cause for concern? *Suggestion:* Refer to the Mollier diagram of Fig. F.3 of App. F.

7.44. A pump operates adiabatically with liquid water entering at T_1 and P_1 with a mass flow rate $\dot{m}$. The discharge pressure is P_2, and the pump efficiency is η. For one of the following sets of operating conditions, determine the power requirement of the pump and the temperature of the water discharged from the pump.

(a) $T_1 = 25$°C, $P_1 = 100$ kPa, $\dot{m} = 20$ kg·s^{-1}, $P_2 = 2000$ kPa, $\eta = 0.75$,
   $\beta = 257.2 \times 10^{-6}$ K^{-1}.
(b) $T_1 = 90$°C, $P_1 = 200$ kPa, $\dot{m} = 30$ kg·s^{-1}, $P_2 = 5000$ kPa, $\eta = 0.70$,
   $\beta = 696.2 \times 10^{-6}$ K^{-1}.
(c) $T_1 = 60$°C, $P_1 = 20$ kPa, $\dot{m} = 15$ kg·s^{-1}, $P_2 = 5000$ kPa, $\eta = 0.75$,
   $\beta = 523.1 \times 10^{-6}$ K^{-1}.
(d) $T_1 = 70$(°F), $P_1 = 1$(atm), $\dot{m} = 50$(lb$_m$)(s)$^{-1}$, $P_2 = 20$(atm), $\eta = 0.70$,
   $\beta = 217.3 \times 10^{-6}$ K^{-1}.
(e) $T_1 = 200$(°F), $P_1 = 15$(psia), $\dot{m} = 80$(lb$_m$)(s)$^{-1}$, $P_2 = 1500$(psia), $\eta = 0.75$,
   $\beta = 714.3 \times 10^{-6}$ K^{-1}.

7.45. What is the ideal work for the pumping process of Ex. 7.10? What is the thermodynamic efficiency of the process? What is S_G? What is W_{lost}? Take $T_\sigma = 300$ K.

7.46. Show that the points on the Joule/Thomson inversion curve [for which $\mu = (\partial T/\partial P)_H = 0$] are also characterized by each of the following:

(a) $\left(\dfrac{\partial Z}{\partial T}\right)_P = 0$;   (b) $\left(\dfrac{\partial H}{\partial P}\right)_T = 0$;   (c) $\left(\dfrac{\partial V}{\partial T}\right)_P = \dfrac{V}{T}$;   (d) $\left(\dfrac{\partial Z}{\partial V}\right)_P = 0$;

(e) $V\left(\dfrac{\partial P}{\partial V}\right)_T + T\left(\dfrac{\partial P}{\partial T}\right)_V = 0$

7.47. According to Prob. 7.3, the thermodynamic sound speed c depends on the *PVT* equation of state. Show how isothermal sound-speed measurements can be used to estimate the second virial coefficient B of a gas. Assume that Eq. (3.36) applies, and that the ratio C_P/C_V is given by its ideal-gas value.

7.48. Real-gas behavior for turbomachinery is sometimes empirically accommodated through the expression $\dot{W} = \langle Z \rangle \dot{W}^{ig}$, where $\dot{W}^{ig}$ is the ideal-gas mechanical power and $\langle Z \rangle$ is some suitably defined average value of the compressibility factor.

    (*a*) Rationalize this expression.
    (*b*) Devise a turbine example incorporating real-gas behavior via residual properties, and determine a numerical value of $\langle Z \rangle$ for the example.

7.49. Operating data are taken on an air turbine. For a particular run, $P_1 = 8$ bar, $T_1 = 600$ K, and $P_2 = 1.2$ bar. However, the recorded outlet temperature is only partially legible; it could be $T_2 = 318$, 348, or 398 K. Which must it be? For the given conditions, assume air to be an ideal gas with constant $C_P = (7/2)R$.

7.50. Liquid benzene at 25°C and 1.2 bar is converted to vapor at 200°C and 5 bar in a two-step steady-flow process: compression by a pump to 5 bar, followed by vaporization in a counterflow heat exchanger. Determine the power requirement of the pump and the duty of the exchanger in kJ·mol^{-1}. Assume a pump efficiency of 70%, and treat benzene vapor as as an ideal gas with constant $C_P = 105$ J·mol^{-1}·K^{-1}.

7.51. Liquid benzene at 25°C and 1.2 bar is converted to vapor at 200°C and 5 bar in a two-step steady-flow process: vaporization in a counterflow heat exchanger at 1.2 bar, followed by compression as a gas to 5 bar. Determine the duty of the exchanger and the power requirement of the compressor in kJ·mol^{-1}. Assume a compressor efficiency of 75%, and treat benzene vapor as as an ideal gas with constant $C_P = 105$ J·mol^{-1}·K^{-1}.

7.52. Of the processes proposed in Probs. 7.50 and 7.51, which would you recommend? Why?

7.53. Liquids (identified below) at 25°C are completely vaporized at 1(atm) in a countercurrent heat exchanger. Saturated steam is the heating medium, available at four pressures: 4.5, 9, 17, and 33 bar. Which variety of steam is most appropriate for each case? Assume a minimum approach ΔT of 10°C for heat exchange.

    (*a*) Benzene;
    (*b*) *n*-Decane;
    (*c*) Ethylene glycol;
    (*d*) *o*-Xylene

7.54. One hundred (100) kmol·h^{-1} of ethylene is compressed from 1.2 bar and 300 K to 6 bar by an electric-motor-driven compressor. Determine the capital cost C of the unit. Treat ethylene as an ideal gas with constant $C_P = 50.6$ J·mol^{-1}·K^{-1}.

Data: η(compressor) = 0.70

$$C(\text{compressor})/\$ = 3040(\dot{W}_S/\text{kW})^{0.952}$$
where $\dot{W}_S \equiv$ *isentropic* power requirement for the compressor.

$$C(\text{motor})/\$ = 380(|\dot{W}_e|/\text{kW})^{0.855}$$
where $\dot{W}_e \equiv$ *delivered* shaft power of motor.

7.55. Four different types of drivers for gas compressors are: electric motors, gas expanders, steam turbines, and internal-combustion engines. Suggest when each might be appropriate. How would you estimate operating costs for each of these drivers? Ignore such add-ons as maintenance, operating labor, and overhead.

7.56. Two schemes are proposed for the reduction in pressure of ethylene gas at 375 K and 18 bar to 1.2 bar in a steady-flow process:

(*a*) Pass it through a throttle valve.
(*b*) Send it through an adiabatic expander of 70% efficiency.

For each proposal, determine the downstream temperature, and the rate of entropy generation in J·mol^{-1}·K^{-1}. What is the power output for proposal (*b*) in kJ·mol^{-1}? Discuss the pros and cons of the two proposals. Do not assume ideal gases.

7.57. A stream of hydrocarbon gas at 500°C is cooled by continuously combining it with a stream of light oil in an adiabatic tower. The light oil enters as a liquid at 25°C; the combined stream leaves as a gas at 200°C.

(*a*) Draw a carefully labeled flow diagram for the process.
(*b*) Let F and D denote respectively the molar flow rates of hot hydrocarbon gas and light oil. Use data given below to determine a numerical value for the oil-to-gas ratio D/F. Explain your analysis.
(*c*) What is the advantage to quenching the hydrocarbon gas with a *liquid* rather than with another (cooler) gas? Explain.

Data: $C_P^v(\text{ave}) = 150$ J·mol^{-1}·K^{-1} for the hydrocarbon gas.

   $C_P^v(\text{ave}) = 200$ J·mol^{-1}·K^{-1} for the oil vapor.

   $\Delta H^{lv}(\text{oil}) = 35{,}000$ J·mol^{-1} at 25°C.

Chapter 8

Production of Power from Heat

In everyday experience, we often speak of "using" energy. For example, one's utility bills are determined by the quantities of electrical energy and chemical energy (e.g. natural gas) "used" in one's home. This may appear to conflict with the conservation of energy expressed by the first law of thermodynamics. However, closer examination shows that when we speak of "using" energy, we generally mean the conversion of energy from a form capable of producing mechanical work into heat and/or transfer of heat from a source at higher temperature to a lower temperature. These processes that "use" energy in fact are ones that generate entropy. Such entropy generation is inevitable. However, we "use" energy most efficiently when we minimize entropy generation.

Except for nuclear power, the sun is the ultimate source of almost all mechanical and electrical power used by humankind. Energy reaches the earth by solar radiation at a tremendous rate, exceeding current total rates of human energy use by orders of magnitude. Of course the same amount of energy radiates into space from the earth, such that the earth's temperature remains nearly constant. However, incoming solar radiation has an effective temperature near 6000 K, twenty times that of the earth. This large temperature difference implies that most of the incoming solar radiation can, in principle, be converted to mechanical or electrical power. Of course, a key role of sunlight is to support the growth of vegetation. Over millions of years a fraction of this organic matter has been transformed into deposits of coal, oil, and natural gas. Combustion of these fossil fuels provided the power needed to enable the Industrial Revolution and to transform civilization. The large-scale power plants that dot the landscape depend on combustion of these fuels, and to a lesser extent on nuclear fission, for the transfer of heat to the working fluid (H_2O) of steam power plants. These are large-scale heat engines that convert part of the heat into mechanical energy. As a consequence of the second law, the thermal efficiencies of these plants seldom exceed 35%.

Water evaporated by sunlight and transported over land by wind is the source of precipitation that ultimately produces hydroelectric power on a significant scale. Solar radiation also energizes atmospheric winds, which in favorable locations are increasingly used to drive large wind turbines for the production of power. As fossil fuels become less readily available and more expensive, and with increasing concern about atmospheric pollution and global warming, development of alternative sources of power has become urgent. An especially attractive means of harnessing solar energy is through photovoltaic cells, which convert radiation directly into electricity. They are safe, simple, durable, and operate at ambient temperatures, but expense has largely limited their use to small-scale special applications. Their future use on a much larger scale seems inevitable, and the photovoltaic industry is growing at a rapid pace. An alternative for harnessing solar energy at large scales is solar-thermal technology, in which sunlight is focused with mirrors to heat a working fluid that drives a heat engine.

A common device for the direct conversion of chemical (molecular) energy into electrical energy, without the intermediate generation of heat, is the electrochemical cell, that is, a battery. A related device is the *fuel cell*, in which reactants are supplied continuously to the electrodes. The prototype is a cell in which hydrogen reacts with oxygen to produce water through electrochemical conversion. The efficiency of conversion of chemical to electical energy is considerably improved over processes that first convert chemical energy into heat by combustion. This technology has potential application in transportation, and may well find other uses.

The *internal*-combustion engine is also a heat engine, wherein high temperatures are attained by conversion of the chemical energy of a fuel directly into internal energy within the work-producing device. Examples are Otto, Diesel, and gas-turbine engines.[1]

In this chapter we briefly analyze:

- Steam power plants in relation to the Carnot, Rankine, and regenerative cycles
- Internal-combustion engines in relation to the Otto, Diesel, and Brayton cycles
- Jet and rocket engines

8.1   THE STEAM POWER PLANT

Figure 8.1 shows a simple steady-state steady-flow cyclic process in which steam generated in a boiler is expanded in an adiabatic turbine to produce work. The discharge stream from the turbine passes to a condenser from which it is pumped adiabatically back to the boiler. The processes that occur as the working fluid flows around the cycle are represented by lines on the TS diagram of Fig. 8.2. The sequence of these lines conforms to a *Carnot* cycle, as described in Chapter 5. The operation as represented is reversible, consisting of two isothermal steps connected by two adiabatic steps.

Step $1 \rightarrow 2$ is isothermal vaporization taking place in a boiler at temperature T_H, wherein heat is transferred to saturated-liquid water at rate $\dot{Q}_H$, producing saturated vapor. Step $2 \rightarrow 3$ is reversible adiabatic expansion of saturated vapor in a turbine producing a two-phase mixture of saturated liquid and vapor at T_C. This isentropic expansion is represented by a vertical line. Step $3 \rightarrow 4$ is an isothermal partial-condensation process at lower temperature T_C, wherein heat is transferred to the surroundings at rate $\dot{Q}_C$. Step $4 \rightarrow 1$ is an isentropic compression in a pump. Represented by a vertical line, it takes the cycle back to its origin, producing saturated-liquid water at point 1. The power produced by the turbine $\dot{W}_{\text{turbine}}$ is much greater than the

---

[1]Details of steam-power plants and internal-combustion engines can be found in E. B. Woodruff, H. B. Lammers, and T. S. Lammers, *Steam Plant Operation*, 9th ed., McGraw-Hill, New York, 2011; C. F. Taylor, *The Internal Combustion Engine in Theory and Practice: Thermodynamics, Fluid Flow, Performance*, 2nd ed., MIT Press, Boston, 1984; and J. Heywood, *Internal Combustion Engine Fundamentals* McGraw-Hill, New York, 1988.

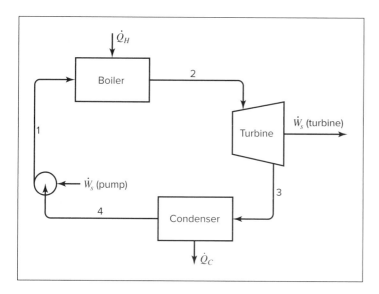

Figure 8.1: Simple steam power plant.

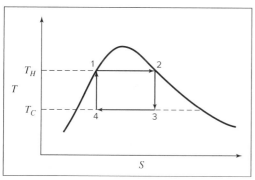

Figure 8.2: Carnot cycle on a *TS* diagram.

power requirement of the pump $\dot{W}_{\text{pump}}$, The net power output is equal to the difference between the rate of heat input in the boiler and the rate of heat rejection in the condenser.

The thermal efficiency of this cycle is:

$$\eta_{\text{Carnot}} = 1 - \frac{T_C}{T_H} \tag{5.7}$$

Clearly, η increases as T_H increases and as T_C decreases. Although the efficiencies of practical heat engines are lowered by irreversibilities, it is still true that their efficiencies are increased when the average temperature at which heat is absorbed in the boiler is increased and when the average temperature at which heat is rejected in the condenser is decreased.

The Rankine Cycle

The thermal efficiency of the Carnot cycle just described and given by Eq. (5.7) could serve as a standard of comparison for actual steam power plants. However, severe practical difficulties attend the operation of equipment intended to carry out steps $2 \rightarrow 3$ and $4 \rightarrow 1$. Turbines that take in saturated steam produce an exhaust with high liquid content, which causes severe erosion problems.[2] Even more difficult is the design of a pump that takes in a mixture of liquid and vapor (point 4) and discharges a saturated liquid (point 1). For these reasons, an alternative cycle is taken as the standard, at least for fossil-fuel-burning power plants. It is called the *Rankine cycle*, and it differs from the cycle of Fig. 8.2 in two major respects. First, the heating step 1 → 2 is carried well beyond vaporization so as to produce a superheated vapor, and second, the cooling step 3 → 4 brings about complete condensation, yielding saturated liquid to be pumped to the boiler. The Rankine cycle therefore consists of the four steps shown in Fig. 8.3 and described as follows:

- 1 → 2 A constant-pressure heating process in a boiler. The step lies along an isobar (the pressure of the boiler) and consists of three sections: heating of subcooled liquid water to its saturation temperature, vaporization at constant temperature and pressure, and superheating of the vapor to a temperature well above its saturation temperature.

- 2 → 3 Reversible, adiabatic (isentropic) expansion of vapor in a turbine to the pressure of the condenser. The step normally crosses the saturation curve, producing a wet exhaust. However, the superheating accomplished in step 1 → 2 shifts the vertical line far enough to the right on Fig. 8.3 that the moisture content is not too large.

- 3 → 4 A constant-pressure, constant-temperature process in a condenser to produce saturated liquid at point 4.

- 4 → 1 Reversible, adiabatic (isentropic) pumping of the saturated liquid to the pressure of the boiler, producing compressed (subcooled) liquid. The vertical line (whose length is exaggerated in Fig 8.3) is very short because the temperature rise associated with compression of a liquid is small.

Power plants actually operate on a cycle that departs from the Rankine cycle due to irreversibilities of the expansion and compression steps. Figure 8.4 illustrates the effects of these irreversibilities on steps 2 → 3 and 4 → 1. The lines are no longer vertical but tend in the direction of increasing entropy. The turbine exhaust is normally still wet, but with sufficiently

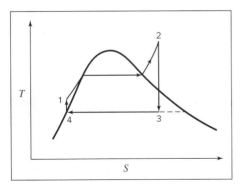

Figure 8.3: Rankine cycle on a *TS* diagram.

---

[2]Nevertheless, nuclear power plants generate saturated steam and operate with turbines designed to eject liquid at various stages of expansion.

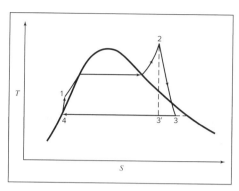

Figure 8.4: Simple practical power cycle.

low moisture content, erosion problems are not serious. Slight subcooling of the condensate in the condenser may occur, but the effect is inconsequential.

The boiler serves to transfer heat from a burning fuel (or from a nuclear reactor or even a solar-thermal heat source) to the cycle, and the condenser transfers heat from the cycle to the surroundings. Neglecting kinetic- and potential-energy changes reduces the energy relations, Eqs. (2.30) and (2.31), to:

$$\dot{Q} = \dot{m}\Delta H \quad (8.1) \qquad Q = \Delta H \quad (8.2)$$

Turbine and pump calculations were treated in detail in Secs. 7.2 and 7.3.

Example 8.1

Steam generated in a power plant at a pressure of 8600 kPa and a temperature of 500°C is fed to a turbine. Exhaust from the turbine enters a condenser at 10 kPa, where it is condensed to saturated liquid, which is then pumped to the boiler.

(a) What is the thermal efficiency of a Rankine cycle operating at these conditions?

(b) What is the thermal efficiency of a practical cycle operating at these conditions if the turbine efficiency and pump efficiency are both 0.75?

(c) If the rating of the power cycle of part (b) is 80,000 kW, what is the steam rate and what are the heat-transfer rates in the boiler and condenser?

Solution 8.1

(a) The turbine operates under the same conditions as the turbine of Ex. 7.6 where, on the basis of 1 kg of steam:

$$(\Delta H)_S = -1274.2 \ \text{kJ·kg}^{-1}$$

Thus $\qquad W_s(\text{isentropic}) = (\Delta H)_S = -1274.2 \ \text{kJ·kg}^{-1}$

Moreover, the enthalpy at the end of isentropic expansion, H_2' in Ex. 7.6, is here:

$$H_3' = 2117.4 \text{ kJ·kg}^{-1}$$

Subscripts refer to Fig. 8.4. The enthalpy of saturated liquid condensate at 10 kPa (and $t^{\text{sat}} = 45.83°C$) is:

$$H_4 = 191.8 \text{ kJ·kg}^{-1}$$

By Eq. (8.2) applied to the condenser,

$$Q(\text{condenser}) = H_4 - H_3' = 191.8 - 2117.4 = -1925.6 \text{ kJ·kg}^{-1}$$

where the minus sign indicates heat flow out of the system.

The pump operates under essentially the same conditions as the pump of Ex. 7.10, where:

$$W_s(\text{isentropic}) = (\Delta H)_S = 8.7 \text{ kJ·kg}^{-1}$$

and $H_1 = H_4 + (\Delta H)_S = 191.8 + 8.7 = 200.5 \text{ kJ·kg}^{-1}$

The enthalpy of superheated steam at 8,600 kPa and 500°C is:

$$H_2 = 3391.6 \text{ kJ·kg}^{-1}$$

By Eq. (8.2) applied to the boiler,

$$Q(\text{boiler}) = H_2 - H_1 = 3391.6 - 200.5 = 3191.1 \text{ kJ·kg}^{-1}$$

The net work of the Rankine cycle is the sum of the turbine work and the pump work:

$$W_s(\text{Rankine}) = -1274.2 + 8.7 = -1265.5 \text{ kJ·kg}^{-1}$$

This result is of course also:

$$W_s(\text{Rankine}) = -Q(\text{boiler}) - Q(\text{condenser})$$
$$= -3191.1 + 1925.6 = -1265.5 \text{ kJ·kg}^{-1}$$

The thermal efficiency of the cycle is:

$$\eta = \frac{-W_s(\text{Rankine})}{Q} = \frac{1265.5}{3191.1} = 0.3966$$

(b) With a turbine efficiency of 0.75, then also from Ex. 7.6:

$$W_s(\text{turbine}) = \Delta H = -955.6 \text{ kJ·kg}^{-1}$$

and $H_3 = H_2 + \Delta H = 3391.6 - 955.6 = 2436.0 \text{ kJ·kg}^{-1}$

For the condenser,

$$Q(\text{condenser}) = H_4 - H_3 = 191.8 - 2436.0 = -2244.2 \text{ kJ·kg}^{-1}$$

By Ex. 7.10 for the pump,

$$W_s(\text{pump}) = \Delta H = 11.6 \text{ kJ·kg}^{-1}$$

The net work of the cycle is therefore:

$$W_s(\text{net}) = -955.6 + 11.6 = -944.0 \text{ kJ·kg}^{-1}$$

and $\qquad H_1 = H_4 + \Delta H = 191.8 + 11.6 = 203.4 \text{ kJ·kg}^{-1}$

Then $\qquad Q(\text{boiler}) = H_2 - H_1 = 3391.6 - 203.4 = 3188.2 \text{ kJ·kg}^{-1}$

The thermal efficiency of the cycle is therefore:

$$\eta = \frac{-W_s(\text{net})}{Q(\text{boiler})} = \frac{944.0}{3188.2} = 0.2961$$

which may be compared with the result of part (*a*).

(*c*) For a power rating of 80,000 kW:

$$\dot{W}_s(\text{net}) = \dot{m}W_s(\text{net})$$

or $\qquad \dot{m} = \dfrac{\dot{W}_s(\text{net})}{W_s(\text{net})} = \dfrac{-80,000 \text{ kJ·s}^{-1}}{-944.0 \text{ kJ·kg}^{-1}} = 84.75 \text{ kg·s}^{-1}$

Then by Eq. (8.1),

$$\dot{Q}(\text{boiler}) = (84.75)(3188.2) = 270.2 \times 10^3 \text{ kJ·s}^{-1}$$

$$\dot{Q}(\text{condenser}) = (84.75)(-2244.2) = -190.2 \times 10^3 \text{ kJ·s}^{-1}$$

Note that

$$\dot{Q}(\text{boiler}) + \dot{Q}(\text{condenser}) = -\dot{W}_s(\text{net})$$

The Regenerative Cycle

The thermal efficiency of a steam power cycle is increased when the pressure and hence the vaporization temperature in the boiler is raised. It is also increased by greater superheating in the boiler. Thus, high boiler pressures and temperatures favor high efficiencies. However, these same conditions increase the capital investment in the plant, because they require heavier construction and more expensive materials of construction. Moreover, these costs increase ever more rapidly as more severe conditions are imposed. Thus, in practice power plants seldom operate at pressures much above 10,000 kPa or temperatures much above 600°C. The thermal efficiency of a power plant increases as pressure and hence temperature in the condenser is reduced. However, condensation temperatures must be higher than the temperature of the cooling medium, usually water, and this is controlled by local conditions of climate and geography. Power plants universally operate with condenser pressures as low as practical.

Most modern power plants operate on a modification of the Rankine cycle that incorporates feedwater heaters. Water from the condenser, rather than being pumped directly back to the boiler, is first heated by steam extracted from the turbine. This is normally done in several stages, with steam taken from the turbine at several intermediate states of expansion. An arrangement with four feedwater heaters is shown in Fig. 8.5. The operating conditions indicated on this figure and described in the following paragraphs are typical, and are the basis for the illustrative calculations of Ex. 8.2.

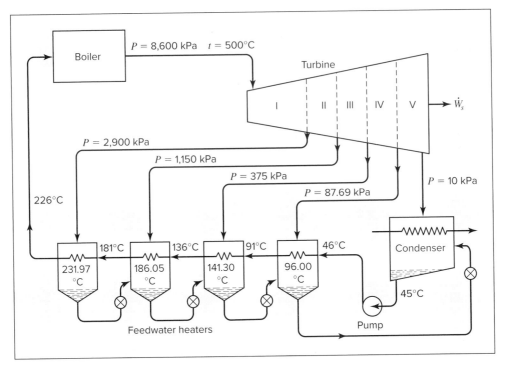

Figure 8.5: Steam power plant with feedwater heating.

The conditions of steam generation in the boiler are the same as in Ex. 8.1: 8600 kPa and 500°C. The exhaust pressure of the turbine, 10 kPa, is also the same. The saturation temperature of the exhaust steam is therefore 45.83°C. Allowing for slight subcooling of the condensate, we fix the temperature of the liquid water from the condenser at 45°C. The feedwater pump, which operates under exactly the conditions of the pump in Ex. 7.10, causes a temperature rise of about 1°C, making the temperature of the feedwater entering the series of heaters equal to 46°C.

The saturation temperature of steam at the boiler pressure of 8600 kPa is 300.06°C, and the temperature to which the feedwater can be raised in the heaters is certainly less. This temperature is a design variable, which is ultimately fixed by economic considerations. However, a value must be chosen before any thermodynamic calculations can be made. We have therefore arbitrarily specified a temperature of 226°C for the feedwater stream entering the boiler. We have also specified that all four feedwater heaters accomplish the same temperature rise. Thus, the total temperature rise of 226 − 46 = 180°C is divided into four 45°C increments. This establishes all intermediate feedwater temperatures at the values shown on Fig. 8.5.

The steam supplied to a given feedwater heater must be at a pressure high enough that its saturation temperature is above that of the feedwater stream leaving the heater. We have here presumed a minimum temperature difference for heat transfer of no less than 5°C, and have chosen extraction steam pressures such that the t^{sat} values shown in the feedwater heaters are at least 5°C greater than the exit temperatures of the feedwater streams. The condensate from each feedwater heater is flashed through a throttle valve to the heater at the next lower pressure, and the collected condensate in the final heater of the series is flashed into the condenser. Thus, all condensate returns from the condenser to the boiler by way of the feedwater heaters.

The purpose of heating the feedwater in this manner is to raise the average temperature at which heat is added to the water in the boiler. This increases the thermal efficiency of the plant, which is said to operate on a *regenerative cycle*.

Example 8.2

Determine the thermal efficiency of the power plant shown in Fig. 8.5, assuming turbine and pump efficiencies of 0.75. If its power rating is 80,000 kW, what is the steam rate from the boiler and what are the heat-transfer rates in the boiler and condenser?

Solution 8.2

Initial calculations are made on the basis of 1 kg of steam entering the turbine from the boiler. The turbine is in effect divided into five sections, as indicated in Fig. 8.5. Because steam is extracted at the end of each section, the flow rate in the turbine decreases from one section to the next. The amounts of steam extracted from the first four sections are determined by energy balances.

This requires enthalpies of the compressed feedwater streams. The effect of pressure at constant temperature on a liquid is given by Eq. (7.25):

$$\Delta H = V(1 - \beta T)\Delta P \qquad \text{(const } T\text{)}$$

For saturated liquid water at 226°C (499.15 K), the steam tables provide:

$$P^{\text{sat}} = 2598.2 \text{ kPa} \quad H = 971.5 \text{ kJ·kg}^{-1} \quad V = 1201 \text{ cm}^3\text{·kg}^{-1}$$

In addition, at this temperature,

$$\beta = 1.582 \times 10^{-3} \text{ K}^{-1}$$

Thus, for a pressure change from the saturation pressure to 8600 kPa:

$$\Delta H = 1201[1 - (1.528 \times 10^{-3})(499.15)]\frac{(8600 - 2598.2)}{10^6} = 1.5 \text{ kJ·kg}^{-1}$$

and $\qquad H = H \text{ (sat. liq.)} + \Delta H = 971.5 + 1.5 = 973.0 \text{ kJ·kg}^{-1}$

Similar calculations yield the enthalpies of the feedwater at other temperatures. All pertinent values are given in the following table.

$t/°C$	226	181	136	91	46
$H/\text{kJ·kg}^{-1}$ for water at t and $P = 8600$ kPa	973.0	771.3	577.4	387.5	200.0

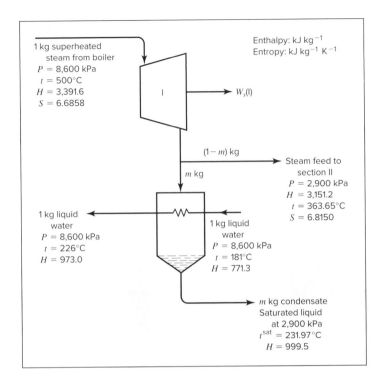

Figure 8.6: Section I of turbine and first feedwater heater.

The first section of the turbine and the first feedwater heater, are shown by Fig. 8.6. The enthalpy and entropy of the steam entering the turbine are found from the tables for superheated steam. The assumption of isentropic expansion of steam in section I of the turbine to 2900 kPa leads to the result:

$$(\Delta H)_S = -320.5 \text{ kJ·kg}^{-1}$$

If we assume that the turbine efficiency is independent of the pressure to which the steam expands, then Eq. (7.16) gives:

$$\Delta H = \eta(\Delta H)_S = (0.75)(-320.5) = -240.4 \text{ kJ·kg}^{-1}$$

By Eq. (7.14),

$$W_s(\text{I}) = \Delta H = -240.4 \text{ kJ}$$

In addition, the enthalpy of steam discharged from this section of the turbine is:

$$H = 3391.6 - 240.4 = 3151.2 \text{ kJ·kg}^{-1}$$

A simple energy balance on the feedwater heater results from the assumption that kinetic- and potential-energy changes are negligible and from the assignments, $\dot{Q} = -\dot{W}_s = 0$. Equation (2.29) then reduces to:

$$\Delta(\dot{m}H)_{\text{fs}} = 0$$

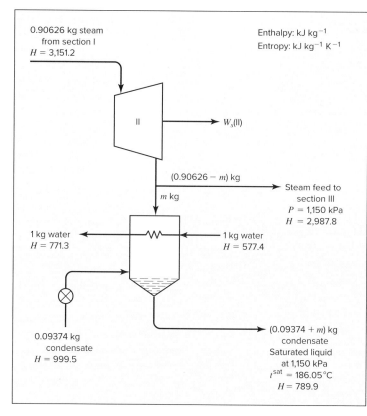

0.90626 kg steam
from section I
$H = 3{,}151.2$

Enthalpy: kJ kg^{-1}
Entropy: kJ kg^{-1} K^{-1}

II

W_s(II)

$(0.90626 - m)$ kg

m kg

Steam feed to
section III
$P = 1{,}150$ kPa
$H = 2{,}987.8$

1 kg water
$H = 771.3$

1 kg water
$H = 577.4$

0.09374 kg
condensate
$H = 999.5$

$(0.09374 + m)$ kg
condensate
Saturated liquid
at 1,150 kPa
$t^{\text{sat}} = 186.05\,°C$
$H = 789.9$

Figure 8.7: Section II of turbine and second feed-water heater.

Thus on the basis of 1 kg of steam entering the turbine (Fig. 8.6):

$$m(999.5 - 3151.2) + (1)(973.0 - 771.3) = 0$$

Then $\quad m = 0.09374$ kg $\quad$ and $\quad 1 - m = 0.90626$ kg

For each kilogram of steam entering the turbine, $1-m$ is the mass of steam flowing into section II of the turbine.

Section II of the turbine and the second feedwater heater are shown in Fig. 8.7. In doing the same calculations as for section I, we assume that each kilogram of steam leaving section II expands from its state *at the turbine entrance* to the exit of section II with an efficiency of 0.75 compared with isentropic expansion. The enthalpy of the steam leaving section II found in this way is:

$$H = 2987.8 \text{ kJ·kg}^{-1}$$

Then on the basis of 1 kg of steam entering the turbine,

$$W_s(\text{II}) = (2987.8 - 3151.2)(0.90626) = -148.08 \text{ kJ}$$

An energy balance on the feedwater heater (Fig. 8.7) gives:

$$(0.09374 + m)(789.9) - (0.09374)(999.5) - m(2987.8) + 1(771.3 - 577.4) = 0$$

and $$m = 0.07971 \text{ kg}$$

Note that throttling the condensate stream does not change its enthalpy.

These results and those of similar calculations for the remaining sections of the turbine are listed in the accompanying table. From the results shown,

$$\sum W_s = -804.0 \text{ kJ} \qquad \text{and} \qquad \sum m = 0.3055 \text{ kg}$$

	$H/\text{kJ·kg}^{-1}$ at section exit	W_s/kJ for section	$t/°\text{C}$ at section exit	State	m/kg of steam extracted
Sec. I	3151.2	−240.40	363.65	Superheated vapor	0.09374
Sec. II	2987.8	−148.08	272.48	Superheated vapor	0.07928
Sec. III	2827.4	−132.65	183.84	Superheated vapor	0.06993
Sec. IV	2651.3	−133.32	96.00	Wet vapor $x = 0.9919$	0.06257
Sec. V	2435.9	−149.59	45.83	Wet vapor $x = 0.9378$	

Thus for every kilogram of steam entering the turbine, the work produced is 804.0 kJ, and 0.3055 kg of steam is extracted from the turbine for the feedwater heaters. The work required by the pump is exactly the work calculated for the pump in Ex. 7.10, that is, 11.6 kJ. The net work of the cycle on the basis of 1 kg of steam generated in the boiler is therefore:

$$W_s(\text{net}) = -804.0 + 11.6 = -792.4 \text{ kJ}$$

On the same basis, the heat added in the boiler is:

$$Q(\text{boiler}) = \Delta H = 3391.6 - 973.0 = 2418.6 \text{ kJ}$$

The thermal efficiency of the cycle is therefore:

$$\eta = \frac{-W_s(\text{net})}{Q(\text{boiler})} = \frac{792.4}{2418.6} = 0.3276$$

This is a significant improvement over the value 0.2961 of Ex. 8.1.

Because $\dot{W}_S(\text{net}) = -80,000 \text{ kJ·s}^{-1}$,

$$\dot{m} = \frac{\dot{W}_s(\text{net})}{W_s(\text{net})} = \frac{-80,000}{-792.4} = 100.96 \text{ kg·s}^{-1}$$

This is the steam rate to the turbine, used to calculate the heat-transfer rate in the boiler:

$$\dot{Q}(\text{boiler}) = \dot{m}\Delta H = (100.96)(2418.6) = 244.2 \times 10^3 \text{ kJ·s}^{-1}$$

The heat-transfer rate to the cooling water in the condenser is:

$$\dot{Q}(\text{condenser}) = -\dot{Q}(\text{boiler}) - \dot{W}_s(\text{net})$$
$$= -244.2 \times 10^3 - (-80.0 \times 10^3)$$
$$= -164.2 \times 10^3 \text{ kJ·s}^{-1}$$

Although the steam generation rate is higher than was found in Ex. 8.1, the heat-transfer rates in the boiler and condenser are appreciably less because their functions are partly taken over by the feedwater heaters.

8.2 INTERNAL-COMBUSTION ENGINES

In a steam power plant, the steam is an inert medium to which heat is transferred from an external source (e.g., burning fuel). It is therefore characterized by large heat-transfer surfaces: (1) for the absorption of heat by the steam at a high temperature in the boiler, and (2) for the rejection of heat from the steam at a relatively low temperature in the condenser. The disadvantage is that when heat must be transferred through walls (as through the metal walls of boiler tubes) the ability of the walls to withstand high temperatures and pressures imposes a limit on the temperature of heat absorption. In an internal-combustion engine, on the other hand, a fuel is burned within the engine itself, and the combustion products serve as the working medium, acting for example on a piston in a cylinder. High temperatures are internal and do not involve heat-transfer surfaces.

Burning of fuel within the internal-combustion engine complicates thermodynamic analysis. Moreover, fuel and air flow steadily into an internal-combustion engine, and combustion products flow steadily out of it; no working medium undergoes a cyclic process, as does steam in a steam power plant. However, for making simple analyses, one imagines cyclic engines with air as the working fluid, equivalent in performance to actual internal-combustion engines. In addition, the combustion step is replaced by the addition to the air of an equivalent amount of heat. In what follows, each internal-combustion engine is introduced by a qualitative description. This is followed by a quantitative analysis of an ideal cycle in which air in its ideal-gas state with constant heat capacities is the working medium.

The Otto Engine

The most common internal-combustion engine, because of its use in automobiles, is the Otto engine.[3] Its cycle consists of four strokes and starts with an intake stroke at essentially constant pressure, during which a piston moving outward draws a fuel/air mixture into a cylinder. This is represented by line $0 \rightarrow 1$ in Fig. 8.8. During the second stroke ($1 \rightarrow 2 \rightarrow 3$), all valves are closed, and the fuel/air mixture is compressed, approximately adiabatically along line segment $1 \rightarrow 2$; the mixture is then ignited, by firing of a spark plug, and combustion occurs so rapidly that the volume remains nearly constant while the pressure rises along line segment $2 \rightarrow 3$. It is during the third stroke ($3 \rightarrow 4 \rightarrow 1$) that work is produced. The high-temperature, high-pressure products of combustion expand, approximately adiabatically along line segment $3 \rightarrow 4$; the exhaust valve then opens and the pressure falls rapidly at nearly constant volume along line segment $4 \rightarrow 1$. During the fourth or exhaust stroke (line $1 \rightarrow 0$), the piston pushes the remaining combustion gases (except for the contents of the clearance volume) from the cylinder. The volume plotted in Fig. 8.8 is the total volume of gas contained in the engine between the piston and the cylinder head.

---

[3]Named for Nikolaus August Otto, the German engineer who demonstrated one of the first practical engines of this type. See http://en.wikipedia.org/wiki/Nikolaus_Otto. Also called a four-stroke spark-ignition engine.

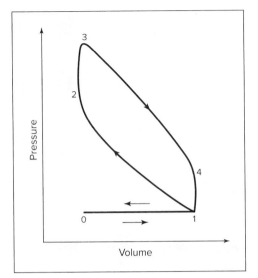
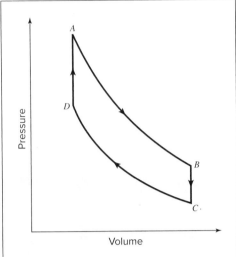

Figure 8.8: Otto engine cycle.   **Figure 8.9**: Air-standard Otto cycle.

The effect of increasing the compression ratio—the ratio of the volumes at the beginning and end of compression from point 1 to point 2—is to increase the efficiency of the
engine, that is, to increase the work produced per unit quantity of fuel. We demonstrate this for
an idealized cycle, called the air-standard Otto cycle, shown in Fig. 8.9. It consists of two adiabatic and two constant-volume steps, which comprise a heat-engine cycle for which the working fluid is air in its ideal-gas state with constant heat capacity. Step CD, a reversible adiabatic
compression, is followed by step DA, where sufficient heat is added to the air at constant
volume to raise its temperature and pressure to the values resulting from combustion in an
actual Otto engine. Then the air is expanded adiabatically and reversibly (step AB) and cooled
at constant volume (step BC) to the initial state at C by heat transfer to the surroundings.

The thermal efficiency η of the air-standard cycle shown in Fig. 8.9 is simply:

$$\eta = \frac{-W(\text{net})}{Q_{DA}} = \frac{Q_{DA} + Q_{BC}}{Q_{DA}} \tag{8.3}$$

For 1 mol of air with constant heat capacities,

$$Q_{DA} = C_V^{ig}(T_A - T_D) \quad \text{and} \quad Q_{BC} = C_V^{ig}(T_C - T_B)$$

Substituting these expressions in Eq. (8.3) gives:

$$\eta = \frac{C_V^{ig}(T_A - T_D) + C_V^{ig}(T_C - T_B)}{C_V^{ig}(T_A - T_D)}$$

or

$$\eta = 1 - \frac{T_B - T_C}{T_A - T_D} \tag{8.4}$$

The thermal efficiency is related in a simple way to the compression ratio, $r \equiv V_C^{ig}/V_D^{ig}$.
Each temperature in Eq. (8.4) is replaced by an appropriate group PV^{ig}/R. Thus,

$$T_B = \frac{P_B V_B^{ig}}{R} = \frac{P_B V_C^{ig}}{R} \qquad T_C = \frac{P_C V_C^{ig}}{R}$$

$$T_A = \frac{P_A V_A^{ig}}{R} = \frac{P_A V_D^{ig}}{R} \qquad T_D = \frac{P_D V_D^{ig}}{R}$$

Substituting into Eq. (8.4) leads to:

$$\eta = 1 - \frac{V_C^{ig}}{V_D^{ig}} \left(\frac{P_B - P_C}{P_A - P_D} \right) = 1 - r \left(\frac{P_B - P_C}{P_A - P_D} \right) \tag{8.5}$$

For the two adiabatic, reversible steps, $PV^\gamma = $ const. Hence:

$$P_B V_C^\gamma = P_A V_D^\gamma \quad \text{(because } V_D = V_A^{ig} \text{ and } V_C^{ig} = V_B^{ig}\text{)}$$
$$P_C V_C^\gamma = P_D V_D^\gamma$$

Division of the first expression by the second yields:

$$\frac{P_B}{P_C} = \frac{P_A}{P_D} \quad \text{whence} \quad \frac{P_B}{P_C} - 1 = \frac{P_A}{P_D} - 1 \quad \text{or} \quad \frac{P_B - P_C}{P_C} = \frac{P_A - P_D}{P_D}$$

Thus,

$$\frac{P_B - P_C}{P_A - P_D} = \frac{P_C}{P_D} = \left(\frac{V_D}{V_C} \right)^\gamma = \left(\frac{1}{r} \right)^\gamma$$

where we have used the relation $P_C V_C^\gamma = P_D V_D^\gamma$. Equation (8.5) now becomes:

$$\eta = 1 - r \left(\frac{1}{r} \right)^\gamma = 1 - \left(\frac{1}{r} \right)^{\gamma-1} \tag{8.6}$$

This equation shows that the thermal efficiency increases rapidly with the compression ratio r at low values of r, but more slowly at high compression ratios. This agrees with the results of actual tests on Otto engines.

Unfortunately, the compression ratio cannot be increased arbitrarily but is limited by pre-ignition of the fuel. For sufficiently high compression ratios, the temperature rise due to compression will ignite the fuel before the compression stroke is complete. This is manifested as "knocking" of the engine. Compression ratios in automobile engines are usually not much above 10. The compression ratio at which pre-ignition occurs depends on the fuel, with the resistance to pre-ignition of fuels indicated by an octane rating. High-octane gasoline does not actually contain more octane than other gasoline, but it does have additives such as alcohols, ethers, and aromatic compounds that increase its resistance to pre-ignition. Iso-octane is arbitrarily assigned an octane number of 100, while n-heptane is assigned an octane number of zero. Other compounds and fuels are assigned octane numbers relative to these standards, based on the compression ratio at which they pre-ignite. Ethanol and methanol have octane numbers well above 100. Racing engines that burn these alcohols can employ compression ratios of 15 or more.

The Diesel Engine

The fundamental difference between the Otto cycle and the Diesel cycle[4] is that in the Diesel cycle the temperature at the end of compression is sufficiently high that combustion is initiated spontaneously. This higher temperature results because of a higher compression ratio that carries the compression step to a higher pressure. The fuel is not injected until the end of the compression step, and then it is added slowly enough that the combustion process occurs at approximately constant pressure.

For the same compression ratio, the Otto engine has a higher efficiency than the Diesel engine. Because preignition limits the compression ratio attainable in the Otto engine, the Diesel engine operates at higher compression ratios, and consequently at higher efficiencies. Compression ratios can exceed 20 in Diesel engines employing indirect fuel injection.

Example 8.3

Sketch the air-standard Diesel cycle on a *PV* diagram, and derive an equation giving the thermal efficiency of this cycle in relation to the compression ratio *r* (ratio of volumes at the beginning and end of the compression step) and the expansion ratio r_e (ratio of volumes at the end and beginning of the adiabatic expansion step).

Solution 8.3

The air-standard Diesel cycle is the same as the air-standard Otto cycle, except that the heat-absorption step (corresponding to the combustion process in the actual engine) is at constant pressure, as indicated by line *DA* in Fig. 8.10.

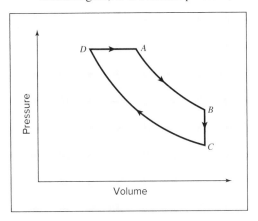

Figure 8.10: Air-standard Diesel cycle.

On the basis of 1 mol of air in its ideal-gas state with with constant heat capacities, the heat quantities absorbed in step *DA* and rejected in step *BC* are:

$$Q_{DA} = C_P^{ig}(T_A - T_D) \qquad \text{and} \qquad Q_{BC} = C_V^{ig}(T_C - T_B)$$

---

[4]Named for Rudolf Diesel, the German engineer who developed one of the first practical compression-ignition engines. See http://en.wikipedia.org/wiki/Rudolf_Diesel.

The thermal efficiency, Eq. (8.3), is:

$$\eta = 1 + \frac{Q_{BC}}{Q_{DA}} = 1 + \frac{C_V^{ig}(T_C - T_B)}{C_P^{ig}(T_A - T_D)} = 1 - \frac{1}{\gamma}\left(\frac{T_B - T_C}{T_A - T_D}\right) \qquad (A)$$

For reversible, adiabatic expansion (step AB) and reversible, adiabatic compression (step CD), Eq. (3.23a) applies:

$$T_A V_A^{\gamma-1} = T_B V_B^{\gamma-1} \qquad \text{and} \qquad T_D V_D^{\gamma-1} = T_C V_C^{\gamma-1}$$

By definition, the compression ratio is $r \equiv V_C^{ig}/V_D^{ig}$, and the expansion ratio is $r_e \equiv V_B^{ig}/V_A^{ig}$. Thus,

$$T_B = T_A \left(\frac{1}{r_e}\right)^{\gamma-1} \qquad T_C = T_D \left(\frac{1}{r}\right)^{\gamma-1}$$

Substituting these equations into Eq. (A) gives:

$$\eta = 1 - \frac{1}{\gamma}\left[\frac{T_A(1/r_e)^{\gamma-1} - T_D(1/r)^{\gamma-1}}{T_A - T_D}\right] \qquad (B)$$

Also $P_A = P_D$, and for the ideal-gas state,

$$P_D V_D^{ig} = RT_D \qquad \text{and} \qquad P_A V_A^{ig} = RT_A$$

Moreover, $V_C^{ig} = V_B^{ig}$, and therefore:

$$\frac{T_D}{T_A} = \frac{V_D^{ig}}{V_A^{ig}} = \frac{V_D^{ig}/V_C^{ig}}{V_A^{ig}/V_B^{ig}} = \frac{r_e}{r}$$

This relation combines with Eq. (B):

$$\eta = 1 - \frac{1}{\gamma}\left[\frac{(1/r_e)^{\gamma-1} - (r_e/r)(1/r)^{\gamma-1}}{1 - r_e/r}\right]$$

or

$$\eta = 1 - \frac{1}{\gamma}\left[\frac{(1/r_e)^{\gamma} - (1/r)^{\gamma}}{1/r_e - 1/r}\right] \qquad (8.7)$$

The Gas-Turbine Engine

The Otto and Diesel engines exemplify direct use of the energy of high-temperature, high-pressure gases acting on a piston within a cylinder; no heat transfer with an external source is required. However, turbines are more efficient than reciprocating engines, and the advantages of internal combustion are combined with those of the turbine in the gas-turbine engine.

The gas turbine is driven by high-temperature gases from a combustion chamber, as indicated in Fig. 8.11. The entering air is compressed (supercharged) to a pressure of several bars before combustion. The centrifugal compressor operates on the same shaft as the turbine, and part of the work of the turbine serves to drive the compressor. The higher the temperature

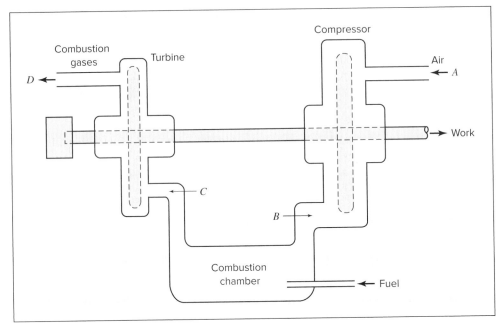

Figure 8.11: Gas-turbine engine.

of the combustion gases entering the turbine, the higher the efficiency of the unit, i.e., the greater the work produced per unit of fuel burned. The limiting temperature is determined by the strength of the metal turbine blades and is generally much lower than the theoretical flame temperature (Ex. 4.7) of the fuel. Sufficient excess air must be supplied to keep the combustion temperature at a safe level.

The idealization of the gas-turbine engine, called the Brayton cycle, is shown on a PV diagram in Fig. 8.12. The working fluid is air in its ideal-gas state with constant heat capacities. Step AB is a reversible adiabatic compression from P_A (atmospheric pressure) to P_B. In step BC heat Q_{BC}, replacing combustion, is added at constant pressure, raising the air temperature. A work-producing isentropic expansion of the air reduces the pressure from P_C to P_D

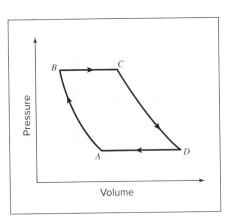

Figure 8.12: Brayton cycle for the gas-turbine engine.

(atmospheric pressure). Step DA is a constant-pressure cooling process that merely completes the cycle. The thermal efficiency of the cycle is:

$$\eta = \frac{-W(\text{net})}{Q_{BC}} = \frac{-(W_{CD} + W_{AB})}{Q_{BC}} \tag{8.8}$$

where each energy quantity is based on 1 mol of air.

The work done as air passes through the compressor is given by Eq. (7.14), and for air in its ideal-gas state with constant heat capacities:

$$W_{AB} = H_B - H_A = C_P^{ig}(T_B - T_A)$$

Similarly, for the heat-addition and turbine processes,

$$Q_{BC} = C_P^{ig}(T_C - T_B) \quad \text{and} \quad W_{CD} = C_P^{ig}(T_C - T_D)$$

Substituting these equations into Eq. (8.8) and simplifying leads to:

$$\eta = 1 - \frac{T_D - T_A}{T_C - T_B} \tag{8.9}$$

Because processes AB and CD are isentropic, the temperatures and pressures are related by Eq. (3.23b):

$$\frac{T_A}{T_B} = \left(\frac{P_A}{P_B}\right)^{(\gamma-1)/\gamma} \tag{8.10}$$

and

$$\frac{T_D}{T_C} = \left(\frac{P_D}{P_C}\right)^{(\gamma-1)/\gamma} = \left(\frac{P_A}{P_B}\right)^{(\gamma-1)/\gamma} \tag{8.11}$$

Solving Eq. (8.10) for T_A and Eq. (8.11) for T_D and substituting the results into Eq. (8.9) reduce it to:

$$\eta = 1 - \left(\frac{P_A}{P_B}\right)^{(\gamma-1)/\gamma} \tag{8.12}$$

Example 8.4

A gas-turbine engine with a compression ratio $P_B/P_A = 6$ operates with air entering the compressor at 25°C. If the maximum permissible temperature in the turbine is 760°C, determine:

(a) The thermal efficiency η of the reversible air-standard cycle for these conditions if $\gamma = 1.4$.

(b) The thermal efficiency for an air-standard cycle for the given conditions if the compressor and turbine operate adiabatically but irreversibly with efficiencies $\eta_c = 0.83$ and $\eta_t = 0.86$.

Solution 8.4

(a) Direct substitution in Eq. (8.12) gives the efficiency:

$$\eta = 1 - (1/6)^{(1.4-1)/1.4} = 1 - 0.60 = 0.40$$

(b) Irreversibilities in both the compressor and turbine reduce the thermal efficiency of the engine because the net work is the difference between the work required by the compressor and the work produced by the turbine. The temperature of air entering the compressor T_A and the temperature of air entering the turbine, the specified maximum for T_C, are the same as for the reversible cycle of part (a). However, the temperature after irreversible compression in the compressor T_B is higher than the temperature after *isentropic* compression T'_B, and the temperature after irreversible expansion in the turbine T_D is higher than the temperature after *isentropic* expansion T'_D.

The thermal efficiency of the engine is given by:

$$\eta = \frac{-(W_{\text{turb}} + W_{\text{comp}})}{Q_{BC}}$$

The two work terms are found from the expressions for isentropic work:

$$W_{\text{turb}} = \eta_t C_P^{ig}(T'_D - T_C)$$

$$W_{\text{comp}} = \frac{C_P^{ig}(T'_B - T_A)}{\eta_c} \tag{A}$$

The heat absorbed to simulate combustion is:

$$Q_{BC} = C_P^{ig}(T_C - T_B)$$

These equations combine to yield:

$$\eta = \frac{\eta_t(T_C - T'_D) - (1/\eta_c)(T'_B - T_A)}{T_C - T_B}$$

An alternative expression for the compression work is:

$$W_{\text{comp}} = C_P^{ig}(T_B - T_A) \tag{B}$$

Equating this and Eq. (A) and using the result to eliminate T_B from the equation for η gives, after simplification:

$$\eta = \frac{\eta_t \eta_c (T_C/T_A - T'_D/T_A) - (T'_B/T_A - 1)}{\eta_c (T_C/T_A - 1) - (T'_B/T_A - 1)} \tag{C}$$

The ratio T_C/T_A depends on given conditions. The ratio T'_B/T_A is related to the pressure ratio by Eq. (8.10). In view of Eq. (8.11), the ratio T'_D/T_A can be expressed as:

$$\frac{T'_D}{T_A} = \frac{T_C T'_D}{T_A T_C} = \frac{T_C}{T_A}\left(\frac{P_A}{P_B}\right)^{(\gamma-1)/\gamma}$$

Substituting these expressions in Eq. (C) yields:

$$\eta = \frac{\eta_t \eta_c (T_C/T_A)(1 - 1/\alpha) - (\alpha - 1)}{\eta_c (T_C/T_A - 1) - (\alpha - 1)} \tag{8.13}$$

where

$$\alpha = \left(\frac{P_B}{P_A}\right)^{(\gamma - 1)/\gamma}$$

One can show by Eq. (8.13) that the thermal efficiency of the gas-turbine engine increases as the temperature of the air entering the turbine (T_C) increases, and as the compressor and turbine efficiencies η_c and η_t increase.

The given efficiency values are here:

$$\eta_t = 0.86 \qquad \text{and} \qquad \eta_c = 0.83$$

Other given data provide:

$$\frac{T_C}{T_A} = \frac{760 + 273.15}{25 + 273.15} = 3.47$$

and

$$\alpha = (6)^{(1.4-1)/1.4} = 1.67$$

Substituting these quantities in Eq. (8.13) gives:

$$\eta = \frac{(0.86)(0.83)(3.47)(1 - 1/1.67) - (1.67 - 1)}{(0.83)(3.47 - 1) - (1.67 - 1)} = 0.235$$

This analysis shows that even with rather high compressor and turbine efficiencies, the thermal efficiency (23.5%) is considerably reduced from the reversible-cycle value of part (a) (40%).

8.3 JET ENGINES; ROCKET ENGINES

In the power cycles so far considered the high-temperature, high-pressure gas expands in a turbine (steam power plant, gas turbine) or in the cylinders of an Otto or Diesel engine with reciprocating pistons. In either case, the power becomes available through a rotating shaft. Another device for expanding the hot gases is a nozzle. Here the power is available as kinetic energy in the jet of exhaust gases leaving the nozzle. The entire power plant, consisting of a compression device and a combustion chamber, as well as a nozzle, is known as a jet engine. Because the kinetic energy of the exhaust gases is directly available for propelling the engine and its attachments, jet engines are most commonly used to power aircraft. There are several types of jet-propulsion engines based on different ways of accomplishing the compression and expansion processes. Because the air striking the engine has kinetic energy (with respect to the engine), its pressure may be increased in a diffuser.

The turbojet engine (usually called simply a jet engine) illustrated in Fig. 8.13 takes advantage of a diffuser to reduce the work of compression. The axial-flow compressor completes

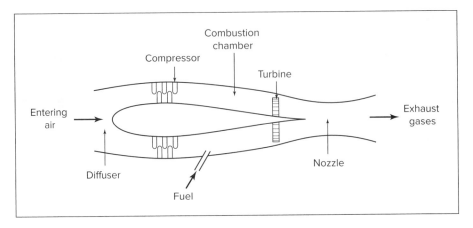

Figure 8.13: The turbojet power plant.

the job of compression, and then fuel is injected and burned in the combustion chamber. Hot combustion-product gases first pass through a turbine where expansion provides just enough power to drive the compressor. The remaining expansion to the exhaust pressure is accomplished in the nozzle. Here, the velocity of the gases with respect to the engine is increased to a level above that of the entering air. This increase in velocity provides a thrust (force) on the engine in the forward direction. If the compression and expansion processes are adiabatic and reversible, the turbojet engine follows the Brayton cycle shown in Fig. 8.12. The only differences are that, physically, the compression and expansion steps are carried out in devices of different types.

A rocket engine differs from a jet engine in that the oxidizing agent is carried with the engine. Instead of depending on the surrounding air to support combustion, the rocket is self-contained. This means that the rocket can operate in a vacuum such as in outer space. In fact, the performance is better in a vacuum because none of the thrust is required to overcome friction forces.

In rockets burning liquid fuels, the oxidizing agent (e.g., liquid oxygen or nitrogen tetroxide) is pumped from tanks into the combustion chamber. Simultaneously, fuel (e.g., hydrogen, kerosene, or monomethylhydrazine) is pumped into the chamber and burned. The combustion takes place at a constant high pressure and produces high-temperature product gases that are expanded in a nozzle, as indicated in Fig. 8.14.

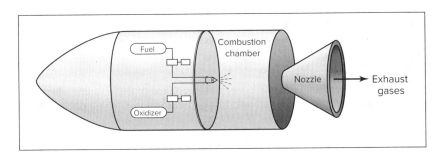

Figure 8.14: Liquid-fuel rocket engine.

In rockets burning solid fuels, the fuel (e.g., organic polymers) and oxidizer (e.g., ammonium perchlorate) are contained together in a solid matrix and stored at the forward end of the combustion chamber. Combustion and expansion are much the same as in a jet engine (Fig. 8.13), but a solid-fuel rocket requires no compression work, and in a liquid-fuel rocket the compression work is small because the fuel and oxidizer are pumped as liquids.

8.4 SYNOPSIS

After thorough study of this chapter, including working through example and end-of-chapter problems, one should be able to:

- Qualitatively describe the idealized Carnot, Rankine, Otto, Diesel, and Brayton cycles and sketch each of them on a *PV* or *TS* diagram.

- Carry out a thermodynamic analysis of a steam power plant, including a plant operating on a regenerative cycle, as in Ex. 8.1 and 8.2.

- Analyze an air-standard Otto cycle or Diesel cycle for a given compression ratio.

- Compute efficiency and heat and work flows for an air-standard Brayton cycle for given combustor temperature and compressor and turbine efficiencies.

- Explain, in simple terms, how jet and rocket engines generate thrust.

8.5 PROBLEMS

8.1. The basic cycle for a steam power plant is shown in Fig. 8.1. The turbine operates adiabatically with inlet steam at 6800 kPa and 550°C, and the exhaust steam enters the condenser at 50°C with a quality of 0.96. Saturated liquid water leaves the condenser and is pumped to the boiler. Neglecting pump work and kinetic- and potential-energy changes, determine the thermal efficiency of the cycle and the turbine efficiency.

8.2. A Carnot engine with H_2O as the working fluid operates on the cycle shown in Fig. 8.2. The H_2O circulation rate is 1 kg·s^{-1}. For $T_H = 475$ K and $T_C = 300$ K, determine:

    (*a*) The pressures at states 1, 2, 3, and 4.
    (*b*) The quality x^v at states 3 and 4.
    (*c*) The rate of heat addition.
    (*d*) The rate of heat rejection.
    (*e*) The mechanical power for each of the four steps.
    (*f*) The thermal efficiency η of the cycle.

8.3. A steam power plant operates on the cycle of Fig. 8.4. For one of the following sets of operating conditions, determine the steam rate, the heat-transfer rates in the boiler and condenser, and the thermal efficiency of the plant.

(a) $P_1 = P_2 = 10{,}000$ kPa; $T_2 = 600°C$; $P_3 = P_4 = 10$ kPa; η(turbine) $= 0.80$;
   η(pump) $= 0.75$; power rating $= 80{,}000$ kW.
(b) $P_1 = P_2 = 7000$ kPa; $T_2 = 550°C$; $P_3 = P_4 = 20$ kPa; η(turbine) $= 0.75$;
   η(pump) $= 0.75$; power rating $= 100{,}000$ kW.
(c) $P_1 = P_2 = 8500$ kPa; $T_2 = 600°C$; $P_3 = P_4 = 10$ kPa; η(turbine) $= 0.80$;
   η(pump) $= 0.80$; power rating $= 70{,}000$ kW.
(d) $P_1 = P_2 = 6500$ kPa; $T_2 = 525°C$; $P_3 = P_4 = 101.33$ kPa; η(turbine) $= 0.78$;
   η(pump) $= 0.75$; power rating $= 50{,}000$ kW.
(e) $P_1 = P_2 = 950$(psia); $T_2 = 1000$(°F); $P_3 = P_4 = 14.7$(psia); η(turbine) $= 0.78$;
   η(pump) $= 0.75$; power rating $= 50{,}000$ kW.
(f) $P_1 = P_2 = 1125$(psia); $T_2 = 1100$(°F); $P_3 = P_4 = 1$(psia); η(turbine) $= 0.80$;
   η(pump) $= 0.75$; power rating $= 80{,}000$ kW.

8.4. Steam enters the turbine of a power plant operating on the Rankine cycle (Fig. 8.3) at 3300 kPa and exhausts at 50 kPa. To show the effect of superheating on the performance of the cycle, calculate the thermal efficiency of the cycle and the quality of the exhaust steam from the turbine for turbine-inlet steam temperatures of 450, 550, and 650°C.

8.5. Steam enters the turbine of a power plant operating on the Rankine cycle (Fig. 8.3) at 600°C and exhausts at 30 kPa. To show the effect of boiler pressure on the performance of the cycle, calculate the thermal efficiency of the cycle and the quality of the exhaust steam from the turbine for boiler pressures of 5000, 7500, and 10,000 kPa.

8.6. A steam power plant employs two adiabatic turbines in series. Steam enters the first turbine at 650°C and 7000 kPa and discharges from the second turbine at 20 kPa. The system is designed for equal power outputs from the two turbines, based on a turbine efficiency of 78% for *each* turbine. Determine the temperature and pressure of the steam in its intermediate state between the two turbines. What is the overall efficiency of the two turbines together with respect to isentropic expansion of the steam from the initial to the final state?

8.7. A steam power plant operating on a regenerative cycle, as illustrated in Fig. 8.5, includes just one feedwater heater. Steam enters the turbine at 4500 kPa and 500°C and exhausts at 20 kPa. Steam for the feedwater heater is extracted from the turbine at 350 kPa, and in condensing raises the temperature of the feedwater to within 6°C of its condensation temperature at 350 kPa. If the turbine and pump efficiencies are both 0.78, what is the thermal efficiency of the cycle, and what fraction of the steam entering the turbine is extracted for the feedwater heater?

8.8. A steam power plant operating on a regenerative cycle, as illustrated in Fig. 8.5, includes just one feedwater heater. Steam enters the turbine at 650(psia) and 900(°F) and exhausts at 1(psia). Steam for the feedwater heater is extracted from the turbine at 50(psia) and in condensing raises the temperature of the feedwater to within 11(°F) of its condensation temperature at 50(psia). If the turbine and pump efficiencies are both 0.78, what is the thermal efficiency of the cycle, and what fraction of the steam entering the turbine is extracted for the feedwater heater?

8.9. A steam power plant operating on a regenerative cycle, as illustrated in Fig. 8.5, includes two feedwater heaters. Steam enters the turbine at 6500 kPa and 600°C and exhausts at 20 kPa. Steam for the feedwater heaters is extracted from the turbine at pressures such that the feedwater is heated to 190°C in two equal increments of temperature rise, with 5°C approaches to the steam-condensation temperature in each feedwater heater. If the turbine and pump efficiencies are each 0.80, what is the thermal efficiency of the cycle, and what fraction of the steam entering the turbine is extracted for each feedwater heater?

8.10. A power plant operating on heat recovered from the exhaust gases of internal combustion engines uses isobutane as the working medium in a modified Rankine cycle in which the upper pressure level is above the critical pressure of isobutane. Thus the isobutane does not undergo a change of phase as it absorbs heat prior to its entry into the turbine. Isobutane vapor is heated at 4800 kPa to 260°C and enters the turbine as a supercritical fluid at these conditions. Isentropic expansion in the turbine produces a superheated vapor at 450 kPa, which is cooled and condensed at constant pressure. The resulting saturated liquid enters the pump for return to the heater. If the power output of the modified Rankine cycle is 1000 kW, what are the isobutane flow rate, the heat-transfer rates in the heater and condenser, and the thermal efficiency of the cycle? The vapor pressure of isobutane can be computed from data given in Table B.2 of App. B.

8.11. A power plant operating on heat from a geothermal source uses isobutane as the working medium in a Rankine cycle (Fig. 8.3). Isobutane is heated at 3400 kPa (a pressure just below its critical pressure) to a temperature of 140°C, at which conditions it enters the turbine. Isentropic expansion in the turbine produces superheated vapor at 450 kPa, which is cooled and condensed to saturated liquid and pumped to the heater/boiler. If the flow rate of isobutane is 75 kg·s^{-1}, what is the power output of the Rankine cycle, and what are the heat-transfer rates in the heater/boiler and cooler/condenser? What is the thermal efficiency of the cycle? The vapor pressure of isobutane is given in Table B.2 of App. B.

Repeat these calculations for a cycle in which the turbine and pump each have an efficiency of 80%.

8.12. For comparison of Diesel- and Otto-engine cycles:

(a) Show that the thermal efficiency of the air-standard Diesel cycle can be expressed as

$$\eta = 1 - \left(\frac{1}{r}\right)^{\gamma-1} \frac{r_c^{\gamma} - 1}{\gamma(r_c - 1)}$$

where r is the compression ratio and r_c is the *cutoff ratio*, defined as $r_c = V_A/V_D$. (See Fig. 8.10.)

(b) Show that for the same compression ratio the thermal efficiency of the air-standard Otto engine is greater than the thermal efficiency of the air-standard Diesel cycle. *Hint:* Show that the fraction which multiplies $(1/r)^{\gamma-1}$ in the preceding equation for η is greater than unity by expanding r_c^{γ} in a Taylor series with the remainder taken to the first derivative.

(c) If $\gamma = 1.4$, how does the thermal efficiency of an air-standard Otto cycle with a compression ratio of 8 compare with the thermal efficiency of an air-standard Diesel cycle with the same compression ratio and a cutoff ratio of 2? How is the comparison changed if the cutoff ratio is 3?

8.13. An air-standard Diesel cycle absorbs 1500 J·mol^{-1} of heat (step *DA* of Fig. 8.10, which simulates combustion). The pressure and temperature at the beginning of the compression step are 1 bar and 20°C, and the pressure at the end of the compression step is 4 bar. Assuming air to be an ideal gas for which $C_P = (7/2)R$ and $C_V = (5/2)R$, what are the compression ratio and the expansion ratio of the cycle?

8.14. Calculate the efficiency for an air-standard gas-turbine cycle (the Brayton cycle) operating with a pressure ratio of 3. Repeat for pressure ratios of 5, 7, and 9. Take $\gamma = 1.35$.

8.15. An air-standard gas-turbine cycle is modified by the installation of a regenerative heat exchanger to transfer energy from the air leaving the turbine to the air leaving the compressor. In an optimum countercurrent exchanger, the temperature of the air leaving the compressor is raised to that of point *D* in Fig. 8.12, and the temperature of the gas leaving the turbine is cooled to that of point *B* in Fig. 8.12. Show that the thermal efficiency of this cycle is given by

$$\eta = 1 - \frac{T_A}{T_C}\left(\frac{P_B}{P_A}\right)^{(\gamma-1)/\gamma}$$

8.16. Consider an air-standard cycle for the turbojet power plant shown in Fig. 8.13. The temperature and pressure of the air entering the compressor are 1 bar and 30°C. The pressure ratio in the compressor is 6.5, and the temperature at the turbine inlet is 1100°C. If expansion in the nozzle is isentropic and if the nozzle exhausts at 1 bar, what is the pressure at the nozzle inlet (turbine exhaust), and what is the velocity of the air leaving the nozzle?

8.17. Air enters a gas-turbine engine (see Fig. 8.11) at 305 K and 1.05 bar and is compressed to 7.5 bar. The fuel is methane at 300 K and 7.5 bar; compressor and turbine efficiencies are each 80%. For one of the turbine inlet temperatures T_C given below, determine: the molar fuel-to-air ratio, the *net* mechanical power delivered per mole of fuel, and the turbine exhaust temperature T_D. Assume complete combustion of the methane and expansion in the turbine to 1(atm).

(a) $T_C = 1000$ K   (b) $T_C = 1250$ K   (c) $T_C = 1500$ K

8.18. Most electrical energy in the United States is generated in large-scale power cycles through conversion of thermal energy to mechanical energy, which is then converted to electrical energy. Assume a thermal efficiency of 0.35 for conversion of thermal to mechanical energy, and an efficiency of 0.95 for conversion of mechanical to electrical energy. Line losses in the distribution system amount to 20%. If the cost of fuel for the power cycle is \$4.00 GJ^{-1}, estimate the cost of electricity delivered to the customer in \$ per kWh. Ignore operating costs, profits, and taxes. Compare this number with that found on a typical electric bill.

8.19. Liquefied natural gas (LNG) is transported in very large tankers, stored as liquid in equilibrium with its vapor at approximately atmospheric pressure. If LNG is essentially pure methane, the storage temperature then is about 111.4 K, the normal boiling point of methane. The enormous amount of cold liquid can in principle serve as a heat sink for an onboard heat engine. Energy discarded to the LNG serves for its vaporization. If the heat source is ambient air at 300 K, and if the efficiency of a heat engine is 60% of its Carnot value, estimate the vaporization rate in moles vaporized per kJ of power output. For methane, $\Delta H_n^{lv} = 8.206$ kJ·mol^{-1}.

8.20. The oceans in the tropics have substantial surface-to-deep-water temperature gradients. Depending on location, relatively constant temperature differences of 15 to 25°C are observed for depths of 500 to 1000 m. This provides the opportunity for using cold (deep) water as a heat sink and warm (surface) water as a heat source for a power cycle. The technology is known as OTEC (Ocean Thermal Energy Conversion).

(*a*) Consider a location where the surface temperature is 27°C and the temperature at a depth of 750 m is 6°C. What is the efficiency of a Carnot engine operating between these temperature levels?

(*b*) Part of the output of a power cycle must be used to pump the cold water to the surface, where the cycle hardware resides. If the inherent efficiency of a real cycle is 0.6 of the Carnot value, and if 1/3 of the generated power is used for moving cold water to the surface, what is the actual efficiency of the cycle?

(*c*) The choice of working fluid for the cycle is critical. Suggest some possibilities. Here you may want to consult a handbook, such as *Perry's Chemical Engineers' Handbook*.

8.21. Air-standard power cycles are conventionally displayed on *PV* diagrams. An alternative is the *PT* diagram. Sketch air-standard cycles on *PT* diagrams for the following:

(*a*) Carnot cycle
(*b*) Otto cycle
(*c*) Diesel cycle
(*d*) Brayton cycle

Why would a *PT* diagram not be helpful for depicting power cycles involving liquid/vapor phase changes?

8.22. A steam plant operates on the cycle of Fig. 8.4. The pressure levels are 10 kPa and 6000 kPa, and steam leaves the turbine as saturated vapor. The pump efficiency is 0.70, and the turbine efficiency is 0.75. Determine the thermal efficiency of the plant.

8.23. Devise a general scheme for analyzing four-step air-standard power cycles. Model each step of the cycle as a polytropic process described by

$$PV^\delta = \text{constant}$$

which implies that

$$TP^{(1-\delta)/\delta} = \text{constant}$$

with a specified value of δ. Decide which states to fix, partially or completely, by values of T and/or P. Analysis here means determination of T and P for initial and final states of each step, Q and W for each step, and the thermal efficiency of the cycle. The analysis should also include a sketch of the cycle on a PT diagram.

Chapter 9

Refrigeration and Liquefaction

The word *refrigeration* implies the maintenance of a temperature below that of the surroundings. It is best known for its use in the air conditioning of buildings and in the preservation of foods and chilling of beverages. Examples of large-scale commercial processes requiring refrigeration are the manufacture of ice and solid CO_2, the dehydration and liquefaction of gases, and the separation of air into oxygen and nitrogen.

It is not our purpose here to deal with details of equipment design, which are left to specialized books.[1] Rather, we consider:

- The model refrigerator, operating on a reverse Carnot cycle
- Refrigeration via the vapor-compression cycle, as in the common household refrigerator and air conditioner
- The choice of refrigerants as influenced by their properties
- Refrigeration based on vapor absorption, an alternative to vapor compression
- Heating or cooling by heat pumps with heat extracted from or rejected to the surroundings
- The liquefaction of gases by refrigeration

9.1 THE CARNOT REFRIGERATOR

As indicated in Sec. 5.2, a refrigerator is a heat pump that absorbs heat from a region at a temperature below that of the surroundings and rejects heat to the surroundings. It operates with the highest possible efficiency on a Carnot refrigeration cycle, the reverse of the Carnot engine cycle, as shown by Fig. 5.1(b). The two isothermal steps provide heat absorption Q_C at the lower temperature T_C and heat rejection Q_H at the higher temperature T_H. The cycle is completed by two adiabatic steps between these two temperatures. The cycle requires the

[1]*ASHRAE Handbook: Refrigeration*, 2014; *Fundamentals*, 2013; *HVAC Systems and Equipment*, 2016; *HVAC Applications*, 2015; American Society of Heating, Refrigerating and Air-Conditioning Engineers, Inc., Atlanta; Shan K. Wang, *Handbook of Air Conditioning and Refrigeration, 2nd Ed.*, McGraw-Hill, New York, 2000.

addition of net work W to the system. Because all steps of the cycle are reversible, it gives the minimum possible work required for a given refrigeration effect.

The working fluid operates in a cycle for which ΔU is zero. The first law for the cycle is therefore:

$$W = -(Q_C + Q_H) \tag{9.1}$$

where we note that Q_C is a positive number and Q_H (larger in absolute value) is negative. The measure of the effectiveness of a refrigerator is its *coefficient of performance* ω, defined as:

$$\omega \equiv \frac{\text{heat absorbed at the lower temperature}}{\text{net work}} = \frac{Q_C}{W} \tag{9.2}$$

Equation (9.1) may be divided by Q_C and then combined with Eq. (5.4):

$$-\frac{W}{Q_C} = 1 + \frac{Q_H}{Q_C} \qquad \frac{-W}{Q_C} = 1 - \frac{T_H}{T_C} = \frac{T_C - T_H}{T_C}$$

Equation (9.2) becomes:
$$\boxed{\omega = \frac{T_C}{T_H - T_C}} \tag{9.3}$$

For example, refrigeration at a temperature level of 5°C in surroundings at 30°C, gives:

$$\omega = \frac{5 + 273.15}{(30 + 273.15) - (5 + 273.15)} = 11.13$$

Equation (9.3) applies only to a refrigerator operating on a Carnot cycle, which yields the maximum possible value of ω for a refrigerator operating between given values of T_H and T_C.

It shows clearly that the refrigeration effect per unit of work decreases as the temperature of heat absorption T_C decreases and as the temperature of heat rejection T_H increases.

9.2   THE VAPOR-COMPRESSION CYCLE

The vapor-compression refrigeration cycle is represented in Fig. 9.1, along with a *TS* diagram showing the four steps of the process. A liquid refrigerant evaporating at constant T and P absorbs heat (line $1 \rightarrow 2$), producing the refrigeration effect. The vapor produced is compressed via dashed line $2 \rightarrow 3'$ for isentropic compression (Fig. 7.6), and via line $2 \rightarrow 3$, sloping in the direction of increasing entropy, for an actual compression process, reflecting inherent irreversibilities. At this higher T and P, it is cooled and condensed (line $3 \rightarrow 4$) with rejection of heat to the surroundings. Liquid from the condenser expands (line $4 \rightarrow 1$) to its original pressure. In principle, this can be carried out in a turbine from which work is obtained. However, for practical reasons it is usually accomplished by throttling through a partly open control valve. The pressure drop in this irreversible process results from fluid friction in the valve. As shown in Sec. 7.1, the throttling process occurs at constant enthalpy.

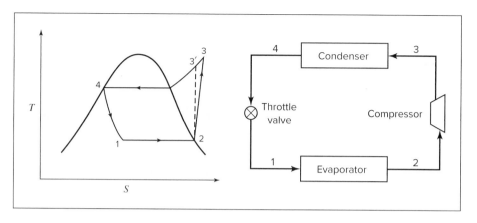

Figure 9.1: Vapor-compression refrigeration cycle.

On the basis of a unit mass of fluid, the equations for the heat absorbed in the evaporator and the heat rejected in the condenser are:

$$Q_C = H_2 - H_1 \quad \text{and} \quad Q_H = H_4 - H_3$$

These equations follow from Eq. (2.31) when the small changes in potential and kinetic energy are neglected. The work of compression is simply: $W = H_3 - H_2$, and by Eq. (9.2), the coefficient of performance is:

$$\omega = \frac{H_2 - H_1}{H_3 - H_2} \tag{9.4}$$

To design the evaporator, compressor, condenser, and auxiliary equipment one must know the rate of circulation of refrigerant $\dot{m}$. This is determined from the rate of heat absorption in the evaporator[2] by the equation:

$$\dot{m} = \frac{\dot{Q}_C}{H_2 - H_1} \tag{9.5}$$

The vapor-compression cycle of Fig. 9.1 is shown on a *PH* diagram in Fig. 9.2, a diagram commonly used in the description of refrigeration processes, because it shows the required enthalpies directly. Although the evaporation and condensation processes are represented by constant-pressure paths, small pressure drops do occur because of fluid friction.

For given T_C and T_H, vapor-compression refrigeration results in lower values of ω than the Carnot cycle because of irreversibilities in expansion and compression. The following example provides an indication of typical values for coefficients of performance.

---

[2]In the United States refrigeration equipment is commonly rated in *tons of refrigeration*; a ton of refrigeration is defined as heat absorption at the rate of 12,000(Btu)(hr)$^{-1}$ or 12,660 kJ per hour. This corresponds approximately to the rate of heat removal required to freeze 1(ton) of water, initially at 32(°F), per day.

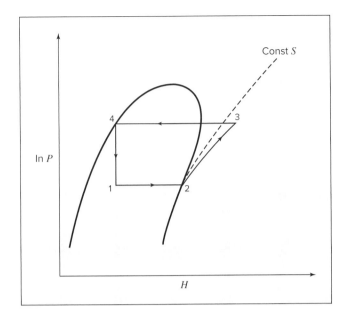

Figure 9.2: Vapor-compression refrigeration cycle on a *PH* diagram.

Example 9.1

A refrigerated space is maintained at −20°C, and cooling water is available at 21°C. Refrigeration capacity is 120,000 kJ·h⁻¹. The evaporator and condenser are of sufficient size that a 5°C minimum-temperature difference for heat transfer can be realized in each. The refrigerant is 1,1,1,2-tetrafluoroethane (HFC-134a), for which data are given in Table 9.1 and Fig. F.2 (App. F).

> (a) What is the value of ω for a Carnot refrigerator?
>
> (b) Calculate ω and $\dot{m}$ for a vapor-compression cycle (Fig. 9.2) if the compressor efficiency is 0.80.

Solution 9.1

> (a) Allowing 5°C temperature differences, the evaporator temperature is −25°C = 248.15 K, and the condenser temperature is 26°C = 299.15 K. Thus, by Eq. (9.3) for a Carnot refrigerator,
>
> $$\omega = \frac{248.15}{299.15 - 248.15} = 4.87$$
>
> (b) With HFC-134a as the refrigerant, enthalpies for states 2 and 4 of Fig. 9.2 are read directly from Table 9.1. The entry at −25°C indicates that HFC-134a vaporizes in the evaporator at a pressure of 1.064 bar. Its properties as a saturated vapor at these conditions are:
>
> $$H_2 = 383.45 \text{ kJ·kg}^{-1} \qquad\qquad S_2 = 1.746 \text{ kJ·kg}^{-1}\cdot\text{K}^{-1}$$

The entry at 26°C in Table 9.1 shows that HFC-134a condenses at 6.854 bar; its enthalpy as a saturated liquid at these conditions is:

$$H_4 = 235.97 \text{ kJ} \cdot \text{kg}^{-1}$$

If the compression step is reversible and adiabatic (isentropic) from saturated vapor at state 2 to superheated vapor at state 3',

$$S'_3 = S_2 = 1.746 \text{ kJ} \cdot \text{kg}^{-1} \cdot \text{K}^{-1}$$

This entropy value and the condenser pressure of 6.854 bar are sufficient to specify the thermodynamic state at point 3'. One could find the other properties at this state using Fig. F.2, following a curve of constant entropy from the saturation curve to the condenser pressure. However, more precise results can be obtained using an electronic resource such as the NIST WebBook. Varying the temperature at a fixed pressure of 6.854 bar shows that the entropy is 1.746 kJ·kg^{-1}·K^{-1} at $T = 308.1$ K. The corresponding enthalpy is:

$$H'_3 = 421.97 \text{ kJ} \cdot \text{kg}^{-1}$$

and the enthalpy change is:

$$(\Delta H)_S = H'_3 - H_2 = 421.97 - 383.45 = 38.52 \text{ kJ} \cdot \text{kg}^{-1}$$

By Eq. (7.17) for a compressor efficiency of 0.80, the actual enthalpy change for step $2 \rightarrow 3$ is:

$$H_3 - H_2 = \frac{(\Delta H)_S}{\eta} = \frac{38.52}{0.80} = 48.15 \text{ kJ} \cdot \text{kg}^{-1}$$

Because the throttling process of step $1 \rightarrow 4$ is isenthalpic, $H_1 = H_4$. The coefficient of performance as given by Eq. (9.4) therefore becomes:

$$\omega = \frac{H_2 - H_4}{H_3 - H_2} = \frac{383.45 - 235.97}{48.15} = 3.06$$

and the HFC-134a circulation rate as given by Eq. (9.5) is:

$$\dot{m} = \frac{\dot{Q}_C}{H_2 - H_4} = \frac{120{,}000}{383.45 - 235.97} = 814 \text{ kg} \cdot \text{h}^{-1}$$

9.3 THE CHOICE OF REFRIGERANT

As shown in Sec. 5.2, the efficiency of a Carnot heat engine is independent of the working medium of the engine. Similarly, the coefficient of performance of a Carnot refrigerator is independent of the refrigerant. The irreversibilities inherent in vapor-compression cycles cause the coefficient of performance of practical refrigerators to depend to a limited extent on the refrigerant. Nevertheless, such characteristics as toxicity, flammability, cost, corrosion properties, and vapor pressure are of greater importance in the choice of refrigerant. Moreover, environmental concerns strongly constrain the range of compounds that can be considered for use as refrigerants. So that air cannot leak into the refrigeration system, the vapor pressure of the refrigerant at the evaporator temperature should be greater than atmospheric pressure. On the other hand, the vapor pressure at the condenser temperature should not be unduly high because of the initial cost and operating expense of high-pressure equipment. These many requirements limit the choice of refrigerants to relatively few fluids.

Ammonia, methyl chloride, carbon dioxide, propane, and other hydrocarbons can serve as refrigerants, particularly in industrial applications. Halogenated hydrocarbons came into common use as refrigerants in the 1930s. Fully halogenated chlorofluorocarbons were the most common refrigerants for several decades. However, these stable molecules were found to persist in the atmosphere for many years, allowing them to reach the stratosphere before finally decomposing by reactions that severely deplete stratospheric ozone. As a result, their production and use is now banned. Certain less-than-fully halogenated hydrocarbons, which cause relatively little ozone depletion, and hydrofluorocarbons that cause no ozone depletion now serve as replacements in many applications. A primary example is 1,1,1,2-tetrafluoroethane (HFC-134a).[3] Unfortunately, these refrigerants have extremely high global warming potentials, hundreds to thousands of times greater than CO_2, and for that reason are now being banned in many countries. New hydrofluorocarbon refrigerants with lower global warming potential, such as 2,3,3,3-tetrafluoropropene (HFO-1234yf), are beginning to replace first-generation hydrofluorocarbon refrigerants such as R134a.

A pressure/enthalpy diagram for 1,1,1,2-tetrafluoroethane (HFC-134a) is shown in Fig. F.2 of App. F; Table 9.1 provides saturation data for the same refrigerant. Tables and diagrams for a variety of other refrigerants are readily available.[4]

Cascade Cycles

Limits placed on the operating pressures of the evaporator and condenser of a refrigeration system also limit the temperature difference $T_H - T_C$ over which a simple vapor-compression cycle can operate. With T_H fixed by the temperature of the surroundings, a lower limit is placed on the temperature level of refrigeration. This can be overcome by the operation of two or more refrigeration cycles employing different refrigerants in a *cascade*. A two-stage cascade is shown in Fig. 9.3.

Here, the two cycles operate so that the heat absorbed in the interchanger by the refrigerant of the higher-temperature cycle 2 serves to condense the refrigerant in the lower-temperature cycle 1. The two refrigerants are chosen such that each cycle operates at a reasonable pressure. For example, assume the following operating temperatures (Fig. 9.3):

$$T_H = 30°C \qquad T'_C = -16°C \qquad T'_H = -10°C \qquad T_C = -50°C$$

If tetrafluoroethane (HFC-134a) is the refrigerant in cycle 2, then the intake and discharge pressures for the compressor are about 1.6 bar and 7.7 bar, and the pressure ratio is

---

[3]The abbreviated designation is nomenclature of the American Society of Heating, Refrigerating, and Air-Conditioning Engineers.

[4]*ASHRAE Handbook: Fundamentals,* Chap. 30, 2013; R. H. Perry and D. Green, *Perry's Chemical Engineers' Handbook,* 8th ed., Sec. 2.20, 2008. "Thermophysical Properties of Fluid Systems" in *NIST Chemistry WebBook,* http://webbook.nist.gov.

Table 9.1: Properties of Saturated 1,1,1,2-Tetrafluoroethane (R134A)[†]

$T(°C)$	P (bar)	Volume m³·kg⁻¹		Enthalpy kJ·kg⁻¹		Entropy kJ·kg⁻¹·K⁻¹	
		V^l	V^v	H^l	H^v	S^l	S^v
−40	0.512	0.000705	0.361080	148.14	374.00	0.796	1.764
−35	0.661	0.000713	0.284020	154.44	377.17	0.822	1.758
−30	0.844	0.000720	0.225940	160.79	380.32	0.849	1.752
−25	1.064	0.000728	0.181620	167.19	383.45	0.875	1.746
−20	1.327	0.000736	0.147390	173.64	386.55	0.900	1.741
−18	1.446	0.000740	0.135920	176.23	387.79	0.910	1.740
−16	1.573	0.000743	0.125510	178.83	389.02	0.921	1.738
−14	1.708	0.000746	0.116050	181.44	390.24	0.931	1.736
−12	1.852	0.000750	0.107440	184.07	391.46	0.941	1.735
−10	2.006	0.000754	0.099590	186.70	392.66	0.951	1.733
−8	2.169	0.000757	0.092422	189.34	393.87	0.961	1.732
−6	2.343	0.000761	0.085867	191.99	395.06	0.971	1.731
−4	2.527	0.000765	0.079866	194.65	396.25	0.980	1.729
−2	2.722	0.000768	0.074362	197.32	397.43	0.990	1.728
0	2.928	0.000772	0.069309	200.00	398.60	1.000	1.727
2	3.146	0.000776	0.064663	202.69	399.77	1.010	1.726
4	3.377	0.000780	0.060385	205.40	400.92	1.020	1.725
6	3.620	0.000785	0.056443	208.11	402.06	1.029	1.724
8	3.876	0.000789	0.052804	210.84	403.20	1.039	1.723
10	4.146	0.000793	0.049442	213.58	404.32	1.049	1.722
12	4.430	0.000797	0.046332	216.33	405.43	1.058	1.721
14	4.729	0.000802	0.043451	219.09	406.53	1.068	1.720
16	5.043	0.000807	0.040780	221.87	407.61	1.077	1.720
18	5.372	0.000811	0.038301	224.66	408.69	1.087	1.719
20	5.717	0.000816	0.035997	227.47	409.75	1.096	1.718
22	6.079	0.000821	0.033854	230.29	410.79	1.106	1.717
24	6.458	0.000826	0.031858	233.12	411.82	1.115	1.717
26	6.854	0.000831	0.029998	235.97	412.84	1.125	1.716
28	7.269	0.000837	0.028263	238.84	413.84	1.134	1.715
30	7.702	0.000842	0.026642	241.72	414.82	1.144	1.715
35	8.870	0.000857	0.023033	249.01	417.19	1.167	1.713
40	10.166	0.000872	0.019966	256.41	419.43	1.191	1.711
45	11.599	0.000889	0.017344	263.94	421.52	1.214	1.709
50	13.179	0.000907	0.015089	271.62	423.44	1.238	1.70
55	14.915	0.000927	0.013140	279.47	425.15	1.261	1.705
60	16.818	0.000950	0.011444	287.50	426.63	1.285	1.702
65	18.898	0.000975	0.009960	295.76	427.82	1.309	1.699
70	21.168	0.001004	0.008653	304.28	428.65	1.333	1.696
75	23.641	0.001037	0.007491	313.13	429.03	1.358	1.691
80	26.332	0.001077	0.006448	322.39	428.81	1.384	1.685

[†] Data in this table are from E. W. Lemmon, M. O. McLinden and D. G. Friend, "Thermophysical Properties of Fluid Systems" in NIST Chemistry WebBook, NIST Standard Reference Database Number 69, Eds. P. J. Linstrom and W. G. Mallard, National Institute of Standards and Technology, Gaithersburg MD, 20899, http://webbook.nist.gov.

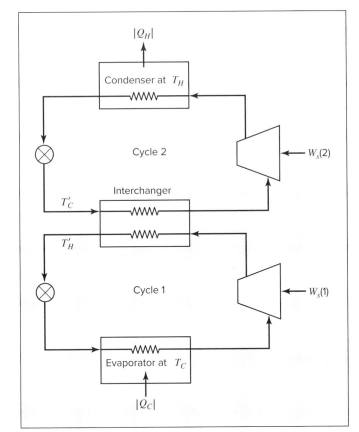

Figure 9.3: A two-stage cascade refrigeration system.

about 4.9. If difluoromethane (R32) is the refrigerant in cycle 1, these pressures are about 1.1 and 5.8 bar, and the pressure ratio is about 5.3. These are all reasonable values. On the other hand, for a single cycle operating between $-50°C$ and $30°C$ with HFC-134a as refrigerant, the intake pressure to the condenser is about 0.29 bar, well below atmospheric pressure. Moreover, for a discharge pressure of about 7.7 bar the pressure ratio is 26, too high to achieve with a single-stage compressor.

9.4   ABSORPTION REFRIGERATION

In vapor-compression refrigeration the work of compression is usually supplied by an electric motor. But the source of the electric energy is most likely a heat engine (central power plant) used to drive a generator. Thus the work for refrigeration comes ultimately from a heat engine. This suggests a combination of cycles wherein work produced by an engine cycle is used internally by a coupled refrigeration cycle in a device that takes in heat both at elevated temperature T_H and at cold temperature T_C and rejects heat to the surroundings at T_S. Before considering practical absorption-refrigeration processes, we first treat a combination of Carnot cycles that produce this effect.

The work required by a Carnot refrigerator absorbing heat at temperature T_C and rejecting heat at the temperature of the surroundings T_S follows from Eqs. (9.2) and (9.3):

$$W = \frac{T_S - T_C}{T_C} Q_C$$

where T_S replaces T_H and Q_C is the heat absorbed. If a source of heat is available at a temperature $T_H > T_S$, then this work may be obtained from a Carnot engine operating between T_H and T_S. The heat required Q_H for the production of work is found from Eq. (5.5), where we have substituted T_S for T_C and changed the sign of W, because W in Eq. (5.5) refers to a Carnot engine, but here it refers to the refrigerator:

$$\frac{-W}{Q_H} = \frac{T_S}{T_H} - 1 \qquad \text{or} \qquad Q_H = W \frac{T_H}{T_H - T_S}$$

Elimination of W gives:
$$\frac{Q_H}{Q_C} = \frac{T_H}{T_H - T_S} \frac{T_S - T_C}{T_C} \tag{9.6}$$

The value of Q_H/Q_C given by this equation is of course a minimum because Carnot cycles cannot be achieved in practice. It provides a limiting value for absorption refrigeration.

A schematic diagram for a typical absorption refrigerator is shown in Fig. 9.4. Note that just as in the preceding derivation, W is eliminated, and both Q_H and Q_C enter the system, with heat discarded only to the surroundings. The essential difference between a vapor-compression and an absorption refrigerator is in the different means employed for compression. The section of the absorption unit to the right of the dashed line in Fig. 9.4 is the same as in a vapor-compression refrigerator, but the section to the left accomplishes compression by what amounts to a heat engine. Refrigerant as vapor from the evaporator is absorbed in a relatively nonvolatile liquid solvent at the pressure of the evaporator and at relatively low temperature. The heat given off in the process is discarded to the surroundings at T_S. The liquid solution from the absorber,

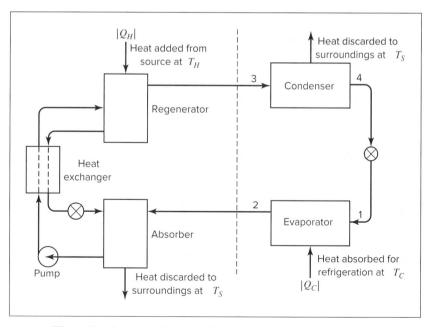

Figure 9.4: Schematic diagram of an absorption-refrigeration unit.

which contains a relatively high concentration of refrigerant, passes to a pump, which raises the pressure of the liquid to that of the condenser. Heat from the higher temperature source at T_H is transferred to the compressed liquid solution, raising its temperature and evaporating the refrigerant from the solvent. Vapor passes from the regenerator to the condenser, and solvent, which now contains a relatively low concentration of refrigerant, returns to the absorber by way of a heat exchanger, which serves to conserve energy and adjust stream temperatures toward optimum values. Low-pressure steam is the usual source of heat for the regenerator.

The most commonly used absorption-refrigeration system operates with water as the refrigerant and a lithium bromide solution as the absorbent. This system is obviously limited to refrigeration temperatures above the freezing point of water. It is treated in detail by Perry and Green.[5] For lower temperatures, ammonia can serve as refrigerant with water as the solvent. An alternative system uses methanol as refrigerant and polyglycolethers as absorbent.

Consider refrigeration at a temperature level of $-10°C$ ($T_C = 263.15$ K) with a heat source of condensing steam at atmospheric pressure ($T_H = 373.15$ K). For a surroundings temperature of $30°C$ ($T_S = 303.15$ K), the minimum possible value of Q_H/Q_C is found from Eq. (9.6):

$$\frac{Q_H}{Q_C} = \left(\frac{373.15}{373.15 - 303.15}\right)\left(\frac{303.15 - 263.15}{263.15}\right) = 0.81$$

For an actual absorption refrigerator, the value would be on the order of three times this result.

9.5   THE HEAT PUMP

The heat pump, a reversed heat engine, is a device for heating houses and commercial buildings during the winter and cooling them during the summer. In the winter it operates so as to absorb heat from the surroundings and reject heat into the building. Refrigerant evaporates in coils placed underground or in the outside air; vapor compression is followed by condensation, heat being transferred to air or water, which is used to heat the building. Compression must be to a pressure such that the condensation temperature of the refrigerant is higher than the required temperature level of the building. The operating cost of the installation is the cost of electric power to run the compressor. If the unit has a coefficient of performance $Q_C/W = 4$, the heat available to heat the house Q_H is equal to five times the energy input to the compressor. Any economic advantage of the heat pump as a heating device depends on the cost of electricity in comparison with the cost of fuels such as oil and natural gas.

The heat pump also serves for air conditioning during the summer. The flow of refrigerant is simply reversed, and heat is absorbed from the building and rejected through underground coils or to the outside air.

Example 9.2

A house has a winter heating requirement of 30 kJ·s^{-1} and a summer cooling requirement of 60 kJ·s^{-1}. Consider a heat-pump installation to maintain the house

---

[5]R. H. Perry and D. Green, *op. cit.*, pp. 11–90 to 11–94.

temperature at 20°C in winter and 25°C in summer. This requires circulation of the refrigerant through interior exchanger coils at 30°C in winter and 5°C in summer. Underground coils provide the heat source in winter and the heat sink in summer. For a year-round ground temperature of 15°C, the heat-transfer characteristics of the coils necessitate refrigerant temperatures of 10°C in winter and 25°C in summer. What are the minimum power requirements for winter heating and summer cooling?

Solution 9.2

The minimum power requirements are provided by a Carnot heat pump. For winter heating, the house coils are at the higher-temperature level T_H, and the heat requirement is $Q_H = 30$ kJ·s^{-1}. Application of Eq. (5.4) gives:

$$Q_C = -Q_H \frac{T_C}{T_H} = 30 \left(\frac{10 + 273.15}{30 + 273.15} \right) = 28.02 \text{ kJ·s}^{-1}$$

This is the heat absorbed in the ground coils. By Eq. (9.1),

$$W = -Q_H - Q_C = 30 - 28.02 = 1.98 \text{ kJ·s}^{-1}$$

Thus the power requirement is 1.98 kW.

For summer cooling, $Q_C = 60$ kJ·s^{-1}, and the house coils are at the lower-temperature level T_C. Combining Eqs. (9.2) and (9.3) and solving for W:

$$W = Q_C \frac{T_H - T_C}{T_C} = 60 \left(\frac{25 - 5}{5 + 273.15} \right) = 4.31 \text{ kJ·s}^{-1}$$

The power requirement here is therefore 4.31 kW. Actual power requirements for practical heat pumps are are likely to be more than twice this lower limit.

9.6 LIQUEFACTION PROCESSES

Liquefied gases are used for a variety of purposes. For example, liquid propane in cylinders serves as a domestic fuel, liquid oxygen is carried in rockets, natural gas is liquefied for ocean transport, and liquid nitrogen provides low-temperature refrigeration. Gas mixtures (e.g., air) are liquefied for separation into their component species by distillation.

Liquefaction results when a gas is cooled to a temperature in the two-phase region. This may be accomplished in several ways:

1. By heat exchange at constant pressure.

2. By an expansion process from which work is obtained.

3. By a throttling process.

The first method requires a heat sink at a temperature lower than that to which the gas is cooled, and is most commonly used to precool a gas prior to its liquefaction by the other two methods. An external refrigerator is required for a gas temperature below that of the surroundings.

The three methods are illustrated in Fig. 9.5. The constant-pressure process (1) approaches the two-phase region (and liquefaction) most closely for a given drop in temperature. The throttling process (3) does not result in liquefaction unless the initial state is at a low enough temperature and high enough pressure for the constant-enthalpy process to cut into the two-phase region. This is indeed the situation for an initial state at A', but not at A, where the temperature is the same but the pressure is lower than at A. The change of state from A to A' may be accomplished by compression of the gas to the pressure at B and constant-pressure cooling to A'. Reference to a *PH* diagram for air[6] shows that at a temperature of 160 K, the pressure must be greater than about 80 bar for any liquefaction to occur along a path of constant enthalpy. Thus, if air is compressed to at least 80 bar and cooled below 160 K, it can be partially liquefied by throttling. An efficient process for cooling the gas is by countercurrent heat exchange with that portion of the gas which does not liquefy in the throttling process.

Figure 9.5: Cooling processes on a *TS* diagram.

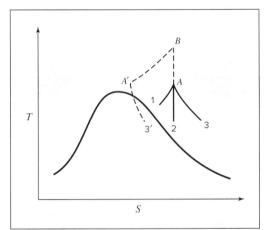

Liquefaction by isentropic expansion along process (2) occurs from lower pressures (for given temperature) than by throttling. For example, continuation of process (2) from initial state A ultimately results in liquefaction.

The throttling process (3) is commonly employed in small-scale commercial liquefaction plants. The temperature of the gas must decrease during expansion, and this indeed occurs with most gases at usual conditions of temperature and pressure. The exceptions are hydrogen and helium, which increase in temperature upon throttling unless the initial temperature is below about 100 K for hydrogen and 20 K for helium. Liquefaction of these gases by throttling requires initial cooling to temperatures lower than obtained by method 1 or 2.

The Linde liquefaction process, which depends solely on throttling expansion, is shown in Fig. 9.6. After compression, the gas is precooled to ambient temperature. It may be even further cooled by refrigeration. The lower the temperature of the gas entering the throttle valve, the greater the fraction of gas that is liquefied. For example, a refrigerant evaporating in the cooler at −40°C provides a lower temperature at the valve than if water at 20°C is the cooling medium.

[6]R. H. Perry and D. Green, *op. cit.*, Fig. 2–5, p. 2–215.

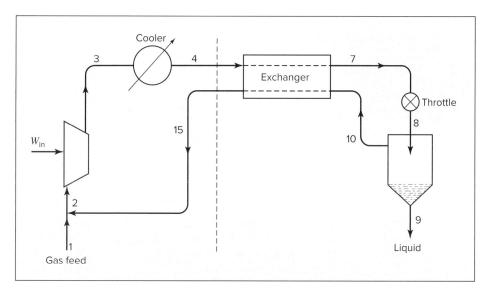

Figure 9.6: Linde liquefaction process.

A more efficient liquefaction process would replace the throttle valve with an expander, but operating such a device into the two-phase region is impractical. The Claude process, shown in Fig. 9.7, is based in part on this idea. Gas at an intermediate temperature is extracted from the heat-exchange system and passed through an expander from which it exhausts as a saturated or slightly superheated vapor. The remaining gas is further cooled and throttled through a valve to produce liquefaction as in the Linde process. The unliquefied portion, which is saturated vapor, mixes with the expander exhaust and returns for recycle through the heat-exchanger system.

An energy balance, Eq. (2.30), applied to that part of the process lying to the right of the dashed vertical line yields:

$$\dot{m}_9 H_9 + \dot{m}_{15} H_{15} - \dot{m}_4 H_4 = \dot{W}_{out}$$

If the expander operates adiabatically, $\dot{W}_{out}$ as given by Eq. (7.13) is:

$$\dot{W}_{out} = \dot{m}_{12}(H_{12} - H_5)$$

Moreover, by a mass balance, $\dot{m}_{15} = \dot{m}_4 - \dot{m}_9$. The energy balance, after division by $\dot{m}_4$, therefore becomes:

$$\frac{\dot{m}_9}{\dot{m}_4} H_9 + \frac{\dot{m}_4 - \dot{m}_9}{\dot{m}_4} H_{15} - H_4 = \frac{\dot{m}_{12}}{\dot{m}_4}(H_{12} - H_5)$$

With the definitions, $z \equiv \dot{m}_9/\dot{m}_4$ and $x \equiv \dot{m}_{12}/\dot{m}_4$, solution of this equation for z yields:

$$z = \frac{x(H_{12} - H_5) + H_4 - H_{15}}{H_9 - H_{15}} \tag{9.7}$$

In this equation z is the fraction of the stream entering the heat-exchanger system that is liquefied, and x is the fraction of this stream that is drawn off between the heat exchangers and

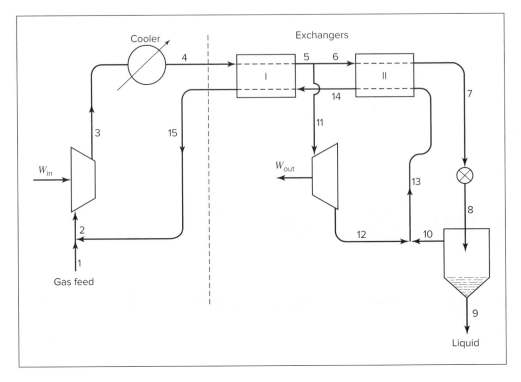

Figure 9.7: Claude liquefaction process.

passed through the expander. This latter quantity (x) is a design variable and must be specified before Eq. (9.7) can be solved for z. Note that the Linde process results when $x = 0$, and in this event Eq. (9.7) reduces to:

$$z = \frac{H_4 - H_{15}}{H_9 - H_{15}} \tag{9.8}$$

Thus the Linde process is a limiting case of the Claude process, obtained when none of the high-pressure gas stream is sent to an expander.

Equations (9.7) and (9.8) suppose that no heat flows into the system from the surroundings. This can never be exactly true, and heat leakage may be significant when temperatures are very low, even with well-insulated equipment.

Example 9.3

Natural gas, assumed here to be pure methane, is liquefied in a Claude process. Compression is to 60 bar and precooling is to 300 K. The expander and throttle exhaust to a pressure of 1 bar. Recycle methane at this pressure leaves the exchanger system (point 15, Fig. 9.7) at 295 K. Assume no heat leaks into the system from the surroundings, an expander efficiency of 75%, and an expander exhaust of saturated vapor. For a draw-off to the expander of 25% of the methane entering the exchanger system ($x = 0.25$), what fraction z of the methane is liquefied, and what is the temperature of the high-pressure stream entering the throttle valve?

Solution 9.3

Data for methane are available in the NIST WebBook,[7] from which the following values were obtained:

$$H_4 = 855.3 \text{ kJ·kg}^{-1} \qquad \text{(at 300 K and 60 bar)}$$
$$H_{15} = 903.0 \text{ kJ·kg}^{-1} \qquad \text{(at 295 K and 1 bar)}$$

For saturated liquid and vapor, at a pressure of 1 bar:

$$T^{\text{sat}} = 111.5 \text{ K}$$
$$H_9 = -0.6 \text{ kJ·kg}^{-1} \qquad \text{(saturated liquid)}$$
$$H_{12} = 510.6 \text{ kJ·kg}^{-1} \qquad \text{(saturated vapor)}$$
$$S_{12} = 4.579 \text{ kJ·kg}^{-1}\text{·K}^{-1} \quad \text{(saturated vapor)}$$

The enthalpy at the draw-off point between exchangers I and II, H_5, is required for solution of Eq. (9.7). The expander efficiency η is known, as is H_{12}, the enthalpy of the expander exhaust. The calculation of H_5 ($=H_{11}$), the expander inlet enthalpy, is less straightforward than the usual calculation of the exhaust enthalpy from the entrance enthalpy. The equation defining expander efficiency can be written:

$$\Delta H = H_{12} - H_5 = \eta (\Delta H)_S = \eta(H'_{12} - H_5)$$

Solution for H_{12} yields:

$$H_{12} = H_5 + \eta(H'_{12} - H_5) \qquad (A)$$

where H'_{12} is the enthalpy at 1 bar as the result of *isentropic* expansion from point 5. This enthalpy is readily found once the conditions at point 5 are known. Thus a trial calculation is indicated, wherein the first step is to assume temperature T_5. This leads to values for H_5 and S_5, from which H'_{12} can be found. All quantities in Eq. (A) are then known, and their substitution into this equation shows whether or not it is satisfied. If not, a new value is chosen for T_5, and the process continues until Eq. (A) is satisfied. For example, at 60 bar and 260 K, the enthalpy and entropy are 745.27 kJ·kg^{-1} and 4.033 kJ·kg^{-1}·K^{-1}, respectively. The saturated liquid and vapor at 1 bar have $S^l = -0.005$ and $S^v = 4.579$, respectively. Using these values, isentropic expansion from 260 K and 60 bar to 1 bar would give a vapor fraction of 0.8808. This would give:

$$H'_{12} = H_9 + 0.8808(H_{12} - H_9) = 449.6 \text{ kJ·kg}^{-1}$$

[7] E. W. Lemmon, M. O. McLinden and D. G. Friend, *op. cit.*, http://webbook.nist.gov.

Using this value in Eq. (A) yields $H_{12} = 508.8$ kJ·kg^{-1}, which is below the known value of $H_{12} = 510.6$ kJ·kg^{-1}. Thus, T_5 must be higher than the assumed value of 260 K. Repeating this process (in an automated fashion using a spreadsheet) for other values of T_5 shows that Eq. (A) is satisfied for:

$$T_5 = 261.2 \text{ K} \quad H_5 = 748.8 \text{ kJ·kg}^{-1} \text{ (at 60 bar)}$$

Substitution of values into Eq. (9.7) now yields:

$$z = \frac{0.25(510.6 - 748.8) + 855.3 - 903.0}{-0.6 - 903.0} = 0.1187$$

Thus 11.9% of the methane entering the exchanger system is liquefied.

The temperature at point 7 depends on its enthalpy, which is found from energy balances on the exchanger system. Thus, for exchanger I,

$$\dot{m}_4(H_5 - H_4) + \dot{m}_{15}(H_{15} - H_{14}) = 0$$

With $\dot{m}_{15} = \dot{m}_4 - \dot{m}_9$ and $\dot{m}_9/\dot{m}_4 = z$, this equation may be rearranged to give:

$$H_{14} = \frac{H_5 - H_4}{1 - z} + H_{15} = \frac{748.8 - 855.3}{1 - 0.1187} + 903.0$$

Then,

$$H_{14} = 782.2 \text{ kJ·kg}^{-1} \quad T_{14} = 239.4 \text{ K(at 1 bar)}$$

where T_{14} is found by evaluating H for methane at 1 bar and varying the temperature to match the known H_{14}.

For exchanger II,

$$\dot{m}_7(H_7 - H_5) + \dot{m}_{14}(H_{14} - H_{12}) = 0$$

With $\dot{m}_7 = \dot{m}_4 - \dot{m}_{12}$ and $\dot{m}_{14} = \dot{m}_4 - \dot{m}_9$ and with the definitions of z and x, this equation upon rearrangement becomes:

$$H_7 = H_5 - \frac{1 - z}{1 - x}(H_{14} - H_{12}) = 748.8 - \frac{1 - 0.1187}{1 - 0.25}(782.2 - 510.6)$$

Then

$$H_7 = 429.7 \text{ kJ·kg}^{-1} \quad T_7 = 199.1 \text{ K(at 60 bar)}$$

As the value of x increases, T_7 decreases, eventually approaching the saturation temperature in the separator, and requiring an exchanger II of infinite area. Thus x is limited on the high side by the cost of the exchanger system.

The other limit is for $x = 0$, the Linde system, for which by Eq. (9.8),

$$z = \frac{855.3 - 903.0}{-0.6 - 903.0} = 0.0528$$

In this case only 5.3% of the gas entering the throttle valve emerges as liquid. The temperature of the gas at point 7 is again found from its enthalpy, calculated by the energy balance:

$$H_7 = H_4 - (1 - z)(H_{15} - H_{10})$$

Substitution of known values yields:

$$H_7 = 855.3 - (1 - 0.0528)(903.0 - 510.6) = 483.6 \text{ kJ·kg}^{-1}$$

The corresponding temperature of the methane entering the throttle valve is $T_7 = 202.1$ K.

9.7 SYNOPSIS

After thorough study of this chapter, including working through example and end-of-chapter problems, one should be able to:

- Compute the coefficient of performance for a Carnot refrigeration cycle and recognize that this represents an upper limit for any real refrigeration process.

- Carry out a thermodynamic analysis of a vapor compression refrigeration cycle like that illustrated in Fig. 9.1.

- Describe a practical absorption refrigeration process and explain why its use might be advantageous.

- Sketch a cascade refrigeration system, explain why one might use such a system, and understand how to approach the selection of refrigerants for such a system.

- Carry out a thermodynamic analysis of a Linde or Claude liquefaction process, such as that presented in Ex. 9.3.

9.8 PROBLEMS

9.1. An easy way to rationalize definitions of cycle performance is to think of them as:

$$\text{Measure of performance} = \frac{\text{What you get}}{\text{What you pay for}}$$

Thus, for an engine, thermal efficiency is $\eta = |W|/|Q_H|$; for a refrigerator, the coefficient of performance is $\omega = |Q_C|/|W|$. Define a coefficient of performance ϕ for a heat pump. What is ϕ for a *Carnot* heat pump?

9.2. The contents of the freezer in a home refrigerator are maintained at $-20°C$. The kitchen temperature is $20°C$. If heat leaks amount to 125,000 kJ per day, and if electricity costs \$0.08/kWh, estimate the yearly cost of running the refrigerator. Assume a coefficient of performance equal to 60% of the Carnot value.

9.3. Consider the startup of a refrigerator. Initially, the contents are at the same temperature as the surroundings: $T_{C_0} = T_H$, where T_H is the (constant) surroundings temperature. With the passage of time, owing to work input, the contents' temperature is reduced from T_{C_0} to its design value T_C. Modeling the process as a Carnot refrigerator operating between an infinite hot reservoir and a *finite* cold reservoir of total heat capacity C^t, determine an expression for the minimum work required to decrease the contents temperature from T_{C_0} to T_C.

9.4. A Carnot refrigerator has tetrafluoroethane as the working fluid. The cycle is the same as that shown by Fig. 8.2, except the directions are reversed. For $T_C = -12°C$ and $T_H = 40°C$, determine:

(*a*) The pressures at states 1, 2, 3, and 4.
(*b*) The quality x^v at states 3 and 4.
(*c*) The heat addition per kg of fluid.
(*d*) The heat rejection per kg of fluid.
(*e*) The mechanical power per kg of fluid for each of the four steps.
(*f*) The coefficient of performance ω for the cycle.

9.5. Which is the more effective way to increase the coefficient of performance of a Carnot refrigerator: to increase T_C with T_H constant, or to decrease T_H with T_C constant? For a real refrigerator, does either of these strategies make sense?

9.6. In comparing the performance of a real cycle with that of a Carnot cycle, one has in principle a choice of temperatures to use for the Carnot calculation. Consider a vapor-compression refrigeration cycle in which the average fluid temperatures in the condenser and evaporator are T_H and T_C, respectively. Corresponding to T_H and T_C, the heat transfer occurs with respect to surroundings at temperature $T_{\sigma H}$ and $T_{\sigma C}$. Which provides the more conservative estimate of ω_{Carnot}: a calculation based on T_H and T_C, or one based on $T_{\sigma H}$ and $T_{\sigma C}$?

9.7. A Carnot engine is coupled to a Carnot refrigerator so that all of the work produced by the engine is used by the refrigerator in extraction of heat from a heat reservoir at $0°C$ at the rate of 35 kJ·s^{-1}. The source of energy for the Carnot engine is a heat reservoir at $250°C$. If both devices discard heat to the surroundings at $25°C$, how much heat does the engine absorb from its heat-source reservoir?

If the actual coefficient of performance of the refrigerator is $\omega = 0.6\,\omega_{Carnot}$ and if the thermal efficiency of the engine is $\eta = 0.6\eta_{Carnot}$, how much heat does the engine absorb from its heat-source reservoir?

9.8. A refrigeration system requires 1.5 kW of power for a refrigeration rate of 4 kJ·s^{-1}.

(*a*) What is the coefficient of performance?
(*b*) How much heat is rejected in the condenser?
(*c*) If heat rejection is at $40°C$, what is the lowest temperature the system can possibly maintain?

9.9. A vapor-compression refrigeration system operates on the cycle of Fig. 9.1. The refrigerant is tetrafluoroethane (Table 9.1, Fig. F.2). For one of the following sets of operating conditions, determine the circulation rate of the refrigerant, the heat-transfer rate in the condenser, the power requirement, the coefficient of performance of the cycle, and the coefficient of performance of a Carnot refrigeration cycle operating between the same temperature levels.

(*a*) Evaporation $T = 0°C$; condensation $T = 26°C$; η(compressor) $= 0.79$; refrigeration rate $= 600$ kJ·s^{-1}.

(*b*) Evaporation $T = 6°C$; condensation $T = 26°C$; η(compressor) $= 0.78$; refrigeration rate $= 500$ kJ·s^{-1}.

(*c*) Evaporation $T = -12°C$; condensation $T = 26°C$; η(compressor) $= 0.77$; refrigeration rate $= 400$ kJ·s^{-1}.

(*d*) Evaporation $T = -18°C$; condensation $T = 26°C$; η(compressor) $= 0.76$; refrigeration rate $= 300$ kJ·s^{-1}.

(*e*) Evaporation $T = -25°C$; condensation $T = 26°C$; η(compressor) $= 0.75$; refrigeration rate $= 200$ kJ·s^{-1}.

9.10. A vapor-compression refrigeration system operates on the cycle of Fig. 9.1. The refrigerant is water. Given that the evaporation $T = 4°C$, the condensation $T = 34°C$, η(compressor) $= 0.76$, and the refrigeration rate $= 1200$ kJ·s^{-1}, determine the circulation rate of the refrigerant, the heat-transfer rate in the condenser, the power requirement, the coefficient of performance of the cycle, and the coefficient of performance of a Carnot refrigeration cycle operating between the same temperature levels.

9.11. A refrigerator with tetrafluoroethane (Table 9.1, Fig. F.2) as refrigerant operates with an evaporation temperature of $-25°C$ and a condensation temperature of $26°C$. Saturated liquid refrigerant from the condenser flows through an expansion valve into the evaporator, from which it emerges as saturated vapor.

(*a*) For a cooling rate of 5 kJ·s^{-1}, what is the circulation rate of the refrigerant?

(*b*) By how much would the circulation rate be reduced if the throttle valve were replaced by a turbine in which the refrigerant expands isentropically?

(*c*) Suppose the cycle of (*a*) is modified by the inclusion of a countercurrent heat exchanger between the condenser and the throttle valve in which heat is transferred to vapor returning from the evaporator. If liquid from the condenser enters the exchanger at $26°C$ and if vapor from the evaporator enters the exchanger at $-25°C$ and leaves at $20°C$, what is the circulation rate of the refrigerant?

(*d*) For each of (*a*), (*b*), and (*c*), determine the coefficient of performance for isentropic compression of the vapor.

9.12. A vapor-compression refrigeration system is conventional except that a countercurrent heat exchanger is installed to subcool the liquid from the condenser by heat exchange with the vapor stream from the evaporator. The minimum temperature difference for heat transfer is $5°C$. Tetrafluoroethane is the refrigerant (Table 9.1, Fig. F.2), evaporating at $-6°C$ and condensing at $26°C$. The heat load on the evaporator is

2000 kJ·s^{-1}. If the compressor efficiency is 75%, what is the power requirement? How does this result compare with the power required by the compressor if the system operates without the heat exchanger? How do the refrigerant circulation rates compare for the two cases?

9.13. Consider the vapor-compression refrigeration cycle of Fig. 9.1 with tetrafluoroethane as refrigerant (Table 9.1, Fig. F.2). If the evaporation temperature is $-12°C$, show the effect of condensation temperature on the coefficient of performance by making calculations for condensation temperatures of 16, 28, and 40°C.

(*a*) Assume isentropic compression of the vapor.
(*b*) Assume a compressor efficiency of 75%.

9.14. A heat pump is used to heat a house in the winter and to cool it in the summer. During the winter, the outside air serves as a low-temperature heat source; during the summer, it acts as a high-temperature heat sink. The heat-transfer rate through the walls and roof of the house is 0.75 kJ·s^{-1} for each °C of temperature difference between the inside and outside of the house, summer and winter. The heat-pump motor is rated at 1.5 kW. Determine the minimum outside temperature for which the house can be maintained at 20°C during the winter and the maximum outside temperature for which the house can be maintained at 25°C during the summer.

9.15. Dry methane is supplied by a compressor and precooling system to the cooler of a Linde liquid-methane system (Fig. 9.6) at 180 bar and 300 K. The low-pressure methane leaves the cooler at a temperature 6°C lower than the temperature of the incoming high-pressure methane. The separator operates at 1 bar, and the product is saturated liquid at this pressure. What is the maximum fraction of the methane entering the cooler that can be liquefied? The NIST Chemistry WebBook (http://webbook.nist.gov/chemistry/fluid/) is a source of data for methane.

9.16. Rework the preceding problem for methane entering at 200 bar and precooled to 240 K by external refrigeration.

9.17. An advertisement is noted in a rural newspaper for a dairy-barn unit that combines a milk cooler with a water heater. Milk must, of course, be refrigerated, and hot water is required for washing purposes. The usual barn is equipped with a conventional air-cooled electric refrigerator and an electric-resistance water heater. The new unit is said to provide both the necessary refrigeration and the required hot water at a cost for electricity about the same as the cost of running just the refrigerator in the usual installation. To assess this claim, compare two refrigeration units: The advertised unit takes 15 kJ·s^{-1} from a milk cooler at $-2°C$, and discards heat through a condenser at 65°C to raise the temperature of water from 13 to 63°C. The conventional unit takes the same amount of heat from the same milk cooler at $-2°C$ and discards heat through an air-cooled condenser at 50°C; in addition, the same amount of water is heated electrically from 13 to 63°C. Estimate the *total* electric power requirements for the two cases, assuming that the actual work in both is 50% greater than required by Carnot refrigerators operating between the given temperatures.

9.18. A two-stage cascade refrigeration system (see Fig. 9.3) operates between $T_C = 210$ K and $T_H = 305$ K. Intermediate temperatures are $T'_C = 255$ K and $T'_H = 260$ K. Coefficients of performance ω of each stage are 65% of the corresponding values for a Carnot refrigerator. Determine ω for the real cascade, and compare it with that for a Carnot refrigerator operating between T_C and T_H.

9.19. Do a parametric study for the Claude liquefaction process treated in Sec. 9.6 and Ex. 9.3. In particular, show numerically the effect of changing the draw-off ratio x on other process variables. The NIST Chemistry WebBook (http://webbook.nist.gov/chemistry/fluid/) is a source of data for methane.

9.20. The condenser of a home refrigerator is commonly underneath the appliance; thus, the condensing refrigerant exchanges heat with household air, which has an average temperature of about 21°C. It is proposed to reconfigure a refrigerator so that the condenser is *outside* the home, where the average yearly temperature is about 10°C. Discuss the pros and cons of this proposal. Assume a freezer temperature of −18°C, and an actual coefficient of performance 60% that of a Carnot refrigerator.

9.21. A common misconception is that the coefficient of performance of a refrigerator must be less than unity. In fact, this is rarely the case. To see why, consider a real refrigerator for which $\omega = 0.6\,\omega_{\text{Carnot}}$. What condition must be satisfied in order for $\omega < 1$? Assume that T_H is fixed.

9.22. A furnace fails in a home in the winter. Mercifully, the electric power remains on. The resident engineer tells her spouse not to worry; they'll move into the kitchen, where the heat discarded from the refrigerator may provide for a temporarily comfortable living space. However (the engineer is reminded), the kitchen loses heat to the outdoors. Use the following data to determine the allowable rate of heat loss (kW) from the kitchen for the engineer's proposal to make sense.

Data: Desired kitchen temperature = 290 K.

Refrigerator freezer temperature = 250 K.

Average mechanical power input to refrigerator = 0.40 kW.

Performance: Actual $\omega = 65\%$ of Carnot ω.

9.23. Fifty (50) kmol·h^{-1} of liquid toluene at 1.2 bar is cooled from 100 to 20°C. A vapor-compression refrigeration cycle is used for the purpose. Ammonia is the working fluid. Condensation in the cycle is effected by an air-cooled fin/fan heat exchanger for which the air temperature may be assumed essentially constant at 20°C. Determine:

(*a*) The low and high pressure levels (bar) in the refrigeration cycle.
(*b*) The circulation rate of ammonia (mol·s^{-1}).

Assume 10°C minimum approach temperature differences for heat exchange. Data for ammonia:

$$\Delta H_n^{lv} = 23.34 \text{ kJ mol}^{-1}$$

$$\ln P^{sat} = 45.327 - \frac{4104.67}{T} - 5.146 \ln T + 615.0\,\frac{P^{sat}}{T^2}$$

where P^{sat} is in bars and T is in kelvins.

Chapter 10

The Framework of Solution Thermodynamics

Our purpose in this chapter is to lay the theoretical foundation for applications of thermodynamics to gas mixtures and liquid solutions. Throughout the chemical, energy, microelectronics, personal care, and pharmaceutical industries, multicomponent fluid mixtures undergo composition changes brought about by mixing and separation processes, the transfer of species from one phase to another, and chemical reaction. Thus, measures of composition become essential variables, along with temperature and pressure, which we already considered in detail in Chap. 6. This adds substantially to the complexity of tabulating and correlating thermodynamic properties, and leads to the introduction of a menagerie of new variables and relationships among them. Applying these relationships to practical problems, such as phase equilibrium calculations, requires that we first map out this "thermodynamic zoo." Thus, in the present chapter, we:

- Develop a *fundamental property relation* that is applicable to open phases of variable composition.

- Define the *chemical potential*, a fundamental new property that facilitates treatment of phase and chemical-reaction equilibria.

- Introduce *partial properties*, a class of thermodynamic properties defined mathematically to distribute total mixture properties among individual species as they exist in a mixture. These are composition-dependent and distinct from the molar properties of pure species.

- Develop property relations for the ideal-gas-state mixture, which provide the basis for treatment of real-gas mixtures.

- Define yet another useful property, the *fugacity*. Related to the chemical potential, it lends itself to mathematical formulation of both phase- and chemical-reaction-equilibrium problems.

- Introduce a useful class of solution properties, known as *excess properties*, in conjunction with an idealization of solution behavior called the *ideal-solution model*, which serves as a reference for real-solution behavior.

Measures of Composition

The three most common measures of composition in thermodynamics are mass fraction, mole fraction, and molar concentration. Mass or mole fraction is defined as the ratio of the mass or number of moles of a particular chemical species in a mixture to the total mass or number of moles of mixture:

$$x_i \equiv \frac{m_i}{m} = \frac{\dot{m}_i}{\dot{m}} \qquad \text{or} \qquad x_i \equiv \frac{n_i}{n} = \frac{\dot{n}_i}{\dot{n}}$$

Molar concentration is defined as the ratio of the mole fraction of a particular chemical species in a mixture or solution to the molar volume of the mixture or solution:

$$C_i \equiv \frac{x_i}{V}$$

This quantity has units of moles of i per unit volume. For flow processes, convenience suggests its expression as a ratio of rates. Multiplying and dividing by molar flow rate $\dot{n}$ gives:

$$C_i \equiv \frac{\dot{n}_i}{q}$$

where $\dot{n}_i$ is molar flow rate of species i, and q is volumetric flow rate.

The molar mass of a mixture or solution is, by definition, the mole-fraction-weighted sum of the molar masses of all species present:

$$\mathcal{M} \equiv \sum_i x_i \mathcal{M}_i$$

In the present chapter, we develop the framework of solution thermodynamics using mole fractions as composition variables. For nonreacting systems, virtually all of the same development can be done using mass fractions, yielding identical definitions and equations. Thus, we may take x_i to represent either a mole fraction or mass fraction in nonreacting systems. In reacting systems, it is nearly always best to work in terms of mole fractions.

10.1 FUNDAMENTAL PROPERTY RELATION

Equation (6.7) relates the total Gibbs energy of any closed system to its *canonical* variables, temperature and pressure:

$$d(nG) = (nV)dP - (nS)dT \qquad\qquad (6.7)$$

where n is the total number of moles of the system. It applies to a single-phase fluid in a closed system wherein no chemical reactions occur. For such a system the composition is necessarily constant, and therefore:

$$\left[\frac{\partial (nG)}{\partial P} \right]_{T,\, n} = nV \qquad \text{and} \qquad \left[\frac{\partial (nG)}{\partial T} \right]_{P,\, n} = -nS$$

The subscript n indicates that the numbers of moles of *all* chemical species are held constant.

For the more general case of a single-phase, *open* system, material may pass into and out of the system, and nG becomes a function of the numbers of moles of the chemical species present. It remains a function of T and P, and we can therefore write the functional relation:

$$nG = g(P, T, n_1, n_2, \ldots, n_i, \ldots)$$

where n_i is the number of moles of species i. The total differential of nG is then:

$$d(nG) = \left[\frac{\partial(nG)}{\partial P}\right]_{T, n} dP + \left[\frac{\partial(nG)}{\partial T}\right]_{P, n} dT + \sum_i \left[\frac{\partial(nG)}{\partial n_i}\right]_{P, T, n_j} dn_i$$

The summation is over all species present, and subscript n_j indicates that all mole numbers except the ith are held constant. The derivative in the final term is given its own symbol and name. Thus, by **definition** the *chemical potential* of species i in the mixture is:

$$\mu_i \equiv \left[\frac{\partial(nG)}{\partial n_i}\right]_{P, T, n_j} \tag{10.1}$$

With this definition and with the first two partial derivatives replaced by (nV) and $-(nS)$, the preceding equation becomes:

$$d(nG) = (nV)dP - (nS)dT + \sum_i \mu_i \, dn_i \tag{10.2}$$

Equation (10.2) is the fundamental property relation for single-phase fluid systems of variable mass and composition. It is the foundation upon which the structure of solution thermodynamics is built. For the special case of one mole of solution, $n = 1$ and $n_i = x_i$:

$$dG = VdP - SdT + \sum_i \mu_i dx_i \tag{10.3}$$

Implicit in this equation is the functional relationship of the molar Gibbs energy to its *canonical* variables, here: T, P, and $\{x_i\}$:

$$G = G(T, P, x_1, x_2, \ldots, x_i, \ldots)$$

Equation (6.11) for a constant-composition solution is a special case of Eq. (10.3). Although the mole numbers n_i of Eq. (10.2) are independent variables, the mole fractions x_i in Eq. (10.3) are not, because $\sum_i x_i = 1$. This precludes certain mathematical operations which depend upon independence of the variables. Nevertheless, Eq. (10.3) does imply:

$$V = \left(\frac{\partial G}{\partial P}\right)_{T, x} \tag{10.4} \qquad S = -\left(\frac{\partial G}{\partial T}\right)_{P, x} \tag{10.5}$$

Other solution properties come from definitions; e.g., the enthalpy, from $H = G + TS$. Thus, by Eq. (10.5),

$$H = G - T\left(\frac{\partial G}{\partial T}\right)_{P, x}$$

When the Gibbs energy is expressed as a function of its canonical variables, it serves as a *generating function,* providing the means for the calculation of all other thermodynamic properties by simple mathematical operations (differentiation and elementary algebra), and it implicitly represents *complete* property information.

This is a more general statement of the conclusion drawn in Sec. 6.1, now extended to systems of variable composition.

10.2 THE CHEMICAL POTENTIAL AND EQUILIBRIUM

Practical applications of the chemical potential will become clearer in later chapters that treat chemical and phase equilibria. However, at this point one can already appreciate its role in these analyses. For a *closed*, single-phase *PVT* system containing chemically reactive species, Eqs. (6.7) and (10.2) must both be valid, the former simply because the system is closed and the second because of its generality. In addition, for a closed system, all differentials dn_i in Eq. (10.2) must result from chemical reaction. Comparison of these two equations shows that they can both be valid only if:

$$\sum_i \mu_i \, dn_i = 0$$

This equation therefore represents a general criterion for chemical-reaction equilibrium in a single-phase closed *PVT* system, and is the basis for the development of working equations for the solution of reaction-equilibrium problems.

With respect to phase equilibrium, we note that for a *closed* nonreacting system consisting of two phases in equilibrium, each individual phase is *open* to the other, and mass transfer between phases may occur. Equation (10.2) applies separately to each phase:

$$d(nG)^\alpha = (nV)^\alpha \, dP - (nS)^\alpha \, dT + \sum_i \mu_i^\alpha \, dn_i^\alpha$$

$$d(nG)^\beta = (nV)^\beta \, dP - (nS)^\beta \, dT + \sum_i \mu_i^\beta \, dn_i^\beta$$

where superscripts α and β identify the phases. For the system to be in thermal and mechanical equilibrium, T and P must be uniform.

The change in the total Gibbs energy of the two-phase system is the sum of the equations for the separate phases. When each total-system property is expressed by an equation of the form,

$$nM = (nM)^\alpha + (nM)^\beta$$

the sum is: $\quad d(nG) = (nV)dP - (nS)dT + \sum_i \mu_i^\alpha dn_i^\alpha + \sum_i \mu_i^\beta dn_i^\beta$

Because the two-phase system is closed, Eq. (6.7) is also valid. Comparison of the two equations shows that at equilibrium:

$$\sum_i \mu_i^\alpha dn_i^\alpha + \sum_i \mu_i^\beta dn_i^\beta = 0$$

The changes dn_i^α and dn_i^β result from mass transfer between the phases; mass conservation therefore requires:

$$dn_i^\alpha = -dn_i^\beta \qquad \text{and} \qquad \sum_i (\mu_i^\alpha - \mu_i^\beta) dn_i^\alpha = 0$$

Quantities dn_i^α are independent and arbitrary, and the only way the left side of the second equation can in general be zero is for each term in parentheses separately to be zero. Hence,

$$\mu_i^\alpha = \mu_i^\beta \qquad (i = 1, 2, \ldots, N)$$

where N is the number of species present in the system. Successive application of this result to pairs of phases permits its generalization to multiple phases; for π phases:

$$\boxed{\mu_i^\alpha = \mu_i^\beta = \cdots = \mu_i^\pi} \qquad (i = 1, 2, \ldots, N) \qquad (10.6)$$

A similar but more comprehensive derivation shows (as we have supposed) that for equilibrium T and P must be the same in all phases.

Thus, multiple phases at the same *T* and *P* are in equilibrium when the chemical potential of each species is the same in all phases.

The application of Eq. (10.6) in later chapters to specific phase-equilibrium problems requires *models* of solution behavior, which provide expressions for G and μ_i as functions of temperature, pressure, and composition. The simplest of these, the ideal-gas state mixture and the ideal solution, are treated in Secs. 10.4 and 10.8, respectively.

10.3 PARTIAL PROPERTIES

The definition of the chemical potential by Eq. (10.1) as the mole-number derivative of nG suggests that other derivatives of this kind may prove useful in solution thermodynamics. Thus, we **define** the partial molar property $\bar{M}_i$ of species i in solution as:

$$\boxed{\bar{M}_i \equiv \left[\frac{\partial(nM)}{\partial n_i} \right]_{P, T, n_j}} \qquad (10.7)$$

Sometimes called a *response function,* it is a *measure* of the response of total property nM to the addition of an infinitesimal amount of species i to a finite amount of solution, at constant T and P.

The generic symbols M and $\bar{M}_i$ may express solution properties on a unit-mass basis as well as on a molar basis. Equation (10.7) retains the same form, with n, the number of moles, replaced by m, representing mass, and yielding partial *specific* properties rather than partial *molar* properties. To accommodate either, one may speak simply of partial properties.

Interest here centers on solutions, for which molar (or unit-mass) properties are represented by the plain symbol M. Partial properties are denoted by an overbar, with a subscript to identify the species; the symbol is therefore $\bar{M}_i$. In addition, properties of the individual species as they exist in the *pure state at the T and P of the solution* are identified by only a subscript, and the symbol is M_i. In summary, the three kinds of properties used in solution thermodynamics are distinguished using the following notation:

Solution properties	M,	for example: V, U, H, S, G
Partial properties	$\bar{M}_i$,	for example: $\bar{V}_i, \bar{U}_i, \bar{H}_i, \bar{S}_i, \bar{G}_i$
Pure-species properties	M_i,	for example: V_i, U_i, H_i, S_i, G_i

Comparison of Eq. (10.1) with Eq. (10.7) written for the Gibbs energy shows that the chemical potential and the partial molar Gibbs energy are identical; i.e.,

$$\mu_i \equiv \bar{G}_i \tag{10.8}$$

Example 10.1

The partial molar volume is defined as:

$$\bar{V}_i \equiv \left[\frac{\partial(nV)}{\partial n_i}\right]_{P,\,T,\,n_j} \tag{A}$$

What physical interpretation can be given to this equation?

Solution 10.1

Suppose an open beaker containing an equimolar mixture of ethanol and water occupies a total volume nV at room temperature T and atmospheric pressure P. Add to this solution a drop of pure water, also at T and P, containing Δn_w moles, and mix it thoroughly into the solution, allowing sufficient time for heat exchange to return the contents of the beaker to the initial temperature. One might expect that the volume of solution increases by an amount equal to the volume of the water added, i.e., by $V_w \Delta n_w$, where V_w is the molar volume of pure water at T and P. If this were true, the total volume change would be:

$$\Delta(nV) = V_w \Delta n_w$$

However, experimental observations show that the actual volume change is somewhat less. Evidently, the *effective* molar volume of water in the final solution is less than the molar volume of pure water at the same T and P. We may therefore write:

$$\Delta(nV) = \tilde{V}_w \Delta n_w \tag{B}$$

where $\tilde{V}_w$ represents the effective molar volume of water in the final solution. Its experimental value is given by:

$$\tilde{V}_w = \frac{\Delta(nV)}{\Delta n_w} \tag{C}$$

In the process described, a drop of water is mixed with a substantial amount of solution, and the result is a small but measurable change in composition of the solution.

For the effective molar volume of the water to be considered a property of the original equimolar solution, the process must be taken to the limit of an infinitesimal drop. Whence, $\Delta n_w \to 0$, and Eq. (*C*) becomes:

$$\tilde{V}_w = \lim_{\Delta n_w \to 0} \frac{\Delta(nV)}{\Delta n_w} = \frac{d(nV)}{dn_w}$$

Because T, P, and n_a (the number of moles of alcohol) are constant, this equation is more appropriately written:

$$\tilde{V}_w = \left[\frac{\partial(nV)}{\partial n_w}\right]_{P, T, n_a}$$

Comparison with Eq. (*A*) shows that in this limit $\tilde{V}_w$ is the partial molar volume $\bar{V}_w$ of the water in the equimolar solution, i.e., the rate of change of the total solution volume with n_w at constant T, P, and n_a for a specific composition. Written for the addition of dn_w moles of water to the solution, Eq. (*B*) is then:

$$d(nV) = \bar{V}_w dn_w \qquad\qquad (D)$$

When $\bar{V}_w$ is considered the molar property of water as it exists in solution, the total volume change $d(nV)$ is merely this molar property multiplied by the number of moles dn_w of water added.

    If dn_w moles of water is added to a volume of *pure* water, then we have every reason to expect the volume change of the system to be:

$$d(nV) = V_w dn_w \qquad\qquad (E)$$

where V_w is the molar volume of pure water at T and P. Comparison of Eqs. (*D*) and (*E*) indicates that $\bar{V}_w = V_w$ when the "solution" is pure water.

Equations Relating Molar and Partial Molar Properties

The definition of a partial molar property, Eq. (10.7), provides the means for calculation of partial properties from solution-property data. Implicit in this definition is another, equally important, equation that allows the reverse, i.e., calculation of solution properties from knowledge of the partial properties. The derivation of this equation starts with the observation that the total thermodynamic properties of a homogeneous phase are functions of T, P, and the numbers of moles of the individual species that comprise the phase.[1] Thus for property M, we can write nM as a function which we could call $\mathbb{M}$:

$$nM = \mathbb{M}(T, P, n_1, n_2, \ldots, n_i, \ldots)$$

The total differential of nM is:

$$d(nM) = \left[\frac{\partial(nM)}{\partial P}\right]_{T, n} dP + \left[\frac{\partial(nM)}{\partial T}\right]_{P, n} dT + \sum_i \left[\frac{\partial(nM)}{\partial n_i}\right]_{P, T, n_j} dn_i$$

---

[1]Mere functionality does not make a set of variables into *canonical* variables. These are the canonical variables only for $M \equiv G$.

where subscript n indicates that *all* mole numbers are held constant, and subscript n_j that all mole numbers *except* n_i are held constant. Because the first two partial derivatives on the right are evaluated at constant n and because the partial derivative of the last term is given by Eq. (10.7), this equation has the simpler form:

$$d(nM) = n\left(\frac{\partial M}{\partial P}\right)_{T,x} dP + n\left(\frac{\partial M}{\partial T}\right)_{P,x} dT + \sum_i \bar{M}_i \, dn_i \qquad (10.9)$$

where subscript x denotes differentiation at constant composition. Because $n_i = x_i n$,

$$dn_i = x_i \, dn + n \, dx_i$$

Moreover,

$$d(nM) = n \, dM + M \, dn$$

When dn_i and $d(nM)$ are replaced in Eq. (10.9), it becomes:

$$n \, dM + M \, dn = n\left(\frac{\partial M}{\partial P}\right)_{T,x} dP + n\left(\frac{\partial M}{\partial T}\right)_{P,x} dT + \sum_i \bar{M}_i(x_i \, dn + n \, dx_i)$$

The terms containing n are collected and separated from those containing dn to yield:

$$\left[dM - \left(\frac{\partial M}{\partial P}\right)_{T,x} dP - \left(\frac{\partial M}{\partial T}\right)_{P,x} dT - \sum_i \bar{M}_i dx_i\right] n + \left[M - \sum_i x_i \bar{M}_i\right] dn = 0$$

In application, one is free to choose a system of any size, as represented by n, and to choose any variation in its size, as represented by dn. Thus n and dn are independent and arbitrary. The only way that the left side of this equation can then, in general, be zero is for *each* term in brackets to be zero. Therefore,

$$dM = \left(\frac{\partial M}{\partial P}\right)_{T,x} dP + \left(\frac{\partial M}{\partial T}\right)_{P,x} dT + \sum_i \bar{M}_i dx_i \qquad (10.10)$$

and

$$\boxed{M = \sum_i x_i \bar{M}_i} \qquad (10.11)$$

Multiplication of Eq. (10.11) by n yields the alternative expression:

$$\boxed{nM = \sum_i n_i \bar{M}_i} \qquad (10.12)$$

Equation (10.10) is in fact just a special case of Eq. (10.9), obtained by setting $n = 1$, which also makes $n_i = x_i$. Equations (10.11) and (10.12) on the other hand are new and vital. Known as *summability relations*, they allow the calculation of mixture properties from partial properties, playing a role opposite to that of Eq. (10.7), which provides for the calculation of partial properties from mixture properties.

One further important equation follows directly from Eqs. (10.10) and (10.11). Differentiation of Eq. (10.11), a general expression for *M*, yields a general expression for *dM*:

$$dM = \sum_i x_i d\bar{M}_i + \sum_i \bar{M}_i dx_i$$

Combining this equation with Eq. (10.10) yields the *Gibbs/Duhem*[2] *equation*:

$$\left(\frac{\partial M}{\partial P}\right)_{T,x} dP + \left(\frac{\partial M}{\partial T}\right)_{P,x} dT - \sum_i x_i \, d\bar{M}_i = 0 \qquad (10.13)$$

This equation must be satisfied for all changes occurring in a homogeneous phase. For the important special case of changes in composition at constant *T* and *P*, it simplifies to:

$$\sum_i x_i d\bar{M}_i = 0 \qquad \text{(const } T, P) \qquad (10.14)$$

Eq. 10.14 shows that the partial molar properties cannot all vary independently. This constraint is analogous to the constraint on mole fractions, which are not all independent because they must sum to one. Similarly, the mole-fraction-weighted sum of the partial molar properties must yield the overall solution property (Eq. 10.11), and this constrains the variation in partial molar properties with composition.

A Rationale for Partial Properties

Central to applied solution thermodynamics, the partial-property concept implies that a solution property represents a "whole," i.e., the sum of its parts as represented by partial properties $\bar{M}_i$ of the constituent species. This is the implication of Eq. (10.11), and it is a proper interpretation provided one understands that the defining equation for $\bar{M}_i$ Eq. (10.7) is an apportioning formula which *arbitrarily* assigns to each species *i* its share of the solution property.[3]

The constituents of a solution are in fact intimately intermixed, and owing to molecular interactions they cannot have private properties of their own. Nevertheless, partial properties, as defined by Eq. (10.7), have all the characteristics of properties of the individual species as they exist in solution. Thus for practical purposes they may be *assigned* as property values to the individual species.

Partial properties, like solution properties, are functions of composition. In the limit as a solution becomes pure in species *i*, both *M* and $\bar{M}_i$ approach the pure-species property M_i. Mathematically,

$$\lim_{x_i \to 1} M = \lim_{x_i \to 1} \bar{M}_i = M_i$$

---

[2]Pierre-Maurice-Marie Duhem (1861–1916), French physicist. See http://en.wikipedia.org/wiki/Pierre Duhem.

[3]Other apportioning equations, which make different allocations of the solution property, are possible and are, in principle, equally valid.

For a partial property of a species that approaches its infinite-dilution limit, i.e., a partial property value of a species as its mole fraction approaches zero, we can make no general statements. Values come from experiment or from models of solution behavior. Because it is an important quantity, we do give it a symbol, and by definition we write:

$$\bar{M}_i^\infty \equiv \lim_{x_i \to 0} \bar{M}_i$$

The essential equations of this section are thus summarized as follows:

Definition:
$$\bar{M}_i \equiv \left[\frac{\partial(nM)}{\partial n_i} \right]_{P, T, n_j} \tag{10.7}$$

which yields partial properties from total properties.

Summability:
$$M = \sum_i x_i \bar{M}_i \tag{10.11}$$

which yields total properties from partial properties.

Gibbs/Duhem:
$$\sum_i x_i d\bar{M}_i = \left(\frac{\partial M}{\partial P} \right)_{T, x} dP + \left(\frac{\partial M}{\partial T} \right)_{P, x} dT \tag{10.13}$$

which shows that the partial properties of species making up a solution are *not* independent of one another.

Partial Properties in Binary Solutions

An equation for a partial property as a function of composition can always be derived from an equation for the solution property by direct application of Eq. (10.7). For binary systems, however, an alternative procedure may be more convenient. Written for a binary solution, the summability relation, Eq. (10.11), becomes:

$$M = x_1 \bar{M}_1 + x_2 \bar{M}_2 \tag{A}$$

Whence,
$$dM = x_1 d\bar{M}_1 + \bar{M}_1 dx_1 + x_2 d\bar{M}_2 + \bar{M}_2 dx_2 \tag{B}$$

When M is known as a function of x_1 at constant T and P, the appropriate form of the Gibbs/Duhem equation is Eq. (10.14), expressed here as:

$$x_1 \, d\bar{M}_1 + x_2 \, d\bar{M}_2 = 0 \tag{C}$$

Because $x_1 + x_2 = 1$, it follows that $dx_1 = -dx_2$. Eliminating dx_2 in favor of dx_1 in Eq. (B) and combining the result with Eq. (C) gives:

$$\frac{dM}{dx_1} = \bar{M}_1 - \bar{M}_2 \tag{D}$$

Two equivalent forms of Eq. (A) result from the elimination separately of x_1 and x_2:

$$M = \bar{M}_1 - x_2(\bar{M}_1 - \bar{M}_2) \quad \text{and} \quad M = x_1(\bar{M}_1 - \bar{M}_2) + \bar{M}_2$$

In combination with Eq. (*D*) these become:

$$\bar{M}_1 = M + x_2 \frac{dM}{dx_1} \quad (10.15) \qquad \bar{M}_2 = M - x_1 \frac{dM}{dx_1} \quad (10.16)$$

Thus for binary systems, the partial properties are readily calculated directly from an expression for the solution property as a function of composition at constant *T* and *P*. The corresponding equations for multicomponent systems are much more complex. They are given in detail by Van Ness and Abbott.[4]

Equation (*C*), the Gibbs/Duhem equation, may be written in derivative forms:

$$x_1 \frac{d\bar{M}_1}{dx_1} + x_2 \frac{d\bar{M}_2}{dx_1} = 0 \quad (E) \qquad \frac{d\bar{M}_1}{dx_1} = -\frac{x_2}{x_1} \frac{d\bar{M}_2}{dx_1} \quad (F)$$

Clearly, when $\bar{M}_1$ and $\bar{M}_2$ are plotted vs. x_1, the slopes must be of opposite sign. Moreover,

$$\lim_{x_1 \to 1} \frac{d\bar{M}_1}{dx_1} = 0 \quad (\text{Provided } \lim_{x_1 \to 1} \frac{d\bar{M}_2}{dx_1} \text{ is finite})$$

Similarly,

$$\lim_{x_2 \to 1} \frac{d\bar{M}_2}{dx_1} = 0 \quad (\text{Provided } \lim_{x_2 \to 1} \frac{d\bar{M}_1}{dx_1} \text{ is finite})$$

Thus, plots of $\bar{M}_1$ and $\bar{M}_2$ vs. x_1 become horizontal as each species approaches purity.

Finally, given an expression for $\bar{M}_1(x_1)$, integration of Eq. (*E*) or Eq. (*F*) yields an expression for $\bar{M}_2(x_1)$ that satisfies the Gibbs/Duhem equation. This means that expressions cannot be specified independently for both $\bar{M}_1(x_1)$ and $\bar{M}_2(x_1)$.

Example 10.2

Describe a graphical interpretation of Eqs. (10.15) and (10.16).

Solution 10.2

Figure 10.1(*a*) shows a representative plot of *M* vs. x_1 for a binary system. The tangent line shown extends across the figure, intersecting the edges (at $x_1 = 1$ and $x_1 = 0$) at points labeled I_1 and I_2. As is evident from the figure, two equivalent expressions can be written for the slope of this tangent line:

$$\frac{dM}{dx_1} = \frac{M - I_2}{x_1} \qquad \text{and} \qquad \frac{dM}{dx_1} = I_1 - I_2$$

---

[4] H. C. Van Ness and M. M. Abbott, *Classical Thermodynamics of Nonelectrolyte Solutions: With Applications to Phase Equilibria*, pp. 46–54, McGraw-Hill, New York, 1982.

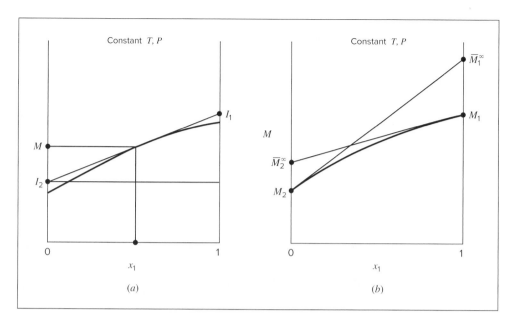

Figure 10.1: (*a*) Graphical construction of Example 10.2. (*b*) Infinite-dilution values of partial properties.

The first equation is solved for I_2; it combines with the second to give I_1:

$$I_2 = M - x_1\frac{dM}{dx_1} \quad \text{and} \quad I_1 = M + (1 - x_1)\frac{dM}{dx_1}$$

Comparisons of these expressions with Eqs. (10.16) and (10.15) show that:

$$I_1 = \bar{M}_1 \quad \text{and} \quad I_2 = \bar{M}_2$$

Thus, the tangent intercepts give directly the values of the two partial properties. These intercepts of course shift as the point of tangency moves along the curve, and the limiting values are indicated by the constructions shown in Fig. 10.1(*b*). For the tangent line drawn at $x_1 = 0$ (pure species 2), $\bar{M}_2 = M_2$, and at the opposite intercept, $\bar{M}_1 = \bar{M}_1^\infty$. Similar comments apply to the tangent drawn at $x_1 = 1$ (pure species 1). In this case $\bar{M}_1 = M_1$ and $\bar{M}_2 = \bar{M}_2^\infty$.

Example 10.3

The need arises in a laboratory for 2000 cm³ of an antifreeze solution consisting of 30 mol-% methanol in water. What volumes of pure methanol and of pure water at 25°C must be mixed to form the 2000 cm³ of antifreeze, also at 25°C? Partial molar volumes for methanol and water in a 30 mol-% methanol solution and their pure-species molar volumes, both at 25°C, are:

Methanol(1) : $\bar{V}_1 = 38.632$ cm$^3 \cdot$mol^{-1} $V_1 = 40.727$ cm$^3 \cdot$mol^{-1}
Water(2) : $\bar{V}_2 = 17.765$ cm$^3 \cdot$mol^{-1} $V_2 = 18.068$ cm$^3 \cdot$mol^{-1}

Solution 10.3

The summability relation, Eq. (10.11), is written for the molar volume of the binary antifreeze solution, and known values are substituted for the mole fractions and partial molar volumes:

$$V = x_1 \bar{V}_1 + x_2 \bar{V}_2 = (0.3)(38.632) + (0.7)(17.765) = 24.025 \text{ cm}^3 \cdot \text{mol}^{-1}$$

Because the required total volume of solution is $V^t = 2000$ cm^3, the total number of moles required is:

$$n = \frac{V^t}{V} = \frac{2000}{24.025} = 83.246 \text{ mol}$$

Of this, 30% is methanol, and 70% is water:

$$n_1 = (0.3)(83.246) = 24.974 \qquad n_2 = (0.7)(83.246) = 58.272 \text{ mol}$$

The volume of each pure species is $V_1^t = n_i V_i$; thus,

$$V_1^t = (24.974)(40.727) = 1017 \text{ cm}^3 \qquad V_2^t = (58.272)(18.068) = 1053 \text{ cm}^3$$

Example 10.4

The enthalpy of a binary liquid system of species 1 and 2 at fixed T and P is represented by the equation:

$$H = 400x_1 + 600x_2 + x_1 x_2 (40x_1 + 20x_2)$$

where H is in J$\cdot$mol^{-1}. Determine expressions for $\bar{H}_1$ and $\bar{H}_2$ as functions of x_1, numerical values for the pure-species enthalpies H_1 and H_2, and numerical values for the partial enthalpies at infinite dilution $\bar{H}_1^\infty$ and $\bar{H}_2^\infty$.

Solution 10.4

Replacing x_2 by $1 - x_1$ in the given equation for H and simplifying gives:

$$H = 600 - 180x_1 - 20x_1^3 \qquad\qquad (A)$$

and

$$\frac{dH}{dx_1} = -180 - 60x_1^2$$

By Eq. (10.15),

$$\bar{H}_1 = H + x_2 \frac{dH}{dx_1}$$

Then,

$$\bar{H}_1 = 600 - 180x_1 - 20x_1^3 - 180x_2 - 60x_1^2 x_2$$

Replacing x_2 with $1 - x_1$ and simplifying:

$$\bar{H}_1 = 420 - 60x_1^2 + 40x_1^3 \tag{B}$$

By Eq. (10.16),

$$\bar{H}_2 = H - x_1 \frac{dH}{dx_1} = 600 - 180x_1 - 20x_1^3 + 180x_1 + 60x_1^3$$

or

$$\bar{H}_2 = 600 + 40x_1^3 \tag{C}$$

One can equally well start with the given equation for H. Because dH/dx_1 is a *total* derivative, x_2 is not a constant. Also, $x_2 = 1 - x_1$; therefore $dx_2/dx_1 = -1$. Differentiation of the given equation for H therefore yields:

$$\frac{dH}{dx_1} = 400 - 600 + x_1 x_2 (40 - 20) + (40x_1 + 20x_2)(-x_1 + x_2)$$

Replacing x_2 with $1 - x_1$ reproduces the expression previously obtained.

A numerical value for H_1 results by substitution of $x_1 = 1$ in either Eq. (A) or (B). Both equations yield $H_1 = 400$ J·mol⁻¹. Similarly, H_2 is found from either Eq. (A) or (C) when $x_1 = 0$. The result is $H_2 = 600$ J·mol⁻¹. The infinite-dilution values $\bar{H}_1^\infty$ and $\bar{H}_2^\infty$ are found from Eqs. (B) and (C) when $x_1 = 0$ in Eq. (B) and $x_1 = 1$ in Eq. (C). The results are: $\bar{H}_1^\infty = 420$ J·mol⁻¹ and $\bar{H}_2^\infty = 640$ J·mol⁻¹

Exercise: Show that the partial properties as given by Eqs. (B) and (C) combine by summability to give Eq. (A) and that they conform to all requirements of the Gibbs/Duhem equation.

Relations Among Partial Properties

We now derive several additional useful relationships among partial properties. By Eq. (10.8), $\mu_i \equiv \bar{G}_i$, and Eq. (10.2) may be written:

$$d(nG) = (nV)dP - (nS)dT + \sum_i \bar{G}_i dn_i \tag{10.17}$$

Application of the criterion of exactness, Eq. (6.13), yields the Maxwell relation,

$$\left(\frac{\partial V}{\partial T} \right)_{P,\, n} = -\left(\frac{\partial S}{\partial P} \right)_{T,\, n} \tag{6.17}$$

plus the two additional equations:

$$\left(\frac{\partial \bar{G}_i}{\partial P} \right)_{T,\, n} = \left[\frac{\partial (nV)}{\partial n_i} \right]_{P,\, T,\, n_j} \qquad \left(\frac{\partial \bar{G}_i}{\partial T} \right)_{P,\, n} = -\left[\frac{\partial (nS)}{\partial n_i} \right]_{P,\, T,\, n_j}$$

where subscript n indicates constancy of all n_i, and therefore of composition, and subscript n_j indicates that all mole numbers except the ith are held constant. We recognize the terms on the right-hand side of these equations as the partial volume and partial entropy, and thus we can rewrite them more simply as:

$$\left(\frac{\partial \bar{G}_i}{\partial P}\right)_{T,x} = \bar{V}_i \quad (10.18) \qquad \left(\frac{\partial \bar{G}_i}{\partial T}\right)_{P,x} = -\bar{S}_i \quad (10.19)$$

These equations allow us to calculate the effects of P and T on the partial Gibbs energy (or chemical potential). They are the partial-property analogs of Eqs. (10.4) and (10.5). Many additional relationships among partial properties can be derived in the same ways that relationships among pure species properties were derived in earlier chapters. More generally, one can prove the following:

> **Every equation that provides a *linear* relation among thermodynamic properties of a *constant-composition* solution has as its counterpart an equation connecting the corresponding partial properties of each species in the solution.**

An example is based on the equation that defines enthalpy: $H = U + PV$. For n moles,

$$nH = nU + P(nV)$$

Differentiation with respect to n_i at constant T, P, and n_j yields:

$$\left[\frac{\partial(nH)}{\partial n_i}\right]_{P,T,n_j} = \left[\frac{\partial(nU)}{\partial n_i}\right]_{P,T,n_j} + P\left[\frac{\partial(nV)}{\partial n_i}\right]_{P,T,n_j}$$

By the definition of partial properties, Eq. (10.7), this becomes:

$$\bar{H}_i = \bar{U}_i + P\bar{V}_i$$

which is the partial-property analog of Eq. (2.10).

In a constant-composition solution, $\bar{G}_i$ is a function of T and P, and therefore:

$$d\bar{G}_i = \left(\frac{\partial \bar{G}_i}{\partial T}\right)_{P,x} dT + \left(\frac{\partial \bar{G}_i}{\partial P}\right)_{T,x} dP$$

By Eqs. (10.18) and (10.19),

$$d\bar{G}_i = -\bar{S}_i dT + \bar{V}_i dP$$

This may be compared with Eq. (6.11). These examples illustrate the parallelism that exists between equations for a constant-composition solution and the corresponding equations for the partial properties of the species in solution. We can therefore write simply by analogy many equations that relate partial properties.

10.4   THE IDEAL-GAS-STATE MIXTURE MODEL

Despite its limited ability to describe actual mixture behavior, the ideal-gas-state mixture model provides a conceptual basis upon which to build the structure of solution thermodynamics. It is a useful property model because it:

- Has a molecular basis.
- Approximates reality in the well-defined limit of zero pressure.
- Is analytically simple.

At the molecular level, the ideal-gas state represents a collection of molecules that do not interact and occupy no volume. This idealization is approached for real molecules in the limit of zero pressure (which implies zero density) because both the energies of intermolecular interactions and the volume fraction occupied by the molecules go to zero with increasing separation of the molecules. Although they do not interact with one another, molecules in the ideal-gas state do have internal *structure;* it is differences in molecular structure that give rise to differences in ideal-gas-state heat capacities (Sec. 4.1), enthalpies, entropies, and other properties.

Molar volumes in the ideal-gas state are $V^{ig} = RT/P$ [Eq. (3.7)] regardless of the nature of the gas. Thus for the ideal-gas state, whether of pure or mixed gases, the molar volume is the same for given T and P. The partial molar volume of species i in the ideal-gas-state mixture is found from Eq. (10.7) applied to the volume; superscript ig denotes the ideal-gas state:

$$\bar{V}_i^{ig} = \left[\frac{\partial(n V^{ig})}{\partial n_i}\right]_{T, P, n_j} = \left[\frac{\partial(nRT/P)}{\partial n_i}\right]_{T, P, n_j} = \frac{RT}{P}\left(\frac{\partial n}{\partial n_i}\right)_{n_j} = \frac{RT}{P}$$

where the final equality depends on the equation $n = n_i + \sum_j n_j$. This result means that for the ideal-gas state at given T and P the partial molar volume, the pure-species molar volume, and the mixture molar volume are identical:

$$\bar{V}_i^{ig} = V_i^{ig} = V^{ig} = \frac{RT}{P} \tag{10.20}$$

We **define** the *partial pressure* of species i in the ideal-gas-state mixture (p_i) as the pressure that species i would exert if it alone occupied the molar volume of the mixture. Thus,[5]

$$p_i \equiv \frac{y_i RT}{V^{ig}} = y_i P \quad (i = 1, 2, \ldots, N)$$

where y_i is the mole fraction of species i. The partial pressures obviously sum to the total pressure.

Because the ideal-gas-state mixture model presumes molecules of zero volume that do not interact, the thermodynamic properties (other than molar volume) of the constituent

---

[5]Note that this definition does *not* make the partial pressure a partial molar property.

species are independent of one another, and each species has its own set of private properties. This is the basis for the following statement of *Gibbs's theorem*:

A partial molar property (other than volume) of a constituent species in an ideal-gas-state mixture is equal to the corresponding molar property of the species in the pure ideal-gas state at the mixture temperature but at a pressure equal to its partial pressure in the mixture.

This is expressed mathematically for generic partial property $\bar{M}_i^{ig} \neq \bar{V}_i^{ig}$ by the equation:

$$\bar{M}_i^{ig}(T, P) = M_i^{ig}(T, p_i) \tag{10.21}$$

Enthalpy in the ideal-gas state is independent of pressure; therefore

$$\bar{H}_i^{ig}(T, P) = H_i^{ig}(T, p_i) = H_i^{ig}(T, P)$$

More simply,

$$\bar{H}_i^{ig} = H_i^{ig} \tag{10.22}$$

where H_i^{ig} is the pure-species value at the *mixture T.* An analogous equation applies for U^{ig} and other properties that are *independent of pressure.*

Entropy in the ideal-gas state *does* depend on pressure, as expressed by Eq. (6.24), restricted to constant temperature:

$$dS_i^{ig} = -Rd \ln P \qquad (\text{const } T)$$

This provides the basis for computing the entropy difference between a gas at its partial pressure in the mixture and at the total pressure of the mixture. Integration from p_i to P gives:

$$S_i^{ig}(T, P) - S_i^{ig}(T, p_i) = -R \ln \frac{P}{p_i} = -R \ln \frac{P}{y_i P} = R \ln y_i$$

Whence,

$$S_i^{ig}(T, p_i) = S_i^{ig}(T, P) - R \ln y_i$$

Comparing this with Eq. (10.21), written for the entropy, yields:

$$\bar{S}_i^{ig}(T, P) = S_i^{ig}(T, P) - R \ln y_i$$

or

$$\bar{S}_i^{ig} = S_i^{ig} - R \ln y_i \tag{10.23}$$

where S_i^{ig} is the pure-species value at the mixture T and P.

For the Gibbs energy in the ideal-gas-state mixture, $G^{ig} = H^{ig} - TS^{ig}$; the parallel relation for partial properties is:

$$\bar{G}_i^{ig} = \bar{H}_i^{ig} - T\bar{S}_i^{ig}$$

In combination with Eqs. (10.22) and (10.23) this becomes:

$$\bar{G}_i^{ig} = H_i^{ig} - TS_i^{ig} + RT \ln y_i$$

or

$$\mu_i^{ig} \equiv \bar{G}_i^{ig} = G_i^{ig} + RT \ln y_i \tag{10.24}$$

Differentiation of this equation in accord with Eqs. (10.18) and (10.19) confirms the results expressed by Eqs. (10.20) and (10.23).

The summability relation, Eq. (10.11), with Eqs. (10.22), (10.23), and (10.24) yields:

$$\boxed{H^{ig} = \sum_i y_i H_i^{ig}} \tag{10.25}$$

$$\boxed{S^{ig} = \sum_i y_i S_i^{ig} - R \sum_i y_i \ln y_i} \tag{10.26}$$

$$\boxed{G^{ig} = \sum_i y_i G_i^{ig} + RT \sum_i y_i \ln y_i} \tag{10.27}$$

Equations analogous to Eq. (10.25) may be written for both C_P^{ig} and V^{ig}. The former appears as Eq. (4.7), but the latter reduces to an identity because of Eq. (10.20).

When Eq. (10.25) is written,

$$H^{ig} - \sum_i y_i H_i^{ig} = 0$$

the difference on the left is the enthalpy change associated with a process in which appropriate amounts of the pure species at T and P are mixed to form one mole of mixture at the same T and P. For the ideal-gas state, this *enthalpy change of mixing* is zero.

When Eq. (10.26) is rearranged as:

$$S^{ig} - \sum_i y_i S_i^{ig} = R \sum_i y_i \ln \frac{1}{y_i}$$

the left side is the *entropy change of mixing* for the ideal-gas state. Because $1/y_i > 1$, this quantity is always positive, in agreement with the second law. The mixing process is inherently irreversible, so the mixing process must increase the total entropy of the system and surroundings together. For ideal-gas-state mixing at constant T and P, using Eq. (10.25) with an energy balance shows that no heat transfer will occur between the system and surroundings. Therefore, the total entropy change of system plus surroundings is only the entropy change of mixing.

An alternative expression for the chemical potential μ_i^{ig} results when G_i^{ig} in Eq. (10.24) is replaced by an expression giving its T and P dependence. This comes from Eq. (6.11) written for the ideal-gas state at constant T:

$$dG_i^{ig} = V_i^{ig} dP = \frac{RT}{P} dP = RT \, d \ln P \quad (\text{const } T)$$

Integration gives:

$$G_i^{ig} = \Gamma_i(T) + RT \ln P \tag{10.28}$$

where $\Gamma_i(T)$, the integration constant at constant T, is a species-dependent function of temperature only.[6] Equation (10.24) is now written:

$$\mu_i^{ig} \equiv \bar{G}_i^{ig} = \Gamma_i(T) + RT \ln(y_i P) \qquad (10.29)$$

where the argument of the logarithm is the partial pressure. Application of the summability relation, Eq. (10.11), produces an expression for the Gibbs energy for the ideal-gas-state mixture:

$$G^{ig} \equiv \sum_i y_i \Gamma_i(T) + RT \sum_i y_i \ln(y_i P) \qquad (10.30)$$

These equations, remarkable in their simplicity, provide a full description of ideal-gas-state behavior. Because T, P, and $\{y_i\}$ are the canonical variables for the Gibbs energy, all other thermodynamic properties for the ideal-gas model can be generated from them.

10.5 FUGACITY AND FUGACITY COEFFICIENT: PURE SPECIES

As is evident from Eq. (10.6), the chemical potential μ_i provides the fundamental criterion for phase equilibrium. This is true as well for chemical-reaction equilibria. However, it exhibits characteristics that discourage its direct use. The Gibbs energy, and hence μ_i, is defined in relation to internal energy and entropy. Because absolute values of internal energy are unknown, the same is true for μ_i. Moreover, Eq. (10.29) shows that μ_i^{ig} approaches negative infinity when either P or y_i approaches zero. This is true not only for the ideal-gas state, but for any gas. Although these characteristics do not preclude the use of chemical potentials, the application of equilibrium criteria is facilitated by the introduction of the *fugacity*,[7] a property that takes the place of μ_i but does not exhibit its less desirable characteristics.

The origin of the fugacity concept resides in Eq. (10.28), valid only for pure species i in the ideal-gas state. For a real fluid, we write an analogous equation that **defines** f_i, the *fugacity* of pure species i:

$$G_i \equiv \Gamma_i(T) + RT \ln f_i \qquad (10.31)$$

This new property f_i, with units of pressure, replaces P in Eq. (10.28). Clearly, if Eq. (10.28) is viewed as a special case of Eq. (10.31), then:

$$f_i^{ig} = P \qquad (10.32)$$

---

[6]A dimensional ambiguity is evident with Eq. (10.28) and with analogous equations to follow in that P has units, whereas ln P must be dimensionless. This difficulty is more apparent than real, because the Gibbs energy is always expressed on a relative scale, absolute values being unknown. Thus in application only *differences* in Gibbs energy appear, leading to *ratios* of quantities with units of pressure in the argument of the logarithm. The only requirement is that consistency of pressure units be maintained.

[7]This quantity originated with Gilbert Newton Lewis (1875–1946), American physical chemist, who also developed the concepts of the partial property and the ideal solution. See http://en.wikipedia.org/wiki/Gilbert N. Lewis.

and the fugacity of pure species i in the ideal-gas state is necessarily equal to its pressure. Subtraction of Eq. (10.28) from Eq. (10.31), both written for the same T and P, gives:

$$G_i - G_i^{ig} = RT \ln \frac{f_i}{P}$$

By the definition of Eq. (6.41), $G_i - G_i^{ig}$ is the *residual Gibbs energy*, G_i^{ig}; thus,

$$G_i^R = RT \ln \frac{f_i}{P} = RT \ln \phi_i \qquad (10.33)$$

where the dimensionless ratio f_i/P has been **defined** as another new property, the *fugacity coefficient*, given by symbol ϕ_i:

$$\phi_i \equiv \frac{f_i}{P} \qquad (10.34)$$

These equations apply to pure species i in any phase at any condition. However, as a special case they must be valid for the ideal-gas state, for which $G_i^R = 0$, $\phi_i = 1$, and Eq. (10.28) is recovered from Eq. (10.31). Moreover, we may write Eq. (10.33) for $P = 0$ and combine it with Eq. (6.45):

$$\lim_{P \to 0} \left(\frac{G_i^R}{RT} \right) = \lim_{P \to 0} \ln \phi_i = J$$

As explained in connection with Eq. (6.48), the value of J is immaterial and is set equal to zero. Whence,

$$\lim_{P \to 0} \ln \phi_i = \lim_{P \to 0} \ln \left(\frac{f_i}{P} \right) = 0$$

and

$$\lim_{P \to 0} \phi_i = \lim_{P \to 0} \frac{f_i}{P} = 1$$

The identification of $\ln \phi_i$ with G_i^R/RT by Eq. (10.33) permits its evaluation by the integral of Eq. (6.49):

$$\ln \phi_i = \int_0^P (Z_i - 1) \frac{dP}{P} \qquad \text{(const } T) \qquad (10.35)$$

Fugacity coefficients (and therefore fugacities) for pure gases are evaluated by this equation from PVT data or from a volume-explicit equation of state.

For example, when the compressibility factor is given by Eq. (3.36), written here with subscripts to indicate that it is applied to a pure substance:

$$Z_i - 1 = \frac{B_{ii} P}{RT}$$

Because the second virial coefficient B_{ii} is a function of temperature only for a pure species, substitution into Eq. (10.35) gives:

$$\ln \phi_i = \frac{B_{ii}}{RT} \int_0^P dP \quad (\text{const } T)$$

and

$$\ln \phi_i = \frac{B_{ii} P}{RT} \tag{10.36}$$

Vapor/Liquid Equilibrium for Pure Species

Equation (10.31), which defines the fugacity of pure species i, may be written for species i as a saturated vapor and as a saturated liquid at the same temperature:

$$G_i^v = \Gamma_i(T) + RT \ln f_i^v \quad (10.37) \qquad G_i^l = \Gamma_i(T) + RT \ln f_i^l \quad (10.38)$$

By difference,

$$G_i^v - G_i^l = RT \ln \frac{f_i^v}{f_i^l}$$

This equation applies to the change of state from saturated liquid to saturated vapor, at temperature T and at the vapor pressure P_i^{sat}. According to Eq. (6.83), $G_i^v - G_i^l = 0$; therefore:

$$f_i^v = f_i^l = f_i^{\text{sat}} \tag{10.39}$$

where f_i^{sat} indicates the value for either saturated liquid or saturated vapor. Coexisting phases of saturated liquid and saturated vapor are in equilibrium; Eq. (10.39) therefore expresses a fundamental principle:

> **For a pure species, coexisting liquid and vapor phases are in equilibrium when they have the same temperature, pressure, and fugacity.[8]**

An alternative formulation is based on the corresponding fugacity coefficients:

$$\phi_i^{\text{sat}} = \frac{f_i^{\text{sat}}}{P_i^{\text{sat}}} \tag{10.40}$$

Whence,

$$\phi_i^v = \phi_i^l = \phi_i^{\text{sat}} \tag{10.41}$$

This equation, expressing the equality of fugacity coefficients, is an equally valid criterion of vapor/liquid equilibrium for pure species.

---

[8]The word *fugacity* is based on a Latin root meaning to flee or escape, also the basis for the word *fugitive*. Thus fugacity has been interpreted to mean "escaping tendency." When the escaping tendency is the same for the two phases, they are in equilibrium. When the escaping tendency of a species is higher in one phase than another, that species will tend to transfer to the phase where its fugacity is lower.

Fugacity of a Pure Liquid

The fugacity of pure species i as a compressed (subcooled) liquid may be calculated as the product of the saturation pressure with three ratios that are each relatively easy to evaluate:

$$f_i^l(P) = \underbrace{\frac{f_i^v(P_i^{\mathrm{sat}})}{P_i^{\mathrm{sat}}}}_{(A)} \underbrace{\frac{f_i^l(P_i^{\mathrm{sat}})}{f_i^v(P_i^{\mathrm{sat}})}}_{(B)} \underbrace{\frac{f_i^l(P)}{f_i^l(P_i^{\mathrm{sat}})}}_{(C)} P_i^{\mathrm{sat}}$$

All terms are at the temperature of interest. Inspection reveals that cancellation of numerators and denominators produces a mathematical identity.

Ratio (A) is the vapor-phase fugacity coefficient of pure vapor i at its vapor/liquid saturation pressure, designated ϕ_i^{sat}. It is given by Eq. (10.35), written,

$$\ln \phi_i^{\mathrm{sat}} = \int_0^{P_i^{\mathrm{sat}}} (Z_i^v - 1) \frac{dP}{P} \qquad (\text{const } T) \qquad (10.42)$$

As shown by Eq. (10.39), expressing the equality of liquid and vapor fugacities at equilibrium, ratio (B) is unity. Ratio (C) reflects the effect of pressure on the fugacity of pure liquid i. This effect is generally small. The basis for its calculation is Eq. (6.11), integrated at constant T to give:

$$G_i - G_i^{\mathrm{sat}} = \int_{P_i^{\mathrm{sat}}}^{P} V_i^l \, dP$$

Another expression for this difference results when Eq. (10.31) is written for both G_i and G_i^{sat}; subtraction then yields:

$$G_i - G_i^{\mathrm{sat}} = RT \ln \frac{f_i}{f_i^{\mathrm{sat}}}$$

The two expressions for $G_i - G_i^{\mathrm{sat}}$ are set equal:

$$\ln \frac{f_i}{f_i^{\mathrm{sat}}} = \frac{1}{RT} \int_{P_i^{\mathrm{sat}}}^{P} V_i^l \, dP$$

Ratio (C) is then:

$$\frac{f_i^l(P)}{f_i^l(P_i^{\mathrm{sat}})} = \exp\left(\frac{1}{RT} \int_{P_i^{\mathrm{sat}}}^{P} V_i^l \, dP \right)$$

Substituting for the three ratios in the initial equation yields:

$$f_i = \phi_i^{\mathrm{sat}} P_i^{\mathrm{sat}} \exp\left(\frac{1}{RT} \int_{P_i^{\mathrm{sat}}}^{P} V_i^l \, dP \right) \qquad (10.43)$$

Because V_i^l, the liquid-phase molar volume, is a very weak function of P at temperatures well below T_c, an excellent approximation is often obtained by taking V_i^l to be constant at the value for saturated liquid. In this case,

$$f_i = \phi_i^{\mathrm{sat}} P_i^{\mathrm{sat}} \exp \frac{V_i^l(P - P_i^{\mathrm{sat}})}{RT} \qquad (10.44)$$

The exponential is known as a Poynting[9] factor. To evaluate the fugacity of a compressed liquid from Eq. (10.44), the following data are required:

- Values of Z_i^v for calculation of ϕ_i^{sat} by Eq. (10.42). These may come from an equation of state, from experiment, or from a generalized correlation.
- The liquid-phase molar volume V_i^l, usually the value for saturated liquid.
- A value for P_i^{sat}.

If Z_i^v is given by Eq. (3.36), the simplest form of the virial equation, then:

$$Z_i^v - 1 = \frac{B_{ii}P}{RT} \quad \text{and} \quad \phi_i^{sat} = \exp\frac{B_{ii}P_i^{sat}}{RT}$$

and Eq. (10.44) becomes:

$$f_i = P_i^{sat}\exp\frac{B_{ii}P_i^{sat} + V_i^l(P - P_i^{sat})}{RT} \tag{10.45}$$

In the following example, data from the steam tables form the basis for calculation of the fugacity and fugacity coefficient of both vapor and liquid water as a function of pressure.

Example 10.5

For H_2O at a temperature of 300°C and for pressures up to 10,000 kPa (100 bar) calculate values of f_i and ϕ_i from data in the steam tables and plot them vs. P.

Solution 10.5

Equation (10.31) is written twice: first, for a state at pressure P; second, for a low-pressure reference state, denoted by $*$, both for temperature T:

$$G_i = \Gamma_i(T) + RT \ln f_i \quad \text{and} \quad G_i^* = \Gamma_i(T) + RT \ln f_i^*$$

Subtraction eliminates $\Gamma_i(T)$, and yields:

$$\ln\frac{f_i}{f_i^*} = \frac{1}{RT}(G_i - G_i^*)$$

By definition $G_i = H_i - TS_i$ and $G_i^* = H_i^* - TS_i^*$; substitution gives:

$$\ln\frac{f_i}{f_i^*} = \frac{1}{R}\left[\frac{H_i - H_i^*}{T} - (S_i - S_i^*)\right] \tag{A}$$

---

[9]John Henry Poynting (1852–1914), British physicist. See http://en.wikipedia.org/wiki/John_Henry_Poynting.

The lowest pressure for which data at 300°C are given in the steam tables is 1 kPa. Steam at these conditions is for practical purposes in its ideal-gas state, for which $f_i^* = P^* = 1$ kPa. Data for this state provide the following reference values:

$$H_i^* = 3076.8 \text{ J·g}^{-1} \quad S_i^* = 10.3450 \text{ J·g}^{-1}\text{·K}^{-1}$$

Equation (A) may now be applied to states of superheated steam at 300°C for various values of P from 1 kPa to the saturation pressure of 8592.7 kPa. For example, at $P = 4000$ kPa and 300°C:

$$H_i = 2962.0 \text{ J·g}^{-1} \quad S_i = 6.3642 \text{ J·g}^{-1}\text{·K}^{-1}$$

Values of H and S must be multiplied by the molar mass of water (18.015 g·mol^{-1}) to put them on a molar basis for substitution into Eq. (A):

$$\ln\frac{f_i}{f^*} = \frac{18.015}{8.314}\left[\frac{2962.0 - 3076.8}{573.15} - (6.3642 - 10.3450)\right] = 8.1917$$

and

$$f_i/f^* = 3611.0$$
$$f_i = (3611.0)(f^*) = (3611.0)(1 \text{ kPa}) = 3611.0 \text{ kPa}$$

Thus the fugacity coefficient at 4000 kPa is:

$$\phi_i = \frac{f_i}{P} = \frac{3611.0}{4000} = 0.9028$$

Similar calculations at other pressures lead to the values plotted in Fig. 10.2 at pressures up to the saturation pressure $P_i^{\text{sat}} = 8592.7$ kPa. At this pressure,

$$\phi_i^{\text{sat}} = 0.7843 \quad \text{and} \quad f_i^{\text{sat}} = 6738.9 \text{ kPa}$$

According to Eqs. (10.39) and (10.41), the saturation values are unchanged by condensation. Although the plots are therefore continuous, they do show discontinuities in slope. Values of f_i and ϕ_i for liquid water at higher pressures are found by application of Eq. (10.44), with V_i^l equal to the molar volume of saturated liquid water at 300°C:

$$V_i^l = (1.403)(18.015) = 25.28 \text{ cm}^3\text{·mol}^{-1}$$

At 10,000 kPa, for example, Eq. (10.44) becomes:

$$f_i = (0.7843)(8592.7) \exp\frac{(25.28)(10,000 - 8592.7)}{(8314)(573.15)} = 6789.8 \text{ kPa}$$

The fugacity coefficient of liquid water at these conditions is:

$$\phi_i = f_i/P = 6789.8/10,000 = 0.6790$$

Such calculations allow the completion of Fig. 10.2, where the solid lines show how f_i and ϕ_i vary with pressure.

The curve for f_i starts at the origin and deviates increasingly from the dashed line for the ideal-gas state ($f_i^{ig} = P$) as the pressure rises. At P_i^{sat} there is a

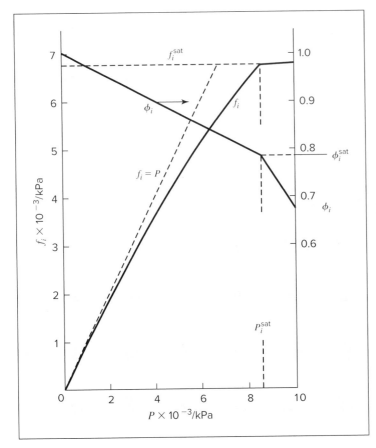

Figure 10.2: Fugacity and fugacity coefficient of steam at 300°C.

discontinuity in slope, and the curve then rises very slowly with increasing pressure, indicating that the fugacity of liquid water at 300°C is a weak function of pressure. This behavior is characteristic of a liquid at a temperature well below its critical temperature. The fugacity coefficient ϕ_i decreases steadily from its zero-pressure value of unity as the pressure rises. Its rapid decrease in the liquid region is a consequence of the near-constancy of the fugacity itself.

10.6 FUGACITY AND FUGACITY COEFFICIENT: SPECIES IN SOLUTION

The definition of the fugacity of a species in solution is parallel to the definition of the pure-species fugacity. For species i in a mixture of real gases or in a solution of liquids, the equation analogous to Eq. (10.29), the ideal-gas-state expression, is:

$$\mu_i \equiv \Gamma_i(T) + RT \ln \hat{f}_i \tag{10.46}$$

where $\hat{f}_i$ is the fugacity of species i in solution, replacing the partial pressure y_iP. This definition of $\hat{f}_i$ does not make it a partial molar property, and it is therefore identified by a circumflex rather than by an overbar.

A direct application of this definition indicates its potential utility. Equation (10.6), the equality of μ_i in every phase, is the fundamental criterion for phase equilibrium. Because all phases in equilibrium are at the same temperature, an alternative and equally general criterion follows immediately from Eq. (10.46):

$$\boxed{\hat{f}_i^{\alpha} = \hat{f}_i^{\beta} = \cdots = \hat{f}_i^{\pi}} \qquad (i = 1, 2, \ldots, N) \qquad (10.47)$$

Thus, multiple phases at the same *T* and *P* are in equilibrium when the fugacity of each constituent species is the same in all phases.

This is the criterion of equilibrium most often applied in the solution of phase-equilibrium problems.

For the specific case of multicomponent vapor/liquid equilibrium, Eq. (10.47) becomes:

$$\hat{f}_i^l = \hat{f}_i^v \ (i = 1, 2, \ldots, N) \qquad (10.48)$$

Equation (10.39) results as a special case when this relation is applied to the vapor/liquid equilibrium of *pure* species i.

The definition of a residual property is given in Sec. 6.2:

$$M^R \equiv M - M^{ig} \qquad (6.41)$$

where M is the molar (or unit-mass) value of a thermodynamic property and M^{ig} is the value that the property would have in its ideal-gas state with the same composition at the same T and P. The defining equation for a *partial residual property* $\bar{M}_i^R$ follows from this equation. Multiplied by n mol of mixture, it becomes:

$$n M^R = nM - n M^{ig}$$

Differentiation with respect to n_i at constant T, P, and n_j gives:

$$\left[\frac{\partial(n M^R)}{\partial n_i} \right]_{P, T, n_j} = \left[\frac{\partial(n M)}{\partial n_i} \right]_{P, T, n_j} - \left[\frac{\partial(n M^{ig})}{\partial n_i} \right]_{P, T, n_j}$$

Reference to Eq. (10.7) shows that each term has the form of a partial molar property. Thus,

$$\bar{M}_i^R = \bar{M}_i - \bar{M}_i^{ig} \qquad (10.49)$$

Because residual properties measure departures from ideal-gas-state values, their most logical and common application is to gas-phase properties, but they are also valid for describing liquid-phase properties. Written for the residual Gibbs energy, Eq. (10.49) becomes:

$$\boxed{\bar{G}_i^R = \bar{G}_i - \bar{G}_i^{ig}} \qquad (10.50)$$

an equation which defines the *partial residual Gibbs energy*.

Subtracting Eq. (10.29) from Eq. (10.46), both written for the same T and P, yields:

$$\mu_i - \mu_i^{ig} = RT \ln \frac{\hat{f}_i}{y_i P}$$

This result combined with Eq. (10.50) and the identity $\mu_i \equiv \bar{G}_i$ gives:

$$\boxed{\bar{G}_i^R = RT \ln \hat{\phi}_i}$$
(10.51)

where by **definition,**

$$\boxed{\hat{\phi}_i \equiv \frac{\hat{f}_i}{y_i P}}$$
(10.52)

The dimensionless ratio $\hat{\phi}_i$ is called the *fugacity coefficient of species i in solution.* Although most commonly applied to gases, the fugacity coefficient may also be used for liquids, and in this case mole fraction y_i is replaced by x_i, the symbol traditionally used for mole fractions in the liquid phase. Because Eq. (10.29) for the ideal-gas state is a special case of Eq. (10.46):

$$\hat{f}_i^{ig} = y_i P$$
(10.53)

Thus the fugacity of species i in an ideal-gas-state mixture is equal to its partial pressure. Moreover, $\hat{\phi}_i^{ig} = 1$, and for the ideal-gas state, $\bar{G}_i^R = 0$.

The Fundamental Residual-Property Relation

The fundamental property relation given by Eq. (10.2) is put into an alternative form through the mathematical identity [also used to generate Eq. (6.37)]:

$$d\left(\frac{nG}{RT}\right) \equiv \frac{1}{RT}d(nG) - \frac{nG}{RT^2}dT$$

In this equation $d(nG)$ is eliminated by Eq. (10.2) and G is replaced by its definition, $H - TS$. The result, after algebraic reduction, is:

$$\boxed{d\left(\frac{nG}{RT}\right) = \frac{nV}{RT}dP - \frac{nH}{RT^2}dT + \sum_i \frac{\bar{G}_i}{RT}dn_i}$$
(10.54)

All terms in Eq. (10.54) have the units of moles; moreover, in contrast to Eq. (10.2), the enthalpy rather than the entropy appears on the right side. Equation (10.54) is a general relation expressing nG/RT as a function of *all* of its canonical variables, T, P, and the mole numbers. It reduces to Eq. (6.37) for the special case of 1 mol of a constant-composition phase. Equations (6.38) and (6.39) follow from either equation, and equations for the other thermodynamic properties then come from appropriate defining equations. Knowledge of G/RT as a function of its canonical variables allows evaluation of all other thermodynamic properties, and therefore implicitly contains complete property information. Unfortunately, we do not have the experimental means by which to exploit this characteristic. That is, we cannot directly measure G/RT as a function of T, P, and composition. However, we can obtain complete thermodynamic information by combining calorimetric and volumetric data. In this regard, an analogous equation relating residual properties proves useful.

Because Eq. (10.54) is general, it may be written for the special case of the ideal-gas state:

$$d\left(\frac{nG^{ig}}{RT}\right) = \frac{nV^{ig}}{RT}dP - \frac{nH^{ig}}{RT^2}dT + \sum_i \frac{\bar{G}_i^{ig}}{RT}dn_i$$

In view of the definitions of residual properties [Eqs. (6.41) and (10.50)], subtracting this equation from Eq. (10.54) gives:

$$d\left(\frac{nG^R}{RT}\right) = \frac{nV^R}{RT}dP - \frac{nH^R}{RT^2}dT + \sum_i \frac{\bar{G}_i^R}{RT}dn_i \tag{10.55}$$

Equation (10.55) is the *fundamental residual-property relation.* Its derivation from Eq. (10.2) parallels the derivation in Chap. 6 that led from Eq. (6.11) to Eq. (6.42). Indeed, Eqs. (6.11) and (6.42) are special cases of Eqs. (10.2) and (10.55), valid for 1 mol of a constant-composition fluid. An alternative form of Eq. (10.55) follows by introduction of the fugacity coefficient as given by Eq. (10.51):

$$d\left(\frac{nG^R}{RT}\right) = \frac{nV^R}{RT}dP - \frac{nH^R}{RT^2}dT + \sum_i \ln \hat{\phi}_i \, dn_i \tag{10.56}$$

Equations so general as Eqs. (10.55) and (10.56) are most useful for practical application in restricted forms. Division of Eqs. (10.55) and (10.56), first, by dP with restriction to constant T and composition, and second, by dT and restriction to constant P and composition leads to:

$$\frac{V^R}{RT} = \left[\frac{\partial(G^R/RT)}{\partial P}\right]_{T,x} \tag{10.57} \qquad \frac{H^R}{RT} = -T\left[\frac{\partial(G^R/RT)}{\partial T}\right]_{P,x} \tag{10.58}$$

These equations are restatements of Eqs. (6.43) and (6.44) wherein the restriction of the derivatives to constant composition is shown explicitly. They lead to Eqs. (6.46), (6.48), and (6.49) for the calculation of residual properties from volumetric data. Moreover, Eq. (10.57) is the basis for the direct derivation of Eq. (10.35), which yields fugacity coefficients from volumetric data. It is through the residual properties that this kind of experimental information enters into the practical application of thermodynamics.

In addition, from Eq. (10.56),

$$\ln \hat{\phi}_i = \left[\frac{\partial(nG^R/RT)}{\partial n_i}\right]_{P,T,n_j} \tag{10.59}$$

This equation demonstrates that the logarithm of the fugacity coefficient of a species in solution is a partial property with respect to G^R/RT.

Example 10.6

Develop a general equation to calculate ln $\hat{\phi}_i$ values from compressibility-factor data.

Solution 10.6

For n mol of a constant-composition mixture, Eq. (6.49) becomes:

$$\frac{nG^R}{RT} = \int_0^P (nZ - n)\frac{dP}{P}$$

In accord with Eq. (10.59) this equation may be differentiated with respect to n_i at constant T, P, and n_j to yield:

$$\ln \hat{\phi}_i = \int_0^P \left[\frac{\partial(nZ - n)}{\partial n_i}\right]_{P, T, n_j} \frac{dP}{P}$$

Because $\partial(nZ)/\partial n_i = \bar{Z}_i$ and $\partial n/\partial n_i = 1$, this reduces to:

$$\ln \hat{\phi}_i = \int_0^P (\bar{Z}_i - 1)\frac{dP}{P} \qquad (10.60)$$

where integration is at constant temperature and composition. This equation is the partial-property analog of Eq. (10.35). It allows the calculation of $\hat{\phi}_i$ values from PVT data.

Fugacity Coefficients from the Virial Equation of State

Values of $\hat{\phi}_i$ for species i in solution are readily found from equations of state. The simplest form of the virial equation provides a useful example. Written for a gas mixture it is exactly the same as for a pure species:

$$Z = 1 + \frac{BP}{RT} \qquad (3.36)$$

The mixture second virial coefficient B is a function of temperature and composition. Its *exact* composition dependence is given by statistical mechanics, which makes the virial equation preeminent among equations of state where it is applicable, i.e., to gases at low to moderate pressures. The equation giving this composition dependence is:

$$\boxed{B = \sum_i \sum_j y_i y_j B_{ij}} \qquad (10.61)$$

where y_i and y_j represent mole fractions in a gas mixture. The indices i and j identify species, and both run over all species present in the mixture. The virial coefficient B_{ij} characterizes bimolecular interactions between molecules of species i and species j, and therefore $B_{ij} = B_{ji}$. The double summation accounts for all possible bimolecular interactions.

For a binary mixture $i = 1, 2$ and $j = 1, 2$; the expansion of Eq. (10.61) then gives:

$$B = y_1 y_1 B_{11} + y_1 y_2 B_{12} + y_2 y_1 B_{21} + y_2 y_2 B_{22}$$

or

$$B = y_1^2 B_{11} + 2y_1 y_2 B_{12} + y_2^2 B_{22} \qquad (10.62)$$

Two types of virial coefficients have appeared: B_{11} and B_{22}, for which the successive sub-scripts are the same, and B_{12}, for which the two subscripts are different. The first type is a pure-species virial coefficient; the second is a mixture property, known as a *cross coefficient*. Both are functions of temperature only. Expressions such as Eqs. (10.61) and (10.62) relate mixture coefficients to pure-species and cross coefficients. They are called *mixing rules*.

Equation (10.62) allows the derivation of expressions for $\ln \hat{\phi}_1$ and $\ln \hat{\phi}_2$ for a binary gas mixture that obeys Eq. (3.36). Written for n mol of gas mixture, it becomes:

$$nZ = n + \frac{nBP}{RT}$$

Differentiation with respect to n_1 gives:

$$\bar{Z}_1 \equiv \left[\frac{\partial(nZ)}{\partial n_1}\right]_{P, T, n_2} = 1 + \frac{P}{RT}\left[\frac{\partial(nB)}{\partial n_1}\right]_{T, n_2}$$

Substitution for $\bar{Z}_1$ in Eq. (10.60) yields:

$$\ln \hat{\phi}_1 = \frac{1}{RT}\int_0^P \left[\frac{\partial(nB)}{\partial n_1}\right]_{T, n_2} dP = \frac{P}{RT}\left[\frac{\partial(nB)}{\partial n_1}\right]_{T, n_2}$$

where the integration is elementary because B is not a function of pressure. All that remains is evaluation of the derivative.

Equation (10.62) for the second virial coefficient may be written:

$$B = y_1(1 - y_2)B_{11} + 2y_1 y_2 B_{12} + y_2(1 - y_1)B_{22}$$
$$= y_1 B_{11} - y_1 y_2 B_{11} + 2y_1 y_2 B_{12} + y_2 B_{22} - y_1 y_2 B_{22}$$

or

$$B = y_1 B_{11} + y_2 B_{22} + y_1 y_2 \delta_{12} \quad \text{with} \quad \delta_{12} \equiv 2B_{12} - B_{11} - B_{22}$$

Multiplying by n and substituting $y_i = n_i/n$ gives,

$$nB = n_1 B_{11} + n_2 B_{22} + \frac{n_1 n_2}{n}\delta_{12}$$

By differentiation:

$$\left[\frac{\partial(nB)}{\partial n_1}\right]_{T, n_2} = B_{11} + \left(\frac{1}{n} - \frac{n_1}{n^2}\right)n_2\delta_{12}$$
$$= B_{11} + (1 - y_1)y_2\delta_{12} = B_{11} + y_2^2\delta_{12}$$

Therefore,

$$\ln \hat{\phi}_1 = \frac{P}{RT}(B_{11} + y_2^2\delta_{12}) \tag{10.63a}$$

Similarly,

$$\ln \hat{\phi}_2 = \frac{P}{RT}(B_{22} + y_1^2\delta_{12}) \tag{10.63b}$$

Equations (10.63) are readily extended for application to multicomponent gas mixtures; the general equation is:[10]

$$\ln \hat{\phi}_k = \frac{P}{RT}\left[B_{kk} + \frac{1}{2}\sum_i \sum_j y_i y_j (2\delta_{ik} - \delta_{ij})\right] \tag{10.64}$$

where the dummy indices i and j run over all species, and

$$\delta_{ik} \equiv 2B_{ik} - B_{ii} - B_{kk} \qquad\qquad \delta_{ij} \equiv 2B_{ij} - B_{ii} - B_{jj}$$

with

$$\delta_{ii} = 0, \delta_{kk} = 0, \text{ etc.}, \qquad\qquad \text{and } \delta_{ki} = \delta_{ik}, \text{ etc.}$$

Example 10.7

Determine the fugacity coefficients as given by Eqs. (10.63) for nitrogen and methane in a $N_2(1)/CH_4(2)$ mixture at 200 K and 30 bar if the mixture contains 40 mol-% N_2. Experimental virial-coefficient data are as follows:

$$B_{11} = -35.2 \quad B_{22} = -105.0 \quad B_{12} = -59.8 \text{ cm}^3 \cdot \text{mol}^{-1}$$

Solution 10.7

By definition, $\delta_{12} = 2B_{12} - B_{11} - B_{22}$. Whence,

$$\delta_{12} = 2(-59.8) + 35.2 + 105.0 = 20.6 \text{ cm}^3 \cdot \text{mol}^{-1}$$

Substitution of numerical values in Eqs. (10.63) yields:

$$\ln \hat{\phi}_1 = \frac{30}{(83.14)(200)}\left[-35.2 + (0.6)^2(20.6)\right] = -0.0501$$

$$\ln \hat{\phi}_2 = \frac{30}{(83.14)(200)}\left[-105.0 + (0.4)^2(20.6)\right] = -0.1835$$

Whence,

$$\hat{\phi}_1 = 0.9511 \quad \text{and} \quad \hat{\phi}_2 = 0.8324$$

Note that the second virial coefficient of the mixture as given by Eq. (10.62) is $B = -72.14 \text{ cm}^3 \cdot \text{mol}^{-1}$, and that substitution in Eq. (3.36) yields a mixture compressibility factor, $Z = 0.870$.

---

[10]H. C. Van Ness and M. M. Abbott, *Classical Thermodynamics of Nonelectrolyte Solutions: With Applications to Phase Equilibria*, pp. 135–140, McGraw-Hill, New York, 1982.

10.7 GENERALIZED CORRELATIONS FOR THE FUGACITY COEFFICIENT

Fugacity Coefficients for Pure Species

The generalized methods developed in Sec. 3.7 for the compressibility factor Z and in Sec. 6.4 for the residual enthalpy and entropy of pure gases are applied here to the fugacity coefficient. Equation (10.35) is put into generalized form by substitution of the relations,

$$P = P_c P_r \qquad dP = P_c dP_r$$

Hence,

$$\ln \phi_i = \int_0^{P_r} (Z_i - 1) \frac{dP_r}{P_r} \tag{10.65}$$

where integration is at constant T_r. Substitution for Z_i by Eq. (3.53) yields:

$$\ln \phi = \int_0^{P_r} (Z^0 - 1) \frac{dP_r}{P_r} + \omega \int_0^{P_r} Z^1 \frac{dP_r}{P_r}$$

where for simplicity subscript i is omitted. This equation may be written in alternative form:

$$\ln \phi = \ln \phi^0 + \omega \ln \phi^1 \tag{10.66}$$

where

$$\ln \phi^0 \equiv \int_0^{P_r} (Z^0 - 1) \frac{dP_r}{P_r} \quad \text{and} \quad \ln \phi^1 \equiv \int_0^{P_r} Z^1 \frac{dP_r}{P_r}$$

The integrals in these equations may be evaluated numerically or graphically for various values of T_r and P_r from the data for Z^0 and Z^1 given in Tables D.1 through D.4 (App. D). Another method, and the one adopted by Lee and Kesler to extend their correlation to fugacity coefficients, is based on an equation of state.

Equation (10.66) may also be written,

$$\phi = (\phi^0)(\phi^1)^\omega \tag{10.67}$$

and we have the option of providing correlations for ϕ^0 and ϕ^1 rather than for their logarithms. This is the choice made here, and Tables D.13 through D.16 present values for these quantities as derived from the Lee/Kesler correlation as functions of T_r and P_r, thus providing a three-parameter generalized correlation for fugacity coefficients. Tables D.13 and D.15 for ϕ^0 can be used alone as a two-parameter correlation which does not incorporate the refinement introduced by the acentric factor.

Example 10.8

Estimate from Eq. (10.67) a value for the fugacity of 1-butene vapor at 200°C and 70 bar.

Solution 10.8

At these conditions, with $T_c = 420.0$ K, $P_c = 40.43$ bar from Table B.1, we have:

$$T_r = 1.127 \quad P_r = 1.731 \quad \omega = 0.191$$

By interpolation in Tables D.15 and D.16 at these conditions,

$$\phi^0 = 0.627 \quad \text{and} \quad \phi^1 = 1.096$$

Equation (10.67) then gives:

$$\phi = (0.627)(1.096)^{0.191} = 0.638$$

and

$$f = \phi P = (0.638)(70) = 44.7 \text{ bar}$$

A useful generalized correlation for $\ln \phi$ results when the simplest form of the virial equation is valid. Equations (3.57) and (3.59) combine to give:

$$Z - 1 = \frac{P_r}{T_r}(B^0 + \omega B^1)$$

Substitution into Eq. (10.65) and integration yield:

$$\ln \phi = \frac{P_r}{T_r}(B^0 + \omega B^1)$$

or

$$\phi = \exp\left[\frac{P_r}{T_r}(B^0 + \omega B^1)\right] \tag{10.68}$$

This equation, used in conjunction with Eqs. (3.61) and (3.62), provides reliable values of ϕ for any nonpolar or slightly polar gas when applied at conditions where Z is approximately linear in pressure. Figure 3.13 again serves as a guide to its applicability.

Named functions HRB(TR,PR,OMEGA) and SRB(TR,PR,OMEGA) for the evaluation of H^R/RT_c and S^R/R by the generalized virial-coefficient correlation are described in Sec. 6.4. Similarly, we introduce here a function named PHIB(TR,PR,OMEGA) for the evaluation of ϕ[11]:

$$\phi = \text{PHIB(TR,PR,OMEGA)}$$

It combines Eq. (10.68) with Eqs. (3.61) and (3.62) to evaluate the fugacity coefficient for given reduced temperature, reduced pressure, and acentric factor. For example, the value of ϕ for carbon dioxide at the conditions of Example 6.4, Step 3, is denoted as:

$$\text{PHIB(0.963,0.203,0.224)} = 0.923$$

---

[11]Sample programs and spreadsheets for the evaluation of these functions are available in the Online Learning Center at http://highered.mheducation.com:80/sites/1259696529.

Extension to Mixtures

The generalized correlation just described is for *pure* gases only. The remainder of this section shows how the virial equation may be generalized to allow the calculation of fugacity coefficients $\hat{\phi}_i$ for species in gas *mixtures*.

The general expression for calculation of $\ln \hat{\phi}_k$ from second-virial-coefficient data is given by Eq. (10.64). Values of the pure-species virial coefficients B_{kk}, B_{ii}, etc., are found from the generalized correlation represented by Eqs. (3.58), (3.59), (3.61), and (3.62). The cross coefficients B_{ik}, B_{ij}, etc., are found from an extension of the same correlation. For this purpose, Eq. (3.59) is rewritten in the more general form:[12]

$$\hat{B}_{ij} = B^0 + \omega_{ij} B^1 \tag{10.69a}$$

where

$$\hat{B}_{ij} \equiv \frac{B_{ij} P_{cij}}{R T_{cij}} \tag{10.69b}$$

and B^0 and B^1 are the same functions of T_r as given by Eqs. (3.61) and (3.62). The combining rules proposed by Prausnitz et al. for the calculation of ω_{ij}, T_{cij}, and P_{cij} are:

$\omega_{ij} = \dfrac{\omega_i + \omega_j}{2}$ (10.70)	$T_{cij} = (T_{ci} T_{cj})^{1/2} (1 - k_{ij})$ (10.71)

$P_{cij} = \dfrac{Z_{cij} R T_{cij}}{V_{cij}}$ (10.72)	$Z_{cij} = \dfrac{Z_{ci} + Z_{cj}}{2}$ (10.73)

$$V_{cij} = \left(\frac{V_{ci}^{1/3} + V_{cj}^{1/3}}{2} \right)^3 \tag{10.74}$$

In Eq. (10.71), k_{ij} is an empirical interaction parameter specific to a particular i-j molecular pair. When $i = j$ and for chemically similar species, $k_{ij} = 0$. Otherwise, it is a small positive number evaluated from minimal *PVT* data or, in the absence of data, set equal to zero. When $i = j$, all equations reduce to the appropriate values for a pure species. When $i \neq j$, these equations define a set of interaction parameters that, while they do not have any fundamental physical significance, do provide useful estimates of fugacity coefficients in moderately non-ideal gas mixtures. Reduced temperature is given for each ij pair by $T_{rij} \equiv T/T_{cij}$. For a mixture, values of B_{ij} from Eq. (10.69b) substituted into Eq. (10.61) yield the mixture second virial coefficient B, and substituted into Eq. (10.64) [Eqs. (10.63) for a binary] they yield values of $\ln \hat{\phi}_i$.

The primary virtue of the generalized correlation for second virial coefficients presented here is simplicity; more accurate, but more complex, correlations appear in the literature.[13]

---

[12]J. M. Prausnitz, R. N. Lichtenthaler, and E. G. de Azevedo, *Molecular Thermodynamics of Fluid-Phase Equilibria*, 3rd ed., pp. 133 and 160, Prentice-Hall, Englewood Cliffs, NJ, 1998.

[13]C. Tsonopoulos, *AIChE J.*, vol. 20, pp. 263–272, 1974, vol. 21, pp. 827–829, 1975, vol. 24, pp. 1112–1115, 1978; C. Tsonopoulos, *Adv. in Chemistry Series* 182, pp. 143–162, 1979; J. G. Hayden and J. P. O'Connell, *Ind. Eng. Chem. Proc. Des. Dev.*, vol. 14, pp. 209–216, 1975; D. W. McCann and R. P. Danner, *Ibid.*, vol. 23, pp. 529–533, 1984; J. A. Abusleme and J. H. Vera, *AIChE J.*, vol. 35, pp. 481–489, 1989; L. Meng, Y. Y. Duan, and X. D. Wang, *Fluid Phase Equilib.*, vol. 260, pp. 354–358, 2007.

Example 10.9

Estimate $\hat{\phi}_1$ and $\hat{\phi}_2$ by Eqs. (10.63) for an equimolar mixture of methyl ethyl ketone(1)/toluene(2) at 50°C and 25 kPa. Set all $k_{ij} = 0$.

Solution 10.9

The required data are as follows:

ij	T_{cij} K	P_{cij} bar	V_{cij} cm³·mol⁻¹	Z_{cij}	ω_{ij}
11	535.5	41.5	267.	0.249	0.323
22	591.8	41.1	316.	0.264	0.262
12	563.0	41.3	291.	0.256	0.293

where values in the last row have been calculated by Eqs. (10.70) through (10.74). The values of T_{rij}, together with B^0, B^1, and B_{ij} calculated for each ij pair by Eqs. (3.65), (3.66), and (10.69), are as follows:

ij	T_{rij}	B^0	B^1	B_{ij} cm³·mol⁻¹
11	0.603	−0.865	−1.300	−1387.
22	0.546	−1.028	−2.045	−1860.
12	0.574	−0.943	−1.632	−1611.

Calculating δ_{12} according to its definition gives:

$$\delta_{12} = 2B_{12} - B_{11} - B_{22} = (2)(-1611) + 1387 + 1860 = 25 \text{ cm}^3\cdot\text{mol}^{-1}$$

Equation (10.63) then yields:

$$\ln \hat{\phi}_1 = \frac{P}{RT}(B_{11} + y_2^2 \delta_{12}) = \frac{25}{(8314)(323.15)}[-1387 + (0.5)^2(25)] = -0.0128$$

$$\ln \hat{\phi}_2 = \frac{P}{RT}(B_{22} + y_1^2 \delta_{12}) = \frac{25}{(8314)(323.15)}[-1860 + (0.5)^2(25)] = -0.0172$$

Whence,

$$\hat{\phi}_1 = 0.987 \qquad \text{and} \qquad \hat{\phi}_2 = 0.983$$

These results are representative of values obtained for vapor phases at typical conditions of low-pressure vapor/liquid equilibrium.

10.8 THE IDEAL-SOLUTION MODEL

The chemical potential as given by the ideal-gas-state mixture model,

$$\boxed{\mu_i^{ig} \equiv \bar{G}_i^{ig} = G_i^{ig}(T, P) + RT \ln y_i}$$

(10.24)

contains a final term that gives it the simplest possible composition dependence. Indeed, this functional form could reasonably serve as the basis for the composition dependence of chemical potential in dense gases and liquids. In it, the composition dependence arises only from the entropy increase due to random intermixing of molecules of different species. This entropy increase due to mixing is the same in any random mixture, and thus can be expected to be present in liquids and dense gases as well. However, the pure-species behavior implied by the term $G_i^{ig}(T, P)$ is unrealistic except for the ideal-gas state. A natural extension of Eq. (10.24) therefore replaces $G_i^{ig}(T, P)$ with $G_i(T, P)$, the Gibbs energy of pure i in its *real physical state* of gas, liquid, or solid. Thus, we **define** an *ideal* solution as one for which:

$$\mu_i^{id} \equiv \bar{G}_i^{id} = G_i(T, P) + RT \ln x_i \qquad (10.75)$$

where superscript *id* denotes an ideal-solution property. Mole fraction is here represented by x_i to reflect the fact that this approach is most often applied to liquids. However, a consequence of this definition is that an ideal-gas-state mixture is a special case, namely, an ideal solution of gases in the ideal-gas state, for which x_i in Eq. (10.75) is replaced by y_i.

All other thermodynamic properties for an ideal solution follow from Eq. (10.75). The partial volume results from differentiation with respect to pressure at constant temperature and composition in accord with Eq. (10.18):

$$\bar{V}_i^{id} = \left(\frac{\partial \bar{G}_i^{id}}{\partial P}\right)_{T,x} = \left(\frac{\partial G_i}{\partial P}\right)_T$$

By Eq. (10.4), $(\partial G_i/\partial P)_T = V_i$; whence,

$$\bar{V}_i^{id} = V_i \qquad (10.76)$$

Similarly, as a result of Eq. (10.19),

$$\bar{S}_i^{id} = -\left(\frac{\partial \bar{G}_i^{id}}{\partial T}\right)_{P,x} = -\left(\frac{\partial G_i}{\partial T}\right)_P - R \ln x_i$$

By Eq. (10.5),

$$\bar{S}_i^{id} = S_i - R \ln x_i \qquad (10.77)$$

Because $\bar{H}_i^{id} = \bar{G}_i^{id} + T\bar{S}_i^{id}$, substitutions by Eqs. (10.75) and (10.77) yield:

$$\bar{H}_i^{id} = G_i + RT \ln x_i + TS_i - RT \ln x_i$$

or

$$\bar{H}_i^{id} = H_i \qquad (10.78)$$

The summability relation, Eq. (10.11), applied to the special case of an ideal solution, is written:

$$M^{id} = \sum_i x_i \bar{M}_i^{id}$$

Application to Eqs. (10.75) through (10.78) yields:

$$G^{id} = \sum_i x_i G_i + RT \sum_i x_i \ln x_i \quad (10.79) \qquad S^{id} = \sum_i x_i S_i - R \sum_i x_i \ln x_i \quad (10.80)$$

$$V^{id} = \sum_i x_i V_i \qquad (10.81) \qquad H^{id} = \sum_i x_i H_i \qquad (10.82)$$

Note the similarity of these equations to Eq. (10.25), (10.26), and (10.27) for ideal-gas-state mixtures. For all properties of an ideal solution, the composition dependence is, by definition, the same as that of ideal-gas-state mixtures, for which this composition dependence is well defined and arises entirely from the entropy increase of random mixing. However, the temperature and pressure dependence are not those of ideal gases but are given by mole-fraction-weighted averages of the pure species properties.

If in Example 10.3 the solution formed by mixing methanol(1) and water(2) were assumed ideal, the final volume would be given by Eq. (10.81), and the V-vs.-x_1 relation would be a straight line connecting the pure-species volumes, V_2 at $x_1 = 0$ with V_1 at $x_1 = 1$. For the specific calculation at $x_1 = 0.3$, the use of V_1 and V_2 in place of partial volumes yields:

$$V_1^t = 983 \qquad V_2^t = 1017 \text{ cm}^3$$

Both values are about 3.4% low.

The Lewis/Randall Rule

The composition dependence of the fugacity of a species in an ideal solution is particularly simple. Recall Eqs. (10.46) and (10.31):

$$\mu_i \equiv \Gamma_i(T) + RT \ln \hat{f}_i \quad (10.46) \qquad G_i \equiv \Gamma_i(T) + RT \ln f_i \quad (10.31)$$

Subtraction yields the *general* equation:

$$\mu_i = G_i + RT \ln (\hat{f}_i / f_i)$$

For the special case of an ideal solution,

$$\mu_i^{id} \equiv \bar{G}_i^{id} = G_i + RT \ln (\hat{f}_i^{id} / f_i)$$

Comparison with Eq. (10.75) gives:

$$\boxed{\hat{f}_i^{id} = x_i f_i} \qquad (10.83)$$

This equation, known as the *Lewis/Randall rule*, applies to each species in an ideal solution at all conditions of temperature, pressure, and composition. It shows that the fugacity of each species in an ideal solution is proportional to its mole fraction; the proportionality constant is the fugacity of *pure* species i in the same physical state as the solution and at the same T and P.

Division of both sides of Eq. (10.83) by Px_i and substitution of $\hat{\phi}_i^{id}$ for $\hat{f}_i^{id}/x_i P_i$ [Eq. (10.52)] and of ϕ_i for f_i/P [Eq. (10.34)] gives an alternative form:

$$\hat{\phi}_i^{id} = \phi_i \tag{10.84}$$

Thus the fugacity coefficient of species i in an ideal solution is equal to the fugacity coefficient of *pure* species i in the same physical state as the solution and at the same T and P. Phases comprised of liquids whose molecules are of similar size and similar chemical nature approximate ideal solutions. Mixtures of isomers closely conform to these conditions. Mixtures of adjacent members of homologous series are also examples.

10.9 EXCESS PROPERTIES

The residual Gibbs energy and the fugacity coefficient are directly related to experimental *PVT* data by Eqs. (6.49), (10.35), and (10.60). Where such data can be adequately correlated by equations of state, thermodynamic-property information is readily provided by residual properties. However, *liquid* solutions are often more easily dealt with through properties that measure their departures, not from ideal-gas-state behavior, but from ideal-solution behavior. Thus the mathematical formalism of *excess* properties is analogous to that of the residual properties, but with ideal-solution behavior rather than ideal-gas-state behavior as a basis.

    If M represents the molar (or unit-mass) value of any extensive thermodynamic property (e.g., V, U, H, S, G, etc.), then an excess property M^E is **defined** as the difference between the actual property value of a solution and the value it would have as an ideal solution at the same temperature, pressure, and composition. Thus,

$$\boxed{M^E \equiv M - M^{id}} \tag{10.85}$$

For example,

$$G^E \equiv G - G^{id} \qquad H^E \equiv H - H^{id} \qquad S^E \equiv S - S^{id}$$

Moreover,

$$G^E = H^E - TS^E \tag{10.86}$$

which follows from Eq. (10.85) and Eq. (6.4), the definition of G.

    The definition of M^E is analogous to the definition of a residual property as given by Eq. (6.41). Indeed, excess properties have a simple relation to residual properties, found by subtracting Eq. (6.41) from Eq. (10.85):

$$M^E - M^R = -(M^{id} - M^{ig})$$

As already noted, ideal-gas-state mixtures are ideal solutions of pure gases in the ideal-gas state. Equations (10.79) through (10.82) therefore become expressions for M^{ig} when M_i is replaced by M_i^{ig}. Equation (10.82) becomes Eq. (10.25), Eq. (10.80) becomes Eq. (10.26), and Eq. (10.79) becomes Eq. (10.27). The two sets of equations, for M^{id} and M^{ig}, therefore provide a general relation for the difference:

$$M^{id} - M^{ig} = \sum_i x_i M_i - \sum_i x_i M_i^{ig} = \sum_i x_i M_i^R$$

This leads immediately to:

$$M^E = M^R - \sum_i x_i M_i^R \tag{10.87}$$

Note that excess properties have no meaning for pure species, whereas residual properties exist for pure species as well as for mixtures.

The partial-property relation analogous to Eq. (10.49) is:

$$\bar{M}_i^E = \bar{M}_i - \bar{M}_i^{id} \tag{10.88}$$

where $\bar{M}_i^E$ is a partial excess property. The fundamental excess-property relation is derived in exactly the same way as the fundamental residual-property relation and leads to analogous results. Equation (10.54), written for the special case of an ideal solution, is subtracted from Eq. (10.54) itself, yielding:

$$d\left(\frac{nG^E}{RT}\right) = \frac{nV^E}{RT}dP - \frac{nH^E}{RT^2}dT + \sum_i \frac{\bar{G}_i^E}{RT}dn_i \tag{10.89}$$

This is the *fundamental excess-property relation,* analogous to Eq. (10.55), the fundamental residual-property relation.

The exact analogy that exists between properties M, residual properties M^R, and excess properties M^E is indicated by Table 10.1. All of the equations that appear are basic property relations, although only Eqs. (10.4) and (10.5) have been shown explicitly before.

Example 10.10

(a) If C_P^E is a constant, independent of T, find expressions for G^E, S^E, and H^E as functions of T.

(b) From the equations developed in part (a), find values for G^E, S^E, and H^E for an equimolar solution of benzene(1)/n-hexane(2) at 323.15 K, given the following excess-property values for an equimolar solution at 298.15 K:

$$C_P^E = -2.86 \text{ J·mol}^{-1}\text{·K}^{-1} \qquad H^E = 897.9 \text{ J·mol}^{-1} \qquad G^E = 384.5 \text{ J·mol}^{-1}$$

Solution 10.10

(a) Let $C_P^E = a$, where a is a constant. From the last column of Table 10.1:

$$C_P^E = -T\left(\frac{\partial^2 G^E}{\partial T^2}\right)_{P,x} \qquad \text{whence} \qquad \left(\frac{\partial^2 G^E}{\partial T^2}\right)_{P,x} = -\frac{a}{T}$$

Integration yields:

$$\left(\frac{\partial G^E}{\partial T}\right)_{P,x} = -a\ln T + b$$

Table 10.1: Summary of Equations for the Gibbs Energy and Related Properties

M in Relation to G	M^R in Relation to G^R	M^E in Relation to G^E
$V = (\partial G/\partial P)_{T,x}$ (10.4)	$V^R = (\partial G^R/\partial P)_{T,x}$	$V^E = (\partial G^E/\partial P)_{T,x}$
$S = -(\partial G/\partial T)_{P,x}$ (10.5)	$S^R = -(\partial G^R/\partial T)_{P,x}$	$S^E = -(\partial G^E/\partial T)_{P,x}$
$H = G + TS$ $= G - T(\partial G/\partial T)_{P,x}$ $= -RT^2\left[\dfrac{\partial(G/RT)}{\partial T}\right]_{P,x}$	$H^R = G^R + TS^R$ $= G^R - T(\partial G^R/\partial T)_{P,x}$ $= -RT^2\left[\dfrac{\partial(G^R/RT)}{\partial T}\right]_{P,x}$	$H^E = G^E + TS^E$ $= G^E - T(\partial G^E/\partial T)_{P,x}$ $= -RT^2\left[\dfrac{\partial(G^E/RT)}{\partial T}\right]_{P,x}$
$C_P = (\partial H/\partial T)_{P,x}$ $= -T(\partial^2 G/\partial T^2)_{P,x}$	$C_P^R = (\partial H^R/\partial T)_{P,x}$ $= -T(\partial^2 G^R/\partial T^2)_{P,x}$	$C_P^E = (\partial H^E/\partial T)_{P,x}$ $= -T(\partial^2 G^E/\partial T^2)_{P,x}$

where b is a constant of integration. A second integration gives:

$$G^E = -a(T \ln T - T) + bT + c \qquad\qquad (A)$$

where c is another integration constant. With $S^E = -(\partial G^E / \partial T)_{P, x}$ (Table 10.1),

$$S^E = a \ln T - b \qquad\qquad (B)$$

Because $H^E = G^E + TS^E$, combination of Eqs. (A) and (B) yields:

$$H^E = aT + c \qquad\qquad (C)$$

(b) Let $C_{P_0}^E$, H_0^E, and G_0^E represent the given values at $T_0 = 298.15$ K. But C_P^E is constant, and therefore $a = C_{P_0}^E = -2.86$ J·mol^{-1}·K^{-1}.

By Eq. (A),

$$c = H_0^E - aT_0 = 1750.6$$

By Eq. (C),

$$b = \frac{G_0^E + a(T_0 \ln T_0 - T_0) - c}{T_0} = -18.0171$$

Substitution of known values into Eqs. (A), (B), and (C) for $T = 323.15$ yields:

$$G^E = 344.4 \text{ J·mol}^{-1} \quad S^E = 1.492 \text{ J·mol}^{-1}\cdot\text{K}^{-1} \quad H^E = 826.4 \text{ J·mol}^{-1}$$

The Nature of Excess Properties

Peculiarities of liquid-mixture behavior are dramatically revealed in the excess properties. Those of primary interest are G^E, H^E, and S^E. The excess Gibbs energy comes from experiment through reduction of vapor/liquid equilibrium data (Chap. 13), and H^E is determined by mixing experiments (Chap. 11). The excess entropy is not measured directly but is found from Eq. (10.86), written:

$$S^E = \frac{H^E - G^E}{T}$$

Excess properties are often strong functions of temperature, but at normal temperatures they are not strongly influenced by pressure. Their composition dependence is illustrated in Fig. 10.3 for six binary liquid mixtures at 50°C and approximately atmospheric pressure. To present S^E with the same units and on the same scale as H^E and G^E, the product TS^E is shown rather than S^E itself. Although the systems exhibit a diversity of behavior, they have common features:

1. All excess properties become zero as either species approaches purity.

2. Although G^E vs. x_1 is approximately parabolic in shape, both H^E and TS^E exhibit individualistic composition dependencies.

3. When an excess property M^E has a single sign (as does G^E in all six cases), the extreme value of M^E (maximum or minimum) often occurs near the equimolar composition.

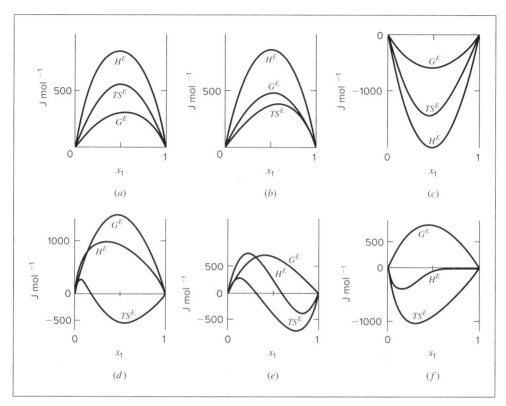

Figure 10.3: Excess properties at 50°C for six binary liquid systems: (*a*) chloroform(1)/*n*-heptane(2); (*b*) acetone(1)/methanol(2); (*c*) acetone(1)/chloroform(2); (*d*) ethanol(1)/*n*-heptane(2); (*e*) ethanol(1)/ chloroform(2); (*f*) ethanol(1)/water(2).

Feature 1 is a consequence of the definition of an excess property, Eq. (10.85); as any x_i approaches unity, both M and M^{id} approach M_i, the corresponding property of pure species *i*. Features 2 and 3 are generalizations based on observation and admit exceptions (note, for example, the behavior of H^E for the ethanol/water system).

10.10 SYNOPSIS

After studying this chapter, including the end-of-chapter problems, one should be able to:

- Define, in words and in symbols, chemical potential and partial properties
- Compute partial properties from mixture properties
- Recall that two phases are in equilibrium, at fixed T and P, when the chemical potential of each species is the same in both phases
- Apply the summability relation to compute mixture properties from partial properties

- Recognize that partial properties cannot vary independently, but are constrained by the Gibbs/Duhem equation
- Understand that all linear relationships among thermodynamic properties of pure species also apply to partial properties of species in a mixture
- Compute partial properties of a mixture in the ideal-gas state
- Define and use fugacity and fugacity coefficients for pure species and for species in a mixture, and recognize that these are defined for convenience in solving phase-equilibrium problems
- Compute fugacities and fugacity coefficients of pure species from PVT data or generalized correlations
- Estimate fugacity coefficients of species in gas mixtures using second virial coefficient correlations
- Recognize that the ideal-solution model provides a reference for description of liquid mixtures
- Relate partial properties in an ideal solution to corresponding pure species properties
- Define and use excess properties of species in a mixture

10.11 PROBLEMS

10.1. What is the change in entropy when 0.7 m^3 of CO_2 and 0.3 m^3 of N_2, each at 1 bar and 25°C, blend to form a gas mixture at the same conditions? Assume ideal gases.

10.2. A vessel, divided into two parts by a partition, contains 4 mol of nitrogen gas at 75°C and 30 bar on one side and 2.5 mol of argon gas at 130°C and 20 bar on the other. If the partition is removed and the gases mix adiabatically and completely, what is the change in entropy? Assume nitrogen to be an ideal gas with $C_V = (5/2)R$ and argon to be an ideal gas with $C_V = (3/2)R$.

10.3. A stream of nitrogen flowing at a rate of 2 kg·s^{-1} and a stream of hydrogen flowing at a rate of 0.5 kg·s^{-1} mix adiabatically in a steady-flow process. If the gases are assumed ideal, what is the rate of entropy increase as a result of the process?

10.4. What is the ideal work for the separation of an equimolar mixture of methane and ethane at 175°C and 3 bar in a steady-flow process into product streams of the pure gases at 35°C and 1 bar if $T_\sigma = 300$ K?

10.5. What is the work required for the separation of air (21-mol-% oxygen and 79-mol-% nitrogen) at 25°C and 1 bar in a steady-flow process into product streams of pure oxygen and nitrogen, also at 25°C and 1 bar, if the thermodynamic efficiency of the process is 5% and if $T_\sigma = 300$ K?

10.6. What is the partial molar temperature? What is the partial molar pressure? Express results in relation to the T and P of the mixture.

10.7. Show that:

(a) The "partial molar mass" of a species in solution is equal to its molar mass.
(b) A partial *specific* property of a species in solution is obtained by division of the partial *molar* property by the molar mass of the species.

10.8. If the molar density of a binary mixture is given by the empirical expression:

$$\rho = a_0 + a_1 x_1 + a_2 x_1^2$$

find the corresponding expressions for $\bar{V}_1$ and $\bar{V}_2$

10.9. For a ternary solution at constant T and P, the composition dependence of molar property M is given by:

$$M = x_1 M_1 + x_2 M_2 + x_3 M_3 + x_1 x_2 x_3 C$$

where M_1, M_2, and M_3 are the values of M for pure species 1, 2, and 3, and C is a parameter independent of composition. Determine expressions for $\bar{M}_1, \bar{M}_2$, and $\bar{M}_3$ by application of Eq. (10.7). As a partial check on your results, verify that they satisfy the summability relation, Eq. (10.11). For this correlating equation, what are the $\bar{M}_i$ at infinite dilution?

10.10. A *pure-component pressure* p_i for species i in a gas mixture may be defined as the pressure that species i would exert if it alone occupied the mixture volume. Thus,

$$p_i \equiv \frac{y_i Z_i RT}{V}$$

where y_i is the mole fraction of species i in the gas mixture, Z_i is evaluated at p_i and T, and V is the molar volume of the gas mixture. Note that p_i as defined here is not a partial pressure $y_i P$, except for an ideal gas. Dalton's "law" of additive pressures states that the total pressure exerted by a gas mixture is equal to the sum of the pure-component pressures of its constituent species: $P = \sum_i p_i$. Show that Dalton's "law" implies that $Z = \sum_i y_i Z_i$, where Z_i is the compressibility factor of pure species i evaluated at the mixture temperature but at its pure-component pressure.

10.11. If for a binary solution one starts with an expression for M (or M^R or M^E) as a function of x_1 and applies Eqs. (10.15) and (10.16) to find $\bar{M}_1$ and $\bar{M}_2$ (or $\bar{M}_1^R$ and $\bar{M}_2^R$ or $\bar{M}_1^E$ and $\bar{M}_2^E$) and then combines these expressions by Eq. (10.11), the initial expression for M is regenerated. On the other hand, if one starts with expressions for $\bar{M}_1$ and $\bar{M}_2$, combines them in accord with Eq. (10.11), and then applies

Eqs. (10.15) and (10.16), the initial expressions for $\bar{M}_1$ and $\bar{M}_2$ are regenerated if and only if the initial expressions for these quantities meet a specific condition. What is the condition?

10.12. With reference to Ex. 10.4,

(a) Apply Eq. (10.7) to Eq. (A) to verify Eqs. (B) and (C).
(b) Show that Eqs. (B) and (C) combine in accord with Eq. (10.11) to regenerate Eq. (A).
(c) Show that Eqs. (B) and (C) satisfy Eq. (10.14), the Gibbs/Duhem equation.
(d) Show that at constant T and P,

$$(d\bar{H}_1/dx_1)_{x_1=1} = (d\bar{H}_2/dx_1)_{x_1=0} = 0$$

(e) Plot values of H, $\bar{H}_1$, and $\bar{H}_2$, calculated by Eqs. (A), (B), and (C), vs. x_1. Label points $H_1, H_2, \bar{H}_1^\infty$, and $\bar{H}_2^\infty$, and show their values.

10.13. The molar volume ($\text{cm}^3 \cdot \text{mol}^{-1}$) of a binary liquid mixture at T and P is given by:

$$V = 120x_1 + 70x_2 + (15x_1 + 8x_2)x_1 x_2$$

(a) Find expressions for the partial molar volumes of species 1 and 2 at T and P.
(b) Show that when these expressions are combined in accord with Eq. (10.11) the given equation for V is recovered.
(c) Show that these expressions satisfy Eq. (10.14), the Gibbs/Duhem equation.
(d) Show that $(d\bar{V}_1/dx_1)_{x_1=1} = (d\bar{V}_2/dx_1)_{x_1=0} = 0$.
(e) Plot values of V, $\bar{V}_1$, and $\bar{V}_2$ calculated by the given equation for V and by the equations developed in (a) vs. x_1. Label points V_1, V_2, $\bar{V}_1^\infty$, and $\bar{V}_2^\infty$ and show their values.

10.14. For a particular binary liquid solution at constant T and P, the molar enthalpies of mixtures are represented by the equation

$$H = x_1(a_1 + b_1 x_1) + x_2(a_2 + b_2 x_2)$$

where the a_i and b_i are constants. Because the equation has the form of Eq. (10.11), it might be that $\bar{H}_i = a_i + b_i x_i$. Show whether this is true.

10.15 Analogous to the conventional partial property $\bar{M}_i$, one can define a constant-T, V partial property $\tilde{M}_i$:

$$\tilde{M}_i \equiv \left[\frac{\partial(nM)}{\partial n_i}\right]_{T, V, n_j}$$

Show that $\tilde{M}_i$ and $\bar{M}_i$ are related by the equation:

$$\tilde{M}_i = \bar{M}_i + (V - \bar{V}_i)\left(\frac{\partial M}{\partial V}\right)_{T, x}$$

Demonstrate that the $\tilde{M}_i$ satisfy a summability relation, $M = \sum_i x_i \tilde{M}_i$.

10.16 From the following compressibility-factor data for CO_2 at 150°C, prepare plots of the fugacity and fugacity coefficient of CO_2 vs. P for pressures up to 500 bar. Compare results with those found from the generalized correlation represented by Eq. (10.68).

P/bar	Z
10	0.985
20	0.970
40	0.942
60	0.913
80	0.885
100	0.869
200	0.765
300	0.762
400	0.824
500	0.910

10.17 For SO_2 at 600 K and 300 bar, determine good estimates of the fugacity and of G^R/RT.

10.18 Estimate the fugacity of isobutylene as a gas:

(*a*) At 280°C and 20 bar;
(*b*) At 280°C and 100 bar.

10.19 Estimate the fugacity of one of the following:

(*a*) Cyclopentane at 110°C and 275 bar. At 110°C the vapor pressure of cyclopentane is 5.267 bar.
(*b*) 1-Butene at 120°C and 34 bar. At 120°C the vapor pressure of 1-butene is 25.83 bar.

10.20 Justify the following equations:

$$\left(\frac{\partial \ln \hat{\phi}_i}{\partial P}\right)_{T,x} = \frac{\bar{V}_i^R}{RT} \qquad \left(\frac{\partial \ln \hat{\phi}_i}{\partial T}\right)_{P,x} = -\frac{\bar{H}_i^R}{RT^2}$$

$$\frac{G^R}{RT} = \sum_i x_i \ln \hat{\phi}_i \qquad \sum_i x_i d \ln \hat{\phi}_i = 0 \quad (\text{const } T, P)$$

10.21 From data in the steam tables, determine a good estimate for f/f^{sat} for liquid water at 150°C and 150 bar, where f^{sat} is the fugacity of saturated liquid at 150°C.

10.22 For one of the following, determine the ratio of the fugacity in the final state to that in the initial state for steam undergoing the isothermal change of state:

(*a*) From 9000 kPa and 400°C to 300 kPa.
(*b*) From 1000(psia) and 800(°F) to 50(psia).

10.23 Estimate the fugacity of one of the following liquids at its normal-boiling-point temperature and 200 bar:

(*a*) *n*-Pentane;
(*b*) Isobutylene;
(*c*) 1-Butene.

10.24 Assuming that Eq. (10.68) is valid for the vapor phase and that the molar volume of saturated liquid is given by Eq. (3.68), prepare plots of *f* vs. *P* and of ϕ vs. *P* for one of the following:

(*a*) Chloroform at 200°C for the pressure range from 0 to 40 bar. At 200°C the vapor pressure of chloroform is 22.27 bar.
(*b*) Isobutane at 40°C for the pressure range from 0 to 10 bar. At 40°C the vapor pressure of isobutane is 5.28 bar.

10.25 For the system ethylene(1)/propylene(2) as a gas, estimate $\hat{f}_1$, $\hat{f}_2$, $\hat{\phi}_1$, and $\hat{\phi}_2$ at $t = 150$°C, $P = 30$ bar, and $y_1 = 0.35$:

(*a*) Through application of Eqs. (10.63).
(*b*) Assuming that the mixture is an ideal solution.

10.26 Rationalize the following expression, valid at sufficiently low pressures, for estimating the fugacity coefficient: $\ln \phi \approx Z - 1$.

10.27 For the system methane(1)/ethane(2)/propane(3) as a gas, estimate $\hat{f}_1$, $\hat{f}_2$, $\hat{f}_3$, $\hat{\phi}_1$, $\hat{\phi}_2$, and $\hat{\phi}_3$ at $t = 100$°C, $P = 35$ bar, $y_1 = 0.21$, and $y_2 = 0.43$:

(*a*) Through application of Eq. (10.64).
(*b*) Assuming that the mixture is an ideal solution.

10.28 Given below are values of G^E /J·mol^{-1}, H^E /J·mol^{-1}, and C_P^E/J·mol^{-1}·K^{-1} for some equimolar binary liquid mixtures at 298.15 K. Estimate values of G^E, H^E, and S^E at 328.15 K for one of the equimolar mixtures by two procedures: (I) Use all the data; (II) Assume $C_P^E = 0$. Compare and discuss your results for the two procedures.

(*a*) Acetone/chloroform: $G^E = -622$, $H^E = -1920$, $C_P^E = 4.2$.
(*b*) Acetone/*n*-hexane: $G^E = 1095$, $H^E = 1595$, $C_P^E = 3.3$.
(*c*) Benzene/isooctane: $G^E = 407$, $H^E = 984$, $C_P^E = -2.7$.
(*d*) Chloroform/ethanol: $G^E = 632$, $H^E = -208$, $C_P^E = 23.0$.
(*e*) Ethanol/*n*-heptane: $G^E = 1445$, $H^E = 605$, $C_P^E = 11.0$.
(*f*) Ethanol/water: $G^E = 734$, $H^E = -416$, $C_P^E = 11.0$.
(*g*) Ethyl acetate/*n*-heptane: $G^E = 759$, $H^E = 1465$, $C_P^E = -8.0$.

10.29 The data in Table 10.2 are experimental values of V^E for binary liquid mixtures of 1,3-dioxolane(1) and isooctane(2) at 298.15 K and 1(atm).

(a) Determine from the data numerical values of parameters a, b, and c in the correlating equation:

$$V^E = x_1 x_2 (a + b x_1 + c x_1^2)$$

(b) Determine from the results of part (a) the maximum value of V^E. At what value of x_1 does this occur?

(c) Determine from the results of part (a) expressions for $\bar{V}_1^E$ and $\bar{V}_2^E$. Prepare a plot of these quantities vs. x_1, and discuss its features.

Table 10.2: Excess Volumes for 1,3-Dioxolane(1)/Isooctane(2) at 298.15 K

x_1	V^E /10^{-3} cm^3·mol^{-1}
0.02715	87.5
0.09329	265.6
0.17490	417.4
0.32760	534.5
0.40244	531.7
0.56689	421.1
0.63128	347.1
0.66233	321.7
0.69984	276.4
0.72792	252.9
0.77514	190.7
0.79243	178.1
0.82954	138.4
0.86835	98.4
0.93287	37.6
0.98233	10.0

R. Francesconi et al., *Int. DATA Ser., Ser. A,* Vol. 25, No. 3, p. 229, 1997.

10.30 For an equimolar vapor mixture of propane(1) and *n*-pentane(2) at 75°C and 2 bar, estimate Z, H^R, and S^R. Second virial coefficients, in cm^3·mol^{-1} are:

t/°C	B_{11}	B_{22}	B_{12}
50	−331	−980	−558
75	−276	−809	−466
100	−235	−684	−399

Equations (3.36), (6.55), (6.56), and (10.62) are pertinent.

10.31 Use the data of Prob. 10.30 to determine $\hat{\phi}_1$ and $\hat{\phi}_2$ as functions of composition for binary vapor mixtures of propane(1) and *n*-pentane(2) at 75°C and 2 bar. Plot the results on a single graph. Discuss the features of this plot.

10.32 For a binary gas mixture described by Eqs. (3.36) and (10.62), prove that:

$$G^E = \delta_{12} P y_1 y_2 \qquad\qquad S^E = -\frac{d\delta_{12}}{dT} P y_1 y_2$$

$$H^E = \left(\delta_{12} - T\frac{d\delta_{12}}{dT}\right) P y_1 y_2 \qquad C_P^E = -T\frac{d^2\delta_{12}}{dT^2} P y_1 y_2$$

See also Eq. (10.87), and note that $\delta_{12} = 2B_{12} - B_{11} - B_{22}$.

10.33 The data in Table 10.3 are experimental values of H^E for binary liquid mixtures of 1,2-dichloroethane(1) and dimethyl carbonate(2) at 313.15 K and 1(atm).

(a) Determine from the data numerical values of parameters a, b, and c in the correlating equation:

$$H^E = x_1 x_2 \left(a + b x_1 + c x_1^2\right)$$

(b) Determine from the results of part (a) the minimum value of H^E. At what value of x_1 does this occur?

(c) Determine from the results of part (a) expressions for $\bar{H}_1^E$ and $\bar{H}_2^E$. Prepare a plot of these quantities vs. x_1, and discuss its features.

Table 10.3 : H^E Values for 1,2-Dichloroethane(1)/Dimethylcarbonate(2) at 313.15 K

x_1	H^E /J·mol^{-1}
0.0426	−23.3
0.0817	−45.7
0.1177	−66.5
0.1510	−86.6
0.2107	−118.2
0.2624	−144.6
0.3472	−176.6
0.4158	−195.7
0.5163	−204.2
0.6156	−191.7
0.6810	−174.1
0.7621	−141.0
0.8181	−116.8
0.8650	−85.6
0.9276	−43.5
0.9624	−22.6

R. Francesconi et al., *Int. Data Ser., Ser. A,*
Vol. 25, No. 3, p. 225, 1997.

10.34 Make use of Eqs. (3.36), (3.61), (3.62), (6.54), (6.55), (6.56), (6.70), (6.71), (10.62), and (10.69)–(10.74), to estimate V, H^R, S^R, and G^R for one of the following binary vapor mixtures:

(a) Acetone(1)/1,3-butadiene(2) with mole fractions $y_1 = 0.28$ and $y_2 = 0.72$ at $t = 60°C$ and $P = 170$ kPa.

(b) Acetonitrile(1)/diethyl ether(2) with mole fractions $y_1 = 0.37$ and $y_2 = 0.63$ at $t = 50°C$ and $P = 120$ kPa.

(c) Methyl chloride(1)/ethyl chloride(2) with mole fractions $y_1 = 0.45$ and $y_2 = 0.55$ at $t = 25°C$ and $P = 100$ kPa.

(d) Nitrogen(1)/ammonia(2) with mole fractions $y_1 = 0.83$ and $y_2 = 0.17$ at $t = 20°C$ and $P = 300$ kPa.

(e) Sulfur dioxide(1)/ethylene(2) with mole fractions $y_1 = 0.32$ and $y_2 = 0.68$ at $t = 25°C$ and $P = 420$ kPa.

Note: Set $k_{ij} = 0$ in Eq. (10.71).

10.35 Laboratory A reports the following results for equimolar values of G^E for liquid mixtures of benzene(1) with 1-hexanol(2):

$$G^E = 805 \text{ J·mol}^{-1} \text{ at } T = 298 \text{ K} \qquad G^E = 785 \text{ J·mol}^{-1} \text{ at } T = 323 \text{ K}$$

Laboratory B reports the following result for the equimolar value of H^E for the same system:

$$H^E = 1060 \text{ J·mol}^{-1} \text{ at } T = 313 \text{ K}$$

Are the results from the two laboratories thermodynamically consistent with one another? Explain.

10.36 The following expressions have been proposed for the partial molar properties of a particular binary mixture:

$$\bar{M}_1 = M_1 + A x_2 \qquad \bar{M}_2 = M_2 + A x_1$$

Here, parameter A is a constant. Can these expressions possibly be correct? Explain.

10.37 Two (2) kmol·hr^{-1} of liquid n-octane (species 1) are continuously mixed with 4 kmol·hr^{-1} of liquid *iso*-octane (species 2). The mixing process occurs at constant T and P; mechanical power requirements are negligible.

(a) Use an energy balance to determine the rate of heat transfer.

(b) Use an entropy balance to determine the rate of entropy generation (W·K^{-1}).

State and justify all assumptions.

10.38 Fifty (50) mol·s^{-1} of enriched air (50 mol-% N_2, 50 mol-% O_2) are produced by continuously combining air (79 mol-% N_2, 21 mol-% O_2) with a stream of pure oxygen.

All streams are at the constant conditions $T = 25°C$ and $P = 1.2(atm)$. There are no moving parts.

(*a*) Determine the rates of air and oxygen $(mol·s^{-1})$.
(*b*) What is the rate of heat transfer for the process?
(*c*) What is the rate of entropy generation $\dot{S}_G$ $(W·K^{-1})$?

    State all assumptions.
    Suggestion: Treat the overall process as a combination of demixing and mixing steps.

10.39 A simple expression for M^E of a *symmetrical* binary system is $M^E = Ax_1x_2$. However, countless other empirical expressions can be proposed which exhibit symmetry. How suitable would the two following expressions be for general application?

(*a*) $M^E = Ax_1^2x_2^2$;
(*b*) $M^E = A\sin(\pi x_1)$

    Suggestion: Look at the implied partial properties $\bar{M}_1^E$ and $\bar{M}_2^E$.

10.40 For a multicomponent mixture containing any number of species, prove that

$$\bar{M}_i = M + \left(\frac{\partial M}{\partial x_i}\right)_{T, P} - \sum_k x_k \left(\frac{\partial M}{\partial x_k}\right)_{T, P}$$

where the summation is over *all* species. Show that for a binary mixture this result reduces to Eqs. (10.15) and (10.16).

10.41 The following empirical two-parameter expression has been proposed for correlation of excess properties of symmetrical liquid mixtures:

$$M^E = Ax_1x_2\left(\frac{1}{x_1 + Bx_2} + \frac{1}{x_2 + Bx_1}\right)$$

Here, quantities A and B are parameters that depend at most on T.

(*a*) Determine from the given equation the implied expressions for $\bar{M}_1^E$ and $\bar{M}_2^E$.
(*b*) Show that the results of part (*a*) satisfy **all** necessary constraints for partial excess properties.
(*c*) Determine from the results of part (*a*) expressions for $\left(\bar{M}_1^E\right)^\infty$ and $\left(\bar{M}_2^E\right)^\infty$.

10.42 Commonly, if M^E for a binary system has a single sign, then the partial properties $\bar{M}_1^E$ and $\bar{M}_2^E$ have the same sign as M^E over the entire composition range. There are occasions, however, where the $\bar{M}_i^E$ may change sign even though M^E has a single sign. In fact, it is the *shape* of the M^E vs. x_1 curve that determines whether the $\bar{M}_i^E$ change sign. Show that a sufficient condition for $\bar{M}_1^E$ and $\bar{M}_2^E$ to have single signs is that the *curvature* of M^E vs. x_1 have a single sign over the entire composition range.

10.43 An engineer claims that the volume expansity of an ideal solution is given by

$$\beta^{id} = \sum_i x_i \beta_i$$

Is this claim valid? If so, show why. If not, find a correct expression for β^{id}.

10.44 Following are data for G^E and H^E (both in J·mol^{-1}) for equimolar mixtures of the same organic liquids. Use *all* of the data to estimate values of G^E, H^E, and TS^E for the equimolar mixture at 25°C.

- At $T = 10°C$: $G^E = 544.0$, $H^E = 932.1$
- At $T = 30°C$: $G^E = 513.2$, $H^E = 893.4$
- At $T = 50°C$: $G^E = 494.2$, $H^E = 845.9$

  Suggestion: Assume C_P^E is constant and use material developed in Example 10.10.

Chapter 11

Mixing Processes

Homogeneous mixtures of different chemical species, particularly liquids, are formed in many ways in both natural and industrial processes. Moreover, "unmixing" processes are required for the separation of mixtures into their constituent species and for the purification of individual chemicals. The thermodynamic properties of mixtures are required for the analysis of these mixing and unmixing processes. It is therefore our purpose in this brief chapter to:

- Define a *standard* mixing process and develop the property changes that accompany it
- Relate properties of a mixture to the properties of its constituents as pure species
- Relate property changes of mixing to excess properties
- Treat in detail the heat effects of mixing and "unmixing" processes

11.1 PROPERTY CHANGES OF MIXING

Mixing processes are carried out in many ways, and each process results in a particular change of state, depending on initial and final conditions of temperature and pressure. For the rational study of mixtures, we must define a *standard* mixing process, much as we have done for standard property changes of chemical reaction. Experimental feasibility suggests mixing at constant T and P. Thus, we take as a standard mixing process one in which appropriate amounts of pure chemical species at T and P are mixed to yield a uniform mixture of specified composition, also at T and P. The pure species are said to be in their *standard states*, and their properties in this state are the pure-species properties V_i, H_i, S_i, etc.

The standard mixing process is represented schematically in Fig. 11.1 for mixing pure species 1 and 2.[1] In this hypothetical device, n_1 moles of pure species 1 are mixed with n_2 moles of pure species 2 to form a homogeneous solution of composition $x_1 = n_1/(n_1 + n_2)$. The observable phenomena accompanying the mixing process are expansion (or contraction)

---

[1]This conceptual illustration is not a practical device for making such measurements. In practice, volume changes of mixing can be measured directly using calibrated glassware (e.g., graduated cylinders) and heats of mixing are measured calorimetrically.

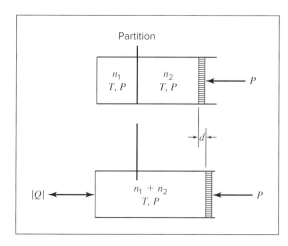

Figure 11.1: Schematic diagram of experimental binary mixing process. Two pure species, both at T and P, are initially separated by a partition, withdrawal of which allows mixing. As mixing occurs, expansion or contraction of the system is accompanied by movement of the piston so that the pressure is constant.

and a temperature change. Expansion is accommodated by movement of the piston so as to maintain pressure P, and temperature change is compensated by heat transfer to restore temperature T.

When mixing is complete, the total volume change of the system (as measured by piston displacement d) is:

$$\Delta V^t = (n_1 + n_2)V - n_1 V_1 - n_2 V_2$$

Because the process occurs at constant pressure, the heat transfer Q is equal to the total enthalpy change of the system:

$$Q = \Delta H^t = (n_1 + n_2)H - n_1 H_1 - n_2 H_2$$

Division of these equations by $n_1 + n_2$ gives:

$$\Delta V \equiv V - x_1 V_1 - x_2 V_2 = \frac{\Delta V^t}{n_1 + n_2}$$

and

$$\Delta H \equiv H - x_1 H_1 - x_2 H_2 = \frac{Q}{n_1 + n_2}$$

Thus the *volume change of mixing* ΔV and the *enthalpy change of mixing* ΔH are found from the measured quantities ΔV^t and Q. Because of its experimental association with Q, ΔH is usually called the *heat of mixing*.

Equations analogous to those for ΔV and ΔH can be written for any property, and they may also be generalized to apply to the mixing of any number of species:

$$\boxed{\Delta M \equiv M - \sum_i x_i M_i} \qquad (11.1)$$

where M can represent any intensive thermodynamic property of the mixture, for example, U, C_P, S, G, or Z. Equation (11.1) is thus the **defining** equation for a class of thermodynamic properties, known as *property changes of mixing*.

Property changes of mixing are functions of T, P, and composition and are directly related to excess properties. This result follows by combination of Eq. (10.85) with each of Eqs. (10.79) through (10.82), which provide expressions for the properties of an ideal solution:

$$M^E \equiv M - M^{id} \tag{10.85}$$

$$V^E = V - \sum_i x_i V_i \tag{11.2}$$

$$H^E = H - \sum_i x_i H_i \tag{11.3}$$

$$S^E = S - \sum_i x_i S_i + R \sum_i x_i \ln x_i \tag{11.4}$$

$$G^E = G - \sum_i x_i G_i - RT \sum_i x_i \ln x_i \tag{11.5}$$

The first two terms on the right side of each equation represent a property change of mixing as defined by Eq. (11.1). Equations (11.2) through (11.5) can therefore be written:

$V^E = \Delta V$	(11.6)	$H^E = \Delta H$	(11.7)
$S^E = \Delta S + R \sum_i x_i \ln x_i$	(11.8)	$G^E = \Delta G - RT \sum_i x_i \ln x_i$	(11.9)

For the special case of an ideal solution, each excess property is zero, and these equations become:

$\Delta V^{id} = 0$	(11.10)	$\Delta H^{id} = 0$	(11.11)
$\Delta S^{id} = -R \sum_i x_i \ln x_i$	(11.12)	$\Delta G^{id} = RT \sum_i x_i \ln x_i$	(11.13)

These equations are alternative forms of Eqs. (10.79) through (10.82). They apply as well to the ideal-gas-state mixture as a special case of an ideal solution.

Equations (11.6) through (11.9) show that excess properties and property changes of mixing are readily calculated one from the other. Although historically property changes of mixing were introduced first, because of their direct relation to experiment, excess properties fit more readily into the theoretical framework of solution thermodynamics. Because of their direct measurability, ΔV and ΔH are the property changes of mixing of major interest. Moreover, they are identical with the corresponding excess properties V^E and H^E.

Figure 11.2 shows excess enthalpies H^E for the ethanol/water system as a function of composition for several temperatures ranging from 30°C to 110°C. For the lower temperatures the behavior is *exothermic*, with heat removal required for isothermal mixing. For the higher temperatures the behavior is *endothermic*, with heat addition required for isothermal mixing. At intermediate temperatures, regions of both exothermic and endothermic behavior appear. Data of this sort are often represented by polynomial equations in mole fraction, multiplied by $x_1 x_2$ to ensure that the excess property goes to zero for both pure components.

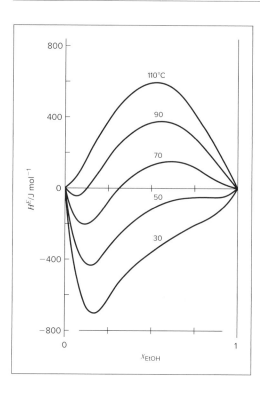

Figure 11.2: Excess enthalpies for ethanol/water. For this system H^E changes rapidly with temperature, going from negative at lower temperatures to positive at the higher temperatures, exhibiting the full range of behavior for excess-property data.

Figure 11.3 illustrates the composition dependence of ΔG, ΔH, and $T\Delta S$ for six binary liquid systems at 50°C and atmospheric pressure. The related quantities G^E, H^E, and TS^E were shown for the same systems in Fig. 10.3. As with excess properties, property changes of mixing exhibit diverse behavior, but most systems have common features:

1. Each ΔM is zero for a pure species.

2. The Gibbs energy change of mixing ΔG is always negative.

3. The entropy change of mixing ΔS is positive.

Feature 1 follows from Eq. (11.1). Feature 2 is a consequence of the requirement that the Gibbs energy be a minimum for an equilibrium state at specified T and P, as discussed in Sec. 12.4. Feature 3 reflects the fact that negative entropy changes of mixing are *unusual*; it is *not* a consequence of the second law of thermodynamics, which merely forbids negative entropy changes of mixing for systems *isolated* from their surroundings. For constant T and P, ΔS is observed to be negative for certain special classes of mixtures, none of which is represented in Fig. 11.3.

For a binary system the partial excess properties are given by Eqs. (10.15) and (10.16) with $M = M^E$. Thus,

$$\bar{M}_1^E = M^E + (1 - x_1)\frac{dM^E}{dx_1} \quad (11.14) \qquad \bar{M}_2^E = M^E - x_1\frac{dM^E}{dx_1} \quad (11.15)$$

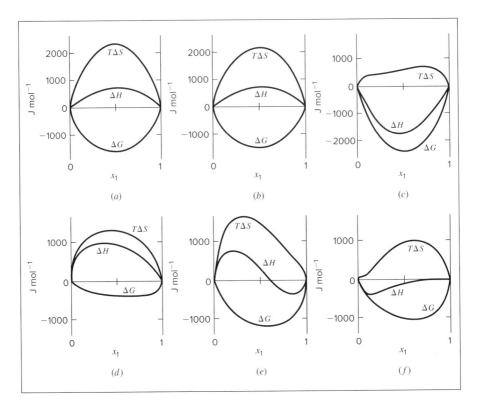

Figure 11.3: Property changes of mixing at 50°C for six binary liquid systems:
(a) chloroform(1)/n-heptane(2); (b) acetone(1)/methanol(2); (c) acetone(1)/chloroform(2);
(d) ethanol(1)/n-heptane(2); (e) ethanol(1)/chloroform(2); (f) ethanol(1)/water(2).

Example 11.1

The heat of mixing for liquid species 1 and 2 at fixed T and P, measured calorimetrically, is represented in some appropriate units by the equation

$$\Delta H = x_1 x_2 (40x_1 + 20x_2)$$

Determine expressions for $\bar{H}_1^E$ and $\bar{H}_2^E$ as functions of x_1.

Solution 11.1

We first recognize that the heat of mixing is equal to the excess enthalpy of the mixture [Eq. (11.7)], so we have

$$H^E = x_1 x_2 (40x_1 + 20x_2)$$

Elimination of x_2 in favor of x_1 and differentiation of the result provides the two equations:

$H^E = 20x_1 - 20x_1^3$ (A)	$\dfrac{dH^E}{dx_1} = 20 - 60x_1^2$ (B)

Substitution into both Eqs. (11.14) and (11.15) with H^E replacing M^E leads to:

$$\bar{H}_1^E = 20 - 60x_1^2 + 40x_1^3 \quad \text{and} \quad \bar{H}_2^E = 40x_1^3$$

Example 11.2

Property changes of mixing and excess properties are related. Show how Figs. 10.3 and 11.3 are generated from correlated data for $\Delta H(x)$ and $G^E(x)$.

Solution 11.2

With $\Delta H(x)$ and $G^E(x)$ given, Eqs. (11.7) and (10.86) provide:

$$H^E = \Delta H \quad \text{and} \quad S^E = \frac{H^E - G^E}{T}$$

These allow completion of Fig. 10.3. Property changes of mixing ΔS and ΔG follow from S^E and G^E by applying Eqs. (11.8) and (11.9):

$$\Delta S = S^E - R \sum_i x_i \ln x_i \qquad \Delta G = G^E + RT \sum_i x_i \ln x_i$$

These permit the completion of Fig. 11.3.

11.2 HEAT EFFECTS OF MIXING PROCESSES

The heat of mixing, defined in accord with Eq. (11.1), is:

$$\Delta H = H - \sum_i x_i H_i \tag{11.16}$$

Equation (11.16) gives the enthalpy change when pure species are mixed at constant T and P to form one mole (or a unit mass) of solution. Data are most commonly available for binary systems, for which Eq. (11.16) solved for H becomes:

$$H = x_1 H_1 + x_2 H_2 + \Delta H \tag{11.17}$$

This equation provides for the calculation of the enthalpies of binary mixtures from enthalpy data for pure species 1 and 2 and heats of mixing.

Data for heats of mixing are usually available for only a limited number of temperatures. If the heat capacities of the pure species and of the mixture are known, heats of mixing are calculated for other temperatures by a method analogous to the calculation of standard heats of reaction at elevated temperatures from the value at 25°C.

Heats of mixing are similar in many respects to heats of reaction. When a chemical reaction occurs, the energy of the products is different from the energy of the reactants at the same T and P because of the chemical rearrangement of the constituent atoms. When a mixture is formed, a similar energy change occurs because of changes in the interactions between molecules. That is, heats of reaction arise from changes in *intramolecular interactions* while heats of mixing arise from changes in *intermolecular interactions*. Intramolecular interactions (chemical bonds) are generally much stronger than intermolecular interactions (arising from electrostatic interactions, van der Waals forces, etc.) and as a result, heats of reaction are usually much larger in magnitude than heats of mixing. Large values of heats of mixing are observed when the intermolecular interactions in the solution are much different than in the pure components. Examples include systems with hydrogen bonding interactions and systems containing electrolytes that dissociate in solution.

Enthalpy/Concentration Diagrams

An *enthalpy/concentration* (H–x) *diagram* is a useful way to represent enthalpy data for binary solutions. It plots enthalpy as a function of composition (mole fraction or mass fraction of one species) for a series of isotherms, all at a fixed pressure (usually 1 standard atmosphere). Equation (11.17) solved for H is directly applicable to each isotherm:

$$H = x_1 H_1 + x_2 H_2 + \Delta H$$

Values of H for the solution depend not only on heats of mixing, but also on enthalpies H_1 and H_2 of the pure species. Once these are established for each isotherm, H is fixed for all solutions because ΔH has unique and measurable values for all compositions and temperatures. Because absolute enthalpies are unknown, arbitrary zero-point conditions are chosen for the enthalpies of the pure species, and this establishes the *basis* of the diagram. The preparation of a complete H–x diagram with many isotherms is a major task, and relatively few have been published.[2] Figure 11.4 presents a simplified H–x diagram for sulfuric acid(1)/water(2) mixtures, with only three isotherms.[3] The basis for this diagram is $H = 0$ for the pure species at 298.15 K, and both the composition and enthalpy are on a mass basis.

A useful feature of an enthalpy/concentration diagram is that all solutions formed by adiabatic mixing of two other solutions are represented by points lying on a straight line connecting the points that represent the initial solutions. This is shown as follows.

Let subscripts a and b identify two initial binary solutions, consisting of n_a and n_b moles or unit masses, respectively. Let subscript c identify the final solution obtained by adiabatic mixing of solutions a and b, for which $\Delta H^t = Q = 0$, and the total energy balance is:

$$(n_a + n_b)H_c = n_a H_a + n_b H_b$$

On an enthalpy/concentration diagram these solutions are represented by points designated a, b, and c, and our purpose is to show that point c lies on the straight line that passes through points a and b. For a line that passes through both points a and b,

$$H_a = m x_{1a} + k \qquad \text{and} \qquad H_b = m x_{1b} + k$$

---

[2]For examples, see D. Green and R. H. Perry, eds., *Perry's Chemical Engineers' Handbook*, 8th ed., pp. 2–220, 2–267, 2–285, 2–323, 2–403, 2–409, McGraw-Hill, New York, 2008. Only the first is in SI units.

[3]Constructed using data from F. Zeleznik, *J. Phys. Chem. Ref. Data*, vol. 20, pp. 1157–1200. 1991. Note that much of the 250 K isotherm is actually below the freezing curve and therefore represents a metastable liquid solution.

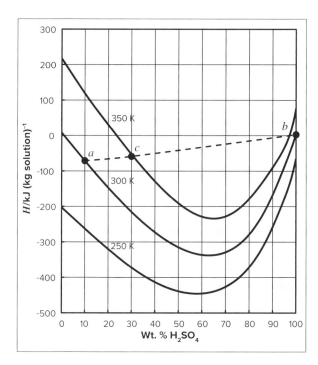

Figure 11.4: H–x diagram for $H_2SO_4(1)/H_2O(2)$. The basis for the diagram is $H = 0$ for the pure species at 298.15 K.

Substituting these expressions into the preceding material balance yields:

$$(n_a + n_b)H_c = n_a(mx_{1a} + k) + n_b(mx_{1b} + k)$$
$$= m(n_a x_{1a} + n_b x_{1b}) + (n_a + n_b)k$$

By a material balance for species 1,

$$(n_a + n_b)x_{1c} = n_a x_{1a} + n_b x_{1b}$$

Combining this equation with the preceding one gives, after reduction:

$$H_c = mx_{1c} + k$$

showing that point c lies on the same straight line as points a and b. This feature can be used to graphically estimate the final temperature when two solutions are mixed adiabatically. Such a graphical estimation is illustrated in Fig. 11.4 for the process of mixing 10 wt.% H_2SO_4 at 300 K (point a) with pure H_2SO_4 at 300 K (point b) in a 3.5:1 ratio, to yield a 30 wt% H_2SO_4 solution (point c). The graphical construction shows that after adiabatic mixing, the temperature will be nearly 350 K.

Because the pure-species enthalpies H_1 and H_2 are arbitrary, when only a single isotherm is considered, they may be set equal to zero, in which case Eq. (11.17) becomes:

$$H = \Delta H = H^E$$

An H^E–x diagram then serves as an enthalpy/concentration diagram for a single temperature. There are many such single-temperature diagrams in the literature, and they are usually accompanied by an equation that represents the curve. An example is the diagram of Fig. 11.5, showing data for sulfuric acid/water at 25°C. Again, $x_{H_2SO_4}$ is the *mass* fraction of sulfuric acid and H^E is on a *unit-mass* basis.

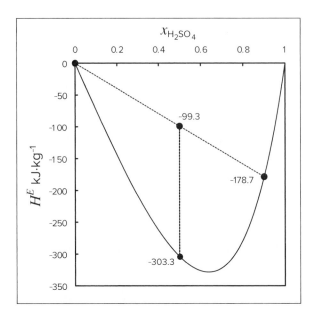

Figure 11.5: Excess enthalpies for $H_2SO_4(1)/H_2O(2)$ at 25°C.

The isotherm shown in Fig. 11.5 is represented by the equation

$$H^E = \left(-735.3 - 824.5x_1 + 195.2x_1^2 - 914.6x_1^3\right)x_1(1 - x_1) \tag{A}$$

The numbers adjacent to points on the graph come from this equation. The form of this equation is typical of models for excess properties: a polynomial in composition multiplied by x_1 and x_2. The x_1x_2 product ensures that the excess property goes to zero for both pure components. Alone, it would take the form of a symmetric parabola with a maximum or minimum at $x_1 = 0.5$. Multiplication by a polynomial in x_1 scales and biases the symmetric parabola.

A simple problem is to find the quantity of heat that must be removed to restore the initial temperature (25°C) when pure water is mixed continuously with a 90% aqueous solution of sulfuric acid to dilute it to 50%. The calculation may be made with the usual material and energy balances. We take as a basis 1 kg of 50% acid produced. If m_a is the mass of 90% acid, a mass balance on the acid is $0.9m_a = 0.5$, from which $m_a = 0.5556$. The energy balance for this process, assuming negligible kinetic- and potential-energy changes, is the difference between the final and initial enthalpies:

$$Q = H_f^E - (H_a^E)(m_a) = -303.3 - (-178.7)(0.5556) = -204.0 \text{ kJ}$$

where the enthalpy values are shown on Fig. 11.5.

An alternative procedure is represented by the two straight lines on Fig. 11.5, the first for adiabatic mixing of pure water with the 90% aqueous solution of acid to form a 50% solution. The enthalpy of this solution lies at $x_{H_2SO4} = 0.5$ on the line connecting the points representing the unmixed species. This line is a direct proportionality, from which $H^E = (-178.7/0.9)(0.5) = -99.3$. The temperature at this point is well above 25°C, and the vertical line represents cooling to 25°C, for which $Q = \Delta H^E = -303.3 - (-99.3) = -204.0 \text{ kJ}$. Because the initial step is adiabatic, this cooling step gives the total heat transfer for the process, indicating that 204 kJ·kg^{-1} are removed from the system. The temperature rise upon adiabatic mixing cannot be obtained directly from a diagram like this with a single isotherm, but it can be estimated if one knows the approximate heat capacity of the final solution.

Equation (A) can be used in Eq. (11.14) and (11.15) to generate values of the partial excess enthalpies $\bar{H}^E_{H_2SO_4}$ and $\bar{H}^E_{H_2O}$ at 25°C. This produces the results shown in Fig. 11.6, where all values are on a unit-mass basis. The two curves are far from symmetric, a consequence of the skewed nature of the H^E curve. At high concentrations of H_2SO_4, $\bar{H}^E_{H_2O}$ reaches high values, in fact the infinite-dilution value approaches the latent heat of water. This is the reason that when water is added to pure sulfuric acid, a very high rate of heat removal is required for isothermal mixing. Under usual circumstances the heat-transfer rate is far from adequate, and the resulting temperature rise causes local boiling and sputtering. This problem does not arise when acid is added to water, because $\bar{H}^E_{H_2SO_4}$ is less than a third the infinite-dilution value of water.

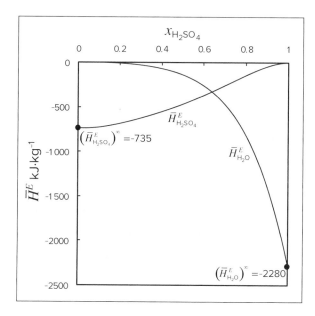

Figure 11.6: Partial excess enthalpies for $H_2SO_4(1)/H_2O(2)$ at 25°C.

Heats of Solution

When solids or gases are dissolved in liquids, the heat effect is called a *heat of solution* and is based on the dissolution of *1 mol of solute*. If species 1 is the solute, then x_1 is the moles of solute per mole of solution. Because ΔH is the heat effect per mole of solution, $\Delta H/x_1$ is the heat effect per mole of solute. Thus,

$$\widetilde{\Delta H} = \frac{\Delta H}{x_1}$$

where $\widetilde{\Delta H}$ is the heat of solution on the basis of one mole of *solute*.

Solution processes are conveniently represented by *physical-change* equations analogous to chemical-reaction equations. When 1 mol of LiCl(s) is mixed with 12 mol of H_2O, the process is represented by:

$$LiCl(s) + 12H_2O(l) \rightarrow LiCl(12H_2O)$$

The designation $LiCl(12H_2O)$ represents a solution of 1 mol of LiCl dissolved in 12 mol of H_2O. The heat of solution for this process at 25°C and 1 bar is $\widetilde{\Delta H} = -33,614$ J. This means that the enthalpy of 1 mol of LiCl in 12 mol of H_2O is 33,614 J less than the combined enthalpies of 1 mol of pure LiCl(*s*) and 12 mol of pure $H_2O(l)$. Equations for physical changes such as this are readily combined with equations for chemical reactions. This is illustrated in the following example, which incorporates the dissolution process just described.

Example 11.3

Calculate the heat of formation of LiCl in 12 mol of H_2O at 25°C.

Solution 11.3

The process implied by the problem statement results in the formation from its constituent elements of 1 mol of LiCl *in solution* in 12 mol of H_2O. The equation representing this process is obtained as follows:

$$
\begin{array}{lll}
\text{Li} + \tfrac{1}{2}\text{Cl}_2 \rightarrow \text{LiCl}(s) & \Delta H^\circ_{298} = -408{,}610 \text{ J} \\
\underline{\text{LiCl}(s) + 12\text{H}_2\text{O}(l) \rightarrow \text{LiCl}(12\text{H}_2\text{O})} & \underline{\widetilde{\Delta H}_{298} = -33{,}614 \text{ J}} \\
\text{Li} + \tfrac{1}{2}\text{Cl}_2 + 12\text{H}_2\text{O}(l) \rightarrow \text{LiCl}(12\text{H}_2\text{O}) & \Delta H^\circ_{298} = -442{,}224 \text{ J}
\end{array}
$$

The first reaction describes a chemical change resulting in the formation of LiCl(*s*) from its elements, and the enthalpy change accompanying this reaction is the standard heat of formation of LiCl(*s*) at 25°C. The second reaction represents the physical change resulting in the dissolution of 1 mol of LiCl(*s*) in 12 mol of $H_2O(l)$, and the enthalpy change is a heat of solution. The overall enthalpy change, −442,224 J, is the heat of formation of LiCl *in* 12 mol of H_2O. This figure does *not* include the heat of formation of the H_2O.

Often heats of solution are not reported directly but must be determined from heats of formation by the reverse of the calculation just illustrated. Typical are data for the heats of formation of 1 mol of LiCl:[4]

LiCl(*s*)	−408,610 J
LiCl · H_2O(*s*)	−712,580 J
LiCl · $2H_2O$(*s*)	−1,012,650 J
LiCl · $3H_2O$(*s*)	−1,311,300 J
LiCl in 3 mol H_2O	− 429,366 J
LiCl in 5 mol H_2O	−436,805 J
LiCl in 8 mol H_2O	−440,529 J
LiCl in 10 mol H_2O	−441,579 J
LiCl in 12 mol H_2O	−442,224 J
LiCl in 15 mol H_2O	−442,835 J

[4] "The NBS Tables of Chemical Thermodynamic Properties," *J. Phys. Chem. Ref. Data*, vol. 11, suppl. 2, pp. 2–291 and 2–292, 1982.

Heats of solution are readily calculated from these data. The reaction representing the dissolution of 1 mol of LiCl(s) in 5 mol of $H_2O(l)$ is obtained as follows:

$$Li + \tfrac{1}{2}Cl_2 + 5H_2O(l) \rightarrow LiCl(5H_2O) \qquad \Delta H^\circ_{298} = -436{,}805 \text{ J}$$

$$\frac{LiCl(s) \rightarrow Li + \tfrac{1}{2}Cl_2 \qquad\qquad\qquad \Delta H^\circ_{298} = 408{,}610 \text{ J}}{}$$

$$LiCl(s) + 5H_2O(l) \rightarrow LiCl(5H_2O) \qquad \widetilde{\Delta H}_{298} = -28{,}195 \text{ J}$$

This calculation can be carried out for each quantity of H_2O for which data are given. The results are then conveniently represented graphically by a plot of $\widetilde{\Delta H}$, the heat of solution per mole of solute, vs. $\tilde n$, the moles of solvent per mole of solute. The composition variable, $\tilde n \equiv n_2/n_1$, is related to x_1:

$$\tilde n = \frac{x_2(n_1 + n_2)}{x_1(n_1 + n_2)} = \frac{1 - x_1}{x_1} \qquad \text{whence} \quad x_1 = \frac{1}{1 + \tilde n}$$

The following equations therefore relate ΔH, the heat of mixing based on 1 mol of solution, and $\widetilde{\Delta H}$, the heat of solution based on 1 mol of solute:

$$\widetilde{\Delta H} = \frac{\Delta H}{x_1} = \Delta H(1 + \tilde n) \quad \text{or} \quad \Delta H = \frac{\widetilde{\Delta H}}{1 + \tilde n}$$

Figure 11.7 shows plots of $\widetilde{\Delta H}$ vs. $\tilde n$ for LiCl(s) and HCl(g) dissolved in water at 25°C. Data in this form are readily applied to the solution of practical problems.

Because water of hydration in solids is an integral part of a chemical compound, the heat of formation of a hydrated salt includes the heat of formation of the water of hydration. The dissolution of 1 mol of $LiCl \cdot 2H_2O(s)$ in 8 mol of H_2O produces a solution containing 1 mol LiCl in 10 mol of H_2O, represented by $LiCl(10H_2O)$. The equations which sum to give this process are:

$$Li + \tfrac{1}{2}Cl_2 + 10H_2O(l) \rightarrow LiCl(10H_2O) \qquad \Delta H^\circ_{298} = -441{,}579 \text{ J}$$

$$LiCl \cdot 2H_2O(s) \rightarrow Li + \tfrac{1}{2}Cl_2 + 2H_2 + O_2 \qquad \Delta H^\circ_{298} = 1{,}012{,}650 \text{ J}$$

$$\frac{2H_2 + O_2 \rightarrow 2H_2O(l) \qquad\qquad\qquad\qquad \Delta H^\circ_{298} = (2)(-285{,}830) \text{ J}}{}$$

$$LiCl \cdot 2H_2O(s) + 8H_2O(l) \rightarrow LiCl(10H_2O) \qquad \widetilde{\Delta H}_{298} = -589 \text{ J}$$

Example 11.4

A single-effect evaporator operating at atmospheric pressure concentrates a 15% (by weight) LiCl solution to 40%. The feed enters the evaporator at a rate of 2 kg·s⁻¹ at 25°C. The normal boiling point of a 40% LiCl solution is about 132°C, and its specific heat is estimated as 2.72 kJ·kg⁻¹·°C⁻¹. What is the heat-transfer rate in the evaporator?

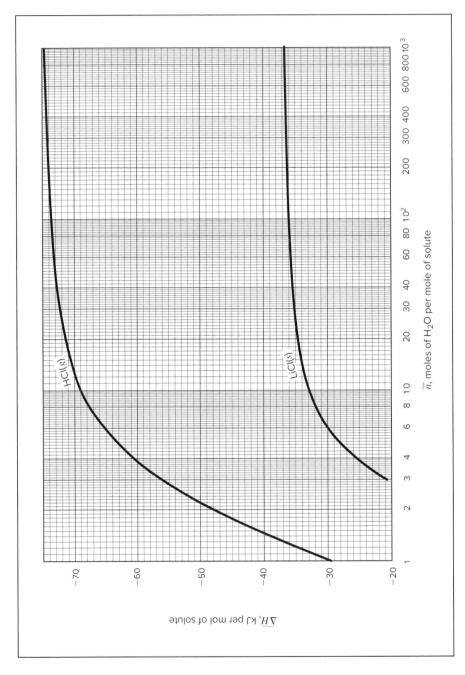

Figure 11.7: Heats of solution at 25°C (Based on data from "The NBS Tables of Chemical Thermodynamic Properties," *J. Phys. Chem. Ref. Data*, vol. 11, suppl. 2, 1982.)

Solution 11.4

The 2 kg of 15% LiCl solution entering the evaporator each second consists of 0.30 kg LiCl and 1.70 kg H_2O. A material balance shows that 1.25 kg of H_2O is evaporated and that 0.75 kg of 40% LiCl solution is produced. The process is represented by Fig. 11.8.

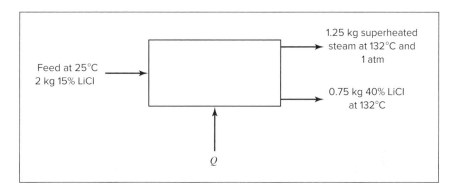

Figure 11.8: Process of Example 11.4.

The energy balance for this flow process is $\Delta H^t = Q$, where ΔH^t is the total enthalpy of the product streams minus the total enthalpy of the feed stream. Thus the problem reduces to finding ΔH^t from available data. Because enthalpy is a state function, the computational path for ΔH^t is immaterial and is selected for convenience and without reference to the actual path followed in the evaporator. The data available are heats of solution of LiCl in H_2O at 25°C (Fig. 11.7), and the calculational path shown in Fig. 11.9 allows their direct use.

The enthalpy changes for the individual steps shown in Fig. 11.9 must add to give the total enthalpy change:

$$\Delta H^t = \Delta H_a^t + \Delta H_b^t + \Delta H_c^t + \Delta H_d^t$$

The individual enthalpy changes are determined as follows:

- ΔH_a^t: This step involves the separation of 2 kg of a 15% LiCl solution into its pure constituents at 25°C. For this "unmixing" process the heat effect is the same as for the corresponding mixing process but is of opposite sign. For 2 kg of 15% LiCl solution, the moles of material entering are:

$$\frac{(0.15)(2000)}{42.39} = 7.077 \text{ mol LiCl} \qquad \frac{(0.85)(2000)}{18.015} = 94.366 \text{ mol } H_2O$$

Thus the solution contains 13.33 mol of H_2O per mole of LiCl. From Fig. 11.7 the heat of solution per mole of LiCl for $\tilde{n} = 13.33$ is $-33,800$ J. For the "unmixing" of 2 kg of solution,

$$\Delta H_a^t = (+33,800)(7.077) = 239,250 \text{ J}$$

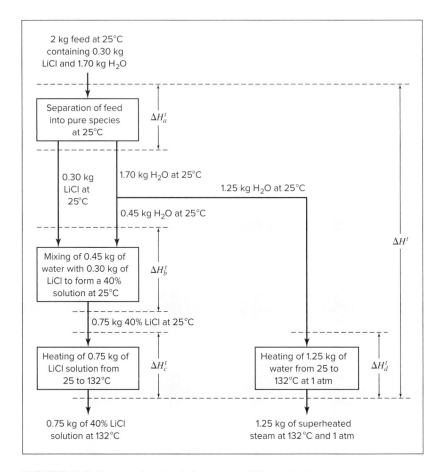

FIGURE 11.9: Computational path for process of Example 11.4.

- ΔH_b^t: This step results in the mixing of 0.45 kg of water with 0.30 kg of LiCl(s) to form a 40% solution at 25°C. This solution comprises:

    $$0.30 \text{ kg} \rightarrow 7.077 \text{ mol LiCl} \qquad \text{and} \qquad 0.45 \text{ kg} \rightarrow 24.979 \text{ mol H}_2\text{O}$$

    Thus the final solution contains 3.53 mol of H_2O per mole of LiCl. From Fig. 11.7 the heat of solution per mole of LiCl for $\tilde{n} = 3.53$ is −23,260 J. Therefore,

    $$\Delta H_b^t = (-23{,}260)(7.077) = -164{,}630 \text{ J}$$

- ΔH_c^t: For this step 0.75 kg of 40% LiCl solution is heated from 25 to 132°C. Because C_P is taken to be constant, $\Delta H_c^t = m\,C_P\,\Delta T$,

    $$\Delta H_c^t = (0.75)(2.72)(132 - 25) = 218.28 \text{ kJ} = 218{,}280 \text{ J}$$

- ΔH_d^t: In this step liquid water is vaporized and heated to 132°C. The enthalpy change is obtained from the steam tables:

    $$\Delta H_d^t = (1.25)(2740.3 - 104.8) = 3294.4 \text{ kJ} = 3{,}294{,}400 \text{ J}$$

Adding the individual enthalpy changes gives:

$$\Delta H = \Delta H_a^t + \Delta H_b^t + \Delta H_c^t + \Delta H_d^t$$
$$= 239{,}250 - 164{,}630 + 218{,}280 + 3{,}294{,}400 = 3{,}587{,}300 \text{ J}$$

The required heat-transfer rate is therefore $3587.3 \text{ kJ} \cdot \text{s}^{-1}$. Although the enthalpy of vaporization of water dominates the overall heat requirement, the heat effect associated with the difference in heat of solution at the initial and final concentration is not negligible.

11.3 SYNOPSIS

After studying this chapter, including the end-of-chapter problems, one should be able to:

- Define, in words and in equations, standard property changes of mixing

- Compute excess properties and partial excess properties from corresponding property changes of mixing

- Interpret and apply enthalpy-concentration diagrams

- Understand conventions for tabulating heats of solution and use them to compute heat effects of mixing and dissolution processes

11.4 PROBLEMS

11.1. At 25°C and atmospheric pressure the volume change of mixing of binary liquid mixtures of species 1 and 2 is given by the equation

$$\Delta V = x_1 x_2 (45 x_1 + 25 x_2)$$

where ΔV is in $\text{cm}^3 \cdot \text{mol}^{-1}$. At these conditions, $V_1 = 110$ and $V_2 = 90 \text{ cm}^3 \cdot \text{mol}^{-1}$. Determine the partial molar volumes $\bar{V}_1$ and $\bar{V}_2$ in a mixture containing 40-mol-% of species 1 at the given conditions.

11.2. The volume change of mixing $(\text{cm}^3 \cdot \text{mol}^{-1})$ for the system ethanol(1)/methyl butyl ether(2) at 25°C is given by the equation

$$\Delta V = x_1 x_2 [-1.026 + 0.0220(x_1 - x_2)]$$

Given that $V_1 = 58.63$ and $V_2 = 118.46 \text{ cm}^3 \cdot \text{mol}^{-1}$, what volume of mixture is formed when 750 cm^3 of pure species 1 is mixed with 1500 cm^3 of species 2 at 25°C? What would be the volume if an ideal solution were formed?

11.3. If $LiCl \cdot 2H_2O(s)$ and $H_2O(l)$ are mixed isothermally at 25°C to form a solution containing 10 mol of water for each mole of LiCl, what is the heat effect per mole of solution?

11.4. If a liquid solution of HCl in water, containing 1 mol of HCl and 4.5 mol of H_2O, absorbs an additional 1 mol of HCl(g) at a constant temperature of 25°C, what is the heat effect?

11.5. What is the heat effect when 20 kg of LiCl(s) is added to 125 kg of an aqueous solution containing 10-wt-% LiCl in an isothermal process at 25°C?

11.6. An LiCl/H_2O solution at 25°C is made by *adiabatically* mixing cool water at 10°C with a 20-mol-% LiCl/H_2O solution at 25°C. What is the composition of the solution formed?

11.7. A 20-mol-% LiCl/H_2O solution at 25°C is made by mixing a 25-mol-% LiCl/H_2O solution at 25°C with chilled water at 5°C. What is the heat effect in joules per mole of final solution?

11.8. A 20-mol-% LiCl/H_2O solution is made by six different mixing processes:

(*a*) Mix LiCl(s) with H_2O(*l*).
(*b*) Mix H_2O(*l*) with a 25-mol-% LiCl/H_2O solution.
(*c*) Mix LiCl·H_2O(s) with H_2O(*l*).
(*d*) Mix LiCl(s) with a 10-mol-% LiCl/H_2O solution.
(*e*) Mix a 25-mol-% LiCl/H_2O solution with a 10-mol-% LiCl/H_2O solution.
(*f*) Mix LiCl·H_2O(s) with a 10-mol-% LiCl/H_2O solution.

Mixing in all cases is isothermal, at 25°C. For each part determine the heat effect in $J·mol^{-1}$ of final solution.

11.9. A stream of 12 $kg·s^{-1}$ of $Cu(NO_3)_2·6H_2O$ and a stream of 15 $kg·s^{-1}$ of water, both at 25°C, are fed to a tank where mixing takes place. The resulting solution passes through a heat exchanger that adjusts its temperature to 25°C. What is the rate of heat transfer in the exchanger?

- For $Cu(NO_3)_2$, $\Delta H^{\circ}_{f_{298}} = -302.9$ kJ.
- For $Cu(NO_3)_2 · 6H_2O$, $\Delta H^{\circ}_{f_{298}} = -2110.8$ kJ.
- The heat of solution of 1 mol of $Cu(NO_3)_2$ in water at 25°C is -47.84 kJ, independent of $\tilde{n}$ for the concentrations of interest here.

11.10. A liquid solution of LiCl in water at 25°C contains 1 mol of LiCl and 7 mol of water. If 1 mol of LiCl·$3H_2O$(s) is dissolved isothermally in this solution, what is the heat effect?

11.11. You need to produce an aqueous LiCl solution by mixing LiCl·$2H_2O$(s) with water. The mixing occurs *both* adiabatically and without change in temperature at 25°C. Determine the mole fraction of LiCl in the final solution.

11.12. Data from the Bureau of Standards (*J. Phys. Chem. Ref. Data*, vol. 11, suppl. 2, 1982) include the following heats of formation for 1 mol of $CaCl_2$ in water at 25°C:

$CaCl_2$ in 10 mol H_2O	-862.74 kJ
$CaCl_2$ in 15 mol H_2O	-867.85 kJ
$CaCl_2$ in 20 mol H_2O	-870.06 kJ
$CaCl_2$ in 25 mol H_2O	-871.07 kJ
$CaCl_2$ in 50 mol H_2O	-872.91 kJ
$CaCl_2$ in 100 mol H_2O	-873.82 kJ
$CaCl_2$ in 300 mol H_2O	-874.79 kJ
$CaCl_2$ in 500 mol H_2O	-875.13 kJ
$CaCl_2$ in 1000 mol H_2O	-875.54 kJ

From these data prepare a plot of $\widetilde{\Delta H}$, the heat of solution at 25°C of $CaCl_2$ in water, vs. $\tilde{n}$, the mole ratio of water to $CaCl_2$.

11.13. A liquid solution contains 1 mol of $CaCl_2$ and 25 mol of water. Using data from Prob. 11.12, determine the heat effect when an additional 1 mol of $CaCl_2$ is dissolved isothermally in this solution.

11.14. Solid $CaCl_2 \cdot 6H_2O$ and liquid water at 25°C are mixed *adiabatically* in a continuous process to form a brine of 15-wt-% $CaCl_2$. Using data from Prob. 11.12, determine the temperature of the brine solution formed. The specific heat of a 15-wt-% aqueous $CaCl_2$ solution at 25°C is 3.28 $kJ \cdot kg^{-1} \cdot °C^{-1}$.

11.15. Consider a plot of $\widetilde{\Delta H}$, the heat of solution based on 1 mol of solute (species 1), vs. $\tilde{n}$, the moles of solvent per mole of solute, at constant T and P. Figure 11.4 is an example of such a plot, except that the plot considered here has a linear rather than logarithmic scale along the abscissa. Let a tangent drawn to the $\widetilde{\Delta H}$ vs. $\tilde{n}$ curve intercept the ordinate at point I.

(*a*) Prove that the slope of the tangent at a particular point is equal to the partial excess enthalpy of the solvent in a solution with the composition represented by $\tilde{n}$; i.e., prove that:

$$\frac{d\widetilde{\Delta H}}{d\tilde{n}} = \bar{H}_2^E$$

(*b*) Prove that the intercept I equals the partial excess enthalpy of the solute in the same solution; i.e., prove that:

$$I = \bar{H}_1^E$$

11.16. Suppose that ΔH for a particular solute(1)/solvent(2) system is represented by the equation:

$$\Delta H = x_1 x_2 (A_{21} x_1 + A_{12} x_2) \qquad (A)$$

Relate the behavior of a plot of $\widetilde{\Delta H}$ vs. $\tilde{n}$ to the features of this equation. Specifically, rewrite Eq. (A) in the form $\widetilde{\Delta H}(\tilde{n})$, and then show that:

(a) $\lim\limits_{\tilde{n}\to 0} \widetilde{\Delta H} = 0$

(b) $\lim\limits_{\tilde{n}\to\infty} \widetilde{\Delta H} = A_{12}$

(c) $\lim\limits_{\tilde{n}\to 0} d\widetilde{\Delta H}/d\tilde{n} = A_{21}$

11.17. If the heat of mixing at temperature t_0 is ΔH_0 and if the heat of mixing of the same solution at temperature t is ΔH, show that the two heats of mixing are related by:

$$\Delta H = \Delta H_0 + \int_{t_0}^{t} \Delta C_P \, dt$$

where ΔC_P is the heat-capacity change of mixing, defined by Eq. (11.1).

Heat of solution data for probs. 11.18 through 11.30 can be obtained from Fig. 11.4.

11.18. What is the heat effect when 75 kg of H_2SO_4 is mixed with 175 kg of an aqueous solution containing 25-wt-% H_2SO_4 in an isothermal process at 300 K?

11.19. For a 50-wt-% aqueous solution of H_2SO_4 at 350 K, what is the excess enthalpy H^E in $kJ \cdot kg^{-1}$?

11.20. A single-effect evaporator concentrates a 20-wt-% aqueous solution of H_2SO_4 to 70-wt-%. The feed rate is 15 $kg \cdot s^{-1}$, and the feed temperature is 300 K. The evaporator is maintained at an absolute pressure of 10 kPa, at which pressure the boiling point of 70-wt-% H_2SO_4 is 102°C. What is the heat-transfer rate in the evaporator?

11.21. What is the heat effect when sufficient $SO_3(l)$ at 25°C is reacted with H_2O at 25°C to give a 50-wt-% H_2SO_4 solution at 60°C?

11.22. A mass of 70 kg of 15-wt-% solution of H_2SO_4 in water at 70°C is mixed at atmospheric pressure with 110 kg of 80-wt-% H_2SO_4 at 38°C. During the process heat in the amount of 20,000 kJ is transferred from the system. Determine the temperature of the product solution.

11.23. An insulated tank, open to the atmosphere, contains 750 kg of 40-wt-% sulfuric acid at 290 K. It is heated to 350 K by injection of saturated steam at 1 bar, which fully condenses in the process. How much steam is required, and what is the final concentration of H_2SO_4 in the tank?

11.24. Saturated steam at 3 bar is throttled to 1 bar and mixed adiabatically with (and condensed by) 45-wt-% sulfuric acid at 300 K in a flow process that raises the temperature of the acid to 350 K. How much steam is required for each pound *mass* of entering acid, and what is the concentration of the hot acid?

11.25. For a 35-wt-% aqueous solution of H_2SO_4 at 300 K, what is the heat of mixing ΔH in $kJ \cdot kg^{-1}$?

11.26. If pure liquid H_2SO_4 at 300 K is added adiabatically to pure liquid water at 300 K to form a 40-wt-% solution, what is the final temperature of the solution?

11.27. A liquid solution containing 1 kg mol H_2SO_4 and 7 kg mol H_2O at 300 K absorbs 0.5 kg mol of $SO_3(g)$, also at 300 K, forming a more concentrated sulfuric acid solution. If the process occurs isothermally, determine the heat transferred.

11.28. Determine the heat of mixing ΔH of sulfuric acid in water and the partial specific enthalpies of H_2SO_4 and H_2O for a solution containing 65-wt-% H_2SO_4 at 300 K.

11.29. It is proposed to cool a stream of 75-wt-% sulfuric acid solution at 330 K by diluting it with chilled water at 280 K. Determine the amount of water that must be added to 1 kg of 75-wt-% acid before cooling below 330 K actually occurs.

11.30. The following liquids, all at atmospheric pressure and 300 K, are mixed: 25 kg of pure water, 40 kg of pure sulfuric acid, and 75 kg of 25-wt-% sulfuric acid.

(a) How much heat is liberated if mixing is isothermal at 300 K?
(b) The mixing process is carried out in two steps: First, the pure sulfuric acid and the 25-wt-% solution are mixed, and the total heat of part (a) is extracted; second, the pure water is added adiabatically. What is the temperature of the intermediate solution formed in the first step?

11.31. A large quantity of very dilute aqueous NaOH solution is neutralized by addition of the stoichiometric amount of a 10-mol-% aqueous HCl solution. Estimate the heat effect per mole of NaOH neutralized if the tank is maintained at 25°C and 1(atm) and the neutralization reaction goes to completion. Data:

- For NaCl, $\lim\limits_{\tilde{n}\to\infty} \widetilde{\Delta H} = 3.88$ kJ·mol^{-1}
- For NaOH, $\lim\limits_{\tilde{n}\to\infty} \widetilde{\Delta H} = -44.50$ kJ·mol^{-1}

11.32. A large quantity of very dilute aqueous HCl solution is neutralized by the addition of the stoichiometric amount of a 10-mol-% aqueous NaOH solution. Estimate the heat effect per mole of HCl neutralized if the tank is maintained at 25°C and 1(atm) and the neutralization reaction goes to completion.

- For NaOH($9H_2O$), $\widetilde{\Delta H} = 45.26$ kJ·mol^{-1}
- For NaCl, $\lim\limits_{\tilde{n}\to\infty} \widetilde{\Delta H} = 3.88$ kJ·mol^{-1}

11.33. (a) Making use of Eqs. (10.15) and (10.16), written for excess properties, show for a binary system that:

$$\bar{M}_1^E = x_2^2\left(X + x_1\frac{dX}{dx_1}\right) \quad \text{and} \quad \bar{M}_2^E = x_1^2\left(X - x_2\frac{dX}{dx_1}\right)$$

where

$$X \equiv \frac{M^E}{x_1 x_2}$$

(b) Plot on a single graph the values of $H^E/(x_1x_2)$, $\bar{H}_1^E$, and $\bar{H}_2^E$ determined from the following heat-of-mixing data for the $H_2SO_4(1)/H_2O(2)$ system at 25°C:

x_1	$-\Delta H/kJ \cdot kg^{-1}$
0.10	73.27
0.20	144.21
0.30	208.64
0.40	262.83
0.50	302.84
0.60	323.31
0.70	320.98
0.80	279.58
0.85	237.25
0.90	178.87
0.95	100.71

x_1 = mass fraction H_2SO_4

Explain with reference to these plots why sulfuric acid is diluted by adding acid to water rather than water to acid.

11.34. A 90-wt-% aqueous H_2SO_4 solution at 25°C is added over a period of 6 hours to a tank containing 4000 kg of pure water also at 25°C. The final concentration of acid in the tank is 50-wt-%. The contents of the tank are cooled continuously to maintain a constant temperature of 25°C. Because the cooling system is designed for a constant rate of heat transfer, this requires the addition of acid at a variable rate. Determine the instantaneous 90-%-acid rate as a function of time, and plot this rate ($kg \cdot s^{-1}$) vs. time. The data of the preceding problem may be fit to a cubic equation expressing $H^E/(x_1x_2)$ as a function of x_1, and the equations of the preceding problem then provide expressions for $\bar{H}_1^E$ and $\bar{H}_2^E$.

11.35. Develop Eq. (11.12) for ΔS^{id} by appropriate application of Eqs. (5.36) and (5.37) to a mixing process.

11.36. Ten thousand (10,000) $kg \cdot h^{-1}$ of an 80-wt-% H_2SO_4 solution in water at 300 K is continuously diluted with chilled water at 280 K to yield a stream containing 50-wt-% H_2SO_4 at 330 K.

(a) What is the mass flow rate of chilled water in $kg \cdot h^{-1}$?

(b) What is the rate of heat transfer in $kJ \cdot h^{-1}$ for the mixing process? Is heat added or removed?

(c) If the mixing occurred *adiabatically*, what would be the temperature of the product stream? Assume here the same inlet conditions and the same product composition as for part (b).

Heat of solution data is available in Fig. 11.4.

Chapter 12

Phase Equilibrium: Introduction

The analysis of phase equilibrium provides the primary motivation for development of the framework of solution thermodynamics and for the creation of models of mixture properties to be used with this framework. As a practical matter, preferential segregation of chemical species in a particular phase provides the basis for nearly all industrial processes for separation and purification of materials, from the distillation of petroleum or alcoholic beverages to the crystallization of pharmaceutical compounds to the capture of carbon dioxide from power plant effluent. As a result, the analysis of phase equilibrium problems is one of the core competencies expected of a chemical engineer. The present chapter focuses on the qualitative description of equilibrium between fluid phases. Chapter 13 then presents the tools required for quantitative analysis of vapor/liquid equilibrium. Chapter 15 treats other types of phase equilibrium in greater detail.

12.1 THE NATURE OF EQUILIBRIUM

Equilibrium is a condition in which no changes occur in the macroscopic properties of an isolated system with time. At equilibrium, all potentials that may cause change are exactly balanced, so no driving force exists for any change in the system. An isolated system consisting of liquid and vapor phases in intimate contact eventually reaches a final state wherein no tendency exists for change to occur within the system. The temperature, pressure, and phase compositions reach final values which thereafter remain fixed. The system is in equilibrium. Nevertheless, at the microscopic level, conditions are not static. The molecules comprising a phase at a given instant are not the same molecules that later occupy the same phase. Molecules constantly pass from one phase to the other. However, the average rate of passage of molecules is the same in both directions, and no net interphase transfer of material occurs. In engineering practice, the assumption of equilibrium is justified when it leads to results of satisfactory accuracy. For example, in the reboiler for a distillation column, equilibrium between vapor and liquid phases is commonly assumed. For finite vaporization rates this is an approximation, but it does not introduce significant error into engineering calculations.

12.2   THE PHASE RULE. DUHEM'S THEOREM

The phase rule for nonreacting systems, presented without proof in Sec. 3.1, results from the application of a rule of algebra. Thus, the number of variables that may be independently fixed in a system at equilibrium is the difference between the total number of variables that characterize the intensive state of the system and the number of independent equations that can be written relating those variables.

The *intensive* state of a PVT system containing N chemical species and π phases in equilibrium is characterized by its temperature T, pressure P, and $N-1$ mole fractions[1] for each phase. The number of these phase-rule variables is $2 + (N-1)(\pi)$. The masses or amounts of the phases are not phase-rule variables because they have no influence on the intensive state of the system.

As will become clear later in this chapter, an independent phase-equilibrium equation may be written connecting intensive variables for each of the N species for each pair of phases present. Thus, the number of independent phase-equilibrium equations is $(\pi-1)(N)$. The difference between the number of phase-rule variables and the number of independent equations connecting them is the number of variables that may be independently fixed. Called the degrees of freedom of the system F, the number is:

$$F = 2 + (N-1)(\pi) - (\pi-1)(N)$$

Upon reduction, this becomes the phase rule:

$$\boxed{F = 2 - \pi + N} \tag{3.1}$$

Duhem's theorem is another rule, similar to the phase rule, that applies to the extensive state of a closed system at equilibrium. When both the extensive state and the intensive state of the system are fixed, the state of the system is said to be *completely determined*, and it is characterized not only by the $2 + (N-1)\pi$ intensive phase-rule variables but also by the π extensive variables represented by the masses (or mole numbers) of the phases. Thus the total number of variables is:

$$2 + (N-1)\pi + \pi = 2 + N\pi$$

For a closed system formed from specified amounts of the chemical species present, a material-balance equation can be written for each of the N chemical species, providing N more equations. These, in addition to the $(\pi-1)N$ phase-equilibrium equations, represent a number of independent equations equal to:

$$(\pi-1)N + N = \pi N$$

The difference between the number of variables and the number of equations is therefore:

$$2 + N\pi - \pi N = 2$$

On the basis of this result, Duhem's theorem is stated as follows:

> **For any closed system formed from known amounts of prescribed chemical species, the equilibrium state is completely determined when any two independent variables are fixed.**

---

[1] Only $N-1$ mole fractions are required, because $\Sigma_i x_i = 1$.

The two independent variables subject to specification may in general be either intensive or extensive. However, the number of *independent intensive* variables is given by the phase rule. Thus when $F = 1$, at least one of the two variables must be extensive, and when $F = 0$, both must be extensive.

12.3 VAPOR/LIQUID EQUILIBRIUM: QUALITATIVE BEHAVIOR

Vapor/liquid equilibrium (VLE) is the state of coexistence of liquid and vapor phases. In this qualitative discussion, we limit consideration to systems comprised of two chemical species because systems of greater complexity cannot be adequately represented graphically.

For a system comprised of two chemical species ($N = 2$), the phase rule becomes $F = 4 - \pi$. Because there must be at least one phase ($\pi = 1$), the maximum number of phase-rule variables that must be specified to fix the intensive state of the system is *three: P, T,* and one mole (or mass) fraction. All equilibrium states of the system can therefore be represented in three-dimensional *P-T*-composition space. Within this space, the states of *pairs* of phases coexisting at equilibrium ($F = 4 - 2 = 2$) define surfaces. A schematic three-dimensional diagram illustrating these surfaces for VLE is shown in Fig. 12.1.

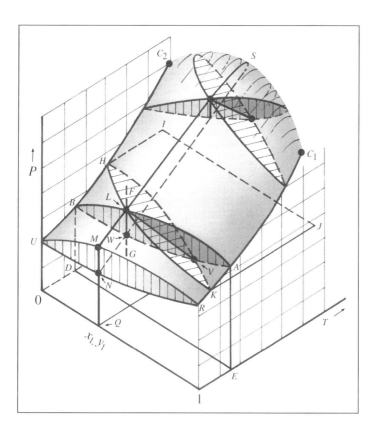

Figure 12.1: *PTxy* diagram for vapor/liquid equilibrium.

This figure shows schematically the P-T-composition surfaces that contain the equilibrium states of saturated vapor and saturated liquid for species 1 and 2 of a binary system. Here, species 1 is the "lighter" or more volatile species. The lower surface contains the saturated-vapor states; it is the P-T-y_1 surface. The upper surface contains the saturated-liquid states; it is the P-T-x_1 surface. These surfaces intersect along the lines $RKAC_1$ and $UBHC_2$, which represent the vapor pressure-vs.-T curves for pure species 1 and 2. Moreover, the lower and upper surfaces form a continuous rounded surface across the top of the diagram between C_1 and C_2, the critical points of pure species 1 and 2; the critical points of the various mixtures of the two species lie along a line on the rounded edge of the surface between C_1 and C_2. This critical locus is defined by the points at which vapor and liquid phases in equilibrium become identical. Because of this geometric feature of an open end (at low T and P) and three closed edges formed by the critical locus and the pure component vapor pressure curves, the pair of surfaces is often called the "phase envelope." No stable single phase exists at points inside the phase envelope. When T, P, and the overall composition correspond to a point inside the envelope, separation into vapor and liquid phases occurs. These have compositions that fall on the lower and upper surface, respectively, at the system T and P.

The subcooled-liquid region lies above the upper surface of Fig. 12.1; the superheated-vapor region lies below the lower surface. The interior space between the two surfaces is the region of coexistence of both liquid and vapor phases. If one starts with a liquid at conditions represented by point F and reduces the pressure at constant temperature and composition along vertical line FG, the first bubble of vapor appears at point L, which lies on the upper surface. Thus, L is called a *bubblepoint*, and the upper surface is called the bubblepoint surface. The state of the vapor bubble in equilibrium with the liquid at L is represented by a point on the lower surface at the temperature and pressure of L. This point is indicated by V. Line LV is an example of a *tie line*, which connects points representing phases in equilibrium.

As the pressure is further reduced along line FG, more liquid vaporizes until at W the process is complete. Thus W lies on the lower surface and represents a state of saturated vapor having the mixture composition. Because W is the point at which the last drops of liquid (dew) disappear, it is called a *dewpoint*, and the lower surface is called the dewpoint surface. Continued reduction of pressure produces expansion of the vapor in the superheated vapor region.

Because of the complexity of three-dimensional diagrams like Fig. 12.1, the detailed characteristics of binary VLE are usually depicted by two-dimensional graphs that display what is seen on various planes that cut the three-dimensional diagram. The three principal planes, each perpendicular to one of the coordinate axes, are illustrated in Fig. 12.1. Thus a vertical plane perpendicular to the temperature axis is outlined as $AEDBLA$. The lines on this plane form a P-x_1-y_1 phase diagram at constant T. If the lines from several such planes are projected on a single parallel plane, a diagram like Fig. 12.2(a) is obtained. It shows P-x_1-y_1 plots for three different temperatures. The one for T_a represents the section of Fig. 12.1 indicated by $AEDBLA$. The horizontal lines are tie lines connecting the compositions of phases in equilibrium. The temperatures T_b and T_d lie between the two pure-species critical temperatures identified by C_1 and C_2 in Fig. 12.1. The curves for these two temperatures therefore do not extend all the way across the diagram. The mixture critical points are denoted by the letter C. Each is a tangent point at which a horizontal line touches the curve. This is so because all tie lines connecting phases in equilibrium are horizontal, and the tie line connecting *identical* phases (the definition of a critical point) must therefore be the last such line to cut the diagram.

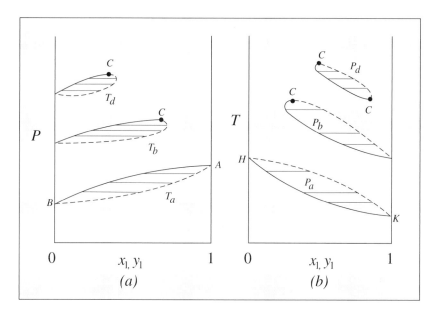

Figure 12.2: (*a*) *Pxy* diagram for three temperatures. (*b*) *Txy* diagram for three pressures.
— Saturated liquid (bubble line); - - - Saturated vapor (dew line)

A horizontal plane passing through Fig. 12.1 perpendicular to the P axis is identified by *KJIHLK*. Viewed from above, the lines on this plane represent a T-x_1-y_1 diagram. When lines for several pressures are projected on a parallel plane, the resulting diagram is that shown in Fig. 12.2(*b*). This figure is analogous to Fig. 12.2(*a*), except that it represents phase behavior for three constant pressures, P_a, P_b, and P_d. The one for P_a represents the section of Fig. 12.1 indicated by *KJIHLK*. Pressure P_b lies between the critical pressures of the two pure species at points C_1 and C_2. Pressure P_d is above the critical pressures of *both* pure species; therefore, the T-x_1-y_1 diagram appears as an "island." Similar P-x_1-y_1 behavior [Fig. 12.2(*a*)] is unusual. Note that on the P-x_1-y_1 plot, the upper curve represents the saturated liquid and the lower curve represents the saturated vapor, but for the T-x_1-y_1, the upper curve represents saturated vapor and the lower curve represents saturated liquid. To avoid confusion, one must keep in mind the fact that vapors form at high T and low P.

Other possible plots include vapor mole fraction y_1 vs. liquid mole fraction x_1 for either the constant-T conditions of Fig. 12.2(*a*) or the constant-P conditions of Fig. 12.2(*b*). Such plots reduce the dimensionality of the representation further by representing the coexisting phases by a single curve, with no information about T or P, rather than as a pair of curves bounding a two-dimensional region. Thus, they convey less information than a T-x_1-y_1 or P-x_1-y_1 plot, but are more convenient for rapidly relating phase compositions at a fixed T or P.

The third plane identified in Fig. 12.1, vertical and perpendicular to the composition axis, passes through points *SLMN* and *Q*. When projected on a parallel plane, the lines from several planes form a diagram such as that shown in Fig. 12.3. This is the *PT* diagram; lines UC_2 and RC_1 are vapor-pressure curves for the pure species, identified by the same letters as in Fig. 12.1. Each interior loop represents the *PT* behavior of saturated liquid and of saturated vapor for a *system of fixed overall composition*. The different loops are for different

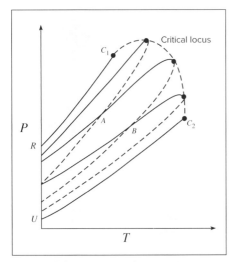

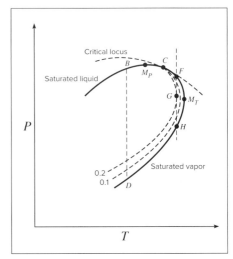

Figure 12.3: *PT* diagram for several compositions.

Figure 12.4: Portion of a *PT* diagram in the critical region.

compositions. Clearly, the *PT* relation for saturated liquid is different from that for saturated vapor of the same composition. This is in contrast to the behavior of a pure species, for which the bubble and dew lines coincide. At points *A* and *B* in Fig. 12.3, saturated-liquid and saturated-vapor lines intersect. At such points a saturated liquid of one composition and a saturated vapor of another composition have the same *T* and *P*, and the two phases are therefore in equilibrium. The tie lines connecting the coinciding points at *A* and at *B* are perpendicular to the *PT* plane, as illustrated by the tie line *LV* in Fig. 12.1.

Critical Points of Binary Mixtures and Retrograde Condensation

The critical point of a binary mixture occurs where the nose of a loop in Fig. 12.3 is tangent to the envelope curve. Put another way, the envelope curve is the critical locus. One can verify this by considering two closely adjacent loops and noting what happens to the point of intersection as their separation becomes infinitesimal. Figure 12.3 shows that the location of the critical point on the nose of the loop varies with composition. For a pure species the critical point is the highest temperature and highest pressure at which vapor and liquid phases can coexist, but for a mixture it is in general neither. Therefore under certain conditions a condensation process occurs as the result of a *reduction* in pressure.

Consider the enlarged nose section of a single *PT* loop shown in Fig. 12.4. The critical point is at *C*. The points of maximum pressure and maximum temperature are identified as M_P and M_T, respectively. The interior dashed curves indicate the fraction of the overall system that is liquid in a two-phase mixture of liquid and vapor. To the left of the critical point *C* a reduction in pressure along a line such as *BD* is accompanied by vaporization of liquid from bubblepoint to dewpoint, as would be expected. However, if the original condition corresponds to point *F*, a state of saturated *vapor*, liquefaction occurs upon reduction of the pressure and reaches a maximum at point *G*, after which vaporization takes place until the dewpoint is reached at point *H*. This phenomenon is called *retrograde condensation*. It can be important in the operation of deep natural-gas wells where the pressure and temperature in the underground formation may be at conditions represented by point *F*. If the pressure at the wellhead is that of point *G*, the product stream from the well is an equilibrium mixture of liquid and vapor. Because the less-volatile species are concentrated in the liquid phase, significant separation is accomplished. Within the underground formation itself, the pressure tends to drop as the gas supply is depleted. If not prevented, this leads to the formation of a liquid phase and a consequent reduction in the production of the well. Repressuring is therefore a common practice; that is, lean gas (gas from which the less-volatile species have been removed) is returned to the underground reservoir to maintain an elevated pressure.

A *PT* diagram for the ethane(1)/*n*-heptane(2) system is shown in Fig. 12.5, and a y_1-x_1 diagram for several pressures for the same system appears in Fig. 12.6. By convention, one typically plots as y_1 and x_1 the mole fractions of the more volatile species in the mixture. The maximum and minimum concentrations of the more volatile species obtainable by distillation at a given pressure are indicated by the points of intersection of the appropriate y_1-x_1 curve with the diagonal, for at these points the vapor and liquid have the same composition. They are in fact mixture critical points, unless $y_1 = x_1 = 0$ or $y_1 = x_1 = 1$. Point *A* in Fig. 12.6 represents the composition of the vapor and liquid phases at the maximum pressure at which the phases can coexist in the ethane/*n*-heptane system. The composition is about 77 mol-% ethane and the pressure is about 87.1 bar. The corresponding point on Fig. 12.5 is labeled *M*. A complete set of consistent phase diagrams for this system has been prepared by Barr-David.[2]

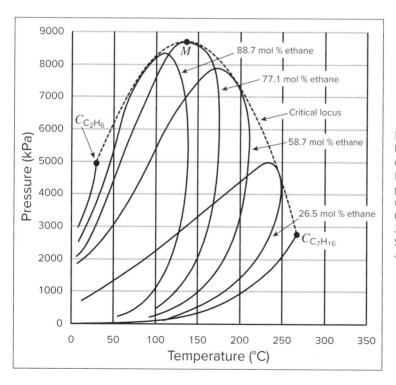

Figure 12.5: Ethane/*n*-heptane *PT* diagram. (Adapted from F. H. Barr-David, "Notes on phase relations of binary mixtures in the region of the critical point, AIChE Journal, Vol 2, Issue 3, September 1956, pp. 426–427.)

[2]F. H. Barr-David, *AIChEJ.*, vol. 2, p. 426, 1956.

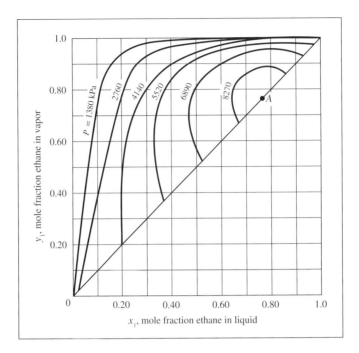

Figure 12.6: Ethane/*n*-heptane *yx* diagram. (Adapted from F. H. Barr-David, "Notes on phase relations of binary mixtures in the region of the critical point, AIChE Journal, Vol 2, Issue 3, September 1956, pp. 426–427.)

The *PT* diagram of Fig. 12.5 is typical for mixtures of nonpolar substances such as hydrocarbons. A *PT* diagram for a very different kind of system, methanol(1)/benzene(2), is shown in Fig. 12.7. The nature of the curves in this figure suggests how difficult it can be to predict phase behavior for species so dissimilar as methanol and benzene, especially at conditions near the mixture critical point.

Low-Pressure Vapor/Liquid Equilibrium Examples

Although VLE in the critical region is of considerable importance in the petroleum and natural-gas industries, most chemical processing is accomplished at much lower pressures. Figures 12.8 and 12.9 display common types of *Pxy* and *Txy* behavior at conditions well removed from the critical region.

Figure 12.8(*a*) shows data for tetrahydrofuran(1)/carbon tetrachloride(2) at 30°C. When the liquid phase behaves as an ideal solution, as defined in Chapter 10, and the vapor phase behaves as an ideal-gas-state mixture, then the system is said to follow Raoult's law. As discussed in Chapter 13, this is the simplest model of vapor/liquid equilibrium. For a system that obeys Raoult's law, the P-x_1 or bubblepoint curve is a straight line connecting the vapor pressures of the pure species. In Fig. 12.8(*a*), the bubblepoint curve lies below the linear P-x_1 relation characteristic of Raoult's-law behavior. When such negative departures from linearity become sufficiently large, relative to the difference between the two pure-species vapor pressures, the P-x_1 curve exhibits a minimum, as illustrated in Fig. 12.8(*b*) for the

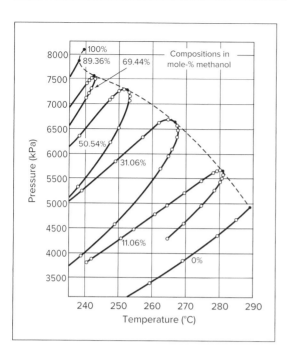

Figure 12.7: Methanol/benzene *PT* diagram. (Adapted from J. M. Skaates and W. B. Kay, "The phase relations of binary systems that form azeotropes," Chemical Engineering Science, Vol 19, Issue 7, July 1964, pp. 431–444.)

chloroform(1)/tetrahydrofuran(2) system at 30°C. This figure shows that the P-y_1 curve also has a minimum at the same point. Thus at this point where $x_1 = y_1$ the dewpoint and bubblepoint curves are tangent to the same horizontal line. A boiling liquid of this composition produces a vapor of exactly the same composition, and the liquid therefore does not change in composition as it evaporates. No separation of such a constant-boiling solution is possible by distillation. The term *azeotrope* is used to describe this state.[3]

The data for furan(1)/carbon tetrachloride(2) at 30°C shown by Fig. 12.8(*c*) provide an example of a system for which the P-x_1 curve lies above the linear P-x_1 relation. The system shown in Fig. 12.8(*d*) for ethanol(1)/toluene(2) at 65°C exhibits positive departures from linearity large enough to cause a *maximum* in the P-x_1 curve. This state is a maximum-pressure azeotrope. Just as for the minimum-pressure azeotrope, the vapor and liquid phases in equilibrium have the identical composition.

Appreciable negative departures from P-x_1 linearity reflect liquid-phase intermolecular attractions that are stronger between unlike than between like pairs of molecules. Conversely, appreciable positive departures result for solutions for which liquid-phase intermolecular interactions between like molecules are stronger than between unlike ones. In this latter case the forces between like molecules may be so strong as to prevent complete miscibility, and the system then forms two separate liquid phases over a range of compositions, as described later in this chapter.

Because distillation processes are carried out more nearly at constant pressure than at constant temperature, T-x_1-y_1 diagrams of data at constant P are of great practical interest. The four such diagrams corresponding to those of Fig. 12.8 are shown for atmospheric pressure in Fig. 12.9. Note that the dewpoint (T-y_1) curves lie above the bubblepoint

---

[3]A compilation of data for such states is given by J. Gmehling, J. Menke, J. Krafczyk, and K. Fischer, *Azeotropic Data*, 2nd ed., John Wiley & Sons, Inc., New York, 2004.

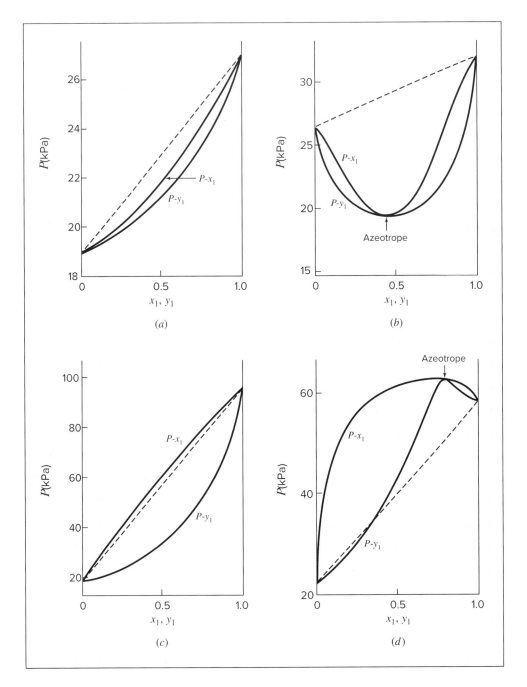

Figure 12.8: *Pxy* diagrams at constant *T:* (*a*) tetrahydrofuran(1)/carbon tetrachloride(2) at 30°C;
(*b*) chloroform(1)/tetrahydrofuran(2) at 30°C; (*c*) furan(1)/carbon tetrachloride(2) at 30°C;
(*d*) ethanol(1)/toluene(2) at 65°C. Dashed lines: *Px* relation for Raoult's law.

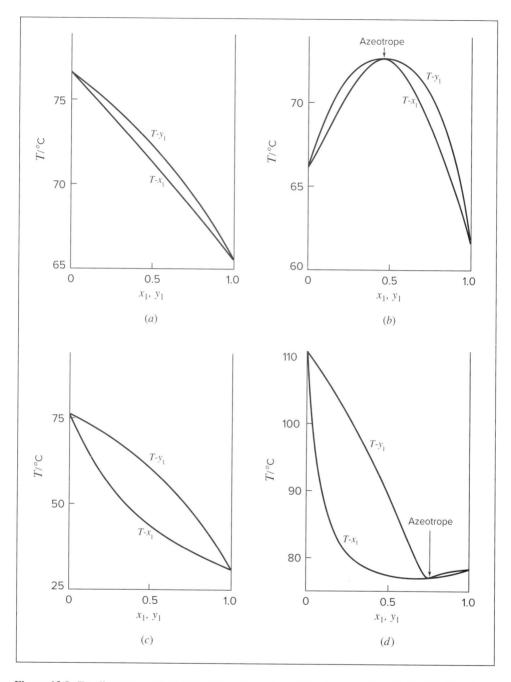

Figure 12.9: *Txy* diagrams at 101.3 kPa: (*a*) tetrahydrofuran(1)/carbon tetrachloride(2); (*b*) chloroform (1)/tetrahydrofuran(2); (*c*) furan(1)/carbon tetrachloride(2); (*d*) ethanol(1)/toluene(2).

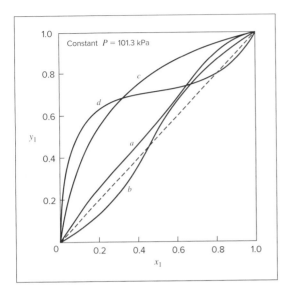

Figure 12.10: yx curves at 101.3 kPa: (a) tetrahydrofuran(1)/carbon tetrachloride(2); (b) chloroform(1)/ tetrahydrofuran(2); (c) furan(1)/ carbon tetrachloride(2); (d) ethanol(1)/ toluene(2).

(T-x_1) curves. Moreover, the minimum-pressure azeotrope of Fig. 12.8(b) appears as a maximum-temperature (or maximum-boiling) azeotrope on Fig. 12.9(b). An analogous correspondence exists between Figs. 12.8(d) and 12.9(d). The y_1-x_1 diagrams at constant P for the same four systems are shown in Fig. 12.10. The point at which a curve crosses the diagonal line of the diagram represents an azeotrope, for at such a point $y_1 = x_1$. Such yx diagrams are useful for qualitative analysis of distillation processes. The greater the separation between the yx curve and the diagonal line, the easier the separation. Examination of Fig. 12.10 immediately shows that complete separation of both tetrahydrofuran/carbon tetrachloride mixtures and furan/carbon tetrachloride mixtures by distillation is possible, and that the separation of the furan/carbon tetrachloride mixture will be much easier to accomplish. Likewise, the diagram shows that the other two systems form azeotropes and cannot be completely separated by distillation at this pressure.

Evaporation of a Binary Mixture at Constant Temperature

The P-x_1-y_1 diagram of Fig. 12.11 describes the behavior of acetonitrile(1)/nitromethane(2) at 75°C. The line labeled P-x_1 represents states of saturated liquid; the subcooled-liquid region lies above this line. The curve labeled P-y_1 represents states of saturated vapor; the superheated-vapor region lies below this curve. Points lying between the saturated-liquid and saturated-vapor lines are in the two-phase region, where saturated liquid and saturated vapor coexist in equilibrium. The P-x_1 and P-y_1 lines meet at the edges of the diagram, where saturated liquid and saturated vapor of the pure species coexist at the vapor pressures P_1^{sat} and P_2^{sat}.

To illustrate the nature of phase behavior in this binary system, we follow the course of a constant-temperature expansion process on the P-x_1-y_1 diagram. We imagine that a subcooled liquid mixture of 60 mol-% acetonitrile and 40 mol-% nitromethane exists in a piston/cylinder arrangement at 75°C. Its state is represented by point a in Fig. 12.11. Withdrawing the piston slowly enough reduces the pressure while maintaining the system at equilibrium at 75°C.

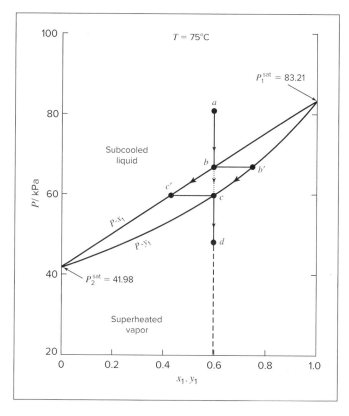

Figure 12.11: Pxy diagram for acetonitrile (1)/ nitromethane (2) at 75°C.

Because the system is closed, the overall composition remains constant during the process, and the states of the system *as a whole* fall on the vertical line descending from point a. When the pressure reaches the value at point b, the system is saturated liquid on the verge of vaporizing. A minuscule further decrease in pressure produces a bubble of vapor, represented by point b'. The two points b and b' together represent the equilibrium state. Point b is a bubblepoint, and the P-x_1 line is the locus of bubblepoints.

As the pressure is further reduced, the amount of vapor increases and the amount of liquid decreases, with the states of the two phases following paths $b'c$ and bc', respectively. The dotted line from point b to point c represents the *overall* states of the two-phase system. Finally, as point c is approached, the liquid phase, represented by point c', has almost disappeared, with only droplets (dew) remaining. Point c is therefore a dewpoint, and the P-y_1 curve is the locus of dewpoints. Once the dew has evaporated, only saturated vapor at point c remains, and further pressure reduction leads to superheated vapor at point d.

During this process, the volume of the system would first remain nearly constant in the subcooled liquid region from point a to point b. From point b to point c, the volume would increase dramatically, but not discontinuously. For a pure substance, the phase transition would occur at a single pressure (the vapor pressure), but for a binary mixture it occurs over a range of pressures. Finally, from point c to point d the volume would be approximately inversely proportional to pressure. Similarly, the heat flow required to maintain constant temperature during pressure reduction would be negligible in the subcooled liquid region and small in the superheated vapor region, but would be substantial between points b and c, where the latent heat of vaporization of the mixture must be supplied.

Evaporation of a Binary Mixture at Constant Pressure

Figure 12.12 is the T-x_1-y_1 diagram for the same system at a constant pressure of 70 kPa. The T-y_1 curve represents states of saturated vapor, with states of superheated vapor lying above it. The T-x_1 curve represents states of saturated liquid, with states of subcooled liquid lying below it. The two-phase region lies between these curves.

    With reference to Fig. 12.12, consider a constant-pressure heating process leading from a state of subcooled liquid at point a to a state of superheated vapor at point d. The path shown on the figure is for a constant overall composition of 60 mol-% acetonitrile. The temperature of the liquid increases as a result of heating from point a to point b, where the first bubble of vapor appears. Thus point b is a bubblepoint, and the T-x_1 curve is the locus of bubblepoints.

    As the temperature is further increased, the amount of vapor increases and the amount of liquid decreases, with the states of the two phases following paths $b'c$ and bc', respectively. The dotted line from point b to point c represents the *overall* states of the two-phase system.

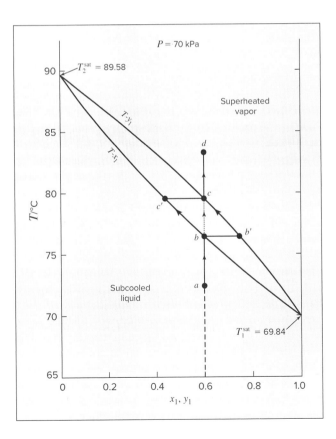

Figure 12.12: *Txy* diagram for acetonitrile(1)/nitromethane (2) at 70 kPa.

Finally, as point c is approached, the liquid phase, represented by point c', has almost disappeared, with only droplets (dew) remaining. Point c is therefore a dewpoint, and the T-y_1 curve is the locus of dewpoints. Once the dew has evaporated, only saturated vapor at point c remains, and further heating leads to superheated vapor at point d.

The change in volume and the heat flows in this process would be similar to those for the constant-temperature evaporation described previously, with a dramatic volume change as the two-phase region is traversed. Above and below the two-phase region, the heat flow and temperature change would be related by the heat capacities of the vapor and liquid, respectively. Within the two-phase region, the apparent heat capacity would be much higher, as it would include both a sensible heat component, required to increase the temperature of both phases, and a larger latent-heat component, required for transfer of material from the liquid to the vapor phase.

12.4 EQUILIBRIUM AND PHASE STABILITY

In the preceding discussion, we have assumed that a single liquid phase was present. Our everyday experience tells us that such an assumption is not always valid; oil-and-vinegar salad dressing provides a prototypical example of its violation. In such cases, the Gibbs energy is lowered by the liquid splitting into two separate phases, and the single phase is said to be unstable. In this section, we demonstrate that the equilibrium state of a closed system at fixed T and P is that which minimizes the Gibbs energy, and we then apply this criterion to the problem of phase stability.

Consider a closed system containing an arbitrary number of species and comprised of an arbitrary number of phases in which the temperature and pressure are spatially uniform (though not necessarily constant in time). The system is initially in a nonequilibrium state with respect to mass transfer between phases and chemical reaction. Changes that occur in the system are necessarily irreversible, and they take the system ever closer to an equilibrium state. We imagine that the system is placed in surroundings such that the system and surroundings are always in thermal and mechanical equilibrium. Heat exchange and expansion work are then accomplished reversibly. Under these circumstances the entropy change of the surroundings is:

$$dS_{surr} = \frac{dQ_{surr}}{T_{surr}} = -\frac{dQ}{T}$$

The final term applies to the system, for which the heat transfer dQ has a sign opposite to that of dQ_{surr}, and the temperature of the system T replaces T_{surr}, because both must have the same value for reversible heat transfer. The second law requires:

$$dS^t + dS_{surr} \geq 0$$

where S^t is the total entropy of the system. Combination of these expressions yields, upon rearrangement:

$$dQ \leq T dS^t \tag{12.1}$$

Application of the first law provides:

$$dU^t = dQ + dW = dQ - P dV^t$$

or

$$dQ = dU^t + P d V^t$$

Combining this equation with Eq. (12.1) gives:

$$dU^t + P dV^t \leq T dS^t$$

or

$$\boxed{dU^t + P dV^t - T dS^t \leq 0} \tag{12.2}$$

Because this relation involves properties only, it must be satisfied for changes in the state of *any* closed system of spatially uniform T and P, without restriction to the conditions of reversibility assumed in its derivation. The inequality applies to every incremental change of the system between nonequilibrium states, and it dictates the direction of change that leads toward equilibrium. The equality holds for changes between equilibrium states (reversible processes).

Equation (12.2) is so general that application to practical problems is difficult; restricted versions are much more useful. For example, by inspection:

$$(dU^t)_{S^t, V^t} \leq 0$$

where the subscripts specify properties held constant. Similarly, for processes that occur at constant U^t and V^t,

$$(dS^t)_{U^t, V^t} \geq 0$$

An *isolated* system is necessarily constrained to constant internal energy and volume, and for such a system it follows directly from the second law that the last equation is valid.

If a process is restricted to occur at constant T and P, then Eq. (12.2) may be written:

$$dU^t_{T, P} + d(PV^t)_{T, P} - d(TS^t)_{T, P} \leq 0$$

or

$$d(U^t + PV^t - TS^t)_{T,P} \leq 0$$

From the definition of the Gibbs energy [Eq. (6.4)],

$$G^t = H^t - TS^t = U^t + PV^t - TS^t$$

Therefore,

$$\boxed{(dG^t)_{T, P} \leq 0} \tag{12.3}$$

Of the possible specializations of Eq. (12.2), this is the most useful, because T and P, which are easily measured and controlled, are more conveniently held constant than are other pairs of variables, such as U^t and V^t.[4]

Equation (12.3) indicates that all irreversible processes occurring at constant T and P proceed in such a direction as to cause a decrease in the Gibbs energy of the system. Therefore:

The equilibrium state of a closed system is that state for which the total Gibbs energy is a minimum with respect to all possible changes at the given *T* and *P*.

---

[4]Although T and P are most easily held constant in experimental work, in molecular simulation studies, other pairs of variables are often more easily held constant.

This criterion of equilibrium provides a general method for determination of equilibrium states. One writes an expression for G^t as a function of the numbers of moles (mole numbers) of the species in the several phases and then finds the set of values for the mole numbers that minimizes G^t, subject to the constraints of mass and element conservation. This procedure can be applied to problems of phase equilibrium, chemical-reaction equilibrium, or combined phase and chemical-reaction equilibrium, and it is most useful for complex equilibrium problems.

Equation (12.3) provides a criterion that must be satisfied by any single phase that is *stable* with respect to the alternative that it split into two phases. It requires that the Gibbs energy of an equilibrium state be the minimum value with respect to all possible changes at the given T and P. Thus, for example, when mixing of two liquids occurs at constant T and P, the total Gibbs energy must decrease, because the mixed state must be the one of lower Gibbs energy with respect to the unmixed state. As a result:

$$G^t \equiv nG < \sum_i n_i G_i \qquad \text{from which} \qquad G < \sum_i x_i G_i$$

or

$$G - \sum_i x_i G_i < 0 \qquad (\text{const } T, P)$$

According to the definition of Eq. (11.1), the quantity on the left is the Gibbs-energy change of mixing. Therefore, $\Delta G < 0$. Thus, as noted in Sec. 11.1, the Gibbs-energy change of mixing must always be negative, and a plot of G vs. x_1 for a binary system must appear as shown by one of the curves of Fig. 12.13. With respect to curve II, however, there is a further consideration. If, when mixing occurs, a system can achieve a lower value of the Gibbs energy by forming *two* phases than by forming a single phase, then the system splits into two phases. This is in fact the situation represented between points α and β on curve II of Fig. 12.13 because the straight dashed line connecting points α and β represents the overall values of G for the range of states consisting of two phases of compositions x_1^α and x_1^β in various proportions. Thus the solid curve shown between points α and β cannot represent stable phases with respect to phase splitting. The equilibrium states between α and β consist of two phases.

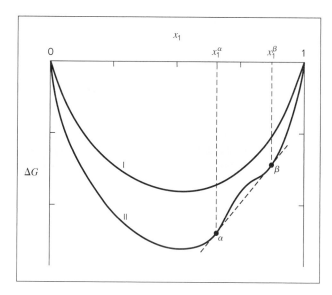

Figure 12.13: Gibbs-energy change of mixing. Curve I, complete miscibility; curve II, two phases between α and β.

These considerations lead to the following criterion of stability for a single-phase binary system for which $\Delta G \equiv G - x_1 G_1 - x_2 G_2$:

> **At fixed temperature and pressure, a single-phase binary mixture is stable if and only if ΔG and its first and second derivatives are continuous functions of x_1, and the second derivative is positive.**

Thus,
$$\frac{d^2 \, \Delta G}{dx_1^2} > 0 \qquad (\text{const } T, P)$$

and
$$\frac{d^2 \, (\Delta G/RT)}{dx_1^2} > 0 \qquad (\text{const } T, P) \tag{12.4}$$

This requirement has a number of consequences. Equation (11.9), rearranged and written for a binary system, becomes:

$$\frac{\Delta G}{RT} = x_1 \, \ln x_1 + x_2 \, \ln x_2 + \frac{G^E}{RT}$$

from which
$$\frac{d(\Delta G/RT)}{dx_1} = \ln x_1 - \ln x_2 + \frac{d(G^E/RT)}{dx_1}$$

and
$$\frac{d^2(\Delta G/RT)}{dx_1^2} = \frac{1}{x_1 x_2} + \frac{d^2(G^E/RT)}{dx_1^2}$$

Hence, stability requires:

$$\frac{d^2 \, (G^E/RT)}{dx_1^2} > -\frac{1}{x_1 x_2} \qquad (\text{const } T, P) \tag{12.5}$$

Liquid/Liquid Equilibrium

For conditions of constant pressure, or when pressure effects are negligible, binary liquid/liquid (LLE) is conveniently displayed on a *solubility diagram*, a plot of T vs. x_1. Figure 12.14 shows binary solubility diagrams of three types. The first, Fig. 12.14(a), shows curves (*binodal curves*) that define an "island." They represent the compositions of coexisting phases: curve UAL for the α phase (rich in species 2), and curve UBL for the β phase (rich in species 1). Equilibrium compositions x_1^α and x_1^β at a particular T are defined by the intersections of a horizontal *tie line* with the binodal curves. At each temperature, these compositions are those for which the curvature of the ΔG vs. x_1 curve changes sign. Between these compositions, it is concave down (negative second derivative) and outside them it is concave up. At these points, the curvature is zero; they are inflection points on the the ΔG vs. x_1 curve. Temperature T_L is a lower *consolute temperature*, or lower *critical solution temperature* (LCST); temperature T_U is an upper *consolute temperature*, or upper *critical solution temperature* (UCST). At temperatures between T_L and T_U, LLE is possible; for $T < T_L$ and $T > T_U$, a single liquid phase is obtained for the full range of compositions. The consolute points are limiting states of two-phase equilibrium for which all properties of the two equilibrium phases are identical.

Actually, the behavior shown on Fig. 12.14(a) is rarely observed; the LLE binodal curves are often interrupted by curves for yet another phase transition. When they intersect the

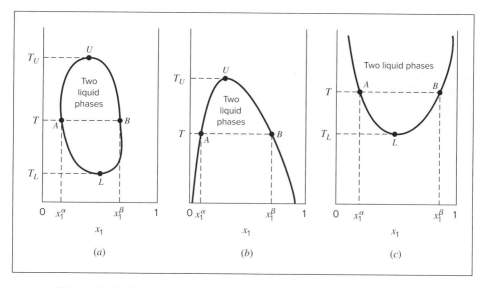

Figure 12.14: Three types of constant-pressure liquid/liquid solubility diagram.

freezing curve, only a UCST exists [Fig. 12.14(*b*)]; when they intersect the VLE bubblepoint curve, only an LCST exists [Fig. 12.14(*c*)]; when they intersect both, no consolute point exists, and yet another behavior is observed.[5]

12.5 VAPOR/LIQUID/LIQUID EQUILIBRIUM

As noted in the previous section, the binodal curves representing LLE can intersect the VLE bubblepoint curve. This gives rise to the phenomenon of vapor/liquid/liquid equilibrium (VLLE). A binary system of two liquid phases and one vapor phase in equilibrium has (by the phase rule) but one degree of freedom. For a given pressure, the temperature and the compositions of all three phases are therefore fixed. On a temperature/composition diagram the points representing the states of the three phases in equilibrium fall on a horizontal line at T^*. In Fig. 12.15, points C and D represent the two liquid phases, and point E represents the vapor phase. If more of either species is added to a system whose overall composition lies between points C and D, and if the three-phase equilibrium pressure is maintained, the phase rule requires that the temperature and the compositions of the phases be unchanged. However, the relative amounts of the phases adjust themselves to reflect the change in overall composition of the system.

At temperatures above T^* in Fig. 12.15, the system may be a single liquid phase, two phases (liquid and vapor), or a single vapor phase, depending on the overall composition.

[5]A comprehensive treatment of LLE is given by J. M. Srensen, T. Magnussen, P. Rasmussen, and Aa. Fredenslund, *Fluid Phase Equilibria*, vol. 2, pp. 297–309, 1979; vol. 3, pp. 47–82, 1979; vol. 4, pp. 151–163, 1980. Large compilations of data include W. Arlt, M. E. A. Macedo, P. Rasmussen, and J. M. Sørensen. *Liquid-Liquid Equilibrium Data Collection*, Chemistry Data Series, vol. V, Parts 1–4, DECHEMA, Frankfurt/Main, 1979–1987, and the IUPAC-NIST solubility database, available online at http://srdata.nist.gov/solubility.

In region α the system is a single liquid, rich in species 2; in region β it is a single liquid, rich in species 1. In region $\alpha - V$, liquid and vapor are in equilibrium. The states of the individual phases fall on lines AC and AE. In region $\beta - V$, liquid and vapor phases, described by lines BD and BE, exist at equilibrium. Finally, in the region designated V, the system is a single vapor phase. Below the three-phase temperature T^*, the system is entirely liquid, with features described in Sec. 12.4; this is the region of LLE.

When a vapor is cooled at constant pressure, it follows a path represented on Fig. 12.15 by a vertical line. Several such lines are shown. If one starts at point k, the vapor first reaches its dewpoint at line BE and then its bubblepoint at line BD, where condensation into single liquid phase β is complete. This is the same process that takes place when the species are completely miscible. If one starts at point n, no condensation of the vapor occurs until temperature T^* is reached. Then condensation occurs entirely at this temperature, producing the two liquid phases represented by points C and D. If one starts at an intermediate point m, the process is a combination of the two just described. After the dewpoint is reached, the vapor, tracing a path along line BE, is in equilibrium with a liquid tracing a path along line BD. However, at temperature T^* the vapor phase is at point E. All remaining condensation therefore occurs at this temperature, producing the two liquids of points C and D.

Figure 12.15 is drawn for a single constant pressure; equilibrium phase compositions, and hence the locations of the lines, change with pressure, but the general nature of the diagram is the same over a range of pressures. For most systems the species become more soluble in one another as the temperature increases, as indicated by lines CG and DH of Fig. 12.15. If this diagram is drawn for successively higher pressures, the corresponding three-phase equilibrium temperatures increase, and lines CG and DH extend further and further until they meet at the liquid/liquid consolute point M, as shown by Fig. 12.16.

As the pressure increases, line CD becomes shorter and shorter (indicated in Fig. 12.16 by lines $C'D'$ and $C''D''$), until at point M it diminishes to a differential length. For still higher pressures (P_4) the temperature is above the critical-solution temperature, and there is but a single liquid phase. The diagram then represents two-phase VLE, and it has the form of Fig. 12.9(d), exhibiting a minimum-boiling azeotrope.

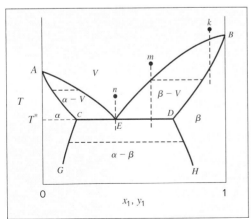

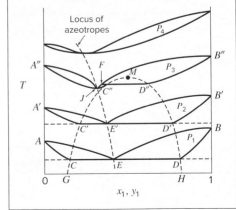

Figure 12.15: *Txy* diagram at constant P for a binary system exhibiting VLLE.

Figure 12.16: *Txy* diagram for several pressures.

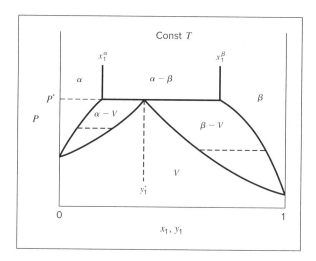

Figure 12.17: *Pxy* diagram at constant *T* for two partially miscible liquids.

For an intermediate range of pressures, the vapor phase in equilibrium with the two liquid phases has a composition that does not lie between the compositions of the two liquids. This is illustrated in Fig. 12.16 by the curves for P_3, which terminate at A'' and B''. The vapor in equilibrium with the two liquids at C'' and D'' is at point F. In addition the system exhibits an azeotrope, as indicated at point J.

Not all systems behave as described in the preceding paragraphs. Sometimes the upper critical-solution temperature is never attained because a vapor/liquid critical temperature is reached first. In other cases the liquid solubilities decrease with an increase in temperature. In this event a lower critical-solution temperature exists, unless solid phases appear first. There are also systems that exhibit both upper and lower critical-solution temperatures.[6]

Figure 12.17 is the phase diagram drawn at *constant T* corresponding to the constant-P diagram of Fig. 12.15. On it we identify the three-phase-equilibrium pressure as P^*, the three-phase-equilibrium vapor composition as y_1^*, and the compositions of the two liquid phases that contribute to the vapor/liquid/liquid equilibrium state as x_1^α and x_1^β. The phase boundaries separating the three liquid-phase regions are solubilities.

Although no two liquids are totally immiscible, this condition is so closely approached in some instances that the assumption of complete immiscibility does not lead to appreciable error for many engineering purposes. The phase characteristics of an immiscible system are illustrated by the temperature/composition diagram of Fig. 12.18(a). This diagram is a special case of Fig. 12.15 wherein phase α is pure species 2 and phase β is pure species 1. Thus lines ACG and BDH of Fig. 12.15 become in Fig. 12.18(a) vertical lines at $x_1 = 0$ and $x_1 = 1$.

In region I, vapor phases with compositions represented by line BE are in equilibrium with pure liquid species 1. Similarly, in region II, vapor phases whose compositions lie along line AE are in equilibrium with pure liquid species 2. Liquid/liquid equilibrium exists in region III, where the two phases are pure liquids of species 1 and 2. If one cools a vapor mixture starting at point m, the constant-composition path is represented by the vertical line shown in the figure. At the dewpoint, where this line crosses line BE, pure liquid species 1 begins to condense. Further reduction in temperature toward T^* causes continued condensation of

---

[6]For a comprehensive discussion of binary fluid-phase behavior, see J. S. Rowlinson and F. L. Swinton, *Liquids and Liquid Mixtures*, 3d ed., Butterworth Scientific, London, 1982.

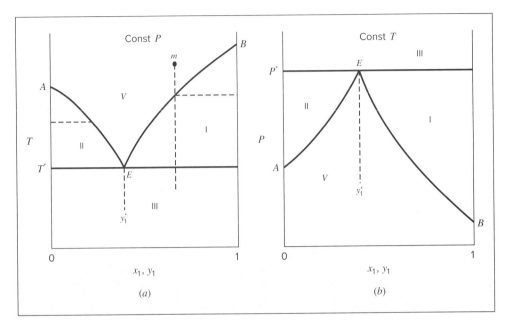

Figure 12.18: Binary system of immiscible liquids. (*a*) *Txy* diagram; (*b*) *Pxy* diagram.

pure species 1; the vapor-phase composition progresses along line *BE* until it reaches point *E*. Here, the remaining vapor condenses at temperature T^*, producing two liquid phases, one of pure species 1 and the other of pure species 2. A similar process, carried out to the left of point *E*, is the same, except that pure species 2 condenses initially. The constant-temperature phase diagram for an immiscible system is represented by Fig. 12.18(*b*).

12.6 SYNOPSIS

After studying this chapter, including the end-of-chapter problems, one should be able to:

- Understand that equilibrium implies absence of driving forces for net changes in the macroscopic state of a system

- State and apply the phase rule and Duhem's theorem for nonreacting systems

- Identify dew point and bubble point surfaces, the critical locus, and pure species vapor-pressure curves that make up a vapor/liquid phase envelope when presented in a *PTxy* diagram like Fig. 12.1

- Interpret and apply *Pxy*, *Txy*, *PT*, and *yx* diagrams representing vapor/liquid equilibrium of binary mixtures

- Sketch the path of an evaporation or condensation process on a *Pxy* or *Txy* diagram

- Understand that minimization of Gibbs energy is a general criterion for equilibrium of a closed system at fixed *T* and *P*

- Recognize that positive curvature of the ΔG vs. x_1 curve is a criterion for phase stability because negative curvature implies that the total Gibbs energy could be lowered via phase splitting

- Define upper consolute point, lower consolute point, high-boiling azeotrope, and low-boiling azeotrope

- Interpret and apply Pxy and Txy diagrams representing vapor/liquid/liquid equilibrium

12.7 PROBLEMS

12.1. Consider a closed vessel of fixed volume containing equal masses of water, ethanol, and toluene at 70°C. Three phases (two liquid and one vapor) are present.

    (*a*) How many variables, in addition to the mass of each component and the temperature, must be specified to fully determine the *intensive* state of the system?

    (*b*) How many variables, in addition to the mass of each component and the temperature, must be specified to fully determine the *extensive* state of the system?

    (*c*) The temperature of the system is increased to 72°C. What, if any, intensive or extensive coordinates of the system remain unchanged?

12.2. Consider a binary (two-species) system in vapor/liquid equilibrium. Enumerate all of the combinations of intensive variables that could be fixed to fully specify the intensive state of the system.

Problems 12.3 through 12.8 refer to the Pxy diagram for ethanol(1)/ethyl acetate(2) at 70°C shown in Fig. 12.19.

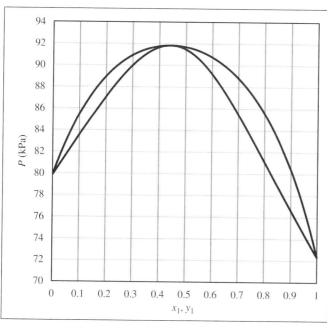

Figure 12.19: Pxy diagram for vapor/liquid equilibrium of ethanol(1)/ethyl acetate (2) at 70°C.

12.3. The pressure above a mixture of ethanol and ethyl acetate at 70°C is measured to be 86 kPa. What are the possible compositions of the liquid and vapor phases?

12.4. The pressure above a mixture of ethanol and ethyl acetate at 70°C is measured to be 78 kPa. What are the possible compositions of the liquid and vapor phases?

12.5. Consider an ethanol(1)/ethyl acetate(2) mixture with $x_1 = 0.70$, initially at 70°C and 100 kPa. Describe the evolution of phases and phase compositions as the pressure is gradually reduced to 70 kPa.

12.6. Consider an ethanol(1)/ethyl acetate(2) mixture with $x_1 = 0.80$, initially at 70°C and 80 kPa. Describe the evolution of phases and phase compositions as the pressure is gradually increased to 100 kPa.

12.7. What is the composition of the azeotrope for the ethanol(1)/ethyl acetate(2) system? Would this be called a high-boiling or low-boiling azeotrope?

12.8. Consider a closed vessel initially containing 1 mol of pure ethyl acetate at 70°C and 86 kPa. Imagine that pure ethanol is slowly added at constant temperature and pressure until the vessel contains 1 mol ethyl acetate and 9 mol ethanol. Describe the evolution of phases and phase compositions during this process. Comment on the practical feasibility of carrying out such a process. What sort of device would be required? How would the total system volume change during this process? At what composition would the system volume reach its maximum value?

Problems 12.9 through 12.14 refer to the *Txy* diagram for ethanol(1)/ethyl acetate(2) shown in Fig. 12.20.

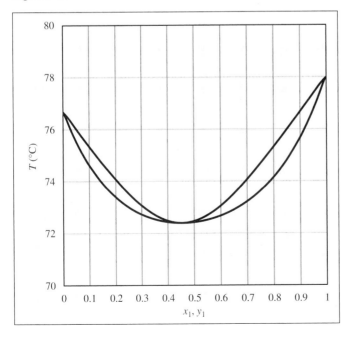

Figure 12.20: *Txy* diagram for vapor/liquid equilibrium of ethanol(1)/ethyl acetate (2) at 100 kPa.

12.9. A mixture of ethanol and ethyl acetate is heated in a closed system at 100 kPa to a temperature of 74°C, and two phases are observed to be present. What are the possible compositions of the liquid and vapor phases?

12.10. A mixture of ethanol and ethyl acetate is heated in a closed system at 100 kPa to a temperature of 77°C, and two phases are observed to be present. What are the possible compositions of the liquid and vapor phases?

12.11. Consider an ethanol(1)/ethyl acetate(2) mixture with $x_1 = 0.70$, initially at 70°C and 100 kPa. Describe the evolution of phases and phase compositions as the temperature is gradually increased to 80°C.

12.12. Consider an ethanol(1)/ethyl acetate(2) mixture with $x_1 = 0.20$, initially at 70°C and 100 kPa. Describe the evolution of phases and phase compositions as the temperature is gradually increased to 80°C.

12.13. Consider an ethanol(1)/ethyl acetate(2) mixture with $x_1 = 0.20$, initially at 80°C and 100 kPa. Describe the evolution of phases and phase compositions as the temperature is gradually reduced to 70°C.

12.14. Consider an ethanol(1)/ethyl acetate(2) mixture with $x_1 = 0.80$, initially at 80°C and 100 kPa. Describe the evolution of phases and phase compositions as the temperature is gradually reduced to 70°C.

12.15. Consider a closed vessel initially containing 1 mol of pure ethyl acetate at 74°C and 100 kPa. Imagine that pure ethanol is slowly added at constant temperature and pressure until the vessel contains 1 mol ethyl acetate and 9 mol ethanol. Describe the evolution of phases and phase compositions during this process. Comment on the practical feasibility of carrying out such a process. What sort of device would be required? How would the total system volume change during this process? At what composition would the system volume reach its maximum value?

Problems 12.16 through 12.21 refer to the Pxy diagram for chloroform(1)/tetrahydrofuran(2) at 50°C shown in Fig. 12.21.

12.16. The pressure above a mixture of chloroform and tetrahydrofuran at 50°C is measured to be 62 kPa. What are the possible compositions of the liquid and vapor phases?

12.17. The pressure above a mixture of chloroform and tetrahydrofuran at 50°C is measured to be 52 kPa. What are the possible compositions of the liquid and vapor phases?

12.18. Consider a chloroform (1)/tetrahydrofuran(2) mixture with $x_1 = 0.80$, initially at 50°C and 70 kPa. Describe the evolution of phases and phase compositions as the pressure is gradually reduced to 50 kPa.

12.19. Consider an chloroform(1)/tetrahydrofuran(2) mixture with $x_1 = 0.90$, initially at 50°C and 50 kPa. Describe the evolution of phases and phase compositions as the pressure is gradually increased to 70 kPa.

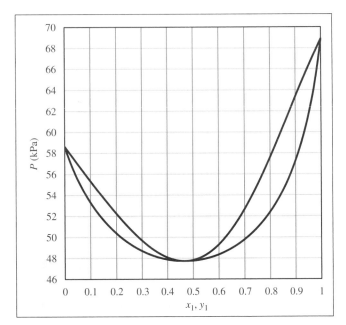

Figure 12.21: *Pxy* diagram for vapor/liquid equilibrium of chloroform(1)/tetrahydrofuran(2) at 50°C.

12.20. What is the composition of the azeotrope for the chloroform(1)/tetrahydrofuran (2) system? Would this be called a high-boiling or low-boiling azeotrope?

12.21. Consider a closed vessel initially containing 1 mol of tetrahydrofuran at 50°C and 52 kPa. Imagine that pure chloroform is slowly added at constant temperature and pressure until the vessel contains 1 mol tetrahydrofuran and 9 mol chloroform. Describe the evolution of phases and phase compositions during this process. Comment on the practical feasibility of carrying out such a process. What sort of device would be required? How would the total system volume change during this process? At what composition would the system volume reach its maximum value?

Problems 12.22 through 12.28 refer to the *Txy* diagram for chloroform(1)/tetrahydrofuran(2) at 120 kPa shown in Fig. 12.22.

12.22. A mixture of chloroform and tetrahydrofuran is heated in a closed system at 120 kPa to a temperature of 75°C, and two phases are observed to be present. What are the possible compositions of the liquid and vapor phases?

12.23. A chloroform and tetrahydrofuran mixture is heated in a closed system at 120 kPa to a temperature of 70°C, and two phases are observed to be present. What are the possible compositions of the liquid and vapor phases?

12.24. Consider a chloroform(1)/tetrahydrofuran(2) mixture with $x_1 = 0.80$, initially at 70°C and 120 kPa. Describe the evolution of phases and phase compositions as the temperature is gradually increased to 80°C.

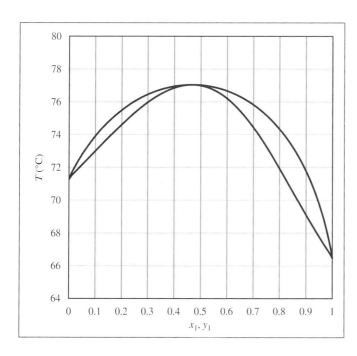

Figure 12.22: Txy diagram for vapor/liquid equilibrium of chloroform(1)/tetrahydrofuran (2) at 120 kPa.

12.25. Consider a chloroform(1)/tetrahydrofuran(2) mixture with $x_1 = 0.20$, initially at 70°C and 120 kPa. Describe the evolution of phases and phase compositions as the temperature is gradually increased to 80°C.

12.26. Consider a chloroform(1)/tetrahydrofuran(2) mixture with $x_1 = 0.10$, initially at 80°C and 120 kPa. Describe the evolution of phases and phase compositions as the temperature is gradually reduced to 70°C.

12.27. Consider a chloroform(1)/tetrahydrofuran(2) mixture with $x_1 = 0.90$, initially at 76°C and 120 kPa. Describe the evolution of phases and phase compositions as the temperature is gradually reduced to 66°C.

12.28. Consider a closed vessel initially containing 1 mol of pure tetrahydrofuran at 74°C and 120 kPa. Imagine that pure chloroform is slowly added at constant temperature and pressure until the vessel contains 1 mol tetrahydrofuran and 9 mol chloroform. Describe the evolution of phases and phase compositions during this process. Comment on the practical feasibility of carrying out such a process. What sort of device would be required? How would the total system volume change during this process? At what composition would the system volume reach its maximum value?

Problems 12.29 through 12.33 refer to the xy diagram provided in Fig. 12.23. This diagram shows xy curves both for ethanol(1)/ethyl acetate(2) and for chloroform(1)/tetrahydrofuran(2), both at a constant pressure of 1 bar. The curves are intentionally unlabeled. Readers should refer to Figs. 12.19 through 12.22 to deduce which curve is for which pair of substances.

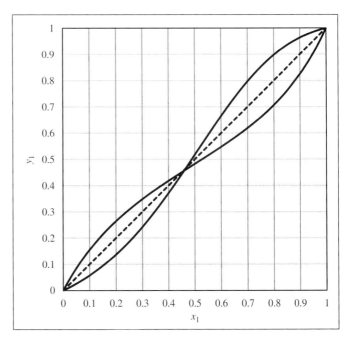

Figure 12.23: xy diagram for ethanol(1)/ethyl acetate(2) and for chloroform(1)/tetrahydrofuran(2), both at a constant pressure of 1 bar. Note that the curves are intentionally unlabeled, but they can be identified based on information presented in Figs. 12.19 through 12.22.

12.29. What is the composition of the vapor phase in equilibrium with a liquid-phase ethanol(1)/ethyl acetate(2) mixture of the following compositions at $P = 1$ bar?

(a) $x_1 = 0.1$
(b) $x_1 = 0.2$
(c) $x_1 = 0.3$
(d) $x_1 = 0.45$
(e) $x_1 = 0.6$
(f) $x_1 = 0.8$
(g) $x_1 = 0.9$

12.30. What is the composition of the liquid phase in equilibrium with a vapor-phase ethanol(1)/ethyl acetate(2) mixture of the following compositions at $P = 1$ bar?
(a) $y_1 = 0.1$
(b) $y_1 = 0.2$
(c) $y_1 = 0.3$
(d) $y_1 = 0.45$
(e) $y_1 = 0.6$
(f) $y_1 = 0.8$
(g) $y_1 = 0.9$

12.31. What is the composition of the vapor phase in equilibrium with a liquid-phase chloroform(1)/tetrahydrofuran(2) mixture of the following compositions at $P = 1$ bar?

(a) $x_1 = 0.1$
(b) $x_1 = 0.2$

(c) $x_1 = 0.3$
(d) $x_1 = 0.45$
(e) $x_1 = 0.6$
(f) $x_1 = 0.8$
(g) $x_1 = 0.9$

12.32. What is the composition of the liquid phase in equilibrium with a vapor-phase chloroform(1)/tetrahydrofuran(2) mixture of the following compositions at $P = 1$ bar?

(a) $y_1 = 0.1$
(b) $y_1 = 0.2$
(c) $y_1 = 0.3$
(d) $y_1 = 0.45$
(e) $y_1 = 0.6$
(f) $y_1 = 0.8$
(g) $y_1 = 0.9$

12.33. Consider a binary liquid mixture for which the excess Gibbs energy is given by $G^E/RT = A x_1 x_2$. What is the minimum value of A for which liquid/liquid equilibrium is possible?

12.34. Consider a binary liquid mixture for which the excess Gibbs energy is given by $G^E/RT = A x_1 x_2 (x_1 + 2x_2)$. What is the minimum value of A for which liquid/liquid equilibrium is possible?

12.35. Consider a binary mixture for which the excess Gibbs energy is given by $G^E/RT = 2.6 x_1 x_2$. For each of the following overall compositions, determine whether one or two liquid phases will be present. If two liquid phases will be present, find their compositions and the amount of each phase present (phase fractions).

(a) $z_1 = 0.2$
(b) $z_1 = 0.3$
(c) $z_1 = 0.5$
(d) $z_1 = 0.7$
(e) $z_1 = 0.8$

12.36. Consider a binary mixture for which the excess Gibbs energy is given by $G^E/RT = 2.1 x_1 x_2 (x_1 + 2x_2)$. For each of the following overall compositions, determine whether one or two liquid phases will be present. If two liquid phases will be present, find their compositions and the amount of each phase present (phase fractions).

(a) $z_1 = 0.2$
(b) $z_1 = 0.3$
(c) $z_1 = 0.5$
(d) $z_1 = 0.7$
(e) $z_1 = 0.8$

Chapter 13

Thermodynamic Formulations for Vapor/Liquid Equilibrium

The objective of this chapter is to apply the framework of solution thermodynamics developed in Chapter 10 to the specific situation of vapor/liquid equilibrium (VLE), as introduced qualitatively in Chapter 12. Because of the practical importance of distillation as a means of separating and purifying chemical species, VLE is the most studied type of phase equilibrium. Approaches developed for analyzing VLE also provide the foundation for most analyses of liquid/liquid equilibrium (LLE), vapor/liquid/liquid equilibrium (VLLE), and combined phase and reaction equilibrium, as considered in Chapter 15.

The analysis of VLE problems in the present chapter begins by developing a general formulation of such problems in terms of vapor-phase fugacity coefficients and liquid-phase activity coefficients. For VLE at low pressure, where the gas phase approaches the ideal-gas state, simplified versions of this formulation are applicable. For those conditions, activity coefficients can be obtained directly from experimental VLE data. These can then be fit to mathematical models. Finally, the models can be used to predict activity coefficients and VLE behavior for situations where experiments have not been performed. Thus, the analyses presented in this chapter allow for the efficient correlation and generalization of the observed behavior of real physical systems.

Specifically, in this chapter, we will:

- Define activity coefficients and relate them to the excess Gibbs energy of a mixture

- Formulate the general criterion for phase equilibrium in terms of vapor-phase fugacity coefficients and liquid-phase activity coefficients (the gamma/phi formulation of VLE)

- Show how this general formulation simplifies to Raoult's law or a modified version of Raoult's law under appropriate conditions

- Perform bubblepoint, dewpoint, and flash calculations using Raoult's law and modified versions thereof

- Illustrate the extraction of activity coefficients and excess Gibbs energy from experimental low-pressure VLE data

- Address the issue of thermodynamic consistency of experimentally derived activity coefficients

- Introduce several excess Gibbs energy and activity coefficient models and the fitting of model parameters to experimental VLE data

- Perform VLE calculations under conditions where the complete gamma/phi formulation is required

- Show that residual properties and excess properties can also be evaluated from cubic equations of state

- Demonstrate the formulation and solution of VLE problems using cubic equations of state

The foundation for VLE calculations was laid in Chapter 10, where Eq. (10.39) was shown to apply for the equilibrium of pure species:

$$f_i^v = f_i^l = f_i^{\text{sat}} \tag{10.39}$$

and Eq. (10.48) was shown to apply to the equilibrium of species in mixtures:

$$\hat{f}_i^v = \hat{f}_i^l \qquad (i = 1, 2, \ldots, N) \tag{10.48}$$

We recall also the definitions of fugacity coefficients, as given by Eqs. (10.34) and (10.52). From the latter we can write for the fugacity coefficient of species i in a vapor phase:

$$\hat{f}_i^v = \hat{\phi}_i^v y_i P \tag{13.1}$$

An analogous equation can be written for the liquid phase, but this phase is often treated differently.

13.1 EXCESS GIBBS ENERGY AND ACTIVITY COEFFICIENTS

In view of Eq. (10.8), $\bar{G}_i = \mu_i$, Eq. (10.46) may be written as:

$$\bar{G}_i = \Gamma_i(T) + RT \ln \hat{f}_i$$

For an ideal solution, in accord with Eq. (10.83), this becomes:

$$\bar{G}_i^{id} = \Gamma_i(T) + RT \ln x_i f_i$$

By difference,

$$\bar{G}_i - \bar{G}_i^{id} = RT \ln \frac{\hat{f}_i}{x_i f_i}$$

Equation (10.88), written for the partial Gibbs energy, shows that the left side of this equation is the partial excess Gibbs energy $\bar{G}_i^E$; the dimensionless ratio $\hat{f}_i / x_i f_i$ appearing on the right is the *activity coefficient* of species i in solution, symbol γ_i. Thus, by **definition,**

$$\boxed{\gamma_i \equiv \frac{\hat{f}_i}{x_i f_i}} \tag{13.2}$$

Whence,
$$\boxed{\bar{G}_i^E = RT \ln \gamma_i} \tag{13.3}$$

These equations establish a thermodynamic foundation for activity coefficients. Comparison with Eq. (10.51) shows that Eq. (13.3) relates $\ln \gamma_i$ to $\bar{G}_i^E$ exactly as Eq. (10.51) relates $\ln \hat{\phi}_i$ to $\bar{G}_i^R$. For an ideal solution, $\bar{G}_i^E = 0$, and therefore $\gamma_i^{id} = 1$.

An alternative form of Eq. (10.89) follows by introduction of the activity coefficient through Eq. (13.3):

$$d\left(\frac{nG^E}{RT}\right) = \frac{nV^E}{RT}\,dP - \frac{nH^E}{RT^2}\,dT + \sum_i \ln \gamma_i \, dn_i \tag{13.4}$$

The generality of this equation precludes its direct practical application. Instead, use is made of restricted forms, which are written by inspection:

$$\frac{V^E}{RT} = \left[\frac{\partial(G^E/RT)}{\partial P}\right]_{T,x} \tag{13.5} \qquad \frac{H^E}{RT} = -T\left[\frac{\partial(G^E/RT)}{\partial T}\right]_{P,x} \tag{13.6}$$

$$\ln \gamma_i = \left[\frac{\partial(nG^E/RT)}{\partial n_i}\right]_{P,T,n_j} \tag{13.7}$$

Equations (13.5) through (13.7) are analogs of Eqs. (10.57) through (10.59) for residual properties. Whereas the fundamental *residual*-property relation derives its usefulness from its direct relation to experimental *PVT* data and equations of state, the fundamental *excess*-property relation is useful because V^E, H^E, and γ_i are all experimentally accessible. Activity coefficients are found from vapor/liquid equilibrium data, as discussed in Sec. 13.5, while V^E and H^E values come from mixing experiments, as discussed in Chap. 11.

Equation (13.7) demonstrates that $\ln \gamma_i$ is a partial property with respect to G^E/RT. It is the analog of Eq. (10.59), which shows the same relation of $\ln \hat{\phi}_i$ to G^R/RT. The partial-property analogs of Eqs. (13.5) and (13.6) are:

$$\left(\frac{\partial \ln \gamma_i}{\partial P}\right)_{T,x} = \frac{\bar{V}_i^E}{RT} \tag{13.8} \qquad \left(\frac{\partial \ln \gamma_i}{\partial T}\right)_{P,x} = -\frac{\bar{H}_i^E}{RT^2} \tag{13.9}$$

These equations allow the calculation of the effect of pressure and temperature on the activity coefficients.

The following forms of the summability and Gibbs/Duhem equations result from the fact that $\ln \gamma_i$ is a partial property with respect *to G^E/RT:*

$$\frac{G^E}{RT} = \sum_i x_i \ln \gamma_i \tag{13.10}$$

$$\sum_i x_i d \ln \gamma_i = 0 \qquad \text{(const } T, P) \tag{13.11}$$

Just as the fundamental property relation of Eq. (10.54) provides complete property information from a canonical equation of state expressing G/RT as a function of T, P, and composition, so the fundamental *residual*-property relation, Eq. (10.55) or (10.56), provides complete *residual*-property information from a *PVT* equation of state, from *PVT* data, or from generalized *PVT* correlations. However, obtaining complete *property* information requires, in addition to *PVT* data, the ideal-gas-state heat capacities of the species that comprise the system. In complete analogy, the fundamental *excess*-property relation, Eq. (13.4), provides complete *excess*-property information, given an equation for G^E/RT as a function of its canonical variables, T, P, and composition. However, this formulation represents less-complete property information than does the residual-property formulation, because it tells us nothing about the properties of the pure constituent chemical species.

13.2 THE GAMMA/PHI FORMULATION OF VLE

Rearranging Eq. (13.2), the definition of the activity coefficient, and writing it for species i in the liquid phase gives:

$$\hat{f}_i^l = x_i \gamma_i^l f_i^l$$

Substitution in Eq. (10.48) for $\hat{f}_i^l$ by this equation and for $\hat{f}_i^v$ by Eq. (13.1) yields:

$$y_i \hat{\phi}_i^v P = x_i \gamma_i^l f_i^l \qquad (i = 1, 2, \ldots, N) \tag{13.12}$$

Transformation of Eq. (13.12) into a working formulation requires the development of suitable expressions for $\hat{\phi}_i^v$, γ_i^l, and f_i^l. Eliminating the pure-species property f_i^l by Eq. (10.44) proves helpful:

$$f_i^l = \phi_i^{\text{sat}} P_i^{\text{sat}} \exp \frac{V_i^l (P - P_i^{\text{sat}})}{RT} \tag{10.44}$$

This equation in combination with Eq. (13.12) yields:

$$\boxed{y_i \Phi_i P = x_i \gamma_i P_i^{\text{sat}} \qquad (i = 1, 2, \ldots, N)} \tag{13.13}$$

where

$$\Phi_i \equiv \frac{\hat{\phi}_i^v}{\phi_i^{\text{sat}}} \exp \left[-\frac{V_i^l (P - P_i^{\text{sat}})}{RT} \right]$$

In Eq. (13.13) γ_i is understood to be a liquid-phase property. Because the Poynting factor, represented by the exponential, at low to moderate pressures differs from unity by only a few parts per thousand, its omission introduces negligible error, and we adopt this simplification to produce the usual working equation:

$$\Phi_i \equiv \frac{\hat{\phi}_i^v}{\phi_i^{\text{sat}}} \tag{13.14}$$

The vapor pressure of pure species i is most commonly given by Eq. (6.90), the Antoine equation:

$$\ln P_i^{\text{sat}} = A_i - \frac{B_i}{T + C_i} \tag{13.15}$$

The gamma/phi formulation of VLE appears in several variations, depending on the treatment of Φ_i and γ_i.

Applications of thermodynamics to vapor/liquid equilibrium calculations encompass the goals of finding the temperature, pressure, and compositions of phases in equilibrium. Indeed, thermodynamics provides the mathematical framework for the systematic correlation, extension, generalization, evaluation, and interpretation of such data. Moreover, it is the means by which the predictions of various theories of molecular physics and statistical mechanics may be applied to practical purposes. None of this can be accomplished without *models* for the behavior of systems in vapor/liquid equilibrium. The two simplest models, already considered, are the ideal-gas state for the vapor phase and the ideal-solution model for the liquid phase. These are combined in the simplest treatment of vapor/liquid equilibrium in what is known as Raoult's law. It is by no means a "law" in the universal sense of the First and Second Laws of Thermodynamics, but it does become valid in a rational limit.

13.3 SIMPLIFICATIONS: RAOULT'S LAW, MODIFIED RAOULT'S LAW, AND HENRY'S LAW

Figure 13.1 shows a vessel in which a vapor mixture and a liquid solution coexist in vapor/liquid equilibrium. If the vapor phase is assumed to be in its ideal-gas state and the liquid phase is assumed to be an ideal solution, both Φ_i and γ_i in Eq. (13.13) approach unity, and this equation reduces to its simplest possible form, Raoult's law:[1]

$$\boxed{y_i P = x_i P_i^{\text{sat}} \qquad (i = 1, 2, \ldots, N)} \qquad (13.16)$$

where x_i is a liquid-phase mole fraction, y_i is a vapor-phase mole fraction, and P_i^{sat} is the vapor pressure of pure species i at the temperature of the system. The product $y_i P$ is the *partial pressure* of species i in the vapor phase. Note that the only thermodynamic function to survive here is the vapor pressure of pure-species i, suggesting its primary importance in VLE calculations.

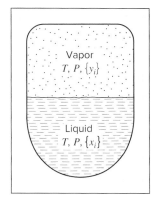

Figure 13.1: Schematic representation of VLE. The temperature T and pressure P are uniform throughout the vessel and can be measured with appropriate instruments. Vapor and liquid samples may be withdrawn for analysis, and this provides experimental values for mole fractions in the vapor $\{y_i\}$ and mole fractions in the liquid $\{x_i\}$.

---

[1]François Marie Raoult (1830–1901), French chemist, see https://en.wikipedia.org/wiki/François-Marie_Raoult.

The ideal-gas-state assumption means that Raoult's law is limited in application to low to moderate pressures. The ideal-solution assumption implies that Raoult's law can have approximate validity only when the species that comprise the system are chemically similar. Just as the ideal-gas state serves as a standard to which real-gas behavior may be compared, the ideal solution represents a standard to which real-solution behavior may be compared. Liquid-phase ideal-solution behavior is promoted when the molecular species are not too different in size and have the same chemical nature. Thus, a mixture of isomers, such as *ortho-*, *meta-*, and *para-*xylene, conforms very closely to ideal-solution behavior. So do mixtures of adjacent members of a homologous series, as for example, *n*-hexane/*n*-heptane, ethanol/propanol, and benzene/toluene. Other examples are acetone/acetonitrile and acetonitrile/nitromethane. Figures 12.11 and 12.12 for the latter system are constructed to represent Raoult's law.

The simple model of VLE represented by Eq. (13.16) provides a realistic description of actual behavior for a relatively small class of systems. Nevertheless, it serves as a standard of comparison for more complex systems. A limitation of Raoult's law is that it can be applied only to species of known vapor pressure, and this requires each species to be "subcritical," i.e., at a temperature below its critical temperature. Raoult's law cannot be applied to situations for which the temperature exceeds the critical temperature of one or more species in the mixture.

Dewpoint and Bubblepoint Calculations with Raoult's Law

Although VLE problems with other combinations of variables are possible, engineering interest centers on dewpoint and bubblepoint calculations, of which there are four types:

> BUBL P: Calculate $\{y_i\}$ and P, given $\{x_i\}$ and T
> DEW P: Calculate $\{x_i\}$ and P, given $\{y_i\}$ and T
> BUBL T: Calculate $\{y_i\}$ and T, given $\{x_i\}$ and P
> DEW T: Calculate $\{x_i\}$ and T, given $\{y_i\}$ and P

In each case the name indicates the quantities to be calculated: either a *BUBL* (vapor) or a *DEW* (liquid) composition *and* either P or T. Thus, one must specify either the vapor-phase or the liquid-phase composition *and* either P or T, thus fixing $1 + (N - 1)$ or N intensive variables, exactly the number of degrees of freedom F required by the phase rule [Eq. (3.1)] for vapor/liquid equilibrium.

Because $\sum_i y_i = 1$, Eq. (13.16) may be summed over all species to yield:

$$P = \sum_i x_i P_i^{\text{sat}} \tag{13.17}$$

This equation finds direct application in bubblepoint calculations, where the vapor-phase composition is unknown. For a binary system with $x_2 = 1 - x_1$,

$$P = P_2^{\text{sat}} + (P_1^{\text{sat}} - P_2^{\text{sat}})x_1 \tag{13.18}$$

and a plot of P vs. x_1 at constant temperature is a straight line connecting P_2^{sat} at $x_1 = 0$ with P_1^{sat} at $x_1 = 1$. The *Pxy* diagram of Fig. 12.11 for acetonitrile(1)/nitromethane(2) shows this linear relation.

For this system at a temperature of 75°C, the pure-species vapor pressures are $P_1^{\text{sat}} = 83.21$ kPa and $P_2^{\text{sat}} = 41.98$ kPa. *BUBL P* calculations are readily carried out by substitution of these values in Eq. (13.18), along with values of x_1. The results allow calcula-tion of the P-x_1 relation. The corresponding values of y_1 are found from Eq. (13.16):

$$y_1 = \frac{x_1 P_1^{sat}}{P}$$

The following table shows the results of calculation. These are the values used to construct the P-x_1-y_1 diagram of Fig. 12.11:

x_1	y_1	P/kPa	x_1	y_1	P/kPa
0.0	0.0000	41.98	0.6	0.7483	66.72
0.2	0.3313	50.23	0.8	0.8880	74.96
0.4	0.5692	58.47	1.0	1.0000	83.21

When P is fixed, the temperature varies along with x_1 and y_1, and the temperature range is bounded by saturation temperatures t_1^{sat} and t_2^{sat}, the temperatures at which the pure species exert vapor pressures equal to P. These temperatures can be calculated from Antoine equation:

$$t_i^{sat} = \frac{B_i}{A_i - \ln P} - C_i$$

Figure 12.12 for acetonitrile(1)/nitromethane(2) at $P = 70$ kPa shows these values as $t_1^{sat} = 69.84°C$ and $t_2^{sat} = 89.58°C$.

The construction of Fig. 12.12 for this system is based on *BUBL T* calculations, which are less direct than *BUBL P* calculations. One cannot solve directly for the temperature because it is buried in the vapor-pressure equations. An iterative or trial-and-error approach is needed in this case. For a binary system and a given value of x_1, Eq. (13.18) must give the specified pressure when the vapor pressures are evaluated at the correct temperature. The most intuitive procedure is simply to make calculations at trial values of T until the correct value of P is gen-erated. The *goal* is the known value for P in Eq. (13.18), and it is found by varying T. Working out a convenient strategy for homing in on the correct final answer using a hand calculator is not difficult. Microsoft Excel's Goal Seek function also does the job quite effectively when varying a single T to find a desired value of P. The Solver function allows this to be done simultaneously for many compositions.[2]

Modified Raoult's Law

Raoult's law results when both γ_i and Φ_i are set equal to unity in Eq. 13.13. For low to mod-erate pressures the latter substitution is usually reasonable. However, modifying Raoult's law to properly evaluate the activity coefficient γ_i, and thus take into account liquid-phase devia-tions from ideal solution behavior, produces a much more reasonable and broadly applicable description of VLE behavior:

$$\boxed{y_i P = x_i \gamma_i P_i^{sat} \qquad (i = 1, 2, \ldots, N)} \qquad (13.19)$$

---

[2]See the Online Learning Center at http://highered.mheducation.com:80/sites/1259696529 for examples.

This equation provides for entirely satisfactory representation of the VLE behavior of a great variety of systems at low to moderate pressures.

Because $\sum_i y_i = 1$, Eq. (13.19) may be summed over all species to yield:

$$P = \sum_i x_i \gamma_i P_i^{\text{sat}} \tag{13.20}$$

Alternatively, Eq. (13.19) may be solved for x_i, in which case summing over all species yields:

$$P = \frac{1}{\sum_i y_i / \gamma_i P_i^{\text{sat}}} \tag{13.21}$$

Bubblepoint and dewpoint calculations made with the modified Raoult's law are only a bit more complex than the same calculations made with Raoult's law. In particular, bubblepoint pressure calculations are straightforward because the specified liquid composition allows immediate evaluation of the activity coefficients. Dewpoint pressure calculations require an iterative solution process because the unknown liquid-phase composition is required to evaluate the activity coefficients. Bubblepoint and dewpoint temperature calculations are further complicated by the temperature dependence of the activity coefficients, along with the temperature dependence of the vapor pressures, but the same iterative or trial-and-error approaches used with Raoult's law calculations can still be employed.

Example 13.1

For the system methanol(1)/methyl acetate(2), the following equations provide a reasonable correlation for the activity coefficients:

$$\ln \gamma_1 = Ax_2^2 \qquad \ln \gamma_2 = Ax_1^2 \qquad \text{where} \qquad A = 2.771 - 0.00523T$$

In addition, the following Antoine equations provide vapor pressures:

$$\ln P_1^{\text{sat}} = 16.59158 - \frac{3643.31}{T - 33.424} \qquad \ln P_2^{\text{sat}} = 14.25326 - \frac{2665.54}{T - 53.424}$$

where T is in kelvins and the vapor pressures are in kPa. Assuming the validity of Eq. (13.19), calculate:

(a) P and $\{y_i\}$ for $T = 318.15$ K and $x_1 = 0.25$.

(b) P and $\{x_i\}$ for $T = 318.15$ K and $y_1 = 0.60$.

(c) T and $\{y_i\}$ for $P = 101.33$ kPa and $x_1 = 0.85$.

(d) T and $\{x_i\}$ for $P = 101.33$ kPa and $y_1 = 0.40$.

(e) The azeotropic pressure and the azeotropic composition for $T = 318.15$ K.

Solution 13.1

In the dewpoint and bubblepoint calculations of parts (a) through (d), the key is the dependence of the activity coefficients on T and x_1. In (a), both values are given, and solution is direct. In (b) only T is given, and solution is by trial with x_1 varied to reproduce the given value of y_1. In (c) only x_1 is given, and T is varied to reproduce the given value of P. In (d) neither T nor x_1 is given, and both are varied alternately, T to yield P and x_1 to yield y_1. In parts (b) through (d), the trial and error calculations are readily automated using Microsoft Excel's Goal Seek function.

(a) A *BUBL P* calculation. For $T = 318.15$ K, the Antoine equations yield $P_1^{sat} = 44.51$ and $P_2^{sat} = 65.64$ kPa. The activity-coefficient correlation provides $A = 1.107$, $\gamma_1 = 1.864$, and $\gamma_2 = 1.072$. By Eq. (13.20), $P = 73.50$ kPa and by Eq. (13.19), $y_1 = 0.282$.

(b) A *DEW P* calculation. With T unchanged from part (a), P_1^{sat}, P_2^{sat}, and A are also unchanged. The unknown liquid-phase composition is varied in trial calculations that evaluate the activity coefficients, P by Eq. (13.21), and y_1 by Eq. (13.19), with the goal of reproducing the given value $y_1 = 0.6$. This leads to final values:

$$P = 62.59 \text{ kPa} \qquad x_1 = 0.8169 \qquad \gamma_1 = 1.0378 \qquad \gamma_2 = 2.0935$$

(c) A *BUBL T* calculation. Solution by trial here varies T until the given value of P is reproduced. A reasonable starting value for T is found from the saturation temperatures of the pure species at the known pressure. The Antoine equation, solved for T, becomes:

$$T_i^{sat} = \frac{B_i}{A_i - \ln P} - C_i$$

Application for $P = 101.33$ kPa leads to: $T_1^{sat} = 337.71$ and $T_2^{sat} = 330.08$ K. An average of these values serves as an initial T: Each trial value of T leads immediately to values for the activity coefficients and to a value for P by Eq. (13.20). The known value of $P = 101.33$ kPa is reproduced when:

$$T = 331.20 \text{ K} \qquad P_1^{sat} = 77.99 \text{ kPa} \qquad P_2^{sat} = 105.35 \text{ kPa}$$
$$A = 1.0388 \qquad \gamma_1 = 1.0236 \qquad \gamma_2 = 2.1182$$

The vapor-phase mole fractions are given by:

$$y_1 = \frac{x_1 \gamma_1 P_1^{sat}}{P} = 0.670 \qquad \text{and} \qquad y_2 = 1 - y_1 = 0.330$$

(d) A *DEW T* calculation. Because $P = 101.33$ kPa, the saturation temperatures are the same as those of part (c), and an average value again serves as an initial value for T. Because the liquid-phase composition is not known, the activity coefficients are initialized as $\gamma_1 = \gamma_2 = 1$. Trial calculations alternately vary T to reproduce the given value of P and then x_1 to reproduce the known value of y_1. The process yields the following final values:

$$T = 326.70 \text{ K} \qquad P_1^{sat} = 64.63 \text{ kPa} \qquad P_2^{sat} = 89.94 \text{ kPa}$$
$$A = 1.0624 \qquad \gamma_1 = 1.3628 \qquad \gamma_2 = 1.2523$$
$$x_1 = 0.4602 \qquad x_2 = 0.5398$$

(e) First determine whether or not an azeotrope exists at the given temperature. This calculation is facilitated by the definition of a quantity called the *relative volatility*:

$$\alpha_{12} \equiv \frac{y_1/x_1}{y_2/x_2} \tag{13.22}$$

at an azeotrope $y_1 = x_1$, $y_2 = x_2$, and $\alpha_{12} = 1$. In general, by Eq. (13.19),

$$\frac{y_i}{x_i} = \frac{\gamma_i P_i^{\text{sat}}}{P}$$

Therefore,

$$\alpha_{12} = \frac{\gamma_1 P_1^{\text{sat}}}{\gamma_2 P_2^{\text{sat}}} \tag{13.23}$$

The correlating equations for the activity coefficients show that when $x_1 = 0$, $\gamma_2 = 1$ and $\gamma_1 = \exp(A)$; when $x_1 = 1$, $\gamma_1 = 1$ and $\gamma_2 = \exp(A)$. Therefore, in these limits,

$$(\alpha_{12})_{x_1=0} = \frac{P_1^{\text{sat}} \exp(A)}{P_2^{\text{sat}}} \qquad \text{and} \qquad (\alpha_{12})_{x_1=1} = \frac{P_1^{\text{sat}}}{P_2^{\text{sat}} \exp(A)}$$

Values of P_1^{sat}, P_2^{sat}, and A are given in part (a) for the temperature of interest. The limiting values of α_{12} are therefore $(\alpha_{12})_{x_1=0} = 2.052$ and $(\alpha_{12})_{x_1=1} = 0.224$. The value at one limit is greater than 1, whereas the value at the other limit is less than 1. Thus, an azeotrope does exist, because α_{12} is a continuous function of x_1 and must pass through the value of 1.0 at some intermediate composition.

For the azeotrope, $\alpha_{12} = 1$, and Eq. (13.23) becomes:

$$\frac{\gamma_1^{\text{az}}}{\gamma_2^{\text{az}}} = \frac{P_2^{\text{sat}}}{P_1^{\text{sat}}} = \frac{65.65}{44.51} = 1.4747$$

The difference between the correlating equations for $\ln \gamma_1$ and $\ln \gamma_2$ provides the general relation:

$$\ln \frac{\gamma_1}{\gamma_2} = A x_2^2 - A x_1^2 = A(x_2 - x_1)(x_2 + x_1) = A(x_2 - x_1) = A(1 - 2x_1)$$

Thus the azeotrope occurs at the value of x_1 for which this equation is satisfied when the activity-coefficient ratio has its azeotrope value of 1.4747, i.e., when

$$\ln \frac{\gamma_1}{\gamma_2} = \ln 1.4747 = 0.388$$

Solution gives $x_1^{\text{az}} = 0.325$. For this value of x_1, $\gamma_1^{\text{az}} = 1.657$. With $x_1^{\text{az}} = y_1^{\text{az}}$, Eq. (13.19) becomes:

$$P^{\text{az}} = \gamma_1^{\text{az}} P_1^{\text{sat}} = (1.657)(44.51)$$

Thus, $P^{\text{az}} = 73.76 \text{ kPa}$ $x_1^{\text{az}} = y_1^{\text{az}} = 0.325$

Activity coefficients are functions of temperature and liquid-phase composition, and ultimately correlations for them are based on experiment. Thus, examination of a set of experimental VLE data and the activity coefficients implied by these data is instructive. Table 13.1 presents such a data set.

Table 13.1: VLE Data for Methyl Ethyl Ketone(1)/Toluene(2) at 50°C

P/kPa	x_1	y_1	$\hat{f}_1^l = y_1 P$	$\hat{f}_2^l = y_2 P$	γ_1	γ_2
12.30 (P_2^{sat})	0.0000	0.0000	0.000	12.300 (P_2^{sat})		1.000
15.51	0.0895	0.2716	4.212	11.298	1.304	1.009
18.61	0.1981	0.4565	8.496	10.114	1.188	1.026
21.63	0.3193	0.5934	12.835	8.795	1.114	1.050
24.01	0.4232	0.6815	16.363	7.697	1.071	1.078
25.92	0.5119	0.7440	19.284	6.636	1.044	1.105
27.96	0.6096	0.8050	22.508	5.542	1.023	1.135
30.12	0.7135	0.8639	26.021	4.099	1.010	1.163
31.75	0.7934	0.9048	28.727	3.023	1.003	1.189
34.15	0.9102	0.9590	32.750	1.400	0.997	1.268
36.09 (P_1^{sat})	1.0000	1.0000	36.090 (P_1^{sat})	0.000	1.000	

The criterion for vapor/liquid equilibrium is that the fugacity of species i is the same in both phases. If the vapor phase is in its ideal-gas state, then the fugacity equals the partial pressure, and

$$\hat{f}_i^l = \hat{f}_i^v = y_i P$$

The liquid-phase fugacity of species i increases from zero at infinite dilution ($x_i = y_i \rightarrow 0$) to P_i^{sat} for pure species i. This is illustrated by the data of Table 13.1 for the methyl ethyl ketone(1)/toluene(2) system at 50°C.[3] The first three columns list experimental P-x_1-y_1 data, and columns 4 and 5 show $\hat{f}_1^l = y_1 P$ and $\hat{f}_2^l = y_2 P$. The fugacities are plotted in Fig. 13.2 as solid lines. The straight dashed lines represent Eq. (10.83), the Lewis/Randall rule, which expresses the composition dependence of the constituent fugacities in an ideal solution:

$$\hat{f}_i^{id} = x_i f_i^l \tag{10.83}$$

Although derived from a particular set of data, Fig. 13.2 illustrates the general nature of the $\hat{f}_1^l$ and $\hat{f}_2^l$ vs. x_1 relationships for a binary liquid solution at constant T. The equilibrium pressure P varies with composition, but its influence on the liquid-phase values of $\hat{f}_1^l$ and $\hat{f}_2^l$ is negligible. Thus a plot at constant T and P would look the same, as indicated in Fig. 13.3 for species i ($i = 1, 2$) in a binary solution at constant T and P.

---

[3]M. Diaz Peña, A. Crespo Colin, and A. Compostizo, J. Chem. Thermodyn., vol. 10, pp. 337–341, 1978.

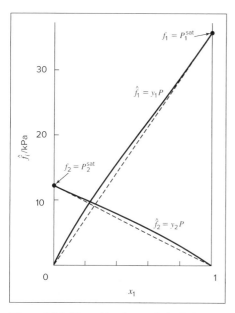

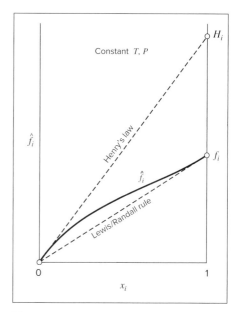

Figure 13.2: Fugacities for methyl ethyl ketone(1)/toluene(2) at 50°C. The dashed lines represent the Lewis/Randall rule.

Figure 13.3: Composition dependence of liquid-phase fugacities for species i in a binary solution.

The lower dashed line in Fig. 13.3, representing the Lewis/Randall rule, is characteristic of ideal-solution behavior. It provides the simplest possible model for the composition dependence of $\hat{f}_i^l$, representing a standard to which actual behavior may be compared. Indeed, the activity coefficient as defined by Eq. (13.2) formalizes this comparison:

$$\gamma_i \equiv \frac{\hat{f}_i^l}{x_i f_i^l} = \frac{\hat{f}_i^l}{\hat{f}_i^{id}}$$

Thus the activity coefficient of a species in solution is the ratio of its actual fugacity to the value given by the Lewis/Randall rule at the same T, P, and composition. For the calculation of experimental values of γ_i, both $\hat{f}_i^l$ and $\hat{f}_i^{id}$ are eliminated in favor of measurable quantities.

$$\gamma_i = \frac{y_i P}{x_i f_i^l} = \frac{y_i P}{x_i P_i^{sat}} \qquad (i = 1, 2, \ldots, N) \qquad (13.24)$$

This is a restatement of Eq. (13.19), the modified Raoult's law, and is adequate for present purposes, allowing easy calculation of activity coefficients from experimental low-pressure VLE data. Values from this equation appear in the last two columns of Table 13.1. Figure 13.4 shows plots of $\ln \gamma_i$ based on experimental measurements for six binary systems at 50°C, illustrating the variety of behavior that is observed. Note in every case that as $x_i \to 1$, $\ln \gamma_i \to 0$ with zero slope. Usually (but not always) the infinite-dilution activity

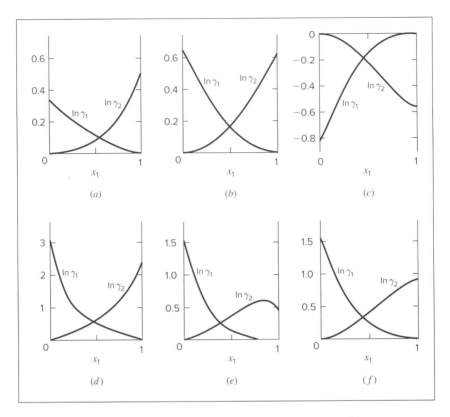

Figure 13.4: Logarithms of the activity coefficients at 50°C for six binary liquid systems: (*a*) chloroform(1)/*n*-heptane(2); (*b*) acetone(1)/methanol(2); (*c*) acetone(1)/chloroform(2); (*d*) ethanol(1)/*n*-heptane(2); (*e*) ethanol(1)/chloroform(2); (*f*) ethanol(1)/water(2).

coefficient is an extreme value. Comparison of these graphs with those of Fig. 10.3 indicates that the $\ln \gamma_i$ generally have the same sign as G^E. That is, positive G^E implies activity coefficients greater than unity and negative G^E implies activity coefficients less than unity, at least over most of the composition range.

Henry's Law

The solid lines in both Figs. 13.2 and 13.3, representing experimental values of $\hat{f}_i^l$, become tangent to the Lewis/Randall-rule lines at $x_i = 1$. This is a consequence of the Gibbs/Duhem equation, as will be shown presently. In the other limit, $x_i \to 0$, $\hat{f}_i^l$ becomes zero. Thus, the ratio $\hat{f}_i^l/x_i$ is indeterminate in this limit, and application of l'Hôpital's rule yields:

$$\lim_{x \to 0} \frac{\hat{f}_i^l}{x_i} = \left(\frac{d\hat{f}_i^l}{dx_i} \right)_{x_i = 0} \equiv \mathcal{H}_i \tag{13.25}$$

Equation (13.25) defines *Henry's constant* $\mathcal{H}_i$ as the limiting slope of the $\hat{f}_i^l$-vs.-x_i curve at $x_i = 0$. As shown by Fig. 13.3, this is the slope of a line drawn tangent to the curve at $x_i = 0$. The equation of this tangent line expresses *Henry's law*.

$$\boxed{\hat{f}_i^l = x_i \mathcal{H}_i}$$

(13.26)

Applicable in the limit as $x_i \rightarrow 0$, it is also of approximate validity for small values of x_i.

For a system of air in equilibrium with liquid water, the mole fraction of water vapor in the air is found from Raoult's law applied to the water with the assumption that the water is essentially pure. Thus, Raoult's law for the water (species 2) becomes $y_2 P = P_2^{sat}$. At 25°C and atmospheric pressure, this equation yields:

$$y_2 = \frac{P_2^{sat}}{P} = \frac{3.166}{101.33} = 0.0312$$

where the pressures are in kPa, and P_2^{sat} comes from the steam tables.

Calculation of the mole fraction of air dissolved in the water is accomplished with Henry's law, applied here for a pressure low enough that the vapor phase is in its ideal-gas state. Values of $\mathcal{H}_i$ come from experiment, and Table 13.2 lists values at 25°C for a few gases dissolved in water. For the air/water system at 25°C and atmospheric pressure, Henry's Law applied to air (species 1) with $y_1 = 1 - 0.0312 = 0.9688$ yields:

$$x_1 = \frac{y_1 P}{\mathcal{H}_1} = \frac{(0.9688)(1.0133)}{72,950} = 1.35 \times 10^{-5}$$

This result justifies the assumption that the water approaches purity.

Table 13.2: Henry's Constants for Gases Dissolved in Water at 25°C

Gas	$\mathcal{H}$/bar
Acetylene	1,350
Air	72,950
Carbon dioxide	1,670
Carbon monoxide	54,600
Ethane	30,600
Ethylene	11,550
Helium	126,600
Hydrogen	71,600
Hydrogen sulfide	550
Methane	41,850
Nitrogen	87,650
Oxygen	44,380

Example 13.2

Assuming that carbonated water contains only $CO_2(1)$ and $H_2O(2)$, determine the compositions of the vapor and liquid phases in a sealed can of "soda" at 25°C if the pressure inside the can is 5 bar.

Solution 13.2

Expecting that the liquid phase will be nearly pure water and the vapor phase will be nearly pure CO_2, we apply Henry's law for CO_2 (species 1) and Raoult's law for water (species 2):

$$y_1 P = x_1 \mathcal{H}_1 \qquad\qquad y_2 P = x_2 P_2^{\text{sat}}$$

With the vapor phase nearly pure CO_2, we obtain the liquid-phase CO_2 mole fraction as

$$x_1 = \frac{y_1 P}{\mathcal{H}_1} \approx \frac{P}{\mathcal{H}_1} = \frac{5}{1670} = 0.0030$$

Similarly, with the liquid phase nearly pure water, we have

$$y_2 = \frac{x_2 P_2^{\text{sat}}}{P} \approx \frac{P_2^{\text{sat}}}{P}$$

From the steam tables, the vapor pressure of water at 25°C is 3.166 kPa, or 0.0317 bar. Thus, $y_2 = 0.0317/5 = 0.0063$. Consistent with our expectations, the liquid is 99.7% water and the vapor is 99.4% CO_2.

Henry's law is related to the Lewis/Randall rule through the Gibbs/Duhem equation, expressed by Eq. (10.14). Written for a binary liquid solution with $\bar{M}_i$ replaced by $\bar{G}_i^l = \mu_i^l$, it becomes:

$$x_1 \, d\mu_1^l + x_2 \, d\mu_2^l = 0 \qquad \text{(const } T, P)$$

Differentiation of Eq. (10.46) at constant T and P yields: $d\mu_i^l = RTd \ln \hat{f}_i^l$. The preceding equation is then

$$x_1 d \ln \hat{f}_1^l + x_2 d \ln \hat{f}_2^l = 0 \qquad \text{(const } T, P)$$

Upon division by dx_1.

$$\boxed{x_1 \frac{d \ln \hat{f}_1^l}{dx_1} + x_2 \frac{d \ln \hat{f}_2^l}{dx_1} = 0 \quad \text{(const } T, P)} \qquad\qquad (13.27)$$

This is a particular form of the Gibbs/Duhem equation. Substitution of $-dx_2$ for dx_1 in the second term produces:

$$x_1 \frac{d \ln \hat{f}_1^l}{dx_1} = x_2 \frac{d \ln \hat{f}_2^l}{dx_2} \qquad \text{or} \qquad \frac{d\hat{f}_1^l/dx_1}{\hat{f}_1^l/x_1} = \frac{d\hat{f}_2^l/dx_2}{\hat{f}_2^l/x_2}$$

In the limit as $x_1 \to 1$ and $x_2 \to 0$,

$$\lim_{x_1 \to 1} \frac{d\hat{f}_1^l/dx_1}{\hat{f}_1^l/x_1} = \lim_{x_2 \to 0} \frac{d\hat{f}_2^l/dx_2}{\hat{f}_2^l/x_2}$$

Because $\hat{f}_1^l = f_1^l$ when $x_1 = 1$, this may be rewritten:

$$\frac{1}{f_1^l}\left(\frac{d\hat{f}_1^l}{dx_1}\right)_{x_1=1} = \frac{\left(d\hat{f}_2^l/dx_2\right)_{x_2=0}}{\lim\limits_{x_2 \to 0}\left(\hat{f}_2^l/x_2\right)}$$

According to Eq. (13.25), the numerator and denominator on the right side of this equation are equal, and therefore:

$$\left(\frac{d\hat{f}_1^l}{dx_1}\right)_{x_1=1} = f_1^l \tag{13.28}$$

This equation is the exact expression of the Lewis/Randall rule as applied to real solutions. It also implies that Eq. (10.83) provides approximately correct values of $\hat{f}_i^l$ when $x_i \approx 1$: $\hat{f}_i^l \approx \hat{f}_i^{id} = x_i f_i^l$.

Henry's law applies to a species as it approaches infinite dilution in a binary solution, and the Gibbs/Duhem equation insures validity of the Lewis/Randall rule for the other species as it approaches purity.

The fugacity shown by Fig. 13.3 is for a species with positive deviations from ideality in the sense of the Lewis/Randall rule. Negative deviations are less common but are also observed; the $\hat{f}_i^l$-vs.-x_i curve then lies below the Lewis/Randall line. In Fig. 13.5 the fugacity of acetone is shown as a function of composition for two different binary liquid solutions at 50°C. When the second species is methanol, acetone exhibits positive deviations from ideality. When the second species is chloroform, the deviations are negative. The fugacity of pure acetone $f_{acetone}$ is of course the same regardless of the identity of the second species. However, Henry's constants, represented by slopes of the two dotted lines, are very different for the two cases.

Example 13.3

A fog consists of spherical water droplets with a radius of about 10^{-6} m. Because of surface tension, the internal pressure within a water droplet is greater than the external pressure. For droplet radius r and surface tension σ, the internal pressure in a water

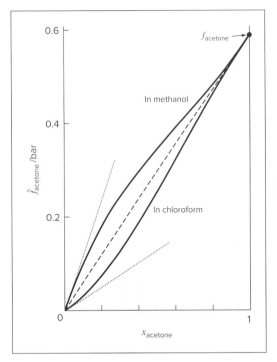

Figure 13.5: Composition dependence of the fugacity of acetone in two binary liquid solutions at 50°C.

droplet is greater than the external pressure by $\delta P = 2\sigma/r$. This increases the fugacity of the water in a droplet above that of bulk water at the same conditions. Thus, it tends to cause the fog to evaporate. However, the fugacity increase may be countered by a decrease in temperature or by the dissolution of atmospheric pollutants. To stabilize the fog with respect to bulk water at the same temperature, determine:

(a) The minimum temperature decrease required.

(b) The minimum concentration of atmospheric pollutants required in the fog droplets.

Solution 13.3

(a) At 25°C the surface tension of pure water is 0.0694 N·m^{-1}, and

$$\delta P = \frac{(2)(0.0694)}{10^{-6}} = 0.1338 \times 10^6 \text{ Pa} \quad \text{or} \quad 1.388 \text{ bar}$$

Two general equations apply to the isothermal change in Gibbs energy resulting from a change in pressure. The first results from differentiation of Eq. (10.31), the defining equation for fugacity, and the second from restriction of Eq. (6.11) to constant temperature:

$$dG_i = RTd \ln f_i \quad \text{(const } T) \quad \text{and} \quad dG_i = V_i dP \quad \text{(const } T)$$

In combination these equations yield:

$$d \ln f_i = \frac{V_i}{RT} dP \quad (\text{const } T)$$

Because the molar volume of water is virtually unaffected by pressure, integration gives:

$$\delta \ln f_{H_2O} = \frac{V_{H_2O}}{RT} \delta P \quad (\text{const } T)$$

For a molar volume of water of 18 cm$^3 \cdot$mol^{-1}, and a pressure change of 1.388 bar,

$$\delta \ln f_{H_2O} = \frac{(18 \text{ cm}^3 \cdot \text{mol})(1.388 \text{ bar})}{(83.14 \text{ cm}^3 \cdot \text{bar} \cdot \text{mol}^{-1} \cdot \text{K}^{-1})\,(298 \text{ K})} = 0.00101$$

This is the amount by which the fugacity of water in a fog droplet exceeds the fugacity of bulk water at the same temperature. Thus the fog is not stable, and it dissipates by evaporation. However, a sufficient decrease in temperature of the fog reduces droplet fugacity enough to stabilize the fog.

An equation giving the effect of temperature on fugacity is developed from Eq. (10.31) by subtracting from it the ideal-gas-state form of the same equation (for which $f_i = P$). Thus

$$G_i - G_i^{ig} = RT \ln f_i - RT \ln P$$

Solution for $\ln f_i$ gives:

$$\ln f_i = \frac{G_i}{RT} - \frac{G_i^{ig}}{RT} + \ln P$$

Differentiation of this general equation with respect to T at constant P yields:

$$\left(\frac{\partial \ln f_i}{\partial T} \right)_P = \left(\frac{\partial (G_i/RT)}{\partial T} \right)_P - \left(\frac{\partial G_i^{ig}/RT}{\partial T} \right)_P$$

Substituting for the partial derivatives on the right by Eq. (6.39) gives:

$$\left(\frac{\partial \ln f_i}{\partial T} \right)_P = \frac{-H_i}{RT^2} + \frac{H_i^{ig}}{RT^2} = \frac{-H_i^R}{RT^2}$$

The residual enthalpy here is approximately the negative of the latent heat of vaporization of water at 25°C; thus

$$-H_i^R = 2442.3 \text{ J} \cdot \text{g}^{-1} \times 18 \text{ g} \cdot \text{mol}^{-1} = 43{,}960 \text{ J} \cdot \text{mol}^{-1}. \text{ Then}$$

$$\left(\frac{\partial \ln f_{H_2O}}{\partial T} \right)_P = \frac{43{,}960 \text{ J} \cdot \text{mol}^{-1}}{(8.314 \text{ J} \cdot \text{mol}^{-1} \cdot \text{K}^{-1})(298 \text{ K})^2} = 0.0595 \text{ K}^{-1}$$

For a small temperature difference, $\delta \ln f_{H_2O} \approx 0.0595\,\delta T$. To find the δT that counters the pressure-induced fugacity increase in a water droplet, we set this fugacity difference equal to -0.00101, the *negative* of the fugacity increase. Thus

$$0.0595\,\delta T = -0.00101 \qquad \text{and} \qquad \delta T = -0.017\ \text{K}$$

This is the minimum temperature decrease of the fog required for stability with respect to bulk water.

(b) The fugacity of water in fog droplets is also lowered by the dissolution of impurities, such as atmospheric pollutants. The droplets then become solutions with water as the major constituent, and their fugacity is $\hat{f}_{H_2O}$. Because the mole fraction of water is near unity, the Lewis/Randall rule provides an excellent approximation, $\hat{f}_{H_2O} = f_{H_2O} x_{H_2O}$. The fugacity change of the water resulting from the dissolution of impurities is then:

$$\delta f_{H_2O} = \hat{f}_{H_2O} - f_{H_2O} = f_{H_2O} x_{H_2O} - f_{H_2O} = f_{H_2O}\left(x_{H_2O} - 1\right) = -f_{H_2O} x_{\text{impurity}}$$

or

$$x_{\text{impurity}} = \frac{-\delta f_{H_2O}}{f_{H_2O}} \approx -\delta \ln f_{H_2O}$$

This quantity is again the negative of the increase caused by surface tension. Thus,

$$x_{\text{impurity}} = 0.00101$$

and impurities of only 0.1% stabilize the fog with respect to bulk water.

13.4   CORRELATIONS FOR LIQUID-PHASE ACTIVITY COEFFICIENTS

Liquid-phase activity coefficients γ_i play a vital role in the gamma/phi formulation of VLE. They are directly related to the excess Gibbs energy. In general, G^E/RT is a function of T, P, and composition, but for liquids at low to moderate pressures it is a very weak function of P. Therefore its pressure dependence is usually neglected, and for applications *at constant T*, excess Gibbs energy is treated as a function of composition alone:

$$\frac{G^E}{RT} = g(x_1, x_2, \ldots, x_n) \qquad (\text{const } T)$$

The Redlich/Kister Expansion

For binary systems (species 1 and 2) the quantity most often selected to be represented by an equation is $G \equiv G^E/(x_1 x_2 RT)$, which may be expressed as a power series in x_1:

$$Y \equiv \frac{G^E}{x_1 x_2 RT} = a + bx_1 + cx_1^2 + \cdots \qquad (\text{const } T)$$

Because $x_2 = 1 - x_1$, mole fraction x_1 serves as the single independent variable. An equivalent power series with certain advantages is known as the Redlich/Kister expansion:[4]

$$Y = A_0 + \sum_{n=1}^{a} A_n z^n \qquad (13.29)$$

where, by definition, $z \equiv x_1 - x_2 = 2x_1 - 1$, a is the order of the power series, and parameters A_n are functions of temperature.

Expressions for the activity coefficients are found from Eqs. (10.15) and (10.16) with G^E/RT replacing M^E.

$$\frac{\bar{G}_1^E}{RT} = \frac{G^E}{RT} + x_2 \frac{d(G^E/RT)}{dx_1} \qquad (13.30) \qquad \frac{\bar{G}_2^E}{RT} = \frac{G^E}{RT} - x_1 \frac{d(G^E/RT)}{dx_1} \qquad (13.31)$$

By Eq. (13.3), $\ln \gamma_i = \bar{G}_i^E/RT$. Moreover, $G^E/RT = x_1 x_2 Y$, and

$$\frac{d(G^E/RT)}{dx_1} = x_1 x_2 \frac{dY}{dx_1} + Y(x_2 - x_1)$$

Making these substitutions in Eqs. (13.30) and (13.31) leads to:

$$\ln \gamma_1 = x_2^2 \left(Y + x_1 \frac{dY}{dx_1} \right) \qquad (13.32) \qquad \ln \gamma_2 = x_1^2 \left(Y - x_2 \frac{dY}{dx_1} \right) \qquad (13.33)$$

where Y is given by Eq. (13.29) and

$$\frac{dY}{dx_1} = \sum_{n=1}^{a} n A_n z^{n-1} \qquad (13.34)$$

For infinite-dilution values, Eqs. (13.32) and (13.33) yield:

$$\ln \gamma_1^\infty = Y(x_1 = 0, x_2 = 1, z = -1) = A_0 + \sum_{n=1}^{a} A_n(-1)^n \qquad (13.35)$$

$$\ln \gamma_2^\infty = Y(x_1 = 1, x_2 = 0, z = 1) = A_0 + \sum_{n=1}^{a} A_n \qquad (13.36)$$

In application, different truncations of these series are appropriate, and truncations with $a \leq 5$ are frequent in the literature.

[4]O. Redlich, A. T. Kister, and C. E. Turnquist, *Chem. Eng. Progr. Symp. Ser. No.* 2, vol. 48, pp. 49–61, 1952.

When all parameters are zero, $\ln \gamma_1 = 0$, $\ln \gamma_2 = 0$, $\gamma_1 = \gamma_2 = 1$. These are the values for an ideal solution, and they represent a limiting case where the excess Gibbs energy is zero. If all parameters except A_0 are zero, $Y = A_0$, and Eqs. (13.32) and (13.33) reduce to:

$$\ln \gamma_1 = A_0 x_2^2 \qquad (13.37) \qquad \ln \gamma_2 = A_0 x_1^2 \qquad (13.38)$$

The symmetrical nature of these relations is evident. Infinite-dilution values of the activity coefficients are $\ln \gamma_1^\infty = \ln \gamma_2^\infty = A_0$.

Most widely found is the two-parameter truncation:

$$Y = A_0 + A_1(x_1 - x_2) = A_0 + A_1(2x_1 - 1)$$

in which Y is linear in x_1. An alternate form of this equation results from the definitions $A_0 + A_1 = A_{21}$ and $A_0 - A_1 = A_{12}$. Eliminating parameters A_0 and A_1 in favor of A_{21} and A_{12}, we obtain:

$$\frac{G^E}{RT} = (A_{21}x_1 + A_{12}x_2)x_1x_2 \qquad (13.39)$$

$$\ln \gamma_1 = x_2^2 [A_{12} + 2(A_{21} - A_{12})x_1] \qquad (13.40)$$

$$\ln \gamma_2 = x_1^2 [A_{21} + 2(A_{12} - A_{21})x_2] \qquad (13.41)$$

These are known as the **Margules[5] equations.** For the limiting conditions of infinite dilution, they imply:

$$\ln \gamma_1^\infty = A_{12} \qquad \text{and} \qquad \ln \gamma_2^\infty = A_{21}$$

The van Laar Equation

Another well-known equation results when the reciprocal expression x_1x_2RT/G^E is expressed as a linear function of x_1:

$$\frac{x_1x_2}{G^E/RT} = A' + B'(x_1 - x_2) = A' + B'(2x_1 - 1)$$

This may also be written:

$$\frac{x_1x_2}{G^E/RT} = A'(x_1 + x_2) + B'(x_1 - x_2) = (A' + B')x_1 + (A' - B')x_2$$

When new parameters are defined by the equations $A' + B' = 1/A_{21}'$ and $A' - B' = 1/A_{12}'$, an equivalent form is obtained:

$$\frac{x_1x_2}{G^E/RT} = \frac{x_1}{A_{21}'} + \frac{x_2}{A_{12}'} = \frac{A_{12}'x_1 + A_{21}'x_2}{A_{12}'A_{21}'}$$

or

$$\frac{G^E}{x_1x_2RT} = \frac{A_{12}'A_{21}'}{A_{12}'x_1 + A_{21}'x_2} \qquad (13.42)$$

[5]Max Margules (1856–1920), Austrian meteorologist and physicist; see http://en.wikipedia.org/wiki/Max_Margules.

The activity coefficients implied by this equation are:

$$\ln \gamma_1 = A'_{12}\left(1 + \frac{A'_{12}x_1}{A'_{21}x_2}\right)^{-2} \qquad (13.43) \qquad \ln \gamma_2 = A'_{21}\left(1 + \frac{A'_{21}x_2}{A'_{12}x_1}\right)^{-2} \qquad (13.44)$$

These are the van Laar[6] equations. When $x_1 = 0$, $\ln \gamma_1^\infty = A'_{12}$; when $x_2 = 0$, $\ln \gamma_2^\infty = A'_{21}$.

The Redlich/Kister expansion and the van Laar equations are special cases of a general treatment based on rational functions, i.e., on equations for $G^E/(x_1 x_2 RT)$ given by ratios of polynomials.[7] They provide great flexibility in the fitting of VLE data for binary systems. However, they have scant theoretical foundation, and therefore they fail to provide a rational basis for extension to multicomponent systems. Moreover, they do not incorporate an explicit temperature dependence of their parameters, though this can be supplied on an *ad hoc* basis.

Local-Composition Models

Theoretical developments in the molecular thermodynamics of liquid-solution behavior are often based on the concept of *local composition*. Within a liquid solution, local compositions, different from the overall mixture composition, are presumed to account for the short-range order and nonrandom molecular orientations that result from differences in molecular size and intermolecular forces. The concept was introduced by G. M. Wilson in 1964 with the publication of a model of solution behavior since known as the Wilson equation.[8] The success of this equation in the correlation of VLE data prompted the development of alternative local-composition models, most notably the NRTL (**N**on-**R**andom-**T**wo-**L**iquid) equation of Renon and Prausnitz[9] and the UNIQUAC (**UNI**versal **QUA**si-**C**hemical) equation of Abrams and Prausnitz.[10] A further significant development, based on the UNIQUAC equation, is the UNIFAC method,[11] in which activity coefficients are calculated from contributions of the various groups making up the molecules of a solution.

Wilson Equation. Like the Margules and van Laar equations, the Wilson equation contains just two parameters for a binary system (Λ_{12} and Λ_{21}). It is written:

$$\frac{G^E}{RT} = -x_1 \ln(x_1 + x_2 \Lambda_{12}) - x_2 \ln(x_2 + x_1 \Lambda_{21}) \qquad (13.45)$$

[6]Johannes Jacobus van Laar (1860–1938), Dutch physical chemist; see http://en.wikipedia.org/wiki/Johannes_van_Laar.

[7]H. C. Van Ness and M. M. Abbott, *Classical Thermodynamics of Nonelectrolyte Solutions: With Applications to Phase Equilibria,* Sec. 5–7, McGraw-Hill, New York, 1982.

[8]G. M. Wilson, *J. Am. Chem. Soc,* vol. 86, pp. 127–130, 1964.

[9]H. Renon and J. M. Prausnitz, *AIChE J.,* vol. 14, p. 135–144, 1968.

[10]D. S. Abrams and J. M. Prausnitz, *AIChE J.,* vol. 21, p. 116–128, 1975.

[11]UNIQUAC Functional-Group Activity Coefficients; proposed by Aa. Fredenslund, R. L. Jones, and J. M. Prausnitz, *AIChE J.,* vol. 21, p. 1086–1099, 1975; given detailed treatment in the monograph Aa. Fredenslund, J. Gmehling, and P. Rasmussen, *Vapor-Liquid Equilibrium Using UNIFAC,* Elsevier, Amsterdam, 1977.

$$\ln \gamma_1 = -\ln(x_1 + x_2\Lambda_{12}) + x_2\left(\frac{\Lambda_{12}}{x_1 + x_2\Lambda_{12}} - \frac{\Lambda_{21}}{x_2 + x_1\Lambda_{21}}\right) \qquad (13.46)$$

$$\ln \gamma_2 = -\ln(x_2 + x_1\Lambda_{21}) - x_1\left(\frac{\Lambda_{12}}{x_1 + x_2\Lambda_{12}} - \frac{\Lambda_{21}}{x_2 + x_1\Lambda_{21}}\right) \qquad (13.47)$$

For infinite dilution, these equations become:

$$\ln \gamma_1^\infty = -\ln \Lambda_{12} + 1 - \Lambda_{21} \qquad \text{and} \qquad \ln \gamma_2^\infty = -\ln \Lambda_{21} + 1 - \Lambda_{12}$$

Note that Λ_{12} and Λ_{21} must always be positive numbers.

NRTL Equation. This equation contains three parameters for a binary system and is written:

$$\frac{G^E}{x_1 x_2 RT} = \frac{G_{21}\tau_{21}}{x_1 + x_2 G_{21}} + \frac{G_{12}\tau_{12}}{x_2 + x_1 G_{12}} \qquad (13.48)$$

$$\ln \gamma_1 = x_2^2\left[\tau_{21}\left(\frac{G_{21}}{x_1 + x_2 G_{21}}\right)^2 + \frac{G_{12}\tau_{12}}{(x_2 + x_1 G_{12})^2}\right] \qquad (13.49)$$

$$\ln \gamma_2 = x_1^2\left[\tau_{12}\left(\frac{G_{12}}{x_2 + x_1 G_{12}}\right)^2 + \frac{G_{21}\tau_{21}}{(x_1 + x_2 G_{21})^2}\right] \qquad (13.50)$$

Here, $G_{12} = \exp(-\alpha\tau_{12}) \qquad G_{21} = \exp(-\alpha\tau_{21})$

and $\tau_{12} = \dfrac{b_{12}}{RT} \qquad \tau_{21} = \dfrac{b_{21}}{RT}$

where α, b_{12}, and b_{21}, parameters specific to a particular pair of species, are independent of composition and temperature. The infinite-dilution values of the activity coefficients are given by the equations:

$$\ln \gamma_1^\infty = \tau_{21} + \tau_{12}\exp(-\alpha\tau_{12}) \qquad \text{and} \qquad \ln \gamma_2^\infty = \tau_{12} + \tau_{21}\exp(-\alpha\tau_{21})$$

The UNIQUAC equation and the UNIFAC method are models of greater complexity and are treated in App. G.

Multicomponent Systems

The local-composition models have limited flexibility in the fitting of data, but they are adequate for most engineering purposes. Moreover, they are implicitly generalizable to multicomponent systems without the introduction of any parameters beyond those required to describe the constituent binary systems. For example, the Wilson equation for multicomponent systems is:

$$\frac{G^E}{RT} = -\sum_i x_i \ln\left(\sum_j x_j\Lambda_{ij}\right) \qquad (13.51)$$

$$\ln \gamma_i = 1 - \ln\left(\sum_j x_j\Lambda_{ij}\right) - \sum_k \frac{x_k\Lambda_{ki}}{\sum_j x_j\Lambda_{kj}} \qquad (13.52)$$

where $\Lambda_{ij} = 1$ for $i = j$, etc. All indices refer to the same species, and summations are over *all* species. For each *ij* pair there are two parameters, because $\Lambda_{ij} \neq \Lambda_{ji}$. For a ternary system the three *ij* pairs are associated with the parameters $\Lambda_{12}, \Lambda_{21}; \Lambda_{13}, \Lambda_{31};$ and $\Lambda_{23}, \Lambda_{32}.$

The temperature dependence of the parameters is given by:

$$\Lambda_{ij} = \frac{V_j}{V_i} \exp \frac{-a_{ij}}{RT} \qquad (i \neq j) \tag{13.53}$$

where V_j and V_i are the molar volumes at temperature T of pure liquids j and i, and a_{ij} is a constant independent of composition and temperature. Thus the Wilson equation, like all other local-composition models, has built into it an *approximate* temperature dependence for the parameters. Moreover, all parameters are found from data for binary (in contrast to multicomponent) systems. This makes parameter determination for the local-composition models a task of manageable proportions.

13.5 FITTING ACTIVITY COEFFICIENT MODELS TO VLE DATA

In Table 13.3 the first three columns repeat the P-x_1-y_1 data of Table 13.1 for the system methyl ethyl ketone(1)/toluene(2). These data points are also shown as circles on Fig. 13.6(a). Values of $\ln \gamma_1$ and $\ln \gamma_2$ are listed in columns 4 and 5, and are shown by the open squares and triangles of Fig. 13.6(b). They are combined for a binary system in accord with Eq. (13.10):

$$\boxed{\frac{G^E}{RT} = x_1 \ln \gamma_1 + x_2 \ln \gamma_2} \tag{13.54}$$

The values of G^E/RT so calculated are then divided by x_1x_2 to provide values of $G^E/(x_1x_2 RT)$; the two sets of numbers are listed in columns 6 and 7 of Table 13.3 and appear as solid circles on Fig. 13.6(b).

The four thermodynamic functions $\ln \gamma_1, \ln \gamma_2, G^E/RT$, and $G^E/(x_1x_2RT)$, are properties of the liquid phase. Figure 13.6(b) shows how their experimental values vary with composition for a particular binary system at a specified temperature. This figure is characteristic of systems for which:

$$\gamma_i \geq 1 \qquad \text{and} \qquad \ln \gamma_i \geq 0 \quad (i = 1, 2)$$

In such cases the liquid phase shows *positive deviations* from Raoult's-law behavior. This is seen also in Fig. 13.6(a), where the P-x_1 data points all lie above the dashed straight line, which represents Raoult's law.

Because the activity coefficient of a species in solution becomes unity as the species becomes pure, each $\ln \gamma_i$ ($i = 1, 2$) tends to zero as $x_i \to 1$. This is evident in Fig. 13.6(b). At the other limit, where $x_i \to 0$ and species i becomes infinitely dilute, $\ln \gamma_i$ approaches a finite limit, namely, $\ln \gamma_i^\infty$. In the limit as $x_1 \to 0$, the dimensionless excess Gibbs energy G^E/RT as given by Eq. (13.54) becomes:

$$\lim_{x_1 \to 0} \frac{G^E}{RT} = (0) \ln \gamma_1^\infty + (1)(0) = 0$$

Table 13.3: VLE Data for Methyl Ethyl Ketone(1)/Toluene(2) at 50°C

P/kPa	x_1	y_1	$\ln \gamma_1$	$\ln \gamma_2$	G^E/RT	$G^E/(x_1 x_2 RT)$
12.30 (P_2^{sat})	0.0000	0.0000		0.000	0.000	
15.51	0.0895	0.2716	0.266	0.009	0.032	0.389
18.61	0.1981	0.4565	0.172	0.025	0.054	0.342
21.63	0.3193	0.5934	0.108	0.049	0.068	0.312
24.01	0.4232	0.6815	0.069	0.075	0.072	0.297
25.92	0.5119	0.7440	0.043	0.100	0.071	0.283
27.96	0.6096	0.8050	0.023	0.127	0.063	0.267
30.12	0.7135	0.8639	0.010	0.151	0.051	0.248
31.75	0.7934	0.9048	0.003	0.173	0.038	0.234
34.15	0.9102	0.9590	−0.003	0.237	0.019	0.227
36.09 (P_1^{sat})	1.0000	1.0000	0.000		0.000	

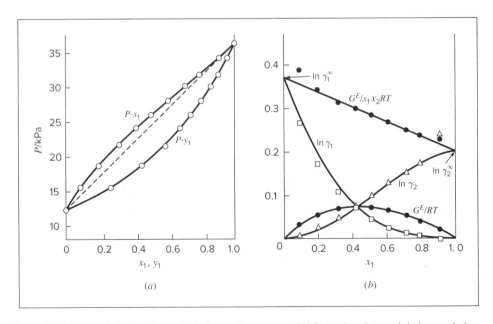

Figure 13.6: The methyl ethyl ketone(1)/toluene(2) system at 50°C. (*a*) *Pxy* data and their correlation. (*b*) Liquid-phase properties and their correlation.

The same result is obtained for $x_2 \to 0$ ($x_1 \to 1$). The value of G^E/RT (and G^E) is therefore zero at both $x_1 = 0$ and $x_1 = 1$.

The quantity $G^E/(x_1 x_2 RT)$ becomes indeterminate both at $x_1 = 0$ and $x_1 = 1$, because G^E is zero in both limits, as is the product $x_1 x_2$. For $x_1 \to 0$, l'Hôpital's rule yields:

$$\lim_{x_1 \to 0} \frac{G^E}{x_1 x_2 RT} = \lim_{x_1 \to 0} \frac{G^E/RT}{x_1} = \lim_{x_1 \to 0} \frac{d(G^E/RT)}{dx_1} \qquad (A)$$

Differentiation of Eq. (13.54) with respect to x_1 provides the derivative of the final member:

$$\frac{d(G^E/RT)}{dx_1} = x_1 \frac{d\ln\gamma_1}{dx_1} + \ln\gamma_1 + x_2 \frac{d\ln\gamma_2}{dx_1} - \ln\gamma_2 \qquad (B)$$

The minus sign preceding the last term comes from $dx_2/dx_1 = -1$, a consequence of the equation, $x_1 + x_2 = 1$. The Gibbs/Duhem equation, Eq. (13.11), written for a binary system, is divided by dx_1 to give:

$$\boxed{x_1 \frac{d\ln\gamma_1}{dx_1} + x_2 \frac{d\ln\gamma_2}{dx_1} = 0 \qquad (\text{const } T, P)} \qquad (13.55)$$

Substitution into Eq. (B) reduces it to:

$$\frac{d(G^E/RT)}{dx_1} = \ln\frac{\gamma_1}{\gamma_2} \qquad (13.56)$$

Applied to the composition limit at $x_1 = 0$, this equation yields:

$$\lim_{x_1 \to 0} \frac{d(G^E/RT)}{dx_1} = \lim_{x_1 \to 0} \ln\frac{\gamma_1}{\gamma_2} = \ln\gamma_1^\infty$$

By Eq. (A), $\displaystyle \lim_{x_1 \to 0} \frac{G^E}{x_1 x_2 RT} = \ln\gamma_1^\infty$ Similarly, $\displaystyle \lim_{x_1 \to 1} \frac{G^E}{x_1 x_2 RT} = \ln\gamma_2^\infty$

Thus the limiting values of $G^E/(x_1 x_2 RT)$ are equal to the infinite-dilution limits of $\ln\gamma_1$ and $\ln\gamma_2$. This result is illustrated in Fig. 13.6(b).

These results depend on Eq. (13.11), which is valid for constant T and P. Although the data of Table 13.3 are for constant T, but variable P, negligible error is introduced through Eq. (13.11) because liquid-phase activity coefficients are very nearly independent of P for systems at low to moderate pressures.

Equation (13.11) has further influence on the nature of Fig. 13.6(b). Rewritten as,

$$\frac{d\ln\gamma_1}{dx_1} = -\frac{x_2}{x_1}\frac{d\ln\gamma_2}{dx_1}$$

it requires the slope of the $\ln\gamma_1$ curve to be everywhere of opposite sign to the slope of the $\ln\gamma_2$ curve. Furthermore, when $x_2 \to 0$ (and $x_1 \to 1$), the slope of the $\ln\gamma_1$ curve is zero. Similarly, when $x_1 \to 0$, the slope of the $\ln\gamma_2$ curve is zero. Thus, each $\ln\gamma_i$ ($i = 1, 2$) curve terminates at zero with zero slope at $x_i = 1$.

Data Reduction

Of the sets of points shown in Fig. 13.6(b), those for $G^E/(x_1 x_2 RT)$ most closely conform to a simple mathematical relation. Thus a straight line provides a reasonable approximation to this set of points, and mathematical expression is given to this linear relation by the equation:

$$\frac{G^E}{x_1 x_2 RT} = A_{21}x_1 + A_{12}x_2$$

where A_{21} and A_{12} are constants in any particular application. This is the Margules equation as given by Eq. (13.39), with corresponding expressions for the activity coefficients given by Eqs. (13.40) and (13.41).

For the methyl ethyl ketone/toluene system considered here, the curves of Fig. 13.6(b) for G^E/RT, $\ln \gamma_1$, and $\ln \gamma_2$ represent Eqs. (13.39), (13.40), and (13.41) with:

$$A_{12} = 0.372 \qquad \text{and} \qquad A_{21} = 0.198$$

These are values of the intercepts at $x_1 = 0$ and $x_1 = 1$ of the straight line drawn to represent the $G^E/(x_1 x_2 RT)$ data points.

A set of VLE data has here been *reduced* to a simple mathematical equation for the dimensionless excess Gibbs energy:

$$\frac{G^E}{RT} = (0.198 x_1 + 0.372 x_2) x_1 x_2$$

This equation concisely stores the information of the data set. Indeed, the Margules equations for $\ln \gamma_1$ and $\ln \gamma_2$ allow construction of a correlation of the original P-x_1-y_1 data set. Equation (13.19) is written for species 1 and 2 of a binary system as:

$$y_1 P = x_1 \gamma_1 P_1^{\text{sat}} \quad \text{and} \quad y_2 P = x_2 \gamma_2 P_2^{\text{sat}}$$

Addition gives,
$$\boxed{P = x_1 \gamma_1 P_1^{\text{sat}} + x_2 \gamma_2 P_2^{\text{sat}}} \qquad (13.57)$$

Whence,
$$\boxed{y_1 = \frac{x_1 \gamma_1 P_1^{\text{sat}}}{x_1 \gamma_1 P_1^{\text{sat}} + x_2 \gamma_2 P_2^{\text{sat}}}} \qquad (13.58)$$

Values of γ_1 and γ_2 from Eqs. (13.40) and (13.41) with A_{12} and A_{21} as determined for the methyl ethyl ketone(1)/toluene(2) system are combined with the experimental values of P_1^{sat} and P_2^{sat} to calculate P and y_1 by Eqs. (13.57) and (13.58) at various values of x_1. The results are shown by the solid lines of Fig. 13.6(a), which represent the calculated P-x_1 and P-y_1 relations. They clearly provide an adequate correlation of the experimental data points.

A second set of P-x_1-y_1 data, for chloroform(1)/1,4-dioxane(2) at 50°C,[12] is given in Table 13.4, along with values of pertinent thermodynamic functions. Figures 13.7(a) and 13.7(b) display all of the experimental values as points. This system shows negative deviations from Raoult's-law behavior; γ_1 and γ_2 are less than unity, and values of $\ln \gamma_1$, $\ln \gamma_2$, G^E/RT, and $G^E/(x_1 x_2 RT)$ are negative. Moreover, the P-x_1 data points in Fig. 13.7(a) all lie below the dashed line representing Raoult's-law behavior. Again the data points for $G^E/(x_1 x_2 RT)$ are reasonably well correlated by the Margules equation, in this case with parameters:

$$A_{12} = -0.72 \qquad \text{and} \qquad A_{21} = -1.27$$

Values of G^E/RT, $\ln \gamma_1$, $\ln \gamma_2$, P, and y_1 calculated by Eqs. (13.39), (13.40), (13.41), (13.57), and (13.58) provide the curves shown for these quantities in Figs. 13.7(a) and 13.7(b). Again, the experimental P-x_1-y_1 data are adequately correlated. Although the correlations provided by

[12]M. L. McGlashan and R. P. Rastogi, *Trans. Faraday Soc,* vol. 54, p. 496, 1958.

Table 13.4: VLE Data for Chloroform(1)/1,4-Dioxane(2) at 50°C

$P/$kPa	x_1	y_1	$\ln \gamma_1$	$\ln \gamma_2$	G^E/RT	$G^E/(x_1 x_2 RT)$
15.79 (P_2^{sat})	0.0000	0.0000		0.000	0.000	
17.51	0.0932	0.1794	−0.722	0.004	−0.064	−0.758
18.15	0.1248	0.2383	−0.694	−0.000	−0.086	−0.790
19.30	0.1757	0.3302	−0.648	−0.007	−0.120	−0.825
19.89	0.2000	0.3691	−0.636	−0.007	−0.133	−0.828
21.37	0.2626	0.4628	−0.611	−0.014	−0.171	−0.882
24.95	0.3615	0.6184	−0.486	−0.057	−0.212	−0.919
29.82	0.4750	0.7552	−0.380	−0.127	−0.248	−0.992
34.80	0.5555	0.8378	−0.279	−0.218	−0.252	−1.019
42.10	0.6718	0.9137	−0.192	−0.355	−0.245	−1.113
60.38	0.8780	0.9860	−0.023	−0.824	−0.120	−1.124
65.39	0.9398	0.9945	−0.002	−0.972	−0.061	−1.074
69.36 (P_1^{sat})	1.0000	1.0000	0.000		0.000	

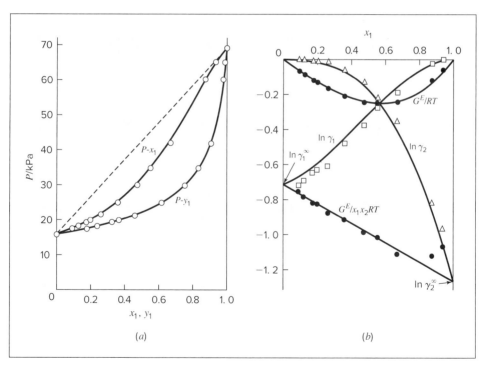

Figure 13.7: The chloroform(1)/1,4-dioxane(2) system at 50°C. (*a*) *Pxy* data and their correlation. (*b*) Liquid-phase properties and their correlation.

the Margules equations for the two sets of VLE data presented here are satisfactory, they are not perfect. The two possible reasons are, first, that the Margules equations are not precisely suited to the data set; second, that the P-x_1-y_1 data themselves are systematically in error such that they do not conform to the requirements of the Gibbs/Duhem equation.

We have presumed in applying the Margules equations that the deviations of the experimental points for $G^E/(x_1x_2RT)$ from the straight lines drawn to represent them result from random error in the data. Indeed, the straight lines do provide excellent correlations of all but a few data points. Only toward edges of a diagram are there significant deviations, and these have been discounted because the error bounds widen rapidly as the edges of such a diagram are approached. In the limits as $x_1 \to 0$ and $x_1 \to 1$, $G^E/(x_1x_2RT)$ becomes indeterminate; experimentally this means that the values are subject to unlimited error and are not measurable. However, the possibility exists that the correlation would be improved were the $G^E/(x_1x_2RT)$ points represented by an appropriate curve, rather than a straight line. Finding the correlation that best represents the data is a trial procedure.

Thermodynamic Consistency

The Gibbs/Duhem equation, Eq. (13.55), imposes a constraint on activity coefficients that may not be satisfied by a set of experimental values derived from P-x_1-y_1 data. The experimental values of $\ln \gamma_1$ and $\ln \gamma_2$ combine by Eq. (13.54) to give values of G^E/RT. This addition process is independent of the Gibbs/Duhem equation. On the other hand, the Gibbs/Duhem equation is implicit in Eq. (13.7), and activity coefficients derived from this equation necessarily obey the Gibbs/Duhem equation. These derived activity coefficients cannot possibly be consistent with the experimental values unless the experimental values also satisfy the Gibbs/Duhem equation. Nor can a P-x_1-y_1 correlation calculated by Eqs. (13.57) and (13.58) be consistent with such experimental values. If the experimental data are inconsistent with the Gibbs/Duhem equation, they are necessarily incorrect as the result of systematic error in the data. Because correlating equations for G^E/RT impose consistency on derived activity coefficients, no correlation exists that can precisely reproduce P-x_1-y_1 data that are inconsistent.

Our purpose now is to develop a simple test for the *consistency* with respect to the Gibbs/Duhem equation of a P-x_1-y_1 data set. Equation (13.54) is written with experimental values, calculated by Eq. (13.24), and denoted by an asterisk:

$$\left(\frac{G^E}{RT}\right)^* = x_1 \ln \gamma_1^* + x_2 \ln \gamma_2^*$$

Differentiation gives:

$$\frac{d(G^E/RT)^*}{dx_1} = x_1 \frac{d\ln \gamma_1^*}{dx_1} + \ln \gamma_1^* + x_2 \frac{d\ln \gamma_2^*}{dx_1} - \ln \gamma_2^*$$

or

$$\frac{d(G^E/RT)^*}{dx_1} = \ln \frac{\gamma_1^*}{\gamma_2^*} + x_1 \frac{d\ln \gamma_1^*}{dx_1} + x_2 \frac{d\ln \gamma_2^*}{dx_1}$$

This equation is subtracted from Eq. (13.56), written for *derived* property values, i.e., those given by a correlation, such as the Margules equations:

$$\frac{d(G^E/RT)}{dx_1} - \frac{d(G^E/RT)^*}{dx_1} = \ln\frac{\gamma_1}{\gamma_2} - \ln\frac{\gamma_1^*}{\gamma_2^*} - \left(x_1\frac{d\ln\gamma_1^*}{dx_1} + x_2\frac{d\ln\gamma_2^*}{dx_1}\right)$$

The differences between like terms are *residuals*, which may be represented by a δ notation. The preceding equation then becomes:

$$\frac{d\,\delta(G^E/RT)}{dx_1} = \delta\ln\frac{\gamma_1}{\gamma_2} - \left(x_1\frac{d\ln\gamma_1^*}{dx_1} + x_2\frac{d\ln\gamma_2^*}{dx_1}\right)$$

If a data set is reduced so as to make the residuals in G^E/RT scatter about zero, then the derivative $d\,\delta(G^E/RT)/dx_1$ is effectively zero, reducing the preceding equation to:

$$\boxed{\delta\ln\frac{\gamma_1}{\gamma_2} = x_1\frac{d\ln\gamma_1^*}{dx_1} + x_2\frac{d\ln\gamma_2^*}{dx_1}} \qquad (13.59)$$

The right side of this equation is exactly the quantity that Eq. (13.55), the Gibbs/Duhem equation, requires to be zero for consistent data. The residual on the left therefore provides a direct measure of deviation from the Gibbs/Duhem equation. The extent to which a data set departs from consistency is measured by the degree to which these residuals fail to scatter about zero.[13]

Example 13.4

VLE data for diethyl ketone(1)/*n*-hexane(2) at 65°C as reported by Maripuri and Ratcliff[14] are given in the first three columns of Table 13.5. Reduce this set of data.

Solution 13.4

The last three columns of Table 13.5 present the experimental values, $\ln\gamma_1^*$, $\ln\gamma_2^*$, and $(G^E/(x_1x_2RT))^*$, calculated from the data by Eqs. (13.24) and (13.54). All values are shown as points on Figs. 13.8(*a*) and 13.8(*b*). The object here is to find an equation for G^E/RT that provides a suitable correlation of the data. Although

---

[13]This test and other aspects of VLE data reduction are treated by H. C. Van Ness, *J. Chem. Thermodyn.*, vol. 27, pp. 113–134, 1995; *Pure & Appl. Chem.*, vol. 67, pp. 859–872, 1995. See also, P. T. Eubank, B. G. Lamonte, and J. F. Javier Alvarado, *J. Chem. Eng. Data*, vol. 45, pp. 1040–1048, 2000.

[14]V. C. Maripuri and G. A. Ratcliff, *J. Appl. Chem. Biotechnol.*, vol. 22, pp. 899–903, 1972.

Table 13.5: VLE Data for Diethyl Ketone(1)/n-Hexane(2) at 65°C

P/kPa	x_1	y_1	$\ln\gamma_1^*$	$\ln\gamma_2^*$	$\left(\dfrac{G^E}{x_1x_2RT}\right)^*$
90.15 (P_2^{sat})	0.000	0.000		0.000	
91.78	0.063	0.049	0.901	0.033	1.481
88.01	0.248	0.131	0.472	0.121	1.114
81.67	0.372	0.182	0.321	0.166	0.955
78.89	0.443	0.215	0.278	0.210	0.972
76.82	0.508	0.248	0.257	0.264	1.043
73.39	0.561	0.268	0.190	0.306	0.977
66.45	0.640	0.316	0.123	0.337	0.869
62.95	0.702	0.368	0.129	0.393	0.993
57.70	0.763	0.412	0.072	0.462	0.909
50.16	0.834	0.490	0.016	0.536	0.740
45.70	0.874	0.570	0.027	0.548	0.844
29.00 (P_1^{sat})	1.000	1.000	0.000		

the data points of Fig. 13.8(b) for $(G^E/(x_1x_2RT))^*$ show scatter, they are adequate to define a straight line:

$$\frac{G^E}{x_1x_2RT} = 0.70x_1 + 1.35x_2$$

This is the Margules equation with $A_{21} = 0.70$ and $A_{12} = 1.35$. Derived values of $\ln\gamma_1$ and $\ln\gamma_2$ are calculated by Eqs. (13.40) and Eqs. (13.41), and derived values of P and y_1 all come from Eqs. (13.57) and (13.58). These results, plotted as solid lines of Figs. 13.8(a) and 13.8(b), clearly do not represent a good correlation of the data.

The difficulty is that the data are not *consistent* with the Gibbs/Duhem equation. That is, the sets of experimental values, $\ln\gamma_1^*$ and $\ln\gamma_2^*$, shown in Table 13.5 are not in accord with Eq. (13.55). However, the values of $\ln\gamma_1$ and $\ln\gamma_2$ derived from the correlation necessarily obey this equation; the experimental and derived values therefore cannot possibly agree, and the resulting correlation cannot provide a precise representation of the complete set of P-x_1-y_1 data.

Application of the test for consistency represented by Eq. (13.59) requires calculation of the residuals $\delta(G^E/RT)$ and $\delta\ln(\gamma_1/\gamma_2)$, values of which are plotted vs. x_1 in Fig. 13.9. The residuals $\delta(G^E/RT)$ distribute themselves about zero,[15] as is required by the test, but the residuals $\delta\ln(\gamma_1/\gamma_2)$, which show the extent to which the data fail to satisfy the Gibbs/Duhem equation, clearly do not. Average absolute values of this residual less than 0.03 indicate data of a high degree of consistency; average absolute values of less than 0.10 are probably acceptable. The data set

---

[15]The simple procedure used here to find a correlation for G^E/RT would be slightly improved by a regression procedure that determines the values of A_{21} and A_{12} that minimize the sum of squares of the *residuals* $\delta(G^E/RT)$.

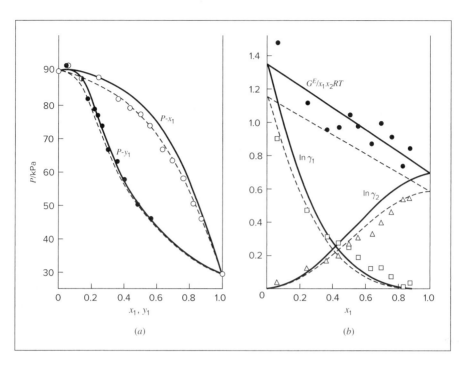

Figure 13.8: The diethyl ketone(1)/*n*-hexane(2) system at 65°C. (*a*) *Pxy* data and their correlations. (*b*) Liquid-phase properties and their correlation.

considered here shows an average absolute deviation of about 0.15 and must therefore contain significant error. Although one cannot be certain where the error lies, the values of y_1 are usually most suspect.

The method just described produces a correlation that is unnecessarily divergent from the experimental values. An alternative is to process just the P-x_1 data; this is possible because the P-x_1-y_1 data set includes more information than necessary. The procedure requires a computer but in principle is simple enough. Assuming that the Margules equation is appropriate to the data, one merely searches for values of the parameters A_{12} and A_{21} that yield pressures by Eq. (13.57) that are as close as possible to the measured values. The method is applicable regardless of the correlating equation assumed and is known as *Barker's method.*[16] Applied to the present data set, it yields the parameters:

$$A_{21} = 0.596 \quad \text{and} \quad A_{12} = 1.153$$

Use of these parameters in Eqs. (13.39), (13.40), (13.41), (13.57), and (13.58) produces the results described by the dashed lines of Figs. 13.8(*a*) and 13.8(*b*). The correlation cannot be precise, but it clearly provides a better overall representation of the experimental P-x_1-y_1 data. Note, however, that it necessarily provides a

[16]J. A. Barker, *Austral. J. Chem.,* vol. 6, pp. 207–210, 1953.

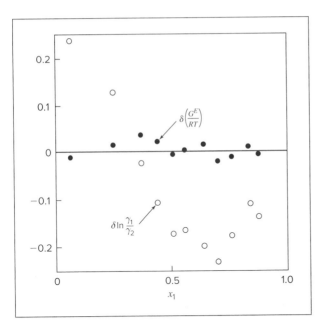

Figure 13.9: Consistency test of data for diethyl ketone(1)/*n*-hexane(2) at 65°C.

worse fit to the experimentally derived ln γ_1, ln γ_2, and $(G^E/(x_1 x_2 RT))$. This fitting procedure ignores the vapor-phase composition data from which those experimentally derived activity coefficients were determined.

Incorporation of Vapor-Phase Fugacity Coefficients

Restriction to low pressures where the vapor phase can be assumed to be in the ideal-gas state is not always possible or even desirable. Practical processes are often operated at elevated pressure to increase throughput or capacity. In that case, experimental data may be taken at elevated pressure to increase their relevance to process conditions.

At moderate pressures Eq. (3.36), the two-term virial expansion in P, is usually adequate for property calculations. The fugacity coefficients required in Eq. (13.14) are then given by Eq. (10.64), here written:

$$\hat{\phi}_i^v = \exp \frac{P}{RT}\left[B_{ii} + \frac{1}{2}\sum_j \sum_k y_j y_k (2\delta_{ji} - \delta_{jk}) \right] \tag{13.60}$$

where $\delta_{ji} \equiv 2B_{ji} - B_{jj} - B_{ii}$ $\delta_{jk} \equiv 2B_{jk} - B_{jj} - B_{kk}$

with $\delta_{ii} = 0$, $\delta_{jj} = 0$, etc., and $\delta_{ij} = \delta_{ji}$, etc. Values of the virial coefficients come from a generalized correlation, as represented for example by Eqs. (10.69) through (10.74). The fugacity coefficient for pure i as a saturated vapor ϕ_i^{sat} is obtained from Eq. (13.60) with δ_{ji} and δ_{jk} set equal to zero:

$$\phi_i^{\text{sat}} = \exp \frac{B_{ii} P_i^{\text{sat}}}{RT} \tag{13.61}$$

Combination of Eqs. (13.14), (13.60), and (13.61) gives:

$$\Phi_i = \exp\frac{B_{ii}(P - P_i^{sat}) + \frac{1}{2}P\sum_j\sum_k y_j y_k(2\delta_{ji} - \delta_{jk})}{RT} \tag{13.62}$$

For a binary system comprised of species 1 and 2, this becomes:

$$\Phi_1 = \exp\frac{B_{11}(P - P_1^{sat}) + Py_2^2\delta_{12}}{RT} \tag{13.63}$$

$$\Phi_2 = \exp\frac{B_{22}(P - P_2^{sat}) + Py_1^2\delta_{12}}{RT} \tag{13.64}$$

The inclusion of Φ_1 and Φ_2 evaluated by Eqs. (13.63) and (13.64) in the reduction of a set of moderate-pressure VLE data is straightforward, because in this situation the values of T, P, and y_1 at each data point are known. After evaluating Φ_i, at each data point, the activity coefficients are calculated from:

$$\gamma_i = \frac{y_i\Phi_i P}{x_i P_i^{sat}} \tag{13.65}$$

These experimentally derived activity coefficients are then combined via Eq. (13.54), and fitting to an excess Gibbs energy model proceeds as usual.

On the other hand, inclusion of the vapor-phase fugacity coefficients introduces significant new complications to bubblepoint calculations and flash calculations because in those cases the vapor-phase composition is unknown. When vapor-phase fugacity coefficients were not included, bubblepoint pressure calculations could be made directly, without requiring any iterative solution procedure. However, with the inclusion of the vapor-phase fugacity coefficients, all types of bubblepoint, dewpoint, and flash calculations require iterative solution.

Extrapolation of Data to Higher Temperatures

A vast store of liquid-phase excess-property data for binary systems at temperatures near 30°C and somewhat higher is available in the literature. Effective use of these data to extend G^E correlations to higher temperatures is critical to employing them for engineering design calculations. The key relations for the temperature dependence of excess properties are Eq. (10.89), written for constant P and x as:

$$d\left(\frac{G^E}{RT}\right) = -\frac{H^E}{RT^2}dT \qquad (\text{const } P, x)$$

and the excess-property analog of Eq. (2.20):

$$dH^E = C_P^E dT \qquad (\text{const } P, x)$$

Integration of the first of these equations from T_0 to T gives:

$$\frac{G^E}{RT} = \left(\frac{G^E}{RT}\right)_{T_0} - \int_{T_0}^{T}\frac{H^E}{RT^2}dT \tag{13.66}$$

Similarly, the second equation may be integrated from T_1 to T:

$$H^E = H_1^E + \int_{T_1}^{T} C_P^E \, dT \tag{13.67}$$

In addition.

$$d C_P^E = \left(\frac{\partial C_P^E}{\partial T} \right)_{P,x} dT$$

Integration from T_2 to T yields:

$$C_P^E = C_{P_2}^E + \int_{T_2}^{T} \left(\frac{\partial C_P^E}{\partial T} \right)_{P,x} dT$$

Combining this equation with Eqs. (13.66) and (13.67) leads to:

$$\frac{G^E}{RT} = \left(\frac{G^E}{RT} \right)_{T_0} - \left(\frac{H^E}{RT} \right)_{T_1} \left(\frac{T}{T_0} - 1 \right) \frac{T_1}{T}$$

$$- \frac{C_{P_2}^E}{R} \left[\ln \frac{T}{T_0} - \left(\frac{T}{T_0} - 1 \right) \frac{T_1}{T} \right] - J \tag{13.68}$$

where

$$J \equiv \int_{T_0}^{T} \frac{1}{RT^2} \int_{T_1}^{T} \int_{T_2}^{T} \left(\frac{\partial C_P^E}{\partial T} \right)_{P,x} dT \, dT \, dT$$

This general equation makes use of excess Gibbs-energy data at temperature T_0, excess enthalpy (heat-of-mixing) data at T_1, and excess heat-capacity data at T_2.

Evaluation of integral J requires information with respect to the temperature dependence of C_P^E. Because of the relative paucity of excess-heat-capacity data, the usual assumption is that this property is constant, independent of T. In this event, integral J is zero, and the closer T_0 and T_1 are to T, the less the influence of this assumption. When no information is available with respect to C_P^E, and excess enthalpy data are available at only a single temperature, the excess heat capacity must be assumed to be zero. In this case only the first two terms on the right side of Eq. (13.68) are retained, and it more rapidly becomes imprecise as T increases.

Because the parameters of two-parameter correlations of G^E data are directly related to infinite-dilution values of the activity coefficients, our primary interest in Eq. (13.68) is its application to binary systems at infinite dilution of one of the constituent species. For this purpose, we divide Eq. (13.68) by the product $x_1 x_2$. For C_P^E independent of T (and thus with $J = 0$), it then becomes:

$$\frac{G^E}{x_1 x_2 RT} = \left(\frac{G^E}{x_1 x_2 RT} \right)_{T_0} - \left(\frac{H^E}{x_1 x_2 RT} \right)_{T_1} \left(\frac{T}{T_0} - 1 \right) \frac{T_1}{T} - \frac{C_P^E}{x_1 x_2 R} \left[\ln \frac{T}{T_0} - \left(\frac{T}{T_0} - 1 \right) \frac{T_1}{T} \right]$$

As shown in Sec. 13.5,

$$\left(\frac{G^E}{x_1 x_2 RT} \right)_{x_i = 0} = \ln \gamma_i^\infty$$

The preceding equation applied at infinite dilution of species i may therefore be written:

$$\ln \gamma_i^\infty = (\ln \gamma_i^\infty)_{T_0} - \left(\frac{H^E}{x_1 x_2 RT}\right)_{T_1, x_i = 0} \left(\frac{T}{T_0} - 1\right)\frac{T_1}{T}$$
$$- \left(\frac{C_P^E}{x_1 x_2 R}\right)_{x_i = 0} \left[\ln\frac{T}{T_0} - \left(\frac{T}{T_0} - 1\right)\frac{T_1}{T}\right] \tag{13.69}$$

Data for the ethanol(1)/water(2) binary system provide a specific illustration. At a base temperature T_0 of 363.15 K (90°C), the VLE data of Pemberton and Mash[17] yield accurate values for infinite-dilution activity coefficients:

$$(\ln \gamma_1^\infty)_{T_0} = 1.7720 \quad \text{and} \quad (\ln \gamma_2^\infty)_{T_0} = 0.9042$$

Correlation of the excess enthalpy data of J. A. Larkin[18] at 110°C yields the values:

$$\left(\frac{H^E}{x_1 x_2 RT}\right)_{T_1, x_1 = 0} = -0.0598 \quad \text{and} \quad \left(\frac{H^E}{x_1 x_2 RT}\right)_{T_1, x_2 = 0} = 0.6735$$

Correlations of the excess enthalpy for the temperature range from 50 to 110°C lead to infinite-dilution values of $C_P^E/(x_1 x_2 R)$, which are nearly constant and equal to

$$\left(\frac{C_P^E}{x_1 x_2 R}\right)_{x_1 = 0} = 13.8 \quad \text{and} \quad \left(\frac{C_P^E}{x_1 x_2 R}\right)_{x_2 = 0} = 7.2$$

Equation (13.69) may be directly applied with these data to estimate $\ln \gamma_1^\infty$ and $\ln \gamma_2^\infty$ for temperatures greater than 90°C. The van Laar equations [Eqs. (13.43) and (13.44)] are appropriate here, with parameters directly related to the infinite-dilution activity coefficients:

$$A'_{12} = \ln \gamma_1^\infty \quad \text{and} \quad A'_{21} = \ln \gamma_2^\infty$$

These data allow prediction of VLE by an equation of state at 90°C and at two higher temperatures, 423.15 and 473.15 K (150 and 200°C), for which measured VLE data are given by Barr-David and Dodge.[19] Pemberton and Mash report pure-species vapor pressures at 90°C for both ethanol and water, but the data of Barr-David and Dodge (at 150 and 200°C) do not include these values. They are therefore calculated from reliable correlations. Results of calculations based on the Peng/Robinson equation of state are given in Table 13.6. Shown for the three temperatures are values of the van Laar parameters A'_{12} and A'_{21}, the pure-species vapor pressures P_1^{sat} and P_2^{sat}, the equation-of-state parameters b_i and q_i, and root-mean-square (RMS) deviations between computed and experimental values for P and y_1.

[17]R. C. Pemberton and C. J. Mash, *Int. DATA Series, Ser. B*, vol. 1, p. 66, 1978.

[18]As reported in *Heats of Mixing Data Collection*, Chemistry Data Series, vol. III, part 1, pp. 457–459, DECHEMA, Frankfurt/Main, 1984.

[19]F. H. Barr-David and B. F. Dodge, *J. Chem. Eng. Data*, vol. 4, pp. 107–121, 1959.

Table 13.6: VLE Results for Ethanol(1)/Water(2)

$T\,^\circ C$	A'_{12}	A'_{21}	P_1^{sat} bar	P_2^{sat} bar	q_1	q_2	RMS % δP	RMS δy_1
90	1.7720	0.9042	1.5789	0.7012	12.0364	15.4551	0.29	*****
150	1.7356	0.7796	9.825	4.760	8.8905	12.2158	2.54	0.005
200	1.5204	0.6001	29.861	15.547	7.0268	10.2080	1.40	0.005

$$b_1 = 54.0645 \qquad b_2 = 18.9772$$

***** Vapor-phase compositions not measured.

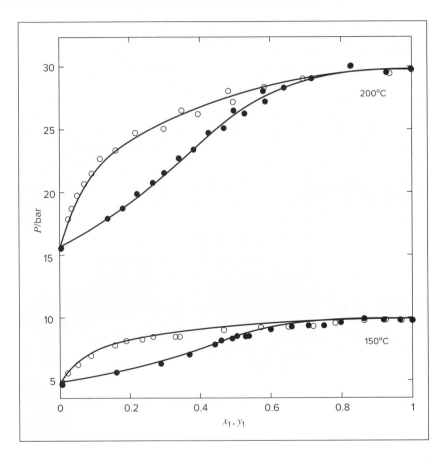

Figure 13.10: *Pxy* diagram for ethanol(1)/water(2). The lines represent predicted values; the points are experimental values.

   The small value of RMS % δP shown for 90°C indicates both the suitability of the van Laar equation for correlation of the VLE data and the capability of the equation of state to reproduce the data. A direct fit of these data with the van Laar equation by the gamma/phi procedure yields RMS % $\delta P = 0.19.$[20] The results at 150 and 200°C are based only on vapor-pressure data for the pure species and on mixture data at lower temperatures. The quality of prediction is indicated by the P-x-y diagram of Fig. 13.13, which reflects the uncertainty of the data as well.

13.6   RESIDUAL PROPERTIES BY CUBIC EQUATIONS OF STATE

In Sec. 6.3 we treated the calculation of residual properties by the virial equations of state, as well as from generalized correlations, but we did not extend the treatment to cubic equations of state at that time. The key differentiating feature of cubic equations of state is their ability to treat both vapor and liquid phase properties. This capability is most valuable in the context of VLE calculations, as considered in the present chapter. Thus, in this section we first treat the computation of residual properties from cubic equations of state and then, in Sec. 13.7, show how these can be used to carry out phase equilibrium calculations.

Results of some generality follow from the generic cubic equation of state:

$$P = \frac{RT}{V - b} - \frac{a(T)}{(V + \epsilon b)(V + \sigma b)} \tag{3.41}$$

Equations (6.58) through (6.60) are compatible with pressure-explicit equations of state. We need only recast Eq. (3.41) to yield Z with density ρ as the independent variable. We therefore divide Eq. (3.41) by ρRT and substitute $V = 1/\rho$. With q given by Eq. (3.47), $q \equiv a(T)/bRT$, the result after alegbraic reduction is:

$$Z = \frac{1}{1 - \rho b} - q\frac{\rho b}{(1 + \epsilon\rho b)(1 + \sigma\rho b)}$$

The two quantities needed for evaluation of the integrals in Eqs. (6.58) through (6.60), $Z - 1$ and $(\partial Z/\partial T)_\rho$, are readily obtained from this equation:

$$Z - 1 = \frac{\rho b}{1 - \rho b} - q\frac{\rho b}{(1 + \epsilon\rho b)(1 + \sigma\rho b)}$$

$$\left(\frac{\partial Z}{\partial T}\right)_\rho = -\left(\frac{dq}{dT}\right)\frac{\rho b}{(1 + \epsilon\rho b)(1 + \sigma\rho b)} \tag{13.70}$$

The integrals of Eqs. (6.58) through (6.60) are now evaluated as follows:

$$\int_0^\rho (Z - 1)\frac{d\rho}{\rho} = \int_0^\rho \frac{\rho b}{1 - \rho b}\frac{d(\rho b)}{\rho b} - q\int_0^\rho \frac{d(\rho b)}{(1 + \epsilon\rho b)(1 + \sigma\rho b)}$$

$$\int_0^\rho \left(\frac{\partial Z}{\partial T}\right)_\rho \frac{d\rho}{\rho} = -\frac{dq}{dT}\int_0^\rho \frac{d(\rho b)}{(1 + \epsilon\rho b)(1 + \sigma\rho b)}$$

[20]As reported in *Vapor-Liquid Equilibrium Data Collection,* Chemistry Data Series, vol. 1, part 1a, p. 145, DECHEMA, Frankfurt/Main, 1981.

These two equations simplify to:

$$\int_0^\rho (Z-1)\frac{d\rho}{\rho} = -\ln(1-\rho b) - qI \qquad \int_0^\rho \left(\frac{\partial Z}{\partial T}\right)_\rho \frac{d\rho}{\rho} = -\frac{dq}{dT}I$$

where by definition, $I \equiv \displaystyle\int_0^\rho \frac{d(\rho b)}{(1+\varepsilon\rho b)(1+\sigma\rho b)}$ (const T)

The generic equation of state presents two cases for the evaluation of this integral:

Case I: $\varepsilon \neq \sigma$ $\qquad\qquad I = \dfrac{1}{\sigma-\varepsilon}\ln\left(\dfrac{1+\sigma\rho b}{1+\varepsilon\rho b}\right)$ $\qquad\qquad$ (13.71)

Application of this and subsequent equations is simpler when ρ is eliminated in favor of Z. The definitions of β by Eq. (3.46), $\beta \equiv bP/RT$, and of $Z \equiv P/\rho RT$, combine to give $\rho b = \beta/Z$. Then:

$$I = \frac{1}{\sigma-\varepsilon}\ln\left(\frac{Z+\sigma\beta}{Z+\varepsilon\beta}\right) \qquad\qquad (13.72)$$

Case II: $\varepsilon = \sigma$ $\qquad\qquad I = \dfrac{\rho b}{1+\varepsilon\rho b} = \dfrac{\beta}{Z+\varepsilon\beta}$

The van der Waals equation is the only one considered here to which Case II applies, and this equation, with $\varepsilon = 0$, reduces to $I = \beta/Z$.

With evaluation of the integrals, Eqs. (6.58) through (6.60) reduce to:

$$\frac{G^R}{RT} = Z - 1 - \ln[(1-\rho b)Z] - qI \qquad\qquad (13.73)$$

or $\qquad\qquad \boxed{\dfrac{G^R}{RT} = Z - 1 - \ln(Z-\beta) - qI}$ $\qquad\qquad$ (13.74)

$$\frac{H^R}{RT} = Z - 1 + T\left(\frac{dq}{dT}\right)I = Z - 1 + T_r\left(\frac{dq}{dT_r}\right)I$$

and $\qquad\qquad \dfrac{S^R}{R} = \ln(Z-\beta) + \left(q + T_r\dfrac{dq}{dT_r}\right)I$

The quantity $T_r\dfrac{dq}{dT_r}$ is readily found from Eq. (3.51):

$$T_r\frac{dq}{dT_r} = \left[\frac{d\ln\alpha(T_r)}{d\ln T_r} - 1\right]q$$

Substitution for this quantity in the preceding two equations yields:

$$\boxed{\frac{H^R}{RT} = Z - 1 + \left[\frac{d\ln\alpha(T_r)}{d\ln T_r} - 1\right]qI} \qquad\qquad (13.75)$$

$$\boxed{\frac{S^R}{R} = \ln(Z - \beta) + \frac{d \ln\alpha(T_r)}{d \ln T_r} q I}$$ (13.76)

Before applying these equations one must find Z by solution of the equation of state itself, typically written in the form of Eq. (3.48) or Eq. (3.49) for a vapor or liquid phase, respectively.

Example 13.5

Find values for the residual enthalpy H^R and the residual entropy S^R of *n*-butane gas at 500 K and 50 bar as given by the Redlich/Kwong equation.

Solution 13.5

For the given conditions:

$$T_r = \frac{500}{425.1} = 1.176 \qquad\qquad P_r = \frac{50}{37.96} = 1.317$$

By Eq. (3.50), with Ω for the Redlich/Kwong equation from Table 3.1,

$$\beta = \Omega \frac{P_r}{T_r} = \frac{(0.08664)(1.317)}{1.176} = 0.09703$$

With values for Ψ and Ω, and with the expression $\alpha(T_r) = T_r^{-1/2}$ from Table 3.1, Eq. (3.51) yields:

$$q = \frac{\Psi\alpha(T_r)}{\Omega T_r} = \frac{0.42748}{(0.08664)(1.176)^{1.5}} = 3.8689$$

Substitution of these values of β and q, along with $\varepsilon = 0$, and $\sigma = 1$ into Eq. (3.48) reduces it to:

$$Z = 1 + 0.09703 - (3.8689)(0.09703)\frac{Z - 0.09703}{Z(Z + 0.09703)}$$

Iterative solution of this equation yields $Z = 0.6850$. Then:

$$I = \ln\frac{Z + \beta}{Z} = 0.13247$$

Given that $\ln\alpha(T_r) = -\frac{1}{2}\ln T_r$, we have $\frac{d \ln\alpha(T_r)}{d \ln T_r} = -\frac{1}{2}$, and Eqs. (13.75) and (13.76) become:

$$\frac{H^R}{RT} = 0.6850 - 1 + (-0.5 - 1)(3.8689)(0.13247) = -1.0838$$

$$\frac{S^R}{R} = \ln(0.6850 - 0.09703) - (0.5)(3.8689)(0.13247) = -0.78735$$

Whence,
$$H^R = (8.314)(500)(-1.0838) = -4505 \text{ J·mol}^{-1}$$
$$S^R = (8.314)(-0.78735) = -6.546 \text{ J·mol}^{-1}\text{·K}^{-1}$$

These results are compared with those of other calculations in Table 13.7.

Table 13.7: Values for Z, H^R, and S^R for n-butane at 500 K and 50 Bar

Method	Z	H^R J·mol^{-1}	S^R J·mol^{-1}·K^{-1}
vdW Eqn.	0.6608	−3937	−5.424
RK Eqn.	0.6850	−4505	−6.546
SRK Eqn.	0.7222	−4824	−7.413
PR Eqn.	0.6907	−4988	−7.426
Lee/Kesler[†]	0.6988	−4966	−7.632
Handbook[‡]	0.7060	−4760	−7.170

[†]Described in Sec. 6.7.
[‡]Values derived from numbers in Table 2–240, p. 2–223, *Chemical Engineers' Handbook,* 7th ed., D. Green and R. H. Perry (eds.), McGraw-Hill, New York, 1997.

13.7   VLE FROM CUBIC EQUATIONS OF STATE

As shown in Sec. 10.6, phases at the same T and P are in equilibrium when the fugacity of each species is the same in all phases. For VLE, this requirement is written:

$$\hat{f}_i^v = \hat{f}_i^l \qquad (i = 1, 2, \ldots, N) \qquad (10.48)$$

An alternative form results from the introduction of the fugacity coefficient, defined by Eq. (10.52):

$$y_i \hat{\phi}_i^v P = x_i \hat{\phi}_i^l P$$

or
$$\boxed{y_i \hat{\phi}_i^v = x_i \hat{\phi}_i^l \qquad (i = 1, 2, \ldots, N)} \qquad (13.77)$$

Applications of this equation with fugacity coefficients evaluated using cubic equations of state are presented in the following subsections.

Vapor Pressures for a Pure Species

Although vapor pressures for a pure species P_i^{sat} are subject to experimental measurement, they are also implicit in a cubic equation of state. Indeed, the simplest application of cubic equations of state for VLE calculations is to find the vapor pressure of a pure species at given temperature T.

The subcritical PV isotherm of Fig. 3.9 labeled $T_2 < T_c$ is reproduced here as Fig. 13.11. Generated by a cubic equation of state, it consists of three segments. The very steep segment on the left (rs) is characteristic of liquids; in the limit as $P \to \infty$, the molar volume V

approaches the constant b [Eq. 3.41]. The segment on the right (tu) with gentle downward slope is characteristic of vapors; in the limit as $P \to 0$ molar volume V approaches infinity. The middle segment (st), containing both a minimum (note here that $P < 0$) and a maximum, provides a smooth transition from liquid to vapor, but has no physical meaning. The actual transition from liquid to vapor occurs at the vapor pressure along a horizontal line like that which connects points M and W.

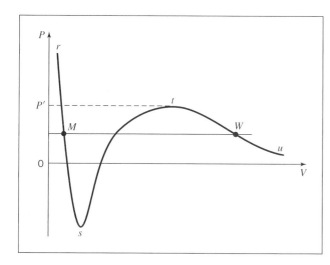

Figure 13.11: Isotherm for $T < T_c$ *on a PV diagram for a pure fluid.*

For pure species i, Eq. (13.77) reduces to Eq. (10.41), $\phi_i^v = \phi_i^l$, which may be written:

$$\ln \phi_i^l - \ln \phi_i^v = 0 \tag{13.78}$$

The fugacity coefficient of a pure liquid or vapor is a function of its temperature and pressure. For a *saturated* liquid or vapor, the equilibrium pressure is P_i^{sat}, and Eq. (13.78) implicitly expresses the functional relation,

$$g(T, P_i^{\text{sat}}) = 0 \qquad \text{or} \qquad P_i^{\text{sat}} = f(T)$$

An isotherm generated by a cubic equation of state, as represented in Fig. 13.11, has three volume roots for a specified pressure between $P = 0$ and $P = P'$. The smallest root lies on the left line segment and is a liquid-like volume, e.g., at point M. The largest root lies on the right line segment and is a vapor-like volume, e.g., at point W.

> **If these points lie at the vapor pressure, then M represents *saturated* liquid, W represents *saturated* vapor, and they exist in phase equilibrium.**

The root lying on the middle line segment has no physical significance.

Two widely used cubic equations of state, developed specifically for VLE calculations, are the Soave/Redlich/Kwong (SRK) equation[21] and the Peng/Robinson (PR) equation.[22] Both are special cases of the generic cubic equation of state, Eq. (3.41). Equation-of-state

[21]G. Soave, *Chem. Eng. Sci.*, vol. 27, pp. 1197–1203, 1972.

[22]D.-Y. Peng and D. B. Robinson, *Ind. Eng. Chem. Fundam.*, vol. 15, pp. 59–64, 1976.

parameters are independent of phase, and in accord with Eqs. (3.44) through (3.47) they are given by:

$a_i(T) = \Psi \dfrac{\alpha(T_{r_i})R^2 T_{c_i}^2}{P_{c_i}}$	(13.79)	$b_i = \Omega \dfrac{RT_{c_i}}{P_{c_i}}$	(13.80)
$\beta_i \equiv \dfrac{b_i P}{RT}$	(13.81)	$q_i \equiv \dfrac{a_i(T)}{b_i RT}$	(13.82)

Written for pure species i as a vapor, Eq. (3.48) becomes:

$$Z_i^v = 1 + \beta_i - q_i \beta_i \frac{Z_i^v - \beta_i}{\left(Z_i^v + \epsilon\beta_i\right)\left(Z_i^v + \sigma\beta_i\right)} \tag{13.83}$$

For pure species i as a liquid, Eq. (3.49) is written:

$$Z_i^l = \beta_i + \left(Z_i^l + \epsilon\beta_i\right)\left(Z_i^l + \sigma\beta_i\right)\left(\frac{1 + \beta_i - Z_i^l}{q_i \beta_i}\right) \tag{13.84}$$

The pure numbers ϵ, σ, Ψ, and Ω and expressions for $\alpha(T_{ri})$ are specific to the equation of state and are given in Table 3.1 for several prototypical cubic equations of state.

In Sec. 13.6 and Sec. 10.5, we developed the following two relationships:

$$\frac{G_i^R}{RT} = Z_i - 1 - \ln(Z_i - \beta_i) - q_i I_i \tag{13.74}$$

$$\frac{G_i^R}{RT} = \ln\phi_i \tag{10.33}$$

Together these yield: $\ln\phi_i = Z_i - 1 - \ln(Z_i - \beta_i) - q_i I_i$ (13.85)

Values for $\ln\phi_i$ are therefore implied by the equation of state. In Eq. (13.85), q_i is given by Eq. (13.82) and I_i by Eq. (13.72). For given T and P, the vapor-phase value, Z_i^v at point W of Fig. 13.11, is given by Eq. (13.83), and the liquid-phase value Z_i^l at point M, by Eq. (13.84). Values for $\ln\phi_i^l$ and $\ln\phi_i^v$ are then found by Eq. (13.85). When they satisfy Eq. (13.77), then P is the vapor pressure P_i^{sat} at temperature T, and M and W represent the states of saturated liquid and vapor implied by the equation of state. Solution may be by trial, by iteration, or by an appropriate nonlinear algebraic equation solution algorithm. The eight equations and eight unknowns are listed in Table 13.8.

The calculation of pure-species vapor pressures as just described may be reversed to allow evaluation of an equation-of-state parameter from a known vapor pressure P_i^{sat} at temperature T. Thus, Eq. (13.85) may be written for each phase of pure-species i and combined in accord with Eq. (13.77). Solving the resulting expression for q_i yields:

$$q_i = \frac{Z_i^v - Z_i^l + \ln\dfrac{Z_i^l - \beta_i}{Z_i^v - \beta_i}}{I_i^v - I_i^l} \tag{13.86}$$

Table 13.8: Equations for Calculating Vapor Pressures

The unknowns are: P_i^{sat}, β_i, Z_i^l, Z_i^v, I_i^l, I_i^v, $\ln \phi_i^l$, and $\ln \phi_i^v$

$$\beta_i \equiv \frac{b_i P_i^{\text{sat}}}{RT}$$

$$Z_i^l = \beta_i + (Z_i^l + \varepsilon\beta_i)(Z_i^l + \sigma\beta_i)\left(\frac{1 + \beta_i - Z_i^l}{q_i \beta_i}\right)$$

$$Z_i^v = 1 + \beta_i - q_i\beta_i \frac{Z_i^v - \beta_i}{\left(Z_i^v + \varepsilon\beta_i\right)\left(Z_i^v + \sigma\beta_i\right)}$$

$$I_i^l = \frac{1}{\sigma - \varepsilon}\ln\frac{Z_i^l + \sigma\beta_i}{Z_i^l + \varepsilon\beta_i} \qquad I_i^v = \frac{1}{\sigma - \varepsilon}\ln\frac{Z_i^v + \sigma\beta_i}{Z_i^v + \varepsilon\beta_i}$$

$$\ln \phi_i^l = Z_i^l - 1 - \ln(Z_i^l - \beta_i) - q_i I_i^l$$

$$\ln \phi_i^v = Z_i^v - 1 - \ln(Z_i^v - \beta_i) - q_i I_i^v$$

$$\ln \phi_i^v = \ln \phi_i^l$$

where $\beta_i \equiv b_i P_i^{\text{sat}}/RT$. For the PR and SRK equations, I_i is given by Eq. (13.72) written for pure species i:

$$I_i = \frac{1}{\sigma - \varepsilon}\ln\frac{Z_i + \sigma\beta_i}{Z_i + \varepsilon\beta_i}$$

This equation yields I_i^v with Z_i^v from Eq. (13.83), and I_i^l with Z_i^l from Eq. (13.79). However, the equations for Z_i^v and Z_i^l contain q_i, the quantity sought. Thus, a solution procedure is required that yields values for eight unknowns given eight equations. An initial value of q_i is provided by the generalized correlation of Eqs. (13.79), (13.80), and (13.82).

Mixture VLE

The fundamental assumption when an equation of state is written for mixtures is that it has exactly the same form as when written for pure species. Thus for mixtures, Eqs. (13.83) and (13.84), written without subscripts, become:

Vapor: $$Z^v = 1 + \beta^v - q^v\beta^v \frac{Z^v - \beta^v}{(Z^v + \varepsilon\beta^v)(Z^v + \sigma\beta^v)} \qquad (13.87)$$

Liquid: $$Z^l = \beta^l + (Z^l + \varepsilon\beta^l)(Z^l + \sigma\beta^l)\left(\frac{1 + \beta^l - Z^l}{q^l\beta^l}\right) \qquad (13.88)$$

Here, β^l, β^v, q^l, and q^v are for mixtures, with definitions:

$$\beta^p \equiv \frac{b^p P}{RT} \qquad (p = l, v) \quad (13.89) \qquad q^p \equiv \frac{a^p}{b^p RT} \qquad (p = l, v) \quad (13.90)$$

The complication is that *mixture* parameters a^p and b^p, and therefore β^p and q^p, are *functions of composition*. Systems in vapor/liquid equilibrium consist in general of two phases with different compositions. The *PV* isotherms generated by an equation of state for these two fixed compositions are represented in Fig. 13.12 by two similar lines: the solid line for the liquid-phase composition and the dashed line for the vapor-phase composition. They are displaced from one another because the equation-of-state parameters are different for the two compositions. However, each line includes three *segments* as described in connection with the isotherm of Fig. 13.11. Thus, we distinguish between the *composition* that characterizes a complete line and the *phases,* all of the same composition, that are associated with the segments of an isotherm.

Figure 13.12: Two *PV* isotherms at the same *T* for mixtures of two different compositions. The solid line is for a liquid-phase composition; the dashed line is for a vapor-phase composition. Point *B* represents a bubblepoint with the liquid-phase composition; point *D* represents a dewpoint with the vapor-phase composition. When these points lie at the same *P* (as shown) they represent phases in equilibrium.

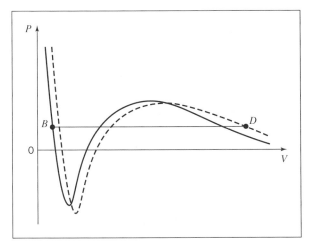

Each line contains a bubblepoint on its left segment representing *saturated* liquid and a dewpoint of the same composition on its right segment representing *saturated* vapor[23]. Because these points for a given line are for the same composition, they do not represent phases in equilibrium and do not lie at the same pressure. (See Fig. 12.3, where for a given constant-composition loop and a given *T*, saturated liquid and saturated vapor are at different pressures.)

For a *BUBL P* calculation, the temperature and liquid composition are known, and this fixes the location of the *PV* isotherm for the composition of the liquid phase (solid line). The *BUBL P* calculation then finds the composition for a second (dashed) line that contains a dewpoint *D* on its vapor segment that lies at the pressure of the bubblepoint *B* on the liquid segment of the solid line. This pressure is then the phase-equilibrium pressure,

---

[23]Note that bubblepoint *B* and dewpoint *D* in Fig. 13.12 are on *different* lines.

and the composition for the dashed line is that of the equilibrium vapor. This equilibrium condition is shown by Fig. 13.12, where bubblepoint B and dewpoint D lie at the same P on isotherms for the same T but representing the different compositions of liquid and vapor in equilibrium.

Because no established theory prescribes the form of the composition dependence of the equation-of-state parameters, empirical *mixing rules* have been proposed to relate mixture parameters to pure-species parameters. The simplest realistic expressions are a linear mixing rule for parameter b and a quadratic mixing rule for parameter a:

$$b = \sum_i x_i b_i \quad (13.91) \qquad a = \sum_i \sum_j x_i x_j a_{ij} \quad (13.92)$$

with $a_{ij} = a_{ji}$. The general mole-fraction variable x_i is used here because these mixing rules are applied to both liquid and vapor mixtures. The a_{ij} are of two types: pure-species parameters (repeated subscripts, e.g., a_{11}) and interaction parameters (unlike subscripts, e.g., a_{12}). Parameter b_i is for pure species i. The interaction parameters a_{ij} are often evaluated from pure-species parameters by *combining rules,* e.g., a geometric-mean rule:

$$a_{ij} = (a_i a_j)^{1/2} \tag{13.93}$$

These equations, known as van der Waals prescriptions, provide for the evaluation of mixture parameters solely from parameters for the pure constituent species. Although they are satisfactory only for mixtures comprised of simple and chemically similar molecules, they allow straightforward calculations that illustrate how complex VLE problems can be solved.

Also useful for application of equations of state to mixtures are *partial* equation-of-state parameters, defined by:

$$\bar{a}_i \equiv \left[\frac{\partial(na)}{\partial n_i} \right]_{T,n_j} \quad (13.94) \qquad \bar{b}_i \equiv \left[\frac{\partial(nb)}{\partial n_i} \right]_{T,n_j} \quad (13.95) \qquad \bar{q}_i \equiv \left[\frac{\partial(nq)}{\partial n_i} \right]_{T,n_j} \quad (13.96)$$

Because equation-of-state parameters are, at most, functions of temperature and composition, these definitions are in accord with Eq. (10.7). They are general equations, valid regardless of the particular mixing or combining rules adopted for the composition dependence of mixture parameters.

Values of $\hat{\phi}_i^l$ and $\hat{\phi}_i^v$ are implicit in an equation of state, and with Eq. (13.77) they allow calculation of *mixture* VLE. The same basic principle applies as for pure-species VLE, but the calculations are more complex. With $\hat{\phi}_i^l$ a function of T, P, and $\{x_i\}$, and $\hat{\phi}_i^v$ a function of T, P, and $\{y_i\}$, Eq. (13.72) represents N relations among the $2N$ variables: T, P, $N - 1$ liquid-phase mole fractions (x_i) and $N - 1$ vapor-phase mole fractions (y_i). Thus, specification of N of these variables, usually either T or P *and* either the liquid- or vapor-phase composition, allows us to solve for the remaining N variables by *BUBL P, DEW P, BUBL T,* and *DEW T* calculations.

Fugacity Coefficients from the Generic Cubic Equation of State

Cubic equations of state give Z as a function of the independent variables T and ρ (or V). For VLE calculations, expressions for the fugacity coefficient $\hat{\phi}_i$ must be given by an equation suited to these variables. The derivation of such an equation starts with Eq. (10.56), written for a mixture with V^R replaced by Eq. (6.40), $V^R = RT(Z-1)/P$:

$$d\left(\frac{nG^R}{RT}\right) = \frac{n(Z-1)}{P}dP - \frac{nH^R}{RT^2}dT + \sum_i \ln\hat{\phi}_i\,dn_i$$

Division by dn_i and restriction to constant T, $n/\rho\,(=nV)$, and $n_j\,(j \neq i)$ leads to:

$$\ln\hat{\phi}_i = \left[\frac{\partial(nG^R/RT)}{\partial n_i}\right]_{T,\,n/\rho,\,n_j} - \frac{n(Z-1)}{P}\left(\frac{\partial P}{\partial n_i}\right)_{T,\,n/\rho,\,n_j} \tag{13.97}$$

For simplicity of notation, the partial derivatives in the following development are written without subscripts, and they are understood to be at constant T, n/ρ, and n_j. Thus, with $P = (nZ)RT/(n/\rho)$,

$$\frac{\partial P}{\partial n_i} = \frac{RT}{n/\rho}\frac{\partial(nZ)}{\partial n_i} = \frac{P}{nZ}\frac{\partial(nZ)}{\partial n_i} \tag{13.98}$$

Combination of Eqs. (13.97) and (13.98) yields:

$$\ln\hat{\phi}_i = \frac{\partial(nG^R/RT)}{\partial n_i} - \left(\frac{Z-1}{Z}\right)\frac{\partial(nZ)}{\partial n_i} = \frac{\partial(nG^R/RT)}{\partial n_i} - \frac{\partial(nZ)}{\partial n_i} + \frac{1}{Z}\left(n\frac{\partial Z}{\partial n_i} + Z\right)$$

Equation (13.73), written for the mixture and multiplied by n, is differentiated to give the first term on the right:

$$\frac{nG^R}{RT} = nZ - n - n\ln[(1 - \rho b)Z] - (nq)I$$

$$\frac{\partial(nG^R/RT)}{\partial n_i} = \frac{\partial(nZ)}{\partial n_i} - 1 - \ln[(1 - \rho b)Z] - n\left[\frac{\partial\ln(1-\rho b)}{\partial n_i} + \frac{\partial\ln Z}{\partial n_i}\right] - nq\frac{\partial I}{\partial n_i} - I\bar{q}_i$$

where use has been made of Eq. (13.96). The equation for $\ln\hat{\phi}_i$ now becomes:

$$\ln\hat{\phi}_i = \frac{\partial(nZ)}{\partial n_i} - 1 - \ln[(1 - \rho b)Z] - n\frac{\partial\ln(1-\rho b)}{\partial n_i}$$

$$-\frac{n}{Z}\frac{\partial Z}{\partial n_i} - nq\frac{\partial I}{\partial n_i} - I\bar{q}_i - \frac{\partial(nZ)}{\partial n_i} + \frac{1}{Z}\left(n\frac{\partial Z}{\partial n_i} + Z\right)$$

This reduces to: $\ln\hat{\phi}_i = \dfrac{n}{1-\rho b}\dfrac{\partial(\rho b)}{\partial n_i} - nq\dfrac{\partial I}{\partial n_i} - \ln[(1 - \rho b)Z] - \bar{q}_i I$

All that remains is evaluation of the two partial derivatives. The first is:

$$\frac{\partial(\rho b)}{\partial n_i} = \frac{\partial\left(\dfrac{nb}{n/\rho}\right)}{\partial n_i} = \frac{\rho}{n}\bar{b}_i$$

The second follows from differentiation of Eq. (13.71). After algebraic reduction this yields:

$$\frac{\partial I}{\partial n_i} = \frac{\partial(\rho b)}{\partial n_i}\frac{1}{(1+\sigma\rho b)(1+\varepsilon\rho b)} = \frac{\bar{b}_i}{nb}\frac{\rho b}{(1+\sigma\rho b)(1+\varepsilon\rho b)}$$

Substitution of these derivatives in the preceding equation for $\ln\hat{\phi}_i$ reduces it to:

$$\ln\hat{\phi}_i = \frac{\bar{b}_i}{b}\left[\frac{\rho b}{1-\rho b} - q\frac{\rho b}{(1+\varepsilon\rho b)(1+\sigma\rho b)}\right] - \ln[(1-\rho b)Z] - \bar{q}_i I$$

Reference to Eq. (13.70) shows that the term in the first set of square brackets is $Z-1$. Therefore,

$$\ln\hat{\phi}_i = \frac{\bar{b}_i}{b}(Z-1) - \ln[(1-\rho b)Z] - \bar{q}_i I$$

Moreover, $\qquad \beta \equiv \dfrac{bP}{RT} \qquad$ and $\qquad Z \equiv \dfrac{P}{\rho RT}; \qquad$ whence $\qquad \rho b = \dfrac{\beta}{Z}$

Thus, $\qquad\qquad\qquad \ln\hat{\phi}_i = \dfrac{\bar{b}_i}{b}(Z-1) - \ln(Z-\beta) - \bar{q}_i I$

Because experience has shown that Eq. (13.91) is an acceptable mixing rule for parameter b, it is adopted here. Whence,

$$nb = \sum_i n_i b_i$$

and $\qquad\qquad \bar{b}_i \equiv \left[\dfrac{\partial(nb)}{\partial n_i}\right]_{T,n_j} = \left[\dfrac{\partial(n_i b_i)}{\partial n_i}\right]_{T,n_j} + \sum_j\left[\dfrac{\partial(n_j b_j)}{\partial n_i}\right]_{T,n_j} = b_i$

The equation for $\ln\hat{\phi}_i$ is therefore written:

$$\boxed{\ln\hat{\phi}_i = \frac{b_i}{b}(Z-1) - \ln(Z-\beta) - \bar{q}_i I} \qquad\qquad (13.99)$$

where I is evaluated by Eq. (13.72). For the special case of pure species i, this becomes:

$$\boxed{\ln\phi_i = Z_i - 1 - \ln(Z_i - \beta_i) - q_i I_i} \qquad\qquad (13.100)$$

Application of these equations requires prior evaluation of Z at the conditions of interest by an equation of state.

Parameter q is defined in relation to parameters a and b by Eq. (13.90). The relation of partial parameter $\bar{q}_i$ to $\bar{a}_i$ and $\bar{b}_i$ is found by differentiation of this equation, written:

$$nq = \frac{n(na)}{RT(nb)}$$

Whence, $\qquad \bar{q}_i \equiv \left[\dfrac{\partial(nq)}{\partial n_i}\right]_{T,n_j} = q\left(1 + \dfrac{\bar{a}_i}{a} - \dfrac{\bar{b}_i}{b}\right) = q\left(1 + \dfrac{\bar{a}_i}{a} - \dfrac{b_i}{b}\right) \qquad (13.101)$

Any two of the three partial parameters form an independent pair, and any one of them can be found from the other two.[24]

---

[24] Because q, a, and b are not linearly related, $\bar{q}_i \neq \bar{a}_i / \bar{b}_i RT$

Example 13.6

A vapor mixture of $N_2(1)$ and $CH_4(2)$ at 200 K and 30 bar contains 40 mol-% N_2. Determine the fugacity coefficients of nitrogen and methane in the mixture by Eq. (13.99) and the Redlich/Kwong equation of state.

Solution 13.6

For the Redlich/Kwong equation, $\varepsilon = 0$ and $\sigma = 1$, and Eq. (13.87) becomes:

$$Z = 1 + \beta - q\beta \frac{Z - \beta}{Z(Z + \beta)} \tag{A}$$

where β and q are given by Eqs. (13.89) and (13.90). Superscripts are here omitted because all calculations are for a vapor phase. The mixing rules most commonly used with the Redlich/Kwong equation for parameters $a(T)$ and b are given by Eqs. (13.91) through (13.93). For a binary mixture they become:

$$a = y_1^2 a_1 + 2 y_1 y_2 \sqrt{a_1 a_2} + y_2^2 a_2 \tag{B}$$

$$b = y_1 b_1 + y_2 b_2 \tag{C}$$

In Eq. (B), a_1 and a_2 are pure-species parameters given by Eq. (13.79) written for the Redlich/Kwong equation:

$$a_i = 0.42748 \frac{T_{r_i}^{-1/2} (83.14)^2 T_{c_i}^2}{P_{c_i}} \text{ bar·cm}^6\text{·mol}^{-2} \tag{D}$$

In Eq. (C), b_1 and b_2 are pure-species parameters, given by Eq. (13.80):

$$b_i = 0.08664 \frac{83.14 \, T_{c_i}}{P_{c_i}} \text{ cm}^3\text{·mol}^{-1} \tag{E}$$

Critical constants for nitrogen and methane from Table B.1 of App. B and calculated values for b_i and a_i from Eqs. (D) and (E) are:

	T_{c_i}/K	T_{r_i}	P_{c_i}/bar	b_i	$10^{-5} a_i$
$N_2(1)$	126.2	1.5848	34.00	26.737	10.995
$CH_4(2)$	190.6	1.0493	45.99	29.853	22.786

Mixture parameters by Eqs. (B), (C), and (13.90) are:

$$a = 17.560 \times 10^5 \text{ bar·cm}^6\text{·mol}^{-2} \qquad b = 28.607 \text{ cm}^3\text{·mol}^{-1} \qquad q = 3.6916$$

Equation (A) becomes:

$$Z = 1 + \beta - 3.6916 \frac{\beta(Z - \beta)}{Z(Z + \beta)} \qquad \text{with} \qquad \beta = 0.051612$$

where β comes from Eq. (13.89). Solution yields $Z = 0.85393$. Moreover, Eq. (13.72) reduces to:

$$I = \ln\frac{Z+\beta}{Z} = 0.05868$$

Application of Eq. (13.94) to Eq. (*B*) yields:

$$\bar{a}_1 = \left[\frac{\partial(na)}{\partial n_1}\right]_{T,n_2} = 2y_1 a_1 + 2y_2\sqrt{a_1 a_2} - a$$

$$\bar{a}_2 = \left[\frac{\partial(na)}{\partial n_2}\right]_{T,n_1} = 2y_2 a_2 + 2y_1\sqrt{a_1 a_2} - a$$

By Eq. (13.95) applied to Eq. (*C*),

$$\bar{b}_1 = \left[\frac{\partial(nb)}{\partial n_1}\right]_{T,n_2} = b_1 \qquad\qquad \bar{b}_2 = \left[\frac{\partial(nb)}{\partial n_2}\right]_{T,n_1} = b_2$$

Whence, by Eq. (13.101):

$$\bar{q}_1 = q\left(\frac{2y_1 a_1 + 2y_2\sqrt{a_1 a_2}}{a} - \frac{b_1}{b}\right) \tag{F}$$

$$\bar{q}_2 = q\left(\frac{2y_2 a_2 + 2y_1\sqrt{a_1 a_2}}{a} - \frac{b_2}{b}\right) \tag{G}$$

Substitution of numerical values into these equations and into Eq. (13.99) leads to the following results:

	$\bar{q}_i$	$\ln\hat{\phi}_i$	$\hat{\phi}_i$
$N_2(1)$	2.39194	−0.05664	0.94493
$CH_4(2)$	4.55795	−0.19966	0.81901

The values of $\hat{\phi}_i$ agree reasonably well with those found in Ex. 10.7.

Equations (13.77) and (13.99) provide the basis for VLE calculations for mixtures, but they incorporate a number of mixture parameters (e.g., a^l, b^v) and thermodynamic functions (e.g., Z^l, Z^v) that may initially be unknown. The calculation therefore becomes one of solving simultaneous equations equal in number to the number of unknowns. The equations available fall into several classes: mixing- and combining-rule and parameter equations for the liquid phase, the same equations for the vapor phase, and equilibrium and related equations. All of these parameters and equations have already been presented, but they are classified for convenience in Table 13.9. The presumption is that all pure-species parameters (e.g., a_i and b_i) are known and that either the temperature T or pressure P and either the liquid-phase *or* vapor-phase composition is specified. In addition to the N primary unknowns (T or P and either liquid or vapor phase composition), Table 13.9 enumerates $12 + 4N$ auxiliary variables and a total of $12 + 5N$ equations.

Table 13.9: VLE Calculations Based on Equations of State

A. Mixing and combining rules. Liquid phase.

$$a^l = \sum_i \sum_j x_i x_j (a_i a_j)^{1/2}$$

$$b^l = \sum_i x_i b_i$$

2 equations; 2 variables: a^l, b^l

B. Mixing and combining rules. Vapor phase.

$$a^v = \sum_i \sum_j y_i y_j (a_i a_j)^{1/2}$$

$$b^v = \sum_i y_i b_i$$

2 equations; 2 variables: a^v, b^v

C. Dimensionless parameters. Liquid phase.

$$\beta^l = b^l P/(RT) \qquad q^l = a^l/(b^l RT)$$

$$\bar{q}_i^l = q^l \left(1 + \frac{\bar{a}_i^l}{a^l} - \frac{b_i}{b^l} \right) \qquad (i = 1, 2, \ldots, N)$$

$2 + N$ equations; $2 + N$ variables: b^l, q^l, $\{\bar{q}_i^l\}$

D. Dimensionless parameters. Vapor phase.

$$\beta^v = b^v P/(RT) \qquad q^v = a^v/(b^v RT)$$

$$\bar{q}_i^v = q^v \left(1 + \frac{\bar{a}_i^v}{a^v} - \frac{b_i}{b^v} \right) \qquad (i = 1, 2, \ldots, N)$$

$2 + N$ equations; $2 + N$ variables: b^v, q^v, $\{\bar{q}_i^v\}$

E. Equilibrium and related equations.

$$y_i \hat{\phi}_i^v = x_i \hat{\phi}_i^l \qquad (i = 1, 2, \ldots, N)$$

$$\ln \hat{\phi}_i^l = \frac{b_i}{b^l}(Z^l - 1) - \ln(Z^l - \beta^l) - \bar{q}_i^l \, I^l \qquad (i = 1, 2, \ldots, N)$$

$$\ln \hat{\phi}_i^v = \frac{b_i}{b^v}(Z^v - 1) - \ln(Z^v - \beta^v) - \bar{q}_i^v \, I^v \qquad (i = 1, 2, \ldots, N)$$

$$Z^p = 1 + \beta^p - q^p \beta^p \frac{Z^p - \beta^p}{(Z + \varepsilon \beta^p)(Z + \sigma \beta^p)} \qquad (p = v, l)$$

$$I^p = \frac{1}{\sigma - \varepsilon} \ln \left(\frac{Z^p + \sigma \beta^p}{Z^p + \varepsilon \beta^p} \right) \qquad (p = v, l)$$

$4 + 3N$ equations; $4 + 2N$ variables: $\{\hat{\phi}_i^v\}$, $\{\hat{\phi}_i^l\}$, I^l, Z^l, I^v, Z^v

A direct and somewhat intuitive solution procedure makes use of Eq. (13.77), rewritten as $y_i = K_i x_i$. Because $\sum_i y_i = 1$,

$$\sum_i K_i x_i = 1 \tag{13.102}$$

where K_i, the K-value, is given by:

$$K_i = \frac{\hat{\phi}_i^l}{\hat{\phi}_i^v} \tag{13.103}$$

Thus for bubblepoint calculations, where the liquid-phase composition is known, the problem is to find the set of K-values that satisfies Eq. (13.102).

Example 13.7

Develop the *Pxy* diagram at 37.78°C for the methane(1)/*n*-butane(2) binary system. Base calculations on the Soave/Redlich/Kwong equation with mixing rules given by Eqs. (13.91) through (13.93). Experimental data at this temperature for comparison are given by Sage et al.[25]

Solution 13.7

The procedure here is to do a *BUBL P* calculation for each experimental data point. For each calculation, estimated values of P and y_1 are required to initiate iteration. These estimates are here provided by the experimental data. Where no such data are available, several trials may be required to find values for which the iterative procedure converges.

Pure-species parameters a_i and b_i are found from Eqs. (13.79) and (13.80) with constants and an expression for $\alpha(T_r)$ from Table 3.1. For a temperature of 310.93 K [37.78°C] and with critical constants and ω_i from Table B.1, calculations provide the following pure-species values:

	T_{c_i}/K	T_{r_i}	ω_i	$\alpha(T_r)$	P_{c_i}/bar	b_i	$10^{-6}a_i$
CH$_4$(1)	190.6	1.6313	0.012	0.7425	45.99	29.853	1.7331
n-C$_4$H$_{10}$(2)	425.1	0.7314	0.200	1.2411	37.96	80.667	17.458

The units of b_i are cm^3·mol^{-1}, and those of a_i are bar·cm^6·mol^{-2}.

Note that the temperature of interest is greater than the critical temperature of methane. The *Pxy* diagram will therefore be of the type shown by Fig. 12.2(*a*) for temperature T_b. The equations for $\alpha(T_r)$ given in Table 3.1 are based on vapor-pressure data, which extend only to the critical temperature. However, they may be applied to temperatures modestly above the critical temperature.

[25]B. H. Sage, B. L. Hicks, and W. N. Lacey, *Industrial and Engineering Chemistry*, vol. 32, pp. 1085–1092, 1940.

The mixing rules adopted here are the same as in Ex. 13.6, where Eqs. (B), (C), (F), and (G) give mixture parameters for the vapor phase. When applied to the liquid phase, x_i replaces y_i as the mole-fraction variable:

$$a^l = x_1^2 a_1 + 2x_1 x_2 \sqrt{a_1 a_2} + x_2^2 a_2 \qquad\qquad b^l = x_1 b_1 + x_2 b_2$$

$$\bar{q}_1^l = q^l \left(\frac{2x_1 a_1 + 2x_2 \sqrt{a_1 a_2}}{a^l} - \frac{b_1}{b^l} \right) \qquad \bar{q}_2^l = q^l \left(\frac{2x_2 a_2 + 2x_1 \sqrt{a_1 a_2}}{a^l} - \frac{b_2}{b^l} \right)$$

where q^l is given by Eq. (13.90).

For the SRK equation, $\varepsilon = 0$ and $\sigma = 1$; Eqs. (13.83) and (13.84) reduce to:

$$Z^l = \beta^l + Z^l (Z^l + \beta^l) \left(\frac{1 + \beta^l - Z^l}{q^l \beta^l} \right) \qquad Z^v = 1 + \beta^v - q^v \beta^v \frac{Z^v - \beta^v}{Z^v (Z^v + \beta^v)}$$

where β^l, β^v, q^l, and q^v are given by Eqs. (13.89) and (13.90). The first set of *BUBL P* calculations is made for the assumed pressure. With the given liquid-phase composition and assumed vapor-phase composition, values for Z^l and Z^v are determined by the preceding equations, and fugacity coefficients $\hat{\phi}_i^l$ and $\hat{\phi}_i^v$ then follow from Eq. (13.99). Values of K_1 and K_2 come from Eq. (13.103). The constraint $y_1 + y_2 = 1$ has not been imposed, and Eq. (13.102) is unlikely to be satisfied. In this event, $K_1 x_1 + K_2 x_2 \neq 1$, and a new vapor composition for the next iteration is given by the normalizing equation:

$$y_1 = \frac{K_1 x_1}{K_1 x_1 + K_2 x_2} \quad \text{with} \quad y_2 = 1 - y_1$$

This new vapor composition allows reevaluation of $\{\hat{\phi}_i^v\}$, $\{K_i\}$, and $\{K_i x_i\}$. If the sum $K_1 x_1 + K_2 x_2$ has changed, a new vapor composition is found and the sequence of calculations is repeated. Continued iteration leads to stable values of all quantities. If the sum $K_1 x_1 + K_2 x_2$ is not unity, the assumed pressure is incorrect, and it must be adjusted according by some rational scheme. When $\sum_i K_i x_i > 1$, P is too low; when $\sum_i K_i x_i < 1$, P is too high. The entire iterative procedure is then repeated with a new pressure P. The last calculated values of y_i are used for the initial estimate of $\{y_i\}$. The process continues until $K_1 x_1 + K_2 x_2 = 1$. Of course the same result can be obtained by any other method capable of solving the set of equations given in Table 13.9, with P, y_1, and y_2 as the unknowns.

The results of all calculations are shown by the solid lines of Fig. 13.13. Experimental values appear as points. The root-mean-square percentage difference between experimental and calculated pressures is 3.9%, and the root-mean-square deviation between experimental and calculated y values is 0.013. These results, based on the simple mixing rules of Eqs. (13.91) and (13.92), are representative for systems that exhibit modest and well-behaved deviations from ideal-solution behavior, e.g., for systems comprised of hydrocarbons and cryogenic fluids.

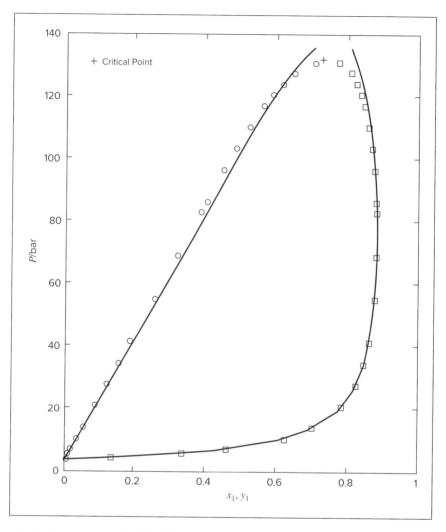

Figure 13.13: *Pxy* diagram at 37.8°C (100°F) for methane(1)/*n*-butane(2). Lines represent values from *BUBL P* calculations with the SRK equation; points are experimental values.

13.8 FLASH CALCULATIONS

In previous sections, we have focused on bubblepoint and dewpoint calculations, which are common calculations in practice, and which provide a basis for constructing phase diagrams for VLE. Perhaps an even more important application of VLE is the flash calculation. The name originates from the fact that a liquid at a pressure equal to or greater than its bubblepoint pressure "flashes" or partially evaporates when the pressure is reduced below its bubblepoint pressure, producing a two-phase system of vapor and liquid in equilibrium. We consider here only the *P, T*-flash, which refers to any calculation of the quantities and compositions of the vapor and liquid phases making up a two-phase system in equilibrium at known T, P, and overall composition. This poses a problem known to be determinate on the basis of Duhem's theorem, because two independent variables (T and P) are specified for a system of fixed over-all composition, that is, a system formed from given masses of nonreacting chemical species.

Consider a system containing one mole of nonreacting chemical species with an overall composition represented by the set of mole fractions $\{z_i\}$. Let $\mathcal{L}$ be the moles of liquid, with mole fractions $\{x_i\}$, and let $\mathcal{V}$ be the moles of vapor, with mole fractions $\{y_i\}$. The material balance equations are:

$$\mathcal{L} + \mathcal{V} = 1$$
$$z_i = x_i \mathcal{L} + y_i \mathcal{V} \qquad (i = 1, 2, \ldots, N)$$

Combining these equations to eliminate $\mathcal{L}$ gives:

$$z_i = x_i(1 - \mathcal{V}) + y_i \mathcal{V} \qquad (13.104)$$

The K-value, as defined in the previous section ($K_i \equiv y_i/x_i$), is a convenient construct for use in flash calculations. Substituting $x_i = y_i/K_i$ into Eq. (13.104) and solving for y_i yields:

$$y_i = \frac{z_i K_i}{1 + \mathcal{V}(K_i - 1)} \qquad (i = 1, 2, \ldots, N) \qquad (13.105)$$

Because $\sum_i y_i = 1$, summing this expression over all species gives us a single equation in which, for known K-values, the only unknown is $\mathcal{V}$.

$$\sum_i \frac{z_i K_i}{1 + \mathcal{V}(K_i - 1)} = 1 \qquad (13.106)$$

One general approach to solving a P, T-flash problem is to find the value of $\mathcal{V}$, between 0 and 1, that satisfies this equation. Note that $\mathcal{V} = 1$ is always a trivial solution to this equation. Having done so, the vapor-phase mole fractions are then obtained from Eq. (13.105), the liquid-phase mole fractions are obtained from $x_i = y_i/K_i$, and $\mathcal{L}$ is given by $\mathcal{L} = 1 - \mathcal{V}$. When Raoult's law can be applied, the K-values are constant and this is straightforward, as shown in the following example. Historically, K-values for light hydrocarbons were often taken from a set of charts constructed by DePriester, hence called DePriester charts. These similarly provided a set of constant K-values for use in the preceding calculations.[26]

Example 13.8

The system acetone(1)/acetonitrile(2)/nitromethane(3) at 80°C and 110 kPa has the overall composition $z_1 = 0.45$, $z_2 = 0.35$, $z_3 = 0.20$. Assuming that Raoult's law is appropriate to this system, determine $\mathcal{L}$, $\mathcal{V}$, $\{x_i\}$, and $\{y_i\}$. The vapor pressures of the pure species at 80°C are:

$$P_1^{sat} = 195.75 \qquad P_2^{sat} = 97.84 \qquad P_3^{sat} = 50.32 \text{ kPa}$$

[26]C. L. DePriester, Chem. Eng. Progr. Symp. Ser. No. 7, vol. 49, p. 42, 1953.

Solution 13.8

First, do a *BUBL P* calculation with $\{z_i\} = \{x_i\}$ to determine P_{bubl}:

$$P_{bubl} = x_1 P_1^{sat} + x_2 P_2^{sat} + x_3 P_3^{sat}$$
$$P_{bubl} = (0.45)(195.75) + (0.35)(97.84) + (0.20)(50.32) = 132.40 \text{ kPa}$$

Next, do a *DEW P* calculation with $\{z_i\} = \{y_i\}$ to find P_{dew}:

$$P_{dew} = \frac{1}{y_1/P_1^{sat} + y_2/P_2^{sat} + y_3/P_3^{sat}} = 101.52 \text{ kPa}$$

Because the given pressure lies between P_{bubl} and P_{dew}, the system is in the two-phase region, and a flash calculation is possible.

From Raoult's law, Eq. (13.16), we have $K_i = y_i/x_i = P_i^{sat}/P$, from which:

$$K_1 = 1.7795 \qquad K_2 = 0.8895 \qquad K_3 = 0.4575$$

Substituting known values into Eq. (13.106) gives:

$$\frac{(0.45)(0.7795)}{1 + 0.7795\mathcal{V}} + \frac{(0.35)(0.8895)}{1 - 0.1105\mathcal{V}} + \frac{(0.20)(0.4575)}{1 - 0.5425\mathcal{V}} = 1$$

Trial-and-error or iterative solution for $\mathcal{V}$ followed by evaluation of the other unknowns yields:

$$\mathcal{V} = 0.7364 \text{ mol} \qquad \mathcal{L} = 1 - \mathcal{V} = 0.2636 \text{ mol}$$

$$y_1 = \frac{(0.45)(1.7795)}{1 + (0.7795)(0.7634)} = 0.5087 \qquad y_2 = 0.3389 \qquad y_3 = 0.1524$$

$$x_1 = \frac{y_1}{K_1} = \frac{0.5087}{1.7795} = 0.2859 \qquad x_2 = 0.3810 \qquad x_3 = 0.3331$$

Reassuringly, $\sum_i x_i = \sum_i y_i = 1$.

The procedure of the preceding example is applicable regardless of the number of species present. However, for the simple case of a binary flash calculation using Raoult's law, explicit solution is possible. In that case, we have:

$$P = x_1 P_1^{sat} + x_2 P_2^{sat} = x_1 P_1^{sat} + (1 - x_1)P_2^{sat}$$

Solving for x_1 gives:

$$x_1 = \frac{P - P_2^{sat}}{P_1^{sat} - P_2^{sat}}$$

With x_1 known, the remaining variables follow immediately from Raoult's law [Eq. (13.16)] and the overall mass balances [Eq. (13.104)].

When the K-values are not constant, the general approach remains the same as in Ex. 13.8, but an additional level of iterative solution is required. In the preceding treatment, we solved for y_i to obtain Eq. (13.105). If, instead, we eliminate y_i using $y_i = K_i x_i$, we obtain an alternative expression:

$$x_i = \frac{z_i}{1 + \mathcal{V}(K_i - 1)} \qquad (i = 1, 2, \ldots, N) \qquad (13.107)$$

Because both sets of mole fractions must sum to unity, $\sum_i x_i = \sum_i y_i = 1$. Thus, if we sum Eq. (13.105) over all species and subtract unity from this sum, the difference F_y is zero:

$$F_y = \sum_i \frac{z_i K_i}{1 + \mathcal{V}(K_i - 1)} - 1 = 0 \qquad (13.108)$$

Similar treatment of Eq. (13.107) yields the difference F_x, which is also zero:

$$F_x = \sum_i \frac{z_i}{1 + \mathcal{V}(K_i - 1)} - 1 = 0 \qquad (13.109)$$

A P, T-flash problem can be solved by finding a value of $\mathcal{V}$ that makes either F_y or F_x equal to zero for known T, P, and overall composition. A more convenient function for applying a *general* solution procedure[27] is the difference $F \equiv F_y - F_x$:

$$F = \sum_i \frac{z_i(K_i - 1)}{1 + \mathcal{V}(K_i - 1)} = 0 \qquad (13.110)$$

The advantage of this function is apparent from its derivative:

$$\frac{dF}{d\mathcal{V}} = -\sum_i \frac{z_i(K_i - 1)\,2}{[1 + \mathcal{V}(K_i - 1)]^2} \qquad (13.111)$$

Because $dF/d\mathcal{V}$ is always negative, the F vs. $\mathcal{V}$ relation is monotonic. This, in turn, makes this form of the equation exceptionally well-suited for solution by Newton's method (App. H). Equation H.1 for the nth iteration of Newton's method becomes:

$$F + \left(\frac{dF}{d\mathcal{V}}\right)\Delta \mathcal{V} = 0 \qquad (13.112)$$

where $\Delta \mathcal{V} \equiv \mathcal{V}_{n+1} - \mathcal{V}_n$, and F and $(dF/d\mathcal{V})$ are found by Eq. (13.110) and (13.111). In these equations, for the general gamma/phi formulation of VLE, the K-values come from Eq. (13.13), written:

$$K_i = \frac{y_i}{x_i} = \frac{\gamma_i P_i^{\text{sat}}}{\Phi_i P} \qquad (i = 1, 2, \ldots, N) \qquad (13.113)$$

with the Φ_i given by Eq. (13.14). The K-values contain all of the thermodynamic information and are related in a complex way to T, P, $\{y_i\}$, and $\{x_i\}$. Because solution is for $\{y_i\}$ and $\{x_i\}$, the P, T-flash calculation inevitably requires iterative solution. This remains the case, even at low pressure where we can assume $\Phi_i = 1$, because the activity coefficients still depend upon the unknown $\{x_i\}$.

One typically proceeds by performing a BUBL P calculation and a DEW P calculation prior to the flash calculation. If the given pressure is below P_{dew} for the specified T and $\{z_i\}$,

---

[27]H. H. Rachford, Jr., and J. D. Rice, *J. Petrol. Technol.*, vol. 4(10), sec. 1, p. 19 and sec. 2, p. 3, October 1952.

then the system exists as superheated vapor, and no flash calculation is possible. Similarly, if the given pressure is above P_{bubl} for the specified T and $\{z_i\}$, then the system is a subcooled liquid, and no flash calculation is possible. If the specified P falls between P_{dew} and P_{bubl} for the specified T and $\{z_i\}$, then the system exists as an equilibrium mixture of vapor and liquid, and we can proceed with the flash calculation. The results of the preliminary DEW P and BUBL P calculations then provide useful initial estimates of $\{y_i\}$, $\{\hat{\phi}_i\}$, and $\mathcal{V}$. For the dewpoint, $\mathcal{V} = 1$, with calculated values of P_{dew}, $\gamma_{i,dew}$, and $\hat{\phi}_{i,dew}$; for the bubblepoint, $\mathcal{V} = 0$, with calculated values of P_{bubl}, $\gamma_{i,bubl}$, and $\hat{\phi}_{i,bubl}$. The simplest procedure is to interpolate linearly between dewpoint and bubblepoint with respect to P:

$$\mathcal{V} = \frac{P_{bubl} - P}{P_{bubl} - P_{dew}}$$

$$\gamma_i = \gamma_{i,dew} + (\gamma_{i,bubl} - \gamma_{i,dew})\frac{P - P_{dew}}{P_{bubl} - P_{dew}}$$

$$\hat{\phi}_i = \hat{\phi}_{i,dew} + (\hat{\phi}_{i,bubl} - \hat{\phi}_{i,dew})\frac{P - P_{dew}}{P_{bubl} - P_{dew}}$$

With these initial values of the γ_i and $\hat{\phi}_i$, one can now calculate initial values of K_i from Eq. (13.113). Using these values with Eqs. (13.110) and (13.111) one applies Newton's method, iterating on Eq. (13.112) to obtain a solution for $\mathcal{V}$. One then proceeds as in Ex. 13.8 to compute $\mathcal{L}$, $\{x_i\}$, and $\{y_i\}$. The computed compositions are used to obtain new estimates of $\{\gamma_i\}$ and $\{\hat{\phi}_i\}$, from which new K-values are computed. The procedure is repeated until the change in $\{x_i\}$ and $\{y_i\}$ from one iteration to the next is negligible. The same basic procedure can be applied with K-values computed by application of a cubic equation of state to both phases, as exemplified by Eq. (13.103).

13.9 SYNOPSIS

After studying this chapter, including the end-of-chapter problems, one should be able to:

- Understand the relationship between excess Gibbs energy and activity coefficients
- Explain and interpret each of the following five types of VLE calculations:
  - Bubblepoint pressure (BUBL P) calculations
  - Dewpoint pressure (DEW P) calculations
  - Bubblepoint temperature (BUBL T) calculations
  - Dewpoint temperature (DEW T) calculations
  - P, T-flash calculations
- Carry out each of the five types of VLE calculations using each of the following VLE formulations:
  - Raoult's law
  - Modified Raoult's law, with activity coefficients

- The full gamma/phi formulation
- A cubic equation of state applied to both the liquid and vapor phases
- State and apply Henry's law
- Compute liquid phase fugacities, activity coefficients, and excess Gibbs energy from low-pressure VLE data
- Fit excess Gibbs energy to models including the Margules equation, the van Laar equation, and the Wilson equation
- Evaluate the thermodynamic consistency of a set of low-pressure binary VLE data
- Fit activity coefficient models, including the Margules equation, the van Laar equation, and the Wilson equation directly to P vs. x_1 data
- Compute activity coefficients and excess properties from
  - The Margules equations
  - The van Laar equation
  - The Wilson equation
  - The NRTL equation
- Compute residual properties and fugacities for pure species and mixtures from a cubic equation of state, and use these in VLE calculations

13.10 PROBLEMS

Solutions to some of the problems of this chapter require vapor pressures as a function of temperature. Table B.2, Appendix B, lists parameter values for the Antoine equation, from which these can be computed.

13.1 Assuming the validity of Raoult's law, do the following calculations for the benzene(1)/toluene(2) system:

 (a) Given $x_1 = 0.33$ and $T = 100°C$, find y_1 and P.
 (b) Given $y_1 = 0.33$ and $T = 100°C$, find x_1 and P.
 (c) Given $x_1 = 0.33$ and $P = 120$ kPa, find y_1 and T.
 (d) Given $y_1 = 0.33$ and $P = 120$ kPa, find x_1 and T.
 (e) Given $T = 105°C$ and $P = 120$ kPa, find x_1 and y_1.
 (f) For part (e), if the overall mole fraction of benzene is $z_1 = 0.33$, what molar fraction of the two-phase system is vapor?
 (g) Why is Raoult's law likely to be an excellent VLE model for this system at the stated (or computed) conditions?

13.2. Assuming Raoult's law to be valid, prepare a Pxy diagram for a temperature of 90°C and a txy diagram for a pressure of 90 kPa for one of the following systems:

 (a) Benzene(1)/ethylbenzene(2).
 (b) 1-Chlorobutane(1)/chlorobenzene(2).

13.3. Assuming Raoult's law to apply to the system n-pentane(1)/n-heptane(2),

    (*a*) What are the values of x_1 and y_1 at $t = 55°C$ and $P = \frac{1}{2}(P_1^{sat} + P_2^{sat})$? For these conditions plot the fraction of system that is vapor $\mathcal{V}$ vs. overall composition z_1.

    (*b*) For $t = 55°C$ and $z_1 = 0.5$, plot P, x_1, and y_1 vs. $\mathcal{V}$.

13.4. Rework Prob. 13.3 for one of the following:

    (*a*) $t = 65°C$; (*b*) $t = 75°C$; (*c*) $t = 85°C$; (*d*) $t = 95°C$.

13.5. Prove: An equilibrium liquid/vapor system described by Raoult's law cannot exhibit an azeotrope.

13.6. Of the following binary liquid/vapor systems, which can be approximately modeled by Raoult's law? For those that cannot, why not? Table B.1 (App. B) may be useful.

    (*a*) Benzene/toluene at 1(atm).
    (*b*) n-Hexane/n-heptane at 25 bar.
    (*c*) Hydrogen/propane at 200 K.
    (*d*) Iso-octane/n-octane at 100°C.
    (*e*) Water/n-decane at 1 bar.

13.7. A single-stage liquid/vapor separation for the benzene(1)/ethylbenzene(2) system must produce phases of the following equilibrium compositions. For one of these sets, determine T and P in the separator. What additional information is needed to compute the relative amounts of liquid and vapor leaving the separator? Assume that Raoult's law applies.

    (*a*) $x_1 = 0.35$, $y_1 = 0.70$.
    (*b*) $x_1 = 0.35$, $y_1 = 0.725$.
    (*c*) $x_1 = 0.35$, $y_1 = 0.75$.
    (*d*) $x_1 = 0.35$, $y_1 = 0.775$.

13.8. Do all four parts of Prob. 13.7, and compare the results. The required temperatures and pressures vary significantly. Discuss possible processing implications of the various temperature and pressure levels.

13.9. A mixture containing equimolar amounts of benzene(1), toluene(2), and ethylbenzene(3) is flashed to conditions T and P. For one of the following conditions, determine the equilibrium mole fractions $\{x_i\}$ and $\{y_i\}$ of the liquid and vapor phases formed and the molar fraction $\mathcal{V}$ of the vapor formed. Assume that Raoult's law applies.

    (*a*) $T = 110°C$, $P = 90$ kPa.
    (*b*) $T = 110°C$, $P = 100$ kPa.
    (*c*) $T = 110°C$, $P = 110$ kPa.
    (*d*) $T = 110°C$, $P = 120$ kPa.

13.10. Do all four parts of Prob. 13.9, and compare the results. Discuss any trends that appear.

13.11. A binary mixture of mole fraction z_1 is flashed to conditions T and P. For one of the following, determine the equilibrium mole fractions x_1 and y_1 of the liquid and vapor phases formed, the molar fraction V of the vapor formed, and the fractional recovery R of species 1 in the vapor phase (defined as the ratio for species 1 of moles in the vapor to moles in the feed). Assume that Raoult's law applies.

    (*a*) Acetone(1)/acetonitrile(2), $z_1 = 0.75$, $T = 340$ K, $P = 115$ kPa.
    (*b*) Benzene(1)/ethylbenzene(2), $z_1 = 0.50$, $T = 100°C$, $P = 0.75(atm)$.
    (*c*) Ethanol(1)/1-propanol(2), $z_1 = 0.25$, $T = 360$ K, $P = 0.80(atm)$.
    (*d*) 1-Chlorobutane(1)/chlorobenzene(2), $z_1 = 0.50$, $T = 125°C$, $P = 1.75$ bar.

13.12. Humidity, relating to the quantity of moisture in atmospheric air, is accurately given by equations derived from the ideal-gas law and Raoult's law for H_2O.

    (*a*) The *absolute humidity h* is defined as the mass of water vapor in a unit mass of dry air. Show that it is given by:

$$h = \frac{\mathcal{M}_{H_2O}}{\mathcal{M}_{air}} \frac{p_{H_2O}}{P - p_{H_2O}}$$

    where $\mathcal{M}$ represents a molar mass and p_{H_2O} is the partial pressure of the water vapor, i.e., $p_{H_2O} = y_{H_2O} P$.
    (*b*) The *saturation humidity* h^{sat} is defined as the value of h when air is in equilibrium with a large body of pure water. Show that it is given by:

$$h^{sat} = \frac{\mathcal{M}_{H_2O}}{\mathcal{M}_{air}} \frac{p_{H_2O}^{sat}}{P - p_{H_2O}^{sat}}$$

    where $p_{H_2O}^{sat}$ is the vapor pressure of water at the ambient temperature.
    (*c*) The percentage humidity is defined as the ratio of h to its saturation value, expressed as a percentage. On the other hand, the relative humidity is defined as the ratio of the partial pressure of water vapor in air to its vapor pressure, expressed as a percentage. What is the relation between these two quantities?

13.13. A concentrated binary solution containing mostly species 2 (but $x_2 \neq 1$) is in equilibrium with a vapor phase containing both species 1 and 2. The pressure of this two-phase system is 1 bar; the temperature is 25°C. At this temperature, $\mathcal{H}_1 = 200$ bar and $P_2^{sat} = 0.10$ bar. Determine good estimates of x_1 and y_1. State and justify all assumptions.

13.14. Air, even more than carbon dioxide, is inexpensive and nontoxic. Why is it not the gas of choice for making soda water and (cheap) champagne effervescent? Table 13.2 may provide useful data.

13.15. Helium-laced gases are used as breathing media for deep-sea divers. Why? Table 13.2 may provide useful data.

13.16. A binary system of species 1 and 2 consists of vapor and liquid phases in equilibrium at temperature T. The *overall* mole fraction of species 1 in the system is $z_1 = 0.65$. At temperature T, $\ln \gamma_1 = 0.67 x_2^2$; $\ln \gamma_2 = 0.67 x_1^2$; $P_1^{sat} = 32.27$ kPa; and $P_2^{sat} = 73.14$ kPa. Assuming the validity of Eq. (13.19),

(a) Over what range of pressures can this system exist as two phases at the given T and z_1?
(b) For a liquid-phase mole fraction $x_1 = 0.75$, what is the pressure P and what molar fraction V of the system is vapor?
(c) Show whether or not the system exhibits an azeotrope.

13.17. For the system ethyl ethanoate(1)/n-heptane(2) at 343.15 K, $\ln\gamma_1 = 0.95\,x_2^2$; $\ln\gamma_2 = 0.95\,x_1^2$; $P_1^{sat} = 79.80$ kPa; and $P_2^{sat} = 40.50$ kPa. Assuming the validity of Eq. (13.19),

(a) Make a BUBL P calculation for $T = 343.15$ K, $x_1 = 0.05$.
(b) Make a DEW P calculation for $T = 343.15$ K, $y_1 = 0.05$.
(c) What are the azeotrope composition and pressure at $T = 343.15$ K?

13.18. A liquid mixture of cyclohexanone(1)/phenol(2) for which $x_1 = 0.6$ is in equilibrium with its vapor at 144°C. Determine the equilibrium pressure P and vapor composition y_1 from the following information:

- $\ln\gamma_1 = A\,x_2^2$ $\ln\gamma_2 = A\,x_1^2$
- At 144°C, $P_1^{sat} = 75.20$ kPa and $P_2^{sat} = 31.66$ kPa
- The system forms an azeotrope at 144°C for which $x_1^{az} = y_1^{az} = 0.294$.

13.19. A binary system of species 1 and 2 consists of vapor and liquid phases in equilibrium at temperature T, for which $\ln\gamma_1 = 1.8\,x_2^2$, $\ln\gamma_2 = 1.8\,x_1^2$, $P_1^{sat} = 1.24$ bar, and $P_2^{sat} = 40.50$ kPa. Assuming the validity of Eq. (13.19),

(a) For what range of values of the overall mole fraction z_1 can this two-phase system exist with a liquid mole fraction $x_1 = 0.65$?
(b) What are the pressure P and vapor mole fraction y_1 within this range?
(c) What are the pressure and composition of the azeotrope at temperature T?

13.20. For the acetone(1)/methanol(2) system, a vapor mixture for which $z_1 = 0.25$ and $z_2 = 0.75$ is cooled to temperature T in the two-phase region and flows into a separation chamber at a pressure of 1 bar. If the composition of the liquid product is to be $x_1 = 0.175$, what is the required value of T, and what is the value of y_1? For liquid mixtures of this system, to a good approximation, $\ln\gamma_1 = 0.64\,x_2^2$ and $\ln\gamma_2 = 0.64\,x_1^2$.

13.21. The following is a rule of thumb: For a binary system in VLE at low pressure, the equilibrium vapor-phase mole fraction y_1 corresponding to an equimolar liquid mixture is approximately

$$y_1 = \frac{P_1^{sat}}{P_1^{sat} + P_2^{sat}}$$

where P_i^{sat} is a pure-species vapor pressure. Clearly, this equation is valid if Raoult's law applies. Prove that it is also valid for VLE described by Eq. (13.19), with $\ln\gamma_1 = A\,x_2^2$ and $\ln\gamma_2 = A\,x_1^2$.

13.22. A process stream contains light species 1 and heavy species 2. A relatively pure liquid stream containing mostly 2 is desired, obtained by a single-stage liquid/vapor separation. Specifications of the equilibrium composition are: $x_1 = 0.002$ and $y_1 = 0.950$. Use data given below to determine T (K) and P (bar) for the separator. Assume that

Eq. (13.19) applies; the calculated P should be used to validate this assumption. Data: For the liquid phase,

$$\ln\gamma_1 = 0.93\,x_2^2$$
$$\ln\gamma_2 = 0.93\,x_1^2$$
$$\ln P_i^{\text{sat}}/\text{bar} = A_i - \frac{B_i}{T/\text{K}}$$

$A_1 = 10.08,\ B_1 = 2572.0,\ A_2 = 11.63,\ B_2 = 6254.0$

13.23. If a system exhibits VLE, at least one of the K-values must be greater than 1.0 and at least one must be less than 1.0. Offer a proof of this observation.

13.24. Flash calculations are simpler for binary systems than for the general multicomponent case because the equilibrium compositions for a binary are independent of the overall composition. Show that, for a binary system in VLE,

$$x_1 = \frac{1 - K_2}{K_1 - K_2} \qquad y_1 = \frac{K_1(1 - K_2)}{K_1 - K_2}$$
$$\mathcal{V} = \frac{z_1(K_1 - K_2) - (1 - K_2)}{(K_1 - 1)(1 - K_2)}$$

13.25. The NIST Chemistry WebBook reports critically evaluated Henry's constants for selected chemicals in water at 25°C. Henry's constants from this source, denoted here by k_{Hi}, appear in the VLE equation written for the solute in the form:

$$m_i = k_{Hi}y_iP$$

where m_i is the liquid-phase molality of solute species i, expressed as mol i/kg solvent.

(a) Determine an algebraic relation connecting k_{Hi} to $\mathcal{H}_i$, Henry's constant in Eq. (13.26). Assume that x_i is "small."
(b) The NIST Chemistry WebBook provides a value of 0.034 mol·kg^{-1}·bar^{-1} for k_{Hi} of CO_2 in H_2O at 25°C. What is the implied value of $\mathcal{H}_i$ in bar? Compare this with the value given in Table 13.2, which came from a different source.

13.26. (a) A feed containing equimolar amounts of acetone(1) and acetonitrile(2) is throttled to pressure P and temperature T. For what pressure range (atm) will two phases (liquid and vapor) be formed for $T = 50$°C? Assume that Raoult's law applies.
(b) A feed containing equimolar amounts of acetone(1) and acetonitrile(2) is throttled to pressure P and temperature T. For what temperature range (°C) will two phases (liquid and vapor) be formed for $P = 0.5$(atm)? Assume that Raoult's law applies.

13.27. A binary mixture of benzene(1) and toluene(2) is flashed to 75 kPa and 90°C. Analysis of the effluent liquid and vapor streams from the separator yields: $x_1 = 0.1604$ and $y_1 = 0.2919$. An operator remarks that the product streams are "off-spec," and you are asked to diagnose the problem.

(a) Verify that the exiting streams are not in binary equilibrium.
(b) Verify that an air leak into the separator could be the cause.

13.28. Ten (10) kmol·hr^{-1} of hydrogen sulfide gas is burned with the stoichiometric amount of pure oxygen in a special unit. Reactants enter as gases at 25°C and 1(atm). Products leave as two streams in equilibrium at 70°C and 1(atm): a phase of pure liquid water, and a saturated vapor stream containing H_2O and SO_2.

(*a*) What is the composition (mole fractions) of the product vapor stream?
(*b*) What are the rates (kmol·hr^{-1}) of the two product streams?

13.29. Physiological studies show the neutral comfort level (NCL) of moist air corresponds to an absolute humidity of about 0.01 kg H_2O per kg of dry air.

(*a*) What is the vapor-phase mole fraction of H_2O at the NCL?
(*b*) What is the partial pressure of H_2O at the NCL? Here, and in part (*c*), take $P =$ 1.01325 bar.
(*c*) What is the dewpoint temperature (°F) at the NCL?

13.30. An industrial dehumidifier accepts 50 kmol·hr^{-1} of moist air with a dewpoint of 20°C. Conditioned air leaving the dehumidifier has a dewpoint temperature of 10°C. At what rate (kg·hr^{-1}) is liquid water removed in this steady-flow process? Assume P is constant at 1(atm).

13.31. Vapor/liquid-equilibrium azeotropy is impossible for binary systems rigorously described by Raoult's law. For real systems (those with $\gamma_i \neq 1$), azeotropy is inevitable at temperatures where the P_i^{sat} are equal. Such a temperature is called a *Bancroft point*. Not all binary systems exhibit such a point. With Table B.2 of App. B as a resource, identify three binary systems with Bancroft points, and determine the T and P coordinates. *Ground rule*: A Bancroft point must lie in the temperature ranges of validity of the Antoine equations.

13.32. The following is a set of VLE data for the system methanol(1)/water(2) at 333.15 K:

P/kPa	x_1	y_1	P/kPa	x_1	y_1
19.953	0.0000	0.0000	60.614	0.5282	0.8085
39.223	0.1686	0.5714	63.998	0.6044	0.8383
42.984	0.2167	0.6268	67.924	0.6804	0.8733
48.852	0.3039	0.6943	70.229	0.7255	0.8922
52.784	0.3681	0.7345	72.832	0.7776	0.9141
56.652	0.4461	0.7742	84.562	1.0000	1.0000

Extracted from K. Kurihara et al., *J. Chem. Eng. Data*, vol. 40, pp. 679–684, 1995.

(*a*) Basing calculations on Eq. (13.24), find parameter values for the Margules equation that provide the best fit of G^E/RT to the data, and prepare a *Pxy* diagram that compares the experimental points with curves determined from the correlation.
(*b*) Repeat (*a*) for the van Laar equation.
(*c*) Repeat (*a*) for the Wilson equation.

(d) Using Barker's method, find parameter values for the Margules equation that provide the best fit of the P–x_1 data. Prepare a diagram showing the residuals δP and δy_1 plotted vs. x_1.

(e) Repeat (d) for the van Laar equation.

(f) Repeat (d) for the Wilson equation.

13.33. If Eq. (13.24) is valid for isothermal VLE in a binary system, show that:

$$\left(\frac{dP}{dx_1}\right)_{x_1=0} \geq -P_2^{\text{sat}} \qquad \left(\frac{dP}{dx_1}\right)_{x_1=1} \leq P_1^{\text{sat}}$$

13.34. The following is a set of VLE data for the system acetone(1)/methanol(2) at 55°C:

P/kPa	x_1	y_1	P/kPa	x_1	y_1
68.728	0.0000	0.0000	97.646	0.5052	0.5844
72.278	0.0287	0.0647	98.462	0.5432	0.6174
75.279	0.0570	0.1295	99.811	0.6332	0.6772
77.524	0.0858	0.1848	99.950	0.6605	0.6926
78.951	0.1046	0.2190	100.278	0.6945	0.7124
82.528	0.1452	0.2694	100.467	0.7327	0.7383
86.762	0.2173	0.3633	100.999	0.7752	0.7729
90.088	0.2787	0.4184	101.059	0.7922	0.7876
93.206	0.3579	0.4779	99.877	0.9080	0.8959
95.017	0.4050	0.5135	99.799	0.9448	0.9336
96.365	0.4480	0.5512	96.885	1.0000	1.0000

Extracted from D. C. Freshwater and K. A. Pike, *J. Chem. Eng. Data*, vol. 12, pp. 179–183, 1967.

(a) Basing calculations on Eq. (13.24), find parameter values for the Margules equation that provide the best fit of G^E/RT to the data, and prepare a Pxy diagram that compares the experimental points with curves determined from the correlation.

(b) Repeat (a) for the van Laar equation.

(c) Repeat (a) for the Wilson equation.

(d) Using Barker's method, find parameter values for the Margules equation that provide the best fit of the P–x_1 data. Prepare a diagram showing the residuals δP and δy_1 plotted vs. x_1.

(e) Repeat (d) for the van Laar equation.

(f) Repeat (d) for the Wilson equation.

13.35. The excess Gibbs energy for binary systems consisting of liquids not too dissimilar in chemical nature is represented to a reasonable approximation by the equation:

$$G^E/RT = A x_1 x_2$$

where A is a function of temperature only. For such systems, it is often observed that the ratio of the vapor pressures of the pure species is nearly constant over a considerable temperature range. Let this ratio be r, and determine the range of values of A, expressed as a function of r, for which no azeotrope can exist. Assume the vapor phase to be an ideal gas.

13.36. For the ethanol(1)/chloroform(2) system at 50°C, the activity coefficients show interior extrema with respect to composition [see Fig. 13.4(*e*)].

(*a*) Prove that the van Laar equation cannot represent such behavior.
(*b*) The two-parameter Margules equation can represent this behavior, but only for particular ranges of the ratio A_{21}/A_{12}. What are they?

13.37. VLE data for methyl *tert*-butyl ether(1)/dichloromethane(2) at 308.15 K are as follows:

P/kPa	x_1	y_1	P/kPa	x_1	y_1
85.265	0.0000	0.0000	59.651	0.5036	0.3686
83.402	0.0330	0.0141	56.833	0.5749	0.4564
82.202	0.0579	0.0253	53.689	0.6736	0.5882
80.481	0.0924	0.0416	51.620	0.7676	0.7176
76.719	0.1665	0.0804	50.455	0.8476	0.8238
72.422	0.2482	0.1314	49.926	0.9093	0.9002
68.005	0.3322	0.1975	49.720	0.9529	0.9502
65.096	0.3880	0.2457	49.624	1.0000	1.0000

Extracted from F. A. Mato, C. Berro, and A. Péneloux, *J. Chem. Eng. Data*, vol. 36, pp. 259–262, 1991.

The data are well correlated by the three-parameter Margules equation [an extension of Eq. (13.39)]:

$$\frac{G^E}{RT} = (A_{21}x_1 + A_{12}x_2 - Cx_1x_2)x_1x_2$$

Implied by this equation are the expressions:

$$\ln \gamma_1 = x_2^2 \left[A_{12} + 2(A_{21} - A_{12} - C)x_1 + 3Cx_1^2 \right]$$
$$\ln \gamma_2 = x_1^2 \left[A_{21} + 2(A_{12} - A_{21} - C)x_2 + 3Cx_2^2 \right]$$

(*a*) Basing calculations on Eq. (13.24), find the values of parameters A_{12}, A_{21}, and C that provide the best fit of G^E/RT to the data.
(*b*) Prepare a plot of $\ln \gamma_1$, $\ln \gamma_2$, and $G^E/(x_1x_2RT)$ vs. x_1 showing both the correlation and experimental values.
(*c*) Prepare a Pxy diagram [see Fig. 13.8(*a*)] that compares the experimental data with the correlation determined in (*a*).
(*d*) Prepare a consistency-test diagram like Fig. 13.9.
(*e*) Using Barker's method, find the values of parameters A_{12}, A_{21}, and C that provide the best fit of the P–x_1 data. Prepare a diagram showing the residuals δP and δy_1 plotted vs. x_1.

13.38. Equations analogous to Eqs. (10.15) and (10.16) apply for excess properties. Because $\ln \gamma_i$ is a partial property with respect to G^E/RT, these analogous equations can be written for $\ln \gamma_1$ and $\ln \gamma_2$ in a binary system.

(a) Write these equations, and apply them to Eq. (13.42) to show that Eqs. (13.43) and (13.44) are indeed obtained.
(b) The alternative procedure is to apply Eq. (13.7). Show that by doing so Eqs. (13.43) and (13.44) are again reproduced.

13.39. The following is a set of activity-coefficient data for a binary liquid system as determined from VLE data:

x_1	γ_1	γ_2	x_1	γ_1	γ_2
0.0523	1.202	1.002	0.5637	1.120	1.102
0.1299	1.307	1.004	0.6469	1.076	1.170
0.2233	1.295	1.006	0.7832	1.032	1.298
0.2764	1.228	1.024	0.8576	1.016	1.393
0.3482	1.234	1.022	0.9388	1.001	1.600
0.4187	1.180	1.049	0.9813	1.003	1.404
0.5001	1.129	1.092			

Inspection of these experimental values suggests that they are *noisy*, but the question is whether they are *consistent*, and therefore possibly on average correct.

(a) Find experimental values for G^E/RT and plot them along with the experimental values of $\ln \gamma_1$ and $\ln \gamma_2$ on a single graph.
(b) Develop a valid correlation for the composition dependence of G^E/RT and show lines on the graph of part (a) that represent this correlation for all three of the quantities plotted there.
(c) Apply the consistency test described in Ex. 13.4 to these data, and draw a conclusion with respect to this test.

13.40. Following are VLE data for the system acetonitrile(1)/benzene(2) at 45°C:

P/kPa	x_1	y_1	P/kPa	x_1	y_1
29.819	0.0000	0.0000	36.978	0.5458	0.5098
31.957	0.0455	0.1056	36.778	0.5946	0.5375
33.553	0.0940	0.1818	35.792	0.7206	0.6157
35.285	0.1829	0.2783	34.372	0.8145	0.6913
36.457	0.2909	0.3607	32.331	0.8972	0.7869
36.996	0.3980	0.4274	30.038	0.9573	0.8916
37.068	0.5069	0.4885	27.778	1.0000	1.0000

Extracted from I. Brown and F. Smith, *Austral. J. Chem.*, vol. 8, p. 62, 1955.

The data are well correlated by the three-parameter Margules equation (see Prob. 13.37).

(a) Basing calculations on Eq. (13.24), find the values of parameters A_{12}, A_{21}, and C that provide the best fit of G^E/RT to the data.

(b) Prepare a plot of $\ln \gamma_1$, $\ln \gamma_2$, and $G^E/x_1x_2 RT$ vs. x_1 showing both the correlation and experimental values.

(c) Prepare a Pxy diagram [see Fig. 13.8(a)] that compares the experimental data with the correlation determined in (a).

(d) Prepare a consistency-test diagram like Fig. 13.9.

(e) Using Barker's method, find the values of parameters A_{12}, A_{21}, and C that provide the best fit of the P–x_1 data. Prepare a diagram showing the residuals δP and δy_1 plotted vs. x_1.

13.41. An unusual type of low-pressure VLE behavior is that of *double azeotropy*, in which the dew and bubble curves are S-shaped, thus yielding at different compositions both a minimum-pressure and a maximum-pressure azeotrope. Assuming that Eq. (13.57) applies, determine under what circumstances double azeotropy is likely to occur.

13.42. Rationalize the following rule of thumb, appropriate for an *equimolar* binary liquid mixture:

$$\frac{G^E}{RT}(\text{equimolar}) \approx \frac{1}{8}\ln(\gamma_1^\infty \gamma_2^\infty)$$

Problems 13.43 through 13.54 require parameter values for the Wilson or NRTL equation for liquid-phase activity coefficients. Table 13.10 gives parameter values for both equations. Antoine equations for vapor pressure are given in Table B.2, Appendix B.

13.43. For one of the binary systems listed in Table 13.10, based on Eq. (13.19) and the Wilson equation, prepare a Pxy diagram for $t = 60°C$.

13.44. For one of the binary systems listed in Table 13.10, based on Eq. (13.19) and the Wilson equation, prepare a txy diagram for $P = 101.33$ kPa.

13.45. For one of the binary systems listed in Table 13.10, based on Eq. (13.19) and the NRTL equation, prepare a Pxy diagram for $t = 60°C$.

13.46. For one of the binary systems listed in Table 13.10, based on Eq. (13.19) and the NRTL equation, prepare a txy diagram for $P = 101.33$ kPa.

13.47. For one of the binary systems listed in Table 13.10, based on Eq. (13.19) and the Wilson equation, make the following calculations:

(a) BUBL P: $t = 60°C$, $x_1 = 0.3$.

(b) DEW P: $t = 60°C$, $y_1 = 0.3$.

(c) P, T – flash: $t = 60°C$, $P = \frac{1}{2}(P_{\text{bubble}} + P_{\text{dew}})$, $z_1 = 0.3$.

(d) If an azeotrope exists at $t = 60°C$, find P^{az} and $x_1^{\text{az}} = y_1^{\text{az}}$.

13.48. Work Prob. 13.47 for the NRTL equation.

Table 13.10: Parameter Values for the Wilson and NRTL Equations

Parameters a_{12}, a_{21}, b_{12}, and b_{21} have units of cal·mol^{-1}, and V_1 and V_2 have units of cm^3·mol^{-1}.

System	V_1 V_2	Wilson Equation a_{12}	a_{21}	NRTL Equation b_{12}	b_{21}	α
Acetone(1) Water(2)	74.05 18.07	291.27	1448.01	631.05	1197.41	0.5343
Methanol(1) Water(2)	40.73 18.07	107.38	469.55	−253.88	845.21	0.2994
1-Propanol(1) Water(2)	75.14 18.07	775.48	1351.90	500.40	1636.57	0.5081
Water(1) 1,4-Dioxane(2)	18.07 85.71	1696.98	−219.39	715.96	548.90	0.2920
Methanol(1) Acetonitrile(2)	40.73 66.30	504.31	196.75	343.70	314.59	0.2981
Acetone(1) Methanol(2)	74.05 40.73	−161.88	583.11	184.70	222.64	0.3084
Methyl acetate(1) Methanol(2)	79.84 40.73	−31.19	813.18	381.46	346.54	0.2965
Methanol(1) Benzene(2)	40.73 89.41	1734.42	183.04	730.09	1175.41	0.4743
Ethanol(1) Toluene(2)	58.68 106.85	1556.45	210.52	713.57	1147.86	0.5292

Values are those recommended by Gmehling et al., *Vapor-Liquid Equilibrium Data Collection*, Chemistry Data Series, vol. I, parts 1a, 1b, 2c, and 2e, DECHEMA, Frankfurt/Main, 1981–1988.

13.49. For one of the binary systems listed in Table 13.10, based on Eq. (13.19) and the Wilson equation, make the following calculations:

(a) *BUBL T*: $P = 101.33$ kPa, $x_1 = 0.3$.
(b) *DEW T*: $P = 101.33$ kPa, $y_1 = 0.3$.
(c) P, T − flash: $P = 101.33$ kPa, $T = \frac{1}{2}(T_{bubble} + T_{dew})$, $z_1 = 0.3$.
(d) If an azeotrope exists at $P = 101.33$ kPa, find T^{az} and $x_1^{az} = y_1^{az}$.

13.50. Work Prob. 13.49 for the NRTL equation.

13.51. For the acetone(1)/methanol(2)/water(3) system, based on Eq. (13.19) and the Wilson equation, make the following calculations:

(a) *BUBL P*: $t = 65°C$, $x_1 = 0.3$, $x_2 = 0.4$.
(b) *DEW P*: $t = 65°C$, $y_1 = 0.3$, $y_2 = 0.4$.
(c) P, T − flash: $t = 65°C$, $P = \frac{1}{2}(P_{bubble} + P_{dew})$, $z_1 = 0.3$, $z_2 = 0.4$.

13.52. Work Prob. 13.51 for the NRTL equation.

13.53. For the acetone(1)/methanol(2)/water(3) system, based on Eq. (13.19) and the Wilson equation, make the following calculations:

(a) BUBL $T:P = 101.33$ kPa, $x_1 = 0.3, x_2 = 0.4$.
(b) DEW $T:P = 101.33$ kPa, $y_1 = 0.3, y_2 = 0.4$.
(c) $P, T -$ flash: $P = 101.33$ kPa, $T = \frac{1}{2}(T_{\text{bubble}} + T_{\text{dew}})$, $z_1 = 0.3$, $z_2 = 0.2$.

13.54. Work Prob. 13.53 for the NRTL equation.

13.55. The following expressions have been reported for the activity coefficients of species 1 and 2 in a binary liquid mixture at given T and P:
$$\ln \gamma_1 = x_2^2(0.273 + 0.096 \, x_1) \quad \ln \gamma_2 = x_1^2(0.273 - 0.096 \, x_2)$$

(a) Determine the implied expression for G^E/RT.
(b) *Generate* expressions for $\ln \gamma_1$ and $\ln \gamma_2$ from the result of (a).
(c) Compare the results of (b) with the reported expressions for $\ln \gamma_1$ and $\ln \gamma_2$. Discuss any discrepancy. Can the reported expressions possibly be correct?

13.56. Possible correlating equations for $\ln \gamma_1$ in a binary liquid system are given here. For one of these cases, determine by integration of the Gibbs/Duhem equation [Eq. (13.11)] the corresponding equation for $\ln \gamma_2$. What is the corresponding equation for G^E/RT? Note that by its definition, $\gamma_i = 1$ for $x_i = 1$.

(a) $\ln \gamma_1 = A x_2^2$;
(b) $\ln \gamma_1 = x_2^2(A + B x_2)$;
(c) $\ln \gamma_1 = x_2^2(A + B x_2 + C x_2^2)$;

13.57. A storage tank contains a heavy organic liquid. Chemical analysis shows the liquid to contain 600 ppm (molar basis) of water. It is proposed to reduce the water concentration to 50 ppm by boiling the contents of the tank at constant atmospheric pressure. Because the water is lighter than the organic, the vapor will be rich in water; continuous removal of the vapor serves to reduce the water content of the system. Estimate the percentage loss of organic (molar basis) in the boil-off process. Comment on the reasonableness of the proposal.

Suggestion: Designate the system water(1)/organic(2) and do unsteady-state molar balances for water and for water + organic. State all assumptions.

Data: T_{n_2} = normal boiling point of organic = 130°C.
$\gamma_1^\infty = 5.8$ for water in the liquid phase at 130°C.

13.58. Binary VLE data are commonly measured at constant T or at constant P. Isothermal data are *much* preferred for determination of a correlation for G^E for the liquid phase. Why?

13.59. Consider the following model for G^E/RT of a binary mixture:

$$\frac{G^E}{x_1 x_2 RT} = \left(x_1 A_{21}^k + x_2 A_{12}^k\right)^{1/k}$$

This equation in fact represents a *family* of two-parameter expressions for G^E/RT; specification of k leaves A_{12} and A_{21} as the free parameters.

(*a*) Find general expressions for $\ln \gamma_1$ and $\ln \gamma_2$, for any k.
(*b*) Show that $\ln\gamma_1^\infty = A_{12}$ and $\ln \gamma_2^\infty = A_{21}$, for any k.
(*c*) Specialize the model to the cases where k equals $-\infty$, -1, 0, $+1$, and $+\infty$. Two of the cases should generate familiar results. What are they?

13.60. A breathalyzer measures volume-% ethanol in gases exhaled from the lungs. Calibration relates it to volume-% ethanol in the bloodstream. Use VLE concepts to develop an approximate relation between the two quantities. Numerous assumptions are required; state and justify them where possible.

13.61. Table 13.10 gives values of parameters for the Wilson equation for the acetone(1)/methanol(2) system. Estimate values of $\ln\gamma_1^\infty$ and $\ln\gamma_2^\infty$ at 50°C. Compare with the values suggested by Fig. 13.4(*b*). Repeat the exercise with the NRTL equation.

13.62. For a binary system derive the expression for H^E implied by the Wilson equation for G^E/RT. Show that the implied excess heat capacity C_P^E is necessarily *positive*. Recall that the Wilson parameters depend on T, in accord with Eq. (13.53).

13.63. A single P-x_1-y_1 data point is available for a binary system at 25°C. Estimate from the data:

(*a*) The total pressure and vapor-phase composition at 25°C for an equimolar liquid mixture.
(*b*) Whether azeotropy is likely at 25°C.

Data: At 25°C, $P_1^{sat} = 183.4$ and $P_2^{sat} = 96.7$ kPa
      For $x_1 = 0.253$, $y_1 = 0.456$ and $P = 139.1$ kPa

13.64. A single P–x_1 data point is available for a binary system at 35°C. Estimate from the data:

(*a*) The corresponding value of y_1.
(*b*) The total pressure at 35°C for an equimolar liquid mixture.
(*c*) Whether azeotropy is likely at 35°C.

Data: At 25°C, $P_1^{sat} = 120.2$ and $P_2^{sat} = 73.9$ kPa
      For $x_1 = 0.389$, $P = 108.6$ kPa

13.65. The excess Gibbs energy for the system chloroform(1)/ethanol(2) at 55°C is well represented by the Margules equation, $G^E/RT = (1.42\, x_1 + 0.59\, x_2)x_1 x_2$. The vapor pressures of chloroform and ethanol at 55°C are $P_1^{sat} = 82.37$ and $P_2^{sat} = 37.31$ kPa.

(*a*) Assuming the validity of Eq. (13.19), make *BUBL P* calculations at 55°C for liquid-phase mole fractions of 0.25, 0.50, and 0.75.

(*b*) For comparison, repeat the calculations using Eqs. (13.13) and (13.14) with virial coefficients: $B_{11} = -963$, $B_{22} = -1523$, and $B_{12} = 52$ cm^3·mol^{-1}.

13.66. Find expressions for $\hat{\phi}_1$ and $\hat{\phi}_2$ for a binary gas mixture described by Eq. (3.38). The mixing rule for B is given by Eq. (10.62). The mixing rule for C is given by the general equation:

$$C = \sum_i \sum_j \sum_k y_i y_j y_k C_{ijk}$$

where Cs with the same subscripts, regardless of order, are equal. For a binary mixture, this becomes:

$$C = y_1^3 C_{111} + 3 y_1^2 y_2 C_{112} + 3 y_1 y_2^2 C_{122} + y_2^3 C_{222}$$

13.67. A system formed of methane(1) and a light oil(2) at 200 K and 30 bar consists of a vapor phase containing 95 mol-% methane and a liquid phase containing oil and dissolved methane. The fugacity of the methane is given by Henry's law, and at the temperature of interest Henry's constant is $H_1 = 200$ bar. Stating any assumptions, estimate the equilibrium mole fraction of methane in the liquid phase. The second virial coefficient of pure methane at 200 K is -105 cm^3·mol^{-1}.

13.68. Use Eq. (13.13) to reduce one of the following isothermal data sets, and compare the result with that obtained by application of Eq. (13.19). Recall that reduction means developing a numerical expression for G^E/RT as a function of composition.

(*a*) Methylethylketone(1)/toluene(2) at 50°C: Table 13.1.
(*b*) Acetone(1)/methanol(2) at 55°C: Prob. 13.34.
(*c*) Methyl *tert*-butyl ether(1)/dichloromethane(2) at 35°C: Prob. 13.37.
(*d*) Acetonitrile(1)/benzene(2) at 45°C: Prob. 13.40.

Second-virial-coefficient data are as follows:

	Part (*a*)	Part (*b*)	Part (*c*)	Part (*d*)
B_{11}/cm^3·mol^{-1}	-1840	-1440	-2060	-4500
B_{12}/cm^3·mol^{-1}	-1800	-1150	-860	-1300
B_{22}/cm^3·mol^{-1}	-1150	-1040	-790	-1000

13.69. For one of the following substances, determine P^{sat}/bar from the Redlich/Kwong equation at two temperatures: $T = T_n$ (the normal boiling point), and $T = 0.85T_c$. For the second temperature, compare your result with a value from the literature (e.g., Perry's Chemical Engineers' Handbook). Discuss your results.

(*a*) Acetylene; (*b*) Argon; (*c*) Benzene; (*d*) *n*-Butane; (*e*) Carbon monoxide;
(*f*) *n*-Decane; (*g*) Ethylene; (*h*) *n*-Heptane; (*i*) Methane; (*j*) Nitrogen

13.70. Work Prob. 13.69 for one of the following: (a) The Soave/Redlich/Kwong equation; (b) the Peng/Robinson equation.

13.71. Departures from Raoult's law are primarily from liquid-phase nonidealities ($\gamma_i \neq 1$). But vapor-phase nonidealities ($\hat{\phi}_i \neq 1$) also contribute. Consider the special case where the liquid phase is an ideal solution, and the vapor phase a nonideal gas mixture described by Eq. (3.36). Show that departures from Raoult's law at constant temperature are likely to be negative. State clearly any assumptions and approximations.

13.72. Determine a numerical value for the acentric factor ω implied by:
   (a) The van der Waals equation;
   (b) The Redlich/Kwong equation.

13.73. The relative volatility α_{12} is commonly used in applications involving binary VLE. In particular (see Ex. 13.1), it serves as a basis for assessing the possibility of binary azeotropy. (a) Develop an expression for α_{12} based on Eqs. (13.13) and (13.14). (b) Specialize the expression to the composition limits $x_1 = y_1 = 0$ and $x_1 = y_1 = 1$. Compare with the result obtained from modified Raoult's law, Eq. (13.19). The difference between the results reflects the effects of vapor-phase nonidealities. (c) Further specialize the results of part (b) to the case where the vapor phase is an ideal solution of real gases.

13.74. Although isothermal VLE data are preferred for extraction of activity coefficients, a large body of good isobaric data exists in the literature. For a binary isobaric T-x_1-y_1 data set, one can extract point values of γ_i via Eq. (13.13):

$$\gamma_i(x, T_k) = \frac{y_i \Phi_i(T_k, P, y) P}{x_i P_i^{\text{sat}}(T_k)}$$

Here, the variable list for γ_i recognizes a primary dependence on x and T; pressure dependence is normally negligible. The notation T_k emphasizes that temperature varies with data point across the composition range, and the calculated activity coefficients are at different temperatures. However, the usual goal of VLE data reduction and correlation is to develop an appropriate expression for G^E/RT at a single temperature T. A procedure is needed to correct each activity coefficient to such a T chosen near the average for the data set. If a correlation for $H^E(x)$ is available at or near this T, show that the values of γ_i corrected to T can be estimated by the expression:

$$\gamma_i(x, T) = \gamma_i(x, T_k) \exp\left[\frac{-\bar{H}_i^E}{RT}\left(\frac{T}{T_k} - 1\right)\right]$$

13.75. What are the relative contributions of the various terms in the gamma/phi expression for VLE? One way to address the question is through calculation of the activity coefficients for a single binary VLE data point via Eq. (13.19):

$$\gamma_i = \underbrace{\frac{y_i P}{x_i P_i^{\text{sat}}}}_{(A)} \cdot \underbrace{\frac{\hat{\phi}_i}{\phi_i^{\text{sat}}}}_{(B)} \cdot \underbrace{\frac{f_i^{\text{sat}}}{f_i}}_{(C)}$$

Term (A) is the value that would follow from modified Raoult's law; term (B) accounts for vapor-phase nonidealities; term (C) is the Poynting factor [see Eq. (10.44)]. Use the following single-point data for the butanenitrile(1)/benzene(2) system at 318.15 K to evaluate all terms for $i = 1$ and $i = 2$. Discuss the results.

VLE data: $P = 0.20941$ bar, $x_1 = 0.4819$, $y_1 = 0.1813$.

Ancillary data: $P_1^{\text{sat}} = 0.07287$ and $P_2^{\text{sat}} = 0.29871$ bar

$$B_{11} = -7993, \ B_{22} = -1247, \ B_{12} = -2089 \text{ cm}^3 \cdot \text{mol}^{-1}$$
$$V_1^l = 90, \ V_2^l = 92 \text{ cm}^3 \cdot \text{mol}^{-1}$$

13.76. Generate P-x_1-y_1 diagrams at 100°C for one of the systems identified below. Base activity coefficients on the Wilson equation, Eqs. (13.45) to (13.47). Use two procedures: (i) modified Raoult's law, Eq. (13.19), and (ii) the gamma/phi approach, Eq. (13.13), with Φ_i given by Eq. (13.14). Plot the results for both procedures on the same graph. Compare and discuss them.

Data sources: For P_i^{sat} use Table B.2. For vapor-phase nonidealities, use material from Chap. 3; assume that the vapor phase is an (approximately) ideal solution. Estimated parameters for the Wilson equation are given for each system.

(*a*) Benzene(1)/carbon tetrachloride(2): $\Lambda_{12} = 1.0372, \Lambda_{21} = 0.8637$
(*b*) Benzene(1)/cyclohexane(2): $\Lambda_{12} = 1.0773, \Lambda_{21} = 0.7100$
(*c*) Benzene(1)/*n*-heptane(2): $\Lambda_{12} = 1.2908, \Lambda_{21} = 0.5011$
(*d*) Benzene(1)/*n*-hexane(2): $\Lambda_{12} = 1.3684, \Lambda_{21} = 0.4530$
(*e*) Carbon tetrachloride(1)/cyclohexane(2): $\Lambda_{12} = 1.1619, \Lambda_{21} = 0.7757$
(*f*) Carbon tetrachloride(1)/*n*-heptane(2): $\Lambda_{12} = 1.5410, \Lambda_{21} = 0.5197$
(*g*) Carbon tetrachloride(1)/*n*-hexane(2): $\Lambda_{12} = 1.2839, \Lambda_{21} = 0.6011$
(*h*) Cyclohexane(1)/*n*-heptane(2): $\Lambda_{12} = 1.2996, \Lambda_{21} = 0.7046$
(*i*) Cyclohexane(1)/*n*-hexane(2): $\Lambda_{12} = 1.4187, \Lambda_{21} = 0.5901$

Chapter 14

Chemical-Reaction Equilibria

The transformation of raw materials into products of greater value by means of chemical reaction is a major industry, and a vast array of commercial products is obtained by chemical synthesis. Sulfuric acid, ammonia, ethylene, propylene, phosphoric acid, chlorine, nitric acid, urea, benzene, methanol, ethanol, and ethylene glycol are examples of chemicals produced at scales of many billions of kg per year worldwide. These in turn are used in the large-scale manufacture of fibers, paints, detergents, plastics, rubber, paper, and fertilizers, to name a few. Other products produced by chemical reaction range from pharmaceuticals to the array of inorganic materials that undergird the microelectronics and telecommunications industries. While these are smaller in production volume than the commodity chemicals we have mentioned, their economic and societal impacts are also enormous. Clearly, the chemical engineer must be familiar with chemical reactor design, analysis, and operation.

Both the rate and the equilibrium conversion of a chemical reaction depend on the temperature, pressure, and composition of reactants. Often, a reasonable reaction rate is achieved only with a suitable catalyst. For example, the rate of oxidation of sulfur dioxide to sulfur trioxide, carried out with a vanadium pentoxide catalyst, becomes appreciable at about 300°C and increases at higher temperatures. On the basis of rate alone, one would operate the reactor at the highest practical temperature. However, the equilibrium conversion of sulfur dioxide to sulfur trioxide falls as temperature rises, decreasing from about 90% at 520°C to about 50% at 680°C. These values represent maximum possible conversions regardless of catalyst or reaction rate. It is clear that both equilibrium and rate must be considered in the exploitation of chemical reactions for commercial purposes. Although reaction *rates* are not susceptible to thermodynamic treatment, reaction *equilibria* are. Therefore, the central goal of this chapter is to relate the equilibrium composition of reacting systems to their temperature, pressure, and initial composition.

Many industrial reactions are not carried to equilibrium; reactor design is often based on reaction rate or on other considerations such as rates of heat and mass transfer. However, the choice of operating conditions may still be influenced by equilibrium considerations. Moreover, the equilibrium state provides reference against which to measure improvements in a process. Similarly, chemical equilibrium considerations often determine whether investigation of a new process is worthwhile. For example, if thermodynamic analysis indicates that a yield of only 20% is possible at equilibrium and if a 50% yield is necessary for a process to be economically attractive, there is no purpose in proceeding with further analysis. On the other hand, if the equilibrium yield is 80%, further studies of reaction rates and other aspects of the process may be warranted.

Reaction stoichiometry is treated in Sec. 14.1, where we relate composition of a reacting mixture to a single reaction coordinate variable for each chemical reaction that occurs. Criteria for chemical reaction equilibrium are then introduced in Sec. 14.2. The equilibrium constant is defined in Sec. 14.3, and its temperature dependence and evaluation are considered in Secs. 14.4 and 14.5. The relationship between the equilibrium constant and composition is developed in Sec. 14.6. The calculation of equilibrium conversions for single reactions is taken up in Sec. 14.7. In Sec. 14.8, the phase rule is reconsidered in the context of reacting systems. Multiple reaction equilibrium is treated in Sec. 14.9.[1] Finally, in Sec. 14.10 the fuel cell is given an introductory treatment.

14.1 THE REACTION COORDINATE

The general chemical reaction, as written in Sec. 4.6 is:

$$|\nu_1|A_1 + |\nu_2|A_2 + \cdots \rightarrow |\nu_3|A_3 + |\nu_4|A_4 + \cdots \tag{14.1}$$

where $|\nu_i|$ is a stoichiometric coefficient and A_i represents a chemical formula. The symbol ν_i itself is called a stoichiometric *number*, and by the sign convention of Sec. 4.6 it is:

positive (+) for a product   and   negative (−) for a reactant

Thus for the reaction, $CH_4 + H_2O \rightarrow CO + 3H_2$

the stoichiometric numbers are:

$$\nu_{CH_4} = -1 \qquad \nu_{H_2O} = -1 \qquad \nu_{CO} = 1 \qquad \nu_{H_2} = 3$$

The stoichiometric number for a species that does not participate in the reaction, that is, an inert species, is zero.

As the reaction represented by Eq. (14.1) progresses, the *changes* in the numbers of moles of species present are in direct proportion to the stoichiometric numbers. Thus for the preceding reaction, if 0.5 mol of CH_4 disappears by reaction, 0.5 mol of H_2O also disappears; simultaneously 0.5 mol of CO and 1.5 mol of H_2 are formed. Applied to a differential amount of reaction, this principle provides the equations:

$$\frac{dn_2}{\nu_2} = \frac{dn_1}{\nu_1} \qquad \frac{dn_3}{\nu_3} = \frac{dn_1}{\nu_1} \qquad \text{etc.}$$

The list continues to include all species. Comparison of these equations yields:

$$\frac{dn_1}{\nu_1} = \frac{dn_2}{\nu_2} = \frac{dn_3}{\nu_3} = \frac{dn_4}{\nu_4} = \cdots$$

---

[1]For a comprehensive treatment of chemical-reaction equilibria, see W. R. Smith and R. W. Missen, *Chemical Reaction Equilibrium Analysis*, John Wiley & Sons, New York, 1982.

All terms being equal, they can be identified collectively by a single quantity representing an amount of reaction. Thus $d\varepsilon$, a single variable representing the extent to which the reaction has proceeded, is **defined** by the equation:

$$\frac{dn_1}{\nu_1} = \frac{dn_2}{\nu_2} = \frac{dn_3}{\nu_3} = \frac{dn_4}{\nu_4} = \cdots \equiv d\varepsilon \qquad (14.2)$$

The general relation connecting the differential change dn_i with $d\varepsilon$ is therefore:

$$dn_i = \nu_i \, d\varepsilon \qquad (i = 1, 2, \ldots, N) \qquad (14.3)$$

This new variable ε, called the *reaction coordinate*, characterizes the extent or degree to which a reaction has taken place.[2] Only *changes* in ε with respect to changes in a mole number are defined by Eq. (14.3). The definition of ε itself depends for a specific application on setting it equal to *zero* for the initial state of the system prior to reaction. Thus, integration of Eq. (14.3) from an initial unreacted state where $\varepsilon = 0$ and $n_i = n_{i_0}$ to a state reached after an arbitrary amount of reaction gives:

$$\int_{n_{i_0}}^{n_i} dn_i = \nu_i \int_0^\varepsilon d\varepsilon$$

or $\qquad\qquad\qquad n_i = n_{i_0} + \nu_i \varepsilon \qquad (i = 1, 2, \ldots, N) \qquad (14.4)$

Summation over all species yields:

$$n = \sum_i n_i = \sum_i n_{i_0} + \varepsilon \sum_i \nu_i$$

or $\qquad\qquad\qquad\qquad n = n_0 + \nu\varepsilon$

where $\qquad\qquad n \equiv \sum_i n_i \qquad\qquad n_0 \equiv \sum_i n_{i_0} \qquad\qquad \nu \equiv \sum_i \nu_i$

Thus the mole fractions y_i of the species present are related to ε by:

$$y_i = \frac{n_i}{n} = \frac{n_{i_0} + \nu_i \varepsilon}{n_0 + \nu\varepsilon} \qquad (14.5)$$

Application of this equation is illustrated in the following examples.

Example 14.1

The following reaction occurs in a system initially consisting of 2 mol CH_4, 1 mol H_2O, 1 mol CO, and 4 mol H_2:

$$CH_4 + H_2O \rightarrow CO + 3H_2$$

Determine expressions for the mole fractions y_i as functions of ε.

---

[2]The reaction coordinate ε has been given various names, including degree of advancement, degree of reaction, extent of reaction, and progress variable.

Solution 14.1

For the reaction, $\quad \nu = \sum_i \nu_i = -1 - 1 + 1 + 3 = 2$

For the given numbers of moles of species initially present,

$$n_0 = \sum_i n_{i_0} = 2 + 1 + 1 + 4 = 8$$

Equation (14.5) now yields:

$$y_{CH_4} = \frac{2 - \varepsilon}{8 + 2\varepsilon} \qquad y_{H_2O} = \frac{1 - \varepsilon}{8 + 2\varepsilon}$$

$$y_{CO} = \frac{1 + \varepsilon}{8 + 2\varepsilon} \qquad y_{H_2} = \frac{4 + 3\varepsilon}{8 + 2\varepsilon}$$

Example 14.2

Consider a vessel that initially contains only n_0 mol of water vapor. If decomposition occurs according to the reaction

$$H_2O \rightarrow H_2 + \tfrac{1}{2}O_2$$

find expressions that relate the number of moles and the mole fraction of each chemical species to the reaction coordinate ε.

Solution 14.2

For the given reaction, $\nu = -1 + 1 + \frac{1}{2} = \frac{1}{2}$. Application of Eqs. (14.4) and (14.5) yields:

$$n_{H_2O} = n_0 - \varepsilon \qquad y_{H_2O} = \frac{n_0 - \varepsilon}{n_0 + \frac{1}{2}\varepsilon}$$

$$n_{H_2} = \varepsilon \qquad y_{H_2} = \frac{\varepsilon}{n_0 + \frac{1}{2}\varepsilon}$$

$$n_{O_2} = \tfrac{1}{2}\varepsilon \qquad y_{O_2} = \frac{\frac{1}{2}\varepsilon}{n_0 + \frac{1}{2}\varepsilon}$$

The fractional decomposition of water vapor is:

$$\frac{n_0 - n_{H_2O}}{n_0} = \frac{n_0 - (n_0 - \varepsilon)}{n_0} = \frac{\varepsilon}{n_0}$$

Thus when $n_0 = 1$, ε is directly related to the fractional decomposition of the water vapor.

The ν_i are pure numbers without units; Eq. (14.3) therefore requires ε to be expressed in moles. This leads to the concept of a *mole of reaction*, meaning a change in ε of one mole. When $\Delta\varepsilon = 1$ mol, the reaction proceeds to such an extent that the change in mole number of each reactant and product is equal to its stoichiometric number.

Multireaction Stoichiometry

When two or more independent reactions proceed simultaneously, a second subscript, here indicated by j, serves as the reaction index. A separate reaction coordinate ε_j then applies to each reaction. The stoichiometric numbers are doubly subscripted to identify their association with both a species and a reaction. Thus $\nu_{i,j}$ designates the stoichiometric number of species i in reaction j. Because the number of moles of a species n_i may change because of several reactions, the general equation analogous to Eq. (14.3) includes a sum:

$$dn_i = \sum_j \nu_{i,j} d\varepsilon_j \qquad (i = 1, 2, \ldots, N)$$

Integration from $n_i = n_{i_0}$ and $\varepsilon_j = 0$ to arbitrary n_i and ε_j gives:

$$n_i = n_{i_0} + \sum_j \nu_{i,j} \varepsilon_j \qquad (i = 1, 2, \ldots, N) \qquad (14.6)$$

Summing over all species yields:

$$n = \sum_i n_{i_0} + \sum_i \sum_j \nu_{i,j} \varepsilon_j = n_0 + \sum_j \left(\sum_i \nu_{i,j} \right) \varepsilon_j$$

The definition of a total stoichiometric number ν ($\equiv \sum_i \nu_i$) for a single reaction has its counterpart here in the definition:

$$\nu_j \equiv \sum_i \nu_{i,j} \qquad \text{whence} \qquad n = n_0 + \sum_j \nu_j \varepsilon_j$$

Combination of this last equation with Eq. (14.6) gives the mole fraction:

$$\boxed{y_i = \frac{n_{i_0} + \sum_j \nu_{i,j} \varepsilon_j}{n_0 + \sum_j \nu_j \varepsilon_j} \qquad (i = 1, 2, \ldots, N)} \qquad (14.7)$$

Example 14.3

Consider a system in which the following reactions occur:

$$CH_4 + H_2O \rightarrow CO + 3H_2 \qquad (1)$$

$$CH_4 + 2H_2O \rightarrow CO_2 + 4H_2 \qquad (2)$$

where the numbers (1) and (2) indicate the value of j, the reaction index. If 2 mol CH_4 and 3 mol H_2O are initially present, determine expressions for the y_i as functions of ε_1 and ε_2.

Solution 14.3

The stoichiometric numbers $\nu_{i,j}$ can be arrayed as follows:

$i =$	CH_4	H_2O	CO	CO_2	H_2	
j						ν_j
1	-1	-1	1	0	3	2
2	-1	-2	0	1	4	2

Application of Eq. (14.7) now gives:

$$y_{CH_4} = \frac{2 - \varepsilon_1 - \varepsilon_2}{5 + 2\varepsilon_1 + 2\varepsilon_2} \qquad\qquad y_{CO} = \frac{\varepsilon_1}{5 + 2\varepsilon_1 + 2\varepsilon_2}$$

$$y_{H_2O} = \frac{3 - \varepsilon_1 - 2\varepsilon_2}{5 + 2\varepsilon_1 + 2\varepsilon_2} \qquad\qquad y_{CO_2} = \frac{\varepsilon_2}{5 + 2\varepsilon_1 + 2\varepsilon_2}$$

$$y_{H_2} = \frac{3\varepsilon_1 + 4\varepsilon_2}{5 + 2\varepsilon_1 + 2\varepsilon_2}$$

The composition of the system is a function of independent variables ε_1 and ε_2.

14.2 APPLICATION OF EQUILIBRIUM CRITERIA TO CHEMICAL REACTIONS

In Sec. 12.4 we proved that the total Gibbs energy of a closed system at constant T and P must decrease during an irreversible process and that the condition for equilibrium is reached when G^t attains its minimum value. At this equilibrium state,

$$(dG^t)_{T,P} = 0 \tag{12.3}$$

Thus if a mixture of chemical species is not in chemical equilibrium, any reaction that occurs at constant T and P must decrease the total Gibbs energy of the system. The significance of this for a single chemical reaction is illustrated in Fig. 14.1, which shows a schematic diagram of G^t vs. ε, the reaction coordinate. Because ε is the single variable that characterizes the progress of the reaction, and therefore the composition of the system, the total Gibbs energy at constant T and P is determined by ε. The arrows along the curve in Fig. 14.1 indicate the directions of changes in $(G^t)_{T,P}$ that are possible on account of reaction. The reaction coordinate has its equilibrium value ε_e at the minimum of the curve. The meaning of Eq. (12.3) is that differential displacements of the chemical reaction can occur at the equilibrium state without causing changes in the total Gibbs energy of the system.

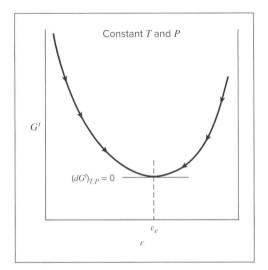

Figure 14.1: Schematic plot of the total Gibbs energy as a function of the reaction coordinate.

Figure 14.1 indicates the two distinctive features of the equilibrium state for given temperature and pressure:

- The total Gibbs energy G^t is a minimum.   • Its differential is zero.

Each of these may serve as a criterion of equilibrium. Thus, we may write an expression for G^t as a function of ε and seek the value of ε that minimizes G^t, or we may differentiate the expression, equate it to zero, and solve for ε. The latter procedure is almost always used for single reactions (Fig. 14.1), and it leads to the method of equilibrium constants, as described in the following sections. Although the method of equilibrium constants can also be applied to multiple reactions, direct minimization of G^t is often more convenient. The direct minimization approach is considered in Sec. 14.9.

Although the equilibrium expressions are *developed* for closed systems at constant T and P, they are not restricted in *application* to systems that are actually closed and reach equilibrium states along paths of constant T and P. Once an equilibrium state is reached, no further changes occur, and the system continues to exist in this state at fixed T and P. How this state was actually attained does not matter. Once it is known that an equilibrium state exists at given T and P, the criteria apply.

14.3 THE STANDARD GIBBS-ENERGY CHANGE AND THE EQUILIBRIUM CONSTANT

The next step in applying the criterion of equilibrium to chemical reactions is to relate the Gibbs energy, which is minimized at equilibrium, to the reaction coordinate. Equation (10.2), the fundamental property relation for single-phase systems, provides an expression for the total differential of the Gibbs energy:

$$d(nG) = (nV)dP - (nS)dT + \sum_i \mu_i \, dn_i \tag{10.2}$$

If changes in the mole numbers n_i occur as the result of a single chemical reaction in a closed system, then by Eq. (14.3) each dn_i may be replaced by the product $\nu_i\, d\varepsilon$. Equation (10.2) then becomes:

$$d(nG) = (nV)dP - (nS)dT + \sum_i \nu_i \mu_i\, d\varepsilon$$

Because nG is a state function, the right side of this equation is an exact differential expression; whence,

$$\sum_i \nu_i \mu_i = \left[\frac{\partial(nG)}{\partial\varepsilon}\right]_{T,P} = \left[\frac{\partial(G^t)}{\partial\varepsilon}\right]_{T,P}$$

Thus the quantity $\sum_i \nu_i \mu_i$ represents, in general, the rate of change of total Gibbs energy of the system with respect to the reaction coordinate at constant T and P. Figure 14.1 shows that this quantity is zero at the equilibrium state. A criterion of chemical-reaction equilibrium is therefore:

$$\boxed{\sum_i \nu_i \mu_i = 0} \tag{14.8}$$

Recall the definition of the fugacity of a species in solution:

$$\mu_i = \Gamma_i(T) + RT \ln \hat{f}_i \tag{10.46}$$

In addition, Eq. (10.31) may be written for pure species i in its *standard state*[3] at the same temperature:

$$G_i^\circ = \Gamma_i(T) + RT \ln f_i^\circ$$

The difference between these two equations is:

$$\mu_i - G_i^\circ = RT \ln \frac{\hat{f}_i}{f_i^\circ} \tag{14.9}$$

Combining Eq. (14.8) with Eq. (14.9) to eliminate μ_i gives for the equilibrium state of a chemical reaction:

$$\sum_i \nu_i [G_i^\circ + RT \ln(\hat{f}_i/f_i^\circ)] = 0$$

or

$$\sum_i \nu_i G_i^\circ + RT \sum_i \ln(\hat{f}_i/f_i^\circ)^{\nu_i} = 0$$

or

$$\ln \prod_i (\hat{f}_i/f_i^\circ)^{\nu_i} = \frac{-\sum_i \nu_i G_i^\circ}{RT}$$

where $\prod_i$ signifies the product over all species i. In exponential form, this equation becomes:

$$\boxed{\prod_i (\hat{f}_i/f_i^\circ)^{\nu_i} = K} \tag{14.10}$$

where the **definition** of K and its logarithm are given by:

---

[3]Standard states are introduced and discussed in Sec. 4.3.

$$K \equiv \exp\left(\frac{-\Delta G°}{RT}\right) \quad (14.11a) \qquad \ln K = \frac{-\Delta G°}{RT} \quad (14.11b)$$

Also by **definition**,
$$\Delta G° \equiv \sum_i \nu_i G_i° \tag{14.12}$$

Because $G_i°$ is a property of pure species i in its standard state at fixed pressure, it depends only on temperature. Thus, Eq. (14.12) implies that $\Delta G°$, and hence K, are also functions only of temperature.

> **Despite its dependence on temperature, K is called the equilibrium constant for the reaction; $\Delta G°$ is called *the standard Gibbs-energy change of reaction.***

The fugacity ratios in Eq. (14.10) provide the connection between the *equilibrium* state of interest and the *standard* states of the individual species, for which data are presumed to be available, as discussed in Sec. 14.5. The standard states are arbitrary, but they must always be at the equilibrium temperature T. The standard states selected need not be the same for all species taking part in a reaction. However, for a particular species the standard state represented by $G_i°$ must be the same state as for the fugacity of that species, $f_i°$.

The function $\Delta G° \equiv \sum_i \nu_i G_i°$ in Eq. (14.12) is the difference between the Gibbs energies of the products and reactants (weighted by their stoichiometric coefficients) when each is in its standard state as a pure substance at the standard-state pressure, but at the *system* temperature. Thus the value of $\Delta G°$ is fixed for a given reaction once the temperature is established, and it is independent of pressure and composition. Other *standard property changes of reaction* are similarly defined. Thus, for the general property M:

$$\Delta M° \equiv \sum_i \nu_i M_i°$$

In accord with this, $\Delta H°$ is defined by Eq. (4.15) and $\Delta C_P°$ by Eq. (4.17). These quantities are functions only of temperature for a given reaction, and they are related to one another by equations analogous to property relations for pure species.

For example, the relation between the standard heat of reaction and the standard Gibbs-energy change of reaction may be developed from Eq. (6.39) written for species i in its standard state:

$$H_i° = -RT^2 \frac{d(G_i°/RT)}{dT}$$

Total derivatives are appropriate here because the properties in the standard state are functions of temperature only. Multiplication of both sides of this equation by ν_i and summation over all species gives:

$$\sum_i \nu_i H_i° = -RT^2 \frac{d\left(\sum_i \nu_i G_i°/RT\right)}{dT}$$

In view of the definitions of Eqs. (4.15) and (14.12), this may be written:

$$\Delta H° = -RT^2 \frac{d(\Delta G°/RT)}{dT} \tag{14.13}$$

14.4 EFFECT OF TEMPERATURE ON THE EQUILIBRIUM CONSTANT

Because the standard-state temperature is that of the equilibrium mixture, the standard property changes of reaction, such as $\Delta G°$ and $\Delta H°$, vary with the equilibrium temperature. The dependence of $\Delta G°$ on T is given by Eq. (14.13), which may be rewritten:

$$\frac{d(\Delta G°/RT)}{dT} = \frac{-\Delta H°}{RT^2}$$

In view of Eq. (14.11b), this becomes:

$$\boxed{\frac{d\ln K}{dT} = \frac{\Delta H°}{RT^2}} \tag{14.14}$$

Equation (14.14) describes the effect of temperature on the equilibrium constant, and hence on the equilibrium conversion. If $\Delta H°$ is negative, i.e., if the reaction is exothermic, the equilibrium constant decreases as the temperature increases. Conversely, K increases with T for an endothermic reaction.

If $\Delta H°$, the standard enthalpy change (heat) of reaction, is assumed to be independent of T, integration of Eq. (14.14) from a particular temperature T' to an arbitrary temperature T leads to the simple result:

$$\ln\frac{K}{K'} = -\frac{\Delta H°}{R}\left(\frac{1}{T} - \frac{1}{T'}\right) \tag{14.15}$$

This approximate equation implies that a plot of $\ln K$ vs. the reciprocal of absolute temperature is a straight line. Figure 14.2, a plot of $\ln K$ vs. $1/T$ for a number of common reactions, illustrates this near linearity. Thus, Eq. (14.15) provides a reasonably accurate relation for the interpolation and extrapolation of equilibrium-constant data.

The *rigorous* development of the effect of temperature on the equilibrium constant is based on the definition of the Gibbs energy, written for a chemical species in its standard state:

$$G_i° = H_i° - TS_i°$$

Multiplication by ν_i and summation over all species gives:

$$\sum_i \nu_i G_i° = \sum_i \nu_i H_i° - T\sum_i \nu_i S_i°$$

Employing the definition of a standard property change of reaction, this is written as:

$$\Delta G° = \Delta H° - T\Delta S° \tag{14.16}$$

The standard heat of reaction is related to temperature:

$$\Delta H° = \Delta H_0° + R\int_{T_0}^{T} \frac{\Delta C_P°}{R}\,dT \tag{4.19}$$

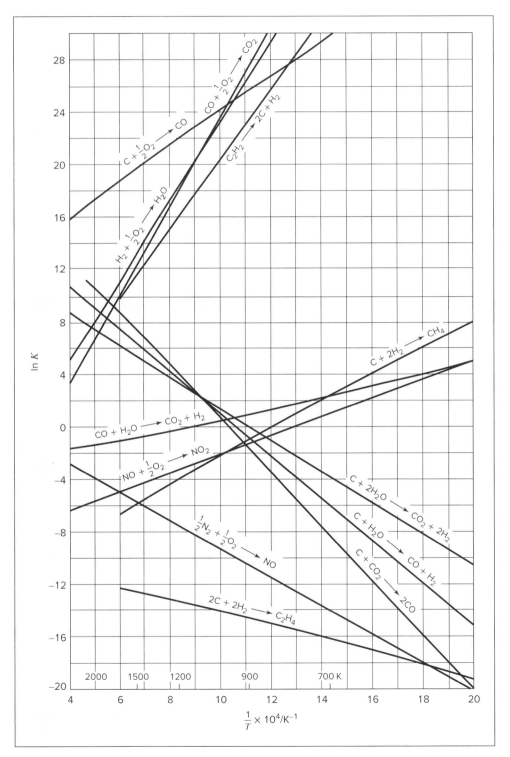

Figure 14.2: Equilibrium constants as a function of temperature.

The temperature dependence of the standard entropy change of reaction is developed similarly. Equation (6.22) is written for the standard-state entropy of species i at the constant standard-state pressure P°:

$$dS_i^\circ = C_{P_i}^\circ \frac{dT}{T}$$

Multiplying by ν_i, summing over all species, and invoking the definition of a standard property change of reaction yields:

$$d\Delta S^\circ = \Delta C_P^\circ \frac{dT}{T}$$

Integration gives:

$$\Delta S^\circ = \Delta S_0^\circ + R \int_{T_0}^{T} \frac{\Delta C_P^\circ}{R} \frac{dT}{T} \tag{14.17}$$

where ΔS° and ΔS_0° are standard entropy changes of reaction at temperature T and at reference temperature T_0, respectively. Equations (14.16), (4.19), and (14.17) are combined to yield:

$$\Delta G^\circ = \Delta H_0^\circ + R \int_{T_0}^{T} \frac{\Delta C_P^\circ}{R} dT - T\Delta S_0^\circ - RT \int_{T_0}^{T} \frac{\Delta C_P^\circ}{R} \frac{dT}{T}$$

We can eliminate ΔS_0° from this equation through the relationship:

$$\Delta S_0^\circ = \frac{\Delta H_0^\circ - \Delta G_0^\circ}{T_0}$$

Doing so produces the following:

$$\Delta G^\circ = \Delta H_0^\circ - \frac{T}{T_0}(\Delta H_0^\circ - \Delta G_0^\circ) + R \int_{T_0}^{T} \frac{\Delta C_P^\circ}{R} dT - RT \int_{T_0}^{T} \frac{\Delta C_P^\circ}{R} \frac{dT}{T}$$

Finally, division by RT yields:

$$\frac{\Delta G^\circ}{RT} = \frac{\Delta G_0^\circ - \Delta H_0^\circ}{RT_0} + \frac{\Delta H_0^\circ}{RT} + \frac{1}{T}\int_{T_0}^{T} \frac{\Delta C_P^\circ}{R} dT - \int_{T_0}^{T} \frac{\Delta C_P^\circ}{R} \frac{dT}{T} \tag{14.18}$$

Recall that by Eq. (14.11b), $\ln K = -\Delta G^\circ / RT$.

When the temperature dependence of the heat capacity of each species is given by Eq. (4.4), the first integral on the right side of Eq. (14.18) is given by Eq. (4.19), represented for computational purposes by:

$$\int_{T_0}^{T} \frac{\Delta C_P^\circ}{R} dT = \text{IDCPH(T0, T; DA, DB, DC, DD)}$$

where D denotes Δ. Similarly, the second integral is given by the analog of Eq. (5.11):

$$\int_{T_0}^{T} \frac{\Delta C_P^\circ}{R} \frac{dT}{T} = \Delta A \ln \frac{T}{T_0} + \left[\Delta B + \left(\Delta C + \frac{\Delta D}{T_0^2 T^2} \right) \left(\frac{T + T_0}{2} \right) \right] (T - T_0) \tag{14.19}$$

The integral is evaluated by a function of exactly the same form as given by Eq. (5.11), and the same computer program therefore serves for evaluation of either integral. The only difference is in the name of the function, here: IDCPS(T0,T;DA,DB,DC,DD). By definition,

$$\int_{T_0}^{T} \frac{\Delta C_P^\circ}{R} \frac{dT}{T} = \text{IDCPS(T0, T; DA, DB, DC, DD)}$$

Thus $\Delta G^\circ / RT$ $(= -\ln K)$ as given by Eq. (14.18) is readily calculated at any temperature from the standard heat of reaction and the standard Gibbs-energy change of reaction at a reference temperature (usually 298.15 K), and from two functions which can be evaluated by standard computational procedures.

The preceding equations may be reorganized so as to factor K into three terms, each representing a basic contribution to its value:

$$K = K_0 K_1 K_2 \tag{14.20}$$

The first factor K_0 represents the equilibrium constant at reference temperature T_0:

$$K_0 \equiv \exp\left(\frac{-\Delta G_0^\circ}{RT_0}\right) \tag{14.21}$$

The second factor K_1 is a multiplier that accounts for the major effect of temperature, such that the product $K_0 K_1$ is the equilibrium constant at temperature T when the heat of reaction is assumed to be independent of temperature:

$$K_1 \equiv \exp\left[\frac{\Delta H_0^\circ}{RT_0}\left(1 - \frac{T_0}{T}\right)\right] \tag{14.22}$$

The third factor K_2 accounts for the much smaller temperature influence resulting from the change of ΔH° with temperature:

$$K_2 \equiv \exp\left(-\frac{1}{T}\int_{T_0}^{T} \frac{\Delta C_P^\circ}{R} dT + \int_{T_0}^{T} \frac{\Delta C_P^\circ}{R} \frac{dT}{T}\right) \tag{14.23}$$

With heat capacities given by Eq. (4.4), the expression for K_2 can be simplified to:

$$K_2 = \exp\left\{ \Delta A\left[\ln\frac{T}{T_0} - \frac{T - T_0}{T}\right] + \frac{1}{2}\Delta B \frac{(T - T_0)^2}{T} \right.$$
$$\left. + \frac{1}{6}\Delta C \frac{(T - T_0)^2 (T + 2T_0)}{T} + \frac{1}{2}\Delta D \frac{(T - T_0)^2}{T^2 T_0^2} \right\} \tag{14.24}$$

14.5 EVALUATION OF EQUILIBRIUM CONSTANTS

Values of ΔG° for many *formation reactions* are tabulated in standard references.[4] The reported values of ΔG_f° are not measured experimentally but are calculated by Eq. (14.16).

---

[4]For example, "TRC Thermodynamic Tables–Hydrocarbons" and "TRC Thermodynamic Tables–Non-hydrocarbons," serial publications of the Thermodynamics Research Center, Texas A & M Univ. System, College Station, Texas; "The NBS Tables of Chemical Thermodynamic Properties," *J. Physical and Chemical Reference Data*, vol. 11, supp. 2, 1982.

The determination of ΔS_f° may be based on the third law of thermodynamics, discussed in Sec. 5.9. Combination of values from Eq. (5.35) for the absolute entropies of the species taking part in the reaction gives the value of ΔS_f°. Entropies (and heat capacities) are also commonly determined by applying statistical mechanics to molecular structure data obtained from spectroscopic measurements or from computational quantum chemistry methods.[5]

Values of $\Delta G_{f_{298}}^\circ$ for a limited number of chemical compounds are listed in Table C.4 of App. C. These are for a temperature of 298.15 K, as are the values of $\Delta H_{f_{298}}^\circ$ listed in the same table. Values of ΔG° for other reactions are calculated from formation-reaction values in exactly the same way that ΔH° values for other reactions are determined from formation-reaction values (Sec. 4.4). In more extensive compilations of data, values of ΔG_f° and ΔH_f° are given for a range of temperatures, rather than only at 298.15 K. Where data are lacking, methods of estimation are available; these have been reviewed by Poling, Prausnitz, and O'Connell.[6]

Example 14.4

Calculate the equilibrium constant for the vapor-phase hydration of ethylene at 145°C and at 320°C from data given in App. C.

Solution 14.4

First determine values for ΔA, ΔB, ΔC, and ΔD for the reaction:

$$C_2H_4(g) + H_2O(g) \rightarrow C_2H_5OH(g)$$

The meaning of Δ is indicated by: $\Delta = (C_2H_5OH) - (C_2H_4) - (H_2O)$. Thus, from the heat-capacity data of Table C.1:

$$\Delta A = 3.518 - 1.424 - 3.470 = -1.376$$

$$\Delta B = (20.001 - 14.394 - 1.450) \times 10^{-3} = 4.157 \times 10^{-3}$$

$$\Delta C = (-6.002 + 4.392 - 0.000) \times 10^{-6} = -1.610 \times 10^{-6}$$

$$\Delta D = (-0.000 - 0.000 - 0.121) \times 10^5 = -0.121 \times 10^5$$

Values of ΔH_{298}° and ΔG_{298}° at 298.15 K for the hydration reaction are found from the heat-of-formation and Gibbs-energy-of-formation data of Table C.4:

$$\Delta H_{298}^\circ = -235,100 - 52,510 - (-241,818) = -45,792 \text{ J·mol}^{-1}$$

$$\Delta G_{298}^\circ = -168,490 - 68,460 - (-228,572) = -8378 \text{ J·mol}^{-1}$$

---

[5]K. S. Pitzer, *Thermodynamics*, 3d ed., chap. 5, McGraw-Hill, New York, 1995.

[6]B. E. Poling, J. M. Prausnitz, and J. P. O'Connell, *The Properties of Gases and Liquids*, 5th ed., chap. 3, McGraw-Hill, New York, 2001.

For $T = 145 + 273.15 = 418.15$ K, values of the integrals in Eq. (14.18) are:

IDCPH(298.15, 418.15; -1.376, 4.157E-3, -1.610E-6, -0.121E+5) = −23.121

IDCPS(298.15, 418.15; -1.376, 4.157E-3, -1.610E-6, -0.121E+5) = −0.0692

Substitution of values into Eq. (14.18) for a reference temperature of 298.15 gives:

$$\frac{\Delta G^{\circ}_{418}}{RT} = \frac{-8378 + 45{,}792}{(8.314)(298.15)} + \frac{-45{,}792}{(8.314)(418.15)} + \frac{-23.121}{418.15} + 0.0692 = 1.9356$$

For $T = 320 + 273.15 = 593.15$ K,

IDCPH(298.15, 593.15; -1.376, 4.157E-3, -1.610E-6, -0.121E+5) = 22.632

IDCPS(298.15, 593.15; -1.376, 4.157E-3, -1.610E-6, -0.121E+5) = 0.0173

Whence,

$$\frac{\Delta G^{\circ}_{593}}{RT} = \frac{-8378 + 45{,}792}{(8.314)(298.15)} + \frac{-45{,}792}{(8.314)(593.15)} + \frac{22.632}{593.15} - 0.0173 = 5.8286$$

Finally,

$$@ \; 418.15 \text{ K}: \quad \ln K = -1.9356 \quad \text{and} \quad K = 1.443 \times 10^{-1}$$
$$@ \; 593.15 \text{ K}: \quad \ln K = -5.8286 \quad \text{and} \quad K = 2.942 \times 10^{-3}$$

Application of Eqs. (14.21), (14.22), and (14.24) provides an alternative solution to this example. By Eq. (14.21),

$$K_0 = \exp\frac{8378}{(8.314)(298.15)} = 29.366$$

Moreover,
$$\frac{\Delta H^{\circ}_0}{RT_0} = \frac{-45{,}792}{(8.314)(298.15)} = -18.473$$

With these values, the following results are readily obtained:

T/K	K_0	K_1	K_2	K
298.15	29.366	1	1	29.366
418.15	29.366	4.985×10^{-3}	0.9860	1.443×10^{-1}
593.15	29.366	1.023×10^{-4}	0.9794	2.942×10^{-3}

Clearly, the influence of K_1 is far greater than that of K_2. This is a typical result and is consistent with the observation that all of the lines on Fig. 14.2 are very nearly linear.

14.6 RELATION OF EQUILIBRIUM CONSTANTS TO COMPOSITION

Gas-Phase Reactions

The standard state for a gas is the ideal-gas state of the pure gas at the standard-state pressure $P°$ of 1 bar. Because the fugacity of an ideal gas is equal to its pressure, $f_i° = P°$ for each species i. Thus for gas-phase reactions $\hat{f}_i/f_i° = \hat{f}_i/P°$, and Eq. (14.10) becomes:

$$\prod_i \left(\frac{\hat{f}_i}{P°} \right)^{\nu_i} = K \tag{14.25}$$

The equilibrium constant K depends only on temperature, but the fugacities are functions of pressure and composition as well. Eq. (14.25) relates K to fugacities of the reacting species as they exist in the real equilibrium mixture. These fugacities reflect the nonidealities of the equilibrium mixture and are functions of temperature, pressure, and composition. This means that for a fixed temperature the composition at equilibrium must change with pressure in such a way that $\prod_i (\hat{f}_i/P°)^{\nu_i}$ remains constant.

The fugacity is related to the fugacity coefficient by Eq. (10.52):

$$\hat{f}_i = \hat{\phi}_i y_i P$$

Substitution of this equation into Eq. (14.25) provides an equilibrium expression displaying the pressure and the composition dependence more explicitly:

$$\boxed{\prod_i (y_i \hat{\phi}_i)^{\nu_i} = \left(\frac{P}{P°} \right)^{-\nu} K} \tag{14.26}$$

where $\nu \equiv \Sigma_i \nu_i$ and $P°$ is the standard-state pressure of 1 bar, *expressed in the same units used for P*. The $\{y_i\}$ may be eliminated in favor of the equilibrium value of the reaction coordinate ε_e. Then, for a fixed temperature Eq. (14.26) relates ε_e to P. In principle, specification of the pressure allows solution for ε_e. However, the problem may be complicated by the dependence of $\hat{\phi}_i$ on composition, i.e., on ε_e. The methods of Secs. 10.6 and 10.7 can be applied to the calculation of $\hat{\phi}_i$ values, for example, by Eq. (10.64). Because of the complexity of the calculations, an iterative procedure, initiated by setting $\hat{\phi}_i = 1$ and formulated for computer solution, is appropriate. Once the initial set $\{y_i\}$ is calculated, the $\{\hat{\phi}_i\}$ are evaluated, and the procedure is repeated to convergence.

If the assumption that the equilibrium mixture is an *ideal solution* is justified, then each $\hat{\phi}_i$ becomes ϕ_i, the fugacity coefficient of pure species i at T and P [Eq. (10.84)]. In this case, Eq. (14.26) becomes:

$$\boxed{\prod_i (y_i \phi_i)^{\nu_i} = \left(\frac{P}{P°} \right)^{-\nu} K} \tag{14.27}$$

Each ϕ_i for a pure species can be evaluated from a generalized correlation once the equilibrium T and P are specified.

For pressures sufficiently low or temperatures sufficiently high, the equilibrium mixture behaves essentially as an ideal gas. In this event, each $\hat{\phi}_i = 1$, and Eq. (14.26) reduces to:

$$\prod_i (y_i)^{\nu_i} = \left(\frac{P}{P^\circ}\right)^{-\nu} K \tag{14.28}$$

In this equation the temperature-, pressure-, and composition-dependent terms are distinct and separate, and solution for any one of ε_e, T, or P, given the other two, is straightforward.

Although Eq. (14.28) holds only for a reaction occurring at a mixture in the ideal-gas state, we can draw some conclusions from it that are true in general:

- According to Eq. (14.14), the effect of temperature on the equilibrium constant K is determined by the sign of ΔH°. Thus when ΔH° is positive, i.e., when the standard reaction is *endothermic*, an increase in T results in an increase in K. Equation (14.28) shows that an increase in K at constant P results in an increase in $\prod_i(y_i)^{\nu_i}$; this implies a shift of the reaction equilibrium toward the products, and thus an increase in ε_e. Conversely, when ΔH° is negative, that is, when the standard reaction is *exothermic*, an increase in T causes a decrease in K and a decrease in $\prod_i(y_i)^{\nu_i}$ at constant P. This implies a shift of the reaction equilibrium toward the reactants, and a decrease in ε_e.

- Eq. (14.28) shows that when the total stoichiometric number ν ($\equiv \sum_i \nu_i$) is negative, an increase in P at constant T causes an increase in $\prod_i(y_i)^{\nu_i}$, implying a shift of the reaction equilibrium toward the products, and an increase in ε_e. If ν is positive, an increase in P at constant T causes a decrease in $\prod_i(y_i)^{\nu_i}$, a shift of the reaction equilibrium toward the reactants, and a decrease in ε_e.

Liquid-Phase Reactions

For a reaction occurring in the liquid phase, we return to:

$$\prod_i (\hat{f}_i/f_i^\circ)^{\nu_i} = K \tag{14.10}$$

The usual standard state for liquids is the pure liquid i at the temperature of the system and at 1 bar, for which the fugacity is f_i°.

According to Eq. (13.2), which defines the activity coefficient,

$$\hat{f}_i = \gamma_i x_i f_i$$

where f_i is the fugacity of pure liquid i at the temperature *and pressure* of the equilibrium mixture. The fugacity ratio can now be expressed:

$$\frac{\hat{f}_i}{f_i^\circ} = \frac{\gamma_i x_i f_i}{f_i^\circ} = \gamma_i x_i \left(\frac{f_i}{f_i^\circ}\right) \tag{14.29}$$

Because the fugacities of liquids are weak functions of pressure, the ratio f_i/f_i° is often taken as unity. However, it is readily evaluated. For pure liquid i, Eq. (10.31) is written twice, first for temperature T and pressure P, and then for the same temperature T but for the standard-state pressure $P°$. The difference between these two equations is:

$$G_i - G_i° = RT \ln \frac{f_i}{f_i°}$$

Integration of Eq. (6.11) at constant temperature T for the change of state of pure liquid i from $P°$ to P yields:

$$G_i - G_i° = \int_{P°}^{P} V_i\, dP$$

As a result,
$$RT \ln \frac{f_i}{f_i°} = \int_{P°}^{P} V_i\, dP$$

Because V_i changes little with pressure for liquids (and solids), integration from $P°$ to P assuming constant V_i is generally an excellent approximation. This integration gives:

$$\ln \frac{f_i}{f_i°} = \frac{V_i(P - P°)}{RT} \tag{14.30}$$

Using Eqs. (14.29) and (14.30), we can now write Eq. (14.10) as:

$$\boxed{\prod_i (x_i \gamma_i)^{\nu_i} = K \exp\left[\frac{(P° - P)}{RT} \sum_i (\nu_i V_i) \right]} \tag{14.31}$$

Except for high pressures, the exponential term is close to unity and may be omitted. Then,

$$\prod_i (x_i \gamma_i)^{\nu_i} = K \tag{14.32}$$

and the only problem is determination of the activity coefficients. One may apply an equation such as the Wilson equation [Eq. (13.45)], or the UNIFAC method [App. G], and the compositions can be found from Eq. (14.32) by a complex iterative computer program. However, the relative ease of experimental investigation for liquid mixtures has worked against the application of Eq. (14.32).

If the equilibrium mixture is an ideal solution, then γ_i is unity, and Eq. (14.32) becomes:

$$\prod_i (x_i)^{\nu_i} = K \tag{14.33}$$

This simple relation is known as the *law of mass action*. Because liquids often form nonideal solutions, Eq. (14.33) can be expected in many cases to yield poor results.

For species known to be present in high concentration, the equation $\hat{f}_i/f_i = x_i$ is usually nearly correct. The reason, as discussed in Sec. 13.3, is that the Lewis/Randall rule [Eq. (10.83)] always becomes valid for a species as its concentration approaches $x_i = 1$. For species at low concentration in aqueous solution, a different procedure has been widely adopted, because in this case the equality of $\hat{f}_i/f_i$ and x_i is usually far from correct. The method is based on the use of a fictitious or hypothetical standard state for the solute, taken as the state that would exist if the solute obeyed Henry's law up to a *molality m* of unity.[7] In this application, Henry's law is expressed as

$$\hat{f}_i = k_i m_i \qquad (14.34)$$

and it is always valid for a species whose concentration approaches zero. This hypothetical state is illustrated in Fig. 14.3. The dashed line drawn tangent to the curve at the origin represents Henry's law, and it is valid in the case shown to a molality much less than unity. However, one can calculate the properties the solute would have if it obeyed Henry's law to a concentration of 1 molal, and this hypothetical state often serves as a convenient standard state for solutes.

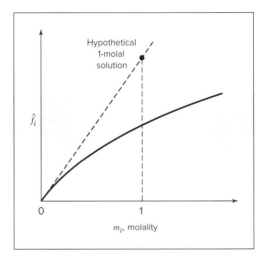

Figure 14.3: Schematic illustration of an alternative standard state for dilute aqueous solutions.

The standard-state fugacity is

$$\hat{f}_i^\circ = k_i m_i^\circ = k_i \times 1 = k_i$$

Hence, for any species at a concentration low enough for Henry's law to hold,

$$\hat{f}_i = k_i m_i = \hat{f}_i^\circ m_i$$

and

$$\frac{\hat{f}_i}{\hat{f}_i^\circ} = m_i \qquad (14.35)$$

The advantage of this standard state is that it provides a very simple relation between fugacity and concentration for cases in which Henry's law is at least approximately valid. Its range does not commonly extend to a concentration of 1 molal. In the rare case where it does, the standard state is a real state of the solute. This standard state is useful only where ΔG° data are available for the standard state of a 1-molal solution. Otherwise, the equilibrium constant cannot be evaluated by Eq. (14.11).

---

[7]Molality is a measure of solute concentration, expressed as moles of solute per kilogram of solvent.

14.7 EQUILIBRIUM CONVERSIONS FOR SINGLE REACTIONS

Suppose a single reaction occurs in a *homogeneous* system, and suppose the equilibrium constant is known. In this event, the calculation of the phase composition at equilibrium is straightforward if the phase is assumed to be in the ideal-gas state [Eq. (14.28)] or if it can be treated as an ideal solution [Eq. (14.27) or (14.33)]. When an assumption of ideality is not reasonable, the problem is still tractable for gas-phase reactions through application of an equation of state and solution by computer. For *heterogeneous* systems, where more than one phase is present, the problem is more complicated and requires the superposition of the criterion for phase equilibrium developed in Sec. 10.6. At equilibrium, there can be no tendency for change to occur, either by mass transfer between phases or by chemical reaction. We present in what follows, mainly by example, procedures useful for carrying out equilibrium calculations, first, for single-phase reactions, and second, for heterogeneous reactions.

Single-Phase Reactions

The following examples illustrate the application of equations developed in the preceding sections.

Example 14.5

The water-gas-shift reaction,

$$CO(g) + H_2O(g) \rightarrow CO_2(g) + H_2(g)$$

is carried out under several sets of conditions enumerated below. Calculate the fraction of steam reacted in each case. Assume the mixture behaves as an ideal gas.

(a) The reactants consist of 1 mol of H_2O vapor and 1 mol of CO. The temperature is 1100 K and the pressure is 1 bar.

(b) Same as (a) except that the pressure is 10 bar.

(c) Same as (a) except that 2 mol of N_2 is included in the reactants.

(d) The reactants are 2 mol of H_2O and 1 mol of CO. Other conditions are the same as in (a).

(e) The reactants are 1 mol of H_2O and 2 mol of CO. Other conditions are the same as in (a).

(f) The initial mixture consists of 1 mol of H_2O, 1 mol of CO, and 1 mol of CO_2. Other conditions are the same as in (a).

(g) Same as (a) except that the temperature is 1650 K.

Solution 14.5

(*a*) For the given reaction at 1100 K, $10^4/T = 9.05$, and from Fig. 14.2, $\ln K = 0$ and $K = 1$. For this reaction $\nu = \sum_i \nu_i = 1 + 1 - 1 - 1 = 0$. Because the reaction mixture can be treated as an ideal gas, Eq. (14.28) applies, and here becomes:

$$\frac{y_{H_2} y_{CO_2}}{y_{CO} y_{H_2O}} = K = 1 \tag{A}$$

By Eq. (14.5),

$$y_{CO} = \frac{1 - \varepsilon_e}{2} \qquad y_{H_2O} = \frac{1 - \varepsilon_e}{2} \qquad y_{CO_2} = \frac{\varepsilon_e}{2} \qquad y_{H_2} = \frac{\varepsilon_e}{2}$$

Substitution of these values into Eq. (*A*) gives:

$$\frac{\varepsilon_e^2}{(1 - \varepsilon_e)^2} = 1 \qquad \text{from which} \qquad \varepsilon_e = 0.5$$

Therefore the fraction of the steam that reacts is 0.5.

(*b*) Because $\nu = 0$, the increase in pressure has no effect on the ideal-gas reaction, and ε_e is still 0.5.

(*c*) The N_2 does not take part in the reaction and serves only as a diluent. It does increase the initial number of moles n_0 from 2 to 4, and the mole fractions are all reduced by a factor of 2. However, Eq. (*A*) is unchanged and reduces to the same expression as before. Therefore, ε_e is again 0.5.

(*d*) In this case the mole fractions at equilibrium are:

$$y_{CO} = \frac{1 - \varepsilon_e}{3} \qquad y_{H_2O} = \frac{2 - \varepsilon_e}{3} \qquad y_{CO_2} = \frac{\varepsilon_e}{3} \qquad y_{H_2} = \frac{\varepsilon_e}{3}$$

and Eq. (*A*) becomes:

$$\frac{\varepsilon_e^2}{(1 - \varepsilon_e)(2 - \varepsilon_e)} = 1 \qquad \text{from which} \qquad \varepsilon_e = 0.667$$

The fraction of steam that reacts is then $0.667/2 = 0.333$.

(*e*) Here the expressions for y_{CO} and y_{H_2O} are interchanged, but this leaves the equilibrium equation the same as in (*d*). Therefore $\varepsilon_e = 0.667$, and the fraction of steam that reacts is 0.667.

(*f*) In this case Eq. (*A*) becomes:

$$\frac{\varepsilon_e(1 + \varepsilon_e)}{(1 - \varepsilon_e)^2} = 1 \qquad \text{from which} \qquad \varepsilon_e = 0.333$$

The fraction of steam reacted is 0.333.

(*g*) At 1650 K, $10^4/T = 6.06$, and from Fig. 14.2, $\ln K = -1.15$ or $K = 0.316$. Therefore Eq. (*A*) becomes:

$$\frac{\varepsilon_e^2}{(1 - \varepsilon_e)^2} = 0.316 \qquad \text{from which} \qquad \varepsilon_e = 0.36$$

The reaction is exothermic, and the extent of reaction decreases with increasing temperature.

Example 14.6

Estimate the maximum conversion of ethylene to ethanol by vapor-phase hydration at 250°C and 35 bar for an initial steam-to-ethylene molar ratio of 5.

Solution 14.6

The calculation of K for this reaction is treated in Ex. 14.4. For a temperature of 250°C or 523.15 K the calculation yields:

$$K = 10.02 \times 10^{-3}$$

The appropriate equilibrium expression is Eq. (14.26). This equation requires evaluation of the fugacity coefficients of the species present in the equilibrium mixture. This can be accomplished with Eq. (10.64). However, the calculations involve iteration because the fugacity coefficients are functions of composition. For purposes of illustration, we assume here that the reaction mixture can be treated as an ideal solution, which eliminates the need for iteration. This calculation would also serve as the first iteration of a more rigorous calculation using mixture fugacity coefficients from Eq. (10.64). With the assumption of ideal solution behavior, Eq. (14.26) reduces to Eq. (14.27), which requires fugacity coefficients of the *pure* gases of the reacting mixture at the equilibrium T and P. Because $\nu = \sum_i \nu_i = -1$, this equation becomes:

$$\frac{y_{\text{EtOH}} \phi_{\text{EtOH}}}{y_{C_2H_4} \phi_{C_2H_4} y_{H_2O} \phi_{H_2O}} = \left(\frac{P}{P^\circ}\right)(10.02 \times 10^{-3}) \tag{A}$$

Computations based on Eq. (10.68) in conjunction with Eqs. (3.61) and (3.62) provide values represented by:

$$\text{PHIB(TR,PR,OMEGA)} = \phi_i$$

The results of these calculations are summarized in the following table:

	T_{c_i}/K	P_{c_i}/bar	ω_i	T_{r_i}	P_{r_i}	B^0	B^1	ϕ_i
C_2H_4	282.3	50.40	0.087	1.853	0.694	−0.074	0.126	0.977
H_2O	647.1	220.55	0.345	0.808	0.159	−0.511	−0.281	0.887
EtOH	513.9	61.48	0.645	1.018	0.569	−0.327	−0.021	0.827

The critical data and values of ω_i are from App. B. The temperature and pressure in all cases are 523.15 K and 35 bar. Substitution of values for ϕ_i and for (P/P°) into Eq. (A) gives:

$$\frac{y_{\text{EtOH}}}{y_{C_2H_4} y_{H_2O}} = \frac{(0.977)(0.887)}{(0.827)}(35)(10.02 \times 10^{-3}) = 0.367 \tag{B}$$

By Eq. (14.5),

$$y_{C_2H_4} = \frac{1 - \varepsilon_e}{6 - \varepsilon_e} \qquad y_{H_2O} = \frac{5 - \varepsilon_e}{6 - \varepsilon_e} \qquad y_{EtOH} = \frac{\varepsilon_e}{6 - \varepsilon_e}$$

Substituting these into Eq. (B) yields:

$$\frac{\varepsilon_e(6 - \varepsilon_e)}{(5 - \varepsilon_e)(1 - \varepsilon_e)} = 0.367 \qquad \text{or} \qquad \varepsilon_e^2 - 6.000\varepsilon_e + 1.342 = 0$$

The solution to this quadratic equation for the smaller root is $\varepsilon_e = 0.233$. Because the larger root is greater than unity, and would therefore correspond to a negative mole fraction of ethylene, it does not represent a physically possible result. The maximum conversion of ethylene to ethanol under the stated conditions is therefore 23.3%. To carry out a more rigorous calculation without assuming that the gases form an ideal solution, one would next evaluate the mixture fugacity coefficients from Eq. (10.64), use the resulting values in Eq. (B), and calculate a new value of ε_e and then iterate until the value of ε_e stops changing. However, this is rarely necessary in practice.

In this reaction, increasing the temperature decreases K and hence the conversion. Increasing the pressure increases the conversion. Equilibrium considerations therefore suggest that the operating pressure be as high as possible (limited by condensation) and the temperature as low as possible. However, even with the best catalyst known, the minimum temperature for a reasonable reaction rate is about 150°C. This is an instance where both equilibrium and reaction rate influence the commercial viability of a reaction process.

The equilibrium conversion is a function of temperature, T, pressure, P, and the steam-to-ethylene ratio in the feed, a. The effects of all three variables are shown in Fig. 14.4. The curves in this figure come from calculations like those illustrated in this example, except that a less precise relation for K as a function of T was used. Comparing the families of curves for different pressures, and the curves for different steam-to-ethylene ratios at a given pressure, illustrates how increasing P or a allows one to reach a given equilibrium ethylene conversion at higher T.

Example 14.7

In a laboratory investigation, acetylene is catalytically hydrogenated to ethylene at 1120°C and 1 bar. If the feed is an equimolar mixture of acetylene and hydrogen, what is the composition of the product stream at equilibrium?

Solution 14.7

The required reaction is obtained by the addition of the two formation reactions written as follows:

$$C_2H_2 \rightarrow 2C + H_2 \quad (I)$$

$$2C + 2H_2 \rightarrow C_2H_4 \quad (II)$$

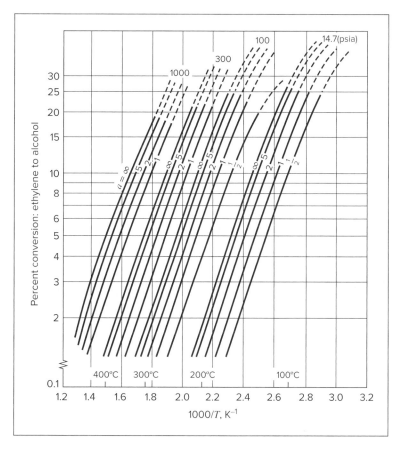

Figure 14.4: Equilibrium conversion of ethylene to ethanol in the vapor phase. Here, a = moles water/ moles ethylene. Dashed lines indicate conditions of water condensation. Data based on equation: $\ln K = 5200/T - 15.0$.

The sum of reactions (I) and (II) is the hydrogenation reaction:

$$C_2H_2 + H_2 \rightarrow C_2H_4$$

Also,
$$\Delta G^\circ = \Delta G_I^\circ + \Delta G_{II}^\circ$$

By Eq. (14.11b),

$$-RT \ln K = -RT \ln K_I - RT \ln K_{II} \qquad \text{or} \qquad K = K_I K_{II}$$

Data for both reactions (I) and (II) are given in Fig. 14.2. For 1120°C (1393.15 K), $10^4/T = 7.18$, the following values are read from the graph:

$$\ln K_I = 12.9 \qquad\qquad K_I = 4.0 \times 10^5$$

$$\ln K_{II} = -12.9 \qquad\qquad K_{II} = 2.5 \times 10^{-6}$$

Therefore, $K = K_1 K_{II} = 1.0$

At this elevated temperature and for a pressure of 1 bar, we can safely assume ideal gases. Application of Eq. (14.28) leads to the expression:

$$\frac{y_{C_2H_4}}{y_{H_2} y_{C_2H_2}} = 1$$

On the basis of one mole of each reactant, Eq. (14.5) gives:

$$y_{H_2} = y_{C_2H_2} = \frac{1 - \varepsilon_e}{2 - \varepsilon_e} \qquad \text{and} \qquad y_{C_2H_4} = \frac{\varepsilon_e}{2 - \varepsilon_e}$$

Therefore, $$\frac{\varepsilon_e(2 - \varepsilon_e)}{(1 - \varepsilon_e)^2} = 1$$

The smaller root of this quadratic expression (the larger is > 1, corresponding to negative mole fractions for the reactants) is: $\varepsilon_e = 0.293$.

The equilibrium composition of the product gas is then:

$$y_{H_2} = y_{C_2H_2} = \frac{1 - 0.293}{2 - 0.293} = 0.414 \qquad \text{and} \qquad y_{C_2H_4} = \frac{0.293}{2 - 0.293} = 0.172$$

Example 14.8

Acetic acid is esterified in the liquid phase with ethanol at 100°C and atmospheric pressure to produce ethyl acetate and water according to the reaction:

$$CH_3COOH(l) + C_2H_5OH(l) \rightarrow CH_3COOC_2H_5(l) + H_2O(l)$$

If initially there is one mole each of acetic acid and ethanol, estimate the mole fraction of ethyl acetate in the reacting mixture at equilibrium.

Solution 14.8

Data for $\Delta H^{\circ}_{f_{298}}$ and $\Delta G^{\circ}_{f_{298}}$ are given for liquid acetic acid, ethanol, and water in Table C.4. For liquid ethyl acetate, the corresponding values are:

$$\Delta H^{\circ}_{f_{298}} = -480,000\,J \qquad \text{and} \qquad \Delta G^{\circ}_{f_{298}} = -332,200\,J$$

The values of ΔH°_{298} and ΔG°_{298} for the reaction are therefore:

$$\Delta H^{\circ}_{298} = -480,000 - 285,830 + 484,500 + 277,690 = -3640\,J$$

$$\Delta G^{\circ}_{298} = -332,200 - 237,130 + 389,900 + 174,780 = -4650\,J$$

By Eq. (14.11b),

$$\ln K_{298} = \frac{-\Delta G^{\circ}_{298}}{RT} = \frac{4650}{(8.314)(298.15)} = 1.8759 \qquad \text{or} \qquad K_{298} = 6.5266$$

For the small temperature change from 298.15 to 373.15 K, Eq. (14.15) is adequate for estimating K. Thus,

$$\ln \frac{K_{373}}{K_{298}} = \frac{-\Delta H^{\circ}_{298}}{R} \left(\frac{1}{373.15} - \frac{1}{298.15} \right)$$

or

$$\ln \frac{K_{373}}{6.5266} = \frac{3640}{8.314} \left(\frac{1}{373.15} - \frac{1}{298.15} \right) = -0.2951$$

and

$$K_{373} = (6.5266)(0.7444) = 4.8586$$

For the given reaction, Eq. (14.5), with x replacing y, yields:

$$x_{\text{AcH}} = x_{\text{EtOH}} = \frac{1 - \varepsilon_e}{2} \qquad x_{\text{EtAc}} = x_{\text{H}_2\text{O}} = \frac{\varepsilon_e}{2}$$

Because the pressure is low, Eq. (14.32) is applicable. In the absence of data for the activity coefficients in this complex system, we assume that the reacting species form an ideal solution. In this case Eq. (14.33) is employed, giving:

$$K = \frac{x_{\text{EtAc}} x_{\text{H}_2\text{O}}}{x_{\text{AcH}} x_{\text{EtOH}}}$$

Thus,

$$4.8586 = \left(\frac{\varepsilon_e}{1 - \varepsilon_e} \right)^2$$

Solution yields:

$$\varepsilon_e = 0.6879 \qquad \text{and} \qquad x_{\text{EtAc}} = 0.6879/2 = 0.344$$

This result is in good agreement with experiment, even though the assumption of ideal solutions may be unrealistic. Carried out in the laboratory, the reaction yields a measured mole fraction of ethyl acetate at equilibrium of about 0.33.

Example 14.9

The gas-phase oxidation of SO_2 to SO_3 is carried out at a pressure of 1 bar with 20% excess air in an adiabatic reactor. Assuming that the reactants enter at 25°C and that equilibrium is attained at the exit, determine the composition and temperature of the product stream from the reactor.

Solution 14.9

The reaction is:

$$SO_2 + \tfrac{1}{2}O_2 \rightarrow SO_3$$

for which, $\quad \Delta H^{\circ}_{298} = -98{,}890 \qquad \Delta G^{\circ}_{298} = -70{,}866 \text{ J mol}^{-1}$

On the basis of one mole of SO_2 entering the reactor,

$$\text{Moles O}_2 \text{ entering} = (0.5)(1.2) = 0.6$$
$$\text{Moles N}_2 \text{ entering} = (0.6)(79/21) = 2.257$$

Application of Eq. (14.4) yields the amounts of the species in the product stream:

$$\text{Moles SO}_2 = 1 - \varepsilon_e$$
$$\text{Moles O}_2 = 0.6 - 0.5\varepsilon_e$$
$$\text{Moles SO}_3 = \varepsilon_e$$
$$\text{Moles N}_2 = 2.257$$

$$\overline{\text{Total Moles} = 3.857 - 0.5\varepsilon_e}$$

Two equations must be written if we are to solve for both ε_e and the temperature. They are an energy balance and an equilibrium equation. For the energy balance, we proceed as in Ex. 4.7:

$$\Delta H^\circ_{298}\, \varepsilon_e + \Delta H^\circ_P = \Delta H = 0 \qquad (A)$$

where all enthalpies are on the basis of 1 mol SO_2 entering the reactor. The enthalpy change of the products as they are heated from 298.15 K to T is:

$$\Delta H^\circ_P = \langle C^\circ_P \rangle_H\, (T - 298.15) \qquad (B)$$

where $\langle C^\circ_P \rangle$ is defined as the mean *total* heat capacity of the product stream:

$$\langle C^\circ_P \rangle_H \equiv \sum_i n_i \langle C^\circ_{P_i} \rangle_H$$

Data from Table C.1 provide $\langle C^\circ_{P_i} \rangle$ values:

$$SO_2: \quad \text{MCPH(298.15, } T; 5.699, 0.801\text{E-}3, 0.0, -1.015\text{E+}5)$$
$$O_2: \quad \text{MCPH(298.15, } T; 3.639, 0.506\text{E-}3, 0.0, -0.227\text{E+}5)$$
$$SO_3: \quad \text{MCPH(298.15, } T; 8.060, 1.056\text{E-}3, 0.0, -2.028\text{E+}5)$$
$$N_2: \quad \text{MCPH(298.15, } T; 3.280, 0.593\text{E-}3, 0.0, 0.040\text{E+}5)$$

Equations (A) and (B) combine to yield:

$$\Delta H^\circ_{298}\, \varepsilon_e + \langle C^\circ_P \rangle_H\, (T - 298.15) = 0$$

Solution for T gives: $\qquad T = \dfrac{-\Delta H^\circ_{298}\, \varepsilon_e}{\langle C^\circ_P \rangle_H} + 298.15 \qquad (C)$

At the conditions of temperature and pressure of the equilibrium state, the assumption of ideal gases is fully justified, and the equilibrium constant is therefore given by Eq. (14.28), which here becomes:

$$K = \left(\frac{\varepsilon_e}{1 - \varepsilon_e} \right) \left(\frac{3.857 - 0.5\varepsilon_e}{0.6 - 0.5\varepsilon_e} \right)^{0.5} \qquad (D)$$

Because $-\ln K = \Delta G^\circ / RT$, Eq. (14.18) can be written:

$$-\ln K = \frac{\Delta G^\circ_0 - \Delta H^\circ_0}{RT_0} + \frac{\Delta H^\circ_0}{RT} + \frac{1}{T}\int_{T_0}^{T} \frac{\Delta C^\circ_P}{R}\, dT - \int_{T_0}^{T} \frac{\Delta C^\circ_P}{R} \frac{dT}{T}$$

Substitution of numerical values yields:

$$\ln K = -11.3054 + \frac{11{,}894.4}{T} - \frac{1}{T}(IDCPH) + IDCPS \qquad (E)$$

$$IDCPH = IDCPH(298.15, T; 0.5415, 0.002E\text{-}3, 0.0, -0.8995E\text{+}5)$$

$$IDCPS = IDCPS(298.15, T; 0.5415, 0.002E\text{-}3, 0.0, -0.8995E\text{+}5)$$

These expressions for the computed values of the integrals show parameters ΔA, ΔB, ΔC, and ΔD as evaluated from data of Table C.1.

An iteration scheme for solution of these equations for ε_e and T that converges fairly rapidly is as follows:

1. Assume a starting value for T.

2. Evaluate IDCPH and IDCPS at this value of T.

3. Solve Eq. (E) for K and Eq. (D) for ε_e, probably by trial.

4. Evaluate $\langle C_P^\circ \rangle_H$ and solve Eq. (C) for T.

5. Find a new value of T as the arithmetic mean of the value just calculated and the initial value; return to step 2.

This scheme converges on the values $\varepsilon_e = 0.77$ and $T = 855.7$ K. For the product,

$$y_{SO_2} = \frac{1 - 0.77}{3.857 - (0.5)(0.77)} = \frac{0.23}{3.472} = 0.0662$$

$$y_{O_2} = \frac{0.6 - (0.5)(0.77)}{3.472} = \frac{0.215}{3.472} = 0.0619$$

$$y_{SO_3} = \frac{0.77}{3.472} = 0.2218 \qquad y_{N_2} = \frac{2.257}{3.472} = 0.6501$$

Reactions in Heterogeneous Systems

When liquid and gas phases are both present in an equilibrium mixture of reacting species, Eq. (10.48), a criterion of vapor/liquid equilibrium, must be satisfied along with the equation of chemical-reaction equilibrium. Suppose, for example, that gas A reacts with liquid water B to form an aqueous solution C. Several methods of treatment exist. The reaction may be considered to occur in the gas phase with transfer of material between phases to maintain phase equilibrium. In this case, the equilibrium constant is evaluated from ΔG° data based on standard states for the species as gases, that is, the ideal-gas states at 1 bar and the reaction temperature. Alternatively, the reaction may be considered to occur in the liquid phase, in which case ΔG° is based on standard states for the species as liquids. Finally, the reaction may be written:

$$A(g) + B(l) \rightarrow C(aq)$$

in which case the $\Delta G°$ value is for mixed standard states: C as a solute in an ideal 1-molal aqueous solution, B as a pure liquid at 1 bar, and A as a pure ideal gas at 1 bar. For this choice of standard states, the relationship between the composition and equilibrium constant, as given by Eq. (14.10), becomes:

$$\frac{\hat{f}_C/f_C°}{(\hat{f}_B/f_B°)(\hat{f}_A/f_A°)} = \frac{m_C}{(\gamma_B x_B)(\hat{f}_A/P°)} = K$$

The second term arises from Eq. (14.35) applied to species C, Eq. (14.29) applied to B with $f_B/f_B° = 1$, and the fact that $f_A° = P°$ for species A in the gas phase. Because K depends on the standard states, this value of K differs from that which would arise from other choices of standard states. However, other choices of standard states would still lead to the same equilibrium composition, provided Henry's law as applied to species C in solution is valid. In practice, a particular choice of standard states may simplify calculations or yield more accurate results because it makes better use of available data. The nature of the calculations required for heterogeneous reactions is illustrated in the following example.

Example 14.10

Estimate the compositions of the liquid and vapor phases when ethylene reacts with water to form ethanol at 200°C and 34.5 bar, conditions that assure the presence of both liquid and vapor phases. The reaction vessel is maintained at 34.5 bar by connection to a source of ethylene at this pressure. Assume no other reactions.

Solution 14.10

According to the phase rule (Sec. 14.8), the system has two degrees of freedom. Specification of both T and P therefore fixes the intensive state of the system, independent of the initial amounts of reactants. Material-balance equations are irrelevant, and we can make no use of equations that relate compositions to the reaction coordinate. Instead, phase-equilibrium relations must provide a sufficient number of equations to allow solution for the unknown compositions.

The most convenient approach to this problem is to regard the chemical reaction as occurring in the vapor phase. Thus,

$$C_2H_4(g) + H_2O(g) \rightarrow C_2H_5OH(g)$$

and the standard states are those of the pure ideal gases at 1 bar. For these standard states, the equilibrium expression is Eq. (14.25), which in this case becomes:

$$K = \frac{\hat{f}_{EtOH}}{\hat{f}_{C_2H_4}\hat{f}_{H_2O}}P° \tag{A}$$

where the standard-state pressure $P°$ is 1 bar (expressed in appropriate units). A general expression for $\ln K$ as a function of T is provided by the results of Ex. 14.4. For 200°C (473.15 K), this equation yields:

$$\ln K = -3.473 \qquad K = 0.0310$$

The task now is to incorporate the phase-equilibrium equations, $\hat{f}_i^v = \hat{f}_i^l$, into Eq. ($A$) and to relate the fugacities to the compositions in such a way that the equations can be readily solved. Equation (A) may be written:

$$K = \frac{\hat{f}_{\text{EtOH}}^v}{\hat{f}_{\text{C}_2\text{H}_4}^v \hat{f}_{\text{H}_2\text{O}}^v} P^\circ = \frac{\hat{f}_{\text{EtOH}}^l}{\hat{f}_{\text{C}_2\text{H}_4}^v \hat{f}_{\text{H}_2\text{O}}^l} P^\circ \tag{B}$$

The liquid-phase fugacities are related to activity coefficients by Eq. (13.2), and the vapor-phase fugacities are related to fugacity coefficients by Eq. (10.52):

$$\hat{f}_i^l = x_i \gamma_i f_i^l \quad (C) \qquad\qquad \hat{f}_i^v = y_i \hat{\phi}_i P \quad (D)$$

Elimination of the fugacities in Eq. (B) by Eqs. (C) and (D) gives:

$$K = \frac{x_{\text{EtOH}} \gamma_{\text{EtOH}} f_{\text{EtOH}}^l P^\circ}{(y_{\text{C}_2\text{H}_4} \hat{\phi}_{\text{C}_2\text{H}_4} P)(x_{\text{H}_2\text{O}} \gamma_{\text{H}_2\text{O}} f_{\text{H}_2\text{O}}^l)} \tag{E}$$

The fugacity f_i^l is for pure liquid i at the temperature and pressure of the system. However, pressure has small effect on the fugacity of a liquid, and to a good approximation, $f_i^l = f_i^{\text{sat}}$; whence by Eq. (10.40),

$$f_i^l = \phi_i^{\text{sat}} P_i^{\text{sat}} \tag{F}$$

In this equation ϕ_i^{sat} is the fugacity coefficient of pure saturated liquid or vapor evaluated at T of the system and at P_i^{sat}, the vapor pressure of pure species i. The assumption that the vapor phase is an ideal solution allows substitution of $\phi_{\text{C}_2\text{H}_4}$ for $\hat{\phi}_{\text{C}_2\text{H}_4}$, where $\phi_{\text{C}_2\text{H}_4}$ is the fugacity coefficient of pure ethylene at the system T and P. With this substitution and that of Eq. (F), Eq. (E) becomes:

$$K = \frac{x_{\text{EtOH}} \gamma_{\text{EtOH}} \phi_{\text{EtOH}}^{\text{sat}} P_{\text{EtOH}}^{\text{sat}} P^\circ}{(y_{\text{C}_2\text{H}_4} \hat{\phi}_{\text{C}_2\text{H}_4} P)(x_{\text{H}_2\text{O}} \gamma_{\text{H}_2\text{O}} \phi_{\text{H}_2\text{O}}^{\text{sat}} P_{\text{H}_2\text{O}}^{\text{sat}})} \tag{G}$$

where the standard-state pressure P° is 1 bar, expressed in the units of P.

In addition to Eq. (G) the following expressions apply. Because $\sum_i y_i = 1$,

$$y_{\text{C}_2\text{H}_4} = 1 - y_{\text{EtOH}} - y_{\text{H}_2\text{O}} \tag{H}$$

Eliminate y_{EtOH} and $y_{\text{H}_2\text{O}}$ from Eq. (H) in favor of x_{EtOH} and $x_{\text{H}_2\text{O}}$ by the vapor/liquid equilibrium relation, $\hat{f}_i^v = \hat{f}_i^l$. Combination with Eqs. (C), (D), and (F) then gives:

$$y_i = \frac{\gamma_i x_i \phi_i^{\text{sat}} P_i^{\text{sat}}}{\phi_i P} \tag{I}$$

where ϕ_i replaces $\hat{\phi}_i$ because of the assumption that the vapor phase is an ideal solution. Equations (H) and (I) yield:

$$y_{\text{C}_2\text{H}_4} = 1 - \frac{x_{\text{EtOH}} \gamma_{\text{EtOH}} \phi_{\text{EtOH}}^{\text{sat}} P_{\text{EtOH}}^{\text{sat}}}{\phi_{\text{EtOH}} P} - \frac{x_{\text{H}_2\text{O}} \gamma_{\text{H}_2\text{O}} \phi_{\text{H}_2\text{O}}^{\text{sat}} P_{\text{H}_2\text{O}}^{\text{sat}}}{\phi_i P} \tag{J}$$

Because ethylene is far more volatile than ethanol or water, we can assume to a good approximation that $x_{C_2H_4} = 0$. Then,

$$x_{H_2O} = 1 - x_{EtOH} \qquad (K)$$

Equations (G), (J), and (K) are the basis for solution of the problem. The primary variables in these equations are: x_{H_2O}, x_{EtOH}, and $y_{C_2H_4}$. Other quantities are either given or determined from correlations of data. The values of P_i^{sat} are:

$$P_{H_2O}^{sat} = 15.55 \qquad P_{EtOH}^{sat} = 30.22 \text{ bar}$$

The quantities ϕ_i^{sat} and ϕ_i are found from the generalized correlation represented by Eq. (10.68) with B^0 and B^1 given by Eqs. (3.61) and (3.62). Computed results are represented by PHIB(TR,PR,OMEGA). With $T = 473.15$K, $P = 34.5$ bar, and critical data and acentric factors from App. B, computations provide:

	$T_{c_i}/$K	$P_{c_i}/$bar	ω_i	T_{r_i}	P_{r_i}	$P_{r_i}^{sat}$	B^0	B^1	ϕ_i	ϕ_i^{sat}
EtOH	513.9	61.48	0.645	0.921	0.561	0.492	−0.399	−0.104	0.753	0.780
H$_2$O	647.1	220.55	0.345	0.731	0.156	0.071	−0.613	−0.502	0.846	0.926
C$_2$H$_4$	282.3	50.40	0.087	1.676	0.685		−0.102	0.119	0.963	

Substitution of values so far determined into Eqs. (G), (J), and (K) reduces these three equations to the following:

$$K = \frac{0.0493 x_{EtOH} \gamma_{EtOH}}{y_{C_2H_4} x_{H_2O} \gamma_{H_2O}} \qquad (L)$$

$$y_{C_2H_4} = 1 - 0.907 x_{EtOH} \gamma_{EtOH} - 0.493 x_{H_2O} \gamma_{H_2O} \qquad (M)$$

$$x_{H_2O} = 1 - x_{EtOH} \qquad (K)$$

The only remaining undetermined thermodynamic properties are γ_{H_2O} and γ_{EtOH}. Because of the highly nonideal behavior of a liquid solution of ethanol and water, these must be determined from experimental data. The required data, found from VLE measurements, are given by Otsuki and Williams.[8] From their results for the ethanol/water system, one can estimate values of γ_{H_2O} and γ_{EtOH} at 200°C. (Pressure has little effect on the activity coefficients of liquids.)

A procedure for solution of the foregoing three equations is as follows:

1. Assume a value for x_{EtOH} and calculate x_{H_2O} by Eq. (K).
2. Determine γ_{H_2O} and γ_{EtOH} from data in the reference cited.
3. Calculate $y_{C_2H_4}$ by Eq. (M).
4. Calculate K by Eq. (L) and compare with the value of 0.0310 determined from standard-reaction data.
5. If the two values agree, the assumed value of x_{EtOH} is correct. If they do not agree, assume a new value of x_{EtOH} and repeat the procedure.

[8]H. Otsuki and F. C. Williams, *Chem. Engr. Progr. Symp. Series No. 6*, vol. 49, pp. 55–67, 1953.

If $x_{EtOH} = 0.06$, then by Eq. (K), $x_{H_2O} = 0.94$, and from the reference cited,

$$\gamma_{EtOH} = 3.34 \qquad \text{and} \qquad \gamma_{H_2O} = 1.00$$

By Eq. (M),

$$y_{C_2H_4} = 1 - (0.907)(3.34)(0.06) - (0.493)(1.00)(0.94) = 0.355$$

The value of K given by Eq. (L) is then:

$$K = \frac{(0.0493)(0.06)(3.34)}{(0.355)(0.94)(1.00)} = 0.0296$$

This result is in close enough agreement with the value, 0.0310, found from standard-reaction data to make further calculations pointless, and the liquid-phase composition is essentially as assumed ($x_{EtOH} = 0.06$, $x_{H_2O} = 0.94$). The remaining vapor-phase compositions ($y_{C_2H_4}$ has already been determined as 0.356) are found by solution of Eq. (I) for y_{H_2O} or y_{EtOH}. All results are summarized in the following table.

	x_i	y_i
EtOH	0.060	0.180
H_2O	0.940	0.464
C_2H_4	0.000	0.356
	$\sum_i x_i = 1.000$	$\sum_i y_i = 1.000$

These results provide reasonable estimates of actual values, provided no other reactions take place.

14.8 PHASE RULE AND DUHEM'S THEOREM FOR REACTING SYSTEMS

The phase rule (applicable to intensive properties) as discussed in Secs. 3.1 and 12.2 for non-reacting systems of π phases and N chemical species is:

$$F = 2 - \pi + N$$

It must be modified for application to systems in which chemical reactions occur. The phase-rule variables are unchanged: temperature, pressure, and $N - 1$ mole fractions in each phase. The total number of these variables is $2 + (N - 1)(\pi)$. The same phase-equilibrium equations apply as before, and they number $(\pi - 1)(N)$. However, Eq. (14.8) provides for each independent reaction an additional relation that must be satisfied at equilibrium. Because the μ_i's are functions of temperature, pressure, and the phase compositions, Eq. (14.8) represents a relation connecting phase-rule variables. If there are r independent chemical reactions at equilibrium within the system, then there is a total of $(\pi - 1)(N) + r$ independent equations relating the phase-rule variables. Taking the difference between the number of variables and the number of equations gives:

$$F = [2 + (N - 1)(\pi)] - [(\pi - 1)(N) + r]$$

or

$$\boxed{F = 2 - \pi + N - r} \tag{14.36}$$

This is the phase rule for reacting systems.

The only remaining problem for application is to determine the number of independent chemical reactions. This can be done systematically as follows:

- Write chemical equations for the formation, from the *constituent elements*, of each chemical compound considered present in the system.

- Combine these equations so as to eliminate from them all elements not considered present *as elements* in the system. A systematic procedure is to select one equation and combine it with each of the others of the set to eliminate a particular element. Then the process is repeated to eliminate another element from the new set of equations. This is done for each element eliminated [see Ex. 14.11(*d*)], and it usually reduces the set by one equation for each element eliminated. However, the simultaneous elimination of two or more elements may occur.

The set of *r* equations resulting from this reduction procedure is a complete set of independent reactions for the *N* species considered present in the system. More than one such set is possible, depending on how the reduction procedure is carried out, but all sets number *r* and are equivalent. The reduction procedure also ensures the following relation:

> *r* ≥ number of compounds present in the system
>   − number of constituent elements *not* present *as elements*

The phase-equilibrium and chemical-reaction-equilibrium equations are the only ones considered in the foregoing treatment as interrelating the phase-rule variables. However, in certain situations *special constraints* may be placed on the system that allow additional equations to be written over and above those considered in the development of Eq. (14.36). If the number of equations resulting from special constraints is *s*, then Eq. (14.36) must be modified to take account of these *s* additional equations. The still more general form of the phase rule that results is:

$$\boxed{F = 2 - \pi + N - r - s} \tag{14.37}$$

Example 14.11 shows how Eqs. (14.36) and (14.37) may be applied to specific systems.

Example 14.11

Determine the number of degrees of freedom *F* for each of the following systems.

(*a*) A system of two miscible nonreacting species that exists as an azeotrope in vapor/liquid equilibrium.

(b) A system prepared by partially decomposing $CaCO_3$ into an evacuated space.

(c) A system prepared by partially decomposing NH_4Cl into an evacuated space.

(d) A system consisting of the gases CO, CO_2, H_2, H_2O, and CH_4 in chemical equilibrium.

Solution 14.11

(a) The system consists of two nonreacting species in two phases. If there were no azeotrope, Eq. (14.36) would apply:

$$F = 2 - \pi + N - r = 2 - 2 + 2 - 0 = 2$$

This is the usual result for binary VLE. However, a special constraint is imposed on the system; it is an azeotrope. This provides an equation, $x_1 = y_1$, not considered in the development of Eq. (14.36). Thus, Eq. (14.37) with $s = 1$ yields $F = 1$. If the system is an azeotrope, then just one phase-rule variable—T, P, or $x_1 (= y_1)$—may be arbitrarily specified.

(b) Here, a single chemical reaction occurs:

$$CaCO_3(s) \rightarrow CaO(s) + CO_2(g)$$

and $r = 1$. Three chemical species are present, and three phases: solid $CaCO_3$, solid CaO, and gaseous CO_2. One might think a special constraint has been imposed by the requirement that the system be prepared in a special way—by decomposing $CaCO_3$. This is not the case, because no equation connecting the phase-rule variables can be written as a result of this requirement. Therefore,

$$F = 2 - \pi + N - r - s = 2 - 3 + 3 - 1 - 0 = 1$$

and there is a single degree of freedom. Thus, $CaCO_3$ exerts a fixed decomposition pressure at fixed T.

(c) The chemical reaction here is:

$$NH_4Cl(s) \rightarrow NH_3(g) + HCl(g)$$

Three species, but only two phases, are present in this case: solid NH_4Cl and a gas mixture of NH_3 and HCl. In addition, a special constraint is imposed by the requirement that the system be formed by the decomposition of NH_4Cl. This means that the gas phase is equimolar in NH_3 and HCl. Thus a special equation, $y_{NH_3} = y_{HCl} (= 0.5)$, connecting the phase-rule variables can be written. Application of Eq. (14.37) gives:

$$F = 2 - \pi + N - r - s = 2 - 2 + 3 - 1 - 1 = 1$$

and the system has but one degree of freedom. This result is the same as that for part (b), and it is a matter of experience that NH_4Cl has a given decomposition pressure at a given temperature. This conclusion is reached quite differently in the two cases.

(d) This system contains five species, all in a single gas phase. There are no special constraints. Only r remains to be determined. The formation reactions for the compounds present are:

$C + \frac{1}{2}O_2 \rightarrow CO$	(A)	$C + O_2 \rightarrow CO_2$	(B)
$H_2 + \frac{1}{2}O_2 \rightarrow H_2O$	(C)	$C + 2H_2 \rightarrow CH_4$	(D)

Systematic elimination of C and O_2, the elements not present in the system, leads to two equations. One such pair of equations is obtained in the following way. Eliminate C from the set of equations by combining Eq. (B), first with Eq. (A) and then with Eq. (D). The two resulting reactions are:

$$\text{From } (B) \text{ and } (A): \quad CO + \tfrac{1}{2}O_2 \rightarrow CO_2 \tag{E}$$

$$\text{From } (B) \text{ and } (D): \quad CH_4 + O_2 \rightarrow 2H_2 + CO_2 \tag{F}$$

Equations (C), (E), and (F) are the new set, and we now eliminate O_2 by combining Eq. (C), first with Eq. (E) and then with Eq. (F). This gives:

$$\text{From } (C) \text{ and } (E): \quad CO_2 + H_2 \rightarrow CO + H_2O \tag{G}$$

$$\text{From } (C) \text{ and } (F): \quad CH_4 + 2H_2O \rightarrow CO_2 + 4H_2 \tag{H}$$

Equations (G) and (H) are an independent set and indicate that $r = 2$. The use of different elimination procedures produces other pairs of equations, but always just two equations.

Application of Eq. (14.37) yields:

$$F = 2 - \pi + N - r - s = 2 - 1 + 5 - 2 - 0 = 4$$

This result means that one is free to specify four phase-rule variables, for example, T, P, and two mole fractions, in an equilibrium mixture of these five chemical species, provided that nothing else is arbitrarily set. In other words, there can be no special constraints, such as the specification that the system be prepared from given amounts of CH_4 and H_2O. This imposes special constraints through material balances that reduce the degrees of freedom to two. (Duhem's theorem; see the following paragraphs.)

Duhem's theorem for nonreacting systems was developed in Sec. 12.2. It states that for any closed system formed initially from given masses of particular chemical species, the

equilibrium state is *completely determined* (extensive as well as intensive properties) by specification of any two independent variables. This theorem gives the difference between the number of independent variables that completely determine the state of the system and the number
of independent equations that can be written connecting these variables:

$$[2 + (N-1)(\pi) + \pi] - [(\pi-1)(N) + N] = 2$$

When chemical reactions occur, a new variable ε_j is introduced into the material-balance equations for each independent reaction. Furthermore, a new equilibrium relation [Eq. (14.8)] can
be written for each independent reaction. Therefore, when chemical-reaction equilibrium is
superimposed on phase equilibrium, r new variables appear and r new equations can be written. The difference between the number of variables and the number of equations therefore is
unchanged, and Duhem's theorem as originally stated holds for reacting systems as well as for
nonreacting systems.

Most chemical-reaction equilibrium problems are posed such that Duhem's theorem
makes them determinate. The usual problem is to find the composition of a system that reaches
equilibrium from an initial state of *fixed amounts of reacting species* when the *two* variables
T and P are specified.

14.9 MULTIREACTION EQUILIBRIA

When the equilibrium state in a reacting system depends on two or more independent chemical reactions, the equilibrium composition can be found by a direct extension of the methods
developed for single reactions. One first determines a set of independent reactions as discussed in Sec. 14.8. With each independent reaction there is associated a reaction coordinate
in accord with the treatment of Sec. 14.1. In addition, a separate equilibrium constant is evaluated for each reaction j, and Eq. (14.10) becomes:

$$\prod_i \left(\frac{\hat{f}_i}{f_i^\circ}\right)^{\nu_{i,j}} = K_j \tag{14.38}$$

with
$$K_j \equiv \exp\left(\frac{-\Delta G_j^\circ}{RT}\right) \qquad (j = 1, 2, \ldots, r)$$

For a gas-phase reaction Eq. (14.38) takes the form:

$$\prod_i \left(\frac{\hat{f}_i}{P^\circ}\right)^{\nu_{i,j}} = K_j \tag{14.39}$$

If the equilibrium mixture is in the ideal-gas state,

$$\prod_i (y_i)^{\nu_{i,j}} = \left(\frac{P}{P^\circ}\right)^{-\nu_j} K_j \tag{14.40}$$

For r independent reactions there are r separate equations of this kind, and the $\{y_i\}$ can be
eliminated by Eq. (14.7) in favor of the r reaction coordinates ε_j. The set of equations is then
solved simultaneously for the r reaction coordinates, as illustrated in the following examples.

Example 14.12

A feedstock of pure *n*-butane is cracked at 750 K and 1.2 bar to produce olefins. Only two reactions have favorable equilibrium conversions at these conditions:

$$C_4H_{10} \rightarrow C_2H_4 + C_2H_6 \quad \text{(I)}$$

$$C_4H_{10} \rightarrow C_3H_6 + CH_4 \quad \text{(II)}$$

If these reactions reach equilibrium, what is the product composition?

With data from App. C and procedures illustrated in Ex. 14.4, the equilibrium constants at 750 K are found to be:

$$K_I = 3.856 \qquad \text{and} \qquad K_{II} = 268.4$$

Solution 14.12

Equations relating the product composition to the reaction coordinates are developed as in Ex. 14.3. With a basis of 1 mol of *n*-butane feed, they here become:

$$y_{C_4H_{10}} = \frac{1 - \varepsilon_I - \varepsilon_{II}}{1 + \varepsilon_I + \varepsilon_{II}}$$

$$y_{C_2H_4} = y_{C_2H_6} = \frac{\varepsilon_I}{1 + \varepsilon_I + \varepsilon_{II}} \qquad y_{C_3H_6} = y_{CH_4} = \frac{\varepsilon_{II}}{1 + \varepsilon_I + \varepsilon_{II}}$$

The equilibrium relations, by Eq. (14.40), are:

$$\frac{y_{C_2H_4} y_{C_2H_6}}{y_{C_4H_{10}}} = \left(\frac{P}{P^\circ}\right)^{-1} K_I \qquad \frac{y_{C_3H_6} y_{CH_4}}{y_{C_4H_{10}}} = \left(\frac{P}{P^\circ}\right)^{-1} K_{II}$$

Combining these equilibrium equations with the mole-fraction equations yields:

$$\frac{\varepsilon_I^2}{(1 - \varepsilon_I - \varepsilon_{II})(1 + \varepsilon_I + \varepsilon_{II})} = \left(\frac{P}{P^\circ}\right)^{-1} K_I \tag{A}$$

$$\frac{\varepsilon_{II}^2}{(1 - \varepsilon_I - \varepsilon_{II})(1 + \varepsilon_I + \varepsilon_{II})} = \left(\frac{P}{P^\circ}\right)^{-1} K_{II} \tag{B}$$

Dividing Eq. (*B*) by Eq. (*A*) and solving for ε_{II} yields:

$$\varepsilon_{II} = \kappa \varepsilon_I \tag{C}$$

where

$$\kappa \equiv \left(\frac{K_{II}}{K_I}\right)^{1/2} \tag{D}$$

Combining Eqs. (*A*) and (*C*) to eliminate ε_{II}, and then solving for ε_I gives:

$$\varepsilon_I = \left[\frac{K_I(P^\circ/P)}{1 + K_I(P^\circ/P)(\kappa + 1)^2}\right]^{1/2} \tag{E}$$

Substitution of numerical values in Eqs. (*D*), (*E*), and (*C*) yields:

$$\kappa = \left(\frac{268.4}{3.856}\right)^{1/2} = 8.343$$

$$\varepsilon_{\mathrm{I}} = \left[\frac{(3.856)(1/1.2)}{1 + (3.856)(1/1.2)(9.343)^2}\right]^{1/2} = 0.1068$$

$$\varepsilon_{\mathrm{II}} = (8.343)(0.1068) = 0.8914$$

The product-gas composition for these reaction coordinates is:

$$y_{C_4H_{10}} = 0.0010 \qquad y_{C_2H_4} = y_{C_2H_6} = 0.0534 \qquad y_{C_3H_6} = y_{CH_4} = 0.4461$$

For this simple reaction scheme, analytical solution is possible. However, this is unusual. In most cases the solution of multireaction-equilibrium problems requires numerical techniques.

Example 14.13

A bed of coal (assumed to be pure carbon) in a coal gasifier is fed with steam and air and produces a gas stream containing H_2, CO, O_2, H_2O, CO_2, and N_2. If the feed to the gasifier consists of 1 mol of steam and 2.38 mol of air, calculate the equilibrium composition of the gas stream at $P = 20$ bar for temperatures of 1000, 1100, 1200, 1300, 1400, and 1500 K. Available data are listed in the following table.

T/K	$\Delta G_f^{\circ}/\mathrm{J \cdot mol^{-1}}$		
	H_2O	CO	CO_2
1000	$-192,420$	$-200,240$	$-395,790$
1100	$-187,000$	$-209,110$	$-395,960$
1200	$-181,380$	$-217,830$	$-396,020$
1300	$-175,720$	$-226,530$	$-396,080$
1400	$-170,020$	$-235,130$	$-396,130$
1500	$-164,310$	$-243,740$	$-396,160$

Solution 14.13

The feed stream to the coal bed consists of 1 mol of steam and 2.38 mol of air, containing:

$$O_2 \colon \ (0.21)(2.38) = 0.5 \text{ mol} \qquad N_2 \colon \ (0.79)(2.38) = 1.88 \text{ mol}$$

The species present at equilibrium are C, H_2, CO, O_2, H_2O, CO_2, and N_2. The formation reactions for the compounds present are:

$$H_2 + \tfrac{1}{2}O_2 \rightarrow H_2O \quad \text{(I)}$$

$$C + \tfrac{1}{2}O_2 \rightarrow CO \quad \text{(II)}$$

$$C + O_2 \rightarrow CO_2 \quad \text{(III)}$$

Because the elements hydrogen, oxygen, and carbon are themselves presumed to be present in the system, this set of three independent reactions is a complete set.

All species are present as gases except carbon, which is a pure solid phase. In the equilibrium expression, Eq. (14.38), the fugacity ratio of the pure carbon, is $\hat{f}_C/f_C^{\circ} = f_C/f_C^{\circ}$, the fugacity of carbon at 20 bar divided by the fugacity of carbon at 1 bar. Because the effect of pressure on the fugacity of a solid is very small, negligible error is introduced by the assumption that this ratio is unity. The fugacity ratio for carbon is then $\hat{f}_C/f_C^{\circ} = 1$, and it may be omitted from the equilibrium expression. With the assumption that the remaining species are ideal gases, Eq. (14.40) is written for the gas phase only, and it provides the following equilibrium expressions for reactions (I) through (III):

$$K_{\mathrm{I}} = \frac{y_{H_2O}}{y_{O_2}^{1/2} y_{H_2}} \left(\frac{P}{P^{\circ}}\right)^{-1/2} \qquad K_{\mathrm{II}} = \frac{y_{CO}}{y_{O_2}^{1/2}} \left(\frac{P}{P^{\circ}}\right)^{-1/2} \qquad K_{\mathrm{III}} = \frac{y_{CO_2}}{y_{O_2}}$$

The reaction coordinates for the three reactions are designated ε_{I}, $\varepsilon_{\mathrm{II}}$, and $\varepsilon_{\mathrm{III}}$. For the initial state,

$$n_{H_2} = n_{CO} = n_{CO_2} = 0 \qquad n_{H_2O} = 1 \qquad n_{O_2} = 0.5 \qquad n_{N_2} = 1.88$$

Moreover, because only the gas-phase species are considered,

$$\nu_{\mathrm{I}} = -\frac{1}{2} \qquad \nu_{\mathrm{II}} = \frac{1}{2} \qquad \nu_{\mathrm{III}} = 0$$

Applying Eq. (14.7) to each species gives:

$$y_{H_2} = \frac{-\varepsilon_{\mathrm{I}}}{3.38 + (\varepsilon_{\mathrm{II}} - \varepsilon_{\mathrm{I}})/2} \qquad\qquad y_{CO} = \frac{\varepsilon_{\mathrm{II}}}{3.38 + (\varepsilon_{\mathrm{II}} - \varepsilon_{\mathrm{I}})/2}$$

$$y_{O_2} = \frac{\tfrac{1}{2}(1 - \varepsilon_{\mathrm{I}} - \varepsilon_{\mathrm{II}}) - \varepsilon_{\mathrm{III}}}{3.38 + (\varepsilon_{\mathrm{II}} - \varepsilon_{\mathrm{I}})/2} \qquad\qquad y_{H_2O} = \frac{1 + \varepsilon_{\mathrm{I}}}{3.38 + (\varepsilon_{\mathrm{II}} - \varepsilon_{\mathrm{I}})/2}$$

$$y_{CO_2} = \frac{\varepsilon_{\mathrm{III}}}{3.38 + (\varepsilon_{\mathrm{II}} - \varepsilon_{\mathrm{I}})/2} \qquad\qquad y_{N_2} = \frac{1.88}{3.38 + (\varepsilon_{\mathrm{II}} - \varepsilon_{\mathrm{I}})/2}$$

Substitution of these expressions for y_i into the equilibrium equations yields:

$$K_{\mathrm{I}} = \frac{(1 + \varepsilon_{\mathrm{I}})(2n)^{1/2}(P/P^{\circ})^{-1/2}}{(1 - \varepsilon_{\mathrm{I}} - \varepsilon_{\mathrm{II}} - 2\varepsilon_{\mathrm{III}})^{1/2}(-\varepsilon_{\mathrm{I}})}$$

$$K_{\mathrm{II}} = \frac{\sqrt{2}\varepsilon_{\mathrm{II}}(P/P^{\circ})^{1/2}}{(1 - \varepsilon_{\mathrm{I}} - \varepsilon_{\mathrm{II}} - 2\varepsilon_{\mathrm{III}})^{1/2} n^{1/2}}$$

$$K_{III} = \frac{2\varepsilon_{III}}{(1 - \varepsilon_I - \varepsilon_{II} - 2\varepsilon_{III})}$$

where

$$n \equiv 3.38 + \frac{\varepsilon_{II} - \varepsilon_I}{2}$$

Numerical values for the K_j calculated by Eq. (14.11) are found to be very large. For example, at 1500 K,

$$\ln K_I = \frac{-\Delta G_I^\circ}{RT} = \frac{164,310}{(8.314)(1500)} = 13.2 \qquad K_I \sim 10^6$$

$$\ln K_{II} = \frac{-\Delta G_{II}^\circ}{RT} = \frac{243,740}{(8.314)(1500)} = 19.6 \qquad K_{II} \sim 10^8$$

$$\ln K_{III} = \frac{-\Delta G_{III}^\circ}{RT} = \frac{396,160}{(8.314)(1500)} = 31.8 \qquad K_{III} \sim 10^{14}$$

With each K_j so large, the quantity $1 - \varepsilon_I - \varepsilon_{II} - 2\varepsilon_{III}$ in the denominator of each equilibrium equation must be nearly zero. This means that the mole fraction of oxygen in the equilibrium mixture is very small. For practical purposes, no oxygen is present.

We therefore reformulate the problem by eliminating O_2 from the formation reactions. For this, Eq. (I) is combined, first with Eq. (II), and then with Eq. (III). This provides two equations:

$$C + CO_2 \rightarrow 2CO \qquad (a)$$

$$H_2O + C \rightarrow H_2 + CO \quad (b)$$

The corresponding equilibrium equations are:

$$K_a = \frac{y_{CO}^2}{y_{CO_2}}\left(\frac{P}{P^\circ}\right) \qquad\qquad K_b = \frac{y_{H_2}y_{CO}}{y_{H_2O}}\left(\frac{P}{P^\circ}\right)$$

The input stream is specified to contain 1 mol H_2, 0.5 mol O_2, and 1.88 mol N_2. Because O_2 has been eliminated from the set of reaction equations, we replace the 0.5 mol of O_2 in the feed by 0.5 mol of CO_2. The presumption is that this amount of CO_2 has been formed by prior reaction of the 0.5 mol O_2 with carbon. Thus the equivalent feed stream contains 1 mol H_2, 0.5 mol CO_2, and 1.88 mol N_2, and the application of Eq. (14.7) to Eqs. (a) and (b) gives:

$$y_{H_2} = \frac{\varepsilon_b}{3.38 + \varepsilon_a + \varepsilon_b} \qquad\qquad y_{CO} = \frac{2\varepsilon_a + \varepsilon_b}{3.38 + \varepsilon_a + \varepsilon_b}$$

$$y_{H_2O} = \frac{1 - \varepsilon_b}{3.38 + \varepsilon_a + \varepsilon_b} \qquad\qquad y_{CO_2} = \frac{0.5 - \varepsilon_a}{3.38 + \varepsilon_a + \varepsilon_b}$$

$$y_{N_2} = \frac{1.88}{3.38 + \varepsilon_a + \varepsilon_b}$$

Because values of y_i must lie between zero and unity, the two expressions on the left and the two on the right show that:

$$0 \leq \varepsilon_b \leq 1 \qquad\qquad -0.5 \leq \varepsilon_a \leq 0.5$$

Combining the expressions for the y_i with the equilibrium equations gives:

$$K_a = \frac{(2\varepsilon_a + \varepsilon_b)^2}{(0.5 - \varepsilon)(3.38 + \varepsilon_a + \varepsilon_b)} \left(\frac{P}{P^\circ}\right) \qquad (A)$$

$$K_b = \frac{\varepsilon_b(2\varepsilon_a + \varepsilon_b)}{(1 - \varepsilon_b)(3.38 + \varepsilon_a + \varepsilon_b)} \left(\frac{P}{P^\circ}\right) \qquad (B)$$

For reaction (a) at 1000 K,

$$\Delta G^\circ_{1000} = 2(-200{,}240) - (-395{,}790) = -4690 \text{ J·mol}^{-1}$$

and by Eq. (14.11),

$$\ln K_a = \frac{4690}{(8.314)(1000)} = 0.5641 \qquad\qquad K_a = 1.758$$

Similarly, for reaction (b),

$$\Delta G^\circ_{1000} = (-200{,}240) - (-192{,}420) = -7820 \text{ J·mol}^{-1}$$

and

$$\ln K_b = \frac{7820}{(8.314)(1000)} = 0.9406 \qquad\qquad K_b = 2.561$$

Equations (A) and (B) with these values for K_a and K_b and with $(P/P^\circ) = 20$ constitute two nonlinear equations in unknowns ε_a and ε_b. An ad hoc iteration scheme can be devised for their solution, but Newton's method for solving an array of nonlinear algebraic equations is attractive. It is described and applied to this example in App. H. The results of calculations for all temperatures are shown in the following table.

T/K	K_a	K_b	ε_a	ε_b
1000	1.758	2.561	-0.0506	0.5336
1100	11.405	11.219	0.1210	0.7124
1200	53.155	38.609	0.3168	0.8551
1300	194.430	110.064	0.4301	0.9357
1400	584.85	268.76	0.4739	0.9713
1500	1514.12	583.58	0.4896	0.9863

Values for the mole fractions y_i of the species in the equilibrium mixture are calculated by the equations already given. The results of all such calculations appear in the following table and are shown graphically in Fig. 14.5.

T/K	y_{H_2}	y_{CO}	y_{H_2O}	y_{CO_2}	y_{N_2}
1000	0.138	0.112	0.121	0.143	0.486
1100	0.169	0.226	0.068	0.090	0.447
1200	0.188	0.327	0.032	0.040	0.413
1300	0.197	0.378	0.014	0.015	0.396
1400	0.201	0.398	0.006	0.005	0.390
1500	0.203	0.405	0.003	0.002	0.387

At the higher temperatures the values of ε_a and ε_b are approaching their upper limiting values of 0.5 and 1.0, indicating that reactions (*a*) and (*b*) are proceeding nearly to completion. In this limit, which is approached even more closely at still higher temperatures, the mole fractions of CO_2 and H_2O approach zero, and for the product species,

$$y_{H_2} = \frac{1}{3.38 + 0.5 + 1.0} = 0.205$$

$$y_{CO} = \frac{1 + 1}{3.38 + 0.5 + 1.0} = 0.410$$

$$y_{N_2} = \frac{1.88}{3.38 + 0.5 + 1.0} = 0.385$$

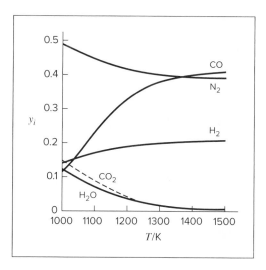

Figure 14.5: Equilibrium compositions of the product gases in Ex. 14.13.

We have here assumed a sufficient depth of coal so that the gases reach equilibrium while in contact with incandescent carbon. This need not be the case; if oxygen and steam are supplied at too high a rate, the reactions may not attain equilibrium or may reach equilibrium after they have left the coal bed. In this event, carbon is not present at equilibrium, and the problem must again be reformulated.

Although Eqs. (*A*) and (*B*) of the preceding example are readily solved, the method of equilibrium constants does not lend itself to standardization that allows a *general* program to be written for computer solution. An alternative criterion of equilibrium, mentioned in Sec. 14.2, is based on the fact that at equilibrium the total Gibbs energy of the system has its minimum value, as illustrated for a single reaction in Fig. 14.1. Applied to multiple reactions, this criterion is the basis for a general scheme of computer solution of multiple reaction equilibria.

The total Gibbs energy of a single-phase system as given by Eq. (10.2) shows that:

$$(G^t)_{T,P} = g(n_1, n_2, n_3, \ldots n_N)$$

The problem is to find the set $\{n_i\}$ that minimizes G^t for specified T and P, subject to the constraints of the material balances. The standard solution to this problem is based on the method of Lagrange multipliers. The procedure for gas-phase reactions is as follows:

1. The first step is to formulate the constraining equations, i.e., the material balances. Although reacting molecular species are not conserved in a closed system, the total number of atoms of each *element* is constant. Let subscript k identify a particular atom. Then define A_k as the total number of atomic masses of the kth element in the system, as determined by the initial constitution of the system. Further, let a_{ik} be the number of atoms of the kth element present in each molecule of chemical species i. The material balance on each element k may then be written:

$$\boxed{\sum_i n_i a_{ik} = A_k \qquad (k = 1, 2, \ldots, w)} \qquad (14.41)$$

or $\qquad\qquad \sum_i n_i a_{ik} - A_k = 0 \qquad (k = 1, 2, \ldots, w)$

   where w is the total number of distinct elements comprising the system.

2. Next, we introduce the Lagrange multipliers λ_k, one for each element, by multiplying each element balance by its λ_k:

$$\lambda_k \left(\sum_i n_i a_{ik} - A_k \right) = 0 \qquad (k = 1, 2, \ldots, w)$$

   These equations are summed over k, giving:

$$\sum_k \lambda_k \left(\sum_i n_i a_{ik} - A_k \right) = 0$$

3. Then a new function F is formed by addition of this last sum to G^t. Thus,

$$F = G^t + \sum_k \lambda_k \left(\sum_i n_i a_{ik} - A_k \right)$$

   This new function is identical to G^t because the summation term is zero. However, the partial derivatives of F and G^t with respect to n_i are different, because the function F incorporates the constraints of the material balances.

4. The minimum value of F (and G^t) occurs when all of the partial derivatives $(\partial F/\partial n_i)_{T,P,n_j}$ are zero. We therefore differentiate the preceding equation, and set the resulting derivative equal to zero:

$$\left(\frac{\partial F}{\partial n_i}\right)_{T,P,n_j} = \left(\frac{\partial G^t}{\partial n_i}\right)_{T,P,n_j} + \sum_k \lambda_k a_{ik} = 0 \qquad (i = 1, 2, \ldots, N)$$

Because the first term on the right is the definition of the chemical potential [see Eq. (10.1)], this equation can be written:

$$\mu_i + \sum_k \lambda_k a_{ik} = 0 \qquad (i = 1, 2, \ldots, N) \tag{14.42}$$

The chemical potential is expressed in terms of fugacity by Eq. (14.9):

$$\mu_i = G_i^\circ + RT \ln(\hat{f}_i/f_i^\circ)$$

For gas-phase reactions and standard states as the pure ideal gases at 1 bar:

$$\mu_i = G_i^\circ + RT \ln(\hat{f}_i/P^\circ)$$

If G_i° is arbitrarily set equal to zero for all *elements* in their standard states, then for compounds $G_i^\circ = \Delta G_{f_i}^\circ$, the standard Gibbs-energy change of formation for species i. In addition, the fugacity is eliminated in favor of the fugacity coefficient by Eq. (10.52), $\hat{f}_i = y_i \hat{\phi}_i P$. With these substitutions, the equation for μ_i becomes:

$$\mu_i = \Delta G_{f_i}^\circ + RT \ln\left(\frac{y_i \hat{\phi}_i P}{P^\circ}\right)$$

Combination with Eq. (14.42) gives:

$$\boxed{\Delta G_{f_i}^\circ + RT \ln\left(\frac{y_i \hat{\phi}_i P}{P^\circ}\right) + \sum_k \lambda_k a_{ik} = 0 \qquad (i = 1, 2, \ldots, N)} \tag{14.43}$$

Note that P° is 1 bar, expressed in the units used for pressure. If species i is an element, $\Delta G_{f_i}^\circ$ is zero.

Equation (14.43) represents N equilibrium equations, one for each chemical species, and Eq. (14.41) represents w material-balance equations, one for each element—a total of $N + w$ equations. The unknowns are the $\{n_i\}$ (note that $y_i = n_i/\sum_i n_i$), of which there are N, and the $\{\lambda_k\}$, of which there are w—a total of $N + w$ unknowns. Thus the number of equations is sufficient for the determination of all unknowns.

The foregoing discussion has presumed that each $\hat{\phi}_i$ is known. If the phase is an ideal gas, then for each species $\hat{\phi}_i = 1$. If the phase is an ideal solution, $\hat{\phi}_i = \phi_i$, and values can at least be estimated. For real gases, $\hat{\phi}_i$ is a function of the $\{y_i\}$, which are being calculated. Thus an iterative procedure is indicated. The calculations are initiated with $\hat{\phi}_i = 1$ for all i. Solution of the equations then provides a preliminary set of $\{y_i\}$. For low pressures or high temperatures this result is usually adequate. Where it is not satisfactory, an equation of state is used together with the calculated $\{y_i\}$ to give a new and more nearly correct set $\{\hat{\phi}_i\}$ for use in Eq. (14.43). Then a new set of $\{y_i\}$ is determined. The process is repeated until successive iterations produce no significant change in $\{y_i\}$. All calculations are well suited to computer solution, including the calculation of $\{\hat{\phi}_i\}$ by equations such as Eq. (10.64).

In the procedure just described, the question of what chemical reactions are involved never enters directly into any of the equations. However, the choice of a set of species is entirely equivalent to the choice of a set of independent reactions among the species. In any event, a set of species or an equivalent set of independent reactions must always be assumed, and different assumptions produce different results.

Example 14.14

Calculate the equilibrium compositions at 1000 K and 1 bar of a gas-phase system containing the species CH_4, H_2O, CO, CO_2, and H_2. In the initial unreacted state there are present 2 mol of CH_4 and 3 mol of H_2O. Values of $\Delta G^\circ_{f_i}$ at 1000 K are:

$$\Delta G^\circ_{f_{CH_4}} = 19{,}720 \text{ J·mol}^{-1} \qquad \Delta G^\circ_{f_{H_2O}} = -192{,}420 \text{ J·mol}^{-1}$$

$$\Delta G^\circ_{f_{CO}} = -200{,}240 \text{ J·mol}^{-1} \qquad \Delta G^\circ_{f_{CO_2}} = -395{,}790 \text{ J·mol}^{-1}$$

Solution 14.14

The required values of A_k are determined from the initial numbers of moles, and the values of a_{ik} come directly from the chemical formulas of the species. These are shown in the following table.

	Element k		
	Carbon	Oxygen	Hydrogen
	A_k = no. of atomic masses of k in the system		
	$A_C = 2$	$A_O = 3$	$A_H = 14$
Species i	a_{ik} = no. of atoms of k per molecule of i		
CH_4	$a_{CH_4,C} = 1$	$a_{CH_4,O} = 0$	$a_{CH_4,H} = 4$
H_2O	$a_{H_2O,C} = 0$	$a_{H_2O,O} = 1$	$a_{H_2O,H} = 2$
CO	$a_{CO,C} = 1$	$a_{CO,O} = 1$	$a_{CO,H} = 0$
CO_2	$a_{CO_2,C} = 1$	$a_{CO_2,O} = 2$	$a_{CO_2,H} = 0$
H_2	$a_{H_2,C} = 0$	$a_{H_2,O} = 0$	$a_{H_2,H} = 2$

At 1 bar and 1000 K the assumption of ideal gases is justified, and each $\hat{\phi}_i$ is unity. Because $P = 1$ bar, $P/P^\circ = 1$, and Eq. (14.43) is written:

$$\frac{\Delta G^\circ_{f_i}}{RT} + \ln \frac{n_i}{\sum_i n_i} + \sum_k \frac{\lambda_k}{RT} a_{ik} = 0$$

The five equations for the five species then become:

$$CH_4 : \quad \frac{19{,}720}{RT} + \ln\frac{n_{CH_4}}{\sum_i n_i} + \frac{\lambda_C}{RT} + \frac{4\lambda_H}{RT} = 0$$

$$H_2O : \quad \frac{-192{,}420}{RT} + \ln\frac{n_{H_2O}}{\sum_i n_i} + \frac{2\lambda_H}{RT} + \frac{\lambda_O}{RT} = 0$$

$$CO : \quad \frac{-200{,}240}{RT} + \ln\frac{n_{CO}}{\sum_i n_i} + \frac{\lambda_C}{RT} + \frac{\lambda_O}{RT} = 0$$

$$CO_2 : \quad \frac{-395{,}790}{RT} + \ln\frac{n_{CO_2}}{\sum_i n_i} + \frac{\lambda_C}{RT} + \frac{2\lambda_O}{RT} = 0$$

$$H_2 : \quad \ln\frac{n_{H_2}}{\sum_i n_i} + \frac{2\lambda_H}{RT} = 0$$

The three atom-balance equations [Eq. (14.41)] and the equation for $\sum_i n_i$ are:

$$C : \quad n_{CH_4} + n_{CO} + n_{CO_2} = 2$$
$$H : \quad 4n_{CH_4} + 2n_{H_2O} + 2n_{H_2} = 14$$
$$O : \quad n_{H_2O} + n_{CO} + 2n_{CO_2} = 3$$
$$\sum_i n_i = n_{CH_4} + n_{H_2O} + n_{CO} + n_{CO_2} + n_{H_2}$$

With $RT = 8314$ J mol^{-1}, simultaneous computer solution of these nine equations[9] produces the following results ($y_i = n_i / \sum_i n_i$):

$$y_{CH_4} = 0.0196$$

$$y_{H_2O} = 0.0980 \qquad\qquad \frac{\lambda_C}{RT} = 0.7635$$

$$y_{CO} = 0.1743 \qquad\qquad \frac{\lambda_O}{RT} = 25.068$$

$$y_{CO_2} = 0.0371 \qquad\qquad \frac{\lambda_H}{RT} = 0.1994$$

$$y_{H_2} = 0.6710$$

The values of λ_k / RT are of no significance, but they are included for completeness.

---

[9]Example computer code for this problem is available in the online learning center at http://highered.mheducation.com:80/sites/1259696529.

14.10   FUEL CELLS

A fuel cell is a device in which a fuel is oxidized electrochemically to produce electric power. It has some characteristics of a battery, in that it consists of two electrodes, separated by an electrolyte. However, the reactants are not stored in the cell but are fed to it continuously, and the products of reaction are continuously withdrawn. The fuel cell is thus not given an initial electric charge, and in operation it does not lose electric charge. It operates as a continuous-flow system as long as fuel and oxygen are supplied, and it produces a steady electric current. Compared to the conventional process of burning a fuel and extracting mechanical work via a heat engine to power a generator, fuel cells provide a more efficient means of converting the chemical energy available by oxidation of fuels into electrical energy. In the context of the present chapter, they provide an interesting example of chemical reaction equilibrium influenced by an external constraint—the electrical circuit driven by the fuel cell.

In a fuel cell, the fuel, e.g., hydrogen, methane, butane, methanol, etc., makes intimate contact with an anode or fuel electrode, and oxygen (usually in air) makes intimate contact with a cathode or oxygen electrode. Half-cell reactions occur at each electrode, and their sum is the overall reaction. Several types of fuel cell exist, each characterized by a particular type of electrolyte.[10]

Cells operating with hydrogen as the fuel are the simplest such devices, and they serve to illustrate basic principles. Schematic diagrams of hydrogen/oxygen fuel cells appear in Fig. 14.6. When the electrolyte is acidic [Fig. 14.6(a)], the half-cell reaction occurring at the hydrogen electrode (anode) is:

$$H_2 \rightarrow 2H^+ + 2e^-$$

and that at the oxygen electrode (cathode) is:

$$\tfrac{1}{2}O_2 + 2e^- + 2H^+ \rightarrow H_2O(g)$$

When the electrolyte is alkaline [Fig. 14.6(b)], the half-cell reaction at the anode is:

$$H_2 + 2OH^- \rightarrow 2H_2O(g) + 2e^-$$

and at the cathode:

$$\tfrac{1}{2}O_2 + 2e^- + H_2O(g) \rightarrow 2OH^-$$

In either case, the sum of the half-cell reactions is the overall reaction of the cell:

$$H_2 + \tfrac{1}{2}O_2 \rightarrow H_2O(g)$$

This of course is the *combustion reaction* of hydrogen, but combustion in the conventional sense of burning does not occur in the cell.

In both cells electrons with negative charge (e^-) are released at the anode, produce an electric current in an external circuit, and are taken up by the reaction occurring at the cathode. The electrolyte does not allow the passage of electrons, but it provides a path for migration

---

[10]Construction details of the various types of fuel cells and extensive explanations of their operation are given by J. Larminie and A. Dicks, *Fuel Cell Systems Explained*, 2nd ed., Wiley, 2003. See also R. O'Hayre, S.-W. Cha, W. Colella, and F. B. Prinz, *Fuel Cell Fundamentals*, 3rd ed., Wiley, 2016.

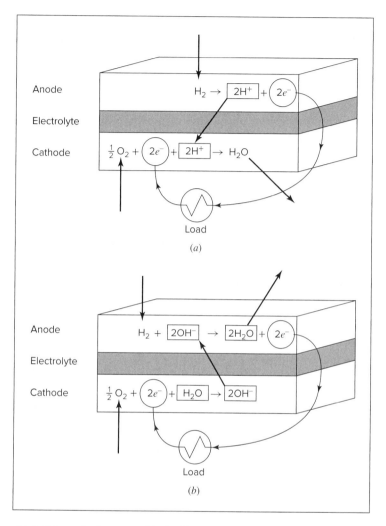

Figure 14.6: Schematic diagrams of fuel cells. (*a*) Acid electrolyte; (*b*) alkaline electrolyte.

of an ion from one electrode to the other. With an acid electrolyte, protons (H^+) migrate from anode to cathode, whereas with an alkaline electrolyte hydroxyl ions (OH^-) migrate from cathode to anode.

For many practical applications the most satisfactory hydrogen/oxygen fuel cell is built around a solid polymer that serves as an acid electrolyte. Because it is very thin and conducts H^+ ions or protons, it is known as a proton-exchange membrane. Each side of the membrane is bonded to a porous electrode loaded with an electrocatalyst such as finely divided platinum. The porous electrodes provide a very large surface area for reaction and accommodate the diffusion of hydrogen and oxygen into the cell and water vapor out of the cell. Cells can be stacked and connected in series to make very compact units with a desired terminal emf. They typically operate at temperatures near 60°C.

Because fuel-cell operation is a steady-flow process, the first law takes the form:

$$\Delta H = Q + W_{elect}$$

where potential- and kinetic-energy terms are omitted as negligible and shaft work has been replaced by electrical work. If the cell operates *reversibly* and *isothermally*,

$$Q = T\Delta S \qquad \text{and} \qquad \Delta H = T\Delta S + W_{elect}$$

The electrical work of a reversible cell is therefore:

$$W_{elect} = \Delta H - T\Delta S = \Delta G \qquad (14.44)$$

where Δ denotes a property change of reaction. The heat transfer to the surroundings required for isothermal operation is:

$$Q = \Delta H - \Delta G \qquad (14.45)$$

With reference to Fig. 14.6(*a*), we note that for each molecule of hydrogen consumed, two electrons pass to the external circuit. On the basis of 1 mol of H_2, the charge (q) transferred between electrodes is:

$$q = 2N_A(-e) \quad \text{coulomb}$$

where $-e$ is the charge on each electron and N_A is Avogadro's number. Because the product $N_A e$ is Faraday's constant F, $q = -2F$.[11] The electrical work is then the product of the charge transferred and the emf (E, expressed in volts) of the cell:

$$W_{elect} = -2FE \text{ joule}$$

The emf of a reversible cell is therefore:

$$E = \frac{-W_{elect}}{2F} = \frac{-\Delta G}{2F} \qquad (14.46)$$

These equations may be applied to a hydrogen/oxygen fuel cell operating at 25°C and 1 bar with pure H_2 and pure O_2 as reactants and pure H_2O vapor as product. If these species are assumed to be ideal gases, then the reaction occurring is the standard formation reaction for $H_2O(g)$ at 298.15 K, for which values from Table C.4 are:

$$\Delta H = \Delta H^{\circ}_{f_{298}} = -241{,}818 \text{ J·mol}^{-1} \qquad \text{and} \qquad \Delta G = \Delta G^{\circ}_{f_{298}} = -228{,}572 \text{ J·mol}^{-1}$$

Equations (14.44) through (14.46) then yield:

$$W_{elect} = -228{,}572 \text{ J·mol}^{-1} \qquad Q = -13{,}246 \text{ J·mol}^{-1} \qquad E = 1.184 \text{ volts}$$

If, as is more commonly the case, air is the source of oxygen, the cell receives O_2 at its partial pressure in air. Because the enthalpy of ideal gases is independent of pressure, the enthalpy change of reaction for the cell is unchanged. However, the Gibbs-energy change of reaction is affected. By Eq. (10.24),

$$G_i^{ig} - \bar{G}_i^{ig} = -RT \ln y_i$$

[11]Faraday's constant is equal to 96,485 C·mol^{-1}.

Therefore, on the basis of 1 mol of H_2O formed,

$$\Delta G = \Delta G_{f_{298}}^{\circ} + (0.5)\left(G_{O_2}^{ig} - \bar{G}_{O_2}^{ig}\right)$$

$$= \Delta G_{f_{298}}^{\circ} + 0.5RT \ln y_{O_2}$$

$$= 228{,}572 - (0.5)(8.314)(298.15)(\ln 0.21) = -226{,}638 \text{ J·mol}^{-1}$$

Equations (14.44) through (14.46) now yield:

$$W_{elect} = -226{,}638 \text{ J·mol}^{-1} \qquad Q = -15{,}180 \text{ J·mol}^{-1} \qquad E = 1.174 \text{ volts}$$

The use of air rather than pure oxygen does not significantly reduce the emf or work output of a reversible cell.

The enthalpy and Gibbs-energy changes of reaction are given as functions of temperature by Eqs. (4.19) and (14.18). For a cell temperature of 60°C (333.15 K), the integrals in these equations are evaluated as:

$$\int_{298.15}^{333.15} \frac{\Delta C_P^{\circ}}{R} dT = \text{IDCPH(298.15, 333.15; } -1.5985, 0.775\text{E-3, 0.0, 0.1515E+5)}$$

$$= -42.0472$$

$$\int_{298.15}^{333.15} \frac{\Delta C_P^{\circ}}{R} \frac{dT}{T} = \text{IDCPS(298.15, 333.15; } -1.5985, 0.775\text{E-3, 0.0, 0.1515E+5)}$$

$$= -0.13334$$

Equations (4.19) and (14.18) then yield:

$$\Delta H_{f_{333}}^{\circ} = -242{,}168 \text{ J·mol}^{-1} \qquad \text{and} \qquad \Delta G_{f_{333}}^{\circ} = -226{,}997 \text{ J·mol}^{-1}$$

With cell operation at 1 bar and oxygen extracted from air, $\Delta H = \Delta H_{f_{333}}^{\circ}$, and

$$\Delta G = -226{,}997 - (0.5)(8.314)(333.15)(\ln 0.21) = -224{,}836 \text{ J·mol}^{-1}$$

Equations (14.44) through (14.46) now yield:

$$W_{elect} = -224{,}836 \text{ J·mol}^{-1} \qquad Q = -17{,}332 \text{ J·mol}^{-1} \qquad E = 1.165 \text{ volts}$$

Thus cell operation at 60°C rather than at 25°C reduces the voltage and work output of a reversible cell by only a small amount.

These calculations for a reversible cell show that the electrical work output is more than 90% of the heat that would be released (ΔH) by actual combustion of the fuel. Were this heat supplied to a Carnot engine operating at practical temperature levels, a much smaller fraction would be converted into work. The reversible operation of a fuel cell implies that the external circuit exactly balances its emf, with the result that its current output is negligible. In actual operation under reasonable load, internal irreversibilities inevitably reduce the emf of the cell and decrease its production of electrical work while increasing the amount of heat transfer to the surroundings. The operating emf of hydrogen/oxygen fuel cells is often 0.6–0.7 volts, and the work output is closer to 50% of the heating value of the fuel. Nevertheless, the irreversibilities of a fuel cell are far less than those inherent in combustion of the fuel and production of work by a practical heat engine. A fuel cell has the additional advantages of simplicity, of clean and quiet operation, and of directly producing electrical energy. Fuels other than hydrogen can also be used in fuel cells, but they require the development of effective catalysts. Methanol, for example, reacts at the anode of a proton-exchange membrane fuel cell according to the equation:

$$CH_3OH + H_2O \rightarrow 6H^+ + 6e^- + CO_2$$

The usual reaction of oxygen to form water vapor occurs at the cathode.

14.11 SYNOPSIS

After studying this chapter, including the end-of-chapter problems, one should be able to:

- Define the reaction coordinate for a single reaction or multiple reaction coordinates for multiple reactions and write species concentrations in terms of the reaction coordinate(s)

- Understand that minimization of the Gibbs energy with respect to allowable changes is a general criterion for equilibrium of chemical reactions

- Define the standard Gibbs energy of reaction and the equilibrium constant

- Compute the standard Gibbs energy of reaction and the equilibrium constant from tables of thermodynamic data

- Understand that the majority of the temperature dependence of reaction equilibria is given by $d(\ln K)/dT = -\Delta H°/(RT^2)$, which implies that plots of $\ln K$ vs. $1/T$ are very nearly linear

- Compute the equilibrium constant for a reaction at arbitrary temperature using heat capacity integrals with standard enthalpies of formation and Gibbs energies of formation (or standard entropies)

- Solve for the equilibrium composition of a mixture of gases undergoing one or more chemical reactions

  - In the ideal-gas state

  - Using pure-component fugacity coefficients for moderate deviations from the ideal-gas state

  - Using mixture fugacity coefficients for conditions where deviations from the ideal-gas state are more significant

- Solve for the equilibrium composition of a liquid mixture undergoing one or more chemical reactions, including nonideal solutions where the treatment must include activity coefficients

- Employ the method of Lagrange multipliers to treat multiple-reaction equilibrium
- Set up the equations for combined phase and reaction equilibrium problems and understand conceptually how to approach a solution to the problem
- Explain the operation of a fuel cell and compute the maximum possible emf and work produced by a given fuel/oxidizer combination

14.12 PROBLEMS

14.1. Develop expressions for the mole fractions of reacting species as functions of the reaction coordinate for:

(a) A system initially containing 2 mol NH_3 and 5 mol O_2 and undergoing the reaction:

$$4NH_3(g) + 5O_2(g) \rightarrow 4NO(g) + 6H_2O(g)$$

(b) A system initially containing 3 mol H_2S and 5 mol O_2 and undergoing the reaction:

$$2H_2S(g) + 3O_2(g) \rightarrow 2H_2O(g) + 2SO_2(g)$$

(c) A system initially containing 3 mol NO_2, 4 mol NH_3, and 1 mol N_2 and undergoing the reaction:

$$6NO_2(g) + 8NH_3(g) \rightarrow 7N_2(g) + 12H_2O(g)$$

14.2. A system initially containing 2 mol C_2H_4 and 3 mol O_2 undergoes the reactions:

$$C_2H_4(g) + \tfrac{1}{2}O_2(g) \rightarrow \langle(CH_2)_2\rangle O(g)$$

$$C_2H_4(g) + 3O_2(g) \rightarrow 2CO_2(g) + 2H_2O(g)$$

Develop expressions for the mole fractions of the reacting species as functions of the reaction coordinates for the two reactions.

14.3. A system formed initially of 2 mol CO_2, 5 mol H_2, and 1 mol CO undergoes the reactions:

$$CO_2(g) + 3H_2(g) \rightarrow CH_3OH(g) + H_2O(g)$$

$$CO_2(g) + H_2(g) \rightarrow CO(g) + H_2O(g)$$

Develop expressions for the mole fractions of the reacting species as functions of the reaction coordinates for the two reactions.

14.4. Consider the water-gas-shift reaction:

$$H_2(g) + CO_2(g) \rightarrow H_2O(g) + CO(g)$$

At high temperatures and low to moderate pressures the reacting species form an ideal-gas mixture. By Eq. (10.27):

$$G = \sum_i y_i G_i + RT \sum_i y_i \ln y_i$$

When the Gibbs energies of the elements in their standard states are set equal to zero, $G_i = \Delta G_{f_i}^\circ$ for each species, and then:

$$G = \sum_i y_i \Delta G_{f_i}^\circ + RT \sum_i y_i \ln y_i \tag{A}$$

At the beginning of Sec. 14.2 we noted that Eq. (12.3) is a criterion of equilibrium. Applied to the water-gas-shift reaction with the understanding that T and P are constant, this equation becomes:

$$dG^t = d(nG) = ndG + Gdn = 0 \qquad n\frac{dG}{d\varepsilon} + G\frac{dn}{d\varepsilon} = 0$$

Here, however, $dn/d\varepsilon = 0$. The equilibrium criterion therefore becomes:

$$\frac{dG}{d\varepsilon} = 0 \tag{B}$$

Once the y_i are eliminated in favor of ε, Eq. (A) relates G to ε. Data for $\Delta G_{f_i}^\circ$ for the compounds of interest are given with Ex. 14.13. For a temperature of 1000 K (the reaction is unaffected by P) and for a feed of 1 mol H_2 and 1 mol CO_2:

(a) Determine the equilibrium value of ε by application of Eq. (B).
(b) Plot G vs. ε, indicating the location of the equilibrium value of ε determined in (a).

14.5. Rework Prob. 14.4 for a temperature of:

(a) 1100 K;   (b) 1200 K;   (c) 1300 K.

14.6. Use the method of equilibrium constants to verify the value of ε found as an answer in one of the following:

(a) Prob. 14.4; (b) Prob. 14.5(a); (c) Prob. 14.5(b); (d) Prob. 14.5(c).

14.7. Develop a general equation for the standard Gibbs-energy change of reaction ΔG° as a function of temperature for one of the reactions given in parts (a), (f), (i), (n), (r), (t), (u), (x), and (y) of Prob. 4.23.

14.8. For ideal gases, exact mathematical expressions can be developed for the effect of T and P on ε_e. For conciseness, let $\prod_i (y_i)^{\nu_i} \equiv K_y$. Then:

$$\left(\frac{\partial \varepsilon_e}{\partial T}\right)_P = \left(\frac{\partial K_y}{\partial T}\right)_P \frac{d\varepsilon_e}{dK_y} \qquad \text{and} \qquad \left(\frac{\partial \varepsilon_e}{\partial P}\right)_T = \left(\frac{\partial K_y}{\partial P}\right)_T \frac{d\varepsilon_e}{dK_y}$$

Use Eqs. (14.28) and (14.14), to show that:

(a) $\left(\dfrac{\partial \varepsilon_e}{\partial T}\right)_P = \dfrac{K}{RT^2} \dfrac{d\varepsilon_e}{dK_y} \Delta H^\circ$

(b) $\left(\dfrac{\partial \varepsilon_e}{\partial P}\right)_T = \dfrac{K_y}{P} \dfrac{d\varepsilon_e}{dK_y} (-\nu)$

(c) $d\varepsilon_e/dK_y$ is always positive. (*Note*: It is equally valid and perhaps easier to show that the reciprocal is positive.)

14.9. For the ammonia synthesis reaction written:

$$\tfrac{1}{2}N_2(g) + \tfrac{3}{2}H_2(g) \rightarrow NH_3(g)$$

with 0.5 mol N_2 and 1.5 mol H_2 as the initial amounts of reactants and with the assumption that the equilibrium mixture is an ideal gas, show that:

$$\varepsilon_e = 1 - \left(1 + 1.299K\frac{P}{P^\circ}\right)^{-1/2}$$

14.10. Peter, Paul, and Mary, members of a thermodynamics class, are asked to find the equilibrium composition at a particular T and P and for given initial amounts of reactants for the following gas-phase reaction:

$$2NH_3 + 3NO \rightarrow 3H_2O + \tfrac{5}{2}N_2 \qquad (A)$$

Each solves the problem correctly in a different way. Mary bases her solution on reaction (A) as written. Paul, who prefers whole numbers, multiplies reaction (A) by 2:

$$4NH_3 + 6NO \rightarrow 6H_2O + 5N_2 \qquad (B)$$

Peter, who usually does things backward, deals with the reaction:

$$3H_2O + \tfrac{5}{2}N_2 \rightarrow 2NH_3 + 3NO \qquad (C)$$

Write the chemical-equilibrium equations for the three reactions, indicate how the equilibrium constants are related, and show why Peter, Paul, and Mary all obtain the same result.

14.11. The following reaction reaches equilibrium at 500°C and 2 bar:

$$4HCl(g) + O_2(g) \rightarrow 2H_2O(g) + 2Cl_2(g)$$

If the system initially contains 5 mol HCl for each mole of oxygen, what is the composition of the system at equilibrium? Assume ideal gases.

14.12. The following reaction reaches equilibrium at 650°C and atmospheric pressure:

$$N_2(g) + C_2H_2(g) \rightarrow 2HCN(g)$$

If the system initially is an equimolar mixture of nitrogen and acetylene, what is the composition of the system at equilibrium? What would be the effect of doubling the pressure? Assume ideal gases.

14.13. The following reaction reaches equilibrium at 350°C and 3 bar:

$$CH_3CHO(g) + H_2(g) \rightarrow C_2H_5OH(g)$$

If the system initially contains 1.5 mol H_2 for each mole of acetaldehyde, what is the composition of the system at equilibrium? What would be the effect of reducing the pressure to 1 bar? Assume ideal gases.

14.14. The following reaction, hydrogenation of styrene to ethylbenzene, reaches equilibrium at 650°C and atmospheric pressure:

$$C_6H_5CH{:}CH_2(g) + H_2(g) \rightarrow C_6H_5.C_2H_5(g)$$

If the system initially contains 1.5 mol H_2 for each mole of styrene ($C_6H_5CH{:}CH_2$), what is the composition of the system at equilibrium? Assume ideal gases.

14.15. The gas stream from a sulfur burner is composed of 15-mol-% SO_2, 20-mol-% O_2, and 65-mol-% N_2. This gas stream at 1 bar and 480°C enters a catalytic converter, where the SO_2 is further oxidized to SO_3. Assuming that the reaction reaches equilibrium, how much heat must be removed from the converter to maintain isothermal conditions? Base your answer on 1 mol of entering gas.

14.16. For the cracking reaction,

$$C_3H_8(g) \rightarrow C_2H_4(g) + CH_4(g)$$

the equilibrium conversion is negligible at 300 K, but it becomes appreciable at temperatures above 500 K. For a pressure of 1 bar, determine:

(*a*) The fractional conversion of propane at 625 K.
(*b*) The temperature at which the fractional conversion is 85%.

14.17. Ethylene is produced by the dehydrogenation of ethane. If the feed includes 0.5 mol of steam (an inert diluent) per mole of ethane and if the reaction reaches equilibrium at 1100 K and 1 bar, what is the composition of the product gas on a water-free basis?

14.18. The production of 1,3-butadiene can be carried out by the dehydrogenation of 1-butene:

$$C_2H_5CH{:}CH_2(g) \rightarrow H_2C{:}CHHC{:}CH_2(g) + H_2(g)$$

Side reactions are suppressed by the introduction of steam. If equilibrium is attained at 950 K and 1 bar and if the reactor product contains 10-mol-% 1,3-butadiene, find:

(*a*) The mole fractions of the other species in the product gas.
(*b*) The mole fraction of steam required in the feed.

14.19. The production of 1,3-butadiene can be carried out by the dehydrogenation of *n*-butane:

$$C_4H_{10}(g) \rightarrow H_2C{:}CHHC{:}CH_2(g) + 2H_2(g)$$

Side reactions are suppressed by the introduction of steam. If equilibrium is attained at 925 K and 1 bar and if the reactor product contains 12-mol-% 1,3-butadiene, find:

(*a*) The mole fractions of the other species in the product gas.
(*b*) The mole fraction of steam required in the feed.

14.20. For the ammonia synthesis reaction,

$$\tfrac{1}{2}N_2(g) + \tfrac{3}{2}H_2 \rightarrow NH_3(g)$$

the equilibrium conversion to ammonia is large at 300 K, but it decreases rapidly with increasing T. However, reaction rates become appreciable only at higher temperatures. For a feed mixture of hydrogen and nitrogen in the stoichiometric proportions,

(a) What is the equilibrium mole fraction of ammonia at 1 bar and 300 K?
(b) At what temperature does the equilibrium mole fraction of ammonia equal 0.50 for a pressure of 1 bar?
(c) At what temperature does the equilibrium mole fraction of ammonia equal 0.50 for a pressure of 100 bar, assuming the equilibrium mixture is an ideal gas?
(d) At what temperature does the equilibrium mole fraction of ammonia equal 0.50 for a pressure of 100 bar, assuming the equilibrium mixture is an ideal solution of gases?

14.21. For the methanol synthesis reaction,

$$CO(g) + 2H_2(g) \rightarrow CH_3OH(g)$$

the equilibrium conversion to methanol is large at 300 K, but it decreases rapidly with increasing T. However, reaction rates become appreciable only at higher temperatures. For a feed mixture of carbon monoxide and hydrogen in the stoichiometric proportions,

(a) What is the equilibrium mole fraction of methanol at 1 bar and 300 K?
(b) At what temperature does the equilibrium mole fraction of methanol equal 0.50 for a pressure of 1 bar?
(c) At what temperature does the equilibrium mole fraction of methanol equal 0.50 for a pressure of 100 bar, assuming the equilibrium mixture is an ideal gas?
(d) At what temperature does the equilibrium mole fraction of methanol equal 0.50 for a pressure of 100 bar, assuming the equilibrium mixture is an ideal solution of gases?

14.22. Limestone ($CaCO_3$) decomposes upon heating to yield quicklime (CaO) and carbon dioxide. At what temperature is the decomposition pressure of limestone 1(atm)?

14.23. Ammonium chloride [$NH_4Cl(s)$] decomposes upon heating to yield a gas mixture of ammonia and hydrochloric acid. At what temperature does ammonium chloride exert a decomposition pressure of 1.5 bar? For $NH_4Cl(s)$, $\Delta H^\circ_{f_{298}} = -314{,}430$ J·mol^{-1} and $\Delta G^\circ_{f_{298}} = -202{,}870$ J·mol^{-1}.

14.24. A chemically reactive system contains the following species in the gas phase: NH_3, NO, NO_2, O_2, and H_2O. Determine a complete set of independent reactions for this system. How many degrees of freedom does the system have?

14.25. The relative compositions of the pollutants NO and NO_2 in air are governed by the reaction,

$$NO + \tfrac{1}{2}O_2 \rightarrow NO_2$$

For air containing 21-mol-% O_2 at 25°C and 1.0133 bar, what is the concentration of NO in parts per million if the total concentration of the two nitrogen oxides is 5 ppm?

14.26. Consider the gas-phase oxidation of ethylene to ethylene oxide at a pressure of 1 bar with 25% excess air. If the reactants enter the process at 25°C, if the reaction proceeds adiabatically to equilibrium, and if there are no side reactions, determine the composition and temperature of the product stream from the reactor.

14.27. Carbon black is produced by the decomposition of methane:

$$CH_4(g) \rightarrow C(s) + 2H_2(g)$$

For equilibrium at 650°C and 1 bar,

(a) What is the gas-phase composition if pure methane enters the reactor, and what fraction of the methane decomposes?
(b) Repeat part (a) if the feed is an equimolar mixture of methane and nitrogen.

14.28. Consider the reactions

$$\tfrac{1}{2}N_2(g) + \tfrac{1}{2}O_2(g) \rightarrow NO(g)$$

$$\tfrac{1}{2}N_2(g) + O_2(g) \rightarrow NO_2(g)$$

If these reactions come to equilibrium after combustion in an internal-combustion engine at 2000 K and 200 bar, estimate the mole fractions of NO and NO_2 present for mole fractions of nitrogen and oxygen in the combustion products of 0.70 and 0.05.

14.29. Oil refineries often have both H_2S and SO_2 to dispose of. The following reaction suggests a means of getting rid of both at once:

$$2H_2S(g) + SO_2(g) \rightarrow 3S(s) + 2H_2O(g)$$

For reactants in the stoichiometric proportion, estimate the percent conversion of each reactant if the reaction comes to equilibrium at 450°C and 8 bar.

14.30. Species N_2O_4 and NO_2 as gases come to equilibrium by the reaction: $N_2O_4 \rightarrow 2NO_2$.

(a) For $T = 350$ K and $P = 5$ bar, calculate the mole fractions of these species in the equilibrium mixture. Assume ideal gases.
(b) If an equilibrium mixture of N_2O_4 and NO_2 at conditions of part (a) flows through a throttle valve to a pressure of 1 bar and through a heat exchanger that restores its initial temperature, how much heat is exchanged, assuming chemical equilibrium is again attained in the final state? Base the answer on an amount of mixture equivalent to 1 mol of N_2O_4, i.e., as though the NO_2 were present as N_2O_4.

14.31. The following isomerization reaction occurs in the *liquid* phase: A → B, where A and B are miscible liquids for which: $G^E/RT = 0.1x_Ax_B$. If $\Delta G^\circ_{298} = -1000$ J, what is the equilibrium composition of the mixture at 25°C? How much error is introduced if one assumes that A and B form an ideal solution?

14.32. Hydrogen gas can be produced by the reaction of steam with "water gas," an equimolar mixture of H_2 and CO obtained by the reaction of steam with coal. A stream of "water gas" mixed with steam is passed over a catalyst to convert CO to CO_2 by the reaction:

$$H_2O(g) + CO(g) \rightarrow H_2(g) + CO_2(g)$$

Subsequently, unreacted water is condensed and carbon dioxide is absorbed, leaving a product that is mostly hydrogen. The equilibrium conditions are 1 bar and 800 K.

(*a*) Is any advantage gained by carrying out the reaction at pressures above 1 bar?
(*b*) Would increasing the equilibrium temperature increase the conversion of CO?
(*c*) For the given equilibrium conditions, determine the molar ratio of steam to "water gas" (H_2 + CO) required to produce a *product* gas containing only 2-mol-% CO after cooling to 20°C, where the unreacted H_2O has been virtually all condensed.
(*d*) Is there any danger that solid carbon will form at the equilibrium conditions by the reaction

$$2CO(g) \rightarrow CO_2(g) + C(s)$$

14.33. The feed gas to a methanol synthesis reactor is composed of 75-mol-% H_2, 15-mol-% CO, 5-mol-% CO_2, and 5-mol-% N_2. The system comes to equilibrium at 550 K and 100 bar with respect to the reactions:

$$2H_2(g) + CO(g) \rightarrow CH_3OH(g) \qquad H_2(g) + CO_2(g) \rightarrow CO(g) + H_2O(g)$$

Assuming ideal gases, determine the composition of the equilibrium mixture.

14.34. "Synthesis gas" can be produced by the catalytic re-forming of methane with steam. The reactions are:

$$CH_4(g) + H_2O(g) \rightarrow CO(g) + 3H_2(g) \qquad CO(g) + H_2O(g) \rightarrow CO_2(g) + H_2(g)$$

Assume equilibrium is attained for both reactions at 1 bar and 1300 K.

(*a*) Would it be better to carry out the reaction at pressures above 1 bar?
(*b*) Would it be better to carry out the reaction at temperatures below 1300 K?
(*c*) Estimate the molar ratio of hydrogen to carbon monoxide in the synthesis gas if the feed consists of an equimolar mixture of steam and methane.
(*d*) Repeat part (*c*) for a steam-to-methane mole ratio in the feed of 2.
(*e*) How could the feed composition be altered to yield a lower ratio of hydrogen to carbon monoxide in the synthesis gas than is obtained in part (*c*)?
(*f*) Is there any danger that carbon will deposit by the reaction $2CO \rightarrow C + CO_2$ under conditions of part (*c*)? Part (*d*)? If so, how could the feed be altered to prevent carbon deposition?

14.35. Consider the gas-phase isomerization reaction: A→B.

    (a) Assuming ideal gases, develop from Eq. (14.28) the chemical-reaction equilibrium equation for the system.

    (b) The result of part (a) should suggest that there is *one* degree of freedom for the equilibrium state. Upon verifying that the phase rule indicates *two* degrees of freedom, explain the discrepancy.

14.36. A low-pressure, gas-phase isomerization reaction, A→B, occurs at conditions such that vapor and liquid phases are present.

    (a) Prove that the equilibrium state is univariant.

    (b) Suppose T is specified. Show how to calculate x_A, y_A, and P. State carefully, and justify, any assumptions.

14.37. Set up the equations required for solution of Ex. 14.14 by the method of equilibrium constants. Verify that your equations yield the same equilibrium compositions as given in the example.

14.38. Reaction-equilibrium calculations may be useful for estimation of the compositions of hydrocarbon feedstocks. A particular feedstock, available as a low-pressure gas at 500 K, is identified as "aromatic C8." It could in principle contain the C_8H_{10} isomers: *o*-xylene (OX), *m*-xylene (MX), *p*-xylene (PX), and ethylbenzene (EB). Estimate how much of each species is present, assuming the gas mixture has come to equilibrium at 500 K and low pressure. The following is a set of independent reactions (why?):

OX → MX (I)	OX → PX (II)	OX → EB (III)

    (a) Write reaction-equilibrium equations for each equation of the set. State clearly any assumptions.

    (b) Solve the set of equations to obtain algebraic expressions for the equilibrium vapor-phase mole fractions of the four species in relation to the equilibrium constants, K_I, K_{II}, K_{III}.

    (c) Use the following data to determine numerical values for the equilibrium constants at 500 K. State clearly any assumptions.

    (d) Determine numerical values for the mole fractions of the four species.

Species	$\Delta H^{\circ}_{f_{298}} / \text{J·mol}^{-1}$	$\Delta G^{\circ}_{f_{298}} / \text{J·mol}^{-1}$
OX(g)	19,000	122,200
MX(g)	17,250	118,900
PX(g)	17,960	121,200
EB(g)	29,920	130,890

14.39. Ethylene oxide as a vapor and water as liquid, both at 25°C and 101.33 kPa, react to form a liquid solution containing ethylene glycol (1,2-ethanediol) at the same conditions:

$$\langle(CH_2)_2\rangle O + H_2O \rightarrow CH_2OH.CH_2OH$$

If the initial molar ratio of ethylene oxide to water is 3.0, estimate the equilibrium conversion of ethylene oxide to ethylene glycol.

At equilibrium the system consists of liquid and vapor in equilibrium, and the intensive state of the system is fixed by the specification of T and P. Therefore, one must first determine the phase compositions, independent of the ratio of reactants. These results may then be applied in the material-balance equations to find the equilibrium conversion.

Choose as standard states for water and ethylene glycol the pure liquids at 1 bar and for ethylene oxide the pure ideal gas at 1 bar. Assume any water present in the liquid phase has an activity coefficient of unity and that the vapor phase is an ideal gas. The partial pressure of ethylene oxide over the liquid phase is given by:

$$p_i/\text{kPa} = 415x_i$$

The vapor pressure of ethylene glycol at 25°C is so low that its concentration in the vapor phase is negligible.

14.40. In chemical-reaction engineering, special measures of product distribution are sometimes used when multiple reactions occur. Two of these are *yield* Y_j and *selectivity* $S_{j/k}$. We adopt the following definitions[12]:

$$Y_j \equiv \frac{\text{moles formed of desired product } j}{\text{moles of } j \text{ that would be formed with no side reactions and with complete consumption of the limiting reactant species}}$$

$$S_{j/k} \equiv \frac{\text{moles formed of desired product } j}{\text{moles formed of undesired product } k}$$

For any particular application, yield and selectivity can be related to component rates and reaction coordinates. For two-reaction schemes the two reaction coordinates can be found from Y_j and $S_{j/k}$, allowing the usual material-balance equations to be written.

Consider the gas-phase reactions:

$$A + B \rightarrow C \quad \text{(I)} \qquad\qquad A + C \rightarrow D \quad \text{(II)}$$

Here, C is the desired product, and D is the undesired by-product. If the feed to a steady-flow reactor contains 10 kmol·h^{-1} of A and 15 kmol·h^{-1} of B, and if $Y_C = 0.40$ and $S_{C/D} = 2.0$, determine complete product rates and the product composition (mole fractions), using reaction coordinates.

---

[12]R. M. Felder, R. W. Rousseau, and L. G. Bullard, *Elementary Principles of Chemical Processes*, 4th ed., Sec. 4.6d, Wiley, New York, 2015.

14.41. The following problems involving chemical-reaction stoichiometry are to be solved through the use of reaction coordinates.

(a) Feed to a gas-phase reactor comprises 50 kmol·h^{-1} of species A, and 50 kmol·h^{-1} of species B. Two independent reactions occur:

$$A + B \rightarrow C \quad (I) \qquad\qquad A + C \rightarrow D \quad (II)$$

Analysis of the gaseous effluent shows mole fractions $y_A = 0.05$ and $y_B = 0.10$.

(i) What is the reactor effluent rate in kmol·h^{-1} ?
(ii) What are the mole fractions y_C and y_D in the effluent?

(b) Feed to a gas-phase reactor comprises 40 kmol·h^{-1} of species A, and 40 kmol·h^{-1} of species B. Two independent reactions occur:

$$A + B \rightarrow C \quad (I) \qquad\qquad A + 2B \rightarrow D \quad (II)$$

Analysis of the gaseous effluent shows mole fractions: $y_C = 0.52$ and $y_D = 0.04$. Determine the rates (kmol·h^{-1}) of all species in the effluent stream.
(c) Feed to a gas-phase reactor is 100 kmol·h^{-1} of pure species A. Two independent reactions occur:

$$A \rightarrow B + C \quad (I) \qquad\qquad A + B \rightarrow D \quad (II)$$

Reaction (I) produces valuable species C and coproduct B. The side reaction (II) produces by-product D. Analysis of the gaseous effluent shows mole fractions $y_C = 0.30$ and $y_D = 0.10$. Determine the rates (kmol·h^{-1}) of all species in the effluent stream.
(d) The feed to a gas-phase reactor is 100 kmol·h^{-1}, containing 40 mol-% species A and 60 mol-% species B. Two independent reactions occur:

$$A + B \rightarrow C \quad (I) \qquad\qquad A + B \rightarrow D + E \quad (II)$$

Analysis of the gaseous effluent shows mole fractions $y_C = 0.25$ and $y_D = 0.20$. Determine:

(i) Rates in kmol·h^{-1} of all species in the effluent stream.
(ii) Mole fractions of all species in the effluent stream.

14.42. The following is an industrial-safety rule of thumb: compounds with large positive ΔG_f° must be handled and stored carefully. Explain.

14.43. Two important classes of reactions are *oxidation* reactions and *cracking* reactions. One class is invariably endothermic; the other, exothermic. Which is which? For which class of reactions (oxidation or cracking) does equilibrium conversion increase with increasing T?

14.44. The standard heat of reaction ΔH° for gas-phase reactions is independent of the choice of standard-state pressure P°. (Why?) However, the numerical value of ΔG° for such reactions does depend on P°. Two choices of P° are conventional: 1 bar (the basis adopted in this text), and 1.01325 bar. Show how to convert ΔG° for gas-phase reactions from values based on $P^\circ = 1$ bar to those based on $P^\circ = 1.01325$ bar.

14.45. Ethanol is produced from ethylene via the gas-phase reaction

$$C_2H_4(g) + H_2O(g) \rightarrow C_2H_5OH(g)$$

Reaction conditions are 400 K and 2 bar.

(*a*) Determine a numerical value for the equilibrium constant K for this reaction at 298.15 K.
(*b*) Determine a numerical value for K for this reaction at 400 K.
(*c*) Determine the composition of the *equilibrium* gas mixture for an equimolar feed containing only ethylene and H_2O. State all assumptions.
(*d*) For the same feed as in part (*c*), but for $P = 1$ bar, would the equilibrium mole fraction of ethanol be higher or lower? Explain.

14.46. A good source for formation data for compounds is the NIST Chemistry WebBook site. Values of ΔH_f°, but not of ΔG_f°, are reported. Instead, values of absolute standard entropies S° are listed for compounds and elements. To illustrate the use of NIST data, let H_2O_2 be the compound of interest. Values provided by the Chemistry WebBook are as follows:

- $\Delta H_f^\circ[H_2O_2(g)] = 136.1064 \text{ J·mol}^{-1}$
- $S^\circ[H_2O_2(g)] = 232.95 \text{ J·mol}^{-1}\cdot\text{K}^{-1}$
- $S^\circ[H_2(g)] = 130.680 \text{ J·mol}^{-1}\cdot\text{K}^{-1}$
- $S^\circ[O_2(g)] = 205.152 \text{ J·mol}^{-1}\cdot\text{K}^{-1}$

All data are for the ideal-gas state at 298.15 K and 1 bar. Determine a value of $\Delta G_{f_{298}}^\circ$ for $H_2O_2(g)$.

14.47. Reagent-grade, liquid-phase chemicals often contain as impurities isomers of the nominal compound, with a consequent effect on the vapor pressure. This can be quantified by phase-equilibrium/reaction-equilibrium analysis. Consider a system containing isomers A and B in vapor/liquid equilibrium, and also in equilibrium with respect to the reaction A $\rightarrow$ B at relatively low pressure.

(*a*) For the reaction in the liquid phase, determine an expression for P (the "mixture vapor pressure") in terms of $P_A^{\text{sat}}, P_B^{\text{sat}}$, and K^l, the reaction equilibrium constant. Check the result for the limits $K^l = 0$ and $K^l = \infty$.
(*b*) For the reaction in the vapor phase, repeat part (*a*). Here, the relevant reaction equilibrium constant is K^v.
(*c*) If equilibrium prevails, then whether the reaction is assumed to occur in one phase or the other makes no difference. Thus the results for parts (*a*) and (*b*) must be equivalent. Use this idea to show the connection between K^l and K^v through the pure-species vapor pressures.

(*d*) Why is the assumption of ideal gases and ideal solutions both reasonable and prudent?

(*e*) Results for parts (*a*) and (*b*) should suggest that P depends on T only. Show that this is in accord with the phase rule.

14.48. Cracking propane is a route to light olefin production. Suppose that *two* cracking reactions occur in a steady-flow reactor:

$$C_3H_8(g) \rightarrow C_3H_6(g) + H_2(g) \qquad (I)$$

$$C_3H_8(g) \rightarrow C_2H_4(g) + CH_4(g) \qquad (II)$$

Calculate the product composition if both reactions go to equilibrium at 1.2 bar and

(*a*) 750 K;   (*b*) 1000 K;   (*c*)   1250 K

14.49. Equilibrium at 425 K and 15 bar is established for the gas-phase isomerization reaction

$$n\text{-}C_4H_{10}(g) \rightarrow iso\text{-}C_4H_{10}(g)$$

Estimate the composition of the equilibrium mixture by two procedures:

(*a*) Assume an ideal-gas mixture.

(*b*) Assume an ideal solution with the equation of state given by Eq. (3.36).

Compare and discuss the results.

Data: For *iso*-butane, $\Delta H^\circ_{f_{298}} = -134{,}180 \text{ J·mol}^{-1}$; $\Delta G^\circ_{f_{298}} = -20{,}760 \text{ J·mol}^{-1}$

Chapter 15

Topics in Phase Equilibria

In chapters 12 and 13, we introduced phase equilibrium in general terms, but we focused primarily on vapor/liquid equilibrium. VLE has long been considered the most important type of phase equilibrium for chemical engineers, due to the prevalence of distillation as a separation method in the chemical industry. However, a wide variety of other phase equilibria are of importance in chemical engineering. This chapter deals more generally with phase equilibria, with consideration given in separate sections to liquid/liquid, vapor/liquid/liquid, solid/liquid, solid/vapor, adsorption, and osmotic equilibria. In each case, we aim to provide a qualitative introduction and a quantitative framework that is sufficient to begin to treat practical problems and that provides a foundation for more specialized study of these topics.

15.1 LIQUID/LIQUID EQUILIBRIUM

The criteria for stability of a single liquid phase and the general features of liquid/liquid equilibrium were introduced in Sec. 12.4. A key result from that section is the following criterion of stability for a single-phase binary system for which the change in Gibbs energy upon mixing is $\Delta G \equiv G - x_1 G_1 - x_2 G_2$:

At fixed temperature and pressure, a single-phase binary mixture is stable if and only if ΔG and its first and second derivatives are continuous functions of x_1, and the second derivative is positive.

Thus,
$$\frac{d^2 \Delta G}{dx_1^2} > 0 \qquad \text{(const } T, P\text{)}$$

and
$$\boxed{\frac{d^2(\Delta G/RT)}{dx_1^2} > 0 \qquad \text{(const } T, P\text{)}} \qquad (12.4)$$

This criterion can be most readily applied in the context of an activity coefficient formulation of the excess Gibbs energy. For a binary mixture Eq. (13.10) is:

$$\frac{G^E}{RT} = x_1 \ln \gamma_1 + x_2 \ln \gamma_2$$

587

and
$$\frac{d(G^E/RT)}{dx_1} = \ln \gamma_1 - \ln \gamma_2 + x_1 \frac{d \ln \gamma_1}{dx_1} + x_2 \frac{d \ln \gamma_2}{dx_1}$$

According to Eq. (13.11), the activity-coefficient form of the Gibbs/Duhem equation, the last two terms sum to zero; whence:

$$\frac{d(G^E/RT)}{dx_1} = \ln \gamma_1 - \ln \gamma_2$$

A second differentiation and a second application of the Gibbs/Duhem equation gives:

$$\frac{d^2(G^E/RT)}{dx_1^2} = \frac{d \ln \gamma_1}{dx_1} - \frac{d \ln \gamma_2}{dx_1} = \frac{1}{x_2} \frac{d \ln \gamma_1}{dx_1}$$

This equation in combination with Eq. (12.5) yields:

$$\frac{d \ln \gamma_1}{dx_1} > -\frac{1}{x_1} \qquad (\text{const } T, P)$$

which is yet another condition for stability. It is equivalent to Eq. (12.4), from which it ultimately derives. Other stability criteria follow directly, e.g.,

$$\frac{d\hat{f}_1}{dx_1} > 0 \qquad \frac{d\mu_1}{dx_1} > 0 \qquad (\text{const } T, P)$$

The last three stability conditions can equally well be written for species 2; thus for *either* species in a binary mixture:

$$\boxed{\frac{d \ln \gamma_i}{dx_i} > -\frac{1}{x_i} \qquad (\text{const } T, P)} \qquad (15.1)$$

$$\boxed{\frac{d\hat{f}_i}{dx_i} > 0 \qquad (\text{const } T, P) \qquad (15.2)} \qquad \boxed{\frac{d\mu_i}{dx_i} > 0 \qquad (\text{const } T, P) \qquad (15.3)}$$

Example 15.1

The stability criteria apply to a *particular* phase. However, there is nothing to preclude their application to problems in phase equilibria, where the phase of interest (e.g., a liquid mixture) is in equilibrium with another phase (e.g., a vapor mixture). Consider binary isothermal vapor/liquid equilibria at pressures low enough that the vapor phase may be considered an ideal-gas mixture. What are the implications of liquid-phase stability to the features of isothermal *Pxy* diagrams such as those in Fig. 12.8?

Solution 15.1

Focus initially on the *liquid* phase. By Eq. (15.2) applied to species 1,

$$\frac{d\hat{f}_1}{dx_1} = \hat{f}_1 \frac{d\ln \hat{f}_1}{dx_1} > 0$$

whence, because $\hat{f}_1$ cannot be negative,

$$\frac{d\ln \hat{f}_1}{dx_1} > 0$$

Similarly, with Eq. (15.2) applied to species 2 and $dx_2 = -dx_1$:

$$\frac{d\ln \hat{f}_2}{dx_1} < 0$$

Combination of the last two inequalities gives:

$$\frac{d\ln \hat{f}_1}{dx_1} - \frac{d\ln \hat{f}_2}{dx_1} > 0 \qquad \text{(const } T, P) \qquad (A)$$

which is the basis for the first part of this analysis. Because $\hat{f}_i^v = y_i P$ for an ideal-gas mixture and because $\hat{f}_i^l = \hat{f}_i^v$ for VLE, the left side of Eq. (A) may be written:

$$\frac{d\ln \hat{f}_1}{dx_1} - \frac{d\ln \hat{f}_2}{dx_1} = \frac{d\ln y_1 P}{dx_1} - \frac{d\ln y_2 P}{dx_1} = \frac{d\ln y_1}{dx_1} - \frac{d\ln y_2}{dx_1}$$

$$= \frac{1}{y_1}\frac{dy_1}{dx_1} - \frac{1}{y_2}\frac{dy_2}{dx_1} = \frac{1}{y_1}\frac{dy_1}{dx_1} + \frac{1}{y_2}\frac{dy_1}{dx_1} = \frac{1}{y_1 y_2}\frac{dy_1}{dx_1}$$

Thus, Eq. (A) yields:

$$\frac{dy_1}{dx_1} > 0 \qquad (B)$$

which is an essential feature of binary VLE. Note that, although P is not constant for isothermal VLE, Eq. (A) is still approximately valid, because its application is to the *liquid* phase, for which properties are insensitive to pressure.

The next part of this analysis draws on the fugacity form of the Gibbs/Duhem equation, Eq. (13.27), applied again to the *liquid* phase:

$$x_1 \frac{d\ln \hat{f}_1}{dx_1} + x_2 \frac{d\ln \hat{f}_2}{dx_1} = 0 \qquad \text{(const } T, P) \qquad (13.27)$$

Note again that the restriction here to constant P is of no significance, because of the insensitivity of liquid-phase properties to pressure. With $\hat{f}_i = y_i P$ for low-pressure VLE,

$$x_1 \frac{d\ln y_1 P}{dx_1} + x_2 \frac{d\ln y_2 P}{dx_1} = 0$$

$$\frac{1}{P}\frac{dP}{dx_1} = \frac{(y_1 - x_1)}{y_1 y_2}\frac{dy_1}{dx_1} \qquad (C)$$

Because by Eq. (B) $dy_1/dx_1 > 0$, Eq. (C) asserts that the sign of dP/dx_1 is the same as the sign of the quantity $y_1 - x_1$.

The last part of this analysis is based only upon mathematics, according to which, at constant T,

$$\frac{dP}{dy_1} = \frac{dP/dx_1}{dy_1/dx_1} \qquad (D)$$

But by Eq. (B), $dy_1/dx_1 > 0$. Thus dP/dy_1 has the same sign as dP/dx_1.

In summary, the stability requirement implies the following for VLE in binary systems at constant temperature:

$$\boxed{\frac{dy_1}{dx_1} > 0 \qquad \frac{dP}{dx_1}, \frac{dP}{dy_1}, \text{ and } (y_1 - x_1) \text{ have the same sign}}$$

At an azeotrope, where $y_1 = x_1$,

$$\frac{dP}{dx_1} = 0 \qquad \text{and} \qquad \frac{dP}{dy_1} = 0$$

Although derived for conditions of low pressure, these results are of general validity, as illustrated by the VLE data shown in Fig. 12.8.

Many pairs of chemical species, were they to mix to form a single liquid phase in a certain composition range, would not satisfy the stability criterion of Eq. (12.4). Such systems therefore split in this composition range into two liquid phases of different compositions. If the phases are at thermodynamic equilibrium, the phenomenon is an example of *liquid/liquid equilibrium* (LLE), which is important for industrial operations such as solvent extraction.

The equilibrium criteria for LLE are the same as for VLE, namely, uniformity of T, P, and of the fugacity $\hat{f}_i$ for each chemical species throughout both phases. For LLE in a system of N species at uniform T and P, we denote the liquid phases by superscripts α and β, and we write the equilibrium criteria as:

$$\hat{f}_i^\alpha = \hat{f}_i^\beta \qquad (i = 1, 2, \ldots, N)$$

With the introduction of activity coefficients, this becomes:

$$x_i^\alpha \gamma_i^\alpha f_i^\alpha = x_i^\beta \gamma_i^\beta f_i^\beta$$

If each pure species exists as liquid at the system temperature, $f_i^\alpha = f_i^\beta = f_i$; whence,

$$\boxed{x_i^\alpha \gamma_i^\alpha = x_i^\beta \gamma_i^\beta \qquad (i = 1, 2, \ldots, N)} \qquad (15.4)$$

In Eq. (15.4), activity coefficients γ_i^α and γ_i^β are derived from the *same function* G^E/RT; thus they are functionally identical, distinguished mathematically only by the mole fractions to which they apply. For a liquid/liquid system containing N chemical species:

$$\gamma_i^\alpha = \gamma_i(x_1^\alpha, x_2^\alpha, \ldots, x_{N-1}^\alpha, T, P) \qquad (15.5a)$$

$$\gamma_i^\beta = \gamma_i(x_1^\beta, x_2^\beta, \ldots, x_{N-1}^\beta, T, P) \qquad (15.5b)$$

According to Eqs. (15.4) and (15.5), N equilibrium equations can be written in $2N$ intensive variables (T, P, and $N - 1$ independent mole fractions for each phase). Solution of the equilibrium equations for LLE therefore requires prior specification of numerical values for N of the intensive variables. This is in accord with the phase rule, Eq. (3.1), from which $F = 2 - \pi + N = 2 - 2 + N = N$. The same result is obtained for VLE with no special constraints on the equilibrium state.

In the general description of LLE, any number of species may be considered, and pressure may be a significant variable. We treat here a simpler (but important) special case, that of *binary* LLE either at constant pressure or at conditions at which the effect of pressure on the activity coefficients may be ignored. With but one independent mole fraction per phase, Eq. (15.4) gives:

$$x_1^\alpha \gamma_1^\alpha = x_1^\beta \gamma_1^\beta \qquad (15.6a) \qquad (1 - x_1^\alpha)\gamma_2^\alpha = \left(1 - x_1^\beta\right)\gamma_2^\beta \qquad (15.6b)$$

where

$$\gamma_i^\alpha = \gamma_i(x_1^\alpha, T) \qquad (15.7a) \qquad \gamma_i^\beta = \gamma_i(x_1^\beta, T) \qquad (15.7b)$$

With two equations and three variables (x_1^α, x_1^β, and T), fixing one of the variables allows solution of Eqs. (15.6) for the remaining two. Because $\ln \gamma_i$, rather than γ_i, is a more natural thermodynamic function, application of Eqs. (15.6) often proceeds from the rearrangements:

$$\ln \frac{\gamma_1^\alpha}{\gamma_1^\beta} = \ln \frac{x_1^\beta}{x_1^\alpha} \qquad (15.8a) \qquad \ln \frac{\gamma_2^\alpha}{\gamma_2^\beta} = \ln \frac{1 - x_1^\beta}{1 - x_1^\alpha} \qquad (15.8b)$$

Example 15.2

Develop equations that apply to the limiting case of binary LLE for which the α phase is very dilute in species 1 and the β phase is very dilute in species 2.

Solution 15.2

For the case described, to a good approximation,

$$\gamma_1^\alpha \simeq \gamma_1^\infty \qquad \gamma_2^\alpha \simeq 1 \qquad \gamma_1^\beta \simeq 1 \qquad \gamma_2^\beta \simeq \gamma_2^\infty$$

Substitution into the equilibrium equations, Eqs. (15.6), gives:

$$x_1^\alpha \gamma_1^\infty \simeq x_1^\beta \qquad \text{and} \qquad 1 - x_1^\alpha \simeq \left(1 - x_1^\beta\right)\gamma_2^\infty$$

and solution for the mole fractions yields the approximate expressions:

$$x_1^\alpha = \frac{\gamma_2^\infty - 1}{\gamma_1^\infty \gamma_2^\infty - 1} \qquad (A) \qquad x_1^\beta = \frac{\gamma_1^\infty (\gamma_2^\infty - 1)}{\gamma_1^\infty \gamma_2^\infty - 1} \qquad (B)$$

Alternatively, solution for the infinite-dilution activity coefficients gives:

$$\gamma_1^\infty = \frac{x_1^\beta}{x_1^\alpha} \quad (C) \qquad \gamma_2^\infty = \frac{1 - x_1^\alpha}{1 - x_1^\beta} \quad (D)$$

Equations (A) and (B) provide order-of-magnitude estimates of equilibrium compositions from two-parameter expressions for G^E/RT, where the γ_i^∞ are usually related to the parameters in a simple way. Equations (C) and (D) serve the opposite function; they provide simple explicit expressions for γ_i^∞ in relation to measurable equilibrium compositions. Equations (C) and (D) show that positive deviations from ideal-solution behavior promote LLE:

$$\gamma_1^\infty \simeq \frac{1}{x_1^\alpha} > 1 \quad \text{and} \quad \gamma_2^\infty \simeq \frac{1}{x_2^\beta} > 1$$

The extreme example of binary LLE is that of *complete immiscibility* of the two species. When $x_1^\alpha = x_2^\beta = 0$, γ_1^β and γ_2^α are unity, and Eqs. (15.6) therefore require:

$$\gamma_1^\alpha = \gamma_2^\beta = \infty$$

Strictly speaking, no two liquids are completely immiscible. However, actual solubilities may be so small (e.g., for some hydrocarbon/water systems) that the idealizations $x_1^\alpha = x_2^\beta = 0$ provide suitable approximations for practical calculations (Ex. 15.7).

Example 15.3

The simplest expression for G^E/RT capable of predicting LLE is:

$$\frac{G^E}{RT} = A x_1 x_2 \qquad (A)$$

Derive the equations resulting from application of this equation to LLE.

Solution 15.3

The activity coefficients implied by the given equation are:

$$\ln \gamma_1 = A x_2^2 = A(1 - x_1)^2 \qquad \text{and} \qquad \ln \gamma_2 = A x_1^2$$

Specializing these two expressions to the α and β phases and combining them with Eqs. (15.8) gives:

$$A\left[(1 - x_1^\alpha)^2 - (1 - x_1^\beta)^2\right] = \ln \frac{x_1^\beta}{x_1^\alpha} \qquad (B)$$

$$A\left[(x_1^\alpha)^2 - (x_1^\beta)^2\right] = \ln \frac{1 - x_1^\beta}{1 - x_1^\alpha} \qquad (C)$$

Given a value of parameter A, one finds equilibrium compositions x_1^α and x_1^β as the solution to Eqs. (B) and (C).

Solubility curves implied by Eq. (A) are symmetrical about $x_1 = 0.5$. This can be inferred from the fact that x_1^α and x_1^β appear in exactly the same form in Eqs. (B) and (C). This symmetry can be expressed by the relationship

$$x_1^\beta = 1 - x_1^\alpha \tag{D}$$

Substituting Eq. (D) into Eqs. (B) and (C) reduces them both to the *same* equation:

$$A(1 - 2x_1) = \ln \frac{1 - x_1}{x_1} \tag{E}$$

This implies that the inferred symmetry about $x_1 = 0.5$ is correct. When $A > 2$, this equation has three real roots: $x_1 = 1/2$, $x_1 = r$, and $x_1 = 1 - r$, where $0 < r < 1/2$. The latter two roots are the *equilibrium* compositions (x_1^α and x_1^β), whereas the first root is a trivial solution. For $A < 2$ only the trivial solution exists; the value $A = 2$ corresponds to a consolute point, where the three roots converge to the value 0.5. Table 15.1 shows values of A as calculated from Eq. (E) for various values of x_1^α ($= 1 - x_1^\beta$). Note particularly the sensitivity of x_1^α to small increases in A from its limiting value of 2.

Table 15.1: Liquid/Liquid Equilibrium Compositions Implied by Eq. (A)

A	x_1^α	A	x_1^α
2.0000	0.50	2.4780	0.15
2.0067	0.45	2.7465	0.10
2.0273	0.40	3.2716	0.05
2.0635	0.35	4.6889	0.01
2.1182	0.30	5.3468	0.005
2.1972	0.25	6.9206	0.001
2.3105	0.20	7.6080	0.0005

The actual *shape* of a solubility curve is determined by the temperature dependence of G^E/RT. Assume the following T dependence of parameter A in Eq. (A):

$$A = \frac{a}{T} + b - c \ln T \tag{F}$$

where a, b, and c are constants. From Table 10.1, we have

$$H^E = -RT^2 \left[\frac{\partial (G^E/RT)}{\partial T} \right]_{P,x}$$

Applying this to Eq. (A) shows that the temperature dependence of Eq. (F) makes the excess enthalpy H^E linear in T, and the excess heat capacity C_P^E independent of T:

$$H^E = R(a + cT)x_1 x_2 \tag{G}$$

$$C_P^E = \left(\frac{\partial H^E}{\partial T}\right)_{P,x} = Rcx_1x_2 \tag{H}$$

The excess enthalpy and the temperature dependence of A are directly related. From Eq. (F),

$$\frac{dA}{dT} = -\frac{1}{T^2}(a + cT)$$

Combination of this equation with Eq. (G) yields:

$$\frac{dA}{dT} = -\frac{H^E}{x_1x_2RT^2}$$

Thus dA/dT is negative for an endothermic system (positive H^E) and positive for an exothermic system (negative H^E). A negative value of dA/dT at a consolute point implies a UCST because A decreases to 2.0 as T *increases*. Conversely, a positive value implies an LCST because A decreases to 2.0 as T *decreases*. Hence a system described by Eqs. (A) and (F) exhibits a UCST if endothermic at the consolute point and an LCST if exothermic at the consolute point. Equation (F) written for a consolute point $(A = 2)$ becomes:

$$T \ln T = \frac{a}{c} - \left(\frac{2 - b}{c}\right)T \tag{I}$$

Depending on the values of a, b, and c, this equation has zero, one, or two temperature roots.

Consider hypothetical binary systems described by Eqs. (A) and (F) and for which LLE obtains in the temperature range 250 to 450 K. Setting $c = 3.0$ makes the excess heat capacity positive, independent of T, for which by Eq. (H) the maximum value (at $x_1 = x_2 = 0.5$) is 6.24 J mol^{-1} K^{-1}. For the first case, let

$$A = \frac{-975}{T} + 22.4 - 3 \ln T$$

Here, Eq. (I) has two roots, corresponding to an LCST and a UCST:

$$T_L = 272.9 \qquad \text{and} \qquad T_U = 391.2 \text{ K}$$

Values of A are plotted vs. T in Fig. 15.1(*a*) and the solubility curve [from Eq. (E)] is shown by Fig. 15.1(*b*). This case—that of a closed solubility loop—is of the type shown by Fig. 12.14(*a*). It requires that H^E *change sign* in the temperature interval for which LLE is posssible.

As a second case, let

$$A = \frac{-540}{T} + 21.1 - 3 \ln T$$

Here, Eq. (I) has only *one* root in the temperature range 250 to 450 K. It is a UCST, $T_U = 346.0$ K, because Eq. (G) yields positive H^E at this temperature. Values of A and the corresponding solubility curve are given by Fig. 15.2.

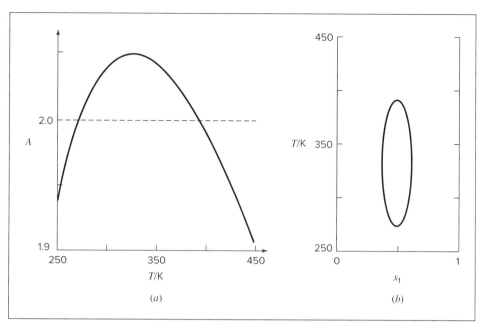

Figure 15.1: (*a*) A vs. T. (*b*) Solubility diagram for a binary system described by $G^E/RT = Ax_1x_2$ with $A = -975/T + 22.4 - 3 \ln T$. This is an example for which H^E changes sign.

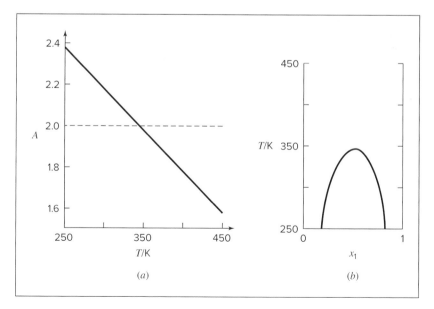

Figure 15.2: (*a*) A vs. T; (*b*) Solubility diagram for a binary system described by $G^E/RT = Ax_1x_2$ with $A = -540/T + 21.1 - 3 \ln T$. In this example H^E is positive and does not change sign.

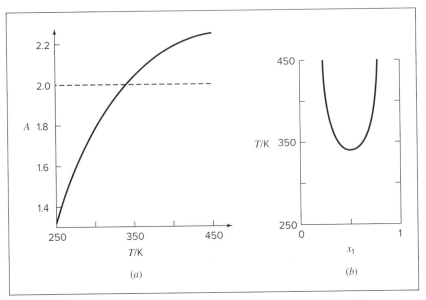

Figure 15.3: (*a*) A vs. T; (*b*) Solubility diagram for a binary system described by $G^E/RT = Ax_1x_2$ with $A = -1500/T + 23.9 - 3\ln T$. In this case H^E is negative and does not change sign.

Finally, let

$$A = \frac{-1500}{T} + 23.9 - 3\ln T$$

This case is similar to the second, there being only one T (339.7 K) that solves Eq. (*I*) for the temperature range considered. However, this is a LCST, because H^E is now negative. Values of A and the solubility curve are shown in Fig. 15.3.

Example 15.3 demonstrates in a "brute-force" way that LLE cannot be predicted by the expression $G^E/RT = Ax_1x_2$ for values of $A < 2$. If the goal is merely to determine under what conditions LLE can occur, but not to find the compositions of the coexisting phases, then one may instead invoke the stability criteria of Sec. 12.4 and determine under what conditions they are satisfied.

Example 15.4

If $G^E/RT = Ax_1x_2$ for a liquid phase, show by stability analysis that LLE is predicted for $A \geq 2$.

Solution 15.4

Application of the stability criterion requires evaluation of the derivative:

$$\frac{d^2(G^E/RT)}{dx_1^2} = \frac{d^2(Ax_1x_2)}{dx_1^2} = -2A$$

Thus, stability requires:
$$2A < \frac{1}{x_1 x_2}$$

When $x_1 = x_2 = 1/2$, the right side of this inequality has its minimum value of 4; thus $A < 2$ yields stability of single-phase mixtures over the entire composition range. Conversely, if $A > 2$, then binary mixtures described by $G^E/RT = A x_1 x_2$ form two liquid phases over some part of the composition range.

Example 15.5

Some expressions for G^E/RT are incapable of representing LLE. An example is the Wilson equation:

$$\frac{G^E}{RT} = -x_1 \ln(x_1 + x_2 \Lambda_{12}) - x_2 \ln(x_2 + x_1 \Lambda_{21}) \qquad (13.45)$$

Show that the stability criteria are satisfied for all values of Λ_{12}, Λ_{21}, and x_1.

Solution 15.5

An equivalent form of inequality (15.1) for species 1 is:

$$\frac{d \ln(x_1 \gamma_1)}{d x_1} > 0 \qquad (A)$$

For the Wilson equation, $\ln \gamma_1$ is given by Eq. (13.46). Addition of $\ln x_1$ to both sides of that equation yields:

$$\ln(x_1 \gamma_1) = -\ln\left(1 + \frac{x_2}{x_1} \Lambda_{12}\right) + x_2 \left(\frac{\Lambda_{12}}{x_1 + x_2 \Lambda_{12}} - \frac{\Lambda_{21}}{x_2 + x_1 \Lambda_{21}}\right)$$

from which:
$$\frac{d \ln(x_1 \gamma_1)}{d x_1} = \frac{x_2 \Lambda_{21}^2}{x_1 (x_1 + x_2 \Lambda_{12})^2} + \frac{\Lambda_{21}^2}{(x_2 + x_1 \Lambda_{21})^2}$$

All quantities on the right side of this equation are positive, and therefore Eq. (A) is satisfied for all x_1 and for all nonzero Λ_{12} and Λ_{21}.[1] Thus inequality (15.1) is always satisfied, and LLE cannot be represented by the Wilson equation.

15.2 VAPOR/LIQUID/LIQUID EQUILIBRIUM (VLLE)

As noted in Sec. 12.4, the binodal curves representing LLE can intersect the VLE bubblepoint curve. This gives rise to the phenomenon of vapor/liquid/liquid equilibrium (VLLE), as illustrated in Fig. 12.15 though 12.18. For a binary system with three phases in equilibrium, the phase rule tells us that only one degree of freedom remains. Thus, at a given T, two binary

---

[1]Both Λ_{12} and Λ_{21} are positive *definite*, because $\Lambda_{12} = \Lambda_{21} = 0$ yields infinite values for γ_1^∞ and γ_2^∞.

liquid phases in equilibrium have a fixed vapor pressure, and at a given P, they have a fixed boiling point temperature.

The compositions of the vapor and liquid phases in equilibrium for partially miscible systems are calculated in the same way as for miscible systems. In the regions where a single liquid is in equilibrium with its vapor, VLE calculations are done as in Chapter 13. Because limited miscibility implies highly nonideal behavior, any general assumption of liquid-phase ideality is excluded. Even a combination of Henry's law, valid for a species at infinite dilution, and Raoult's law, valid for a species as it approaches purity, is not very useful, because each approximates actual behavior for only a very small composition range. Thus G^E is large, and its composition dependence is often not adequately represented by simple equations. Nevertheless, the NRTL and UNIQUAC equations and the UNIFAC method (App. G) often provide suitable correlations for activity coefficients.

Example 15.6

Careful equilibrium measurements for the diethyl ether(1)/water(2) system at 35°C have been reported.[2] Discuss the correlation and behavior of the phase-equilibrium data for this system.

Solution 15.6

The Pxy behavior of this system is shown by Fig. 15.4, where the very rapid rise in pressure with increasing liquid-phase ether concentration in the dilute-ether region is apparent. The three-phase pressure, $P^* = 104.6$ kPa, is reached at an ether mole fraction of only 0.0117. Here, y_1 also increases very rapidly to its three-phase value of $y_1^* = 0.946$. In the dilute-water region, on the other hand, rates of change are quite small, as shown to an expanded scale in Fig. 15.4(b).

The curves in Fig. 15.4 provide an excellent correlation of the VLE data. They result from *BUBL P* calculations carried out with the excess Gibbs energy and activity coefficients from a four-parameter modified Margules equation:

$$\frac{G^E}{x_1 x_2 RT} = A_{21} x_1 + A_{12} x_2 - Q$$

$$\ln \gamma_1 = x_2^2 \left[A_{12} + 2(A_{21} - A_{12})x_1 - Q - x_1 \frac{dQ}{dx_1} \right]$$

$$\ln \gamma_2 = x_1^2 \left[A_{21} + 2(A_{12} - A_{21})x_2 - Q + x_2 \frac{dQ}{dx_1} \right]$$

$$Q = \frac{\alpha_{12} x_1 \alpha_{21} x_2}{\alpha_{12} x_1 + \alpha_{21} x_2} \qquad \frac{dQ}{dx_1} = \frac{\alpha_{12} \alpha_{21} \left(\alpha_{21} x_2^2 - \alpha_{12} x_1^2 \right)}{(\alpha_{12} x_1 + \alpha_{21} x_2)^2}$$

$$A_{21} = 3.35629 \qquad A_{12} = 4.62424 \qquad \alpha_{12} = 3.78608 \qquad \alpha_{21} = 1.81775$$

---

[2]M. A. Villamañán, A. J. Allawi, and H. C. Van Ness, *J. Chem. Eng. Data*, vol. 29, pp. 431–435, 1984.

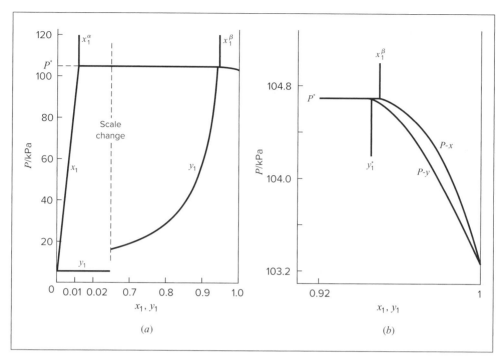

Figure 15.4: (*a*) *Pxy* diagram at 35°C for diethyl ether(1)/water(2); (*b*) Detail of ether-rich region.

The *BUBL P* calculations also require values of Φ_1 and Φ_2, which come from Eqs. (13.63) and (13.64) with virial coefficients:

$$B_{11} = -996 \qquad B_{22} = -1245 \qquad B_{12} = -567 \text{ cm}^3 \cdot \text{mol}^{-1}$$

In addition, the vapor pressures of the pure species at 35°C are:

$$P_1^{\text{sat}} = 103.264 \qquad P_2^{\text{sat}} = 5.633 \text{ kPa}$$

The high degree of nonideality of the liquid phase is indicated by the values of the activity coefficients of the dilute species, which range for diethyl ether from $\gamma_1 = 81.8$ at $x_1^\alpha = 0.0117$ to $\gamma_1^\infty = 101.9$ at $x_1 = 0$ and for water from $\gamma_2 = 19.8$ at $x_1^\beta = 0.9500$ to $\gamma_2^\infty = 28.7$ at $x_1 = 1$.

Thermodynamic insight into the phenomenon of low-pressure VLLE is provided by the modified Raoult's-law expression, Eq. (13.19). For temperature T and the three-phase-equilibrium pressure P^*, the modified Raoult's law applies to each liquid phase:

$$x_i^\alpha \gamma_i^\alpha P_i^{\text{sat}} = y_i^* P^* \qquad \text{and} \qquad x_i^\beta \gamma_i^\beta P_i^{\text{sat}} = y_i^* P^*$$

Implicit in these equations is the LLE requirement of Eq. (15.4). Thus four equations can be written for a binary system:

$x_1^\alpha \gamma_1^\alpha P_1^{\text{sat}} = y_1^* P^*$ (A)	$x_1^\beta \gamma_1^\beta P_1^{\text{sat}} = y_1^* P^*$ (B)
$x_2^\alpha \gamma_2^\alpha P_2^{\text{sat}} = y_2^* P^*$ (C)	$x_2^\beta \gamma_2^\beta P_2^{\text{sat}} = y_2^* P^*$ (D)

All of these equations are correct, but two of them are preferred over the others. Consider the expressions for $y_1^* P^*$:

$$x_1^\alpha \gamma_1^\alpha P_1^{\text{sat}} = x_1^\beta \gamma_1^\beta P_1^{\text{sat}} = y_1^* P^*$$

For the case of two species that approach complete immiscibility,

$$x_1^\alpha \to 0 \qquad \gamma_1^\alpha \to \gamma_1^\infty \qquad x_1^\beta \to 1 \qquad \gamma_1^\beta \to 1$$

Thus,

$$(0)(\gamma_1^\infty) P_1^{\text{sat}} = P_1^{\text{sat}} = y_1^* P^*$$

This equation implies that $\gamma_1^\infty \to \infty$; a similar derivation shows that $\gamma_2^\infty \to \infty$. Thus Eqs. (B) and (C), which include neither γ_1^α nor γ_2^β, are chosen as the more useful expressions. They may be added to give the three-phase pressure:

$$P^* = x_1^\beta \gamma_1^\beta P_1^{\text{sat}} + x_2^\alpha \gamma_2^\alpha P_2^{\text{sat}} \tag{15.9}$$

In addition, the three-phase vapor composition is given by Eq. (B):

$$y_1^* = \frac{x_1^\beta \gamma_1^\beta P_1^{\text{sat}}}{P^*} \tag{15.10}$$

For the diethyl ether(1)/water(2) system at 35°C (Ex. 15.6), the correlation for G^E/RT provides the values:

$$\gamma_1^\beta = 1.0095 \qquad \gamma_2^\alpha = 1.0013$$

These allow calculation of P^* and y_1^* by Eqs. (15.9) and (15.10):

$$P^* = (0.9500)(1.0095)(103.264) + (0.9883)(1.0013)(5.633) = 104.6 \text{ kPa}$$

and

$$y_1^* = \frac{(0.9500)(1.0095)(103.264)}{104.6} = 0.946$$

Although no two liquids are totally immiscible, this condition is so closely approached in some instances that the assumption of complete immiscibility does not lead to appreciable error. The phase characteristics of an immiscible system were illustrated by the temperature/composition diagram of Fig. 12.18. Numerical calculations for immiscible systems are particularly simple, because of the following identities:

$$x_2^\alpha = 1 \qquad \gamma_2^\alpha = 1 \qquad x_1^\beta = 1 \qquad \gamma_1^\beta = 1$$

The three-phase-equilibrium pressure P^* as given by Eq. (15.9) is therefore:

$$P^* = P_1^{\text{sat}} + P_2^{\text{sat}}$$

Substitution of this equation and $x_1^\beta = \gamma_1^\beta = 1$ into Eq. (15.10) gives:

$$y_1^* = \frac{P_1^{\text{sat}}}{P_1^{\text{sat}} + P_2^{\text{sat}}}$$

For region I where vapor is in equilibrium with pure liquid 1, Eq. (13.19) becomes:

$$y_1(\text{I})P = P_1^{\text{sat}} \quad \text{or} \quad y_1(\text{I}) = \frac{P_1^{\text{sat}}}{P}$$

Similarly, for region II where vapor is in equilibrium with pure liquid 2,

$$y_2(\text{II})P = [1 - y_1(\text{II})]P = P_2^{\text{sat}} \quad \text{or} \quad y_1(\text{II}) = 1 - \frac{P_2^{\text{sat}}}{P}$$

Example 15.7

Prepare a table of temperature/composition data for the benzene(1)/water(2) system at a pressure of 101.33 kPa (1 atm) from the vapor-pressure data in the accompanying table.

$t/°C$	$P_1^{\text{sat}}/\text{kPa}$	$P_2^{\text{sat}}/\text{kPa}$	$P_1^{\text{sat}} + P_2^{\text{sat}}$
60	52.22	19.92	72.14
70	73.47	31.16	104.63
75	86.40	38.55	124.95
80	101.05	47.36	148.41
80.1	101.33	47.56	148.89
90	136.14	70.11	206.25
100.0	180.04	101.33	281.37

Solution 15.7

Assume that benzene and water are completely immiscible as liquids. Then the three-phase equilibrium temperature $t*$ is estimated as:

$$P(t^*) = P_1^{\text{sat}} + P_2^{\text{sat}} = 101.33 \text{ kPa}$$

The last column of the table of vapor pressures shows that $t*$ lies between 60 and 70°C, and interpolation yields $t^* = 69.0°C$. At this temperature, again by interpolation: $P_1^{\text{sat}}(t^*) = 71.31$ kPa. Thus,

$$y_1^* = \frac{P_1^{\text{sat}}}{P_1^{\text{sat}} + P_2^{\text{sat}}} = \frac{71.31}{101.33} = 0.704$$

For the two regions of vapor/liquid equilibrium, as labelled in part (a) of Fig. 12.18.

$$y_1(\text{I}) = \frac{P_1^{\text{sat}}}{P} = \frac{P_1^{\text{sat}}}{101.33}$$

and
$$y_1(\text{II}) = 1 - \frac{P_2^{\text{sat}}}{P} = 1 - \frac{P_2^{\text{sat}}}{101.33}$$

Application of these equations for a number of temperatures gives the results summarized in the table that follows.

$t/^\circ$C	$y_1(\text{II})$	$y_1(\text{I})$
100	0.000	-
90	0.308	-
80.1	0.531	1.000
80	0.533	0.997
75	0.620	0.853
70	0.693	0.725
69.0	0.704	0.704

15.3 SOLID/LIQUID EQUILIBRIUM (SLE)

Phase behavior involving the solid and liquid states is the basis of important separation processes (e.g., crystallization) across many facets of chemical and materials engineering. Indeed, a wide variety of binary phase behavior is observed for systems exhibiting solid/solid, solid/liquid, and solid/solid/liquid equilibria. We develop here a rigorous formulation of solid/liquid equilibrium (SLE), and we present as applications analyses of two limiting classes of behavior. Comprehensive treatments can be found elsewhere.[3]

The basis for representing SLE is:

$$\hat{f}_i^l = \hat{f}_i^s \quad \text{(all } i)$$

where uniformity of T and P is understood. As with LLE, each $\hat{f}_i$ is eliminated in favor of an activity coefficient. Thus,

$$x_i \gamma_i^l f_i^l = z_i \gamma_i^s f_i^s \quad \text{(all } i)$$

where x_i and z_i are, respectively, the mole fractions of species i in the liquid and solid solutions. Equivalently,

$$x_i \gamma_i^l = z_i \gamma_i^s \psi_i \quad \text{(all } i) \tag{15.11}$$

where
$$\psi_i \equiv f_i^s / f_i^l \tag{15.12}$$

The right side of this equation, defining ψ_i as the ratio of fugacities at the T and P of the system, may be written in expanded form:

$$\frac{f_i^s(T, P)}{f_i^l(T, P)} = \frac{f_i^s(T, P)}{f_i^s(T_{m_i}, P)} \cdot \frac{f_i^s(T_{m_i}, P)}{f_i^l(T_{m_i}, P)} \cdot \frac{f_i^l(T_{m_i}, P)}{f_i^l(T, P)}$$

---

[3]See, e.g., R. T. DeHoff, *Thermodynamics in Materials Science*, chaps. 9 and 10, McGraw-Hill, New York, 1993. A data compilation is given by H. Knapp, M. Teller, and R. Langhorst, *Solid-Liquid Equilibrium Data Collection*, Chemistry Data Series, vol. VIII, DECHEMA, Frankfurt/Main, 1987.

where T_{mi} is the melting temperature ("freezing point") of pure species i, i.e., the temperature at which pure-species SLE occurs. Thus the second ratio on the right side is *unity* because $f_i^l = f_i^s$ at the melting point of pure species i. Hence,

$$\psi_i = \frac{f_i^s(T, P)}{f_i^s(T_{m_i}, P)} \cdot \frac{f_i^l(T_{m_i}, P)}{f_i^l(T, P)} \tag{15.13}$$

According to Eq. (15.13), evaluation of ψ_i requires expressions for the effect of temperature on fugacity. By Eq. (10.33), with $\phi_i = f_i/P$,

$$\ln \frac{f_i}{P} = \frac{G_i^R}{RT} \qquad \ln f_i = \frac{G_i^R}{RT} + \ln P$$

Whence,

$$\left(\frac{\partial \ln f_i}{\partial T}\right)_P = \left[\frac{\partial (G_i^R/RT)}{\partial T}\right]_P = -\frac{H_i^R}{RT^2}$$

where the second equality comes from Eq. (10.58). Integration of this equation for a *phase* from T_{mi} to T gives:

$$\frac{f_i(T, P)}{f_i(T_{m_i}, P)} = \exp \int_{T_{m_i}}^{T} -\frac{H_i^R}{RT^2} dT \tag{15.14}$$

Equation (15.14) is applied separately to the solid and liquid phases. The resulting expressions are substituted into Eq. (15.13), which is then reduced by the identity:

$$-(H_i^{R,s} - H_i^{R,l}) = -[(H_i^s - H_i^{ig}) - (H_i^l - H_i^{ig})] = H_i^l - H_i^s$$

This yields the exact expression:

$$\psi_i = \exp \int_{T_{m_i}}^{T} \frac{H_i^l - H_i^s}{RT^2} dT \tag{15.15}$$

Evaluation of the integral proceeds as follows:

$$H_i(T) = H_i(T_{m_i}) + \int_{T_{m_i}}^{T} C_{P_i} dT$$

and

$$C_{P_i}(T) = C_{P_i}(T_{m_i}) + \int_{T_{m_i}}^{T} \left(\frac{\partial C_{P_i}}{\partial T}\right)_P dT$$

Hence, for a *phase*,

$$H_i(T) = H_i(T_{m_i}) + C_{P_i}(T_{m_i})(T - T_{m_i}) + \int_{T_{m_i}}^{T} \int_{T_{m_i}}^{T} \left(\frac{\partial C_{P_i}}{\partial T}\right)_P dT dT \tag{15.16}$$

Applying Eq. (15.16) separately to the solid and liquid phases and performing the integration required by Eq. (15.15) yields:

$$\int_{T_{m_i}}^{T} \frac{H_i^l - H_i^s}{RT^2} \, dT = \frac{\Delta H_i^{sl}}{RT_{m_i}} \left(\frac{T - T_{m_i}}{T} \right)$$

$$+ \frac{\Delta C_{P_i}^{sl}}{R} \left[\ln \frac{T}{T_{m_i}} - \left(\frac{T - T_{m_i}}{T} \right) \right] + I$$

(15.17)

where integral I is defined by:

$$I \equiv \int_{T_{m_i}}^{T} \frac{1}{RT^2} \int_{T_{m_i}}^{T} \int_{T_{m_i}}^{T} \left[\frac{\partial (C_{P_i}^l - C_{P_i}^s)}{\partial T} \right]_P dT \, dT \, dT$$

In Eq. (15.17), ΔH_i^{sl} is the enthalpy change of melting ("heat of fusion") and $\Delta C_{P_i}^{sl}$ is the heat-capacity change of melting. Both quantities are evaluated at the melting temperature T_{m_i}.

Equations (15.11), (15.15), and (15.17) provide a formal basis for the solution of problems in solid/liquid equilibria. The full rigor of Eq. (15.17) is rarely maintained. For purposes of development, pressure has been carried through as a thermodynamic variable. However, its effect is rarely included in engineering applications. The triple integral represented by I is a second-order contribution and is normally neglected. The heat-capacity change of melting can be significant, but it is not always available; moreover, inclusion of the term involving $\Delta C_{P_i}^{sl}$ adds little to a qualitative understanding of SLE. With the assumptions that I and $\Delta C_{P_i}^{sl}$ are negligible, Eqs. (15.15) and (15.17) together yield:

$$\psi_i = \exp \frac{\Delta H_i^{sl}}{RT_{m_i}} \left(\frac{T - T_{m_i}}{T} \right)$$

(15.18)

With ψ_i given by Eq. (15.18), all that is required for formulating an SLE problem is a set of statements about the temperature and composition dependence of the activity coefficients γ_i^l and γ_i^s. In the general case, this requires algebraic expressions for $G^E(T, \text{composition})$ for both liquid and solid solutions. Consider two limiting special cases:

I. Assume ideal-solution behavior for both phases, i.e., let $\gamma_i^l = 1$ and $\gamma_i^s = 1$ for all T and compositions.

II. Assume ideal-solution behavior for the liquid phase ($\gamma_i^l = 1$), and complete immiscibility for all species in the solid state (i.e., set $z_i \gamma_i^s = 1$).

These two cases, restricted to binary systems, are considered in the following.

Case I

The two equilibrium equations which follow from Eq. (15.11) are:

$x_1 = z_1 \psi_1$	(15.19a)	$x_2 = z_2 \psi_2$	(15.19a)

where ψ_1 and ψ_2 are given by Eq. (15.18) with $i = 1,2$. Because $x_2 = 1 - x_1$ and $z_2 = 1 - z_1$, Eqs. (15.19) can be solved to give x_1 and z_1 as explicit functions of the ψ_i's and thus of T:

$$x_1 = \frac{\psi_1(1 - \psi_2)}{\psi_1 - \psi_2} \quad (15.20) \qquad z_1 = \frac{1 - \psi_2}{\psi_1 - \psi_2} \quad (15.21)$$

with

$$\psi_1 = \exp \frac{\Delta H_1^{sl}}{R T_{m_1}} \left(\frac{T - T_{m_1}}{T} \right) \quad (15.22a) \qquad \psi_2 = \exp \frac{\Delta H_2^{sl}}{R T_{m_2}} \left(\frac{T - T_{m_2}}{T} \right) \quad (15.22b)$$

Inspection of these results verifies that $x_i = z_i, = 1$ for $T = T_{m_i}$. Moreover, analysis shows that both x_i and z_i vary monotonically with T. Hence, systems described by Eqs. (15.19) exhibit lens-shaped SLE diagrams, as shown in Fig. 15.5(a), where the upper line is the freezing curve and the lower line is the melting curve. The liquid-solution region lies above the freezing curve, and the solid-solution region lies below the melting curve. Examples of systems exhibiting diagrams of this type range from nitrogen/carbon monoxide at cryogenic temperatures to copper/nickel at high temperature. Comparison of this figure with Fig. (12.12) suggests that Case I-SLE behavior is analogous to Raoult's-law behavior for VLE. Comparison of the assumptions leading to Eqs. (15.19) and (13.16) confirms the analogy. As with Raoult's law, Eq. (15.19) rarely describes the behavior of actual systems. However, it is an important limiting case, and it serves as a standard against which observed SLE can be compared.

Case II

The two equilibrium equations resulting from Eq. (15.11) are here:

$$x_1 = \psi_1 \quad (15.23) \qquad x_2 = \psi_2 \quad (15.24)$$

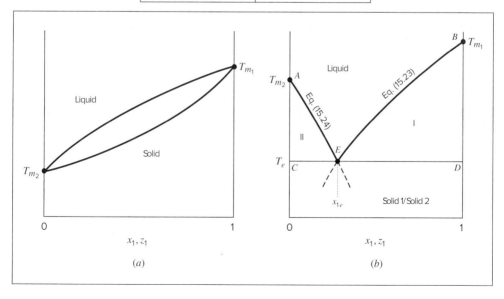

Figure 15.5: *Txz* diagrams. (a) Case I, ideal liquid and solid solutions; (b) Case II, ideal liquid solution; immiscible solids.

where ψ_1 and ψ_2 are given as functions solely of temperature by Eqs. (15.22). Thus x_1 and x_2 are also solely functions of temperature, and Eqs. (15.23) and (15.24) can apply simultaneously only for the particular temperature where $\psi_1 + \psi_2 = 1$ and hence $x_1 + x_2 = 1$. This is the *eutectic temperature* T_e. Thus, three distinct equilibrium situations exist: one where Eq. (15.23) alone applies, one where Eq. (15.24) alone applies, and the special case where they apply together at T_e.

- Equation (15.23) alone applies. By this equation and Eq. (15.22a),

$$x_1 = \exp\frac{\Delta H_1^{sl}}{RT_{m_1}}\left(\frac{T - T_{m_1}}{T}\right) \tag{15.25}$$

This equation is valid only from $T = T_{m_1}$, where $x_1 = 1$, to $T = T_e$, where $x_1 = x_{1e}$, the *eutectic composition*. Equation (15.25) therefore applies where a liquid solution is in equilibrium with pure species 1 as a solid phase. This is represented by region I on Fig. 15.5(b), where liquid solutions with compositions x_1 given by line BE are in equilibrium with pure solid 1.

- Equation (15.24) alone applies. By this equation and Eq. (15.22b), with $x_2 = 1 - x_1$:

$$x_1 = 1 - \exp\frac{\Delta H_2^{sl}}{RT_{m_2}}\left(\frac{T - T_{m_2}}{T}\right) \tag{15.26}$$

This equation is valid only from $T = T_{m_2}$, where $x_1 = 0$, to $T = T_e$, where $x_1 = x_{1e}$, the eutectic composition. Equation (15.26) therefore applies where a liquid solution is in equilibrium with pure species 2 as a solid phase. This is represented by region II on Fig. 15.5(b), where liquid solutions with compositions x_1 given by line AE are in equilibrium with pure solid 2.

- Equations (15.23) and (15.24) apply simultaneously, and are set equal because they must both give the eutectic composition x_{1e}. The resulting expression,

$$\exp\frac{\Delta H_1^{sl}}{RT_{m_1}}\left(\frac{T - T_{m_1}}{T}\right) = 1 - \exp\frac{\Delta H_2^{sl}}{RT_{m_2}}\left(\frac{T - T_{m_2}}{T}\right) \tag{15.27}$$

is satisfied for the single temperature $T = T_e$. Substitution of T_e into either Eq. (15.25) or (15.26) yields the eutectic composition. Coordinates T_e and x_{1e} define a *eutectic state*, a special state of three-phase equilibrium, lying along line CED on Fig. 15.5(b), for which liquid of composition x_{1e} coexists with pure solid 1 and pure solid 2. This is a state of solid/solid/liquid equilibrium. At temperatures below T_e the two pure immiscible solids coexist.

Figure 15.5(b), the phase diagram for Case II, is an exact analog of Fig. 12.18(a) for immiscible liquids because the assumptions upon which its generating equations are based are analogs of the corresponding VLLE assumptions.

15.4   SOLID/VAPOR EQUILIBRIUM (SVE)

At temperatures below its triple point, a pure solid can vaporize. Solid/vapor equilibrium for a pure species is represented on a *PT* diagram by the *sublimation curve* (see Fig. 3.1); here, as for VLE, the equilibrium pressure for a particular temperature is called the (solid/vapor) saturation pressure P^{sat}.

We consider in this section the equilibrium of a pure solid (species 1) with a binary vapor *mixture* containing species 1 and a second species (species 2), assumed insoluble in the solid phase. Because it is usually the major constituent of the vapor phase, species 2 is conventionally called the *solvent* species. Hence species 1 is the *solute* species, and its mole fraction y_1 in the vapor phase is its *solubility* in the solvent. The goal is to develop a procedure for computing y_1 as a function of T and P for vapor solvents.

Only one phase-equilibrium equation can be written for this system, because species 2, by assumption, does not distribute between the two phases. The solid is *pure* species 1. Thus,

$$f_1^s = \hat{f}_1^v$$

Equation (10.44) for a pure liquid is, with minor change of notation, appropriate here:

$$f_1^s = \phi_1^{sat} P_1^{sat} \exp \frac{V_1^s(P - P_1^{sat})}{RT}$$

where P_1^{sat} is the solid/vapor saturation pressure at temperature T, and V_1^s is the molar volume of the solid. For the vapor phase, by Eq. (10.52),

$$\hat{f}_1^v = y_1 \hat{\phi}_1 P$$

Combining the three preceding equations and solving for y_1 gives:

$$y_1 = \frac{P_1^{sat}}{P} F_1 \tag{15.28}$$

where

$$F_1 \equiv \frac{\phi_1^{sat}}{\hat{\phi}_1} \exp \frac{V_1^s(P - P_1^{sat})}{RT} \tag{15.29}$$

Function F_1 reflects vapor-phase nonidealities through ϕ_1^{sat} and $\hat{\phi}_1$ and the effect of pressure on the fugacity of the solid through the exponential Poynting factor. For sufficiently low pressures, both effects are negligible, in which case $F_1 \approx 1$ and $y_1 \approx P_1^{sat}/P$. At moderate and high pressures, vapor-phase nonidealities become important, and for very high pressures even the Poynting factor cannot be ignored. Because F_1 is generally observed to be greater than unity, it is sometimes called an "enhancement factor" because according to Eq. (15.28) it leads to a solid solubility *greater* than that which would be observed in the absence of these pressure-induced effects.

Estimation of Solid Solubility at High Pressure

Solubilities at temperatures and pressures above the critical values of the solvent have important applications for supercritical separation processes. Examples are extraction of caffeine from coffee beans and separation of asphaltenes from heavy petroleum fractions. For a typical solid/vapor equilibrium (SVE) problem, the solid/vapor saturation pressure P_1^{sat} is very small, and the *saturated* vapor is for practical purposes an ideal gas. Hence ϕ_1^{sat} for pure solute vapor at this pressure is close to unity. Moreover, except for very low values of the system pressure P, the solid solubility y_1 is small, and $\hat{\phi}_1$ can be approximated by $\hat{\phi}_1^\infty$, the vapor-phase fugacity coefficient of the solute at infinite dilution. Finally, because P_1^{sat} is very small, the pressure difference $P - P_1^{sat}$ in the Poynting factor is nearly equal to P at any pressure where this factor is important. With these usually reasonable approximations, Eq. (15.29) reduces to:

$$F_1 = \frac{1}{\hat{\phi}_1^\infty} \exp \frac{PV_1^s}{RT} \tag{15.30}$$

an expression suitable for many engineering applications. In this equation, P_1^{sat} and V_1^s are pure-species properties, found in a handbook or estimated from a suitable correlation. Quantity $\hat{\phi}_1^\infty$, on the other hand, must be computed from a *PVT* equation of state—one suitable for vapor mixtures at high pressures.

Cubic equations of state, such as the Soave/Redlich/Kwong (SRK) and Peng/Robinson (PR) equations, are usually satisfactory for this kind of calculation. Equation (13.99) for $\hat{\phi}_i$, developed in Sec. 13.7, is applicable here, but with a slightly modified combining rule for interaction parameter a_{ij} used in the calculation of $\bar{q}_i$. Thus, Eq. (13.93) is replaced by:

$$a_{ij} = (1 - l_{ij})(a_i a_j)^{1/2} \tag{15.31}$$

The additional binary interaction parameter l_{ij} must be found for each ij pair $(i \neq j)$ from experimental data. By convention, $l_{ij} = l_{ji}$ and $l_{ii} = l_{jj} = 0$.

Partial parameter $\bar{a}_i$ is found by application of Eq. (13.94) with a from Eq. (13.92):

$$\bar{a}_i = -a + 2\sum_j y_j a_{ji}$$

Substitution of this expression into Eq. (13.101) yields:

$$\bar{q}_i = q\left(\frac{2\sum_j y_j a_{ji}}{a} - \frac{b_i}{b}\right) \tag{15.32}$$

where b and q are given by Eqs. (13.91) and (13.90).

For species 1 at infinite dilution in a binary system, the "mixture" is pure species 2. In this event, Eqs. (13.99), (15.31), and (15.32) yield an expression for $\hat{\phi}_1^\infty$:

$$\ln \hat{\phi}_1^\infty = \frac{b_1}{b_2}(Z_2 - 1) - \ln(Z_2 - \beta_2) - q_2\left[2(1 - l_{12})\left(\frac{a_1}{a_2}\right)^{1/2} - \frac{b_1}{b_2}\right]I_2 \tag{15.33}$$

where by Eq. (13.72),

$$I_2 = \frac{1}{\sigma - \epsilon}\ln\frac{Z_2 + \sigma\beta_2}{Z_2 + \epsilon\beta_2}$$

Equation (15.33) is used in conjunction with Eqs. (13.81) and (13.83), which provide values of β_2 and Z_2 corresponding to a particular T and P.

As an example, consider the calculation of the solubility of naphthalene(1) in carbon dioxide(2) at 35°C (308.15 K) and pressures up to 300 bar. Strictly speaking, this is not solid/*vapor* equilibrium because the critical temperature of CO_2 is 31.1°C. However, the development of this section remains valid.

The basis is Eq. (15.30), with $\hat{\phi}_1^\infty$ determined from Eq. (15.33) written for the SRK equation of state. For solid naphthalene at 35°C,

$$P_1^{sat} = 2.9 \times 10^{-4} \text{ bar} \qquad \text{and} \qquad V_1^s = 125 \text{ cm}^3\cdot\text{mol}^{-1}$$

Equations (15.33) and (13.83) become the SRK expressions on assignment of the values $\sigma = 1$ and $\varepsilon = 0$. Evaluation of parameters a_1, a_2, b_1, and b_2 requires values for T_c, P_c, and ω, which are found in App. B. Thus Eqs. (13.79) and (13.80) give:

$$a_1 = 7.299 \times 10^7 \text{ bar·cm}^6 \cdot \text{mol}^{-2} \qquad b_1 = 133.1 \text{ cm}^3 \cdot \text{mol}^{-1}$$
$$a_2 = 3.664 \times 10^6 \text{ bar·cm}^6 \cdot \text{mol}^{-2} \qquad b_2 = 29.68 \text{ cm}^3 \cdot \text{mol}^{-1}$$

By Eq. (13.82), $$q_2 = \frac{a_2}{b_2 RT} = 4.819$$

With these values, Eqs. (15.33), (13.81), and (13.83) become:

$$\ln \hat{\phi}_1^\infty = 4.485(Z_2 - 1) - \ln(Z_2 - \beta_2) + [21.61 - 43.02(1 - l_{12})] \ln \frac{Z_2 + \beta_2}{Z_2} \qquad (A)$$

$$\beta_2 = 1.1585 \times 10^{-3} P \qquad (P/\text{bar}) \qquad (B)$$

$$Z_2 = 1 + \beta_2 - 4.819 \beta_2 \frac{Z_2 - \beta_2}{Z_2(Z_2 + \beta_2)} \qquad (C)$$

To find $\hat{\phi}_1^\infty$ for a given l_{12} and P, one first evaluates β_2 using Eq. (B) and solves Eq. (C) for Z_2. Substitution of these values into Eq. (A) gives $\hat{\phi}_1^\infty$. For example, for $P = 200$ bar and $l_{12} = 0$, Eq. (B) gives $\beta_2 = 0.2317$, and solution of Eq. (C) yields $Z_2 = 0.4426$. By Eq. (A), $\hat{\phi}_1^\infty = 4.74 \times 10^{-5}$. This small value leads by Eq. (15.30) to a large enhancement factor F_1.

Tsekhanskaya et al.[4] report solubility data for naphthalene in carbon dioxide at 35°C and high pressures, given as circles on Fig. 15.6. The sharp increase in solubility as the pressure approaches the critical value (73.83 bar for CO_2) is typical of supercritical systems. Shown for comparison are the results of calculations based on Eqs. (15.28) and (15.30), under various assumptions. The lowest curve shows the "ideal solubility" P_1^{sat}/P, for which the enhancement factor F_1 is unity. The dashed curve incorporates the Poynting effect, which is significant at the higher pressures. The topmost curve includes the Poynting effect as well as $\hat{\phi}_1^\infty$, estimated from Eq. (15.33) with SRK constants and with $l_{12} = 0$; this purely predictive result captures the general trends of the data, but it overestimates the solubility at the higher pressures. *Correlation* of the data requires a nonzero value for the interaction parameter; the value $l_{12} = 0.088$ produces the semi-quantitative representation shown on Fig. 15.6 as the second curve from the top.

15.5 EQUILIBRIUM ADSORPTION OF GASES ON SOLIDS

The process by which certain porous solids bind large numbers of molecules to their surfaces is known as adsorption. Not only does it serve as a separation process, but it is also a vital part of catalytic-reaction processes. As a separation process, adsorption is used most often for removal of low-concentration impurities and pollutants from fluid streams. It is also the basis for chromatography. In surface-catalyzed reactions, the initial step is usually adsorption of reactant species; the final step is usually the reverse process, desorption of product species.

[4]Y. V. Tsekhanskaya, M. B. Iomtev, and E. V. Mushkina, *Russian J. Phys. Chem.*, vol. 38, pp. 1173–1176, 1964.

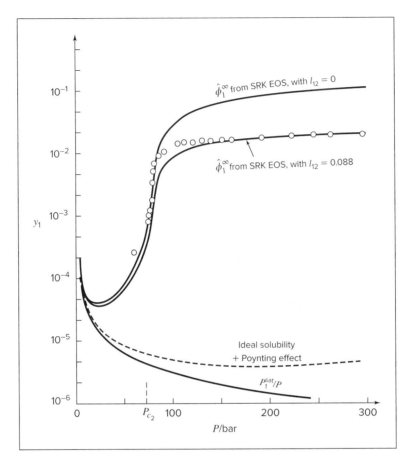

Figure 15.6: Solubility of naphthalene(1) in carbon dioxide(2) at 35°C. Circles are data. Curves are computed from Eqs. (15.28) and (15.30) under various assumptions as labeled.

Because most industrially important reactions are catalytic, adsorption plays a fundamental role in reaction engineering.

The nature of the adsorbing surface is the determining factor in adsorption. To be useful as an *adsorbent*, a solid must present a large surface area per unit mass (specific surface areas up to 1500 m² per gram are not uncommon). This can be achieved with porous solids such as activated carbon, silica gels, aluminas, zeolites, and metal-organic frameworks (MOFs), all of which contain many cavities or pores with diameters as small as a fraction of a nanometer. Surfaces of such solids are necessarily irregular at the length scale of atoms and molecules, and they present *sites* of particular attraction for adsorbing molecules. If the sites are close together, the adsorbed molecules may interact with one another; if they are sufficiently dispersed, the adsorbed molecules may interact only with the sites. Depending upon the strength of the forces binding them to the sites, these *adsorbate* molecules may be mobile or fixed in position. Relatively weak electrostatic and van der Waals interactions favor mobility of adsorbate molecules and result in *physical adsorption*. On the other hand, much stronger quasichemical forces can act to fix molecules to the surface, by *chemisorption*.

Although adsorption may be classified in several ways, the usual distinction is between physical adsorption and chemisorption. Based on the strength of the binding forces, this division is observed experimentally in the magnitudes of the heat of adsorption.

In the adsorption of gases, the number of molecules adsorbed upon a solid surface depends on conditions in the gas phase. For very low pressures, relatively few molecules are adsorbed, and only a fraction of the solid surface is covered. As the gas pressure increases at a given temperature, surface coverage increases. When all sites become occupied, the adsorbed molecules are said to form a *monolayer*. Further increase in pressure promotes *multilayer* adsorption. In some cases, multilayer adsorption may occur on one part of a porous solid when vacant sites remain on another part.

The complexities of solid surfaces, particularly those of high-surface-area porous materials of greatest practical interest, limit molecular-level understanding of the adsorption process. They do not, however, prevent development of an exact thermodynamic description of adsorption equilibrium, applicable to both physical adsorption and chemisorption and equally to monolayer and multilayer adsorption. The thermodynamic framework is independent of any *particular* theoretical or empirical description of material behavior. However, in application such a description is essential, and meaningful results require appropriate models of behavior.

The thermodynamic treatment of gas/adsorbate equilibrium is in many respects analogous to that of vapor/liquid equilibrium. However, the definition of a system to which the equations of thermodynamics apply presents a problem. The force field of the solid adsorbent influences properties in the adjacent gas phase, but its effect decreases rapidly with distance. Thus the properties of the gas change sharply in the immediate neighborhood of the solid surface, but they do not change discontinuously. A region of change exists which contains gradients in the properties of the gas, but the distance into the gas phase over which the solid exerts influence cannot be precisely established.

This problem is circumvented by a construct first devised by J. W. Gibbs. Imagine that the gas-phase properties extend unchanged up to the solid surface. Differences between the actual and the unchanged properties can then be attributed to a mathematical surface, treated as a two-dimensional phase with its own thermodynamic properties. This provides not only a precisely defined surface phase to account for the singularities of the interfacial region, but it also extracts them from the three-dimensional gas phase so that it too may be treated precisely. The solid, despite the influence of its force field, is presumed to be inert and not otherwise to participate in the gas/adsorbate equilibrium. Thus for purposes of thermodynamic analysis, the adsorbate is treated as a two-dimensional phase, inherently an *open* system because it is in equilibrium with the gas phase.

The fundamental property relation for an open *PVT* system is given by Eq. (10.2):

$$d(nG) = (nV)dP - (nS)dT + \sum_i \mu_i \, dn_i$$

An analogous equation may be written for a two-dimensional phase. The key difference is that pressure and molar volume are not appropriate variables for a two-dimensional phase. Pressure is replaced by the *spreading pressure* Π, and the molar volume by the *molar area a*:

$$d(nG) = (na)d\Pi - (nS)dT + \sum_i \mu_i \, dn_i \tag{15.34}$$

This equation is written on the basis of a unit mass, usually a gram or a kilogram, of solid adsorbent. Thus n is the *specific* amount adsorbed, that is, the number of moles of adsorbate *per unit mass of adsorbent*. Moreover, area A is defined as the specific surface area, that is, the area *per unit mass of adsorbent*, a quantity characteristic of a particular adsorbent. The molar area, $a \equiv A/n$, is the surface area per mole of adsorbate.

The spreading pressure is the two-dimensional analog of pressure, having units of force per unit length, akin to surface tension. It can be pictured as the force in the plane of the surface that must be exerted perpendicular to each unit length of edge to keep the surface from spreading, that is, to keep it in mechanical equilibrium. It is not subject to direct experimental measurement, and it must be calculated, significantly complicating the treatment of adsorbed-phase equilibrium.

Because the spreading pressure adds an extra variable, the number of degrees of freedom for gas/adsorbate equilibrium is given by an altered version of the phase rule. For gas/adsorbate equilibrium, $\pi = 2$; therefore,

$$F = N - \pi + 3 = N - 2 + 3 = N + 1$$

Thus for adsorption of a pure species,

$$F = 1 + 1 = 2$$

and two phase-rule variables, e.g., T and P or T and n, must be fixed independently to establish an equilibrium state. Note that the inert solid phase is counted neither as a phase nor as a species.

Recall the summability relation for the Gibbs energy, which follows from Eqs. (10.8) and (10.12):

$$nG = \sum_i n_i \mu_i$$

Differentiation gives: $$d(nG) = \sum_i \mu_i \, dn_i + \sum_i n_i \, d\mu_i$$

Comparison with Eq. (15.34) shows:

$$(nS)dT - (na)d\Pi + \sum_i n_i \, d\mu_i = 0$$

or $$SdT - ad\Pi + \sum_i x_i d\mu_i = 0$$

This is the Gibbs/Duhem equation for the adsorbate. Restricting it to constant temperature produces the *Gibbs adsorption isotherm*:

$$-a \, d\Pi + \sum_i x_i \, d\mu_i = 0 \qquad \text{(const } T) \tag{15.35}$$

The condition of equilibrium between adsorbate and gas presumes the same temperature for the two phases and requires:

$$\mu_i = \mu_i^g$$

where μ_i^g represents the gas-phase chemical potential. For a change in equilibrium conditions,

$$d\mu_i = d\mu_i^g$$

If the gas phase is an *ideal gas* (the usual assumption), then differentiation of Eq. (10.29) at constant temperature yields:

$$d\mu_i^g = RT \, d \ln (y_i \, P)$$

Combining the last two equations with the Gibbs adsorption isotherm gives:

$$-\frac{a}{RT} d\Pi + d \ln P + \sum_i x_i \, d \ln y_i = 0 \qquad (\text{const } T) \qquad (15.36)$$

where x_i and y_i represent adsorbate and gas-phase mole fractions, respectively.

Pure-Gas Adsorption

Basic to the experimental study of pure-gas adsorption are measurements at constant temperature of n, the moles of gas adsorbed, as a function of P, the pressure in the gas phase. Each set of data represents an *adsorption isotherm* for the pure gas on a particular solid adsorbent. Available data are summarized by Valenzuela and Myers.[5] The correlation of such data requires an analytical relation between n and P, and such a relation should be consistent with Eq. (15.36).

Written for a pure chemical species, this equation becomes:

$$\frac{a}{RT} d\Pi = d \ln P \; (\text{const } T) \qquad (15.37)$$

The compressibility-factor analog for an adsorbate is defined by the equation:

$$z \equiv \frac{\Pi a}{RT} \qquad (15.38)$$

Differentiation at constant T yields:

$$dz = \frac{\Pi}{RT} da + \frac{a}{RT} d\Pi$$

Replacing the last term by Eq. (15.37) and eliminating Π/RT in favor of z/a in accord with Eq. (15.38) yields:

$$-d \ln P = z \frac{da}{a} - dz$$

Substituting $a = A/n$ and $da = -A dn/n^2$ gives:

$$-d \ln P = -z \frac{dn}{n} - dz$$

Adding dn/n to both sides of this equation and rearranging,

$$d \ln \frac{n}{P} = (1 - z) \frac{dn}{n} - dz$$

---

[5]D. P. Valenzuela and A. L. Myers, *Adsorption Equilibrium Data Handbook*, Prentice Hall, Englewood Cliffs, NJ, 1989.

Integration from $P = 0$ (where $n = 0$ and $z = 1$) to $P = P$ and $n = n$ yields:

$$\ln\frac{n}{P} - \ln\lim_{P \to 0}\frac{n}{P} = \int_0^n (1 - z)\frac{dn}{n} + 1 - z$$

The limiting value of n/P as $n \to 0$ and $P \to 0$ must be found by extrapolation of experimental data. Applying l'Hôpital's rule to this limit gives:

$$\lim_{P \to 0}\frac{n}{P} = \lim_{P \to 0}\frac{dn}{dP} \equiv k$$

Thus k is defined as the limiting slope of an isotherm as $P \to 0$ and is known as Henry's constant for adsorption. It is a function of temperature only for a given adsorbent and adsorbate, and it is characteristic of the specific interaction between a particular adsorbent and a particular adsorbate.

The preceding equation may therefore be written:

$$\ln\frac{n}{kP} = \int_0^n (1 - z)\frac{dn}{n} + 1 - z$$

or
$$n = kP \ \exp\left[\int_0^n (1 - z)\frac{dn}{n} + 1 - z\right] \qquad (15.39)$$

This general relation between n, the moles adsorbed, and P, the gas-phase pressure, includes z, the adsorbate compressibility factor, which may be represented by an equation of state for the adsorbate. The simplest such equation is the ideal-gas analog, $z = 1$, and in this case Eq. (15.39) yields $n = kP$, which is Henry's law for adsorption.

An equation of state known as the ideal-lattice-gas equation[6] has been developed specifically for an adsorbate:

$$z = -\frac{m}{n} \ln\left(1 - \frac{n}{m}\right)$$

where m is a constant. This equation is based on the presumptions that the surface of the adsorbent is a two-dimensional lattice of energetically equivalent sites, each of which may bind an adsorbate molecule, and that the bound molecules do not interact with each other. The validity of this model is therefore limited to no more than monolayer coverage. Substitution of this equation into Eq. (15.39) and integration leads to the *Langmuir isotherm*:[7]

$$n = \left(\frac{m - n}{m}\right)kP$$

Solution for n yields:
$$n = \frac{mP}{\dfrac{m}{k} + P} \qquad (15.40)$$

---

[6] See, e.g., T. L. Hill, *An Introduction to Statistical Mechanics*, sec. 7–1, Addison-Wesley, Reading, MA, 1960, reprinted by Dover, 1987.

[7] Irving Langmuir (1881–1957), the second American to receive the Nobel Prize in chemistry, awarded for his contributions in the field of surface chemistry. See http://www.nobelprize.org/nobel_prizes/chemistry/laureates/1932/ and http://en.wikipedia.org/wiki/Irving_Langmuir.

Alternatively,
$$n = \frac{kbP}{b + P}$$
(15.41)

where $b \equiv m/k$, and k is Henry's constant. Note that when $P \to 0$, n/P properly approaches k. At the other extreme, where $P \to \infty$, n approaches m, the *saturation* value of the specific amount absorbed, representing full monolayer coverage.

Based on the same assumptions as for the ideal-lattice-gas equation, Langmuir in 1918 derived Eq. (15.40) by noting that at equilibrium the rate of adsorption and the rate of desorption of gas molecules must be the same.[8] For monolayer adsorption, the number of sites may be divided into the fraction occupied θ and the fraction vacant $1 - \theta$. By definition,

$$\theta \equiv \frac{n}{m} \qquad \text{and} \qquad 1 - \theta = \frac{m - n}{m}$$

where m is the value of n for full monolayer coverage. For the assumed conditions, the rate of adsorption is proportional to the rate at which molecules strike unoccupied sites on the surface, which in turn is proportional to both the pressure and the fraction $1 - \theta$ of unoccupied surface sites. The rate of desorption is proportional to the occupied fraction θ of sites. Equating the two rates gives:

$$\kappa P \frac{m - n}{m} = \kappa' \frac{n}{m}$$

where κ and κ' are proportionality (rate) constants. Solving for n and rearranging yields:

$$n = \frac{\kappa m P}{\kappa P + \kappa'} = \frac{mP}{\dfrac{1}{K} + P}$$

where $K \equiv \kappa/\kappa'$, the ratio of the forward and reverse adsorption rate constants, is the conventional adsorption equilibrium constant. The second equality in this equation is equivalent to Eq. (15.40), and it indicates that the adsorption equilibrium constant is equal to Henry's constant divided by m, i.e., $K = k/m$.

Because the assumptions upon which it is based are fulfilled at low surface coverage, the Langmuir isotherm is always valid as $\theta \to 0$ and as $n \to 0$. Even though these assumptions become unrealistic at higher surface coverage, the Langmuir isotherm may provide an approximate overall fit to n vs. P data; however, it does not lead to reasonable values for m.

Substituting $a = A/n$ in Eq. (15.37) gives:

$$\frac{A\, d\Pi}{RT} = n\, d \ln P$$

Integration at constant temperature from $P = 0$ (where $\Pi = 0$) to $P = P$ and $\Pi = \Pi$ yields:

$$\frac{\Pi A}{RT} = \int_0^P \frac{n}{P}\, dP$$
(15.42)

This equation provides the *only* means of evaluating spreading pressure. The integration may be carried out numerically or graphically with experimental data, or the data may be fit to an

---

[8] I. Langmuir, *J. Am. Chem. Soc.*, vol. 40, p. 1361, 1918.

equation for an isotherm. For example, if the integrand n/P is given by Eq. (15.41), the Langmuir isotherm, then:

$$\frac{\Pi A}{RT} = kb \ln \frac{P+b}{b} \tag{15.43}$$

an equation valid for $n \to 0$.

No equation of state is known that leads to an adsorption isotherm which in general fits experimental data over the entire range of n from zero to full monolayer coverage. Isotherms that find practical use are often three-parameter empirical extensions of the Langmuir isotherm. An example is the Toth equation:[9]

$$n = \frac{mP}{(b + P^t)^{1/t}} \tag{15.44}$$

which adds an exponent (t) as the third parameter and reduces to the Langmuir equation for $t = 1$. When the integrand of Eq. (15.42) is expressed by the Toth equation and most other three-parameter equations, its integration requires numerical methods. Moreover, the empirical element of such equations often introduces a singularity that makes them behave improperly in the limit as $P \to 0$. Thus for the Toth equation ($t < 1$) the second derivative d^2n/dP^2 approaches $-\infty$ in this limit, making values of Henry's constant as calculated by this equation too large. Nevertheless, the Toth equation finds frequent practical use as an adsorption isotherm. However, it is not always suitable, and a number of other adsorption isotherms are in use, as discussed by Suzuki.[10] Among them, the Freundlich equation,

$$\theta = \frac{n}{m} = \alpha P^{1/\beta} \qquad (\beta > 1) \tag{15.45}$$

is a two-parameter (α and β) isotherm that often successfully correlates experimental data for low and intermediate values of θ.

Example 15.8

Nakahara et al.[11] report data for ethylene adsorbed on a carbon molecular sieve ($A = 650$ m$^2 \cdot$g^{-1}) at 50°C. The data, shown as filled circles in Fig. 15.7, consist of pairs of values (P, n), where P is the equilibrium gas pressure in kPa and n is moles of adsorbate per kg of adsorbent. Trends shown by the data are typical for physical adsorption on a heterogeneous adsorbent at low-to-moderate surface coverage. Use these data to illustrate numerically the concepts developed for pure-gas adsorption.

---

[9]J. Toth, *Adsorption. Theory, Modelling, and Analysis*, Dekker, New York, 2002.

[10]M. Suzuki, *Adsorption Engineering*, pp. 35–51, Elsevier, Amsterdam, 1990.

[11]T. Nakahara, M. Hirata, and H. Mori, *J. Chem. Eng. Data*, vol. 27, pp. 317–320, 1982.

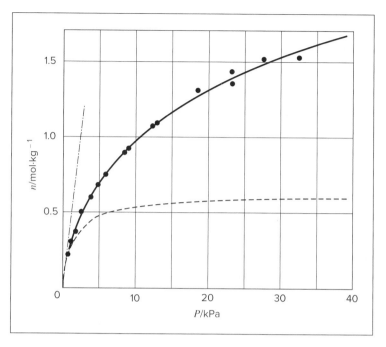

Figure 15.7: Adsorption isotherm for ethylene on a carbon molecular sieve at 50°C. Legend: • experimental data; – • – • – • Henry's law; _____ Toth equation; – – – Langmuir equation $n \to 0$.

Solution 15.8

The solid line in Fig. 15.7 represents a curve-fit to the data by Eq. (15.44), the Toth equation, with parameter values as reported by Valenzuela and Myers (*loc. cit.*):

$$m = 4.7087 \qquad b = 2.1941 \qquad t = 0.3984$$

These imply an apparent value of Henry's constant:

$$k(\text{Toth}) = \lim_{P \to \infty} \frac{n}{P} = \frac{m}{b^{1/t}} = 0.6551 \text{ mol·kg}^{-1}\text{·kPa}^{-1}$$

Although the overall quality of the fit is excellent, the value of Henry's constant is too large, as we will show.

Extraction of Henry's constant from an adsorption isotherm is facilitated when n/P (rather than n) is considered the dependent variable and n (rather than P) the independent variable. The data plotted in this form are shown by Fig. 15.8. On this plot, Henry's constant is the extrapolated intercept:

$$k = \lim_{P \to 0} \frac{n}{P} = \lim_{n \to 0} \frac{n}{P}$$

where the second equality follows from the first because $n \to 0$ as $P \to 0$. Evaluation of the intercept (and hence of k) was done in this case by fitting all of the n/P data by a cubic polynomial in n:

$$\frac{n}{P} = C_0 + C_1 n + C_2 n^2 + C_3 n^3$$

The evaluated parameters are:

$$C_0 = 0.4016 \qquad C_1 = -0.6471 \qquad C_2 = -0.4567 \qquad C_3 = -0.1200$$

Whence, $k = C_0 = 0.4016 \; \text{mol·kg}^{-1} \cdot \text{kPa}^{-1}$

Representation of n/P by the cubic polynomial appears as the solid curve in Fig. 15.8, and the extrapolated intercept ($C_0 = k = 0.4016$) is indicated by an open circle. For comparison, the dotted line is the low-n portion of the n/P curve given by the Toth equation. Here it is apparent that the extrapolated intercept $k(\text{Toth})$, off-scale on this figure, is too high. The Toth equation cannot provide an accurate representation of adsorption behavior at very low values of n or P.

The Langmuir equation, on the other hand, is always suitable for sufficiently small n or P. Rearrangement of Eq. (15.41) gives:

$$\frac{n}{P} = k - \frac{1}{b} n$$

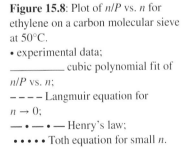

Figure 15.8: Plot of n/P vs. n for ethylene on a carbon molecular sieve at 50°C.
• experimental data;
_____ cubic polynomial fit of n/P vs. n;
– – – – Langmuir equation for $n \to 0$;
— • — • — Henry's law;
• • • • • Toth equation for small n.

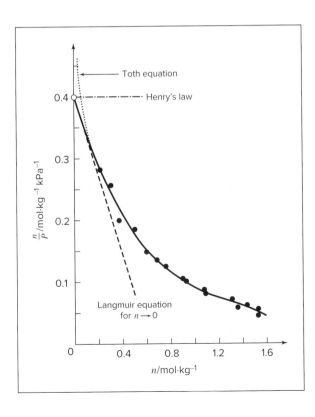

which shows that the Langmuir equation implies a linear variation of n/P with n. Hence, the limiting tangent to the "true" isotherm on a plot of n/P vs. n represents the Langmuir approximation to the isotherm for small n, as shown by the dashed lines in Figs. 15.7 and 15.8. It is given by the equation:

$$\frac{n}{P} = 0.4016 - 0.6471n$$

or, equivalently, by
$$n = \frac{0.6206P}{1.5454 + P}$$

Figures 15.7 and 15.8 show that Henry's law (represented by the dot-dash lines) and the limiting form of the Langmuir equation provide, respectively, in this example upper and lower bounds for the actual isotherm. The Langmuir isotherm when fit to *all* the experimental data yields a curve (not shown) in Fig. 15.7 that fits the data reasonably well, but not as well as the three-parameter Toth expression.

Neither the spreading pressure nor the adsorbate equation of state is required for an empirical correlation of single-species adsorption data. However, a set of (n, P) data *implies* an equation of state for the adsorbed phase, and hence a relationship between the spreading pressure Π and the moles adsorbed. By Eq. (15.42),

$$\frac{\Pi A}{RT} = \int_0^P \frac{n}{P}\, dP = \int_0^n \frac{n}{P}\frac{dP}{dn}\, dn$$

Equation (15.38) may be written:

$$z = \frac{\Pi A}{nRT}$$

Whence,
$$z = \frac{1}{n}\int_0^P \frac{n}{P}\, dP = \frac{1}{n}\int_0^n \frac{n}{P}\frac{dP}{dn}\, dn$$

Finding numerical values for z and Π depends on evaluation of the integral:

$$I \equiv \int_0^P \frac{n}{P}\, dP = \int_0^n \frac{n}{P}\frac{dP}{dn}\, dn$$

Choice of the form depends on whether P or n is the independent variable. The Toth equation gives the integrand n/P as a function of P, and therefore:

$$I(\text{Toth}) = \int_0^P \frac{m\,dP}{(b + P^t)^{1/t}}$$

The cubic polynomial gives n/P as a function of n, whence,

$$I(\text{cubic}) = \int_0^n \left(\frac{C_0 - C_2 n^2 - 2C_3 n^3}{C_0 + C_1 n + C_2 n^2 + C_3 n^3} \right) dn$$

These two expressions permit numerical determination of $z(n)$ and $\Pi(n)$ as a result of correlations presented in this example. Thus, for $n = 1$ mol·kg^{-1} and $A = 650$ m^2·g^{-1}, both the Toth and cubic-polynomial equations yield $z = 1.69$. From this result,

$$\Pi = \frac{nRT}{A} z = \frac{1\,\text{mol·kg}^{-1} \times 83.14\,\text{cm}^3 \cdot \text{bar·mol}^{-1} \cdot \text{K}^{-1} \times 323.15\,\text{K}}{650{,}000\,\text{m}^2 \cdot \text{kg}^{-1}}$$

$$\times\; 1.69 \times 10^{-6}\,\text{m}^3 \cdot \text{cm}^{-3} \times 10^5\,\text{N·m}^{-2} \cdot \text{bar}^{-1}$$

$$= 6.99 \times 10^{-3}\; \text{N·m}^{-1} = 6.99\; \text{mN·m}^{-1} = 6.99\; \text{dyn·cm}^{-1}$$

The adsorptive capacity of an adsorbent depends directly on its specific surface area A, but determination of these large values is not a trivial matter. The means is provided by the adsorption process itself. The basic idea is to measure the quantity of a gas adsorbed at full monolayer coverage and to multiply the number of molecules adsorbed by the area occupied by a single molecule. Two difficulties attend this procedure. First is the problem of detecting the point of full monolayer coverage. Second, one finds measured areas are different with different gases as adsorbates. The latter problem is generally circumvented by the adoption of nitrogen as a standard adsorbate. The procedure is to make measurements of the (physical) adsorption of N_2 at its normal boiling point ($-195.8°C$) for pressures up to its vapor pressure of l(atm). The result is a curve that initially, at low P, looks like that in Fig. 15.7. When monolayer coverage is nearly complete, multilayer adsorption begins, and the curve changes direction, with n increasing ever more rapidly with pressure. Finally, as the pressure approaches l(atm), the vapor pressure of the N_2 adsorbate, the curve becomes nearly vertical because of condensation in the pores of the adsorbent. The problem is to identify the point on the curve that represents full monolayer coverage. The usual procedure is to fit the Brunauer/Emmett/Teller (BET) equation, a two-parameter extension of the Langmuir isotherm to multilayer adsorption, to the n vs. P data. From this, one can determine a value for m.[12] Once m is known, multiplication by Avogadro's number and by the area occupied by one adsorbed N_2 molecule (16.2 Å^2) yields the surface area. The method has its uncertainties, particularly for molecular sieves where the pores may contain unadsorbed molecules. Nevertheless, it is a useful and widely used tool for characterizing and comparing adsorption capacities.

Heat of Adsorption

The Clapeyron equation, derived in Sec. 6.5 for the latent heat of phase transition of pure chemical species, is also applicable to pure-gas adsorption equilibrium. Here, however, the two-phase equilibrium pressure depends not only on temperature but also on surface coverage or the amount adsorbed. Thus the analogous equation for adsorption is written

$$\left(\frac{\partial P}{\partial T}\right)_n = \frac{\Delta H^{av}}{T \Delta V^{av}} \tag{15.46}$$

where subscript n signifies that the derivative is taken at constant amount adsorbed. Superscript av denotes a property change of *desorption*, i.e., the difference between the vapor-phase

---

[12]J. M. Smith, *Chemical Kinetics*, 3d ed., sec. 8–1, McGraw-Hill, New York, 1981.

and the adsorbed-phase property. The quantity $\Delta H^{av} \equiv H^{v} - H^{a}$ is defined as the *isosteric heat of adsorption* and is usually a positive quantity.[13] The heat of adsorption is a useful indication of the strength of the forces binding adsorbed molecules to the surface of the adsorbent, and its magnitude can therefore often be used to distinguish between physical adsorption and chemisorption.

The dependence of heats of adsorption on surface coverage has its basis in the energetic heterogeneity of most solid surfaces. The first sites on a surface to be occupied are those which attract adsorbate molecules most strongly and with the greatest release of energy. Thus the heat of adsorption usually decreases with surface coverage. Once all sites are occupied and multilayer adsorption begins, the dominant forces become those between adsorbate molecules, and for subcritical species the decreasing heat of adsorption approaches the heat of vaporization.

Assumed in the derivation of the Langmuir isotherm is the energetic equivalence of all adsorption sites, implying that the heat of adsorption is independent of surface coverage. This explains in part the inability of the Langmuir isotherm to provide a close fit to most experimental data over a wide range of surface coverage. The Freundlich isotherm, Eq. (15.45), implies a logarithmic decrease in the heat of adsorption with surface coverage.

As in the development of the Clausius/Clapeyron equation (Example 6.6), if for low pressures one assumes that the gas phase is ideal and that the adsorbate is of negligible volume compared with the gas-phase volume, Eq. (15.46) becomes:

$$\left(\frac{\partial \ln P}{\partial T}\right)_{n} = \frac{\Delta H^{av}}{RT^{2}} \tag{15.47}$$

Application of this equation requires the measurement of isotherms, such as the one at 50°C in Fig. 15.7, at several temperatures. Cross plotting yields sets of P vs. T relations at constant n, from which values for the partial derivative of Eq. (15.47) can be obtained. For chemisorption, ΔH^{av} values usually range from 60 to 170 kJ·mol^{-1}. For physical adsorption, they are smaller. For example, measured values at very low coverage for the physical adsorption of nitrogen and *n*-butane on 5A zeolite are 18.0 and 43.1 kJ·mol^{-1}, respectively.[14]

Mixed-Gas Adsorption

Mixed-gas adsorption is treated similarly to the gamma/phi formulation of VLE (Sec. 13.2). With a gas-phase property denoted by superscript g, Eqs. (10.31) and (10.46), which define fugacity, are rewritten:

$$G_{i}^{g} = \Gamma_{i}^{g}(T) + RT \ln f_{i}^{g} \tag{15.48} \qquad \mu_{i}^{g} = \Gamma_{i}^{g}(T) + RT \ln \hat{f}_{i}^{g} \tag{15.49}$$

Note as a result of Eqs. (10.32) and (10.53) that:

$$\lim_{P \to 0} \frac{f_{i}^{g}}{P} = 1 \qquad \text{and} \qquad \lim_{P \to 0} \frac{\hat{f}_{i}^{g}}{y_{i} P} = 1$$

[13]Other heats of adsorption, defined differently, are also in use. However, the isosteric heat is the most common, and it is the one needed for energy balances on adsorption columns.

[14]N. Hashimoto and J. M. Smith, *Ind. Eng. Chem. Fund.*, vol. 12, p. 353, 1973.

For the adsorbate, analogous equations are:

$$G_i = \Gamma_i(T) + RT \ln f_i \qquad (15.50) \qquad \mu_i = \Gamma_i(T) + RT \ln \hat{f}_i \qquad (15.51)$$

with
$$\lim_{\Pi \to 0} \frac{f_i}{\Pi} = 1 \qquad \text{and} \qquad \lim_{\Pi \to 0} \frac{\hat{f}_i}{x_i \Pi} = 1$$

The Gibbs energies as given by Eqs. (15.48) and (15.50) may be equated for pure-gas/adsorbate equilibrium:

$$\Gamma_i^g(T) + RT \ln f_i^g = \Gamma_i(T) + RT \ln f_i$$

Rearrangement gives:

$$\frac{f_i}{f_i^g} = \exp\left[\frac{\Gamma_i^g(T) + \Gamma_i(T)}{RT}\right] \equiv F_i(T) \qquad (15.52)$$

The limiting value of f_i/f_i^g as both P and Π approach zero can be used to evaluate $F_i(T)$:

$$\lim_{\substack{P \to 0 \\ \Pi \to 0}} \frac{f_i}{f_i^g} = \lim_{\substack{P \to 0 \\ \Pi \to 0}} \frac{\Pi}{P} = \lim_{\substack{n_i \to 0 \\ P \to 0}} \frac{n_i}{P} \lim_{\substack{\Pi \to 0 \\ n_i \to 0}} \frac{\Pi}{n_i}$$

The first limit of the last member is Henry's constant k_i; the second limit is evaluated from Eq. (15.48), written $\Pi/n_i = z_i RT/A$; thus,

$$\lim_{\substack{\Pi \to 0 \\ n_i \to 0}} \frac{\Pi}{n_i} = \frac{RT}{A}$$

In combination with Eq. (15.52) these equations give:

$$F_i(T) = \frac{k_i RT}{A} \qquad (15.53) \qquad f_i = \frac{k_i RT}{A} f_i^g \qquad (15.54)$$

Similarly, equating Eqs. (15.49) and (15.51) yields:

$$\Gamma_i^g(T) + RT \ln \hat{f}_i^g = \Gamma_i(T) + RT \ln \hat{f}_i$$

from which
$$\frac{\hat{f}_i}{\hat{f}_i^g} = \exp\left[\frac{\Gamma_i^g(T) + \Gamma_i(T)}{RT}\right] \equiv F_i(T)$$

Then by Eq. (15.53),
$$\hat{f}_i = \frac{k_i RT}{A} \hat{f}_i^g \qquad (15.55)$$

These equations show that equality of fugacities is not a proper criterion for gas/adsorbate equilibrium. This is also evident from the fact that the units of gas-phase fugacities are those of pressure, while the units of adsorbate fugacities are those of spreading pressure. In most applications the fugacities appear as ratios, and the factor $k_i RT/A$ cancels. Nevertheless it is instructive to note that equality of chemical potentials, not fugacities, is the fundamental criterion of phase equilibrium.

An activity coefficient for the constituent species of a mixed-gas adsorbate is defined by the equation:

$$\gamma_i \equiv \frac{\hat{f}_i}{x_i f_i^\circ}$$

where $\hat{f}_i$ and f_i° are evaluated at the same T and *spreading pressure* Π. The degree sign ($^\circ$) denotes values for the equilibrium adsorption of *pure i* at the spreading pressure of the *mixture*. Substitution for the fugacities by Eqs. (15.54) and (15.55) gives:

$$\gamma_i = \frac{\hat{f}_i^g(P)}{x_i f_i^g(P_i^\circ)}$$

The fugacities are evaluated at the pressures indicated in parentheses, where P is the equilibrium mixed-gas pressure and P_i° is the equilibrium pure-gas pressure that produces the same spreading pressure. If the gas-phase fugacities are eliminated in favor of fugacity coefficients [Eqs. (10.34) and (10.52)], then:

$$\gamma_i = \frac{y_i \hat{\phi}_i P}{x_i \phi_i P_i^\circ}$$

or

$$y_i \hat{\phi}_i P = x_i \phi_i P_i^\circ \gamma_i \tag{15.56}$$

The usual assumption is that the gas phase is ideal; the fugacity coefficients are then unity:

$$y_i P = x_i P_i^\circ \gamma_i \tag{15.57}$$

These equations provide the means for calculation of activity coefficients from mixed-gas adsorption data. Alternatively, if γ_i values can be predicted, they allow calculation of adsorbate composition. In particular, if the mixed-gas adsorbate forms an ideal solution, then $\gamma = 1$, and the resulting equation is the adsorption analog of Raoult's law:

$$y_i P = x_i P_i^\circ \tag{15.58}$$

This equation is always valid as $P \to 0$ and within the pressure range for which Henry's law is a suitable approximation.

Equation (15.42) is applicable not only for pure-gas adsorption but also for adsorption of a constant-composition gas mixture. Applied where Henry's law is valid, it yields:

$$\frac{\Pi A}{RT} = kP \tag{15.59}$$

where k is the mixed-gas Henry's constant. For adsorption of pure species i at the same spreading pressure, this becomes:

$$\frac{\Pi A}{RT} = k_i P_i^\circ$$

Combining these two equations with Eq. (15.58) gives:

$$y_i k_i = x_i k$$

Summing over all i,

$$k = \sum_i y_i k_i \tag{15.60}$$

Eliminating k between these two equations yields:

$$x_i = \frac{y_i k_i}{\sum_i y_i k_i} \tag{15.61}$$

This simple equation, requiring only data for pure-gas adsorption, provides adsorbate compositions in the limit as $P \to 0$.

For an ideal adsorbed solution, in analogy with Eq. (10.81) for volumes,

$$a = \sum_i x_i a_i^\circ$$

where a is the molar area for the mixed-gas adsorbate and a_i° is the molar area of the pure-gas adsorbate at the same temperature and spreading pressure. Because $a = A/n$ and $a_i^\circ = A/n_i^\circ$, this equation may be written:

$$\frac{1}{n} = \sum_i \frac{x_i}{n_i^\circ}$$

or
$$n = \frac{1}{\sum_i (x_i / n_i^\circ)} \tag{15.62}$$

where n is the specific amount of mixed-gas adsorbate and n_i° is the specific amount of pure-i adsorbate at the same spreading pressure. The amount of species i in the *mixed-gas adsorbate* is of course $n_i = x_i n$.

The prediction of mixed-gas adsorption equilibria by *ideal-adsorbed-solution theory*[15] is based on Eqs. (15.58) and (15.62). The following is a brief outline of the procedure. Because there are $N + 1$ degrees of freedom, both T and P, as well as the gas-phase composition, must be specified. Solution is for the adsorbate composition and the specific amount adsorbed. Adsorption isotherms for *each pure species* must be known over the pressure range from zero to the value that produces the spreading pressure of the mixed-gas adsorbate. For purposes of illustration we assume Eq. (15.41), the Langmuir isotherm, to apply for each pure species, writing it:

$$n_i^\circ = \frac{k_i b_i P_i^\circ}{b_i + P_i^\circ} \tag{A}$$

The inverse of Eq. (15.43) provides an expression for P_i°, which yields values of P_i° corresponding to the spreading pressure of the mixed-gas adsorbate:

$$P_i^\circ = b_i \left(\exp \frac{\psi}{k_i b_i} - 1 \right) \tag{B}$$

where
$$\psi \equiv \frac{\Pi A}{RT}$$

[15]A. L. Myers and J. M. Prausnitz, *AIChE J.*, vol. 11, pp. 121–127, 1965; D. P. Valenzuela and A. L. Myers, *op. cit.*

The following steps then constitute a solution procedure:

- An initial estimate of ψ is found from the Henry's-law equations. Combining the definition of ψ with Eqs. (15.59) and (15.60) yields:

$$\psi = P\sum_i y_i k_i$$

- With this estimate of ψ, calculate P_i° for each species i by Eq. (B) and n_i° for each species i by Eq. (A).

- One can show that the error in ψ is approximated by:

$$\delta\psi = \frac{P\sum_i \dfrac{y_i}{P_i^\circ} - 1}{P\sum_i \dfrac{y_i}{P_i^\circ n_i^\circ}}$$

Moreover, the approximation becomes increasingly exact as $\delta\psi$ decreases. If $\delta\psi$ is smaller than some preset tolerance (say $\delta\psi < \psi \times 10^{-7}$), the calculation goes to the final step; if not, a new value, $\psi = \psi + \delta\psi$, is determined, and the calculation returns to the preceding step.

- Calculate x_i for each species i by Eq. (15.58):

$$x_i = \frac{y_i P}{P_i^\circ}$$

Calculate the specific amount absorbed by Eq. (15.62).

Use of the Langmuir isotherm has made this computational scheme appear quite simple, because direct solution for P_i° (step 2) is possible. However, most equations for the adsorption isotherm are less tractable, and this calculation must be done numerically. This significantly increases the computational task, but it does not alter the general procedure.

Predictions of adsorption equilibria by ideal-adsorbed-solution theory are usually satisfactory when the specific amount adsorbed is less than a third of the saturation value for monolayer coverage. At higher adsorbed amounts, appreciable negative deviations from ideality are promoted by differences in size of the adsorbate molecules and by adsorbent heterogeneity. One must then have recourse to Eq. (15.57). The difficulty is in obtaining values of the activity coefficients, which are strong functions of both spreading pressure and temperature. This contrasts with activity coefficients for liquid phases, which for most applications are insensitive to pressure. This topic is treated by Talu et al.[16]

15.6 OSMOTIC EQUILIBRIUM AND OSMOTIC PRESSURE

Most of the earth's water resides in the oceans as seawater. For some regions, this is the ultimate source of fresh water for public and commercial use. Conversion of seawater to fresh

[16]O. Talu, J. Li, and A. L. Myers, *Adsorption*, vol. 1, pp. 103–112, 1995.

water requires the separation of more-or-less pure water from an aqueous solution containing dissolved solute species. Historically, this has been achieved by distillation. However, in recent years *reverse osmosis* has outpaced distillation, and the majority of worldwide desalination capacity consists of reverse osmosis facilities. Central to an understanding of osmotic separations are the concepts of osmotic equilibrium and osmotic pressure, the topics of this section.

Consider the idealized physical situation represented by Fig. 15.9. A chamber is divided into two compartments by a rigid semipermeable partition (membrane). The left compartment contains a binary solute(1)/solvent(2) liquid mixture, and the right contains pure solvent; the partition is permeable to solvent species 2 only. Temperature is uniform and constant throughout, but movable pistons permit independent adjustment of the pressures in the two compartments.

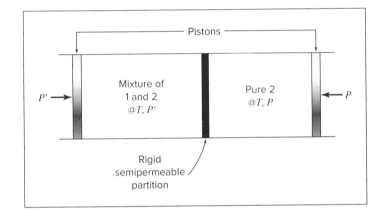

Figure 15.9: Idealized osmotic system.

Suppose that pressure is the same in the two compartments: $P' = P$. This implies inequality of the fugacity $\hat{f}_2$ of the only distributed species (the solvent), for by Eq. (15.2),

$$\frac{d\hat{f}_2}{dx_2} > 0 \qquad (\text{const } T, P)$$

meaning that $\hat{f}_2(T, P' = P, x_2 < 1) < \hat{f}_2(T, P, x_2 = 1) \equiv f_2(T, P)$

Thus, if $P' = P$, the solvent fugacity is smaller in the left compartment than in the right. The difference in solvent fugacities represents a driving force for mass transfer, and solvent diffuses through the partition, from right to left.

Equilibrium is established when pressure P' is increased to an appropriate value P^*, such that

$$\hat{f}_2(T, P' = P^*, x_2 < 1) = f_2(T, P)$$

The pressure difference, $\Pi \equiv P^* - P$, is the *osmotic pressure* of the solution, defined implicitly through the equilibrium equation for species 2, which in abbreviated form is:

$$\hat{f}_2(P + \Pi, x_2) = f_2(P) \qquad (15.63)$$

Equation (15.63) is a basis for developing explicit expressions for osmotic pressure Π. Development is facilitated by the identity:

$$\hat{f}_2(P + \Pi, x_2) \equiv f_2(P) \cdot \frac{\hat{f}_2(P, x_2)}{f_2(P)} \cdot \frac{\hat{f}_2(P + \Pi, x_2)}{\hat{f}_2(P, x_2)} \tag{15.64}$$

The first ratio on the right is, by Eq. (13.2),

$$\frac{\hat{f}_2(P, x_2)}{f_2(P)} = x_2 \gamma_2$$

where γ_2 is the activity coefficient of solvent in the mixture at pressure P. The second ratio is a *Poynting factor*, representing here a pressure effect on the fugacity of a species in solution. An expression for this factor is readily found from Eq. (10.46):

$$\left(\frac{\partial \ln \hat{f}_i}{\partial P} \right)_{T,x} = \frac{1}{RT} \left(\frac{\partial \mu_i}{\partial P} \right)_{T,x}$$

By Eqs. (10.18) and (10.8),

$$\left(\frac{\partial \mu_i}{\partial P} \right)_{T,x} = \bar{V}_i$$

Thus, for solvent species 2,

$$\left(\frac{\partial \ln \hat{f}_2}{\partial P} \right)_{T,x} = \frac{\bar{V}_2}{RT}$$

Whence,

$$\frac{\hat{f}_2(P + \Pi, x_2)}{\hat{f}_2(P, x_2)} = \exp \int_P^{P+\Pi} \frac{\bar{V}_2}{RT} dP$$

Equation (15.64) therefore becomes:

$$\hat{f}(P + \Pi, x_2) = x_2 \gamma_2 \, f_2(P) \exp \int_P^{P+\Pi} \frac{\bar{V}_2}{RT} dP$$

Combination with Eq. (15.63) yields:

$$x_2 \gamma_2 \exp \int_P^{P+\Pi} \frac{\bar{V}_2}{RT} dP = 1$$

or

$$\boxed{\int_P^{P+\Pi} \frac{\bar{V}_2}{RT} dP = -\ln(x_2 \gamma_2)} \tag{15.65}$$

Equation (15.65) is exact; working expressions for Π follow by rational approximation.

If we ignore the effect of pressure on $\bar{V}_2$, the integral becomes $\Pi \bar{V}_2 / RT$. Solution for Π then yields:

$$\Pi = -\frac{RT}{\bar{V}_2} \ln(x_2 \gamma_2) \tag{15.66}$$

If in addition the solution is sufficiently dilute in solute 1,

$$\bar{V}_2 \approx V_2 \qquad \gamma_2 \approx 1 \qquad \text{and} \qquad \ln(x_2 \gamma_2) \approx \ln(1 - x_1) \approx -x_1$$

With these approximations, Eq. (15.66) becomes:

$$\Pi = \frac{x_1 RT}{V_2} \tag{15.67}$$

Equation (15.67) is known as the van't Hoff equation.[17]

Equation (15.65) is valid when species 1 is a nonelectrolyte. If the solute is a strong (completely dissociated) electrolyte containing m ions, then the right side is:

$$-\ln\left(x_2^m \gamma_2\right)$$

and the van't Hoff equation becomes:

$$\Pi = \frac{mx_1 RT}{V_2}$$

Osmotic pressure can be quite large, even for very dilute solutions. Consider an aqueous solution containing mole fraction $x_1 = 0.001$ of a nonelectrolyte solute species at 25°C. Then

$$\Pi = 0.001 \times \frac{1}{18.02} \frac{\text{mol}}{\text{cm}^3} \times 83.14 \frac{\text{bar·cm}^3}{\text{mol·K}} \times 298.15\ \text{K} = 1.38\ \text{bar}$$

With reference to Fig. 15.9, this means that for a pure solvent pressure $P = 1$ bar, the pressure P' on the solution must be 2.38 bar to prevent diffusion of solvent from right to left, i.e., to establish *osmotic equilibrium*.[18] Pressures P' greater than this value make:

$$\hat{f}_2(P', x_2) > f_2(P)$$

and a driving force exists for transfer of water (solvent) from left to right. This observation serves as motivation for the process of *reverse osmosis*, wherein a solvent (commonly water) is separated from a solution by the application of sufficient pressure to provide the driving force needed for solvent transfer through a membrane that, for practical purposes, is permeable only to the solvent. The minimum pressure difference (solution pressure vs. pure-solvent pressure) is the osmotic pressure Π.

In practice, pressure differences significantly greater than Π are used to drive reverse osmosis. For example, seawater has an osmotic pressure of about 25 bar, but working pressures of 50 to 80 bar are employed to enhance the rate of recovery of fresh water. A feature of such separations is that they require mechanical power only for pumping the solution to an appropriate pressure level. This contrasts with distillation schemes, where steam is the usual source of energy. A brief overview of reverse osmosis is given by Perry and Green.[19]

---

[17]Jacobus Henricus van't Hoff (1852–1911), Dutch chemist who won the first Nobel Prize for chemistry in 1901. See http://www.nobelprize.org/nobel_prizes/chemistry/laureates/1901/ and http://en.wikipedia.org/wiki/Jacobus_Henricus_van_'t_Hoff.

[18]Note that, unlike conventional phase equilibrium, pressures are *unequal* for osmotic equilibrium, owing to the special constraints imposed by the rigid semipermeable partition.

[19]R. H. Perry and D. Green, *Perry's Chemical Engineers' Handbook*, 8th ed., pp. 20–36—20–40 and 20–45—20–50, McGraw-Hill, New York, 2008.

15.7 SYNOPSIS

After studying this chapter, including the end-of-chapter problems, one should be able to:

- Understand and interpret LLE, VLLE, SLE, and SVE phase diagrams

- Apply the criterion for stability of a homogeneous phase to determine whether a liquid mixture described by a particular excess Gibbs energy model will split into multiple phases for a particular overall composition

- Solve binary LLE problems using activity coefficient models to describe both liquid phases

- Assess whether a particular excess Gibbs energy model is capable of predicting LLE and, if so, for what range of parameter values

- Construct a *Txy* phase diagram for a system of two immiscible liquids that exhibit VLLE, using pure-species vapor-pressure data

- Construct *Txz* diagrams for limiting cases of binary SLE in which the liquid phase forms an ideal solution and the solid either (I) forms an ideal solution or (II) consists of two pure components

- Analyze the SVE of a pure component solid in equilibrium with a high-pressure vapor or supercritical fluid phase to estimate the solubility of the solid

- Explain the concept of spreading pressure, in the context of adsorption of gases on solids

- Employ and interpret common isotherms for gas adsorption, such as Henry's law for adsorption, the Langmuir isotherm, Toth isotherm, and Freundlich isotherm

- Compute spreading pressure for given conditions using one of the common isotherms

- Interpret heat of adsorption measurements and apply them in the context of the Clapeyron equation for gas adsorption

- Appreciate the complexities of formal thermodynamic treatment of mixed-gas adsorption and solve mixed-gas adsorption problems under idealized conditions

- Explain the concept of osmotic pressure and its relationship to reverse osmosis separation processes

- Compute osmotic pressure for dilute systems of electrolytes and nonelectrolytes

15.8 PROBLEMS

15.1. An absolute upper bound on G^E for stability of an equimolar binary mixture is $G^E = RT \ln 2$. Develop this result. What is the corresponding bound for an equimolar mixture containing N species?

15.2. A binary liquid system exhibits LLE at 25°C. Determine from each of the following sets of miscibility data estimates for parameters A_{12} and A_{21} in the Margules equation at 25°C:

$(a) x_1^\alpha = 0.10, x_1^\beta = 0.90; (b) x_1^\alpha = 0.20, x_1^\beta = 0.90; (c) x_1^\alpha = 0.10, x_1^\beta = 0.80.$

15.3. Work Prob. 15.2 for the van Laar equation.

15.4. Consider a binary vapor-phase mixture described by Eqs. (3.36) and (10.62). Under what (highly unlikely) conditions would one expect the mixture to split into two immiscible vapor phases?

15.5. Figures 15.1, 15.2, and 15.3 are based on Eqs. (A) and (F) of Ex. 15.3 with C_P^E assumed to be *positive* and given by $C_P^E/R = 3x_1x_2$. Graph the corresponding figures for the following cases, in which C_P^E is assumed to be *negative*:

(a) $A = \dfrac{975}{T} - 18.4 + 3\ln T$

(b) $A = \dfrac{540}{T} - 17.1 + 3\ln T$

(c) $A = \dfrac{1500}{T} - 19.9 + 3\ln T$

15.6. It has been suggested that a value for G^E of at least $0.5\,RT$ is required for liquid/liquid phase splitting in a binary system. Offer some justification for this statement.

15.7. Pure liquid species 2 and 3 are for practical purposes immiscible in one another. Liquid species 1 is soluble in both liquid 2 and liquid 3. One mole each of liquids 1, 2, and 3 are shaken together to form an equilibrium mixture of two liquid phases: an α-phase containing species 1 and 2, and a β-phase containing species 1 and 3. What are the mole fractions of species 1 in the α and β phases, if at the temperature of the experiment, the excess Gibbs energies of the phases are given by:

$$\frac{(G^E)^\alpha}{RT} = 0.4x_1^\alpha x_2^\alpha \qquad \text{and} \qquad \frac{(G^E)^\beta}{RT} = 0.8x_1^\beta x_3^\beta$$

15.8. It is demonstrated in Ex. 15.5 that the Wilson equation for G^E is incapable of representing LLE. Show that the simple modification of Wilson's equation given by:

$$G^E/RT = -C[x_1\,\ln(x_1 + x_2\Lambda_{12}) + x_2\,\ln(x_2 + x_1\Lambda_{21})]$$

can represent LLE. Here, C is a constant.

15.9. Vapor sulfur hexafluoride SF_6 at pressures of about 1600 kPa is used as a dielectric in large primary circuit breakers for electric transmission systems. As liquids, SF_6 and H_2O are essentially immiscible, and one must therefore specify a low enough moisture content in the vapor SF_6 so that if condensation occurs in cold weather, a liquid water phase will not form first in the system. For a preliminary determination, assume the vapor phase can be treated as an ideal gas and prepare the phase diagram [like Fig. 12.18(a)] for $H_2O(1)/SF_6(2)$ at 1600 kPa in the composition range up to 1000 parts per million of water (mole basis). The following approximate equations for vapor pressure are adequate:

$$\ln P_1^{\text{sat}}/\text{kPa} = 19.1478 - \frac{5363.70}{T/\text{K}} \qquad \ln P_2^{\text{sat}}/\text{kPa} = 14.6511 - \frac{2048.97}{T/\text{K}}$$

15.10. In Ex. 15.2 a plausibility argument was developed from the LLE *equilibrium* equations to demonstrate that positive deviations from ideal-solution behavior are conducive to liquid/liquid phase splitting.

(*a*) Use one of the binary stability criteria to reach this same conclusion.
(*b*) Is it possible *in principle* for a system exhibiting negative deviations from ideality to form two liquid phases?

15.11. Toluene(l) and water(2) are essentially immiscible as liquids. Determine the dew-point temperatures and the compositions of the first drops of liquid formed when vapor mixtures of these species with mole fractions $z_1 = 0.2$ and $z_1 = 0.7$ are cooled at a constant pressure of 101.33 kPa. What is the bubblepoint temperature and the composition of the last drop of vapor in each case? See Table B.2 for vapor-pressure equations.

15.12. *n*-Heptane(l) and water(2) are essentially immiscible as liquids. A vapor mixture containing 65-mol-% water at 100°C and 101.33 kPa is cooled slowly at constant pressure until condensation is complete. Construct a plot for the process showing temperature vs. the equilibrium mole fraction of heptane in the residual vapor. See Table B.2 for vapor-pressure equations.

15.13. Consider a binary system of species 1 and 2 in which the liquid phase exhibits partial miscibility. In the regions of miscibility, the excess Gibbs energy at a particular temperature is expressed by the equation:

$$G^E/RT = 2.25\,x_1 x_2$$

In addition, the vapor pressures of the pure species are:

$$P_1^{\text{sat}} = 75 \text{ kPa} \qquad \text{and} \qquad P_2^{\text{sat}} = 110 \text{ kPa}$$

Making the usual assumptions for low-pressure VLE, prepare a *Pxy* diagram for this system at the given temperature.

15.14. The system water(1)/*n*-pentane(2)/*n*-heptane(3) exists as a vapor at 101.33 kPa and 100°C with mole fractions $z_1 = 0.45$, $z_2 = 0.30$, $z_3 = 0.25$. The system is slowly cooled at constant pressure until it is completely condensed into a water phase and a hydrocarbon phase. Assuming that the two liquid phases are immiscible, that the vapor phase is an ideal gas, and that the hydrocarbons obey Raoult's law, determine:

(*a*) The dewpoint temperature of the mixture and composition of the first condensate.
(*b*) The temperature at which the second liquid phase appears and its initial composition.
(*c*) The bubblepoint temperature and the composition of the last bubble of vapor.

See Table B.2 for vapor-pressure equations.

15.15. Work the preceding problem for mole fractions $z_1 = 0.32$, $z_2 = 0.45$, $z_3 = 0.23$.

15.16. The Case I behavior for SLE (Sec. 15.4) has an analog for VLE. Develop the analogy.

15.17. An assertion with respect to Case II behavior for SLE (Sec. 15.4) was that the condition $z_i \gamma_i^s = 1$ corresponds to complete immiscibility for all species in the solid state. Prove this.

15.18. Use results of Sec. 15.4 to develop the following (approximate) rules of thumb:

(a) The solubility of a solid in a liquid solvent increases with increasing T.

(b) The solubility of a solid in a liquid solvent is independent of the identity of the solvent species.

(c) Of two solids with roughly the same heat of fusion, that solid with the lower melting point is the more soluble in a given liquid solvent at a given T.

(d) Of two solids with similar melting points, that solid with the smaller heat of fusion is the more soluble in a given liquid solvent at a given T.

15.19. Estimate the solubility of naphthalene(1) in carbon dioxide(2) at a temperature of 80°C at pressures up to 300 bar. Use the procedure described in Sec. 15.4, with $l_{12} = 0.088$. Compare the results with those shown in Fig. 15.6. Discuss any differences. $P_1^{sat} = 0.0102$ bar at 80°C.

15.20. Estimate the solubility of naphthalene(1) in nitrogen(2) at a temperature of 35°C at pressures up to 300 bar. Use the procedure described in Sec. 15.4, with $l_{12} = 0$. Compare the results with those shown in Fig. 15.6 for the naphthalene/CO_2 system at 35°C with $l_{12} = 0$. Discuss any differences.

15.21. The qualitative features of SVE at high pressures shown in Fig. 15.6 are determined by the equation of state for the gas. To what extent can these features be represented by the two-term virial equation in pressure, Eq. (3.36)?

15.22 The UNILAN equation for pure-species adsorption is:

$$n = \frac{m}{2s} \ln \left(\frac{c + P e^s}{c + P e^{-s}} \right)$$

$$z = (1 - bn)^{-1}$$

where m, s, and c are positive empirical constants.

(a) Show that the UNILAN equation reduces to the Langmuir isotherm for $s = 0$. (*Hint*: Apply l'Hôpital's rule.)

(b) Show that Henry's constant k for the UNILAN equation is:

$$k(\text{UNILAN}) = \frac{m}{cs} \sinh s$$

(c) Examine the *detailed* behavior of the UNILAN equation at zero pressure ($P \to 0$, $n \to 0$).

15.23. In Ex. 15.8, Henry's constant for adsorption k, identified as the intercept on a plot of n/P vs. n, was found from a polynomial curve-fit of n/P vs. n. An alternative procedure is based on a plot of $\ln(P/n)$ vs. n. Suppose that the adsorbate equation of state is a power series in n: $z = 1 + Bn + Cn^2 + \cdots$. Show how from a plot (or a polynomial curve-fit) of $\ln(P/n)$ vs. n one can extract values of k and B. [*Hint:* Start with Eq. (15.39).]

15.24. It was assumed in the development of Eq. (15.39) that the gas phase is *ideal*, with $Z = 1$. Suppose for a *real* gas phase that $Z = Z(T, P)$. Determine the analogous expression to Eq. (15.39) appropriate for a real (nonideal) gas phase. [*Hint:* Start with Eq. (15.35).]

15.25. Use results reported in Ex. 15.8 to prepare plots of Π vs. n and z vs. n for ethylene adsorbed on a carbon molecular sieve. Discuss the plots.

15.26. Suppose that the adsorbate equation of state is given by $z = (1 - bn)^{-1}$, where b is a constant. Find the implied adsorption isotherm, and show under what conditions it reduces to the Langmuir isotherm.

15.27. Suppose that the adsorbate equation of state is given by $z = 1 + \beta n$, where β is a function of T only. Find the implied adsorption isotherm, and show under what conditions it reduces to the Langmuir isotherm.

15.28. Derive the result given in the third step of the procedure for predicting adsorption equilibria by ideal-adsorbed-solution theory at the end of Sec. 15.5.

15.29. Consider a ternary system comprising solute species 1 and a mixed solvent (species 2 and 3). Assume that:

$$\frac{G^E}{RT} = A_{12}x_1x_2 + A_{13}x_1x_3 + A_{23}x_2x_3$$

Show that Henry's constant $\mathcal{H}_1$ for species 1 in the mixed solvent is related to Henry's constants $\mathcal{H}_{1,2}$ and $\mathcal{H}_{1,3}$ for species 1 in the pure solvents by:

$$\ln \mathcal{H}_1 = x_2' \ln \mathcal{H}_{1,2} + x_3' \ln \mathcal{H}_{1,3} - A_{23}x_2'x_3'$$

Here x_2' and x_3' are solute-free mole fractions:

$$x_2' \equiv \frac{x_2}{x_2 + x_3} \qquad x_3' \equiv \frac{x_3}{x_2 + x_3}$$

15.30. It is possible in principle for a binary liquid system to show more than one region of LLE for a particular temperature. For example, the solubility diagram might have two side-by-side "islands" of partial miscibility separated by a homogeneous phase. What would the ΔG vs. x_1 diagram at constant T look like for this case? Suggestion: See Fig. 12.13 for a mixture showing *normal* LLE behavior.

15.31. With $\bar{V}_2 = V_2$, Eq. (15.66) for the osmotic pressure may be represented as a power series in x_1:

$$\frac{\Pi V_2}{x_1 RT} = 1 + Bx_1 + Cx_1^2 + \cdots$$

Reminiscent of Eqs. (3.33) and (3.34), this series is called an *osmotic virial expansion*. Show that the second osmotic virial coefficient B is:

$$B = \frac{1}{2}\left[1 - \left(\frac{d^2 \ln \gamma_2}{dx_1^2}\right)_{x_1=0}\right]$$

What is B for an ideal solution? What is B if $G^E = A x_1 x_2$?

15.32. A liquid-process feed stream F contains 99 mol-% of species 1 and 1 mol-% of impurity, species 2. The impurity level is to be reduced to 0.1 mol-% by contacting the feed stream with a stream S of pure liquid solvent, species 3, in a mixer/settler. Species 1 and 3 are essentially immiscible. Owing to "good chemistry," it is expected that species 2 will selectively concentrate in the solvent phase.

   (a) With the equations given below, determine the required solvent-to-feed ratio n_S/n_F.
   (b) What is mole fraction x_2 of impurity in the solvent phase leaving the mixer/settler?
   (c) What is "good" about the chemistry here? With respect to liquid-phase nonidealities, what would be "bad" chemistry for the proposed operation?

   Given: $G_{12}^E/RT = 1.5 x_1 x_2$ $G_{23}^E/RT = -0.8 x_2 x_3$

15.33. At 25°C the solubility of n-hexane in water is 2 ppm (molar basis), and the solubility of water in n-hexane is 520 ppm. Estimate the activity coefficients for the two species in the two phases.

15.34. A binary liquid mixture is only partially miscible at 298 K. If the mixture is to be made homogeneous by increasing the temperature, what must be the sign of H^E?

15.35. The *spinodal curve* for a binary liquid system is the locus of states for which

$$\frac{d^2(\Delta G/RT)}{dx_1^2} = 0 \text{ (const } T, P)$$

Thus it separates regions of stability from instability with respect to liquid/liquid phase splitting. For a given T, there are normally two spinodal compositions (if any). They are the same at a consolute temperature. On curve II of Fig. 12.13 they are a pair of compositions between x_1^α and x_1^β, corresponding to zero curvature.

Suppose a liquid mixture is described by the symmetrical equation

$$\frac{G^E}{RT} = A(T) x_1 x_2$$

   (a) Find an expression for the spinodal compositions as a function of $A(T)$.
   (b) Assume that $A(T)$ is the expression used to generate Fig. 15.2. Plot on a single graph the solubility curve and the spinodal curve. Discuss.

15.36. Two special models of liquid-solution behavior are the *regular solution*, for which $S^E = 0$ everywhere, and the *athermal solution*, for which $H^E = 0$ everywhere.

(a) Ignoring the P-dependence of G^E, show that for a regular solution,

$$\frac{G^E}{RT} = \frac{F_R(x)}{RT}$$

(b) Ignoring the P-dependence of G^E, show that for an athermal solution,

$$\frac{G^E}{RT} = F_A(x)$$

(c) Suppose that G^E/RT is described by the symmetrical equation

$$\frac{G^E}{RT} = A(T)x_1 x_2$$

From parts (a) and (b), we conclude that

$$\frac{G^E}{RT} = \frac{\alpha}{RT}x_1 x_2 \qquad \text{(regular)} \tag{A}$$

$$\frac{G^E}{RT} = \beta x_1 x_2 \qquad \text{(athermal)} \tag{B}$$

where α and β are *constants*. What are the implications of Eqs. (A) and (B) with respect to the shapes of predicted solubility diagrams for LLE? Find from Eq. (A) an expression for the consolute temperature, and show that it must be an *upper* consolute temperature.

Suggestion: See Ex. 15.3 for numerical guidance.

15.37. Many fluids could be used as solvent species for supercritical separation processes (Sec. 15.4). But the two most popular choices seem to be carbon dioxide and water. Why? Discuss the pros and cons of using CO_2 vs. H_2O as a supercritical solvent.

Chapter 16

Thermodynamic Analysis of Processes

The purpose of this chapter is to present a procedure for the analysis of practical processes from a thermodynamic perspective. It is an extension of the *ideal work* and *lost work* concepts presented in Secs. 5.7 and 5.8.

Real irreversible processes are amenable to thermodynamic analysis. The goal of such an analysis is to determine how efficiently energy is used or produced and to show quantitatively the effect of inefficiencies in each step of a process. The cost of energy is of concern in any manufacturing operation, and the first step in any attempt to reduce energy requirements is to determine where and to what extent energy is wasted through process irreversibilities. The treatment here is limited to steady-state flow processes, because of their predominance in industrial practice.

16.1 THERMODYNAMIC ANALYSIS OF STEADY-STATE FLOW PROCESSES

Most industrial processes involving fluids consist of multiple steps, and lost-work calculations are then made for each step separately. By Eq. (5.29),

$$\dot{W}_{\text{lost}} = T_\sigma \dot{S}_G$$

For a single surroundings temperature T_σ, summing over the steps of a process gives:

$$\Sigma \dot{W}_{\text{lost}} = T_\sigma \Sigma \dot{S}_G$$

Dividing the former equation by the latter yields:

$$\frac{\dot{W}_{\text{lost}}}{\Sigma \dot{W}_{\text{lost}}} = \frac{\dot{S}_G}{\Sigma \dot{S}_G}$$

Thus an analysis of the lost work, made by calculation of the fraction that each individual lost-work term represents of the total, is the same as an analysis of the rate of entropy generation, made by expressing each individual entropy-generation term as a fraction of the sum of all entropy-generation terms. Recall that all terms in these equations are positive.

An alternative to the lost-work or entropy-generation analysis is a work analysis. For this, Eq. (5.26) becomes:

$$\Sigma \dot{W}_{\text{lost}} = \dot{W}_s - \dot{W}_{\text{ideal}} \tag{16.1}$$

For a work-requiring process, all of these work quantities are positive and $\dot{W}_s > \dot{W}_{\text{ideal}}$. The preceding equation is then written:

$$\boxed{\dot{W}_s = \dot{W}_{\text{ideal}} + \Sigma \dot{W}_{\text{lost}}} \tag{16.2}$$

A work analysis expresses each individual work term on the right as a fraction of $\dot{W}_s$.

For a work-producing process, $\dot{W}_s$ and $\dot{W}_{\text{ideal}}$ are negative, and $|\dot{W}_{\text{ideal}}| > |\dot{W}_s|$. Equation (16.1) is therefore best written:

$$\boxed{|\dot{W}_{\text{ideal}}| = |\dot{W}_s| + \Sigma \dot{W}_{\text{lost}}} \tag{16.3}$$

A work analysis expresses each individual work term on the right as a fraction of $|\dot{W}_{\text{ideal}}|$. Such an analysis cannot be carried out if a process is so inefficient that W_{ideal} is negative, indicating that the process should produce work, but $\dot{W}_s$ is positive, indicating that the process in fact requires work. A lost-work or entropy-generation analysis is always possible.

Example 16.1

The operating conditions of a practical steam power plant are described in Ex. 8.1, parts (b) and (c). In addition, steam is generated in a furnace/boiler unit where methane is burned completely to CO_2 and H_2O with 25% excess air. The flue gas leaving the furnace has a temperature of 460 K, and $T_\sigma = 298.15$ K. Make a thermodynamic analysis of the power plant.

Solution 16.1

A flow diagram of the power plant is shown in Fig. 16.1. The conditions and properties for key points in the steam cycle, taken from Ex. 8.1, are as follows:

Point	State of steam	$t/°C$	P/kPa	$H/\text{kJ·kg}^{-1}$	$S/\text{kJ·kg}^{-1}\text{·K}^{-1}$
1	Subcooled liquid	45.83	8600	203.4	0.6580
2	Superheated vapor	500	8600	3391.6	6.6858
3	Wet vapor, $x = 0.9378$	45.83	10	2436.0	7.6846
4	Saturated liquid	45.83	10	191.8	0.6493

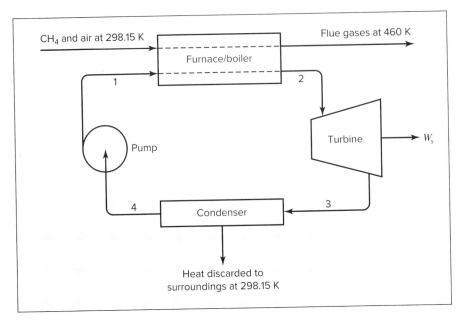

Figure 16.1: Power cycle of Ex. 16.1.

Because the steam undergoes a cyclic process, the only changes that must be considered for calculation of the ideal work are those of the gases passing through the furnace. The reaction occurring is:

$$CH_4 + 2O_2 \rightarrow CO_2 + 2H_2O$$

For this reaction, data from Table C.4 give:

$$\Delta H^\circ_{298} = -393{,}509 + (2)(-241{,}818) - (-74{,}520) = -802{,}625 \text{ J}$$
$$\Delta G^\circ_{298} = -394{,}359 + (2)(-228{,}572) - (-50{,}460) = -801{,}043 \text{ J}$$

Whence, $$\Delta S^\circ_{298} = \frac{\Delta H^\circ_{298} - \Delta G^\circ_{298}}{298.15} = -5.306 \text{ J K}^{-1}$$

On the basis of 1 mol of methane burned with 25% excess air, the air entering the furnace contains:

$$O_2: \quad (2)(1.25) = 2.5 \text{ mol}$$
$$N_2: \quad (2.5)(79/21) = 9.405 \text{ mol}$$

Total: 11.905 mol air

After complete combustion of the methane, the flue gas contains:

CO$_2$:	1 mol	$y_{CO_2} = 0.0775$
H$_2$O:	2 mol	$y_{H_2O} = 0.1550$
O$_2$:	0.5 mol	$y_{O_2} = 0.0387$
N$_2$:	9.405 mol	$y_{N_2} = 0.7288$
Total:	12.905 mol flue gas	$\sum y_i = 1.0000$

The change of state that occurs in the furnace is from methane and air at atmospheric pressure and 298.15 K, the temperature of the surroundings, to flue gas at atmospheric pressure and 460 K. For this change of state, ΔH and ΔS are calculated for the path shown in Fig. 16.2. The assumption of ideal gases is reasonable here and is the basis of calculation for ΔH and ΔS for each of the four steps shown in Fig. 16.2.

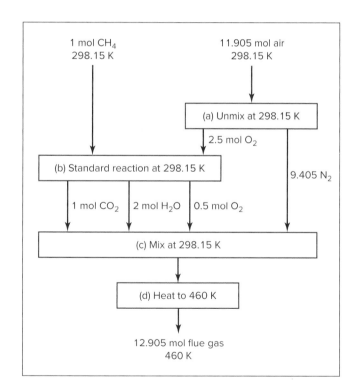

Figure 16.2: Calculation path for combustion process of Ex. 16.1.

Step a: For *unmixing* the entering air, Eqs. (11.11) and (11.12) with changes of sign give:

$$\Delta H_a = 0$$
$$\Delta S_a = nR \sum_i y_i \ln y_i$$
$$= (11.905)(8.314)(0.21 \ln 0.21 + 0.79 \ln 0.79) = -50.870 \text{ J·K}^{-1}$$

Step b: For the standard reaction at 298.15 K,

$$\Delta H_b = \Delta H^\circ_{298} = -802{,}625 \text{ J} \qquad\qquad \Delta S_b = \Delta S^\circ_{298} = -5.306 \text{ J}\cdot\text{K}^{-1}$$

Step c: For *mixing* to form the flue gas,

$$\Delta H_c = 0$$
$$\Delta S_c = -nR \sum_i y_i \ln y_i$$
$$= -(12.905)(8.314)(0.0775 \ln 0.0775 + 0.1550 \ln 0.1550$$
$$+ 0.0387 \ln 0.0387 + 0.7288 \ln 0.7288) = 90.510 \text{ J}\cdot\text{K}^{-1}$$

Step d: For the heating step, the mean heat capacities between 298.15 and 460 K are calculated by Eqs. (4.9) and (5.13) with data from Table C.1. The results in $\text{J}\cdot\text{mol}^{-1}\cdot\text{K}^{-1}$ are summarized as follows:

	$\langle C_p \rangle_H$	$\langle C_p \rangle_S$
CO_2	41.649	41.377
H_2O	34.153	34.106
N_2	29.381	29.360
O_2	30.473	0.997

Each individual heat capacity is multiplied by the number of moles of that species in the flue gas, and the products are summed over all species. This gives total mean heat capacities for the 12.905 mol of mixture:

$$\langle C^t_p \rangle_H = 401.520 \qquad \text{and} \qquad \langle C^t_p \rangle_S = 400.922 \text{ J}\cdot\text{K}^{-1}$$

Then,

$$\Delta H_d = \langle C^t_p \rangle_H (T_2 - T_1) = (401.520)(460 - 298.15) = 64{,}986 \text{ J}$$
$$\Delta S_d = \langle C^t_p \rangle_S \ln \frac{T_2}{T_1} = 400.922 \ln \frac{460}{298.15} = 173.852 \text{ J}\cdot\text{K}^{-1}$$

For the total process on the basis of 1 mol CH_4 burned,

$$\Delta H = \sum_i \Delta H_i = 0 - 802{,}625 + 0 + 64{,}986 = -737{,}639 \text{ J}$$
$$\Delta S = \sum_i \Delta S_i = -50.870 - 5.306 + 90.510 + 173.852 = 208.186 \text{ J}\cdot\text{K}^{-1}$$

Thus, $$\Delta H = -737.64 \text{ kJ} \qquad\qquad \Delta S = 0.2082 \text{ kJ}\cdot\text{K}^{-1}$$

The steam rate found in Ex. 8.1 is $\dot m = 84.75 \text{ kg}\cdot\text{s}^{-1}$. An energy balance for the furnace/boiler unit, where heat is transferred from the combustion gases to the steam, allows calculation of the entering methane rate $\dot n_{CH4}$:

$$(84.75)(3391.6 - 203.4) + \dot n_{CH_4}(-737.64) = 0$$

whence $$\dot n_{CH_4} = 366.30 \text{ mol}\cdot\text{s}^{-1}$$

The ideal work for the process is given by Eq. (5.21):

$$\dot{W}_{\text{ideal}} = 366.30[-737.64 - (298.15)(0.2082)] = -292.94 \times 10^3 \text{ kJ·s}^{-1}$$

or

$$\dot{W}_{\text{ideal}} = -292.94 \times 10^3 \text{ kW}$$

The rate of entropy generation in each of the four units of the power plant is calculated by Eq. (5.17), and the lost work is then given by Eq. (5.29).

- *Furnace/boiler*: We have assumed no heat transfer from the furnace/boiler to the surroundings; therefore $\dot{Q} = 0$. The term $\Delta(S\dot{m})_{\text{fs}}$ is simply the sum of the entropy changes of the two streams multiplied by their rates:

$$\dot{S}_G = (366.30)(0.2082) + (84.75)(6.6858 - 0.6580) = 587.12 \text{ kJ·s}^{-1}\text{·K}^{-1}$$

or

$$\dot{S}_G = 587.12 \text{ kW·K}^{-1}$$

and

$$\dot{W}_{\text{lost}} = T_\sigma \dot{S}_G = (298.15)(587.12) = 175.05 \times 10^3 \text{ kW}$$

- *Turbine*: For adiabatic operation,

$$\dot{S}_G = (84.75)(7.6846 - 6.6858) = 84.65 \text{ kW·K}^{-1}$$

and

$$\dot{W}_{\text{lost}} = (298.15)(84.65) = 25.24 \times 10^3 \text{ kW}$$

- *Condenser*: The condenser transfers heat from the condensing steam to the surroundings at 298.15 K in an amount determined in Ex. 8.1:

$$\dot{Q}(\text{condenser}) = -190.2 \times 10^3 \text{ kJ·s}^{-1}$$

Thus

$$\dot{S}_G = (84.75)(0.6493 - 7.6846) + \frac{190,200}{298.15} = 41.69 \text{ kW·K}^{-1}$$

and

$$\dot{W}_{\text{lost}} = (298.15)(41.69) = 12.32 \times 10^3 \text{ kW}$$

- *Pump*: Because the pump operates adiabatically,

$$\dot{S}_G = (84.75)(0.6580 - 0.6493) = 0.74 \text{ kW·K}^{-1}$$

and

$$\dot{W}_{\text{lost}} = (298.15)(0.74) = 0.22 \times 10^3 \text{ kW}$$

The entropy-generation analysis is:

	kW·K^{-1}	Percent of $\sum \dot{S}_G$
$\dot{S}_G$(furnace/boiler)	587.12	82.2
$\dot{S}_G$(turbine)	84.65	11.9
$\dot{S}_G$(condenser)	41.69	5.8
$\dot{S}_G$(pump)	0.74	0.1
$\sum \dot{S}_G$	714.20	100.0

A work analysis is carried out in accord with Eq. (16.3):

$$|\dot{W}_{\text{ideal}}| = |\dot{W}_s| + \sum \dot{W}_{\text{lost}}$$

The results of this analysis are:

	kW	Percent of $|\dot{W}_{ideal}|$
$|\dot{W}_s|$ (from Ex. 8.1)	80.00×10^3	$27.3(=\eta_t)$
$\dot{W}_{lost}$ (furnace/boiler)	175.05×10^3	59.8
$\dot{W}_{lost}$ (turbine)	25.24×10^3	8.6
$\dot{W}_{lost}$ (condenser)	12.43×10^3	4.2
$\dot{W}_{lost}$ (pump)	0.22×10^3	0.1
$|\dot{W}_{ideal}|$	292.94×10^3	100.0

The thermodynamic efficiency of the power plant is 27.3%, and the major source of inefficiency is the furnace/boiler. The combustion process itself accounts for most of the entropy generation in this unit, and the remainder is the result of heat transfer across finite temperature differences.

Example 16.2

Methane is liquefied in a simple Linde system, as shown in Fig. 16.3. The methane enters the compressor at 1 bar and 300 K, and after compression to 60 bar it is cooled back to 300 K. The product is saturated liquid methane at 1 bar. The unliquefied methane, also at 1 bar, is returned through a heat exchanger, where it is heated to 295 K by the high-pressure methane. A heat leak into the heat exchanger of 5 kJ is assumed for each kilogram of methane entering the compressor. Heat leaks to other parts of the liquefier are assumed negligible. Make a thermodynamic analysis of the process for a surroundings temperature of $T_\sigma = 300$ K.

Solution 16.2

Methane compression from 1 to 60 bar is assumed to be carried out in a three-stage machine with inter- and aftercooling to 300 K and a compressor efficiency of 75%. The actual work of this compression is estimated as 1000 kJ per kilogram of methane. The fraction of the methane that is liquefied z is calculated by an energy balance:

$$H_4 z + H_6(1 - z) - H_2 = Q$$

where Q is the heat leak from the surroundings. Solution for z gives

$$z = \frac{H_6 - H_2 - Q}{H_6 - H_4} = \frac{1188.9 - 1140.0 - 5}{1188.9 - 285.4} = 0.0486$$

This result may be compared with the value of 0.0541 obtained in Ex. 9.3, for the same operating conditions, but with no heat leak.

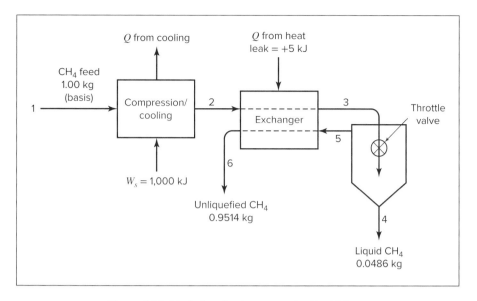

Figure 16.3: Linde liquefaction system for Ex. 16.2.

The properties at the various key points of the process, given in the accompanying table, are either available or are calculated by standard methods. Data used here are from Perry and Green.[1] The basis of all calculations is 1 kg of methane entering the process, and all rates are expressed on this basis.

Point	State of CH$_4$	T/K	P/bar	H/kJ·kg^{-1}	S/kJ·kg^{-1}·K^{-1}
1	Superheated vapor	300.0	1	1198.8	11.629
2	Superheated vapor	300.0	60	1140.0	9.359
3	Superheated vapor	207.1	60	772.0	7.798
4	Saturated liquid	111.5	1	285.4	4.962
5	Saturated vapor	111.5	1	796.9	9.523
6	Superheated vapor	295.0	1	1188.9	11.589

The ideal work depends on the overall changes in the methane passing through the liquefier. Application of Eq. (5.21) gives:

$$\dot{W}_{\text{ideal}} = \Delta(H\dot{m})_{\text{fs}} - T_\sigma \Delta(S\dot{m})_{\text{fs}}$$
$$= [(0.0486)(285.4) + (0.9514)(1188.9) - 1198.8]$$
$$-(300)[(0.0486)(4.962) + (0.9514)(11.589) - 11.629] = 53.8 \text{ kJ}$$

The rate of entropy generation and the lost work for each of the individual steps of the process is calculated by Eqs. (5.28) and (5.29).

[1]R. H. Perry and D. Green, *Perry's Chemical Engineers' Handbook*, 7th ed., pp. 2-251 and 2-253, McGraw-Hill, New York, 1997.

- *Compression/cooling*: Heat transfer for this step is given by an energy balance:

$$\dot{Q} = \Delta H - \dot{W}_s = (H_2 - H_1) - \dot{W}_s$$
$$= (1140.0 - 1999.8) - 1000$$
$$= -1059.8 \text{ kJ}$$

Then,
$$\dot{S}_G = S_2 - S_1 - \frac{\dot{Q}}{T_\sigma}$$
$$= 9.359 - 11.629 + \frac{1059.8}{300}$$
$$= 1.2627 \text{ kJ·kg}^{-1}\cdot\text{K}^{-1}$$
$$\dot{W}_{\text{lost}} = (300)(1.2627) = 378.8 \text{ kJ·kg}^{-1}$$

- *Exchanger*: With $\dot{Q}$ equal to the heat leak,

$$\dot{S}_G = (S_6 - S_5)(1 - z) + (S_3 - S_2)(1) - \frac{\dot{Q}}{T_\sigma}$$

Then,
$$\dot{S}_G = (11.589 - 9.523)(0.9514) + (7.798 - 9.359) - \frac{5}{300}$$
$$= 0.3879 \text{ kJ·kg}^{-1}\cdot\text{K}^{-1}$$
$$\dot{W}_{\text{lost}} = (300)(0.3879) = 116.4 \text{ kJ·kg}^{-1}$$

- *Throttle*: For adiabatic operation of the throttle and separator,

$$\dot{S}_G = S_4 z + S_5(1 - z) - S_3$$
$$= (4.962)(0.0486) + (9.523)(0.9514) - 7.798$$
$$= 1.5033 \text{ kJ·kg}^{-1}\cdot\text{K}^{-1}$$
$$\dot{W}_{\text{lost}} = (300)(1.5033) = 451.0 \text{ kJ·kg}^{-1}$$

The entropy-generation analysis is:

	kJ·kg^{-1}·K^{-1}	Percent of $\Sigma\dot{S}_G$
$\dot{S}_G$(compression/cooling)	1.2627	40.0
$\dot{S}_G$(exchanger)	0.3879	12.3
$\dot{S}_G$(throttle)	1.5033	47.7
$\Sigma\dot{S}_G$	3.1539	100.0

The work analysis, based on Eq. (16.2), is:

	kW·kg^{-1}	Percent of $\dot{W}_s$
$\dot{W}_{\text{ideal}}$	53.8	5.4($=\eta_t$)
$\dot{W}_{\text{lost}}$(compression/cooling)	378.8	37.9
$\dot{W}_{\text{lost}}$(exchanger)	116.4	11.6
$\dot{W}_{\text{lost}}$(throttle)	451.0	45.1
$\dot{W}_s$	1000.0	100.0

The largest loss occurs in the throttling step. Replacing this highly irreversible process with a turbine results in a considerable increase in efficiency. Of course, it would also result in a considerable increase in the capital cost of the equipment.

From the standpoint of energy conservation, the thermodynamic efficiency of a process should be as high as possible and the entropy generation or lost work as low as possible. The final design depends largely on economic considerations, and the cost of energy is an important factor. The thermodynamic analysis of a specific process shows the locations of the major inefficiencies and hence the pieces of equipment or steps in the process that could be altered or replaced to advantage. However, this sort of analysis gives no hint as to the nature of the changes that might be made. It merely shows that the present design is wasteful of energy and that there is room for improvement. One function of the chemical engineer is to try to devise a better process and to use ingenuity to keep operating costs, as well as capital expenditures, low. Each newly devised process may, of course, be analyzed to determine what improvement has been made.

16.2 SYNOPSIS

After studying this chapter, including the end-of-chapter problems, one should be able to

- Carry out a step-by-step thermodynamic analysis of a steady-flow process like those illustrated in Ex. 16.1 and Ex. 16.2

- Identify the contributions of each process step to the overall rate of entropy generation of the process

- Assign the contributions of each process step to the overall lost work of the process, to identify the best opportunities for improving the thermodynamic efficiency of the overall process

16.3 PROBLEMS

16.1. A plant takes in water at 21°C, cools it to 0°C, and freezes it at this temperature, producing 0.5 kg·s^{-1} of ice. Heat rejection is at 21°C. The heat of fusion of water is 333.5 kJ·kg^{-1}.

(*a*) What is $\dot{W}_{ideal}$ for the process?

(*b*) What is the power requirement of a single Carnot heat pump operating between 0 and 21°C? What is the thermodynamic efficiency of this process? What is its irreversible feature?

(*c*) What is the power requirement if an ideal tetrafluoroethane vapor-compression refrigeration cycle is used? *Ideal* here implies isentropic compression, infinite cooling-water rate in the condenser, and minimum heat-transfer driving forces in evaporator and condenser of 0°C. What is the thermodynamic efficiency of this process? What are its irreversible features?

(*d*) What is the power requirement of a tetrafluoroethane vapor-compression cycle for which the compressor efficiency is 75%, the minimum temperature differences in evaporator and condenser are 5°C, and the temperature rise of the cooling water in the condenser is 10°C? Make a thermodynamic analysis of this process.

16.2. Consider a steady-flow process in which the following gas-phase reaction takes place: $CO + \frac{1}{2}O_2 \rightarrow CO_2$. The surroundings are at 300 K.

(*a*) What is W_{ideal} when the reactants enter the process as pure carbon monoxide and as air containing the stoichiometric amount of oxygen, both at 25°C and 1 bar, and the products of complete combustion leave the process at the same conditions?

(*b*) The overall process is exactly the same as in (*a*), but the CO is here burned in an adiabatic reactor at 1 bar. What is W_{ideal} for the process of cooling the flue gases to 25°C? What is the irreversible feature of the overall process? What is its thermodynamic efficiency? What has increased in entropy, and by how much?

16.3. A plant has saturated steam available at 2700 kPa, but there is little use for this steam. Rather, steam at 1000 kPa is required. Also available is saturated steam at 275 kPa. A suggestion is that the 275 kPa steam be compressed to 1000 kPa using the work of expanding the 2700 kPa steam to 1000 kPa. The two streams at 1000 kPa would then be mixed. Determine the rates at which steam at each initial pressure must be supplied to provide enough steam at 1000 kPa so that upon condensation to saturated liquid, heat in the amount of 300 kJ·s^{-1} is released

(*a*) if the process is carried out in a completely reversible manner.

(*b*) if the higher-pressure steam expands in a turbine of 78% efficiency and the lower-pressure steam is compressed in a machine of 75% efficiency. Make a thermodynamic analysis of this process.

16.4. Make a thermodynamic analysis of the refrigeration cycle of Ex. 9.1(*b*).

16.5. Make a thermodynamic analysis of the refrigeration cycle described in one of the parts of Prob. 9.9. Assume that the refrigeration effect maintains a heat reservoir at a temperature 5°C above the evaporation temperature and that T_σ is 5°C below the condensation temperature.

16.6. Make a thermodynamic analysis of the refrigeration cycle described in the first paragraph of Prob. 9.12. Assume that the refrigeration effect maintains a heat reservoir at a temperature 5°C above the evaporation temperature and that T_σ is 5°C below the condensation temperature.

16.7. A colloidal solution enters a single-effect evaporator at 100°C. Water is vaporized from the solution, producing a more concentrated solution and 0.5 kg·s⁻¹ of steam at 100°C. This steam is compressed and sent to the heating coils of the evaporator to supply the heat required for its operation. For a minimum heat-transfer driving force across the evaporator coils of 10°C, for a compressor efficiency of 75%, and for adiabatic operation, what is the state of the steam leaving the heating coils of the evaporator? For a surroundings temperature of 300 K, make a thermodynamic analysis of the process.

16.8. Make a thermodynamic analysis of the process described in Prob. 8.9. $T_\sigma = 27°C$.

16.9. Make a thermodynamic analysis of the process described in Ex. 9.3. $T_\sigma = 295$ K.

Appendix A

Conversion Factors and Values of the Gas Constant

Because standard reference books contain data in diverse units, we include Tables A.1 and A.2 to aid in the conversion of values from one set of units to another. Those units having no connection with the SI system are enclosed in parentheses. The following definitions are noted:

(ft) ≡ U.S. defined foot ≡ 0.3048 m
(in) ≡ U.S. defined inch ≡ 0.0254 m
(gal) ≡ U.S. liquid gallon ≡ 231 (in)^3
(lb_m) ≡ U.S. defined pound *mass* (avoirdupois)
≡ 0.45359237 kg
(lb_f) ≡ force to accelerate 1(lb_m) by 32.1740 (ft)·s^{-2}
(atm) ≡ standard atmospheric pressure ≡ 101,325 Pa
(psia) ≡ pounds *force* per square inch absolute pressure
(torr) ≡ pressure exerted by 1 mm mercury at 0°C and standard gravity
(cal) ≡ thermochemical calorie
(Btu) ≡ international steam table British thermal unit
(lb mole) ≡ mass in pounds *mass* with numerical value equal to the molar mass
(R) ≡ absolute temperature in Rankines

The conversion factors of Table A.1 are referred to a single basic or derived unit of the SI system. Conversions between other pairs of units for a given quantity are made as in the following example:

$$1 \text{ bar} = 0.986923 \text{ (atm)} = 750.061 \text{ (torr)}$$

thus

$$1 \text{ (atm)} = \frac{750.061}{0.986923} = 760.00 \text{ (torr)}$$

Table A.1: Conversion Factors

Quantity	Conversion
Length	$1 \text{ m} = 100 \text{ cm}$ $= 3.28084 \text{ (ft)} = 39.3701 \text{ (in)}$
Mass	$1 \text{ kg} = 10^3 \text{ g}$ $= 2.20462 \text{ (lb}_\text{m})$
Force	$1 \text{ N} = 1 \text{ kg·m·s}^{-2}$ $= 10^5 \text{ (dyne)}$ $= 0.224809 \text{ (lb}_\text{f})$
Pressure	$1 \text{ bar} = 10^5 \text{ kg·m}^{-1}\text{·s}^{-2} = 10^5 \text{ N·m}^{-2}$ $= 10^5 \text{ Pa} = 10^2 \text{ kPa}$ $= 10^6 \text{ (dyne)·cm}^{-2}$ $= 0.986923 \text{ (atm)}$ $= 14.5038 \text{ (psia)}$ $= 750.061 \text{ (torr)}$
Volume	$1 \text{ m}^3 = 10^6 \text{ cm}^3 = 10^3 \text{ liters}$ $= 35.3147 \text{ (ft)}^3$ $= 264.172 \text{ (gal)}$
Density	$1 \text{ g·cm}^{-3} = 10^3 \text{ kg·m}^{-3}$ $= 62.4278 \text{ (lb}_\text{m})\text{(ft)}^{-3}$
Energy	$1 \text{ J} = 1 \text{ kg·m}^2\text{·s}^{-2} = 1 \text{ N·m}$ $= 1 \text{ m}^3\text{·Pa} = 10^{-5} \text{ m}^3\text{·bar} = 10 \text{ cm}^3\text{·bar}$ $= 9.86923 \text{ cm}^3\text{·(atm)}$ $= 10^7 \text{ (dyne)·cm} = 10^7 \text{ (erg)}$ $= 0.239006 \text{ (cal)}$ $= 5.12197 \times 10^{-3} \text{ (ft)}^3\text{(psia)} = 0.737562 \text{ (ft)(lb}_\text{f})$ $= 9.47831 \times 10^{-4} \text{ (Btu)} = 2.77778 \times 10^{-7} \text{ kW·h}$
Power	$1 \text{ kW} = 10^3 \text{ W} = 10^3 \text{ kg·m}^2\text{·s}^{-3} = 10^3 \text{ J·s}^{-1}$ $= 239.006 \text{ (cal)·s}^{-1}$ $= 737.562 \text{ (ft)(lb}_\text{f})\text{·s}^{-1}$ $= 0.947831 \text{ (Btu)·s}^{-1}$ $= 1.34102 \text{ (hp)}$

Table A.2: Values of the Universal Gas Constant

$R = 8.314 \text{ J·mol}^{-1}\text{·K}^{-1} = 8.314 \text{ m}^3\text{·Pa·mol}^{-1}\text{·K}^{-1}$
$= 83.14 \text{ cm}^3\text{·bar·mol}^{-1}\text{·K}^{-1} = 8314 \text{ cm}^3\text{·kPa·mol}^{-1}\text{·K}^{-1}$
$= 82.06 \text{ cm}^3\text{·(atm)·mol}^{-1}\text{·K}^{-1} = 62{,}356 \text{ cm}^3\text{·(torr)·mol}^{-1}\text{·K}^{-1}$
$= 1.987 \text{ (cal)·mol}^{-1}\text{·K}^{-1} = 1.986 \text{ (Btu)(lb mole)}^{-1}\text{(R)}^{-1}$
$= 0.7302 \text{ (ft)}^3\text{(atm)(lb mol)}^{-1}\text{(R)}^{-1} = 10.73 \text{ (ft)}^3\text{(psia)(lb mol)}^{-1}\text{(R)}^{-1}$
$= 1545 \text{ (ft)(lb}_\text{f}\text{)(lb mol)}^{-1}\text{(R)}^{-1}$

Appendix B

Properties of Pure Species

Table B.1 Characteristic Properties of Pure Species

Listed here for various chemical species are values for the molar mass (molecular weight), acentric factor ω, critical temperature T_c, critical pressure P_c, critical compressibility factor Z_c, critical molar volume V_c, and normal boiling point T_n. Abstracted from Project 801, DIPPR®, Design Institute for Physical Property Data of the American Institute of Chemical Engineers, they are reproduced with permission. The current full version of this database includes values for 34 constant properties and 15 temperature-dependent thermodynamic and transport properties for 2278 chemical species, and new species are added regularly.

Table B.2 Constants for the Antoine Equation for Vapor Pressures of Pure Species

Table B.1: Characteristic Properties of Pure Species

	Molar mass	ω	T_c/K	P_c/bar	Z_c	V_c cm³·mol⁻¹	T_n/K
Methane	16.043	0.012	190.6	45.99	0.286	98.6	111.4
Ethane	30.070	0.100	305.3	48.72	0.279	145.5	184.6
Propane	44.097	0.152	369.8	42.48	0.276	200.0	231.1
n-Butane	58.123	0.200	425.1	37.96	0.274	255.	272.7
n-Pentane	72.150	0.252	469.7	33.70	0.270	313.	309.2
n-Hexane	86.177	0.301	507.6	30.25	0.266	371.	341.9
n-Heptane	100.204	0.350	540.2	27.40	0.261	428.	371.6
n-Octane	114.231	0.400	568.7	24.90	0.256	486.	398.8
n-Nonane	128.258	0.444	594.6	22.90	0.252	544.	424.0
n-Decane	142.285	0.492	617.7	21.10	0.247	600.	447.3
Isobutane	58.123	0.181	408.1	36.48	0.282	262.7	261.4
Isooctane	114.231	0.302	544.0	25.68	0.266	468.	372.4

Table B.1 (Continued)

	Molar mass	ω	T_c/K	P_c/bar	Z_c	V_c cm³·mol⁻¹	T_n/K
Cyclopentane	70.134	0.196	511.8	45.02	0.273	258.	322.4
Cyclohexane	84.161	0.210	553.6	40.73	0.273	308.	353.9
Methylcyclopentane	84.161	0.230	532.8	37.85	0.272	319.	345.0
Methylcyclohexane	98.188	0.235	572.2	34.71	0.269	368.	374.1
Ethylene	28.054	0.087	282.3	50.40	0.281	131.	169.4
Propylene	42.081	0.140	365.6	46.65	0.289	188.4	225.5
1-Butene	56.108	0.191	420.0	40.43	0.277	239.3	266.9
cis-2-Butene	56.108	0.205	435.6	42.43	0.273	233.8	276.9
trans-2-Butene	56.108	0.218	428.6	41.00	0.275	237.7	274.0
1-Hexene	84.161	0.280	504.0	31.40	0.265	354.	336.3
Isobutylene	56.108	0.194	417.9	40.00	0.275	238.9	266.3
1,3-Butadiene	54.092	0.190	425.2	42.77	0.267	220.4	268.7
Cyclohexene	82.145	0.212	560.4	43.50	0.272	291.	356.1
Acetylene	26.038	0.187	308.3	61.39	0.271	113.	189.4
Benzene	78.114	0.210	562.2	48.98	0.271	259.	353.2
Toluene	92.141	0.262	591.8	41.06	0.264	316.	383.8
Ethylbenzene	106.167	0.303	617.2	36.06	0.263	374.	409.4
Cumene	120.194	0.326	631.1	32.09	0.261	427.	425.6
o-Xylene	106.167	0.310	630.3	37.34	0.263	369.	417.6
m-Xylene	106.167	0.326	617.1	35.36	0.259	376.	412.3
p-Xylene	106.167	0.322	616.2	35.11	0.260	379.	411.5
Styrene	104.152	0.297	636.0	38.40	0.256	352.	418.3
Naphthalene	128.174	0.302	748.4	40.51	0.269	413.	491.2
Biphenyl	154.211	0.365	789.3	38.50	0.295	502.	528.2
Formaldehyde	30.026	0.282	408.0	65.90	0.223	115.	254.1
Acetaldehyde	44.053	0.291	466.0	55.50	0.221	154.	294.0
Methyl acetate	74.079	0.331	506.6	47.50	0.257	228.	330.1
Ethyl acetate	88.106	0.366	523.3	38.80	0.255	286.	350.2
Acetone	58.080	0.307	508.2	47.01	0.233	209.	329.4
Methyl ethyl ketone	72.107	0.323	535.5	41.50	0.249	267.	352.8
Diethyl ether	74.123	0.281	466.7	36.40	0.263	280.	307.6
Methyl *t*-butyl ether	88.150	0.266	497.1	34.30	0.273	329.	328.4
Methanol	32.042	0.564	512.6	80.97	0.224	118.	337.9
Ethanol	46.069	0.645	513.9	61.48	0.240	167.	351.4
1-Propanol	60.096	0.622	536.8	51.75	0.254	219.	370.4
1-Butanol	74.123	0.594	563.1	44.23	0.260	275.	390.8
1-Hexanol	102.177	0.579	611.4	35.10	0.263	381.	430.6
2-Propanol	60.096	0.668	508.3	47.62	0.248	220.	355.4
Phenol	94.113	0.444	694.3	61.30	0.243	229.	455.0

Table B.1 (Continued)

	Molar mass	ω	T_c/K	P_c/bar	Z_c	V_c cm³·mol⁻¹	T_n/K
Ethylene glycol	62.068	0.487	719.7	77.00	0.246	191.0	470.5
Acetic acid	60.053	0.467	592.0	57.86	0.211	179.7	391.1
n-Butyric acid	88.106	0.681	615.7	40.64	0.232	291.7	436.4
Benzoic acid	122.123	0.603	751.0	44.70	0.246	344.	522.4
Acetonitrile	41.053	0.338	545.5	48.30	0.184	173.	354.8
Methylamine	31.057	0.281	430.1	74.60	0.321	154.	266.8
Ethylamine	45.084	0.285	456.2	56.20	0.307	207.	289.7
Nitromethane	61.040	0.348	588.2	63.10	0.223	173.	374.4
Carbon tetrachloride	153.822	0.193	556.4	45.60	0.272	276.	349.8
Chloroform	119.377	0.222	536.4	54.72	0.293	239.	334.3
Dichloromethane	84.932	0.199	510.0	60.80	0.265	185.	312.9
Methyl chloride	50.488	0.153	416.3	66.80	0.276	143.	249.1
Ethyl chloride	64.514	0.190	460.4	52.70	0.275	200.	285.4
Chlorobenzene	112.558	0.250	632.4	45.20	0.265	308.	404.9
Tetrafluoroethane	102.030	0.327	374.2	40.60	0.258	198.0	247.1
Argon	39.948	0.000	150.9	48.98	0.291	74.6	87.3
Krypton	83.800	0.000	209.4	55.02	0.288	91.2	119.8
Xenon	131.30	0.000	289.7	58.40	0.286	118.0	165.0
Helium 4	4.003	−0.390	5.2	2.28	0.302	57.3	4.2
Hydrogen	2.016	−0.216	33.19	13.13	0.305	64.1	20.4
Oxygen	31.999	0.022	154.6	50.43	0.288	73.4	90.2
Nitrogen	28.014	0.038	126.2	34.00	0.289	89.2	77.3
Air†	28.851	0.035	132.2	37.45	0.289	84.8	
Chlorine	70.905	0.069	417.2	77.10	0.265	124.	239.1
Carbon monoxide	28.010	0.048	132.9	34.99	0.299	93.4	81.7
Carbon dioxide	44.010	0.224	304.2	73.83	0.274	94.0	
Carbon disulfide	76.143	0.111	552.0	79.00	0.275	160.	319.4
Hydrogen sulfide	34.082	0.094	373.5	89.63	0.284	98.5	212.8
Sulfur dioxide	64.065	0.245	430.8	78.84	0.269	122.	263.1
Sulfur trioxide	80.064	0.424	490.9	82.10	0.255	127.	317.9
Nitric oxide (NO)	30.006	0.583	180.2	64.80	0.251	58.0	121.4
Nitrous oxide (N₂O)	44.013	0.141	309.6	72.45	0.274	97.4	184.7
Hydrogen chloride	36.461	0.132	324.7	83.10	0.249	81.	188.2
Hydrogen cyanide	27.026	0.410	456.7	53.90	0.197	139.	298.9
Water	18.015	0.345	647.1	220.55	0.229	55.9	373.2
Ammonia	17.031	0.253	405.7	112.80	0.242	72.5	239.7
Nitric acid	63.013	0.714	520.0	68.90	0.231	145.	356.2
Sulfuric acid	98.080	. . .	924.0	64.00	0.147	177.	610.0

† Pseudoparameters for $y_{N_2} = 0.79$ and $y_{O_2} = 0.21$. See Eqs. (6.78)–(6.80).

Table B.2: Constants for the Antoine Equation for Vapor Pressures of Pure Species

$$\ln P^{\text{sat}}/\text{kPa} = A - \frac{B}{t/°C + C}$$

Latent heat of vaporization at the normal boiling point (ΔH_n), and normal boiling point (t_n)

Name	Formula	Parameters for Antoine Eqn.			Temp. Range °C	ΔH_n kJ/mol	$t_n/°C$
		$A^\dagger$	B	C			
Acetone	C_3H_6O	14.3145	2756.22	228.060	−26—77	29.10	56.2
Acetic acid	$C_2H_4O_2$	15.0717	3580.80	224.650	24—142	23.70	117.9
Acetonitrile*	C_2H_3N	14.8950	3413.10	250.523	−27—81	30.19	81.6
Benzene	C_6H_6	13.7819	2726.81	217.572	6—104	30.72	80.0
iso-Butane	C_4H_{10}	13.8254	2181.79	248.870	−83—7	21.30	−11.9
n-Butane	C_4H_{10}	13.6608	2154.70	238.789	−73—19	22.44	−0.5
1-Butanol	$C_4H_{10}O$	15.3144	3212.43	182.739	37—138	43.29	117.6
2-Butanol*	$C_4H_{10}O$	15.1989	3026.03	186.500	25—120	40.75	99.5
iso-Butanol	$C_4H_{10}O$	14.6047	2740.95	166.670	30—128	41.82	107.8
tert-Butanol	$C_4H_{10}O$	14.8445	2658.29	177.650	10—101	39.07	82.3
Carbon tetrachloride	CCl_4	14.0572	2914.23	232.148	−14—101	29.82	76.6
Chlorobenzene	C_6H_5Cl	13.8635	3174.78	211.700	29—159	35.19	131.7
1-Chlorobutane	C_4H_9Cl	13.7965	2723.73	218.265	−17—79	30.39	78.5
Chloroform	$CHCl_3$	13.7324	2548.74	218.552	−23—84	29.24	61.1
Cyclohexane	C_6H_{12}	13.6568	2723.44	220.618	9—105	29.97	80.7
Cyclopentane	C_5H_{10}	13.9727	2653.90	234.510	−35—71	27.30	49.2
n-Decane	$C_{10}H_{22}$	13.9748	3442.76	193.858	65—203	38.75	174.1
Dichloromethane	CH_2Cl_2	13.9891	2463.93	223.240	−38—60	28.06	39.7
Diethyl ether	$C_4H_{10}O$	14.0735	2511.29	231.200	−43—55	26.52	34.4
1,4-Dioxane	$C_4H_8O_2$	15.0967	3579.78	240.337	20—105	34.16	101.3
n-Eicosane	$C_{20}H_{42}$	14.4575	4680.46	132.100	208—379	57.49	343.6
Ethanol	C_2H_6O	16.8958	3795.17	230.918	3—96	38.56	78.2
Ethylbenzene	C_8H_{10}	13.9726	3259.93	212.300	33—163	35.57	136.2
Ethylene glycol*	$C_2H_6O_2$	15.7567	4187.46	178.650	100—222	50.73	197.3
n-Heptane	C_7H_{16}	13.8622	2910.26	216.432	4—123	31.77	98.4
n-Hexane	C_6H_{14}	13.8193	2696.04	224.317	−19—92	28.85	68.7
Methanol	CH_4O	16.5785	3638.27	239.500	−11—83	35.21	64.7
Methyl acetate	$C_3H_6O_2$	14.2456	2662.78	219.690	−23—78	30.32	56.9
Methyl ethyl ketone	C_4H_8O	14.1334	2838.24	218.690	−8—103	31.30	79.6
Nitromethane*	CH_3NO_2	14.7513	3331.70	227.600	56—146	33.99	101.2
n-Nonane	C_9H_{20}	13.9854	3311.19	202.694	46—178	36.91	150.8
iso-Octane	C_8H_{18}	13.6703	2896.31	220.767	2—125	30.79	99.2
n-Octane	C_8H_{18}	13.9346	3123.13	209.635	26—152	34.41	125.6
n-Pentane	C_5H_{12}	13.7667	2451.88	232.014	−45—58	25.79	36.0
Phenol	C_6H_6O	14.4387	3507.80	175.400	80—208	46.18	181.8
1-Propanol	C_3H_8O	16.1154	3483.67	205.807	20—116	41.44	97.2
2-Propanol	C_3H_8O	16.6796	3640.20	219.610	8—100	39.85	82.2

Table B.2 (Continued)

Name	Formula	Parameters for Antoine Eqn.			Temp. Range °C	ΔH_n kJ/mol	$t_n/°C$
		$A^†$	B	C			
Toluene	C_7H_8	13.9320	3056.96	217.625	13—136	33.18	110.6
Water	H_2O	16.3872	3885.70	230.170	0—200	40.66	100.0
o-Xylene	C_8H_{10}	14.0415	3358.79	212.041	40—172	36.24	144.4
m-Xylene	C_8H_{10}	14.1387	3381.81	216.120	35—166	35.66	139.1
p-Xylene	C_8H_{10}	14.0579	3331.45	214.627	35—166	35.67	138.3

Based primarily on data presented by B. E. Poling, J. M. Prausnitz, and J. P. O'Connell, *The Properties of Gases and Liquids*, 5th ed., App. A, McGraw-Hill, New York, 2001.

*Antoine parameters adapted from J. Gmehling, U. Onken, and W. Arlt, *Vapor-Liquid Equilibrium Data Collection*, Chemistry Data Series, vol. I, parts 1–8, DECHEMA, Frankfurt/Main, 1974–1990.

†Antoine parameters A are adjusted to reproduce the listed values of t_n.

Appendix C

Heat Capacities and Property Changes of Formation

Table C.1: Heat Capacities of Gases in the Ideal-Gas State[†]

Constants in equation $C_P^{ig}/R = A + BT + CT^2 + DT^{-2}$ for T (K) from 298 K to T_{max}

Chemical species		T_{max}	$C_{P_{298}}^{ig}/R$	A	$10^3\,B$	$10^6\,C$	$10^{-5}\,D$
Alkanes:							
Methane	CH_4	1500	4.217	1.702	9.081	−2.164	......
Ethane	C_2H_6	1500	6.369	1.131	19.225	−5.561	......
Propane	C_3H_8	1500	9.011	1.213	28.785	−8.824	......
n-Butane	C_4H_{10}	1500	11.928	1.935	36.915	−11.402	......
iso-Butane	C_4H_{10}	1500	11.901	1.677	37.853	−11.945	......
n-Pentane	C_5H_{12}	1500	14.731	2.464	45.351	−14.111	......
n-Hexane	C_6H_{14}	1500	17.550	3.025	53.722	−16.791	......
n-Heptane	C_7H_{16}	1500	20.361	3.570	62.127	−19.486	......
n-Octane	C_8H_{18}	1500	23.174	4.108	70.567	−22.208	......
1-*Alkenes*:							
Ethylene	C_2H_4	1500	5.325	1.424	14.394	−4.392	......
Propylene	C_3H_6	1500	7.792	1.637	22.706	−6.915	......
1-Butene	C_4H_8	1500	10.520	1.967	31.630	−9.873	......
1-Pentene	C_5H_{10}	1500	13.437	2.691	39.753	−12.447	......
1-Hexene	C_6H_{12}	1500	16.240	3.220	48.189	−15.157	......
1-Heptene	C_7H_{14}	1500	19.053	3.768	56.588	−17.847	......
1-Octene	C_8H_{16}	1500	21.868	4.324	64.960	−20.521	......
Miscellaneous organics:							
Acetaldehyde	C_2H_4O	1000	6.506	1.693	17.978	−6.158	......
Acetylene	C_2H_2	1500	5.253	6.132	1.952	......	−1.299
Benzene	C_6H_6	1500	10.259	−0.206	39.064	−13.301	......
1,3-Butadiene	C_4H_6	1500	10.720	2.734	26.786	−8.882	......
Cyclohexane	C_6H_{12}	1500	13.121	−3.876	63.249	−20.928	......
Ethanol	C_2H_6O	1500	8.948	3.518	20.001	−6.002	......
Ethylbenzene	C_8H_{10}	1500	15.993	1.124	55.380	−18.476	......
Ethylene oxide	C_2H_4O	1000	5.784	−0.385	23.463	−9.296	......
Formaldehyde	CH_2O	1500	4.191	2.264	7.022	−1.877	......
Methanol	CH_4O	1500	5.547	2.211	12.216	−3.450	......
Styrene	C_8H_8	1500	15.534	2.050	50.192	−16.662	......
Toluene	C_7H_8	1500	12.922	0.290	47.052	−15.716	......
Miscellaneous inorganics:							
Air		2000	3.509	3.355	0.575	......	−0.016
Ammonia	NH_3	1800	4.269	3.578	3.020	......	−0.186
Bromine	Br_2	3000	4.337	4.493	0.056	......	−0.154
Carbon monoxide	CO	2500	3.507	3.376	0.557	......	−0.031
Carbon dioxide	CO_2	2000	4.467	5.457	1.045	......	−1.157
Carbon disulfide	CS_2	1800	5.532	6.311	0.805	......	−0.906
Chlorine	Cl_2	3000	4.082	4.442	0.089	......	−0.344
Hydrogen	H_2	3000	3.468	3.249	0.422	......	0.083
Hydrogen sulfide	H_2S	2300	4.114	3.931	1.490	......	−0.232
Hydrogen chloride	HCl	2000	3.512	3.156	0.623	......	0.151
Hydrogen cyanide	HCN	2500	4.326	4.736	1.359	......	−0.725
Nitrogen	N_2	2000	3.502	3.280	0.593	......	0.040
Nitrous oxide	N_2O	2000	4.646	5.328	1.214	......	−0.928
Nitric oxide	NO	2000	3.590	3.387	0.629	......	0.014
Nitrogen dioxide	NO_2	2000	4.447	4.982	1.195	......	−0.792
Dinitrogen tetroxide	N_2O_4	2000	9.198	11.660	2.257	......	−2.787
Oxygen	O_2	2000	3.535	3.639	0.506	......	−0.227
Sulfur dioxide	SO_2	2000	4.796	5.699	0.801	......	−1.015
Sulfur trioxide	SO_3	2000	6.094	8.060	1.056	......	−2.028
Water	H_2O	2000	4.038	3.470	1.450	......	0.121

[†]Selected from H. M. Spencer, *Ind. Eng. Chem.*, vol. 40, pp. 2152–2154, 1948; K. K. Kelley, *U.S. Bur. Mines Bull. 584*, 1960; L. B. Pankratz, *U.S. Bur. Mines Bull. 672*, 1982.

Table C.2: Heat Capacities of Solids[†]

Constants for the equation $C_p/R = A + BT + DT^{-2}$ for T (K) from 298 K to T_{max}

Chemical species	T_{max}	$C_{P_{298}}^{ig}/R$	A	$10^3 B$	$10^{-5} D$
CaO	2000	5.058	6.104	0.443	−1.047
CaCO$_3$	1200	9.848	12.572	2.637	−3.120
Ca(OH)$_2$	700	11.217	9.597	5.435	......
CaC$_2$	720	7.508	8.254	1.429	−1.042
CaCl$_2$	1055	8.762	8.646	1.530	−0.302
C (graphite)	2000	1.026	1.771	0.771	−0.867
Cu	1357	2.959	2.677	0.815	0.035
CuO	1400	5.087	5.780	0.973	−0.874
Fe(α)	1043	3.005	−0.111	6.111	1.150
Fe$_2$O$_3$	960	12.480	11.812	9.697	−1.976
Fe$_3$O$_4$	850	18.138	9.594	27.112	0.409
FeS	411	6.573	2.612	13.286	......
I$_2$	386.8	6.929	6.481	1.502	......
LiCl	800	5.778	5.257	2.476	−0.193
NH$_4$Cl	458	10.741	5.939	16.105	......
Na	371	3.386	1.988	4.688	......
NaCl	1073	6.111	5.526	1.963	......
NaOH	566	7.177	0.121	16.316	1.948
NaHCO$_3$	400	10.539	5.128	18.148	......
S (rhombic)	368.3	3.748	4.114	−1.728	−0.783
SiO$_2$ (quartz)	847	5.345	4.871	5.365	−1.001

[†]Selected from K. K. Kelley, *U.S. Bur. Mines Bull. 584*, 1960; L. B. Pankratz, *U.S. Bur. Mines Bull. 672*, 1982.

Table C.3: Heat Capacities of Liquids[†]

Constants for the equation $C_P/R = A + BT + CT^2$ for T from 273.15 to 373.15 K

Chemical species	$C_{P_{298}}^{ig}/R$	A	$10^3 B$	$10^6 C$
Ammonia	9.718	22.626	−100.75	192.71
Aniline	23.070	15.819	29.03	−15.80
Benzene	16.157	−0.747	67.96	−37.78
1,3-Butadiene	14.779	22.711	−87.96	205.79
Carbon tetrachloride	15.751	21.155	−48.28	101.14
Chlorobenzene	18.240	11.278	32.86	−31.90
Chloroform	13.806	19.215	−42.89	83.01
Cyclohexane	18.737	−9.048	141.38	−161.62
Ethanol	13.444	33.866	−172.60	349.17
Ethylene oxide	10.590	21.039	−86.41	172.28
Methanol	9.798	13.431	−51.28	131.13
n-Propanol	16.921	41.653	−210.32	427.20
Sulfur trioxide	30.408	−2.930	137.08	−84.73
Toluene	18.611	15.133	6.79	16.35
Water	9.069	8.712	1.25	−0.18

[†]Based on correlations presented by J. W. Miller, Jr., G. R. Schorr, and C. L. Yaws, *Chem. Eng.*, vol. 83(23), p. 129, 1976.

Table C.4: Standard Enthalpies and Gibbs Energies of Formation
at 298.15 K†

Joules per mole of the substance formed

Chemical species		State (Note 2)	$\Delta H^{\circ}_{f_{298}}$ (Note 1)	$\Delta G^{\circ}_{f_{298}}$ (Note 1)
Alkanes:				
Methane	CH$_4$	(g)	−74,520	−50,460
Ethane	C$_2$H$_6$	(g)	−83,820	−31,855
Propane	C$_3$H$_8$	(g)	−104,680	−24,290
n-Butane	C$_4$H$_{10}$	(g)	−125,790	−16,570
n-Pentane	C$_5$H$_{12}$	(g)	−146,760	−8,650
n-Hexane	C$_6$H$_{14}$	(g)	−166,920	150
n-Heptane	C$_7$H$_{16}$	(g)	−187,780	8,260
n-Octane	C$_8$H$_{18}$	(g)	−208,750	16,260
1-*Alkenes:*				
Ethylene	C$_2$H$_4$	(g)	52,510	68,460
Propylene	C$_3$H$_6$	(g)	19,710	62,205
1-Butene	C$_4$H$_8$	(g)	−540	70,340
1-Pentene	C$_5$H$_{10}$	(g)	−21,280	78,410
1-Hexene	C$_6$H$_{12}$	(g)	−41,950	86,830
1-Heptene	C$_7$H$_{14}$	(g)	−62,760	
Miscellaneous organics:				
Acetaldehyde	C$_2$H$_4$O	(g)	−166,190	−128,860
Acetic acid	C$_2$H$_4$O$_2$	(l)	−484,500	−389,900
Acetylene	C$_2$H$_2$	(g)	227,480	209,970
Benzene	C$_6$H$_6$	(g)	82,930	129,665
Benzene	C$_6$H$_6$	(l)	49,080	124,520
1,3-Butadiene	C$_4$H$_6$	(g)	109,240	149,795
Cyclohexane	C$_6$H$_{12}$	(g)	−123,140	31,920
Cyclohexane	C$_6$H$_{12}$	(l)	−156,230	26,850
1,2-Ethanediol	C$_2$H$_6$O$_2$	(l)	−454,800	−323,080
Ethanol	C$_2$H$_6$O	(g)	−235,100	−168,490
Ethanol	C$_2$H$_6$O	(l)	−277,690	−174,780
Ethylbenzene	C$_8$H$_{10}$	(g)	29,920	130,890
Ethylene oxide	C$_2$H$_4$O	(g)	−52,630	−13,010
Formaldehyde	CH$_2$O	(g)	−108,570	−102,530
Methanol	CH$_4$O	(g)	−200,660	−161,960
Methanol	CH$_4$O	(l)	−238,660	−166,270
Methylcyclohexane	C$_7$H$_{14}$	(g)	−154,770	27,480
Methylcyclohexane	C$_7$H$_{14}$	(l)	−190,160	20,560
Styrene	C$_8$H$_8$	(g)	147,360	213,900
Toluene	C$_7$H$_8$	(g)	50,170	122,050
Toluene	C$_7$H$_8$	(l)	12,180	113,630

Table C.4 (Continued)

Chemical species		State (Note 2)	$\Delta H^{\circ}_{f_{298}}$ (Note 1)	$\Delta G^{\circ}_{f_{298}}$ (Note 1)
Miscellaneous inorganics:				
Ammonia	NH_3	(g)	$-46,110$	$-16,400$
Ammonia	NH_3	(aq)		$-26,500$
Calcium carbide	CaC_2	(s)	$-59,800$	$-64,900$
Calcium carbonate	$CaCO_3$	(s)	$-1,206,920$	$-1,128,790$
Calcium chloride	$CaCl_2$	(s)	$-795,800$	$-748,100$
Calcium chloride	$CaCl_2$	(aq)		$-8,101,900$
Calcium chloride	$CaCl_2 \cdot 6H_2O$	(s)	$-2,607,900$	
Calcium hydroxide	$Ca(OH)_2$	(s)	$-986,090$	$-898,490$
Calcium hydroxide	$Ca(OH)_2$	(aq)		$-868,070$
Calcium oxide	CaO	(s)	$-635,090$	$-604,030$
Carbon dioxide	CO_2	(g)	$-393,509$	$-394,359$
Carbon monoxide	CO	(g)	$-110,525$	$-137,169$
Hydrochloric acid	HCl	(g)	$-92,307$	$-95,299$
Hydrogen cyanide	HCN	(g)	$135,100$	$124,700$
Hydrogen sulfide	H_2S	(g)	$-20,630$	$-33,560$
Iron oxide	FeO	(s)	$-272,000$	
Iron oxide (hematite)	Fe_2O_3	(s)	$-824,200$	$-742,200$
Iron oxide (magnetite)	Fe_3O_4	(s)	$-1,118,400$	$-1,015,400$
Iron sulfide (pyrite)	FeS_2	(s)	$-178,200$	$-166,900$
Lithium chloride	$LiCl$	(s)	$-408,610$	
Lithium chloride	$LiCl \cdot H_2O$	(s)	$-712,580$	
Lithium chloride	$LiCl \cdot 2H_2O$	(s)	$-1,012,650$	
Lithium chloride	$LiCl \cdot 3H_2O$	(s)	$-1,311,300$	
Nitric acid	HNO_3	(l)	$-174,100$	$-80,710$
Nitric acid	HNO_3	(aq)		$-111,250$
Nitrogen oxides	NO	(g)	$90,250$	$86,550$
	NO_2	(g)	$33,180$	$51,310$
	N_2O	(g)	$82,050$	$104,200$
	N_2O_4	(g)	$9,160$	$97,540$
Sodium carbonate	Na_2CO_3	(s)	$-1,130,680$	$-1,044,440$
Sodium carbonate	$Na_2CO_3 \cdot 10H_2O$	(s)	$-4,081,320$	
Sodium chloride	$NaCl$	(s)	$-411,153$	$-384,138$
Sodium chloride	$NaCl$	(aq)		$-393,133$
Sodium hydroxide	$NaOH$	(s)	$-425,609$	$-379,494$
Sodium hydroxide	$NaOH$	(aq)		$-419,150$
Sulfur dioxide	SO_2	(g)	$-296,830$	$-300,194$
Sulfur trioxide	SO_3	(g)	$-395,720$	$-371,060$
Sulfur trioxide	SO_3	(l)	$-441,040$	
Sulfuric acid	H_2SO_4	(l)	$-813,989$	$-690,003$
Sulfuric acid	H_2SO_4	(aq)		$-744,530$
Water	H_2O	(g)	$-241,818$	$-228,572$
Water	H_2O	(l)	$-285,830$	$-237,129$

[†]From *TRC Thermodynamic Tables—Hydrocarbons*, Thermodynamics Research Center, Texas A & M Univ. System, College Station, TX; "The NBS Tables of Chemical Thermodynamic Properties," *J. Phys. and Chem. Reference Data*, vol. 11, supp. 2, 1982.

Notes

1. The standard property changes of formation ΔH_{f298}° and ΔG_{f298}° are the changes occurring when 1 mol of the listed compound is formed from its elements with each substance in its standard state at 298.15 K (25°C).

2. Standard states: (*a*) Gases (*g*): pure ideal gas at 1 bar and 25°C. (*b*) Liquids (*l*) and solids (*s*): pure substance at 1 bar and 25°C. (*c*) Solutes in aqueous solution *(aq):* Hypothetical ideal 1-molal solution of solute in water at 1 bar and 25°C.

Table C.5: Standard Enthalpies and Gibbs Energies of Formation at 298.15 K for Substances in Dilute Aqueous Solution at Zero Ionic Strength[†]

Joules per mole of the substance formed

Chemical species		ΔH_{f298}°	ΔG_{f298}°
Acetaldehyde	C_2H_4O	−212.2	−139.0
Acetate	$C_2H_2O_2^-$	−486.0	−369.3
Acetic acid	$C_2H_3O_2$	−485.8	−396.5
Acetone	C_3H_6O	−221.7	−159.7
Adenosine	$C_{10}H_{13}N_5O_4$	−621.3	−194.5
Adenosine cation	$C_{10}H_{14}N_5O_4^+$	−637.7	−214.3
Adenosine 5′ diphosphate (ADP)	$C_{10}H_{12}N_5O_{10}P_2^{3-}$	−2626.5	−1906.1
	$C_{10}H_{13}N_5O_{10}P_2^{2-}$	−2620.9	−1947.1
	$C_{10}H_{14}N_5O_{10}P_2^-$	−2638.5	−1972.0
Adenosine 5′ monophosphate (AMP)	$C_{10}H_{12}N_5O_{10}P^{2-}$	−1635.4	−1040.5
	$C_{10}H_{13}N_5O_{10}P^-$	−1630.0	−1078.9
	$C_{10}H_{14}N_5O_7P$	−1648.1	−1101.6
Adenosine 5′ triphosphate (ATP)	$C_{10}H_{12}N_5O_{13}P_3^{4-}$	−3619.2	−2768.1
	$C_{10}H_{13}N_5O_{13}P_3^{3-}$	−3612.9	−2811.5
	$C_{10}H_{14}N_5O_{13}P_3^{2-}$	−3627.9	−2838.2
Alanine	$C_3H_7NO_2$	−554.8	−371.0
Ammonia	NH_3	−80.3	−26.5
Ammonium	NH_4^+	−132.5	−79.3
D-arabinose	$C_5H_{10}O_5$	−1043.8	−742.2
L-asparagine	$C_4H_8N_2O_3$	−766.1	−525.9
L-aspartate	$C_4H_7NO_4$	−943.4	−695.9
Citrate	$C_6H_5O_7^{3-}$	−1515.1	−1162.7
	$C_6H_6O_7^{2-}$	−1518.5	−1199.2
	$C_6H_7O_7^-$	−1520.9	−1226.3
Carbon dioxide	CO_2	−413.8	−386.0
Carbonate	CO_3^{-2}	−677.1	−527.8
Bicarbonate	CHO_3^-	−692.0	−586.8
Carbonic acid	CH_2O_3	−694.9	−606.3

Table C.5 (Continued)

Chemical species		$\Delta H^\circ_{f_{298}}$	$\Delta G^\circ_{f_{298}}$
Carbon monoxide	CO	−121.0	−119.9
Ethanol	C_2H_6O	−288.3	−181.6
Ethyl acetate	$C_4H_8O_2$	−482.0	−337.7
Formate	CHO_2^-	−425.6	−351.0
D-fructose	$C_6H_{12}O_6$	−1259.4	−915.5
D-fructose 6-phosphate	$C_6H_{11}O_9P^{2-}$	−2267.7*	−1760.8
	$C_6H_{12}O_9P^-$	−2265.9*	−1796.6
D-fructose 1,6-biphosphate	$C_6H_{11}O_{12}P_2^{3-}$	−3320.1*	−2639.4
	$C_6H_{12}O_{12}P_2^{2-}$	−3318.3*	−2673.9
Fumarate	$C_4H_2O_4^{2-}$	−777.4	−601.9
	$C_4H_3O_4^-$	−774.5	−628.1
	$C_4H_4O_4$	−774.9	−645.8
D-galactose	$C_6H_{12}O_6$	−1255.2	−908.9
D-glucose	$C_6H_{12}O_6$	−1262.2	−915.9
D-glucose 6-phosphate	$C_6H_{11}O_9P^{2-}$	−2276.4	−1763.9
	$C_6H_{12}O_9P^-$	−2274.6	−1800.6
L-glutamate	$C_5H_8NO_4^-$	−979.9	−697.5
L-glutamine	$C_5H_{10}N_2O_3$	−805.0	−528.0
Glycerol	$C_3H_8O_3$	−676.6	−497.5
Glycine	$C_2H_5NO_2$	−523.0	−379.9
Glycylglycine	$C_4H_8N_2O_3$	−734.3	−520.2
Hydrogen	H_2	−4.2	17.6
Hydrogen peroxide	H_2O_2	−191.2	−134.0
Hydrogen ion (Note 2)	H^+	0.0	0.0
Indole	C_8H_7N	97.5	223.8
Lactate	$C_3H_5O_3^-$	−686.6	−516.7
Lactose	$C_{12}H_{22}O_{11}$	−2233.1	−1567.3
L-leucine	$C_6H_{13}NO_2$	−643.4	−352.3
Maltose	$C_{12}H_{22}O_{11}$	−2238.1	−1574.7
D-mannose	$C_6H_{12}O_6$	−1258.7	−910.0
Methane	CH_4	−89.0	−34.3
Methanol	CH_4O	−245.9	−175.3
Methylammonium	CH_6N^+	−124.9	−39.9
Nitrogen	N_2	−10.5	18.7
Nicotinamide-adenine dinucleotide (ox)	NAD^+ (Note 2)	0.0	0.0
Nicotinamide-adenine dinucleotide (red)	NADH (Note 2)	−31.9	22.7
Nicotinamide-adenine dinucleotide phosphate (ox)	$NADP^+$ (Note 2)	0.0	−835.2

Table C.5 (Continued)

Chemical species		ΔH°_{f298}	ΔG°_{f298}
Nicotinamide-adenine dinucleotide phosphate (red)	NADPH (Note 2)	−29.2	−809.2
Oxygen	O_2	−11.7	16.4
Oxalate	$C_2O_4^{2-}$	−825.1	−673.9
Hydrogen phosphate	HPO_4^{2-}	−1299.0	−1096.1
Dihydrogen phosphate	$H_2PO_4^-$	−1302.6	−1137.3
2-propanol	C_3H_8O	−330.8	−185.2
Pyrophosphate	$P_2O_7^{4-}$	−2293.5	−1919.9
	$HP_2O_7^{3-}$	−2294.9	−1973.9
	$H_2P_2O_7^{2-}$	−2295.4	−2012.2
	$H_3P_2O_7^-$	−2290.4	−2025.1
	$H_4P_2O_7$	−2281.2	−2029.9
Pyruvate	$C_3H_3O_3^-$	−596.2	−472.3
D-ribose	$C_5H_{10}O_5$	−1034.0	−738.8
D-ribose 5-phosphate	$C_5H_9O_8P^{2-}$	−2041.5	−1582.6
	$C_5H_{10}O_8P^-$	−2030.2	−1620.8
D-ribulose	$C_5H_{10}O_5$	−1023.0	−735.9
L-sorbose	$C_6H_{12}O_6$	−1263.3	−912.0
Succinate	$C_4H_4O_4^{2-}$	−908.7	−690.4
	$C_4H_5O_4^-$	−908.8	−722.6
	$C_4H_6O_4$	−912.2	−746.6
Sucrose	$C_{12}H_{22}O_{11}$	−2199.9	−1564.7
L-tryptophan	$C_{11}H_{12}N_2O_2$	−405.2	−114.7
Urea	CH_4N_2O	−317.7	−202.8
L-valine	$C_5H_{11}NO_2$	−612.0	−358.7
D-xylose	$C_5H_{10}O_5$	−1045.9	−750.5
D-xylulose	$C_5H_{10}O_5$	−1029.7	−746.2

*Estimated using data from R. N. Goldberg, Y. B. Tewari, and T. N. Bhat, *Thermodynamics of Enzyme Catalyzed Reactions*, NIST Standard Reference Database 74, http://xpdb.nist.gov/enzyme_thermodynamics.

†From Robert A. Alberty, *Thermodynamics of Biochemical Reactions*, Wiley-Interscience, Hoboken, NJ, USA, 2003. Table 3.2, pp. 52–55 and Table 8.2, p. 151.

Notes

1. The standard property changes of formation ΔH°_{f298} and ΔG°_{f298} are the changes occurring when 1 mol of the listed compound is formed from its elements with each substance in its standard state at 298.15 K (25°C), except as noted in Note 2.

2. Conventions used in this table are that $\Delta G^{\circ}_{f298} = \Delta H^{\circ}_{f298} = 0$ for H^+ and for oxidized nicotinamide-adenine dinucleotide (NAD^-_{ox}). For the latter, and other NAD species, no molecular formula is provided because their properties are computed relative to this convention rather than relative to the elements in their standard states.

Appendix D

The Lee/Kesler Generalized-Correlation Tables

The Lee/Kesler tables are adapted and published by permission from "A Generalized Thermodynamic Correlation Based on Three-Parameter Corresponding States," by Byung Ik Lee and Michael G. Kesler, *AIChE J.*, **21,** 510–527 (1975). The numbers printed in italic type are liquid-phase properties.

TABLES

Table D.1: Values of Z^0

$P_r =$	0.0100	0.0500	0.1000	0.2000	0.4000	0.6000	0.8000	1.0000
T_r								
0.30	*0.0029*	*0.0145*	*0.0290*	*0.0579*	*0.1158*	*0.1737*	*0.2315*	0.2892
0.35	*0.0026*	*0.0130*	*0.0261*	*0.0522*	*0.1043*	*0.1564*	0.2084	0.2604
0.40	*0.0024*	*0.0119*	*0.0239*	*0.0477*	*0.0953*	*0.1429*	*0.1904*	0.2379
0.45	*0.0022*	*0.0110*	*0.0221*	*0.0442*	*0.0882*	*0.1322*	*0.1762*	0.2200
0.50	*0.0021*	*0.0103*	*0.0207*	*0.0413*	*0.0825*	*0.1236*	*0.1647*	0.2056
0.55	0.9804	*0.0098*	*0.0195*	*0.0390*	*0.0778*	*0.1166*	*0.1553*	0.1939
0.60	0.9849	*0.0093*	*0.0186*	*0.0371*	*0.0741*	*0.1109*	*0.1476*	0.1842
0.65	0.9881	0.9377	*0.0178*	*0.0356*	*0.0710*	*0.1063*	*0.1415*	0.1765
0.70	0.9904	0.9504	0.8958	*0.0344*	*0.0687*	*0.1027*	*0.1366*	0.1703
0.75	0.9922	0.9598	0.9165	*0.0336*	*0.0670*	*0.1001*	*0.1330*	0.1656
0.80	0.9935	0.9669	0.9319	0.8539	*0.0661*	*0.0985*	*0.1307*	0.1626
0.85	0.9946	0.9725	0.9436	0.8810	*0.0661*	*0.0983*	*0.1301*	0.1614
0.90	0.9954	0.9768	0.9528	0.9015	0.7800	*0.1006*	*0.1321*	0.1630
0.93	0.9959	0.9790	0.9573	0.9115	0.8059	0.6635	*0.1359*	0.1664
0.95	0.9961	0.9803	0.9600	0.9174	0.8206	0.6967	*0.1410*	0.1705
0.97	0.9963	0.9815	0.9625	0.9227	0.8338	0.7240	0.5580	0.1779
0.98	0.9965	0.9821	0.9637	0.9253	0.8398	0.7360	0.5887	0.1844
0.99	0.9966	0.9826	0.9648	0.9277	0.8455	0.7471	0.6138	0.1959
1.00	0.9967	0.9832	0.9659	0.9300	0.8509	0.7574	0.6355	0.2901
1.01	0.9968	0.9837	0.9669	0.9322	0.8561	0.7671	0.6542	0.4648
1.02	0.9969	0.9842	0.9679	0.9343	0.8610	0.7761	0.6710	0.5146
1.05	0.9971	0.9855	0.9707	0.9401	0.8743	0.8002	0.7130	0.6026
1.10	0.9975	0.9874	0.9747	0.9485	0.8930	0.8323	0.7649	0.6880
1.15	0.9978	0.9891	0.9780	0.9554	0.9081	0.8576	0.8032	0.7443
1.20	0.9981	0.9904	0.9808	0.9611	0.9205	0.8779	0.8330	0.7858
1.30	0.9985	0.9926	0.9852	0.9702	0.9396	0.9083	0.8764	0.8438
1.40	0.9988	0.9942	0.9884	0.9768	0.9534	0.9298	0.9062	0.8827
1.50	0.9991	0.9954	0.9909	0.9818	0.9636	0.9456	0.9278	0.9103
1.60	0.9993	0.9964	0.9928	0.9856	0.9714	0.9575	0.9439	0.9308
1.70	0.9994	0.9971	0.9943	0.9886	0.9775	0.9667	0.9563	0.9463
1.80	0.9995	0.9977	0.9955	0.9910	0.9823	0.9739	0.9659	0.9583
1.90	0.9996	0.9982	0.9964	0.9929	0.9861	0.9796	0.9735	0.9678
2.00	0.9997	0.9986	0.9972	0.9944	0.9892	0.9842	0.9796	0.9754
2.20	0.9998	0.9992	0.9983	0.9967	0.9937	0.9910	0.9886	0.9865
2.40	0.9999	0.9996	0.9991	0.9983	0.9969	0.9957	0.9948	0.9941
2.60	1.0000	0.9998	0.9997	0.9994	0.9991	0.9990	0.9990	0.9993
2.80	1.0000	1.0000	1.0001	1.0002	1.0007	1.0013	1.0021	1.0031
3.00	1.0000	1.0002	1.0004	1.0008	1.0018	1.0030	1.0043	1.0057
3.50	1.0001	1.0004	1.0008	1.0017	1.0035	1.0055	1.0075	1.0097
4.00	1.0001	1.0005	1.0010	1.0021	1.0043	1.0066	1.0090	1.0115

Table D.2: Values of Z^1

$P_r =$	0.0100	0.0500	0.1000	0.2000	0.4000	0.6000	0.8000	1.0000
T_r								
0.30	−0.0008	−0.0040	−0.0081	−0.0161	−0.0323	−0.0484	−0.0645	−0.0806
0.35	−0.0009	−0.0046	−0.0093	−0.0185	−0.0370	−0.0554	−0.0738	−0.0921
0.40	−0.0010	−0.0048	−0.0095	−0.0190	−0.0380	−0.0570	−0.0758	−0.0946
0.45	−0.0009	−0.0047	−0.0094	−0.0187	−0.0374	−0.0560	−0.0745	−0.0929
0.50	−0.0009	−0.0045	−0.0090	−0.0181	−0.0360	−0.0539	−0.0716	−0.0893
0.55	−0.0314	−0.0043	−0.0086	−0.0172	−0.0343	−0.0513	−0.0682	−0.0849
0.60	−0.0205	−0.0041	−0.0082	−0.0164	−0.0326	−0.0487	−0.0646	−0.0803
0.65	−0.0137	−0.0772	−0.0078	−0.0156	−0.0309	−0.0461	−0.0611	−0.0759
0.70	−0.0093	−0.0507	−0.1161	−0.0148	−0.0294	−0.0438	−0.0579	−0.0718
0.75	−0.0064	−0.0339	−0.0744	−0.0143	−0.0282	−0.0417	−0.0550	−0.0681
0.80	−0.0044	−0.0228	−0.0487	−0.1160	−0.0272	−0.0401	−0.0526	−0.0648
0.85	−0.0029	−0.0152	−0.0319	−0.0715	−0.0268	−0.0391	−0.0509	−0.0622
0.90	−0.0019	−0.0099	−0.0205	−0.0442	−0.1118	−0.0396	−0.0503	−0.0604
0.93	−0.0015	−0.0075	−0.0154	−0.0326	−0.0763	−0.1662	−0.0514	−0.0602
0.95	−0.0012	−0.0062	−0.0126	−0.0262	−0.0589	−0.1110	−0.0540	−0.0607
0.97	−0.0010	−0.0050	−0.0101	−0.0208	−0.0450	−0.0770	−0.1647	−0.0623
0.98	−0.0009	−0.0044	−0.0090	−0.0184	−0.0390	−0.0641	−0.1100	−0.0641
0.99	−0.0008	−0.0039	−0.0079	−0.0161	−0.0335	−0.0531	−0.0796	−0.0680
1.00	−0.0007	−0.0034	−0.0069	−0.0140	−0.0285	−0.0435	−0.0588	−0.0879
1.01	−0.0006	−0.0030	−0.0060	−0.0120	−0.0240	−0.0351	−0.0429	−0.0223
1.02	−0.0005	−0.0026	−0.0051	−0.0102	−0.0198	−0.0277	−0.0303	−0.0062
1.05	−0.0003	−0.0015	−0.0029	−0.0054	−0.0092	−0.0097	−0.0032	0.0220
1.10	0.0000	0.0000	0.0001	0.0007	0.0038	0.0106	0.0236	0.0476
1.15	0.0002	0.0011	0.0023	0.0052	0.0127	0.0237	0.0396	0.0625
1.20	0.0004	0.0019	0.0039	0.0084	0.0190	0.0326	0.0499	0.0719
1.30	0.0006	0.0030	0.0061	0.0125	0.0267	0.0429	0.0612	0.0819
1.40	0.0007	0.0036	0.0072	0.0147	0.0306	0.0477	0.0661	0.0857
1.50	0.0008	0.0039	0.0078	0.0158	0.0323	0.0497	0.0677	0.0864
1.60	0.0008	0.0040	0.0080	0.0162	0.0330	0.0501	0.0677	0.0855
1.70	0.0008	0.0040	0.0081	0.0163	0.0329	0.0497	0.0667	0.0838
1.80	0.0008	0.0040	0.0081	0.0162	0.0325	0.0488	0.0652	0.0814
1.90	0.0008	0.0040	0.0079	0.0159	0.0318	0.0477	0.0635	0.0792
2.00	0.0008	0.0039	0.0078	0.0155	0.0310	0.0464	0.0617	0.0767
2.20	0.0007	0.0037	0.0074	0.0147	0.0293	0.0437	0.0579	0.0719
2.40	0.0007	0.0035	0.0070	0.0139	0.0276	0.0411	0.0544	0.0675
2.60	0.0007	0.0033	0.0066	0.0131	0.0260	0.0387	0.0512	0.0634
2.80	0.0006	0.0031	0.0062	0.0124	0.0245	0.0365	0.0483	0.0598
3.00	0.0006	0.0029	0.0059	0.0117	0.0232	0.0345	0.0456	0.0565
3.50	0.0005	0.0026	0.0052	0.0103	0.0204	0.0303	0.0401	0.0497
4.00	0.0005	0.0023	0.0046	0.0091	0.0182	0.0270	0.0357	0.0443

Table D.3: Values of Z^0

$P_r =$	1.0000	1.2000	1.5000	2.0000	3.0000	5.0000	7.0000	10.000
T_r								
0.30	0.2892	0.3479	0.4335	0.5775	0.8648	1.4366	2.0048	2.8507
0.35	0.2604	0.3123	0.3901	0.5195	0.7775	1.2902	1.7987	2.5539
0.40	0.2379	0.2853	0.3563	0.4744	0.7095	1.1758	1.6373	2.3211
0.45	0.2200	0.2638	0.3294	0.4384	0.6551	1.0841	1.5077	2.1338
0.50	0.2056	0.2465	0.3077	0.4092	0.6110	1.0094	1.4017	1.9801
0.55	0.1939	0.2323	0.2899	0.3853	0.5747	0.9475	1.3137	1.8520
0.60	0.1842	0.2207	0.2753	0.3657	0.5446	0.8959	1.2398	1.7440
0.65	0.1765	0.2113	0.2634	0.3495	0.5197	0.8526	1.1773	1.6519
0.70	0.1703	0.2038	0.2538	0.3364	0.4991	0.8161	1.1341	1.5729
0.75	0.1656	0.1981	0.2464	0.3260	0.4823	0.7854	1.0787	1.5047
0.80	0.1626	0.1942	0.2411	0.3182	0.4690	0.7598	1.0400	1.4456
0.85	0.1614	0.1924	0.2382	0.3132	0.4591	0.7388	1.0071	1.3943
0.90	0.1630	0.1935	0.2383	0.3114	0.4527	0.7220	0.9793	1.3496
0.93	0.1664	0.1963	0.2405	0.3122	0.4507	0.7138	0.9648	1.3257
0.95	0.1705	0.1998	0.2432	0.3138	0.4501	0.7092	0.9561	1.3108
0.97	0.1779	0.2055	0.2474	0.3164	0.4504	0.7052	0.9480	1.2968
0.98	0.1844	0.2097	0.2503	0.3182	0.4508	0.7035	0.9442	1.2901
0.99	0.1959	0.2154	0.2538	0.3204	0.4514	0.7018	0.9406	1.2835
1.00	0.2901	0.2237	0.2583	0.3229	0.4522	0.7004	0.9372	1.2772
1.01	0.4648	0.2370	0.2640	0.3260	0.4533	0.6991	0.9339	1.2710
1.02	0.5146	0.2629	0.2715	0.3297	0.4547	0.6980	0.9307	1.2650
1.05	0.6026	0.4437	0.3131	0.3452	0.4604	0.6956	0.9222	1.2481
1.10	0.6880	0.5984	0.4580	0.3953	0.4770	0.6950	0.9110	1.2232
1.15	0.7443	0.6803	0.5798	0.4760	0.5042	0.6987	0.9033	1.2021
1.20	0.7858	0.7363	0.6605	0.5605	0.5425	0.7069	0.8990	1.1844
1.30	0.8438	0.8111	0.7624	0.6908	0.6344	0.7358	0.8998	1.1580
1.40	0.8827	0.8595	0.8256	0.7753	0.7202	0.7761	0.9112	1.1419
1.50	0.9103	0.8933	0.8689	0.8328	0.7887	0.8200	0.9297	1.1339
1.60	0.9308	0.9180	0.9000	0.8738	0.8410	0.8617	0.9518	1.1320
1.70	0.9463	0.9367	0.9234	0.9043	0.8809	0.8984	0.9745	1.1343
1.80	0.9583	0.9511	0.9413	0.9275	0.9118	0.9297	0.9961	1.1391
1.90	0.9678	0.9624	0.9552	0.9456	0.9359	0.9557	1.0157	1.1452
2.00	0.9754	0.9715	0.9664	0.9599	0.9550	0.9772	1.0328	1.1516
2.20	0.9856	0.9847	0.9826	0.9806	0.9827	1.0094	1.0600	1.1635
2.40	0.9941	0.9936	0.9935	0.9945	1.0011	1.0313	1.0793	1.1728
2.60	0.9993	0.9998	1.0010	1.0040	1.0137	1.0463	1.0926	1.1792
2.80	1.0031	1.0042	1.0063	1.0106	1.0223	1.0565	1.1016	1.1830
3.00	1.0057	1.0074	1.0101	1.0153	1.0284	1.0635	1.1075	1.1848
3.50	1.0097	1.0120	1.0156	1.0221	1.0368	1.0723	1.1138	1.1834
4.00	1.0115	1.0140	1.0179	1.0249	1.0401	1.0747	1.1136	1.1773

Table D.4: Values of Z^1

$P_r =$	1.0000	1.2000	1.5000	2.0000	3.0000	5.0000	7.0000	10.000
T_r								
0.30	−0.0806	−0.0966	−0.1207	−0.1608	−0.2407	−0.3996	−0.5572	−0.7915
0.35	−0.0921	−0.1105	−0.1379	−0.1834	−0.2738	−0.4523	−0.6279	−0.8863
0.40	−0.0946	−0.1134	−0.1414	−0.1879	−0.2799	−0.4603	−0.6365	−0.8936
0.45	−0.0929	−0.1113	−0.1387	−0.1840	−0.2734	−0.4475	−0.6162	−0.8608
0.50	−0.0893	−0.1069	−0.1330	−0.1762	−0.2611	−0.4253	−0.5831	−0.8099
0.55	−0.0849	−0.1015	−0.1263	−0.1669	−0.2465	−0.3991	−0.5446	−0.7521
0.60	−0.0803	−0.0960	−0.1192	−0.1572	−0.2312	−0.3718	−0.5047	−0.6928
0.65	−0.0759	−0.0906	−0.1122	−0.1476	−0.2160	−0.3447	−0.4653	−0.6346
0.70	−0.0718	−0.0855	−0.1057	−0.1385	−0.2013	−0.3184	−0.4270	−0.5785
0.75	−0.0681	−0.0808	−0.0996	−0.1298	−0.1872	−0.2929	−0.3901	−0.5250
0.80	−0.0648	−0.0767	−0.0940	−0.1217	−0.1736	−0.2682	−0.3545	−0.4740
0.85	−0.0622	−0.0731	−0.0888	−0.1138	−0.1602	−0.2439	−0.3201	−0.4254
0.90	−0.0604	−0.0701	−0.0840	−0.1059	−0.1463	−0.2195	−0.2862	−0.3788
0.93	−0.0602	−0.0687	−0.0810	−0.1007	−0.1374	−0.2045	−0.2661	−0.3516
0.95	−0.0607	−0.0678	−0.0788	−0.0967	−0.1310	−0.1943	−0.2526	−0.3339
0.97	−0.0623	−0.0669	−0.0759	−0.0921	−0.1240	−0.1837	−0.2391	−0.3163
0.98	−0.0641	−0.0661	−0.0740	−0.0893	−0.1202	−0.1783	−0.2322	−0.3075
0.99	−0.0680	−0.0646	−0.0715	−0.0861	−0.1162	−0.1728	−0.2254	−0.2989
1.00	−0.0879	−0.0609	−0.0678	−0.0824	−0.1118	−0.1672	−0.2185	−0.2902
1.01	−0.0223	−0.0473	−0.0621	−0.0778	−0.1072	−0.1615	−0.2116	−0.2816
1.02	−0.0062	−0.0227	−0.0524	−0.0722	−0.1021	−0.1556	−0.2047	−0.2731
1.05	0.0220	0.1059	0.0451	−0.0432	−0.0838	−0.1370	−0.1835	−0.2476
1.10	0.0476	0.0897	0.1630	0.0698	−0.0373	−0.1021	−0.1469	−0.2056
1.15	0.0625	0.0943	0.1548	0.1667	0.0332	−0.0611	−0.1084	−0.1642
1.20	0.0719	0.0991	0.1477	0.1990	0.1095	−0.0141	−0.0678	−0.1231
1.30	0.0819	0.1048	0.1420	0.1991	0.2079	0.0875	0.0176	−0.0423
1.40	0.0857	0.1063	0.1383	0.1894	0.2397	0.1737	0.1008	0.0350
1.50	0.0854	0.1055	0.1345	0.1806	0.2433	0.2309	0.1717	0.1058
1.60	0.0855	0.1035	0.1303	0.1729	0.2381	0.2631	0.2255	0.1673
1.70	0.0838	0.1008	0.1259	0.1658	0.2305	0.2788	0.2628	0.2179
1.80	0.0816	0.0978	0.1216	0.1593	0.2224	0.2846	0.2871	0.2576
1.90	0.0792	0.0947	0.1173	0.1532	0.2144	0.2848	0.3017	0.2876
2.00	0.0767	0.0916	0.1133	0.1476	0.2069	0.2819	0.3097	0.3096
2.20	0.0719	0.0857	0.1057	0.1374	0.1932	0.2720	0.3135	0.3355
2.40	0.0675	0.0803	0.0989	0.1285	0.1812	0.2602	0.3089	0.3459
2.60	0.0634	0.0754	0.0929	0.1207	0.1706	0.2484	0.3009	0.3475
2.80	0.0598	0.0711	0.0876	0.1138	0.1613	0.2372	0.2915	0.3443
3.00	0.0535	0.0672	0.0828	0.1076	0.1529	0.2268	0.2817	0.3385
3.50	0.0497	0.0591	0.0728	0.0949	0.1356	0.2042	0.2584	0.3194
4.00	0.0443	0.0527	0.0651	0.0849	0.1219	0.1857	0.2378	0.2994

Table D.5: Values of $(H^R)^0/RT_c$

$P_r =$	0.0100	0.0500	0.1000	0.2000	0.4000	0.6000	0.8000	1.0000
T_r								
0.30	−6.045	−6.043	−6.040	−6.034	−6.022	−6.011	−5.999	−5.987
0.35	−5.906	−5.904	−5.901	−5.895	−5.882	−5.870	−5.858	−5.845
0.40	−5.763	−5.761	−5.757	−5.751	−5.738	−5.726	−5.713	−5.700
0.45	−5.615	−5.612	−5.609	−5.603	−5.590	−5.577	−5.564	−5.551
0.50	−5.465	−5.463	−5.459	−5.453	−5.440	−5.427	−5.414	−5.401
0.55	−0.032	−5.312	−5.309	−5.303	−5.290	−5.278	−5.265	−5.252
0.60	−0.027	−5.162	−5.159	−5.153	−5.141	−5.129	−5.116	−5.104
0.65	−0.023	−0.118	−5.008	−5.002	−4.991	−4.980	−4.968	−4.956
0.70	−0.020	−0.101	−0.213	−4.848	−4.838	−4.828	−4.818	−4.808
0.75	−0.017	−0.088	−0.183	−4.687	−4.679	−4.672	−4.664	−4.655
0.80	−0.015	−0.078	−0.160	−0.345	−4.507	−4.504	−4.499	−4.494
0.85	−0.014	−0.069	−0.141	−0.300	−4.309	−4.313	−4.316	−4.316
0.90	−0.012	−0.062	−0.126	−0.264	−0.596	−4.074	−4.094	−4.108
0.93	−0.011	−0.058	−0.118	−0.246	−0.545	−0.960	−3.920	−3.953
0.95	−0.011	−0.056	−0.113	−0.235	−0.516	−0.885	−3.763	−3.825
0.97	−0.011	−0.054	−0.109	−0.225	−0.490	−0.824	−1.356	−3.658
0.98	−0.010	−0.053	−0.107	−0.221	−0.478	−0.797	−1.273	−3.544
0.99	−0.010	−0.052	−0.105	−0.216	−0.466	−0.773	−1.206	−3.376
1.00	−0.010	−0.051	−0.103	−0.212	−0.455	−0.750	−1.151	−2.584
1.01	−0.010	−0.050	−0.101	−0.208	−0.445	−0.721	−1.102	−1.796
1.02	−0.010	−0.049	−0.099	−0.203	−0.434	−0.708	−1.060	−1.627
1.05	−0.009	−0.046	−0.094	−0.192	−0.407	−0.654	−0.955	−1.359
1.10	−0.008	−0.042	−0.086	−0.175	−0.367	−0.581	−0.827	−1.120
1.15	−0.008	−0.039	−0.079	−0.160	−0.334	−0.523	−0.732	−0.968
1.20	−0.007	−0.036	−0.073	−0.148	−0.305	−0.474	−0.657	−0.857
1.30	−0.006	−0.031	−0.063	−0.127	−0.259	−0.399	−0.545	−0.698
1.40	−0.005	−0.027	−0.055	−0.110	−0.224	−0.341	−0.463	−0.588
1.50	−0.005	−0.024	−0.048	−0.097	−0.196	−0.297	−0.400	−0.505
1.60	−0.004	−0.021	−0.043	−0.086	−0.173	−0.261	−0.350	−0.440
1.70	−0.004	−0.019	−0.038	−0.076	−0.153	−0.231	−0.309	−0.387
1.80	−0.003	−0.017	−0.034	−0.068	−0.137	−0.206	−0.275	−0.344
1.90	−0.003	−0.015	−0.031	−0.062	−0.123	−0.185	−0.246	−0.307
2.00	−0.003	−0.014	−0.028	−0.056	−0.111	−0.167	−0.222	−0.276
2.20	−0.002	−0.012	−0.023	−0.046	−0.092	−0.137	−0.182	−0.226
2.40	−0.002	−0.010	−0.019	−0.038	−0.076	−0.114	−0.150	−0.187
2.60	−0.002	−0.008	−0.016	−0.032	−0.064	−0.095	−0.125	−0.155
2.80	−0.001	−0.007	−0.014	−0.027	−0.054	−0.080	−0.105	−0.130
3.00	−0.001	−0.006	−0.011	−0.023	−0.045	−0.067	−0.088	−0.109
3.50	−0.001	−0.004	−0.007	−0.015	−0.029	−0.043	−0.056	−0.069
4.00	−0.000	−0.002	−0.005	−0.009	−0.017	−0.026	−0.033	−0.041

Table D.6: Values of $(H^R)^1/RT_c$

$P_r =$	0.0100	0.0500	0.1000	0.2000	0.4000	0.6000	0.8000	1.0000
T_r								
0.30	−11.098	−11.096	−11.095	−11.091	−11.083	−11.076	−11.069	−11.062
0.35	−10.656	−10.655	−10.654	−10.653	−10.650	−10.646	−10.643	−10.640
0.40	−10.121	−10.121	−10.121	−10.120	−10.121	−10.121	−10.121	−10.121
0.45	−9.515	−9.515	−9.516	−9.517	−9.519	−9.521	−9.523	−9.525
0.50	−8.868	−8.869	−8.870	−8.872	−8.876	−8.880	−8.884	−8.888
0.55	−0.080	−8.211	−8.212	−8.215	−8.221	−8.226	−8.232	−8.238
0.60	−0.059	−7.568	−7.570	−7.573	−7.579	−7.585	−7.591	−7.596
0.65	−0.045	−0.247	−6.949	−6.952	−6.959	−6.966	−6.973	−6.980
0.70	−0.034	−0.185	−0.415	−6.360	−6.367	−6.373	−6.381	−6.388
0.75	−0.027	−0.142	−0.306	−5.796	−5.802	−5.809	−5.816	−5.824
0.80	−0.021	−0.110	−0.234	−0.542	−5.266	−5.271	−5.278	−5.285
0.85	−0.017	−0.087	−0.182	−0.401	−4.753	−4.754	−4.758	−4.763
0.90	−0.014	−0.070	−0.144	−0.308	−0.751	−4.254	−4.248	−4.249
0.93	−0.012	−0.061	−0.126	−0.265	−0.612	−1.236	−3.942	−3.934
0.95	−0.011	−0.056	−0.115	−0.241	−0.542	−0.994	−3.737	−3.712
0.97	−0.010	−0.052	−0.105	−0.219	−0.483	−0.837	−1.616	−3.470
0.98	−0.010	−0.050	−0.101	−0.209	−0.457	−0.776	−1.324	−3.332
0.99	−0.009	−0.048	−0.097	−0.200	−0.433	−0.722	−1.154	−3.164
1.00	−0.009	−0.046	−0.093	−0.191	−0.410	−0.675	−1.034	−2.471
1.01	−0.009	−0.044	−0.089	−0.183	−0.389	−0.632	−0.940	−1.375
1.02	−0.008	−0.042	−0.085	−0.175	−0.370	−0.594	−0.863	−1.180
1.05	−0.007	−0.037	−0.075	−0.153	−0.318	−0.498	−0.691	−0.877
1.10	−0.006	−0.030	−0.061	−0.123	−0.251	−0.381	−0.507	−0.617
1.15	−0.005	−0.025	−0.050	−0.099	−0.199	−0.296	−0.385	−0.459
1.20	−0.004	−0.020	−0.040	−0.080	−0.158	−0.232	−0.297	−0.349
1.30	−0.003	−0.013	−0.026	−0.052	−0.100	−0.142	−0.177	−0.203
1.40	−0.002	−0.008	−0.016	−0.032	−0.060	−0.083	−0.100	−0.111
1.50	−0.001	−0.005	−0.009	−0.018	−0.032	−0.042	−0.048	−0.049
1.60	−0.000	−0.002	−0.004	−0.007	−0.012	−0.013	−0.011	−0.005
1.70	−0.000	−0.000	−0.000	−0.000	0.003	0.009	0.017	0.027
1.80	0.000	0.001	0.003	0.006	0.015	0.025	0.037	0.051
1.90	0.001	0.003	0.005	0.011	0.023	0.037	0.053	0.070
2.00	0.001	0.003	0.007	0.015	0.030	0.047	0.065	0.085
2.20	0.001	0.005	0.010	0.020	0.040	0.062	0.083	0.106
2.40	0.001	0.006	0.012	0.023	0.047	0.071	0.095	0.120
2.60	0.001	0.006	0.013	0.026	0.052	0.078	0.104	0.130
2.80	0.001	0.007	0.014	0.028	0.055	0.082	0.110	0.137
3.00	0.001	0.007	0.014	0.029	0.058	0.086	0.114	0.142
3.50	0.002	0.008	0.016	0.031	0.062	0.092	0.122	0.152
4.00	0.002	0.008	0.016	0.032	0.064	0.096	0.127	0.158

Table D.7: Values of $(H^R)^0/RT_c$

$P_r =$	1.0000	1.2000	1.5000	2.0000	3.0000	5.0000	7.0000	10.000
T_r								
0.30	−5.987	−5.975	−5.957	−5.927	−5.868	−5.748	−5.628	−5.446
0.35	−5.845	−5.833	−5.814	−5.783	−5.721	−5.595	−5.469	−5.278
0.40	−5.700	−5.687	−5.668	−5.636	−5.572	−5.442	−5.311	−5.113
0.45	−5.551	−5.538	−5.519	−5.486	−5.421	−5.288	−5.154	−5.950
0.50	−5.401	−5.388	−5.369	−5.336	−5.279	−5.135	−4.999	−4.791
0.55	−5.252	−5.239	−5.220	−5.187	−5.121	−4.986	−4.849	−4.638
0.60	−5.104	−5.091	−5.073	−5.041	−4.976	−4.842	−4.794	−4.492
0.65	−4.956	−4.949	−4.927	−4.896	−4.833	−4.702	−4.565	−4.353
0.70	−4.808	−4.797	−4.781	−4.752	−4.693	−4.566	−4.432	−4.221
0.75	−4.655	−4.646	−4.632	−4.607	−4.554	−4.434	−4.393	−4.095
0.80	−4.494	−4.488	−4.478	−4.459	−4.413	−4.303	−4.178	−3.974
0.85	−4.316	−4.316	−4.312	−4.302	−4.269	−4.173	−4.056	−3.857
0.90	−4.108	−4.118	−4.127	−4.132	−4.119	−4.043	−3.935	−3.744
0.93	−3.953	−3.976	−4.000	−4.020	−4.024	−3.963	−3.863	−3.678
0.95	−3.825	−3.865	−3.904	−3.940	−3.958	−3.910	−3.815	−3.634
0.97	−3.658	−3.732	−3.796	−3.853	−3.890	−3.856	−3.767	−3.591
0.98	−3.544	−3.652	−3.736	−3.806	−3.854	−3.829	−3.743	−3.569
0.99	−3.376	−3.558	−3.670	−3.758	−3.818	−3.801	−3.719	−3.548
1.00	−2.584	−3.441	−3.598	−3.706	−3.782	−3.774	−3.695	−3.526
1.01	−1.796	−3.283	−3.516	−3.652	−3.744	−3.746	−3.671	−3.505
1.02	−1.627	−3.039	−3.422	−3.595	−3.705	−3.718	−3.647	−3.484
1.05	−1.359	−2.034	−3.030	−3.398	−3.583	−3.632	−3.575	−3.420
1.10	−1.120	−1.487	−2.203	−2.965	−3.353	−3.484	−3.453	−3.315
1.15	−0.968	−1.239	−1.719	−2.479	−3.091	−3.329	−3.329	−3.211
1.20	−0.857	−1.076	−1.443	−2.079	−2.801	−3.166	−3.202	−3.107
1.30	−0.698	−0.860	−1.116	−1.560	−2.274	−2.825	−2.942	−2.899
1.40	−0.588	−0.716	−0.915	−1.253	−1.857	−2.486	−2.679	−2.692
1.50	−0.505	−0.611	−0.774	−1.046	−1.549	−2.175	−2.421	−2.486
1.60	−0.440	−0.531	−0.667	−0.894	−1.318	−1.904	−2.177	−2.285
1.70	−0.387	−0.446	−0.583	−0.777	−1.139	−1.672	−1.953	−2.091
1.80	−0.344	−0.413	−0.515	−0.683	−0.996	−1.476	−1.751	−1.908
1.90	−0.307	−0.368	−0.458	−0.606	−0.880	−1.309	−1.571	−1.736
2.00	−0.276	−0.330	−0.411	−0.541	−0.782	−1.167	−1.411	−1.577
2.20	−0.226	−0.269	−0.334	−0.437	−0.629	−0.937	−1.143	−1.295
2.40	−0.187	−0.222	−0.275	−0.359	−0.513	−0.761	−0.929	−1.058
2.60	−0.155	−0.185	−0.228	−0.297	−0.422	−0.621	−0.756	−0.858
2.80	−0.130	−0.154	−0.190	−0.246	−0.348	−0.508	−0.614	−0.689
3.00	−0.109	−0.129	−0.159	−0.205	−0.288	−0.415	−0.495	−0.545
3.50	−0.069	−0.081	−0.099	−0.127	−0.174	−0.239	−0.270	−0.264
4.00	−0.041	−0.048	−0.058	−0.072	−0.095	−0.116	−0.110	−0.061

Table D.8: Values of $(H^R)^1/RT_c$

$P_r =$	1.0000	1.2000	1.5000	2.0000	3.0000	5.0000	7.0000	10.000
T_r								
0.30	−11.062	−11.055	−11.044	−11.027	−10.992	−10.935	−10.872	−10.781
0.35	−10.640	−10.637	−10.632	−10.624	−10.609	−10.581	−10.554	−10.529
0.40	−10.121	−10.121	−10.121	−10.122	−10.123	−10.128	−10.135	−10.150
0.45	−9.525	−9.527	−9.531	−9.537	−9.549	−9.576	−9.611	−9.663
0.50	−8.888	−8.892	−8.899	−8.909	−8.932	−8.978	−9.030	−9.111
0.55	−8.238	−8.243	−8.252	−8.267	−8.298	−8.360	−8.425	−8.531
0.60	−7.596	−7.603	−7.614	−7.632	−7.669	−7.745	−7.824	−7.950
0.65	−6.980	−6.987	−6.997	−7.017	−7.059	−7.147	−7.239	−7.381
0.70	−6.388	−6.395	−6.407	−6.429	−6.475	−6.574	−6.677	−6.837
0.75	−5.824	−5.832	−5.845	−5.868	−5.918	−6.027	−6.142	−6.318
0.80	−5.285	−5.293	−5.306	−5.330	−5.385	−5.506	−5.632	−5.824
0.85	−4.763	−4.771	−4.784	−4.810	−4.872	−5.000	−5.149	−5.358
0.90	−4.249	−4.255	−4.268	−4.298	−4.371	−4.530	−4.688	−4.916
0.93	−3.934	−3.937	−3.951	−3.987	−4.073	−4.251	−4.422	−4.662
0.95	−3.712	−3.713	−3.730	−3.773	−3.873	−4.068	−4.248	−4.497
0.97	−3.470	−3.467	−3.492	−3.551	−3.670	−3.885	−4.077	−4.336
0.98	−3.332	−3.327	−3.363	−3.434	−3.568	−3.795	−3.992	−4.257
0.99	−3.164	−3.164	−3.223	−3.313	−3.464	−3.705	−3.909	−4.178
1.00	−2.471	−2.952	−3.065	−3.186	−3.358	−3.615	−3.825	−4.100
1.01	−1.375	−2.595	−2.880	−3.051	−3.251	−3.525	−3.742	−4.023
1.02	−1.180	−1.723	−2.650	−2.906	−3.142	−3.435	−3.661	−3.947
1.05	−0.877	−0.878	−1.496	−2.381	−2.800	−3.167	−3.418	−3.722
1.10	−0.617	−0.673	−0.617	−1.261	−2.167	−2.720	−3.023	−3.362
1.15	−0.459	−0.503	−0.487	−0.604	−1.497	−2.275	−2.641	−3.019
1.20	−0.349	−0.381	−0.381	−0.361	−0.934	−1.840	−2.273	−2.692
1.30	−0.203	−0.218	−0.218	−0.178	−0.300	−1.066	−1.592	−2.086
1.40	−0.111	−0.115	−0.128	−0.070	−0.044	−0.504	−1.012	−1.547
1.50	−0.049	−0.046	−0.032	0.008	0.078	−0.142	−0.556	−1.080
1.60	−0.005	0.004	0.023	0.065	0.151	0.082	−0.217	−0.689
1.70	0.027	0.040	0.063	0.109	0.202	0.223	0.028	−0.369
1.80	0.051	0.067	0.094	0.143	0.241	0.317	0.203	−0.112
1.90	0.070	0.088	0.117	0.169	0.271	0.381	0.330	0.092
2.00	0.085	0.105	0.136	0.190	0.295	0.428	0.424	0.255
2.20	0.106	0.128	0.163	0.221	0.331	0.493	0.551	0.489
2.40	0.120	0.144	0.181	0.242	0.356	0.535	0.631	0.645
2.60	0.130	0.156	0.194	0.257	0.376	0.567	0.687	0.754
2.80	0.137	0.164	0.204	0.269	0.391	0.591	0.729	0.836
3.00	0.142	0.170	0.211	0.278	0.403	0.611	0.763	0.899
3.50	0.152	0.181	0.224	0.294	0.425	0.650	0.827	1.015
4.00	0.158	0.188	0.233	0.306	0.442	0.680	0.874	1.097

Table D.9: Values of $(S^R)^0/R$

$P_r =$	0.0100	0.0500	0.1000	0.2000	0.4000	0.6000	0.8000	1.0000
T_r								
0.30	−11.614	−10.008	−9.319	−8.635	−7.961	−7.574	−7.304	−7.099
0.35	−11.185	−9.579	−8.890	−8.205	−7.529	−7.140	−6.869	−6.663
0.40	−10.802	−9.196	−8.506	−7.821	−7.144	−6.755	−6.483	−6.275
0.45	−10.453	−8.847	−8.157	−7.472	−6.794	−6.404	−6.132	−5.924
0.50	−10.137	−8.531	−7.841	−7.156	−6.479	−6.089	−5.816	−5.608
0.55	−0.038	−8.245	−7.555	−6.870	−6.193	−5.803	−5.531	−5.324
0.60	−0.029	−7.983	−7.294	−6.610	−5.933	−5.544	−5.273	−5.066
0.65	−0.023	−0.122	−7.052	−6.368	−5.694	−5.306	−5.036	−4.830
0.70	−0.018	−0.096	−0.206	−6.140	−5.467	−5.082	−4.814	−4.610
0.75	−0.015	−0.078	−0.164	−5.917	−5.248	−4.866	−4.600	−4.399
0.80	−0.013	−0.064	−0.134	−0.294	−5.026	−4.694	−4.388	−4.191
0.85	−0.011	−0.054	−0.111	−0.239	−4.785	−4.418	−4.166	−3.976
0.90	−0.009	−0.046	−0.094	−0.199	−0.463	−4.145	−3.912	−3.738
0.93	−0.008	−0.042	−0.085	−0.179	−0.408	−0.750	−3.723	−3.569
0.95	−0.008	−0.039	−0.080	−0.168	−0.377	−0.671	−3.556	−3.433
0.97	−0.007	−0.037	−0.075	−0.157	−0.350	−0.607	−1.056	−3.259
0.98	−0.007	−0.036	−0.073	−0.153	−0.337	−0.580	−0.971	−3.142
0.99	−0.007	−0.035	−0.071	−0.148	−0.326	−0.555	−0.903	−2.972
1.00	−0.007	−0.034	−0.069	−0.144	−0.315	−0.532	−0.847	−2.178
1.01	−0.007	−0.033	−0.067	−0.139	−0.304	−0.510	−0.799	−1.391
1.02	−0.006	−0.032	−0.065	−0.135	−0.294	−0.491	−0.757	−1.225
1.05	−0.006	−0.030	−0.060	−0.124	−0.267	−0.439	−0.656	−0.965
1.10	−0.005	−0.026	−0.053	−0.108	−0.230	−0.371	−0.537	−0.742
1.15	−0.005	−0.023	−0.047	−0.096	−0.201	−0.319	−0.452	−0.607
1.20	−0.004	−0.021	−0.042	−0.085	−0.177	−0.277	−0.389	−0.512
1.30	−0.003	−0.017	−0.033	−0.068	−0.140	−0.217	−0.298	−0.385
1.40	−0.003	−0.014	−0.027	−0.056	−0.114	−0.174	−0.237	−0.303
1.50	−0.002	−0.011	−0.023	−0.046	−0.094	−0.143	−0.194	−0.246
1.60	−0.002	−0.010	−0.019	−0.039	−0.079	−0.120	−0.162	−0.204
1.70	−0.002	−0.008	−0.017	−0.033	−0.067	−0.102	−0.137	−0.172
1.80	−0.001	−0.007	−0.014	−0.029	−0.058	−0.088	−0.117	−0.147
1.90	−0.001	−0.006	−0.013	−0.025	−0.051	−0.076	−0.102	−0.127
2.00	−0.001	−0.006	−0.011	−0.022	−0.044	−0.067	−0.089	−0.111
2.20	−0.001	−0.004	−0.009	−0.018	−0.035	−0.053	−0.070	−0.087
2.40	−0.001	−0.004	−0.007	−0.014	−0.028	−0.042	−0.056	−0.070
2.60	−0.001	−0.003	−0.006	−0.012	−0.023	−0.035	−0.046	−0.058
2.80	−0.000	−0.002	−0.005	−0.010	−0.020	−0.029	−0.039	−0.048
3.00	−0.000	−0.002	−0.004	−0.008	−0.017	−0.025	−0.033	−0.041
3.50	−0.000	−0.001	−0.003	−0.006	−0.012	−0.017	−0.023	−0.029
4.00	−0.000	−0.001	−0.002	−0.004	−0.009	−0.013	−0.017	−0.021

Table D.10: Values of $(S^R)^1/R$

$P_r =$	0.0100	0.0500	0.1000	0.2000	0.4000	0.6000	0.8000	1.0000
T_r								
0.30	−16.782	−16.774	−16.764	−16.744	−16.705	−16.665	−16.626	−16.586
0.35	−15.413	−15.408	−15.401	−15.387	−15.359	−15.333	−15.305	−15.278
0.40	−13.990	−13.986	−13.981	−13.972	−13.953	−13.934	−13.915	−13.896
0.45	−12.564	−12.561	−12.558	−12.551	−12.537	−12.523	−12.509	−12.496
0.50	−11.202	−11.200	−11.197	−11.092	−11.082	−11.172	−11.162	−11.153
0.55	−0.115	−9.948	−9.946	−9.942	−9.935	−9.928	−9.921	−9.914
0.60	−0.078	−8.828	−8.826	−8.823	−8.817	−8.811	−8.806	−8.799
0.65	−0.055	−0.309	−7.832	−7.829	−7.824	−7.819	−7.815	−7.510
0.70	−0.040	−0.216	−0.491	−6.951	−6.945	−6.941	−6.937	−6.933
0.75	−0.029	−0.156	−0.340	−6.173	−6.167	−6.162	−6.158	−6.155
0.80	−0.022	−0.116	−0.246	−0.578	−5.475	−5.468	−5.462	−5.458
0.85	−0.017	−0.088	−0.183	−0.400	−4.853	−4.841	−4.832	−4.826
0.90	−0.013	−0.068	−0.140	−0.301	−0.744	−4.269	−4.249	−4.238
0.93	−0.011	−0.058	−0.120	−0.254	−0.593	−1.219	−3.914	−3.894
0.95	−0.010	−0.053	−0.109	−0.228	−0.517	−0.961	−3.697	−3.658
0.97	−0.010	−0.048	−0.099	−0.206	−0.456	−0.797	−1.570	−3.406
0.98	−0.009	−0.046	−0.094	−0.196	−0.429	−0.734	−1.270	−3.264
0.99	−0.009	−0.044	−0.090	−0.186	−0.405	−0.680	−1.098	−3.093
1.00	−0.008	−0.042	−0.086	−0.177	−0.382	−0.632	−0.977	−2.399
1.01	−0.008	−0.040	−0.082	−0.169	−0.361	−0.590	−0.883	−1.306
1.02	−0.008	−0.039	−0.078	−0.161	−0.342	−0.552	−0.807	−1.113
1.05	−0.007	−0.034	−0.069	−0.140	−0.292	−0.460	−0.642	−0.820
1.10	−0.005	−0.028	−0.055	−0.112	−0.229	−0.350	−0.470	−0.577
1.15	−0.005	−0.023	−0.045	−0.091	−0.183	−0.275	−0.361	−0.437
1.20	−0.004	−0.019	−0.037	−0.075	−0.149	−0.220	−0.286	−0.343
1.30	−0.003	−0.013	−0.026	−0.052	−0.102	−0.148	−0.190	−0.226
1.40	−0.002	−0.010	−0.019	−0.037	−0.072	−0.104	−0.133	−0.158
1.50	−0.001	−0.007	−0.014	−0.027	−0.053	−0.076	−0.097	−0.115
1.60	−0.001	−0.005	−0.011	−0.021	−0.040	−0.057	−0.073	−0.086
1.70	−0.001	−0.004	−0.008	−0.016	−0.031	−0.044	−0.056	−0.067
1.80	−0.001	−0.003	−0.006	−0.013	−0.024	−0.035	−0.044	−0.053
1.90	−0.001	−0.003	−0.005	−0.010	−0.019	−0.028	−0.036	−0.043
2.00	−0.000	−0.002	−0.004	−0.008	−0.016	−0.023	−0.029	−0.035
2.20	−0.000	−0.001	−0.003	−0.006	−0.011	−0.016	−0.021	−0.025
2.40	−0.000	−0.001	−0.002	−0.004	−0.008	−0.012	−0.015	−0.019
2.60	−0.000	−0.001	−0.002	−0.003	−0.006	−0.009	−0.012	−0.015
2.80	−0.000	−0.001	−0.001	−0.003	−0.005	−0.008	−0.010	−0.012
3.00	−0.000	−0.001	−0.001	−0.002	−0.004	−0.006	−0.008	−0.010
3.50	−0.000	−0.000	−0.001	−0.001	−0.003	−0.004	−0.006	−0.007
4.00	−0.000	−0.000	−0.001	−0.001	−0.002	−0.003	−0.005	−0.006

Table D.11: Values of $(S^R)^0/R$

$P_r =$	1.0000	1.2000	1.5000	2.0000	3.0000	5.0000	7.0000	10.000
T_r								
0.30	−7.099	−6.935	−6.740	−6.497	−6.180	−5.847	−5.683	−5.578
0.35	−6.663	−6.497	−6.299	−6.052	−5.728	−5.376	−5.194	−5.060
0.40	−6.275	−6.109	−5.909	−5.660	−5.330	−4.967	−4.772	−4.619
0.45	−5.924	−5.757	−5.557	−5.306	−4.974	−4.603	−4.401	−4.234
0.50	−5.608	−5.441	−5.240	−4.989	−4.656	−4.282	−4.074	−3.899
0.55	−5.324	−5.157	−4.956	−4.706	−4.373	−3.998	−3.788	−3.607
0.60	−5.066	−4.900	−4.700	−4.451	−4.120	−3.747	−3.537	−3.353
0.65	−4.830	−4.665	−4.467	−4.220	−3.892	−3.523	−3.315	−3.131
0.70	−4.610	−4.446	−4.250	−4.007	−3.684	−3.322	−3.117	−2.935
0.75	−4.399	−4.238	−4.045	−3.807	−3.491	−3.138	−2.939	−2.761
0.80	−4.191	−4.034	−3.846	−3.615	−3.310	−2.970	−2.777	−2.605
0.85	−3.976	−3.825	−3.646	−3.425	−3.135	−2.812	−2.629	−2.463
0.90	−3.738	−3.599	−3.434	−3.231	−2.964	−2.663	−2.491	−2.334
0.93	−3.569	−3.444	−3.295	−3.108	−2.860	−2.577	−2.412	−2.262
0.95	−3.433	−3.326	−3.193	−3.023	−2.790	−2.520	−2.362	−2.215
0.97	−3.259	−3.188	−3.081	−2.932	−2.719	−2.463	−2.312	−2.170
0.98	−3.142	−3.106	−3.019	−2.884	−2.682	−2.436	−2.287	−2.148
0.99	−2.972	−3.010	−2.953	−2.835	−2.646	−2.408	−2.263	−2.126
1.00	−2.178	−2.893	−2.879	−2.784	−2.609	−2.380	−2.239	−2.105
1.01	−1.391	−2.736	−2.798	−2.730	−2.571	−2.352	−2.215	−2.083
1.02	−1.225	−2.495	−2.706	−2.673	−2.533	−2.325	−2.191	−2.062
1.05	−0.965	−1.523	−2.328	−2.483	−2.415	−2.242	−2.121	−2.001
1.10	−0.742	−1.012	−1.557	−2.081	−2.202	−2.104	−2.007	−1.903
1.15	−0.607	−0.790	−1.126	−1.649	−1.968	−1.966	−1.897	−1.810
1.20	−0.512	−0.651	−0.890	−1.308	−1.727	−1.827	−1.789	−1.722
1.30	−0.385	−0.478	−0.628	−0.891	−1.299	−1.554	−1.581	−1.556
1.40	−0.303	−0.375	−0.478	−0.663	−0.990	−1.303	−1.386	−1.402
1.50	−0.246	−0.299	−0.381	−0.520	−0.777	−1.088	−1.208	−1.260
1.60	−0.204	−0.247	−0.312	−0.421	−0.628	−0.913	−1.050	−1.130
1.70	−0.172	−0.208	−0.261	−0.350	−0.519	−0.773	−0.915	−1.013
1.80	−0.147	−0.177	−0.222	−0.296	−0.438	−0.661	−0.799	−0.908
1.90	−0.127	−0.153	−0.191	−0.255	−0.375	−0.570	−0.702	−0.815
2.00	−0.111	−0.134	−0.167	−0.221	−0.625	−0.497	−0.620	−0.733
2.20	−0.087	−0.105	−0.130	−0.172	−0.251	−0.388	−0.492	−0.599
2.40	−0.070	−0.084	−0.104	−0.138	−0.201	−0.311	−0.399	−0.496
2.60	−0.058	−0.069	−0.086	−0.113	−0.164	−0.255	−0.329	−0.416
2.80	−0.048	−0.058	−0.072	−0.094	−0.137	−0.213	−0.277	−0.353
3.00	−0.041	−0.049	−0.061	−0.080	−0.116	−0.181	−0.236	−0.303
3.50	−0.029	−0.034	−0.042	−0.056	−0.081	−0.126	−0.166	−0.216
4.00	−0.021	−0.025	−0.031	−0.041	−0.059	−0.093	−0.123	−0.162

Table D.12: Values of $(S^R)^1/R$

$P_r =$	1.0000	1.2000	1.5000	2.0000	3.0000	5.0000	7.0000	10.000
T_r								
0.30	−16.586	−16.547	−16.488	−16.390	−16.195	−15.837	−15.468	−14.925
0.35	−15.278	−15.251	−15.211	−15.144	−15.011	−14.751	−14.496	−14.153
0.40	−13.896	−13.877	−13.849	−13.803	−13.714	−13.541	−13.376	−13.144
0.45	−12.496	−12.482	−12.462	−12.430	−12.367	−12.248	−12.145	−11.999
0.50	−11.153	−11.143	−11.129	−11.107	−11.063	−10.985	−10.920	−10.836
0.55	−9.914	−9.907	−9.897	−9.882	−9.853	−9.806	−9.769	−9.732
0.60	−8.799	−8.794	−8.787	−8.777	−8.760	−8.736	−8.723	−8.720
0.65	−7.810	−7.807	−7.801	−7.794	−7.784	−7.779	−7.785	−7.811
0.70	−6.933	−6.930	−6.926	−6.922	−6.919	−6.929	−6.952	−7.002
0.75	−6.155	−6.152	−6.149	−6.147	−6.149	−6.174	−6.213	−6.285
0.80	−5.458	−5.455	−5.453	−5.452	−5.461	−5.501	−5.555	−5.648
0.85	−4.826	−4.822	−4.820	−4.822	−4.839	−4.898	−4.969	−5.082
0.90	−4.238	−4.232	−4.230	−4.236	−4.267	−4.351	−4.442	−4.578
0.93	−3.894	−3.885	−3.884	−3.896	−3.941	−4.046	−4.151	−4.300
0.95	−3.658	−3.647	−3.648	−3.669	−3.728	−3.851	−3.966	−4.125
0.97	−3.406	−3.391	−3.401	−3.437	−3.517	−3.661	−3.788	−3.957
0.98	−3.264	−3.247	−3.268	−3.318	−3.412	−3.569	−3.701	−3.875
0.99	−3.093	−3.082	−3.126	−3.195	−3.306	−3.477	−3.616	−3.796
1.00	−2.399	−2.868	−2.967	−3.067	−3.200	−3.387	−3.532	−3.717
1.01	−1.306	−2.513	−2.784	−2.933	−3.094	−3.297	−3.450	−3.640
1.02	−1.113	−1.655	−2.557	−2.790	−2.986	−3.209	−3.369	−3.565
1.05	−0.820	−0.831	−1.443	−2.283	−2.655	−2.949	−3.134	−3.348
1.10	−0.577	−0.640	−0.618	−1.241	−2.067	−2.534	−2.767	−3.013
1.15	−0.437	−0.489	−0.502	−0.654	−1.471	−2.138	−2.428	−2.708
1.20	−0.343	−0.385	−0.412	−0.447	−0.991	−1.767	−2.115	−2.430
1.30	−0.226	−0.254	−0.282	−0.300	−0.481	−1.147	−1.569	−1.944
1.40	−0.158	−0.178	−0.200	−0.220	−0.290	−0.730	−1.138	−1.544
1.50	−0.115	−0.130	−0.147	−0.166	−0.206	−0.479	−0.823	−1.222
1.60	−0.086	−0.098	−0.112	−0.129	−0.159	−0.334	−0.604	−0.969
1.70	−0.067	−0.076	−0.087	−0.102	−0.127	−0.248	−0.456	−0.775
1.80	−0.053	−0.060	−0.070	−0.083	−0.105	−0.195	−0.355	−0.628
1.90	−0.043	−0.049	−0.057	−0.069	−0.089	−0.160	−0.286	−0.518
2.00	−0.035	−0.040	−0.048	−0.058	−0.077	−0.136	−0.238	−0.434
2.20	−0.025	−0.029	−0.035	−0.043	−0.060	−0.105	−0.178	−0.322
2.40	−0.019	−0.022	−0.027	−0.034	−0.048	−0.086	−0.143	−0.254
2.60	−0.015	−0.018	−0.021	−0.028	−0.041	−0.074	−0.120	−0.210
2.80	−0.012	−0.014	−0.018	−0.023	−0.025	−0.065	−0.104	−0.180
3.00	−0.010	−0.012	−0.015	−0.020	−0.031	−0.058	−0.093	−0.158
3.50	−0.007	−0.009	−0.011	−0.015	−0.024	−0.046	−0.073	−0.122
4.00	−0.006	−0.007	−0.009	−0.012	−0.020	−0.038	−0.060	−0.100

Table D.13: Values of ϕ^0

$P_r =$	0.0100	0.0500	0.1000	0.2000	0.4000	0.6000	0.8000	1.0000
T_r								
0.30	0.0002	0.0000	0.0000	0.0000	0.0000	0.0000	0.0000	0.0000
0.35	0.0034	0.0007	0.0003	0.0002	0.0001	0.0001	0.0001	0.0000
0.40	0.0272	0.0055	0.0028	0.0014	0.0007	0.0005	0.0004	0.0003
0.45	0.1321	0.0266	0.0135	0.0069	0.0036	0.0025	0.0020	0.0016
0.50	0.4529	0.0912	0.0461	0.0235	0.0122	0.0085	0.0067	0.0055
0.55	0.9817	0.2432	0.1227	0.0625	0.0325	0.0225	0.0176	0.0146
0.60	0.9840	0.5383	0.2716	0.1384	0.0718	0.0497	0.0386	0.0321
0.65	0.9886	0.9419	0.5212	0.2655	0.1374	0.0948	0.0738	0.0611
0.70	0.9908	0.9528	0.9057	0.4560	0.2360	0.1626	0.1262	0.1045
0.75	0.9931	0.9616	0.9226	0.7178	0.3715	0.2559	0.1982	0.1641
0.80	0.9931	0.9683	0.9354	0.8730	0.5445	0.3750	0.2904	0.2404
0.85	0.9954	0.9727	0.9462	0.8933	0.7534	0.5188	0.4018	0.3319
0.90	0.9954	0.9772	0.9550	0.9099	0.8204	0.6823	0.5297	0.4375
0.93	0.9954	0.9795	0.9594	0.9183	0.8375	0.7551	0.6109	0.5058
0.95	0.9954	0.9817	0.9616	0.9226	0.8472	0.7709	0.6668	0.5521
0.97	0.9954	0.9817	0.9638	0.9268	0.8570	0.7852	0.7112	0.5984
0.98	0.9954	0.9817	0.9638	0.9290	0.8610	0.7925	0.7211	0.6223
0.99	0.9977	0.9840	0.9661	0.9311	0.8650	0.7980	0.7295	0.6442
1.00	0.9977	0.9840	0.9661	0.9333	0.8690	0.8035	0.7379	0.6668
1.01	0.9977	0.9840	0.9683	0.9354	0.8730	0.8110	0.7464	0.6792
1.02	0.9977	0.9840	0.9683	0.9376	0.8770	0.8166	0.7551	0.6902
1.05	0.9977	0.9863	0.9705	0.9441	0.8872	0.8318	0.7762	0.7194
1.10	0.9977	0.9886	0.9750	0.9506	0.9016	0.8531	0.8072	0.7586
1.15	0.9977	0.9886	0.9795	0.9572	0.9141	0.8730	0.8318	0.7907
1.20	0.9977	0.9908	0.9817	0.9616	0.9247	0.8892	0.8531	0.8166
1.30	0.9977	0.9931	0.9863	0.9705	0.9419	0.9141	0.8872	0.8590
1.40	0.9977	0.9931	0.9886	0.9772	0.9550	0.9333	0.9120	0.8892
1.50	1.0000	0.9954	0.9908	0.9817	0.9638	0.9462	0.9290	0.9141
1.60	1.0000	0.9954	0.9931	0.9863	0.9727	0.9572	0.9441	0.9311
1.70	1.0000	0.9977	0.9954	0.9886	0.9772	0.9661	0.9550	0.9462
1.80	1.0000	0.9977	0.9954	0.9908	0.9817	0.9727	0.9661	0.9572
1.90	1.0000	0.9977	0.9954	0.9931	0.9863	0.9795	0.9727	0.9661
2.00	1.0000	0.9977	0.9977	0.9954	0.9886	0.9840	0.9795	0.9727
2.20	1.0000	1.0000	0.9977	0.9977	0.9931	0.9908	0.9886	0.9840
2.40	1.0000	1.0000	1.0000	0.9977	0.9977	0.9954	0.9931	0.9931
2.60	1.0000	1.0000	1.0000	1.0000	1.0000	0.9977	0.9977	0.9977
2.80	1.0000	1.0000	1.0000	1.0000	1.0000	1.0000	1.0023	1.0023
3.00	1.0000	1.0000	1.0000	1.0000	1.0023	1.0023	1.0046	1.0046
3.50	1.0000	1.0000	1.0000	1.0023	1.0023	1.0046	1.0069	1.0093
4.00	1.0000	1.0000	1.0000	1.0023	1.0046	1.0069	1.0093	1.0116

Table D.14: Values of ϕ^1

$P_r =$	0.0100	0.0500	0.1000	0.2000	0.4000	0.6000	0.8000	1.0000
T_r								
0.30	*0.0000*	*0.0000*	*0.0000*	*0.0000*	*0.0000*	*0.0000*	*0.0000*	0.0000
0.35	*0.0000*	*0.0000*	*0.0000*	*0.0000*	*0.0000*	*0.0000*	*0.0000*	0.0000
0.40	*0.0000*	*0.0000*	*0.0000*	*0.0000*	*0.0000*	*0.0000*	*0.0000*	0.0000
0.45	*0.0002*	*0.0002*	*0.0002*	*0.0002*	*0.0002*	*0.0002*	*0.0002*	0.0002
0.50	*0.0014*	*0.0014*	*0.0014*	*0.0014*	*0.0014*	*0.0014*	*0.0013*	0.0013
0.55	0.9705	*0.0069*	*0.0068*	*0.0068*	*0.0066*	*0.0065*	*0.0064*	0.0063
0.60	0.9795	*0.0227*	*0.0226*	*0.0223*	*0.0220*	*0.0216*	*0.0213*	0.0210
0.65	0.9863	0.9311	*0.0572*	*0.0568*	*0.0559*	*0.0551*	*0.0543*	0.0535
0.70	0.9908	0.9528	0.9036	*0.1182*	*0.1163*	*0.1147*	*0.1131*	0.1116
0.75	0.9931	0.9683	0.9332	*0.2112*	*0.2078*	*0.2050*	*0.2022*	0.1994
0.80	0.9954	0.9772	0.9550	0.9057	*0.3302*	*0.3257*	*0.3212*	0.3168
0.85	0.9977	0.9863	0.9705	0.9375	*0.4774*	*0.4708*	*0.4654*	0.4590
0.90	0.9977	0.9908	0.9795	0.9594	0.9141	*0.6323*	*0.6250*	0.6165
0.93	0.9977	0.9931	0.9840	0.9705	0.9354	0.8953	*0.7227*	0.7144
0.95	0.9977	0.9931	0.9885	0.9750	0.9484	0.9183	*0.7888*	0.7797
0.97	1.0000	0.9954	0.9908	0.9795	0.9594	0.9354	0.9078	0.8413
0.98	1.0000	0.9954	0.9908	0.9817	0.9638	0.9440	0.9225	0.8729
0.99	1.0000	0.9954	0.9931	0.9840	0.9683	0.9528	0.9332	0.9036
1.00	1.0000	0.9977	0.9931	0.9863	0.9727	0.9594	0.9440	0.9311
1.01	1.0000	0.9977	0.9931	0.9885	0.9772	0.9638	0.9528	0.9462
1.02	1.0000	0.9977	0.9954	0.9908	0.9795	0.9705	0.9616	0.9572
1.05	1.0000	0.9977	0.9977	0.9954	0.9885	0.9863	0.9840	0.9840
1.10	1.0000	1.0000	1.0000	1.0000	1.0023	1.0046	1.0093	1.0163
1.15	1.0000	1.0000	1.0023	1.0046	1.0116	1.0186	1.0257	1.0375
1.20	1.0000	1.0023	1.0046	1.0069	1.0163	1.0280	1.0399	1.0544
1.30	1.0000	1.0023	1.0069	1.0116	1.0257	1.0399	1.0544	1.0716
1.40	1.0000	1.0046	1.0069	1.0139	1.0304	1.0471	1.0642	1.0815
1.50	1.0000	1.0046	1.0069	1.0163	1.0328	1.0496	1.0666	1.0865
1.60	1.0000	1.0046	1.0069	1.0163	1.0328	1.0496	1.0691	1.0865
1.70	1.0000	1.0046	1.0093	1.0163	1.0328	1.0496	1.0691	1.0865
1.80	1.0000	1.0046	1.0069	1.0163	1.0328	1.0496	1.0666	1.0840
1.90	1.0000	1.0046	1.0069	1.0163	1.0328	1.0496	1.0666	1.0815
2.00	1.0000	1.0046	1.0069	1.0163	1.0304	1.0471	1.0642	1.0815
2.20	1.0000	1.0046	1.0069	1.0139	1.0304	1.0447	1.0593	1.0765
2.40	1.0000	1.0046	1.0069	1.0139	1.0280	1.0423	1.0568	1.0716
2.60	1.0000	1.0023	1.0069	1.0139	1.0257	1.0399	1.0544	1.0666
2.80	1.0000	1.0023	1.0069	1.0116	1.0257	1.0375	1.0496	1.0642
3.00	1.0000	1.0023	1.0069	1.0116	1.0233	1.0352	1.0471	1.0593
3.50	1.0000	1.0023	1.0046	1.0023	1.0209	1.0304	1.0423	1.0520
4.00	1.0000	1.0023	1.0046	1.0093	1.0186	1.0280	1.0375	1.0471

Table D.15: Values of ϕ^0

$P_r =$	1.0000	1.2000	1.5000	2.0000	3.0000	5.0000	7.0000	10.000
T_r								
0.30	0.0000	0.0000	0.0000	0.0000	0.0000	0.0000	0.0000	0.0000
0.35	0.0000	0.0000	0.0000	0.0000	0.0000	0.0000	0.0000	0.0000
0.40	0.0003	0.0003	0.0003	0.0002	0.0002	0.0002	0.0002	0.0003
0.45	0.0016	0.0014	0.0012	0.0010	0.0008	0.0008	0.0009	0.0012
0.50	0.0055	0.0048	0.0041	0.0034	0.0028	0.0025	0.0027	0.0034
0.55	0.0146	0.0127	0.0107	0.0089	0.0072	0.0063	0.0066	0.0080
0.60	0.0321	0.0277	0.0234	0.0193	0.0154	0.0132	0.0135	0.0160
0.65	0.0611	0.0527	0.0445	0.0364	0.0289	0.0244	0.0245	0.0282
0.70	0.1045	0.0902	0.0759	0.0619	0.0488	0.0406	0.0402	0.0453
0.75	0.1641	0.1413	0.1188	0.0966	0.0757	0.0625	0.0610	0.0673
0.80	0.2404	0.2065	0.1738	0.1409	0.1102	0.0899	0.0867	0.0942
0.85	0.3319	0.2858	0.2399	0.1945	0.1517	0.1227	0.1175	0.1256
0.90	0.4375	0.3767	0.3162	0.2564	0.1995	0.1607	0.1524	0.1611
0.93	0.5058	0.4355	0.3656	0.2972	0.2307	0.1854	0.1754	0.1841
0.95	0.5521	0.4764	0.3999	0.3251	0.2523	0.2028	0.1910	0.2000
0.97	0.5984	0.5164	0.4345	0.3532	0.2748	0.2203	0.2075	0.2163
0.98	0.6223	0.5370	0.4529	0.3681	0.2864	0.2296	0.2158	0.2244
0.99	0.6442	0.5572	0.4699	0.3828	0.2978	0.2388	0.2244	0.2328
1.00	0.6668	0.5781	0.4875	0.3972	0.3097	0.2483	0.2328	0.2415
1.01	0.6792	0.5970	0.5047	0.4121	0.3214	0.2576	0.2415	0.2500
1.02	0.6902	0.6166	0.5224	0.4266	0.3334	0.2673	0.2506	0.2582
1.05	0.7194	0.6607	0.5728	0.4710	0.3690	0.2958	0.2773	0.2844
1.10	0.7586	0.7112	0.6412	0.5408	0.4285	0.3451	0.3228	0.3296
1.15	0.7907	0.7499	0.6918	0.6026	0.4875	0.3954	0.3690	0.3750
1.20	0.8166	0.7834	0.7328	0.6546	0.5420	0.4446	0.4150	0.4198
1.30	0.8590	0.8318	0.7943	0.7345	0.6383	0.5383	0.5058	0.5093
1.40	0.8892	0.8690	0.8395	0.7925	0.7145	0.6237	0.5902	0.5943
1.50	0.9141	0.8974	0.8730	0.8375	0.7745	0.6966	0.6668	0.6714
1.60	0.9311	0.9183	0.8995	0.8710	0.8222	0.7586	0.7328	0.7430
1.70	0.9462	0.9354	0.9204	0.8995	0.8610	0.8091	0.7907	0.8054
1.80	0.9572	0.9484	0.9376	0.9204	0.8913	0.8531	0.8414	0.8590
1.90	0.9661	0.9594	0.9506	0.9376	0.9162	0.8872	0.8831	0.9057
2.00	0.9727	0.9683	0.9616	0.9528	0.9354	0.9183	0.9183	0.9462
2.20	0.9840	0.9817	0.9795	0.9727	0.9661	0.9616	0.9727	1.0093
2.40	0.9931	0.9908	0.9908	0.9886	0.9863	0.9931	1.0116	1.0568
2.60	0.9977	0.9977	0.9977	0.9977	1.0023	1.0162	1.0399	1.0889
2.80	1.0023	1.0023	1.0046	1.0069	1.0116	1.0328	1.0593	1.1117
3.00	1.0046	1.0069	1.0069	1.0116	1.0209	1.0423	1.0740	1.1298
3.50	1.0093	1.0116	1.0139	1.0186	1.0304	1.0593	1.0914	1.1508
4.00	1.0116	1.0139	1.0162	1.0233	1.0375	1.0666	1.0990	1.1588

Table D.16: Values of ϕ^1

$P_r =$	1.0000	1.2000	1.5000	2.0000	3.0000	5.0000	7.0000	10.000
T_r								
0.30	0.0000	0.0000	0.0000	0.0000	0.0000	0.0000	0.0000	0.0000
0.35	0.0000	0.0000	0.0000	0.0000	0.0000	0.0000	0.0000	0.0000
0.40	0.0000	0.0000	0.0000	0.0000	0.0000	0.0000	0.0000	0.0000
0.45	0.0002	0.0002	0.0002	0.0002	0.0001	0.0001	0.0001	0.0001
0.50	0.0013	0.0013	0.0013	0.0012	0.0011	0.0009	0.0008	0.0006
0.55	0.0063	0.0062	0.0061	0.0058	0.0053	0.0045	0.0039	0.0031
0.60	0.0210	0.0207	0.0202	0.0194	0.0179	0.0154	0.0133	0.0108
0.65	0.0536	0.0527	0.0516	0.0497	0.0461	0.0401	0.0350	0.0289
0.70	0.1117	0.1102	0.1079	0.1040	0.0970	0.0851	0.0752	0.0629
0.75	0.1995	0.1972	0.1932	0.1871	0.1754	0.1552	0.1387	0.1178
0.80	0.3170	0.3133	0.3076	0.2978	0.2812	0.2512	0.2265	0.1954
0.85	0.4592	0.4539	0.4457	0.4325	0.4093	0.3698	0.3365	0.2951
0.90	0.6166	0.6095	0.5998	0.5834	0.5546	0.5058	0.4645	0.4130
0.93	0.7145	0.7063	0.6950	0.6761	0.6457	0.5916	0.5470	0.4898
0.95	0.7798	0.7691	0.7568	0.7379	0.7063	0.6501	0.6026	0.5432
0.97	0.8414	0.8318	0.8185	0.7998	0.7656	0.7096	0.6607	0.5984
0.98	0.8730	0.8630	0.8492	0.8298	0.7962	0.7379	0.6887	0.6266
0.99	0.9036	0.8913	0.8790	0.8590	0.8241	0.7674	0.7178	0.6546
1.00	0.9311	0.9204	0.9078	0.8872	0.8531	0.7962	0.7464	0.6823
1.01	0.9462	0.9462	0.9333	0.9162	0.8831	0.8241	0.7745	0.7096
1.02	0.9572	0.9661	0.9594	0.9419	0.9099	0.8531	0.8035	0.7379
1.05	0.9840	0.9954	1.0186	1.0162	0.9886	0.9354	0.8872	0.8222
1.10	1.0162	1.0280	1.0593	1.0990	1.1015	1.0617	1.0186	0.9572
1.15	1.0375	1.0520	1.0814	1.1376	1.1858	1.1722	1.1403	1.0864
1.20	1.0544	1.0691	1.0990	1.1588	1.2388	1.2647	1.2474	1.2050
1.30	1.0715	1.0914	1.1194	1.1776	1.2853	1.3868	1.4125	1.4061
1.40	1.0814	1.0990	1.1298	1.1858	1.2942	1.4488	1.5171	1.5524
1.50	1.0864	1.1041	1.1350	1.1858	1.2942	1.4689	1.5740	1.6520
1.60	1.0864	1.1041	1.1350	1.1858	1.2883	1.4689	1.5996	1.7140
1.70	1.0864	1.1041	1.1324	1.1803	1.2794	1.4622	1.6033	1.7458
1.80	1.0839	1.1015	1.1298	1.1749	1.2706	1.4488	1.5959	1.7620
1.90	1.0814	1.0990	1.1272	1.1695	1.2618	1.4355	1.5849	1.7620
2.00	1.0814	1.0965	1.1220	1.1641	1.2503	1.4191	1.5704	1.7539
2.20	1.0765	1.0914	1.1143	1.1535	1.2331	1.3900	1.5346	1.7219
2.40	1.0715	1.0864	1.1066	1.1429	1.2190	1.3614	1.4997	1.6866
2.60	1.0666	1.0814	1.1015	1.1350	1.2023	1.3397	1.4689	1.6482
2.80	1.0641	1.0765	1.0940	1.1272	1.1912	1.3183	1.4388	1.6144
3.00	1.0593	1.0715	1.0889	1.1194	1.1803	1.3002	1.4158	1.5813
3.50	1.0520	1.0617	1.0789	1.1041	1.1561	1.2618	1.3614	1.5101
4.00	1.0471	1.0544	1.0691	1.0914	1.1403	1.2303	1.3213	1.4555

Appendix E

Steam Tables

INTERPOLATION

When a value is required from a table at conditions which lie between listed values, interpolation is necessary. If M, the quantity sought, is a function of a single independent variable X and if linear interpolation is appropriate, as in the tables for saturated steam, then a direct proportionality exists between corresponding differences in M and in X. When M, the value at X, is intermediate between two given values, M_1 at X_1 and M_2 at X_2, then:

$$M = \left(\frac{X_2 - X}{X_2 - X_1}\right)M_1 + \left(\frac{X - X_1}{X_2 - X_1}\right)M_2 \qquad (E.1)$$

For example, the enthalpy of saturated vapor steam at 140.8°C is intermediate between the following values taken from Table E.1:

t	H
$t_1 = 140°C$	$H_1 = 2733.1 \text{ kJ·kg}^{-1}$
$t = 140.8°C$	$H = ?$
$t_2 = 142°C$	$H_2 = 2735.6 \text{ kJ·kg}^{-1}$

Substitution of values into Eq. (E.1) with $M = H$ and $t = X$ yields:

$$H = \frac{1.2}{2}(2733.1) + \frac{0.8}{2}(2735.6) = 2734.1 \text{ kJ·kg}^{-1}$$

When M is a function of two independent variables X and Y and linear interpolation is appropriate, as in the tables for superheated steam, then double linear interpolation is required. Data for quantity M at values of the independent variables X and Y adjacent to the given values are represented as follows:

	X_1	X	X_2
Y_1	$M_{1,1}$		$M_{1,2}$
Y		$M = ?$	
Y_2	$M_{2,1}$		$M_{2,2}$

Double linear interpolation between the given values of M is represented by:

$$M = \left[\left(\frac{X_2 - X}{X_2 - X_1}\right)M_{1,1} + \left(\frac{X - X_1}{X_2 - X_1}\right)M_{1,2}\right]\frac{Y_2 - Y}{Y_2 - Y_1}$$

$$+ \left[\left(\frac{X_2 - X}{X_2 - X_1}\right)M_{2,1} + \left(\frac{X - X_1}{X_2 - X_1}\right)M_{2,2}\right]\frac{Y - Y_1}{Y_2 - Y_1}$$

(E.2)

Example E.1

From data in the steam tables, find:

a. The specific volume of superheated steam at 816 kPa and 512°C.

b. The temperature and specific entropy of superheated steam at $P = 2950$ kPa and $H = 3150.6$ kJ·kg^{-1}.

Solution E.1

(*a*) The following table shows specific volumes from Table E.2 for superheated steam at conditions adjacent to those specified:

P/kPa	t = 500°C	t = 512°C	t = 550°C
800	443.17		472.49
816		V = ?	
825	429.65		458.10

Substitution of values in Eq. (E.2) with $M = V$, $X = t$, and $Y = P$ yields:

$$V = \left[\frac{38}{50}(443.17) + \frac{12}{50}(472.49)\right]\frac{9}{25}$$

$$+ \left[\frac{38}{50}(429.65) + \frac{12}{50}(458.10)\right]\frac{16}{25} = 441.42 \text{ cm}^3\cdot\text{g}^{-1}$$

(*b*) The following table shows enthalpy data from Table E.2 for superheated steam at conditions adjacent to those specified:

P/kPa	t₁ = 350°C	t = ?	t₂ = 375°C
2900	3119.7		3177.4
2950	H_{t_1}	H = 3150.6	H_{t_2}
3000	3117.5		3175.6

Here, the direct use of Eq. (E.2) is not convenient. Rather, for $P = 2950$ kPa, interpolate linearly at $t_1 = 350°C$ for H_{t_1} and at $t_2 = 375°C$ for H_{t_2}, applying Eq. (E.1) twice, first at t_1 and second at t_2, with $M = H$ and $X = P$:

$$H_{t_1} = \frac{50}{100}(3119.7) + \frac{50}{100}(3117.5) = 3118.6$$

$$H_{t_2} = \frac{50}{100}(3177.4) + \frac{50}{100}(3175.6) = 3176.5$$

A third linear interpolation between these values with $M = t$ and $X = H$ in Eq. (E.1) yields:

$$t = \frac{3176.5 - 3150.6}{3176.5 - 3118.6}(350) + \frac{3150.6 - 3118.6}{3176.5 - 3118.6}(375) = 363.82°C$$

Given this temperature, a table of entropy values can now be constructed:

P/kPa	$t = 350°C$	$t = 363.82°C$	$t = 375°C$
2900	6.7654		6.8563
2950		$S = ?$	
3000	6.7471		6.8385

Application of Eq. (E.2) with $M = S$, $X = t$, and $Y = P$ yields:

$$S = \left[\frac{11.18}{25}(6.7654) + \frac{13.82}{25}(6.8563)\right]\frac{50}{100}$$

$$+ \left[\frac{11.18}{25}(6.7471) + \frac{13.82}{25}(6.8385)\right]\frac{50}{100} = 6.8066 \text{ kJ·mol}^{-1}$$

As a check, one can apply Eq. (E.2) with $M = H$, $X = t$, and $Y = P$, confirming that doing so produces $H = 3150.6$ kJ·kg^{-1}.

STEAM TABLES   Page

These tables were generated by computer from programs[1] based on "The 1976 International Formulation Committee Formulation for Industrial Use: A Formulation of the Thermodynamic Properties of Ordinary Water Substance," as published in the *ASME Steam Tables,* 4th ed., App. I, pp. 11–29, The Am. Soc. Mech. Engrs., New York, 1979. These tables served as a world-wide standard for 30 years and are entirely adequate for instructional purposes. However, they have been replaced by the "International Association for the Properties of Water and Steam Industrial Formulation 1997 for the Thermodynamic Properties of Water and Steam." These and other tables are discussed by A. H. Harvey and W. T. Parry, "Keep Your Steam Tables Up to Date," *Chemical Engineering Progress,* vol. 95, no. 11, p. 45, Nov. 1999.

[1]We gratefully acknowledge the contributions of Professor Charles Muckenfuss, of Debra L. Sauke, and of Eugene N. Dorsi, whose efforts produced the computer programs from which these tables derive.

TABLE E.1 Properties of Saturated Steam

V = SPECIFIC VOLUME cm³·g⁻¹ H = SPECIFIC ENTHALPY kJ·kg⁻¹
U = SPECIFIC INTERNAL ENERGY kJ·kg⁻¹ S = SPECIFIC ENTROPY kJ·kg⁻¹·K⁻¹

t (°C)	T (K)	P (kPa)	SPECIFIC VOLUME V sat. liq.	evap.	sat. vap.	INTERNAL ENERGY U sat. liq.	evap.	sat. vap.	ENTHALPY H sat. liq.	evap.	sat. vap.	ENTROPY S sat. liq.	evap.	sat. vap.
0	273.15	0.611	1.000	206300.	206300.	−0.04	2375.7	2375.6	−0.04	2501.7	2501.6	0.0000	9.1578	9.1578
0.01	273.16	0.611	1.000	206200.	206200.	0.00	2375.6	2375.6	0.00	2501.6	2501.6	0.0000	9.1575	9.1575
1	274.15	0.657	1.000	192600.	192600.	4.17	2372.7	2376.9	4.17	2499.2	2503.4	0.0153	9.1158	9.1311
2	275.15	0.705	1.000	179900.	179900.	8.39	2369.9	2378.3	8.39	2496.8	2505.2	0.0306	9.0741	9.1047
3	276.15	0.757	1.000	168200.	168200.	12.60	2367.1	2379.7	12.60	2494.5	2507.1	0.0459	9.0326	9.0785
4	277.15	0.813	1.000	157300.	157300.	16.80	2364.3	2381.1	16.80	2492.1	2508.9	0.0611	8.9915	9.0526
5	278.15	0.872	1.000	147200.	147200.	21.01	2361.4	2382.4	21.01	2489.7	2510.7	0.0762	8.9507	9.0269
6	279.15	0.935	1.000	137800.	137800.	25.21	2358.6	2383.8	25.21	2487.4	2512.6	0.0913	8.9102	9.0014
7	280.15	1.001	1.000	129100.	129100.	29.41	2355.8	2385.2	29.41	2485.0	2514.4	0.1063	8.8699	8.9762
8	281.15	1.072	1.000	121000.	121000.	33.60	2353.0	2386.6	33.60	2482.6	2516.2	0.1213	8.8300	8.9513
9	282.15	1.147	1.000	113400.	113400.	37.80	2350.1	2387.9	37.80	2480.3	2518.1	0.1362	8.7903	8.9265
10	283.15	1.227	1.000	106400.	106400.	41.99	2347.3	2389.3	41.99	2477.9	2519.9	0.1510	8.7510	8.9020
11	284.15	1.312	1.000	99910.	99910.	46.18	2344.5	2390.7	46.19	2475.5	2521.7	0.1658	8.7119	8.8776
12	285.15	1.401	1.000	93830.	93840.	50.38	2341.7	2392.1	50.38	2473.2	2523.6	0.1805	8.6731	8.8536
13	286.15	1.497	1.001	88180.	88180.	54.56	2338.9	2393.4	54.57	2470.8	2525.4	0.1952	8.6345	8.8297
14	287.15	1.597	1.001	82900.	82900.	58.75	2336.1	2394.8	58.75	2468.5	2527.2	0.2098	8.5963	8.8060
15	288.15	1.704	1.001	77980.	77980.	62.94	2333.2	2396.2	62.94	2466.1	2529.1	0.2243	8.5582	8.7826
16	289.15	1.817	1.001	73380.	73380.	67.12	2330.4	2397.6	67.13	2463.8	2530.9	0.2388	8.5205	8.7593
17	290.15	1.936	1.001	69090.	69090.	71.31	2327.6	2398.9	71.31	2461.4	2532.7	0.2533	8.4830	8.7363
18	291.15	2.062	1.001	65090.	65090.	75.49	2324.8	2400.3	75.50	2459.0	2534.5	0.2677	8.4458	8.7135
19	292.15	2.196	1.002	61340.	61340.	79.68	2322.0	2401.7	79.68	2456.7	2536.4	0.2820	8.4088	8.6908
20	293.15	2.337	1.002	57840.	57840.	83.86	2319.2	2403.0	83.86	2454.3	2538.2	0.2963	8.3721	8.6684
21	294.15	2.485	1.002	54560.	54560.	88.04	2316.4	2404.4	88.04	2452.0	2540.0	0.3105	8.3356	8.6462
22	295.15	2.642	1.002	51490.	51490.	92.22	2313.6	2405.8	92.23	2449.6	2541.8	0.3247	8.2994	8.6241
23	296.15	2.808	1.002	48620.	48620.	96.40	2310.7	2407.1	96.41	2447.2	2543.6	0.3389	8.2634	8.6023
24	297.15	2.982	1.003	45920.	45930.	100.6	2307.9	2408.5	100.6	2444.9	2545.5	0.3530	8.2277	8.5806
25	298.15	3.166	1.003	43400.	43400.	104.8	2305.1	2409.9	104.8	2442.5	2547.3	0.3670	8.1922	8.5592
26	299.15	3.360	1.003	41030.	41030.	108.9	2302.3	2411.2	108.9	2440.2	2549.1	0.3810	8.1569	8.5379
27	300.15	3.564	1.003	38810.	38810.	113.1	2299.5	2412.6	113.1	2437.8	2550.9	0.3949	8.1218	8.5168
28	301.15	3.778	1.004	36730.	36730.	117.3	2296.7	2414.0	117.3	2435.4	2552.7	0.4088	8.0870	8.4959
29	302.15	4.004	1.004	34770.	34770.	121.5	2293.8	2415.3	121.5	2433.1	2554.5	0.4227	8.0524	8.4751

t	T/K	P	V^l	V^{lv}	V^v	U^l	U^{lv}	U^v	H^l	H^{lv}	H^v	S^l	S^{lv}	S^v
30	303.15	4.241	1.004	32930.	32930.	125.7	2291.1	2416.7	125.7	2430.7	2556.4	0.4365	8.0180	8.4546
31	304.15	4.491	1.005	31200.	31200.	129.8	2288.3	2418.0	129.8	2428.3	2558.2	0.4503	7.9839	8.4342
32	305.15	4.753	1.005	29570.	29570.	134.0	2285.4	2419.4	134.0	2425.9	2560.0	0.4640	7.9500	8.4140
33	306.15	5.029	1.005	28040.	28040.	138.2	2282.6	2420.8	138.2	2423.6	2561.8	0.4777	7.9163	8.3939
34	307.15	5.318	1.006	26600.	26600.	142.4	2279.7	2422.1	142.4	2421.2	2563.6	0.4913	7.8828	8.3740
35	308.15	5.622	1.006	25240.	25240.	146.6	2276.9	2423.5	146.6	2418.8	2565.4	0.5049	7.8495	8.3543
36	309.15	5.940	1.006	23970.	23970.	150.7	2274.1	2424.8	150.7	2416.4	2567.2	0.5184	7.8164	8.3348
37	310.15	6.274	1.007	22760.	22760.	154.9	2271.3	2426.2	154.9	2414.1	2569.0	0.5319	7.7835	8.3154
38	311.15	6.624	1.007	21630.	21630.	159.1	2268.4	2427.5	159.1	2411.7	2570.8	0.5453	7.7509	8.2962
39	312.15	6.991	1.007	20560.	20560.	163.3	2265.6	2428.9	163.3	2409.3	2572.6	0.5588	7.7184	8.2772
40	313.15	7.375	1.008	19550.	19550.	167.4	2262.8	2430.2	167.5	2406.9	2574.4	0.5721	7.6861	8.2583
41	314.15	7.777	1.008	18590.	18590.	171.6	2259.9	2431.6	171.6	2404.5	2576.2	0.5854	7.6541	8.2395
42	315.15	8.198	1.009	17690.	17690.	175.8	2257.1	2432.9	175.8	2402.1	2577.9	0.5987	7.6222	8.2209
43	316.15	8.639	1.009	16840.	16840.	180.0	2254.3	2434.2	180.0	2399.7	2579.7	0.6120	7.5905	8.2025
44	317.15	9.100	1.009	16040.	16040.	184.2	2251.4	2435.6	184.2	2397.3	2581.5	0.6252	7.5590	8.1842
45	318.15	9.582	1.010	15280.	15280.	188.3	2248.6	2436.9	188.4	2394.9	2583.3	0.6383	7.5277	8.1661
46	319.15	10.09	1.010	14560.	14560.	192.5	2245.7	2438.3	192.5	2392.5	2585.1	0.6514	7.4966	8.1481
47	320.15	10.61	1.011	13880.	13880.	196.7	2242.9	2439.6	196.7	2390.1	2586.9	0.6645	7.4657	8.1302
48	321.15	11.16	1.011	13230.	13230.	200.9	2240.0	2440.9	200.9	2387.7	2588.6	0.6776	7.4350	8.1125
49	322.15	11.74	1.012	12620.	12620.	205.1	2237.2	2442.3	205.1	2385.3	2590.4	0.6906	7.4044	8.0950
50	323.15	12.34	1.012	12040.	12050.	209.2	2234.3	2443.6	209.3	2382.9	2592.2	0.7035	7.3741	8.0776
51	324.15	12.96	1.013	11500.	11500.	213.4	2231.5	2444.9	213.4	2380.5	2593.9	0.7164	7.3439	8.0603
52	325.15	13.61	1.013	10980.	10980.	217.6	2228.6	2446.2	217.6	2378.1	2595.7	0.7293	7.3138	8.0432
53	326.15	14.29	1.014	10490.	10490.	221.8	2225.8	2447.6	221.8	2375.7	2597.5	0.7422	7.2840	8.0262
54	327.15	15.00	1.014	10020.	10020.	226.0	2222.9	2448.9	226.0	2373.2	2599.2	0.7550	7.2543	8.0093
55	328.15	15.74	1.015	9577.9	9578.9	230.2	2220.0	2450.2	230.2	2370.8	2601.0	0.7677	7.2248	7.9925
56	329.15	16.51	1.015	9157.7	9158.7	234.3	2217.2	2451.5	234.4	2368.4	2602.7	0.7804	7.1955	7.9759
57	330.15	17.31	1.016	8758.7	8759.8	238.5	2214.3	2452.8	238.5	2365.9	2604.5	0.7931	7.1663	7.9595
58	331.15	18.15	1.016	8379.8	8380.8	242.7	2211.4	2454.1	242.7	2363.5	2606.2	0.8058	7.1373	7.9431
59	332.15	19.02	1.017	8019.7	8020.8	246.9	2208.6	2455.4	246.9	2361.1	2608.0	0.8184	7.1085	7.9269
60	333.15	19.92	1.017	7677.5	7678.5	251.1	2205.7	2456.8	251.1	2358.6	2609.7	0.8310	7.0798	7.9108
61	334.15	20.86	1.018	7352.1	7353.2	255.3	2202.8	2458.1	255.3	2356.2	2611.4	0.8435	7.0513	7.8948
62	335.15	21.84	1.018	7042.7	7043.7	259.4	2199.9	2459.4	259.5	2353.7	2613.2	0.8560	7.0230	7.8790
63	336.15	22.86	1.019	6748.2	6749.3	263.6	2197.0	2460.7	263.6	2351.3	2614.9	0.8685	6.9948	7.8633
64	337.15	23.91	1.019	6468.0	6469.0	267.8	2194.1	2462.0	267.8	2348.8	2616.6	0.8809	6.9667	7.8477
65	338.15	25.01	1.020	6201.3	6202.3	272.0	2191.2	2463.2	272.0	2346.3	2618.4	0.8933	6.9388	7.8322
66	339.15	26.15	1.020	5947.2	5948.2	276.2	2188.3	2464.5	276.2	2343.9	2620.1	0.9057	6.9111	7.8168
67	340.15	27.33	1.021	5705.2	5706.2	280.4	2185.4	2465.8	280.4	2341.4	2621.8	0.9180	6.8835	7.8015
68	341.15	28.56	1.022	5474.6	5475.6	284.6	2182.5	2467.1	284.6	2338.9	2623.5	0.9303	6.8561	7.7864
69	342.15	29.84	1.022	5254.8	5255.8	288.8	2179.6	2468.4	288.8	2336.4	2625.2	0.9426	6.8288	7.7714

TABLE E.1 Properties of Saturated Steam (Continued)

t (°C)	T (K)	P (kPa)	SPECIFIC VOLUME V			INTERNAL ENERGY U			ENTHALPY H			ENTROPY S		
			sat. liq.	evap.	sat. vap.	sat. liq.	evap.	sat. vap.	sat. liq.	evap.	sat. vap.	sat. liq.	evap.	sat. vap.
70	343.15	31.16	1.023	5045.2	5046.3	292.9	2176.7	2469.7	293.0	2334.0	2626.9	0.9548	6.8017	7.7565
71	344.15	32.53	1.023	4845.4	4846.4	297.1	2173.8	2470.9	297.2	2331.5	2628.6	0.9670	6.7747	7.7417
72	345.15	33.96	1.024	4654.7	4655.7	301.3	2170.9	2472.2	301.4	2329.0	2630.3	0.9792	6.7478	7.7270
73	346.15	35.43	1.025	4472.7	4473.7	305.5	2168.0	2473.5	305.5	2326.5	2632.0	0.9913	6.7211	7.7124
74	347.15	36.96	1.025	4299.0	4300.0	309.7	2165.1	2474.8	309.7	2324.0	2633.7	1.0034	6.6945	7.6979
75	348.15	38.55	1.026	4133.1	4134.1	313.9	2162.1	2476.0	313.9	2321.5	2635.4	1.0154	6.6681	7.6835
76	349.15	40.19	1.027	3974.6	3975.7	318.1	2159.2	2477.3	318.1	2318.9	2637.1	1.0275	6.6418	7.6693
77	350.15	41.89	1.027	3823.3	3824.3	322.3	2156.3	2478.5	322.3	2316.4	2638.7	1.0395	6.6156	7.6551
78	351.15	43.65	1.028	3678.6	3679.6	326.5	2153.3	2479.8	326.5	2313.9	2640.4	1.0514	6.5896	7.6410
79	352.15	45.47	1.029	3540.3	3541.3	330.7	2150.4	2481.1	330.7	2311.4	2642.1	1.0634	6.5637	7.6271
80	353.15	47.36	1.029	3408.1	3409.1	334.9	2147.4	2482.3	334.9	2308.8	2643.8	1.0753	6.5380	7.6132
81	354.15	49.31	1.030	3281.6	3282.6	339.1	2144.5	2483.5	339.1	2306.3	2645.4	1.0871	6.5123	7.5995
82	355.15	51.33	1.031	3160.6	3161.6	343.3	2141.5	2484.8	343.3	2303.8	2647.1	1.0990	6.4868	7.5858
83	356.15	53.42	1.031	3044.8	3045.8	347.5	2138.6	2486.0	347.5	2301.2	2648.7	1.1108	6.4615	7.5722
84	357.15	55.57	1.032	2933.9	2935.0	351.7	2135.6	2487.3	351.7	2298.6	2650.4	1.1225	6.4362	7.5587
85	358.15	57.80	1.033	2827.8	2828.8	355.9	2132.6	2488.5	355.9	2296.1	2652.0	1.1343	6.4111	7.5454
86	359.15	60.11	1.033	2726.1	2727.2	360.1	2129.7	2489.7	360.1	2293.5	2653.6	1.1460	6.3861	7.5321
87	360.15	62.49	1.034	2628.8	2629.8	364.3	2126.7	2490.9	364.3	2290.9	2655.3	1.1577	6.3612	7.5189
88	361.15	64.95	1.035	2535.5	2536.5	368.5	2123.7	2492.2	368.5	2288.4	2656.9	1.1693	6.3365	7.5058
89	362.15	67.49	1.035	2446.0	2447.0	372.7	2120.7	2493.4	372.7	2285.8	2658.5	1.1809	6.3119	7.4928
90	363.15	70.11	1.036	2360.3	2361.3	376.9	2117.7	2494.6	376.9	2283.2	2660.1	1.1925	6.2873	7.4799
91	364.15	72.81	1.037	2278.0	2279.1	381.1	2114.7	2495.8	381.1	2280.6	2661.7	1.2041	6.2629	7.4670
92	365.15	75.61	1.038	2199.2	2200.2	385.3	2111.7	2497.0	385.4	2278.0	2663.4	1.2156	6.2387	7.4543
93	366.15	78.49	1.038	2123.5	2124.5	389.5	2108.7	2498.2	389.6	2275.4	2665.0	1.2271	6.2145	7.4416
94	367.15	81.46	1.039	2050.9	2051.9	393.7	2105.7	2499.4	393.8	2272.8	2666.6	1.2386	6.1905	7.4291
95	368.15	84.53	1.040	1981.2	1982.2	397.9	2102.7	2500.6	398.0	2270.2	2668.1	1.2501	6.1665	7.4166
96	369.15	87.69	1.041	1914.3	1915.3	402.1	2099.7	2501.8	402.2	2267.5	2669.7	1.2615	6.1427	7.4042
97	370.15	90.94	1.041	1850.0	1851.0	406.3	2096.6	2503.0	406.4	2264.9	2671.3	1.2729	6.1190	7.3919
98	371.15	94.30	1.042	1788.3	1789.3	410.5	2093.6	2504.1	410.6	2262.2	2672.9	1.2842	6.0954	7.3796
99	372.15	97.76	1.043	1729.0	1730.0	414.7	2090.6	2505.3	414.8	2259.6	2674.4	1.2956	6.0719	7.3675

100	373.15	101.33	1.044	1672.0	1673.0	419.0	2087.5	2506.5	419.1	2256.9	2676.0	1.3069	6.0485	7.3554
102	375.15	108.78	1.045	1564.5	1565.5	427.4	2081.4	2508.8	427.5	2251.6	2679.1	1.3294	6.0021	7.3315
104	377.15	116.68	1.047	1465.1	1466.2	435.8	2075.3	2511.1	435.9	2246.3	2682.2	1.3518	5.9560	7.3078
106	379.15	125.04	1.049	1373.1	1374.2	444.3	2069.2	2513.4	444.4	2240.9	2685.3	1.3742	5.9104	7.2845
108	381.15	133.90	1.050	1287.9	1288.9	452.7	2063.0	2515.7	452.9	2235.4	2688.3	1.3964	5.8651	7.2615
110	383.15	143.27	1.052	1208.9	1209.9	461.2	2056.8	2518.0	461.3	2230.0	2691.3	1.4185	5.8203	7.2388
112	385.15	153.16	1.054	1135.6	1136.6	469.6	2050.6	2520.2	469.8	2224.5	2694.3	1.4405	5.7758	7.2164
114	387.15	163.62	1.055	1067.5	1068.5	478.1	2044.3	2522.4	478.3	2219.0	2697.2	1.4624	5.7318	7.1942
116	389.15	174.65	1.057	1004.2	1005.2	486.6	2038.1	2524.6	486.7	2213.4	2700.2	1.4842	5.6881	7.1723
118	391.15	186.28	1.059	945.3	946.3	495.0	2031.8	2526.8	495.2	2207.9	2703.1	1.5060	5.6447	7.1507
120	393.15	198.54	1.061	890.5	891.5	503.5	2025.4	2529.0	503.7	2202.2	2706.0	1.5276	5.6017	7.1293
122	395.15	211.45	1.062	839.4	840.5	512.0	2019.1	2531.1	512.2	2196.6	2708.8	1.5491	5.5590	7.1082
124	397.15	225.04	1.064	791.8	792.8	520.5	2012.7	2533.2	520.7	2190.9	2711.6	1.5706	5.5167	7.0873
126	399.15	239.33	1.066	747.3	748.4	529.0	2006.3	2535.3	529.2	2185.2	2714.4	1.5919	5.4747	7.0666
128	401.15	254.35	1.068	705.8	706.9	537.5	1999.9	2537.4	537.8	2179.4	2717.2	1.6132	5.4330	7.0462
130	403.15	270.13	1.070	667.1	668.1	546.0	1993.4	2539.4	546.3	2173.6	2719.9	1.6344	5.3917	7.0261
132	405.15	286.70	1.072	630.8	631.9	554.5	1986.9	2541.4	554.8	2167.8	2722.6	1.6555	5.3507	7.0061
134	407.15	304.07	1.074	596.9	598.0	563.1	1980.4	2543.4	563.4	2161.9	2725.3	1.6765	5.3099	6.9864
136	409.15	322.29	1.076	565.1	566.2	571.6	1973.8	2545.4	572.0	2155.9	2727.9	1.6974	5.2695	6.9669
138	411.15	341.38	1.078	535.3	536.4	580.2	1967.2	2547.4	580.5	2150.0	2730.5	1.7182	5.2293	6.9475
140	413.15	361.38	1.080	507.4	508.5	588.7	1960.6	2549.3	589.1	2144.0	2733.1	1.7390	5.1894	6.9284
142	415.15	382.31	1.082	481.2	482.3	597.3	1953.9	2551.2	597.7	2137.9	2735.6	1.7597	5.1499	6.9095
144	417.15	404.20	1.084	456.6	457.7	605.9	1947.2	2553.1	606.3	2131.8	2738.1	1.7803	5.1105	6.8908
146	419.15	427.09	1.086	433.5	434.6	614.4	1940.5	2554.9	614.9	2125.7	2740.6	1.8008	5.0715	6.8723
148	421.15	451.01	1.089	411.8	412.9	623.0	1933.7	2556.8	623.5	2119.5	2743.0	1.8213	5.0327	6.8539
150	423.15	476.00	1.091	391.4	392.4	631.6	1926.9	2558.6	632.1	2113.2	2745.4	1.8416	4.9941	6.8358
152	425.15	502.08	1.093	372.1	373.2	640.2	1920.1	2560.3	640.8	2106.9	2747.7	1.8619	4.9558	6.8178
154	427.15	529.29	1.095	354.0	355.1	648.9	1913.2	2562.1	649.4	2100.6	2750.0	1.8822	4.9178	6.8000
156	429.15	557.67	1.098	336.9	338.0	657.5	1906.3	2563.8	658.1	2094.2	2752.3	1.9023	4.8800	6.7823
158	431.15	587.25	1.100	320.8	321.9	666.1	1899.3	2565.5	666.8	2087.7	2754.5	1.9224	4.8424	6.7648
160	433.15	618.06	1.102	305.7	306.8	674.8	1892.3	2567.1	675.5	2081.3	2756.7	1.9425	4.8050	6.7475
162	435.15	650.16	1.105	291.3	292.4	683.5	1885.3	2568.8	684.2	2074.7	2758.9	1.9624	4.7679	6.7303
164	437.15	683.56	1.107	277.8	278.9	692.1	1878.2	2570.4	692.9	2068.1	2761.0	1.9823	4.7309	6.7133
166	439.15	718.31	1.109	265.0	266.1	700.8	1871.1	2571.9	701.6	2061.4	2763.1	2.0022	4.6942	6.6964
168	441.15	754.45	1.112	252.9	254.0	709.5	1863.9	2573.4	710.4	2054.7	2765.1	2.0219	4.6577	6.6796

TABLE E.1 Properties of Saturated Steam (Continued)

t (°C)	T (K)	P (kPa)	SPECIFIC VOLUME V sat. liq.	evap.	sat. vap.	INTERNAL ENERGY U sat. liq.	evap.	sat. vap.	ENTHALPY H sat. liq.	evap.	sat. vap.	ENTROPY S sat. liq.	evap.	sat. vap.
170	443.15	792.02	1.114	241.4	242.6	718.2	1856.7	2574.9	719.1	2047.9	2767.1	2.0416	4.6214	6.6630
172	445.15	831.06	1.117	230.6	231.7	727.0	1849.5	2576.4	727.9	2041.1	2769.0	2.0613	4.5853	6.6465
174	447.15	871.60	1.120	220.3	221.5	735.7	1842.2	2577.8	736.7	2034.2	2770.9	2.0809	4.5493	6.6302
176	449.15	913.68	1.122	210.6	211.7	744.4	1834.8	2579.3	745.5	2027.3	2772.7	2.1004	4.5136	6.6140
178	451.15	957.36	1.125	201.4	202.5	753.2	1827.4	2580.6	754.3	2020.2	2774.5	2.1199	4.4780	6.5979
180	453.15	1002.7	1.128	192.7	193.8	762.0	1820.0	2581.9	763.1	2013.1	2776.3	2.1393	4.4426	6.5819
182	455.15	1049.6	1.130	184.4	185.5	770.8	1812.5	2583.2	772.0	2006.0	2778.0	2.1587	4.4074	6.5660
184	457.15	1098.3	1.133	176.5	177.6	779.6	1804.9	2584.5	780.8	1998.8	2779.6	2.1780	4.3723	6.5503
186	459.15	1148.8	1.136	169.0	170.2	788.4	1797.3	2585.7	789.7	1991.5	2781.2	2.1972	4.3374	6.5346
188	461.15	1201.0	1.139	161.9	163.1	797.2	1789.7	2586.9	798.6	1984.2	2782.8	2.2164	4.3026	6.5191
190	463.15	1255.1	1.142	155.2	156.3	806.1	1782.0	2588.1	807.5	1976.7	2784.3	2.2356	4.2680	6.5036
192	465.15	1311.1	1.144	148.8	149.9	814.9	1774.2	2589.2	816.5	1969.3	2785.7	2.2547	4.2336	6.4883
194	467.15	1369.0	1.147	142.6	143.8	823.8	1766.4	2590.2	825.4	1961.7	2787.1	2.2738	4.1993	6.4730
196	469.15	1428.9	1.150	136.8	138.0	832.7	1758.6	2591.3	834.4	1954.1	2788.4	2.2928	4.1651	6.4578
198	471.15	1490.9	1.153	131.3	132.4	841.6	1750.6	2592.3	843.4	1946.4	2789.7	2.3117	4.1310	6.4428
200	473.15	1554.9	1.156	126.0	127.2	850.6	1742.6	2593.2	852.4	1938.6	2790.9	2.3307	4.0971	6.4278
202	475.15	1621.0	1.160	121.0	122.1	859.5	1734.6	2594.1	861.4	1930.7	2792.1	2.3495	4.0633	6.4128
204	477.15	1689.3	1.163	116.2	117.3	868.5	1726.5	2595.0	870.5	1922.8	2793.2	2.3684	4.0296	6.3980
206	479.15	1759.8	1.166	111.6	112.8	877.5	1718.3	2595.8	879.5	1914.7	2794.3	2.3872	3.9961	6.3832
208	481.15	1832.6	1.169	107.2	108.4	886.5	1710.1	2596.6	888.6	1906.6	2795.3	2.4059	3.9626	6.3686
210	483.15	1907.7	1.173	103.1	104.2	895.5	1701.8	2597.3	897.7	1898.5	2796.2	2.4247	3.9293	6.3539
212	485.15	1985.2	1.176	99.09	100.26	904.5	1693.5	2598.0	906.9	1890.2	2797.1	2.4434	3.8960	6.3394
214	487.15	2065.1	1.179	95.28	96.46	913.6	1685.1	2598.7	916.0	1881.8	2797.9	2.4620	3.8629	6.3249
216	489.15	2147.5	1.183	91.65	92.83	922.7	1676.6	2599.3	925.2	1873.4	2798.6	2.4806	3.8298	6.3104
218	491.15	2232.4	1.186	88.17	89.36	931.8	1668.0	2599.8	934.4	1864.9	2799.3	2.4992	3.7968	6.2960
220	493.15	2319.8	1.190	84.85	86.04	940.9	1659.4	2600.3	943.7	1856.2	2799.9	2.5178	3.7639	6.2817
222	495.15	2409.9	1.194	81.67	82.86	950.1	1650.7	2600.8	952.9	1847.5	2800.5	2.5363	3.7311	6.2674
224	497.15	2502.7	1.197	78.62	79.82	959.2	1642.0	2601.2	962.2	1838.7	2800.9	2.5548	3.6984	6.2532
226	499.15	2598.2	1.201	75.71	76.91	968.4	1633.1	2601.5	971.5	1829.8	2801.4	2.5733	3.6657	6.2390
228	501.15	2696.5	1.205	72.92	74.12	977.6	1624.2	2601.8	980.9	1820.8	2801.7	2.5917	3.6331	6.2249

230	503.15	2797.6	1.209	70.24	71.45	986.9	1615.2	2602.1	990.3	1811.7	2802.0	2.6102	3.6006	6.2107
232	505.15	2901.6	1.213	67.68	68.89	996.2	1606.1	2602.3	999.7	1802.5	2802.2	2.6286	3.5681	6.1967
234	507.15	3008.6	1.217	65.22	66.43	1005.4	1597.0	2602.4	1009.1	1793.2	2802.3	2.6470	3.5356	6.1826
236	509.15	3118.6	1.221	62.86	64.08	1014.8	1587.7	2602.5	1018.6	1783.8	2802.3	2.6653	3.5033	6.1686
238	511.15	3231.7	1.225	60.60	61.82	1024.1	1578.4	2602.5	1028.1	1774.2	2802.3	2.6837	3.4709	6.1546
240	513.15	3347.8	1.229	58.43	59.65	1033.5	1569.0	2602.5	1037.6	1764.6	2802.2	2.7020	3.4386	6.1406
242	515.15	3467.2	1.233	56.34	57.57	1042.9	1559.5	2602.4	1047.2	1754.9	2802.0	2.7203	3.4063	6.1266
244	517.15	3589.8	1.238	54.34	55.58	1052.3	1549.9	2602.2	1056.8	1745.0	2801.8	2.7386	3.3740	6.1127
246	519.15	3715.7	1.242	52.41	53.66	1061.8	1540.2	2602.0	1066.4	1735.0	2801.4	2.7569	3.3418	6.0987
248	521.15	3844.9	1.247	50.56	51.81	1071.3	1530.5	2601.8	1076.1	1724.9	2801.0	2.7752	3.3096	6.0848
250	523.15	3977.6	1.251	48.79	50.04	1080.8	1520.6	2601.4	1085.8	1714.7	2800.4	2.7935	3.2773	6.0708
252	525.15	4113.7	1.256	47.08	48.33	1090.4	1510.6	2601.0	1095.5	1704.3	2799.8	2.8118	3.2451	6.0569
254	527.15	4253.4	1.261	45.43	46.69	1100.0	1500.5	2600.5	1105.3	1693.8	2799.1	2.8300	3.2129	6.0429
256	529.15	4396.7	1.266	43.85	45.11	1109.6	1490.4	2600.0	1115.2	1683.2	2798.3	2.8483	3.1807	6.0290
258	531.15	4543.7	1.271	42.33	43.60	1119.3	1480.1	2599.3	1125.0	1672.4	2797.4	2.8666	3.1484	6.0150
260	533.15	4694.3	1.276	40.86	42.13	1129.0	1469.7	2598.6	1134.9	1661.5	2796.4	2.8848	3.1161	6.0010
262	535.15	4848.8	1.281	39.44	40.73	1138.7	1459.2	2597.8	1144.9	1650.4	2795.3	2.9031	3.0838	5.9869
264	537.15	5007.1	1.286	38.08	39.37	1148.5	1448.5	2597.0	1154.9	1639.2	2794.1	2.9214	3.0515	5.9729
266	539.15	5169.3	1.291	36.77	38.06	1158.3	1437.8	2596.1	1165.0	1627.8	2792.8	2.9397	3.0191	5.9588
268	541.15	5335.5	1.297	35.51	36.80	1168.2	1426.9	2595.0	1175.1	1616.3	2791.4	2.9580	2.9866	5.9446
270	543.15	5505.8	1.303	34.29	35.59	1178.1	1415.9	2593.9	1185.2	1604.6	2789.9	2.9763	2.9541	5.9304
272	545.15	5680.2	1.308	33.11	34.42	1188.0	1404.7	2592.7	1195.4	1592.8	2788.2	2.9947	2.9215	5.9162
274	547.15	5858.7	1.314	31.97	33.29	1198.0	1393.4	2591.4	1205.7	1580.8	2786.5	3.0131	2.8889	5.9019
276	549.15	6041.5	1.320	30.88	32.20	1208.0	1382.0	2590.1	1216.0	1568.5	2784.6	3.0314	2.8561	5.8876
278	551.15	6228.7	1.326	29.82	31.14	1218.1	1370.4	2588.6	1226.4	1556.2	2782.6	3.0499	2.8233	5.8731
280	553.15	6420.2	1.332	28.79	30.13	1228.3	1358.7	2587.0	1236.8	1543.6	2780.4	3.0683	2.7903	5.8586
282	555.15	6616.1	1.339	27.81	29.14	1238.5	1346.8	2585.3	1247.3	1530.8	2778.1	3.0868	2.7573	5.8440
284	557.15	6816.6	1.345	26.85	28.20	1248.7	1334.8	2583.5	1257.9	1517.8	2775.7	3.1053	2.7241	5.8294
286	559.15	7021.8	1.352	25.93	27.28	1259.0	1322.6	2581.6	1268.5	1504.6	2773.2	3.1238	2.6908	5.8146
288	561.15	7231.5	1.359	25.03	26.39	1269.4	1310.2	2579.6	1279.2	1491.2	2770.5	3.1424	2.6573	5.7997
290	563.15	7446.1	1.366	24.17	25.54	1279.8	1297.7	2577.5	1290.0	1477.6	2767.6	3.1611	2.6237	5.7848
292	565.15	7665.4	1.373	23.33	24.71	1290.3	1284.9	2575.3	1300.9	1463.8	2764.6	3.1798	2.5899	5.7697
294	567.15	7889.7	1.381	22.52	23.90	1300.9	1272.0	2572.9	1311.8	1449.7	2761.5	3.1985	2.5560	5.7545
296	569.15	8118.9	1.388	21.74	23.13	1311.5	1258.9	2570.4	1322.8	1435.4	2758.2	3.2173	2.5218	5.7392
298	571.15	8353.2	1.396	20.98	22.38	1322.2	1245.6	2567.8	1333.9	1420.8	2754.7	3.2362	2.4875	5.7237

TABLE E.1 Properties of Saturated Steam (Continued)

t (°C)	T (K)	P (kPa)	SPECIFIC VOLUME V			INTERNAL ENERGY U			ENTHALPY H			ENTROPY S		
			sat. liq.	evap.	sat. vap.	sat. liq.	evap.	sat. vap.	sat. liq.	evap.	sat. vap.	sat. liq.	evap.	sat. vap.
300	573.15	8592.7	1.404	20.24	21.65	1333.0	1232.0	2565.0	1345.1	1406.0	2751.0	3.2552	2.4529	5.7081
302	575.15	8837.4	1.412	19.53	20.94	1343.8	1218.3	2562.1	1356.3	1390.9	2747.2	3.2742	2.4182	5.6924
304	577.15	9087.3	1.421	18.84	20.26	1354.8	1204.3	2559.1	1367.7	1375.5	2743.2	3.2933	2.3832	5.6765
306	579.15	9342.7	1.430	18.17	19.60	1365.8	1190.1	2555.9	1379.1	1359.8	2739.0	3.3125	2.3479	5.6604
308	581.15	9603.6	1.439	17.52	18.96	1376.9	1175.6	2552.5	1390.7	1343.9	2734.6	3.3318	2.3124	5.6442
310	583.15	9870.0	1.448	16.89	18.33	1388.1	1161.0	2549.1	1402.4	1327.6	2730.0	3.3512	2.2766	5.6278
312	585.15	10142.1	1.458	16.27	17.73	1399.4	1146.0	2545.4	1414.2	1311.0	2725.2	3.3707	2.2404	5.6111
314	587.15	10420.0	1.468	15.68	17.14	1410.8	1130.8	2541.6	1426.1	1294.1	2720.2	3.3903	2.2040	5.5943
316	589.15	10703.0	1.478	15.09	16.57	1422.3	1115.2	2537.5	1438.1	1276.8	2714.9	3.4101	2.1672	5.5772
318	591.15	10993.4	1.488	14.53	16.02	1433.9	1099.4	2533.3	1450.3	1259.1	2709.4	3.4300	2.1300	5.5599
320	593.15	11289.1	1.500	13.98	15.48	1445.7	1083.2	2528.9	1462.6	1241.1	2703.7	3.4500	2.0923	5.5423
322	595.15	11591.0	1.511	13.44	14.96	1457.5	1066.7	2524.3	1475.1	1222.6	2697.6	3.4702	2.0542	5.5244
324	597.15	11899.2	1.523	12.92	14.45	1469.5	1049.9	2519.4	1487.7	1203.6	2691.3	3.4906	2.0156	5.5062
326	599.15	12213.7	1.535	12.41	13.95	1481.7	1032.6	2514.3	1500.4	1184.2	2684.6	3.5111	1.9764	5.4876
328	601.15	12534.8	1.548	11.91	13.46	1494.0	1014.8	2508.8	1513.4	1164.2	2677.6	3.5319	1.9367	5.4685
330	603.15	12862.5	1.561	11.43	12.99	1506.4	996.7	2503.1	1526.5	1143.6	2670.2	3.5528	1.8962	5.4490
332	605.15	13197.0	1.575	10.95	12.53	1519.1	978.0	2497.0	1539.9	1122.5	2662.3	3.5740	1.8550	5.4290
334	607.15	13538.3	1.590	10.49	12.08	1531.9	958.7	2490.6	1553.4	1100.7	2654.1	3.5955	1.8129	5.4084
336	609.15	13886.7	1.606	10.03	11.63	1544.9	938.9	2483.7	1567.2	1078.1	2645.3	3.6172	1.7700	5.3872
338	611.15	14242.3	1.622	9.58	11.20	1558.1	918.4	2476.4	1581.2	1054.8	2636.0	3.6392	1.7261	5.3653
340	613.15	14605.2	1.639	9.14	10.78	1571.5	897.2	2468.7	1595.5	1030.7	2626.2	3.6616	1.6811	5.3427
342	615.15	14975.5	1.657	8.71	10.37	1585.2	875.2	2460.5	1610.0	1005.7	2615.7	3.6844	1.6350	5.3194
344	617.15	15353.5	1.676	8.286	9.962	1599.2	852.5	2451.7	1624.9	979.7	2604.7	3.7075	1.5877	5.2952
346	619.15	15739.3	1.696	7.870	9.566	1613.5	828.9	2442.4	1640.2	952.8	2593.0	3.7311	1.5391	5.2702
348	621.15	16133.1	1.718	7.461	9.178	1628.1	804.5	2432.6	1655.8	924.8	2580.7	3.7553	1.4891	5.2444

350	623.15	16535.1	1.741	7.058	8.799	1643.0	779.2	2422.2	1671.8	895.9	2567.7	3.7801	1.4375	5.2177
352	625.15	16945.5	1.766	6.654	8.420	1659.4	751.5	2410.8	1689.3	864.2	2553.5	3.8071	1.3822	5.1893
354	627.15	17364.4	1.794	6.252	8.045	1676.3	722.4	2398.7	1707.5	830.9	2538.4	3.8349	1.3247	5.1596
356	629.15	17792.2	1.824	5.850	7.674	1693.4	692.2	2385.6	1725.9	796.2	2522.1	3.8629	1.2654	5.1283
358	631.15	18229.0	1.858	5.448	7.306	1710.8	660.5	2371.4	1744.7	759.9	2504.6	3.8915	1.2037	5.0953
360	633.15	18675.1	1.896	5.044	6.940	1728.8	627.1	2355.8	1764.2	721.3	2485.4	3.9210	1.1390	5.0600
361	634.15	18901.7	1.917	4.840	6.757	1738.0	609.5	2347.5	1774.2	701.0	2475.2	3.9362	1.1052	5.0414
362	635.15	19130.7	1.939	4.634	6.573	1747.5	591.2	2338.7	1784.6	679.8	2464.4	3.9518	1.0702	5.0220
363	636.15	19362.1	1.963	4.425	6.388	1757.3	572.1	2329.3	1795.3	657.8	2453.0	3.9679	1.0338	5.0017
364	637.15	19596.1	1.988	4.213	6.201	1767.4	552.0	2319.4	1806.4	634.6	2440.9	3.9846	0.9958	4.9804
365	638.15	19832.6	2.016	3.996	6.012	1778.0	530.8	2308.8	1818.0	610.0	2428.0	4.0021	0.9558	4.9579
366	639.15	20071.6	2.046	3.772	5.819	1789.1	508.2	2297.3	1830.2	583.9	2414.1	4.0205	0.9134	4.9339
367	640.15	20313.2	2.080	3.540	5.621	1801.0	483.8	2284.8	1843.2	555.7	2399.0	4.0401	0.8680	4.9081
368	641.15	20557.5	2.118	3.298	5.416	1813.8	457.3	2271.1	1857.3	525.1	2382.4	4.0613	0.8189	4.8801
369	642.15	20804.4	2.162	3.039	5.201	1827.8	427.9	2255.7	1872.8	491.1	2363.9	4.0846	0.7647	4.8492
370	643.15	21054.0	2.214	2.759	4.973	1843.6	394.5	2238.1	1890.2	452.6	2342.8	4.1108	0.7036	4.8144
371	644.15	21306.4	2.278	2.446	4.723	1862.0	355.3	2217.3	1910.5	407.4	2317.9	4.1414	0.6324	4.7738
372	645.15	21561.6	2.364	2.075	4.439	1884.6	306.6	2191.2	1935.6	351.4	2287.0	4.1794	0.5446	4.7240
373	646.15	21819.7	2.496	1.588	4.084	1916.0	238.9	2154.9	1970.5	273.5	2244.0	4.2325	0.4233	4.6559
374	647.15	22080.5	2.843	0.623	3.466	1983.9	95.7	2079.7	2046.1	109.5	2156.2	4.3493	0.1692	4.5185
374.15	647.30	22120.0	3.170	0.000	3.170	2037.3	0.0	2037.3	2107.4	0.0	2107.4	4.4429	0.0000	4.4429

TABLE E.2 Properties of Superheated Steam

TEMPERATURE: t°C
(TEMPERATURE: T kelvins)

P/kPa (t^{sat}/°C)		sat. liq.	sat. vap.	75 (348.15)	100 (373.15)	125 (398.15)	150 (423.15)	175 (448.15)	200 (473.15)	225 (498.15)	250 (523.15)
1 (6.98)	V	1.000	129200.	160640.	172180.	183720.	195270.	206810.	218350.	229890.	241430.
	U	29.334	2385.2	2480.8	2516.4	2552.3	2588.5	2624.9	2661.7	2698.8	2736.3
	H	29.335	2514.4	2641.5	2688.6	2736.0	2783.7	2831.7	2880.1	2928.7	2977.7
	S	0.1060	8.9767	9.3828	9.5136	9.6365	9.7527	9.8629	9.9679	10.0681	10.1641
10 (45.83)	V	1.010	14670.	16030.	17190.	18350.	19510.	20660.	21820.	22980.	24130.
	U	191.822	2438.0	2479.7	2515.6	2551.6	2588.0	2624.5	2661.4	2698.6	2736.1
	H	191.832	2584.8	2640.0	2687.5	2735.2	2783.1	2831.2	2879.6	2928.4	2977.4
	S	0.6493	8.1511	8.3168	8.4486	8.5722	8.6888	8.7994	8.9045	9.0049	9.1010
20 (60.09)	V	1.017	7649.8	8000.0	8584.7	9167.1	9748.0	10320.	10900.	11480.	12060.
	U	251.432	2456.9	2478.4	2514.6	2550.9	2587.4	2624.1	2661.0	2698.3	2735.8
	H	251.453	2609.9	2638.4	2686.3	2734.2	2782.3	2830.6	2879.2	2928.0	2977.1
	S	0.8321	7.9094	7.9933	8.1261	8.2504	8.3676	8.4785	8.5839	8.6844	8.7806
30 (69.12)	V	1.022	5229.3	5322.0	5714.4	6104.6	6493.2	6880.8	7267.5	7653.8	8039.7
	U	289.271	2468.6	2477.1	2513.6	2550.2	2586.8	2623.6	2660.7	2698.0	2735.6
	H	289.302	2625.4	2636.8	2685.1	2733.3	2781.6	2830.0	2878.7	2927.6	2976.8
	S	0.9441	7.7695	7.8024	7.9363	8.0614	8.1791	8.2903	8.3960	8.4967	8.5930
40 (75.89)	V	1.027	3993.4	······	4279.2	4573.3	4865.8	5157.2	5447.8	5738.0	6027.7
	U	317.609	2477.1	······	2512.6	2549.4	2586.2	2623.2	2660.3	2697.7	2735.4
	H	317.650	2636.9	······	2683.8	2732.3	2780.9	2829.5	2878.2	2927.2	2976.5
	S	1.0261	7.6709	······	7.8009	7.9268	8.0450	8.1566	8.2624	8.3633	8.4598
50 (81.35)	V	1.030	3240.2	······	3418.1	3654.5	3889.3	4123.0	4356.0	4588.5	4820.5
	U	340.513	2484.0	······	2511.7	2548.6	2585.6	2622.7	2659.9	2697.4	2735.1
	H	340.564	2646.0	······	2682.6	2731.4	2780.1	2828.9	2877.7	2926.8	2976.1
	S	1.0912	7.5947	······	7.6953	7.8219	7.9406	8.0526	8.1587	8.2598	8.3564
75 (91.79)	V	1.037	2216.9	······	2269.8	2429.4	2587.3	2744.2	2900.2	3055.8	3210.9
	U	384.374	2496.7	······	2509.2	2546.7	2584.2	2621.6	2659.0	2696.7	2734.5
	H	384.451	2663.0	······	2679.4	2728.9	2778.2	2827.4	2876.6	2925.8	2975.3
	S	1.2131	7.4570	······	7.5014	7.6300	7.7500	7.8629	7.9697	8.0712	8.1681
100 (99.63)	V	1.043	1693.7	······	1695.5	1816.7	1936.3	2054.7	2172.3	2289.4	2406.1
	U	417.406	2506.1	······	2506.6	2544.7	2582.7	2620.4	2658.1	2695.9	2733.9
	H	417.511	2675.4	······	2676.2	2726.5	2776.3	2825.9	2875.4	2924.9	2974.5
	S	1.3027	7.3598	······	7.3618	7.4923	7.6137	7.7275	7.8349	7.9369	8.0342

P (kPa)	(t_sat)												
101.325	(100.00)	V	1.044	1673.0			1673.0	1792.7	1910.7	2027.7	2143.8	2259.3	2374.5
		U	418.959	2506.5			2506.5	2544.7	2582.6	2620.4	2658.1	2695.9	2733.9
		H	419.064	2676.0			2676.0	2726.4	2776.2	2825.8	2875.3	2924.8	2974.5
		S	1.3069	7.3554			7.3554	7.4860	7.6075	7.7213	7.8288	7.9308	8.0280
125	(105.99)	V	1.049	1374.6				1449.1	1545.6	1641.0	1735.6	1829.6	1923.2
		U	444.224	2513.4				2542.9	2581.2	2619.3	2657.2	2695.2	2733.3
		H	444.356	2685.2				2724.0	2774.4	2824.4	2874.2	2923.9	2973.7
		S	1.3740	7.2847				7.3844	7.5072	7.6219	7.7300	7.8324	7.9300
150	(111.37)	V	1.053	1159.0				1204.0	1285.2	1365.2	1444.4	1523.0	1601.3
		U	466.968	2519.5				2540.9	2579.7	2618.1	2656.3	2694.4	2732.7
		H	467.126	2693.4				2721.5	2772.5	2822.9	2872.9	2922.9	2972.9
		S	1.4336	7.2234				7.2953	7.4194	7.5352	7.6439	7.7468	7.8447
175	(116.06)	V	1.057	1003.34				1028.8	1099.1	1168.2	1236.4	1304.1	1371.3
		U	486.815	2524.7				2538.9	2578.2	2616.9	2655.3	2693.7	2732.1
		H	487.000	2700.3				2719.0	2770.5	2821.3	2871.7	2921.9	2972.0
		S	1.4849	7.1716				7.2191	7.3447	7.4614	7.5708	7.6741	7.7724
200	(120.23)	V	1.061	885.44				897.47	959.54	1020.4	1080.4	1139.8	1198.9
		U	504.489	2529.2				2536.9	2576.6	2615.7	2654.4	2692.9	2731.4
		H	504.701	2706.3				2716.4	2768.5	2819.8	2870.5	2920.9	2971.2
		S	1.5301	7.1268				7.1523	7.2794	7.3971	7.5072	7.6110	7.7096
225	(123.99)	V	1.064	792.97				795.25	850.97	905.44	959.06	1012.1	1064.7
		U	520.465	2533.2				2534.8	2575.1	2614.5	2653.5	2692.2	2730.8
		H	520.705	2711.6				2713.8	2766.5	2818.2	2869.3	2919.9	2970.4
		S	1.5705	7.0873				7.0928	7.2213	7.3400	7.4508	7.5551	7.6540
250	(127.43)	V	1.068	718.44					764.09	813.47	861.98	909.91	957.41
		U	535.077	2536.8					2573.5	2613.3	2652.5	2691.4	2730.2
		H	535.343	2716.4					2764.5	2816.7	2868.0	2918.9	2969.6
		S	1.6071	7.0520					7.1689	7.2886	7.4001	7.5050	7.6042
275	(130.60)	V	1.071	657.04					693.00	738.21	782.55	826.29	869.61
		U	548.564	2540.0					2571.9	2612.1	2651.6	2690.7	2729.6
		H	548.858	2720.7					2762.5	2815.1	2866.8	2917.9	2968.7
		S	1.6407	7.0201					7.1211	7.2419	7.3541	7.4594	7.5590
300	(133.54)	V	1.073	605.56					633.74	675.49	716.35	756.60	796.44
		U	561.107	2543.0					2570.3	2610.8	2650.6	2689.9	2729.0
		H	561.429	2724.7					2760.4	2813.5	2865.5	2916.9	2967.9
		S	1.6716	6.9909					7.0771	7.1990	7.3119	7.4177	7.5176

TABLE E.2 Properties of Superheated Steam (Continued)

TEMPERATURE: t °C
(TEMPERATURE: T kelvins)

P/kPa (t^{sat}/°C)		sat. liq.	sat. vap.	300 (573.15)	350 (623.15)	400 (673.15)	450 (723.15)	500 (773.15)	550 (823.15)	600 (873.15)	650 (923.15)
1 (6.98)	V	1.000	129200.	264500.	287580.	310660.	333730.	356810.	379880.	402960.	426040.
	U	29.334	2385.2	2812.3	2889.9	2969.1	3049.9	3132.4	3216.7	3302.6	3390.3
	H	29.335	2514.4	3076.8	3177.5	3279.7	3383.6	3489.2	3596.5	3705.6	3816.4
	S	0.1060	8.9767	10.3450	10.5133	10.6711	10.8200	10.9612	11.0957	11.2243	11.3476
10 (45.83)	V	1.010	14670.	26440.	28750.	31060.	33370.	35670.	37980.	40290.	42600.
	U	191.822	2438.0	2812.2	2889.8	2969.0	3049.8	3132.3	3216.6	3302.6	3390.3
	H	191.832	2584.8	3076.6	3177.3	3279.6	3383.5	3489.1	3596.5	3705.5	3816.3
	S	0.6493	8.1511	9.2820	9.4504	9.6083	9.7572	9.8984	10.0329	10.1616	10.2849
20 (60.09)	V	1.017	7649.8	13210.	14370.	15520.	16680.	17830.	18990.	20140.	21300.
	U	251.432	2456.9	2812.0	2889.6	2968.9	3049.7	3132.3	3216.5	3302.5	3390.2
	H	251.453	2609.9	3076.4	3177.1	3279.4	3383.4	3489.0	3596.4	3705.4	3816.2
	S	0.8321	7.9094	8.9618	9.1303	9.2882	9.4372	9.5784	9.7130	9.8416	9.9650
30 (69.12)	V	1.022	5229.3	8810.8	9581.2	10350.	11120.	11890.	12660.	13430.	14190.
	U	289.271	2468.6	2811.8	2889.5	2968.7	3049.6	3132.2	3216.5	3302.5	3390.2
	H	289.302	2625.4	3076.1	3176.9	3279.3	3383.3	3488.9	3596.3	3705.4	3816.2
	S	0.9441	7.7695	8.7744	8.9430	9.1010	9.2499	9.3912	9.5257	9.6544	9.7778
40 (75.89)	V	1.027	3993.4	6606.5	7184.6	7762.5	8340.1	8917.6	9494.9	10070.	10640.
	U	317.609	2477.1	2811.6	2889.4	2968.6	3049.5	3132.1	3216.4	3302.4	3390.1
	H	317.650	2636.9	3075.9	3176.6	3279.1	3383.1	3488.8	3596.2	3705.3	3816.1
	S	1.0261	7.6709	8.6413	8.8100	8.9680	9.1170	9.2583	9.3929	9.5216	9.6450
50 (81.35)	V	1.030	3240.2	5283.9	5746.7	6209.1	6671.4	7133.5	7595.5	8057.4	8519.2
	U	340.513	2484.0	2811.5	2889.2	2968.5	3049.4	3132.0	3216.3	3302.3	3390.1
	H	340.564	2646.0	3075.7	3176.6	3279.0	3383.0	3488.7	3596.1	3705.2	3816.0
	S	1.0912	7.5947	8.5380	8.7068	8.8649	9.0139	9.1552	9.2898	9.4185	9.5419
75 (91.79)	V	1.037	2216.9	3520.5	3829.4	4138.0	4446.4	4754.7	5062.8	5370.9	5678.9
	U	384.374	2496.7	2811.0	2888.9	2968.2	3049.2	3131.8	3216.1	3302.2	3389.9
	H	384.451	2663.0	3075.1	3176.1	3278.6	3382.7	3488.4	3595.8	3705.0	3815.9
	S	1.2131	7.4570	8.3502	8.5191	8.6773	8.8265	8.9678	9.1025	9.2312	9.3546
100 (99.63)	V	1.043	1693.7	2638.7	2870.8	3102.5	3334.0	3565.3	3796.5	4027.7	4258.8
	U	417.406	2506.1	2810.6	2888.6	2968.0	3049.0	3131.6	3216.0	3302.0	3389.8
	H	417.511	2675.4	3074.5	3175.6	3278.2	3382.4	3488.1	3595.6	3704.8	3815.7
	S	1.3027	7.3598	8.2166	8.3858	8.5442	8.6934	8.8348	8.9695	9.0982	9.2217

Superheated steam — properties (V, U, H, S) at pressure P (kPa) with saturation temperature in parentheses (°C). The first data column is the saturated value.

P (t^{sat})		Sat								
101.325 (100.00)	V	1.044	1673.0	2604.2	2833.2	3061.9	3290.3	3518.7	3746.9	3975.0	4203.1
	U	418.959	2506.5	2810.6	2888.5	2968.0	3048.9	3131.6	3215.9	3302.0	3389.8
	H	419.064	2676.0	3074.4	3175.6	3278.2	3382.3	3488.1	3595.6	3704.8	3815.7
	S	1.3069	7.3554	8.2105	8.3797	8.5381	8.6873	8.8287	8.9634	9.0922	9.2156
125 (105.99)	V	1.049	1374.6	2109.7	2295.6	2481.2	2666.5	2851.7	3036.8	3221.8	3406.7
	U	444.224	2513.4	2810.2	2888.2	2967.7	3048.7	3131.4	3215.8	3301.9	3389.7
	H	444.356	2685.2	3073.9	3175.2	3277.8	3382.0	3487.9	3595.4	3704.6	3815.5
	S	1.3740	7.2847	8.1129	8.2823	8.4408	8.5901	8.7316	8.8663	8.9951	9.1186
150 (111.37)	V	1.053	1159.0	1757.0	1912.2	2066.9	2221.5	2375.9	2530.2	2684.5	2838.6
	U	466.968	2519.5	2809.7	2887.9	2967.4	3048.5	3131.2	3215.6	3301.7	3389.5
	H	467.126	2693.4	3073.3	3174.7	3277.5	3381.7	3487.6	3595.1	3704.4	3815.3
	S	1.4336	7.2234	8.0280	8.1976	8.3562	8.5056	8.6472	8.7819	8.9108	9.0343
175 (116.06)	V	1.057	1003.34	1505.1	1638.3	1771.1	1903.7	2036.1	2168.4	2300.7	2432.9
	U	486.815	2524.7	2809.3	2887.5	2967.1	3048.3	3131.0	3215.4	3301.6	3389.4
	H	487.000	2700.3	3072.7	3174.2	3277.1	3381.4	3487.3	3594.9	3704.2	3815.1
	S	1.4849	7.1716	7.9561	8.1259	8.2847	8.4341	8.5758	8.7106	8.8394	8.9630
200 (120.23)	V	1.061	885.44	1316.2	1432.8	1549.2	1665.3	1781.2	1897.1	2012.9	2128.6
	U	504.489	2529.2	2808.8	2887.2	2966.9	3048.0	3130.8	3215.3	3301.4	3389.2
	H	504.701	2706.3	3072.1	3173.8	3276.7	3381.1	3487.0	3594.7	3704.0	3815.0
	S	1.5301	7.1268	7.8937	8.0638	8.2226	8.3722	8.5139	8.6487	8.7776	8.9012
225 (123.99)	V	1.064	792.97	1169.2	1273.1	1376.6	1479.9	1583.0	1686.0	1789.0	1891.9
	U	520.465	2533.2	2808.4	2886.9	2966.6	3047.8	3130.6	3215.1	3301.2	3389.1
	H	520.705	2711.6	3071.5	3173.3	3276.3	3380.8	3486.8	3594.4	3703.8	3814.8
	S	1.5705	7.0873	7.8385	8.0088	8.1679	8.3175	8.4593	8.5942	8.7231	8.8467
250 (127.43)	V	1.068	718.44	1051.6	1145.2	1238.5	1331.5	1424.4	1517.2	1609.9	1702.5
	U	535.077	2536.8	2808.0	2886.5	2966.3	3047.6	3130.4	3214.9	3301.1	3389.0
	H	535.343	2716.4	3070.9	3172.8	3275.9	3380.4	3486.5	3594.2	3703.6	3814.6
	S	1.6071	7.0520	7.7891	7.9597	8.1188	8.2686	8.4104	8.5453	8.6743	8.7980
275 (130.60)	V	1.071	657.04	955.45	1040.7	1125.5	1210.2	1294.7	1379.0	1463.3	1547.6
	U	548.564	2540.0	2807.5	2886.2	2966.0	3047.3	3130.2	3214.7	3300.9	3388.8
	H	548.858	2720.7	3070.3	3172.4	3275.5	3380.1	3486.2	3594.0	3703.4	3814.4
	S	1.6407	7.0201	7.7444	7.9151	8.0744	8.2243	8.3661	8.5011	8.6301	8.7538
300 (133.54)	V	1.073	605.56	875.29	953.52	1031.4	1109.0	1186.5	1263.9	1341.2	1418.5
	U	561.107	2543.0	2807.1	2885.8	2965.8	3047.1	3130.0	3214.5	3300.8	3388.7
	H	561.429	2724.7	3069.7	3171.9	3275.2	3379.8	3486.0	3593.7	3703.2	3814.2
	S	1.6716	6.9909	7.7034	7.8744	8.0338	8.1838	8.3257	8.4608	8.5898	8.7135

TABLE E.2 Properties of Superheated Steam (Continued)

TEMPERATURE: t°C
(TEMPERATURE: T kelvins)

P/kPa (t^sat/°C)		sat. liq.	sat. vap.	150 (423.15)	175 (448.15)	200 (473.15)	220 (493.15)	240 (513.15)	260 (533.15)	280 (553.15)	300 (573.15)
325 (136.29)	V	1.076	561.75	583.58	622.41	660.33	690.22	719.81	749.18	778.39	807.47
	U	572.847	2545.7	2568.7	2609.6	2649.6	2681.2	2712.7	2744.0	2775.3	2806.6
	H	573.197	2728.3	2758.4	2811.9	2864.2	2905.6	2946.6	2987.5	3028.2	3069.0
	S	1.7004	6.9640	7.0363	7.1592	7.2729	7.3585	7.4400	7.5181	7.5933	7.6657
350 (138.87)	V	1.079	524.00	540.58	576.90	612.31	640.18	667.75	695.09	722.27	749.33
	U	583.892	2548.2	2567.1	2608.3	2648.6	2680.4	2712.0	2743.4	2774.8	2806.2
	H	584.270	2731.6	2756.3	2810.3	2863.0	2904.5	2945.7	2986.7	3027.6	3068.4
	S	1.7273	6.9392	6.9982	7.1222	7.2366	7.3226	7.4045	7.4828	7.5581	7.6307
375 (141.31)	V	1.081	491.13	503.29	537.46	570.69	596.81	622.62	648.22	673.64	698.94
	U	594.332	2550.6	2565.4	2607.1	2647.7	2679.6	2711.3	2742.8	2774.3	2805.7
	H	594.737	2734.7	2754.1	2808.6	2861.7	2903.4	2944.8	2985.9	3026.9	3067.8
	S	1.7526	6.9160	6.9624	7.0875	7.2027	7.2891	7.3713	7.4499	7.5254	7.5981
400 (143.62)	V	1.084	462.22	470.66	502.93	534.26	558.85	583.14	607.20	631.09	654.85
	U	604.237	2552.7	2563.7	2605.8	2646.7	2678.8	2710.6	2742.2	2773.7	2805.3
	H	604.670	2737.6	2752.0	2807.0	2860.4	2902.3	2943.9	2985.1	3026.2	3067.2
	S	1.7764	6.8943	6.9285	7.0548	7.1708	7.2576	7.3402	7.4190	7.4947	7.5675
425 (145.82)	V	1.086	436.61	441.85	472.47	502.12	525.36	548.30	571.01	593.54	615.95
	U	613.667	2554.8	2562.0	2604.5	2645.7	2678.0	2709.9	2741.6	2773.2	2804.8
	H	614.128	2740.3	2749.8	2805.3	2859.1	2901.2	2942.9	2984.3	3025.5	3066.6
	S	1.7990	6.8739	6.8965	7.0239	7.1407	7.2280	7.3108	7.3899	7.4657	7.5388
450 (147.92)	V	1.088	413.75	416.24	445.38	473.55	495.59	517.33	538.83	560.17	581.37
	U	622.672	2556.7	2560.3	2603.2	2644.7	2677.1	2709.2	2741.0	2772.7	2804.4
	H	623.162	2742.9	2747.7	2803.7	2857.8	2900.2	2942.0	2983.5	3024.8	3066.0
	S	1.8204	6.8547	6.8660	6.9946	7.1121	7.1999	7.2831	7.3624	7.4384	7.5116
475 (149.92)	V	1.091	393.22	393.31	421.14	447.97	468.95	489.62	510.05	530.30	550.43
	U	631.294	2558.5	2558.6	2601.9	2643.7	2676.3	2708.5	2740.4	2772.2	2803.9
	H	631.812	2745.3	2745.5	2802.0	2856.5	2899.1	2941.1	2982.7	3024.1	3065.4
	S	1.8408	6.8365	6.8369	6.9667	7.0850	7.1732	7.2567	7.3363	7.4125	7.4858
500 (151.84)	V	1.093	374.68	......	399.31	424.96	444.97	464.67	484.14	503.43	522.58
	U	639.569	2560.2	......	2600.6	2642.7	2675.5	2707.8	2739.8	2771.7	2803.5
	H	640.116	2747.5	......	2800.3	2855.1	2898.0	2940.1	2981.9	3023.4	3064.8
	S	1.8604	6.8192	......	6.9400	7.0592	7.1478	7.2317	7.3115	7.3879	7.4614

P (kPa) (T sat)		Sat. liq.	Sat. vap.							
525 (153.69)	V	1.095	357.84	……	379.56	404.13	423.28	442.11	460.70	479.11	497.38
	U	647.528	2561.8	……	2599.3	2641.6	2674.6	2707.1	2739.2	2771.2	2803.0
	H	648.103	2749.7	……	2798.6	2853.8	2896.8	2939.2	2981.1	3022.7	3064.1
	S	1.8790	6.8027	……	6.9145	7.0345	7.1236	7.2078	7.2879	7.3645	7.4381
550 (155.47)	V	1.097	342.48	……	361.60	385.19	403.55	421.59	439.38	457.00	474.48
	U	655.199	2563.3	……	2598.0	2640.6	2673.8	2706.4	2738.6	2770.6	2802.6
	H	655.802	2751.7	……	2796.8	2852.5	2895.7	2938.3	2980.3	3022.0	3063.5
	S	1.8970	6.7870	……	6.8900	7.0108	7.1004	7.1849	7.2653	7.3421	7.4158
575 (157.18)	V	1.099	328.41	……	345.20	367.90	385.54	402.85	419.92	436.81	453.56
	U	662.603	2564.8	……	2596.6	2639.6	2672.9	2705.7	2738.0	2770.1	2802.1
	H	663.235	2753.6	……	2795.1	2851.1	2894.6	2937.3	2979.5	3021.3	3062.9
	S	1.9142	6.7720	……	6.8664	6.9880	7.0781	7.1630	7.2436	7.3206	7.3945
600 (158.84)	V	1.101	315.47	……	330.16	352.04	369.03	385.68	402.08	418.31	434.39
	U	669.762	2566.2	……	2595.3	2638.5	2672.1	2705.0	2737.4	2769.6	2801.6
	H	670.423	2755.5	……	2793.3	2849.7	2893.5	2936.4	2978.7	3020.6	3062.3
	S	1.9308	6.7575	……	6.8437	6.9662	7.0567	7.1419	7.2228	7.3000	7.3740
625 (160.44)	V	1.103	303.54	……	316.31	337.45	353.83	369.87	385.67	401.28	416.75
	U	676.695	2567.5	……	2593.9	2637.5	2671.2	2704.2	2736.8	2769.1	2801.2
	H	677.384	2757.2	……	2791.6	2848.4	2892.3	2935.4	2977.8	3019.9	3061.7
	S	1.9469	6.7437	……	6.8217	6.9451	7.0361	7.1217	7.2028	7.2802	7.3544
650 (161.99)	V	1.105	292.49	……	303.53	323.98	339.80	355.29	370.52	385.56	400.47
	U	683.417	2568.7	……	2592.5	2636.4	2670.3	2703.5	2736.2	2768.5	2800.7
	H	684.135	2758.9	……	2789.8	2847.0	2891.2	2934.4	2977.0	3019.2	3061.0
	S	1.9623	6.7304	……	6.8004	6.9247	7.0162	7.1021	7.1835	7.2611	7.3355
675 (163.49)	V	1.106	282.23	……	291.69	311.51	326.81	341.78	356.49	371.01	385.39
	U	689.943	2570.0	……	2591.1	2635.4	2669.5	2702.8	2735.6	2768.0	2800.3
	H	690.689	2760.5	……	2788.0	2845.6	2890.1	2933.5	2976.2	3018.5	3060.4
	S	1.9773	6.7176	……	6.7798	6.9050	6.9970	7.0833	7.1650	7.2428	7.3173
700 (164.96)	V	1.108	272.68	……	280.69	299.92	314.75	329.23	343.46	357.50	371.39
	U	696.285	2571.1	……	2589.7	2634.3	2668.6	2702.1	2735.0	2767.5	2799.8
	H	697.061	2762.0	……	2786.2	2844.2	2888.9	2932.5	2975.4	3017.7	3059.8
	S	1.9918	6.7052	……	6.7598	6.8859	6.9784	7.0651	7.1470	7.2250	7.2997
725 (166.38)	V	1.110	263.77	……	270.45	289.13	303.51	317.55	331.33	344.92	358.36
	U	702.457	2572.2	……	2588.3	2633.2	2667.7	2701.3	2734.3	2767.0	2799.3
	H	703.261	2763.4	……	2784.4	2842.8	2887.7	2931.5	2974.6	3017.0	3059.1
	S	2.0059	6.6932	……	6.7404	6.8673	6.9604	7.0474	7.1296	7.2078	7.2827

TABLE E.2 Properties of Superheated Steam (Continued)

TEMPERATURE: t°C
(TEMPERATURE: T kelvins)

P/kPa (t^sat/°C)		sat. liq.	sat. vap.	325 (598.15)	350 (623.15)	400 (673.15)	450 (723.15)	500 (773.15)	550 (823.15)	600 (873.15)	650 (923.15)
325 (136.29)	V	1.076	561.75	843.68	879.78	951.73	1023.5	1095.0	1166.5	1237.9	1309.2
	U	572.847	2545.7	2845.9	2885.5	2965.5	3046.9	3129.8	3214.4	3300.6	3388.6
	H	573.197	2728.3	3120.1	3171.4	3274.8	3379.5	3485.7	3593.5	3702.9	3814.1
	S	1.7004	6.9640	7.7530	7.8369	7.9965	8.1465	8.2885	8.4236	8.5527	8.6764
350 (138.87)	V	1.079	524.00	783.01	816.57	883.45	950.11	1016.6	1083.0	1149.3	1215.6
	U	583.892	2548.2	2845.6	2885.1	2965.2	3046.6	3129.6	3214.2	3300.5	3388.4
	H	584.270	2731.6	3119.6	3170.9	3274.4	3379.2	3485.4	3593.3	3702.7	3813.9
	S	1.7273	6.9392	7.7181	7.8022	7.9619	8.1120	8.2540	8.3892	8.5183	8.6421
375 (141.31)	V	1.081	491.13	730.42	761.79	824.28	886.54	948.66	1010.7	1072.6	1134.5
	U	594.332	2550.6	2845.2	2884.8	2964.9	3046.4	3129.4	3214.0	3300.3	3388.3
	H	594.737	2734.7	3119.1	3170.5	3274.0	3378.8	3485.1	3593.0	3702.5	3813.7
	S	1.7526	6.9160	7.6856	7.7698	7.9296	8.0798	8.2219	8.3571	8.4863	8.6101
400 (143.62)	V	1.084	462.22	684.41	713.85	772.50	830.92	889.19	947.35	1005.4	1063.4
	U	604.237	2552.7	2844.8	2884.5	2964.6	3046.2	3129.2	3213.8	3300.2	3388.2
	H	604.670	2737.6	3118.5	3170.0	3273.6	3378.5	3484.9	3592.8	3702.3	3813.5
	S	1.7764	6.8943	7.6552	7.7395	7.8994	8.0497	8.1919	8.3271	8.4563	8.5802
425 (145.82)	V	1.086	436.61	643.81	671.56	726.81	781.84	836.72	891.49	946.17	1000.8
	U	613.667	2554.8	2844.4	2884.1	2964.4	3045.9	3129.0	3213.7	3300.0	3388.0
	H	614.128	2740.3	3118.0	3169.5	3273.3	3378.2	3484.6	3592.5	3702.1	3813.4
	S	1.7990	6.8739	7.6265	7.7109	7.8710	8.0214	8.1636	8.2989	8.4282	8.5520
450 (147.92)	V	1.088	413.75	607.73	633.97	686.20	738.21	790.07	841.83	893.50	945.10
	U	622.672	2556.7	2844.0	2883.8	2964.1	3045.7	3128.8	3213.5	3299.8	3387.9
	H	623.162	2742.9	3117.5	3169.1	3272.9	3377.9	3484.3	3592.3	3701.9	3813.2
	S	1.8204	6.8547	7.5995	7.6840	7.8442	7.9947	8.1370	8.2723	8.4016	8.5255
475 (149.92)	V	1.091	393.22	575.44	600.33	649.87	699.18	748.34	797.40	846.37	895.27
	U	631.294	2558.5	2843.6	2883.4	2963.8	3045.4	3128.6	3213.3	3299.7	3387.7
	H	631.812	2745.3	3116.9	3168.6	3272.5	3377.6	3484.0	3592.1	3701.7	3813.0
	S	1.8408	6.8365	7.5739	7.6585	7.8189	7.9694	8.1118	8.2472	8.3765	8.5004
500 (151.84)	V	1.093	374.68	546.38	570.05	617.16	664.05	710.78	757.41	803.95	850.42
	U	639.569	2560.2	2843.2	2883.1	2963.5	3045.2	3128.4	3213.1	3299.5	3387.6
	H	640.116	2747.5	3116.4	3168.1	3272.1	3377.2	3483.8	3591.8	3701.5	3812.8
	S	1.8604	6.8192	7.5496	7.6343	7.7948	7.9454	8.0879	8.2233	8.3526	8.4766

P (kPa)												
525 (153.69)	V	1.095	357.84	520.08	542.66	587.58	632.26	676.80	721.23	765.57	809.85
	U	647.528	2561.8	2842.8	2882.7	2963.2	3045.0	3128.2	3213.0	3299.4	3387.5
	H	648.103	2749.7	3115.9	3167.6	3271.7	3376.9	3483.5	3591.6	3701.3	3812.6
	S	1.8790	6.8027	7.5264	7.6112	7.7719	7.9226	8.0651	8.2006	8.3299	8.4539
550 (155.47)	V	1.097	342.48	496.18	517.76	560.68	603.37	645.91	688.34	730.68	772.96
	U	655.199	2563.3	2842.4	2882.4	2963.0	3044.7	3128.0	3212.8	3299.2	3387.3
	H	655.802	2751.7	3115.3	3167.2	3271.3	3376.6	3483.2	3591.4	3701.1	3812.5
	S	1.8970	6.7870	7.5043	7.5892	7.7500	7.9008	8.0433	8.1789	8.3083	8.4323
575 (157.18)	V	1.099	328.41	474.36	495.03	536.12	576.98	617.70	658.30	698.83	739.28
	U	662.603	2564.8	2842.0	2882.1	2962.7	3044.5	3127.8	3212.6	3299.1	3387.2
	H	663.235	2753.6	3114.8	3166.7	3271.0	3376.3	3482.9	3591.1	3700.9	3812.3
	S	1.9142	6.7720	7.4831	7.5681	7.7290	7.8799	8.0226	8.1581	8.2876	8.4116
600 (158.84)	V	1.101	315.47	454.35	474.19	513.61	552.80	591.84	630.78	669.63	708.41
	U	669.762	2566.2	2841.6	2881.7	2962.4	3044.3	3127.6	3212.4	3298.9	3387.1
	H	670.423	2755.5	3114.3	3166.2	3270.6	3376.0	3482.7	3590.9	3700.7	3812.1
	S	1.9308	6.7575	7.4628	7.5479	7.7090	7.8600	8.0027	8.1383	8.2678	8.3919
625 (160.44)	V	1.103	303.54	435.94	455.01	492.89	530.55	568.05	605.45	642.76	680.01
	U	676.695	2567.5	2841.2	2881.4	2962.1	3044.0	3127.4	3212.2	3298.8	3386.9
	H	677.384	2757.2	3113.7	3165.7	3270.2	3375.6	3482.4	3590.7	3700.5	3811.9
	S	1.9469	6.7437	7.4433	7.5285	7.6897	7.8408	7.9836	8.1192	8.2488	8.3729
650 (161.99)	V	1.105	292.49	418.95	437.31	473.78	510.01	546.10	582.07	617.96	653.79
	U	683.417	2568.7	2840.9	2881.0	2961.8	3043.8	3127.2	3212.1	3298.6	3386.8
	H	684.135	2758.9	3113.2	3165.3	3269.8	3375.3	3482.1	3590.4	3700.3	3811.8
	S	1.9623	6.7304	7.4245	7.5099	7.6712	7.8224	7.9652	8.1009	8.2305	8.3546
675 (163.49)	V	1.106	282.23	403.22	420.92	456.07	491.00	525.77	560.43	595.00	629.51
	U	689.943	2570.0	2840.5	2880.7	2961.6	3043.6	3127.0	3211.9	3298.5	3386.7
	H	690.689	2760.5	3112.6	3164.8	3269.4	3375.0	3481.8	3590.2	3700.1	3811.6
	S	1.9773	6.7176	7.4064	7.4919	7.6534	7.8046	7.9475	8.0833	8.2129	8.3371
700 (164.96)	V	1.108	272.68	388.61	405.71	439.64	473.34	506.89	540.33	573.68	606.97
	U	696.285	2571.1	2840.1	2880.3	2961.3	3043.3	3126.8	3211.7	3298.3	3386.5
	H	697.061	2762.0	3112.1	3164.3	3269.0	3374.7	3481.6	3589.9	3699.9	3811.4
	S	1.9918	6.7052	7.3890	7.4745	7.6362	7.7875	7.9305	8.0663	8.1959	8.3201
725 (166.38)	V	1.110	263.77	375.01	391.54	424.33	456.90	489.31	521.61	553.83	585.99
	U	702.457	2572.2	2839.7	2880.0	2961.0	3043.1	3126.6	3211.5	3298.1	3386.4
	H	703.261	2763.4	3111.5	3163.8	3268.7	3374.1	3481.3	3589.7	3699.7	3811.2
	S	2.0059	6.6932	7.3721	7.4578	7.6196	7.7710	7.9140	8.0499	8.1796	8.3038

TABLE E.2 Properties of Superheated Steam (Continued)

TEMPERATURE: t °C
(TEMPERATURE: T kelvins)

P/kPa (t^{sat}/°C)		sat. liq.	sat. vap.	175 (448.15)	200 (473.15)	220 (493.15)	240 (513.15)	260 (533.15)	280 (553.15)	300 (573.15)	325 (598.15)
750 (167.76)	V	1.112	255.43	260.88	279.05	293.03	306.65	320.01	333.17	346.19	362.32
	U	708.467	2573.3	2586.9	2632.1	2666.8	2700.6	2733.7	2766.4	2798.9	2839.3
	H	709.301	2764.8	2782.5	2841.4	2886.6	2930.6	2973.7	3016.3	3058.5	3111.0
	S	2.0195	6.6817	6.7215	6.8494	6.9429	7.0303	7.1128	7.1912	7.2662	7.3558
775 (169.10)	V	1.113	247.61	251.93	269.63	283.22	296.45	309.41	322.19	334.81	350.44
	U	714.326	2574.3	2585.4	2631.0	2665.9	2699.8	2733.1	2765.9	2798.4	2838.9
	H	715.189	2766.2	2780.7	2840.0	2885.4	2929.8	2972.9	3015.6	3057.9	3110.5
	S	2.0328	6.6705	6.7031	6.8319	6.9259	7.0137	7.0965	7.1751	7.2502	7.3400
800 (170.41)	V	1.115	240.26	243.53	260.79	274.02	286.88	299.48	311.89	324.14	339.31
	U	720.043	2575.3	2584.0	2629.9	2665.0	2699.1	2732.5	2765.4	2797.9	2838.5
	H	720.935	2767.5	2778.8	2838.6	2884.2	2928.6	2972.1	3014.9	3057.3	3109.9
	S	2.0457	6.6596	6.6851	6.8148	6.9094	6.9976	7.0807	7.1595	7.2348	7.3247
825 (171.69)	V	1.117	233.34	235.64	252.48	265.37	277.90	290.15	302.21	314.12	328.85
	U	725.625	2576.2	2582.5	2628.8	2664.1	2698.4	2731.8	2764.8	2797.5	2838.1
	H	726.547	2768.7	2776.9	2837.1	2883.1	2927.6	2971.2	3014.1	3056.6	3109.4
	S	2.0583	6.6491	6.6675	6.7982	6.8933	6.9819	7.0653	7.1443	7.2197	7.3098
850 (172.94)	V	1.118	226.81	228.21	244.66	257.24	269.44	281.37	293.10	304.68	319.00
	U	731.080	2577.1	2581.1	2627.7	2663.2	2697.6	2731.2	2764.3	2797.0	2837.7
	H	732.031	2769.9	2775.1	2835.7	2881.9	2926.6	2970.4	3013.4	3056.0	3108.8
	S	2.0705	6.6388	6.6504	6.7820	6.8777	6.9666	7.0503	7.1295	7.2051	7.2954
875 (174.16)	V	1.120	220.65	221.20	237.29	249.56	261.46	273.09	284.51	295.79	309.72
	U	736.415	2578.0	2579.6	2626.6	2662.3	2696.8	2730.6	2763.7	2796.5	2837.3
	H	737.394	2771.0	2773.1	2834.2	2880.7	2925.6	2969.5	3012.7	3055.3	3108.3
	S	2.0825	6.6289	6.6336	6.7662	6.8624	6.9518	7.0357	7.1152	7.1909	7.2813
900 (175.36)	V	1.121	214.81	· · · · · ·	230.32	242.31	253.93	265.27	276.40	287.39	300.96
	U	741.635	2578.8	· · · · · ·	2625.5	2661.4	2696.1	2729.9	2763.2	2796.1	2836.9
	H	742.644	2772.1	· · · · · ·	2832.7	2879.5	2924.6	2968.7	3012.0	3054.7	3107.7
	S	2.0941	6.6192	· · · · · ·	6.7508	6.8475	6.9373	7.0215	7.1012	7.1771	7.2676
925 (176.53)	V	1.123	209.28	· · · · · ·	223.73	235.46	246.80	257.87	268.73	279.44	292.66
	U	746.746	2579.6	· · · · · ·	2624.3	2660.5	2695.3	2729.3	2762.6	2795.6	2836.5
	H	747.784	2773.2	· · · · · ·	2831.3	2878.3	2923.6	2967.8	3011.2	3054.1	3107.2
	S	2.1055	6.6097	· · · · · ·	6.7357	6.8329	6.9231	7.0076	7.0875	7.1636	7.2543

P (kPa) (T_sat)		Sat. Liq.	Sat. Vap.							
950 (177.67)	V	1.124	204.03		217.48	228.96	240.05	250.86	261.46	271.91	284.81
	U	751.754	2580.4		2623.2	2659.5	2694.6	2728.7	2762.1	2795.1	2836.0
	H	752.822	2774.8		2829.8	2877.0	2922.6	2967.0	3010.5	3053.4	3106.6
	S	2.1166	6.6005		6.7209	6.8187	6.9093	6.9941	7.0742	7.1505	7.2413
975 (178.79)	V	1.126	199.04		211.55	222.79	233.64	244.20	254.56	264.76	277.35
	U	756.663	2581.1		2622.0	2658.6	2693.8	2728.0	2761.5	2794.6	2835.6
	H	757.761	2775.2		2828.3	2875.8	2921.6	2966.1	3009.7	3052.8	3106.1
	S	2.1275	6.5916		6.7064	6.8048	6.8958	6.9809	7.0612	7.1377	7.2286
1000 (179.88)	V	1.127	194.29		205.92	216.93	227.55	237.89	248.01	257.98	270.27
	U	761.478	25819		2620.9	2657.7	2693.0	2727.4	2761.0	2794.2	2835.2
	H	762.605	2776.2		2826.8	2874.6	2920.6	2965.2	3009.0	3052.1	3105.5
	S	2.1382	6.5828		6.6922	6.7911	6.8825	6.9680	7.0485	7.1251	7.2163
1050 (182.02)	V	1.130	185.45		195.45	206.04	216.24	226.15	235.84	245.37	257.12
	U	770.843	2583.3		2618.5	2655.8	2691.5	2726.1	2759.9	2793.2	2834.4
	H	772.029	2778.0		2823.8	2872.1	2918.5	2963.5	3007.5	3050.8	3104.4
	S	2.1588	6.5659		6.6645	6.7647	6.8569	6.9430	7.0240	7.1009	7.1924
1100 (184.07)	V	1.133	177.38		185.92	196.14	205.96	215.47	224.77	233.91	245.16
	U	779.878	2584.5		2616.2	2653.9	2689.9	2724.7	2758.8	2792.2	2833.6
	H	781.124	2779.7		2820.7	2869.6	2916.4	2961.8	3006.0	3049.6	3103.3
	S	2.1786	6.5497		6.6379	6.7392	6.8323	6.9190	7.0005	7.0778	7.1695
1150 (186.05)	V	1.136	169.99		177.22	187.10	196.56	205.73	214.67	223.44	234.25
	U	788.611	2585.8		2613.8	2651.9	2688.3	2723.4	2757.7	2791.3	2832.8
	H	789.917	2781.3		2817.6	2867.1	2914.2	2960.0	3004.5	3048.2	3102.2
	S	2.1977	6.5342		6.6122	6.7147	6.8086	6.8959	6.9779	7.0556	7.1476
1200 (187.96)	V	1.139	163.20		169.23	178.80	187.95	196.79	205.40	213.85	224.24
	U	797.064	2586.9		2611.3	2650.0	2686.7	2722.1	2756.5	2790.3	2832.0
	H	798.430	2782.7		2814.4	2864.5	2912.2	2958.2	3003.0	3046.9	3101.0
	S	2.2161	6.5194		6.5872	6.6909	6.7858	6.8738	6.9562	7.0342	7.1266
1250 (189.81)	V	1.141	156.93		161.88	171.17	180.02	188.56	196.88	205.02	215.03
	U	805.259	2588.0		2608.9	2648.0	2685.1	2720.8	2755.4	2789.3	2831.1
	H	806.685	2784.1		2811.2	2861.9	2910.1	2956.5	3001.5	3045.6	3099.9
	S	2.2338	6.5050		6.5630	6.6680	6.7637	6.8523	6.9353	7.0136	7.1064
1300 (191.61)	V	1.144	151.13		155.09	164.11	172.70	180.97	189.01	196.87	206.53
	U	813.213	2589.0		2606.4	2646.0	2683.5	2719.4	2754.3	2788.4	2830.3
	H	814.700	2785.4		2808.0	2859.3	2908.0	2954.7	3000.0	3044.3	3098.8
	S	2.2510	6.4913		6.5394	6.6457	6.7424	6.8316	6.9151	6.9938	7.0869

TABLE E.2 Properties of Superheated Steam (Continued)

TEMPERATURE: $t°C$
(TEMPERATURE: T kelvins)

P/kPa (t^{sat}/°C)		sat. liq.	sat. vap.	350 (623.15)	375 (648.15)	400 (673.15)	450 (723.15)	500 (773.15)	550 (833.15)	600 (873.15)	650 (923.15)
750 (167.76)	V	1.112	255.43	378.31	394.22	410.05	441.55	472.90	504.15	535.30	566.40
	U	708.467	2573.3	2879.6	2920.1	2960.7	3042.9	3126.3	3211.4	3298.0	3386.2
	H	709.301	2764.8	3163.4	3215.7	3268.3	3374.0	3481.0	3589.5	3699.5	3811.0
	S	2.0195	6.6817	7.4416	7.5240	7.6035	7.7550	7.8981	8.0340	8.1637	8.2880
775 (169.10)	V	1.113	247.61	365.94	381.35	396.69	427.20	457.56	487.81	517.97	548.07
	U	714.326	2574.3	2879.3	2919.8	2960.4	3042.6	3126.1	3211.2	3297.8	3386.1
	H	715.189	2766.2	3162.9	3215.3	3267.9	3373.7	3480.8	3589.2	3699.3	3810.9
	S	2.0328	6.6705	7.4259	7.5084	7.5880	7.7396	7.8827	8.0187	8.1484	8.2727
800 (170.41)	V	1.115	240.26	354.34	369.29	384.16	413.74	443.17	472.49	501.72	530.89
	U	720.043	2575.3	2878.9	2919.5	2960.2	3042.4	3125.9	3211.0	3297.7	3386.0
	H	720.935	2767.5	3162.4	3214.9	3267.5	3373.4	3480.5	3589.0	3699.1	3810.7
	S	2.0457	6.6596	7.4107	7.4932	7.5729	7.7246	7.8678	8.0038	8.1336	8.2579
825 (171.69)	V	1.117	233.34	343.45	357.96	372.39	401.10	429.65	458.10	486.46	514.76
	U	725.625	2576.2	2878.6	2919.1	2959.9	3042.2	3125.7	3210.8	3297.5	3385.8
	H	726.547	2768.7	3161.9	3214.5	3267.1	3373.1	3480.2	3588.8	3698.8	3810.5
	S	2.0583	6.6491	7.3959	7.4786	7.5583	7.7101	7.8533	7.9894	8.1192	8.2436
850 (172.94)	V	1.118	226.81	333.20	347.29	361.31	389.20	416.93	444.56	472.09	499.57
	U	731.080	2577.1	2878.2	2918.8	2959.6	3041.9	3125.5	3210.7	3297.4	3385.7
	H	732.031	2769.9	3161.4	3214.0	3266.7	3372.7	3479.9	3588.5	3698.6	3810.3
	S	2.0705	6.6388	7.3815	7.4643	7.5441	7.6960	7.8393	7.9754	8.1053	8.2296
875 (174.16)	V	1.120	220.65	323.53	337.24	350.87	377.98	404.94	431.79	458.55	485.25
	U	736.415	2578.0	2877.9	2918.5	2959.3	3041.7	3125.3	3210.5	3297.2	3385.6
	H	737.394	2771.0	3161.0	3213.6	3266.3	3372.4	3479.7	3588.3	3698.4	3810.2
	S	2.0825	6.6289	7.3676	7.4504	7.5303	7.6823	7.8257	7.9618	8.0917	8.2161
900 (175.36)	V	1.121	214.81	314.40	327.74	341.01	367.39	393.61	419.73	445.76	471.72
	U	741.635	2578.8	2877.5	2918.2	2959.0	3041.4	3125.1	3210.3	3297.1	3385.4
	H	742.644	2772.1	3160.5	3213.2	3266.0	3372.1	3479.4	3588.1	3698.2	3810.0
	S	2.0941	6.6192	7.3540	7.4370	7.5169	7.6689	7.8124	7.9486	8.0785	8.2030
925 (176.53)	V	1.123	209.28	305.76	318.75	331.68	357.36	382.90	408.32	433.66	458.93
	U	746.746	2579.6	2877.2	2917.9	2958.8	3041.2	3124.9	3210.1	3296.9	3385.3
	H	747.784	2773.2	3160.0	3212.7	3265.6	3371.8	3479.1	3587.8	3698.0	3809.8
	S	2.1055	6.6097	7.3408	7.4238	7.5038	7.6560	7.7995	7.9357	8.0657	8.1902

P (T sat)										
950 (177.67)	V	1.124	204.03	297.57	310.24	322.84	347.87	372.74	397.51	422.19	446.81
	U	751.754	2580.4	2876.8	2917.6	2958.5	3041.0	3124.7	3209.9	3296.7	3385.1
	H	752.822	2774.2	3159.5	3212.3	3265.2	3371.5	3478.8	3587.6	3697.8	3809.6
	S	2.1166	6.6005	7.3279	7.4110	7.4911	7.6433	7.7869	7.9232	8.0532	8.1777
975 (178.79)	V	1.126	199.04	289.81	302.17	314.45	338.86	363.11	387.26	411.32	435.31
	U	756.663	2581.1	2876.5	2917.3	2958.2	3040.7	3124.5	3209.8	3296.6	3385.0
	H	757.761	2775.2	3159.0	3211.9	3264.8	3371.1	3478.6	3587.3	3697.6	3809.4
	S	2.1275	6.5916	7.3154	7.3986	7.4787	7.6310	7.7747	7.9110	8.0410	8.1656
1000 (179.88)	V	1.127	194.29	282.43	294.50	306.49	330.30	353.96	377.52	400.98	424.38
	U	761.478	2581.9	2876.1	2917.0	2957.9	3040.5	3124.3	3209.6	3296.4	3384.9
	H	762.605	2776.2	3158.5	3211.5	3264.4	3370.8	3478.3	3587.1	3697.4	3809.3
	S	2.1382	6.5828	7.3031	7.3864	7.4665	7.6190	7.7627	7.8991	8.0292	8.1537
1050 (182.02)	V	1.130	185.45	268.74	280.25	291.69	314.41	336.97	359.43	381.79	404.10
	U	770.843	2583.3	2875.4	2916.3	2957.4	3040.0	3123.9	3209.2	3296.1	3384.6
	H	772.029	2778.0	3157.6	3210.6	3263.6	3370.2	3477.7	3586.6	3697.0	3808.9
	S	2.1588	6.5659	7.2795	7.3629	7.4432	7.5958	7.7397	7.8762	8.0063	8.1309
1100 (184.07)	V	1.133	177.38	256.28	267.30	278.24	299.96	321.53	342.98	364.35	385.65
	U	779.878	2584.5	2874.7	2915.7	2956.8	3039.6	3123.5	3208.9	3295.8	3384.3
	H	781.124	2779.7	3156.6	3209.7	3262.9	3369.5	3477.2	3586.2	3696.6	3808.5
	S	2.1786	6.5497	7.2569	7.3405	7.4209	7.5737	7.7177	7.8543	7.9845	8.1092
1150 (186.05)	V	1.136	169.99	244.91	255.47	265.96	286.77	307.42	327.97	348.42	368.81
	U	788.611	2585.8	2874.0	2915.1	2956.2	3039.1	3123.1	3208.5	3295.5	3384.1
	H	789.917	2781.3	3155.6	3208.9	3262.1	3368.9	3476.6	3585.7	3696.2	3808.2
	S	2.1977	6.5342	7.2352	7.3190	7.3995	7.5525	7.6966	7.8333	7.9636	8.0883
1200 (187.96)	V	1.139	163.20	234.49	244.63	254.70	274.68	294.50	314.20	333.82	353.38
	U	797.064	2586.9	2873.3	2914.4	2955.7	3038.6	3122.7	3208.2	3295.2	3383.8
	H	798.430	2782.7	3154.6	3208.0	3261.3	3368.2	3476.1	3585.2	3695.8	3807.8
	S	2.2161	6.5194	7.2144	7.2983	7.3790	7.5323	7.6765	7.8132	7.9436	8.0684
1250 (189.81)	V	1.141	156.93	224.90	234.66	244.35	263.55	282.60	301.54	320.39	339.18
	U	805.259	2588.0	2872.5	2913.8	2955.1	3038.1	3122.3	3207.8	3294.9	3383.5
	H	806.685	2784.1	3153.7	3207.1	3260.5	3367.6	3475.5	3584.7	3695.4	3807.5
	S	2.2338	6.5050	7.1944	7.2785	7.3593	7.5128	7.6571	7.7940	7.9244	8.0493
1300 (191.61)	V	1.144	151.13	216.05	225.46	234.79	253.28	271.62	289.85	307.99	326.07
	U	813.213	2589.0	2871.8	2913.2	2954.5	3037.7	3121.9	3207.5	3294.6	3383.2
	H	814.700	2785.4	3152.7	3206.3	3259.7	3366.9	3475.0	3584.3	3695.0	3807.1
	S	2.2510	6.4913	7.1751	7.2594	7.3404	7.4940	7.6385	7.7754	7.9060	8.0309

TABLE E.2 Properties of Superheated Steam (Continued)

TEMPERATURE: t°C
(TEMPERATURE: T kelvins)

P/kPa (t^sat/°C)		sat. liq.	sat. vap.	200 (473.15)	225 (498.15)	250 (523.15)	275 (548.15)	300 (573.15)	325 (598.15)	350 (623.15)	375 (648.15)
1350 (193.35)	V	1.146	145.74	148.79	159.70	169.96	179.79	189.33	198.66	207.85	216.93
	U	820.944	2589.6	2603.9	2653.6	2700.1	2744.4	2787.4	2829.5	2871.5	2912.5
	H	822.491	2786.6	2804.7	2869.2	2929.2	2987.1	3043.0	3097.7	3151.7	3205.4
	S	2.2676	6.4780	6.5165	6.6493	6.7675	6.8750	6.9746	7.0681	7.1566	7.2410
1400 (195.04)	V	1.149	140.72	142.94	153.57	163.55	173.08	182.32	191.35	200.24	209.02
	U	828.465	2590.8	2601.3	2651.7	2698.6	2743.2	2786.4	2828.6	2870.4	2911.9
	H	830.074	2787.8	2801.4	2866.7	2927.6	2985.5	3041.6	3096.5	3150.7	3204.5
	S	2.2837	6.4651	6.4941	6.6285	6.7477	6.8560	6.9561	7.0499	7.1386	7.2233
1450 (196.69)	V	1.151	136.04	137.48	147.86	157.57	166.83	175.79	184.54	193.15	201.65
	U	835.791	2591.6	2598.7	2649.7	2697.1	2742.0	2785.4	2827.8	2869.7	2911.3
	H	837.460	2788.9	2798.1	2864.1	2925.5	2983.9	3040.3	3095.4	3149.7	3203.6
	S	2.2993	6.4526	6.4722	6.6082	6.7286	6.8376	6.9381	7.0322	7.1212	7.2061
1500 (198.29)	V	1.154	131.66	132.38	142.53	151.99	161.00	169.70	178.19	186.53	194.77
	U	842.933	2592.4	2596.1	2647.7	2695.5	2740.8	2784.4	2826.9	2868.9	2910.6
	H	844.663	2789.9	2794.7	2861.5	2923.5	2982.3	3038.9	3094.2	3148.7	3202.8
	S	2.3145	6.4406	6.4508	6.5885	6.7099	6.8196	6.9207	7.0152	7.1044	7.1894
1550 (199.85)	V	1.156	127.55	127.61	137.54	146.77	155.54	164.00	172.25	180.34	188.33
	U	849.901	2593.2	2593.5	2645.8	2694.0	2739.5	2783.4	2826.1	2868.2	2910.0
	H	851.694	2790.8	2791.3	2858.9	2921.5	2980.6	3037.6	3093.1	3147.7	3201.9
	S	2.3292	6.4289	6.4298	6.5692	6.6917	6.8022	6.9038	6.9986	7.0881	7.1733
1600 (201.37)	V	1.159	123.69	......	132.85	141.87	150.42	158.66	166.68	174.54	182.30
	U	856.707	2593.8	......	2643.7	2692.4	2738.3	2782.4	2825.2	2867.5	2909.3
	H	858.561	2791.7	......	2856.3	2919.4	2979.0	3036.2	3091.9	3146.7	3201.0
	S	2.3436	6.4175	......	6.5503	6.6740	6.7852	6.8873	6.9825	7.0723	7.1577
1650 (202.86)	V	1.161	120.05	......	128.45	137.27	145.61	153.64	161.44	169.09	176.63
	U	863.359	2594.5	......	2641.7	2690.9	2737.1	2781.3	2824.4	2866.7	2908.7
	H	865.275	2792.6	......	2853.6	2917.4	2977.3	3034.8	3090.8	3145.7	3200.1
	S	2.3576	6.4065	......	6.5319	6.6567	6.7687	6.8713	6.9669	7.0569	7.1425
1700 (204.31)	V	1.163	116.62	......	124.31	132.94	141.09	148.91	156.51	163.96	171.30
	U	869.866	2595.1	......	2639.6	2689.3	2735.8	2780.3	2823.5	2866.0	2908.0
	H	871.843	2793.4	......	2851.0	2915.3	2975.6	3033.3	3089.6	3144.7	3199.2
	S	2.3713	6.3957	......	6.5138	6.6398	6.7526	6.8557	6.9516	7.0419	7.1277

P (kPa) / (t°C)		Sat. Liq.	Sat. Vap.								
1750 (205.72)	V	1.166	113.38	.....	120.39	128.85	136.82	144.45	151.87	159.12	166.27
	U	876.234	2595.7	.....	2637.6	2687.7	2734.5	2779.3	2822.7	2865.3	2907.4
	H	878.274	2794.1	.....	2848.2	2913.2	2974.0	3032.1	3088.4	3143.7	3198.4
	S	2.3846	6.3853	.....	6.4961	6.6233	6.7368	6.8405	6.9368	7.0273	7.1133
1800 (207.11)	V	1.168	110.32	.....	116.69	124.99	132.78	140.24	147.48	154.55	161.51
	U	882.472	2596.3	.....	2635.5	2686.1	2733.3	2778.2	2821.8	2864.5	2906.7
	H	884.574	2794.8	.....	2845.5	2911.0	2972.3	3030.7	3087.3	3142.7	3197.5
	S	2.3976	6.3751	.....	6.4787	6.6071	6.7214	6.8257	6.9223	7.0131	7.0993
1850 (208.47)	V	1.170	107.41	.....	113.19	121.33	128.96	136.26	143.33	150.23	157.02
	U	888.585	2596.8	.....	2633.3	2684.4	2732.0	2777.2	2820.9	2863.8	2906.1
	H	890.750	2795.5	.....	2842.8	2908.9	2970.6	3029.3	3086.1	3141.7	3196.6
	S	2.4103	6.3651	.....	6.4616	6.5912	6.7064	6.8112	6.9082	6.9993	7.0856
1900 (209.80)	V	1.172	104.65	.....	109.87	117.87	125.35	132.49	139.39	146.14	152.76
	U	894.580	2597.3	.....	2631.2	2682.8	2730.7	2776.2	2820.1	2863.0	2905.4
	H	896.807	2796.1	.....	2840.0	2906.7	2968.8	3027.9	3084.9	3140.7	3195.7
	S	2.4228	6.3554	.....	6.4448	6.5757	6.6917	6.7970	6.8944	6.9857	7.0723
1950 (211.10)	V	1.174	102.031	.....	106.72	114.58	121.91	128.90	135.66	142.25	148.72
	U	900.461	2597.7	.....	2629.0	2681.1	2729.4	2775.1	2819.2	2862.3	2904.8
	H	902.752	2796.7	.....	2837.1	2904.6	2967.1	3026.5	3083.7	3139.7	3194.8
	S	2.4349	6.3459	.....	6.4283	6.5604	6.6772	6.7831	6.8809	6.9725	7.0593
2000 (212.37)	V	1.177	99.536	.....	103.72	111.45	118.65	125.50	132.11	138.56	144.89
	U	906.236	2598.2	.....	2626.9	2679.5	2728.1	2774.0	2818.3	2861.5	2904.1
	H	908.589	2797.2	.....	2834.3	2902.4	2965.4	3025.0	3082.5	3138.6	3193.9
	S	2.4469	6.3366	.....	6.4120	6.5454	6.6631	6.7696	6.8677	6.9596	7.0466
2100 (214.85)	V	1.181	94.890	.....	98.147	105.64	112.59	119.18	125.53	131.70	137.76
	U	917.479	2598.9	.....	2622.4	2676.1	2725.4	2771.9	2816.5	2860.0	2902.8
	H	919.959	2798.2	.....	2828.5	2897.9	2961.9	3022.2	3080.1	3136.6	3192.1
	S	2.4700	6.3187	.....	6.3802	6.5162	6.6356	6.7432	6.8422	6.9347	7.0220
2200 (217.24)	V	1.185	90.652	.....	93.067	100.35	107.07	113.43	119.53	125.47	131.28
	U	928.346	2599.6	.....	2617.9	2672.7	2722.7	2769.7	2814.7	2858.5	2901.5
	H	930.953	2799.1	.....	2822.7	2893.4	2958.5	3019.3	3077.7	3134.5	3190.3
	S	2.4922	6.3015	.....	6.3492	6.4879	6.6091	6.7179	6.8177	6.9107	6.9985
2300 (219.55)	V	1.189	86.769	.....	88.420	95.513	102.03	108.18	114.06	119.77	125.36
	U	938.866	2600.2	.....	2613.3	2669.2	2720.0	2767.6	2812.9	2857.0	2900.2
	H	941.601	2799.8	.....	2816.7	2888.9	2954.7	3016.4	3075.3	3132.4	3188.5
	S	2.5136	6.2849	.....	6.3190	6.4605	6.5835	6.6935	6.7941	6.8877	6.9759

TABLE E.2 Properties of Superheated Steam (Continued)

TEMPERATURE: t/°C
(TEMPERATURE: T kelvins)

P/kPa (t^{sat}/°C)		sat. liq.	sat. vap.	400 (673.15)	425 (698.15)	450 (723.15)	475 (748.15)	500 (773.15)	550 (823.15)	600 (873.15)	650 (923.15)
1350 (193.35)	V	1.146	145.74	225.94	234.88	243.78	252.63	261.46	279.03	296.51	313.93
	U	820.944	2589.9	2953.9	2995.5	3037.2	3079.2	3121.5	3207.1	3294.3	3383.0
	H	822.491	2786.6	3259.0	3312.6	3366.3	3420.2	3474.4	3583.8	3694.5	3806.8
	S	2.2676	6.4780	7.3221	7.4003	7.4759	7.5493	7.6205	7.7576	7.8882	8.0132
1400 (195.04)	V	1.149	140.72	217.72	226.35	234.95	243.50	252.02	268.98	285.85	302.66
	U	828.465	2590.8	2953.4	2994.9	3036.7	3078.7	3121.1	3206.8	3293.9	3382.7
	H	830.074	2787.8	3258.2	3311.8	3365.6	3419.6	3473.9	3583.3	3694.1	3806.4
	S	2.2837	6.4651	7.3045	7.3828	7.4585	7.5319	7.6032	7.7404	7.8710	7.9961
1450 (196.69)	V	1.151	136.04	210.06	218.42	226.72	234.99	243.23	259.62	275.93	292.16
	U	835.791	2591.6	2952.8	2994.4	3036.2	3078.3	3120.7	3206.4	3293.6	3382.4
	H	837.460	2788.9	3257.4	3311.1	3365.0	3419.0	3473.3	3582.9	3693.7	3806.1
	S	2.2993	6.4526	7.2874	7.3658	7.4416	7.5151	7.5865	7.7237	7.8545	7.9796
1500 (198.29)	V	1.154	131.66	202.92	211.01	219.05	227.06	235.03	250.89	266.66	282.37
	U	842.933	2592.4	2952.2	2993.9	3035.8	3077.9	3120.3	3206.0	3293.3	3382.1
	H	844.663	2789.9	3256.6	3310.4	3364.3	3418.4	3472.8	3582.4	3693.3	3805.7
	S	2.3145	6.4406	7.2709	7.3494	7.4253	7.4989	7.5703	7.7077	7.8385	7.9636
1550 (199.85)	V	1.156	127.55	196.24	204.08	211.87	219.63	227.35	242.72	258.00	273.21
	U	849.901	2593.2	2951.7	2993.4	3035.3	3077.4	3119.8	3205.7	3293.0	3381.9
	H	851.694	2790.8	3255.8	3309.7	3363.7	3417.8	3472.2	3581.9	3692.9	3805.3
	S	2.3292	6.4289	7.2550	7.3336	7.4095	7.4832	7.5547	7.6921	7.8230	7.9482
1600 (201.37)	V	1.159	123.69	189.97	197.58	205.15	212.67	220.16	235.06	249.87	264.62
	U	856.707	2593.8	2951.1	2992.9	3034.8	3077.0	3119.4	3205.3	3292.7	3381.6
	H	858.561	2791.7	3255.0	3309.0	3363.0	3417.2	3471.7	3581.4	3692.5	3805.0
	S	2.3436	6.4175	7.2394	7.3182	7.3942	7.4679	7.5395	7.6770	7.8080	7.9333
1650 (202.86)	V	1.161	120.05	184.09	191.48	198.82	206.13	213.40	227.86	242.24	256.55
	U	863.359	2594.5	2950.5	2992.3	3034.3	3076.5	3119.0	3205.0	3292.4	3381.3
	H	865.275	2792.6	3254.2	3308.3	3362.4	3416.7	3471.1	3581.0	3692.1	3804.6
	S	2.3576	6.4065	7.2244	7.3032	7.3794	7.4531	7.5248	7.6624	7.7934	7.9188
1700 (204.31)	V	1.163	116.62	178.55	185.74	192.87	199.97	207.04	221.09	235.06	248.96
	U	869.866	2595.1	2949.9	2991.8	3033.9	3076.1	3118.6	3204.6	3292.1	3381.0
	H	871.843	2793.4	3253.5	3307.6	3361.7	3416.1	3470.6	3580.5	3691.7	3804.3
	S	2.3713	6.3957	7.2098	7.2887	7.3649	7.4388	7.5105	7.6482	7.7793	7.9047

P (t sat)		Sat. liq.	Sat. vap.								
1750	V	1.166	113.38	173.32	180.32	187.26	194.17	201.04	214.71	228.28	241.80
(205.72)	U	876.234	2595.7	2949.3	2991.3	3033.4	3075.5	3118.2	3204.3	3291.8	3380.8
	H	878.274	2794.1	3252.7	3306.9	3361.1	3415.5	3470.0	3580.0	3691.3	3803.9
	S	2.3846	6.3853	7.1955	7.2746	7.3509	7.4248	7.4965	7.6344	7.7656	7.8910
1800	V	1.168	110.32	168.39	175.20	181.97	188.69	195.38	208.68	221.89	235.03
(207.11)	U	882.472	2596.3	2948.8	2990.8	3032.9	3075.2	3117.8	3203.9	3291.5	3380.5
	H	884.574	2794.8	3251.9	3306.1	3360.4	3414.9	3469.5	3579.5	3690.9	3803.6
	S	2.3976	6.3751	7.1816	7.2608	7.3372	7.4112	7.4830	7.6209	7.7522	7.8777
1850	V	1.170	107.41	163.73	170.37	176.96	183.50	190.02	202.97	215.84	228.64
(208.47)	U	888.585	2596.8	2948.2	2990.3	3032.4	3074.8	3117.4	3203.6	3291.1	3380.2
	H	890.750	2795.5	3251.1	3305.4	3359.8	3414.3	3469.0	3579.1	3690.4	3803.2
	S	2.4103	6.3651	7.1681	7.2474	7.3239	7.3980	7.4698	7.6079	7.7392	7.8648
1900	V	1.172	104.65	159.30	165.78	172.21	178.59	184.94	197.57	210.11	222.58
(209.80)	U	894.580	2597.3	2947.6	2989.7	3031.9	3074.3	3117.0	3203.2	3290.8	3380.0
	H	896.807	2796.1	3250.3	3304.7	3359.1	3413.7	3468.4	3578.6	3690.0	3802.8
	S	2.4228	6.3554	7.1550	7.2344	7.3109	7.3851	7.4570	7.5951	7.7265	7.8522
1950	V	1.174	102.031	155.11	161.43	167.70	173.93	180.13	192.44	204.67	216.83
(211.10)	U	900.461	2597.7	2947.0	2989.2	3031.5	3073.9	3116.6	3202.9	3290.5	3379.7
	H	902.752	2796.7	3249.5	3304.0	3358.5	3413.1	3467.8	3578.1	3689.6	3802.5
	S	2.4349	6.3459	7.1421	7.2216	7.2983	7.3725	7.4445	7.5827	7.7142	7.8399
2000	V	1.177	99.536	151.13	157.30	163.42	169.51	175.55	187.57	199.50	211.36
(212.37)	U	906.236	2598.2	2946.4	2988.7	3031.0	3073.5	3116.2	3202.5	3290.2	3379.4
	H	908.589	2797.2	3248.7	3303.3	3357.8	3412.5	3467.3	3577.6	3689.2	3802.1
	S	2.4469	6.3366	7.1296	7.2092	7.2859	7.3602	7.4323	7.5706	7.7022	7.8279
2100	V	1.181	94.890	143.73	149.63	155.48	161.28	167.06	178.53	189.91	201.22
(214.85)	U	917.479	2598.9	2945.3	2987.6	3030.0	3072.6	3115.3	3201.8	3289.6	3378.9
	H	919.959	2798.2	3247.1	3301.8	3356.5	3411.3	3466.2	3576.7	3688.4	3801.4
	S	2.4700	6.3187	7.1053	7.1851	7.2621	7.3365	7.4087	7.5472	7.6789	7.8048
2200	V	1.185	90.652	137.00	142.65	148.25	153.81	159.34	170.30	181.19	192.00
(217.24)	U	928.346	2599.6	2944.1	2986.6	3029.1	3071.7	3114.5	3201.1	3289.0	3378.3
	H	930.953	2799.1	3245.5	3300.4	3355.2	3410.1	3465.1	3575.7	3687.6	3800.7
	S	2.4922	6.3015	7.0821	7.1621	7.2393	7.3139	7.3862	7.5249	7.6568	7.7827
2300	V	1.189	86.769	130.85	136.28	141.65	146.99	152.28	162.80	173.22	183.58
(219.55)	U	938.866	2600.2	2942.9	2985.5	3028.1	3070.8	3113.7	3200.4	3288.3	3377.8
	H	941.601	2799.8	3243.9	3299.0	3353.9	3408.9	3464.0	3574.8	3686.7	3800.0
	S	2.5136	6.2849	7.0598	7.1401	7.2174	7.2922	7.3646	7.5035	7.6355	7.7616

TABLE E.2 Properties of Superheated Steam (Continued)

TEMPERATURE: *t*°C
(TEMPERATURE: *T* kelvins)

P/kPa (*t*^{sat}/°C)		sat. liq.	sat. vap.	225 (498.15)	250 (523.15)	275 (548.15)	300 (573.15)	325 (598.15)	350 (623.15)	375 (648.15)	400 (673.15)
2400 (221.78)	V	1.193	83.199	84.149	91.075	97.411	103.36	109.05	114.55	119.93	125.22
	U	949.066	2600.7	2608.6	2665.6	2717.3	2765.4	2811.1	2855.4	2898.8	2941.7
	H	951.929	2800.4	2810.6	2884.2	2951.1	3013.4	3072.8	3130.4	3186.7	3242.3
	S	2.5343	6.2690	6.2894	6.4338	6.5586	6.6699	6.7714	6.8656	6.9542	7.0384
2500 (223.94)	V	1.197	79.905	80.210	86.985	93.154	98.925	104.43	109.75	114.94	120.04
	U	958.969	2601.2	2603.8	2662.0	2714.5	2763.1	2809.3	2853.9	2897.5	2940.6
	H	961.962	2800.9	2804.3	2879.5	2947.4	3010.4	3070.4	3128.2	3184.8	3240.7
	S	2.5543	6.2536	6.2604	6.4077	6.5345	6.6470	6.7494	6.8442	6.9333	7.0178
2600 (226.04)	V	1.201	76.856	⋯	83.205	89.220	94.830	100.17	105.32	110.33	115.26
	U	968.597	2601.5	⋯	2658.4	2711.7	2760.9	2807.4	2852.3	2896.1	2939.4
	H	971.720	2801.4	⋯	2874.7	2943.6	3007.4	3067.9	3126.1	3183.0	3239.0
	S	2.5736	6.2387	⋯	6.3823	6.5110	6.6249	6.7281	6.8236	6.9131	6.9979
2700 (228.07)	V	1.205	74.025	⋯	79.698	85.575	91.036	96.218	101.21	106.07	110.83
	U	977.968	2601.8	⋯	2654.7	2708.8	2758.6	2805.6	2850.7	2894.8	2938.2
	H	981.222	2801.7	⋯	2869.9	2939.8	3004.4	3065.4	3124.0	3181.2	3237.4
	S	2.5924	6.2244	⋯	6.3575	6.4882	6.6034	6.7075	6.8036	6.8935	6.9787
2800 (230.05)	V	1.209	71.389	⋯	76.437	82.187	87.510	92.550	97.395	102.10	106.71
	U	987.100	2602.1	⋯	2650.9	2705.9	2756.3	2803.7	2849.2	2893.4	2937.0
	H	990.485	2802.0	⋯	2864.9	2936.0	3001.3	3062.8	3121.9	3179.3	3235.8
	S	2.6106	6.2104	⋯	6.3331	6.4659	6.5824	6.6875	6.7842	6.8746	6.9601
2900 (231.97)	V	1.213	68.928	⋯	73.395	79.029	84.226	89.133	93.843	98.414	102.88
	U	996.008	2602.3	⋯	2647.1	2702.9	2754.0	2801.8	2847.6	2892.0	2935.8
	H	999.524	2802.2	⋯	2859.9	2932.1	2998.2	3060.3	3119.7	3177.4	3234.1
	S	2.6283	6.1969	⋯	6.3092	6.4441	6.5621	6.6681	6.7654	6.8563	6.9421
3000 (233.84)	V	1.216	66.626	⋯	70.551	76.078	81.159	85.943	90.526	94.969	99.310
	U	1004.7	2602.4	⋯	2643.2	2700.0	2751.6	2799.9	2846.0	2890.7	2934.6
	H	1008.4	2802.3	⋯	2854.8	2928.2	2995.1	3057.7	3117.5	3175.6	3232.5
	S	2.6455	6.1837	⋯	6.2857	6.4228	6.5422	6.6491	6.7471	6.8385	6.9246
3100 (235.67)	V	1.220	64.467	⋯	67.885	73.315	78.287	82.958	87.423	91.745	95.965
	U	1013.2	2602.5	⋯	2639.2	2697.0	2749.2	2797.9	2844.3	2889.3	2933.4
	H	1017.0	2802.3	⋯	2849.6	2924.2	2991.9	3055.1	3115.4	3173.7	3230.8
	S	2.6623	6.1709	⋯	6.2626	6.4019	6.5227	6.6307	6.7294	6.8212	6.9077

Superheated steam. Values of V (cm³ g⁻¹), U (kJ kg⁻¹), H (kJ kg⁻¹), S (kJ kg⁻¹ K⁻¹).

P (kPa) (t_sat °C)		Sat. liq.		Sat. vap.						
3200 (237.45)	V	1.224	......	62.439	65.380	70.721	75.593	80.158	84.513	88.723	92.829
	U	1021.5	......	2602.5	2635.2	2693.9	2746.8	2796.0	2842.7	2887.9	2932.1
	H	1025.4	......	2802.3	2844.4	2920.2	2988.7	3052.5	3113.2	3171.8	3229.2
	S	2.6786	......	6.1585	6.2398	6.3815	6.5037	6.6127	6.7120	6.8043	6.8912
3300 (239.18)	V	1.227	......	60.529	63.021	68.282	73.061	77.526	81.778	85.883	89.883
	U	1029.7	......	2602.5	2631.1	2690.8	2744.4	2794.0	2841.1	2886.5	2930.9
	H	1033.7	......	2802.3	2839.0	2916.1	2985.5	3049.9	3110.9	3169.9	3227.5
	S	2.6945	......	6.1463	6.2173	6.3614	6.4851	6.5951	6.6952	6.7879	6.8752
3400 (240.88)	V	1.231	......	58.728	60.796	65.982	70.675	75.048	79.204	83.210	87.110
	U	1037.6	......	2602.5	2626.9	2687.7	2741.9	2792.0	2839.4	2885.1	2929.7
	H	1041.8	......	2802.1	2833.6	2912.0	2982.2	3047.2	3108.7	3168.0	3225.9
	S	2.7101	......	6.1344	6.1951	6.3416	6.4669	6.5779	6.6787	6.7719	6.8595
3500 (242.54)	V	1.235	......	57.025	58.693	63.812	68.424	72.710	76.776	80.689	84.494
	U	1045.4	......	2602.4	2622.7	2684.5	2739.5	2790.0	2837.8	2883.7	2928.4
	H	1049.8	......	2802.0	2828.1	2907.8	2979.0	3044.5	3106.5	3166.5	3224.2
	S	2.7253	......	6.1228	6.1732	6.3221	6.4491	6.5611	6.6626	6.7563	6.8443
3600 (244.16)	V	1.238	......	55.415	56.702	61.759	66.297	70.501	74.482	78.308	82.024
	U	1053.1	......	2602.2	2618.4	2681.3	2737.0	2788.0	2836.1	2882.3	2927.2
	H	1057.6	......	2801.7	2822.5	2903.6	2975.6	3041.8	3104.2	3164.2	3222.5
	S	2.7401	......	6.1115	6.1514	6.3030	6.4315	6.5446	6.6468	6.7411	6.8294
3700 (245.75)	V	1.242	......	53.888		59.814	64.282	68.410	72.311	76.055	79.687
	U	1060.6	......	2602.1		2678.0	2734.4	2786.0	2834.4	2880.8	2926.0
	H	1065.2	......	2801.4		2899.3	2972.3	3039.1	3102.0	3162.2	3220.8
	S	2.7547	......	6.1004		6.2841	6.4143	6.5284	6.6314	6.7262	6.8149
3800 (247.31)	V	1.245	......	52.438		57.968	62.372	66.429	70.254	73.920	77.473
	U	1068.0	......	2601.9		2674.7	2731.9	2783.9	2832.7	2879.4	2924.7
	H	1072.7	......	2801.1		2895.0	2968.9	3036.4	3099.7	3160.3	3219.1
	S	2.7689	......	6.0896		6.2654	6.3973	6.5126	6.6163	6.7117	6.8007
3900 (248.84)	V	1.249	......	51.061		56.215	60.558	64.547	68.302	71.894	75.372
	U	1075.3	......	2601.6		2671.4	2729.3	2781.9	2831.0	2877.9	2923.5
	H	1080.1	......	2800.8		2890.6	2965.5	3033.6	3097.4	3158.3	3217.4
	S	2.7828	......	6.0789		6.2470	6.3806	6.4970	6.6015	6.6974	6.7868
4000 (250.33)	V	1.252	......	49.749		54.546	58.833	62.759	66.446	69.969	73.376
	U	1082.4	......	2601.3		2668.0	2726.7	2779.8	2829.3	2876.5	2922.2
	H	1087.4	......	2800.3		2886.1	2962.0	3030.8	3095.1	3156.4	3215.7
	S	2.7965	......	6.0685		6.2288	6.3642	6.4817	6.5870	6.6834	6.7733

TABLE E.2 Properties of Superheated Steam (Continued)

TEMPERATURE: t°C
(TEMPERATURE: T kelvins)

P/kPa (t^{sat}/°C)		sat. liq.	sat. vap.	425 (698.15)	450 (723.15)	475 (748.15)	500 (773.15)	525 (798.15)	550 (823.15)	600 (873.15)	650 (923.15)
2400	V	1.193	83.199	130.44	135.61	140.73	145.82	150.88	155.91	165.92	175.86
(221.78)	U	949.066	2600.7	2984.5	3027.1	3069.9	3112.9	3156.1	3199.6	3287.7	3377.2
	H	951.929	2800.4	3297.5	3352.6	3407.7	3462.9	3518.2	3573.8	3685.9	3799.3
	S	2.5343	6.2690	7.1189	7.1964	7.2713	7.3439	7.4144	7.4830	7.6152	7.7414
2500	V	1.197	79.905	125.07	130.04	134.97	139.87	144.74	149.58	159.21	168.76
(223.94)	U	958.969	2601.2	2983.4	3026.2	3069.0	3112.1	3155.4	3198.9	3287.1	3376.7
	H	961.962	2800.9	3296.1	3351.3	3406.5	3461.7	3517.0	3572.9	3685.1	3798.6
	S	2.5543	6.2536	7.0986	7.1763	7.2513	7.3240	7.3946	7.4633	7.5956	7.7220
2600	V	1.201	76.856	120.11	124.91	129.66	134.38	139.07	143.74	153.01	162.21
(226.04)	U	968.597	2601.5	2982.3	3025.2	3068.1	3111.2	3154.6	3198.2	3286.5	3376.1
	H	971.720	2801.4	3294.6	3349.9	3405.3	3460.6	3516.2	3571.9	3684.3	3797.9
	S	2.5736	6.2387	7.0789	7.1568	7.2320	7.3048	7.3755	7.4443	7.5768	7.7033
2700	V	1.205	74.025	115.52	120.15	124.74	129.30	133.82	138.33	147.27	156.14
(228.07)	U	977.968	2601.8	2981.2	3024.2	3067.2	3110.4	3153.8	3197.5	3285.8	3375.6
	H	981.222	2801.7	3293.1	3348.6	3404.0	3459.5	3515.2	3571.0	3683.5	3797.1
	S	2.5924	6.2244	7.0600	7.1381	7.2134	7.2863	7.3571	7.4260	7.5587	7.6853
2800	V	1.209	71.389	111.25	115.74	120.17	124.58	128.95	133.30	141.94	150.50
(230.05)	U	987.100	2602.1	2980.2	3023.2	3066.3	3109.6	3153.1	3196.8	3285.2	3375.0
	H	990.485	2802.0	3291.7	3347.3	3402.8	3458.4	3514.1	3570.0	3682.6	3796.4
	S	2.6106	6.2104	7.0416	7.1199	7.1954	7.2685	7.3394	7.4084	7.5412	7.6679
2900	V	1.213	68.928	107.28	111.62	115.92	120.18	124.42	128.62	136.97	145.26
(231.97)	U	996.008	2602.3	2979.1	3022.3	3065.5	3108.8	3152.3	3196.1	3284.6	3374.5
	H	999.524	2802.2	3290.2	3346.0	3401.6	3457.3	3513.1	3569.1	3681.8	3795.7
	S	2.6283	6.1969	7.0239	7.1024	7.1780	7.2512	7.3222	7.3913	7.5243	7.6511
3000	V	1.216	66.626	103.58	107.79	111.95	116.08	120.18	124.26	132.34	140.36
(233.84)	U	1004.7	2602.4	2978.0	3021.3	3064.6	3107.9	3151.5	3195.4	3284.0	3373.9
	H	1008.4	2802.3	3288.7	3344.6	3400.4	3456.2	3512.1	3568.1	3681.0	3795.0
	S	2.6455	6.1837	7.0067	7.0854	7.1612	7.2345	7.3056	7.3748	7.5079	7.6349
3100	V	1.220	64.467	100.11	104.20	108.24	112.24	116.22	120.17	128.01	135.78
(235.67)	U	1013.2	2602.5	2976.9	3020.3	3063.7	3107.1	3150.8	3194.7	3283.3	3373.4
	H	1017.0	2802.3	3287.3	3343.3	3399.2	3455.1	3511.0	3567.2	3680.2	3794.3
	S	2.6623	6.1709	6.9900	7.0689	7.1448	7.2183	7.2895	7.3588	7.4920	7.6191

P (t sat)											
3200 (237.45)	V	1.224	62.439	96.859	100.83	104.76	108.65	112.51	116.34	123.95	131.48
	U	1021.5	2602.5	2975.9	3019.3	3062.8	3106.3	3150.0	3193.9	3282.7	3372.8
	H	1025.4	2802.3	3285.8	3342.0	3398.0	3454.0	3510.0	3566.2	3679.3	3793.8
	S	2.6786	6.1585	6.9738	7.0528	7.1290	7.2026	7.2739	7.3433	7.4767	7.6039
3300 (239.18)	V	1.227	60.529	93.805	97.668	101.49	105.27	109.02	112.74	120.13	127.45
	U	1029.7	2602.5	2974.8	3018.3	3061.9	3105.5	3149.2	3193.2	3282.1	3372.3
	H	1033.7	2802.3	3284.3	3340.6	3396.8	3452.8	3509.0	3565.5	3678.5	3792.9
	S	2.6945	6.1463	6.9580	7.0373	7.1136	7.1873	7.2588	7.3282	7.4618	7.5891
3400 (240.88)	V	1.231	58.728	90.930	94.692	98.408	102.09	105.74	109.36	116.54	123.65
	U	1037.6	2602.5	2973.7	3017.4	3061.0	3104.6	3148.4	3192.5	3281.5	3371.7
	H	1041.8	2802.1	3282.8	3339.3	3395.5	3451.7	3507.9	3564.3	3677.7	3792.1
	S	2.7101	6.1344	6.9426	7.0221	7.0986	7.1724	7.2440	7.3136	7.4473	7.5747
3500 (242.54)	V	1.235	57.025	88.220	91.886	95.505	99.088	102.64	106.17	113.15	120.07
	U	1045.4	2602.4	2972.6	3016.4	3060.1	3103.8	3147.7	3191.8	3280.8	3371.2
	H	1049.8	2802.0	3281.3	3338.0	3394.3	3450.6	3506.9	3563.4	3676.9	3791.4
	S	2.7253	6.1228	6.9277	7.0074	7.0840	7.1580	7.2297	7.2993	7.4332	7.5607
3600 (244.16)	V	1.238	55.415	85.660	89.236	92.764	96.255	99.716	103.15	109.96	116.69
	U	1053.1	2602.2	2971.5	3015.4	3059.2	3103.0	3146.9	3191.1	3280.2	3370.6
	H	1057.6	2801.7	3279.8	3336.6	3393.1	3449.5	3505.9	3562.4	3676.1	3790.7
	S	2.7401	6.1115	6.9131	6.9930	7.0698	7.1439	7.2157	7.2854	7.4195	7.5471
3700 (245.75)	V	1.242	53.888	83.238	86.728	90.171	93.576	96.950	100.30	106.93	113.49
	U	1060.6	2602.1	2970.4	3014.4	3058.2	3102.1	3146.1	3190.4	3279.6	3370.1
	H	1065.2	2801.4	3278.4	3335.3	3391.9	3448.4	3504.9	3561.5	3675.2	3790.0
	S	2.7547	6.1004	6.8989	6.9790	7.0559	7.1302	7.2021	7.2719	7.4061	7.5339
3800 (247.31)	V	1.245	52.438	80.944	84.353	87.714	91.038	94.330	97.596	104.06	110.46
	U	1068.0	2601.9	2969.3	3013.4	3057.3	3101.3	3145.4	3189.6	3279.0	3369.5
	H	1072.7	2801.1	3276.8	3333.9	3390.7	3447.2	3503.8	3560.5	3674.4	3789.3
	S	2.7689	6.0896	6.8849	6.9653	7.0424	7.1168	7.1888	7.2587	7.3931	7.5210
3900 (248.84)	V	1.249	51.061	78.767	82.099	85.383	88.629	91.844	95.033	101.35	107.59
	U	1075.3	2601.6	2968.2	3012.4	3056.4	3100.5	3144.6	3188.9	3278.3	3369.0
	H	1080.1	2800.8	3275.3	3332.6	3389.4	3446.1	3502.8	3559.5	3673.6	3788.6
	S	2.7828	6.0789	6.8713	6.9519	7.0292	7.1037	7.1759	7.2459	7.3804	7.5084
4000 (250.33)	V	1.252	49.749	76.698	79.958	83.169	86.341	89.483	92.598	98.763	104.86
	U	1082.4	2601.3	2967.0	3011.4	3055.5	3099.6	3143.8	3188.2	3277.7	3368.4
	H	1087.4	2800.3	3273.8	3331.2	3388.2	3445.0	3501.7	3558.6	3672.8	3787.9
	S	2.7965	6.0685	6.8581	6.9388	7.0163	7.0909	7.1632	7.2333	7.3680	7.4961

TABLE E.2 Properties of Superheated Steam (Continued)

TEMPERATURE: $t°C$
(TEMPERATURE: T kelvins)

P/kPa (t^{sat}/°C)		sat. liq.	sat. vap.	260 (533.15)	275 (548.15)	300 (573.15)	325 (598.15)	350 (623.15)	375 (648.15)	400 (673.15)	425 (698.15)
4100 (251.80)	V	1.256	48.500	50.150	52.955	57.191	61.057	64.680	68.137	71.476	74.730
	U	1089.4	2601.0	2624.6	2664.5	2724.0	2777.7	2827.6	2875.0	2920.9	2965.6
	H	1094.6	2799.9	2830.3	2881.6	2958.5	3028.0	3092.8	3154.4	3214.0	3272.3
	S	2.8099	6.0583	6.1157	6.2107	6.3480	6.4667	6.5727	6.6697	6.7600	6.8450
4200 (253.24)	V	1.259	47.307	48.654	51.438	55.625	59.435	62.998	66.392	69.667	72.856
	U	1096.3	2600.7	2620.4	2661.0	2721.4	2775.6	2825.8	2873.6	2919.7	2964.8
	H	1101.6	2799.4	2824.8	2877.1	2955.0	3025.2	3090.4	3152.4	3212.3	3270.8
	S	2.8231	6.0482	6.0962	6.1929	6.3320	6.4519	6.5587	6.6563	6.7469	6.8323
4300 (254.66)	V	1.262	46.168	47.223	49.988	54.130	57.887	61.393	64.728	67.942	71.069
	U	1103.1	2600.3	2616.2	2657.5	2718.7	2773.4	2824.1	2872.1	2918.4	2963.7
	H	1108.5	2798.5	2819.2	2872.4	2951.4	3022.3	3088.1	3150.4	3210.5	3269.3
	S	2.8360	6.0383	6.0768	6.1752	6.3162	6.4373	6.5450	6.6431	6.7341	6.8198
4400 (256.05)	V	1.266	45.079	45.853	48.601	52.702	56.409	59.861	63.139	66.295	69.363
	U	1109.8	2599.9	2611.8	2653.9	2716.0	2771.3	2822.3	2870.6	2917.1	2962.5
	H	1115.4	2798.3	2813.6	2867.8	2947.8	3019.5	3085.7	3148.4	3208.8	3267.7
	S	2.8487	6.0286	6.0575	6.1577	6.3006	6.4230	6.5315	6.6301	6.7216	6.8076
4500 (257.41)	V	1.269	44.037	44.540	47.273	51.336	54.996	58.396	61.620	64.721	67.732
	U	1116.4	2599.5	2607.4	2650.3	2713.2	2769.1	2820.5	2869.1	2915.8	2961.4
	H	1122.1	2797.7	2807.9	2863.0	2944.2	3016.6	3083.3	3146.4	3207.1	3266.2
	S	2.8612	6.0191	6.0382	6.1403	6.2852	6.4088	6.5182	6.6174	6.7093	6.7955
4600 (258.75)	V	1.272	43.038	43.278	46.000	50.027	53.643	56.994	60.167	63.215	66.172
	U	1122.9	2599.1	2602.9	2646.6	2710.4	2766.9	2818.7	2867.6	2914.5	2960.3
	H	1128.8	2797.0	2802.0	2858.2	2940.5	3013.7	3080.9	3144.4	3205.3	3264.7
	S	2.8735	6.0097	6.0190	6.1230	6.2700	6.3949	6.5050	6.6049	6.6972	6.7838
4700 (260.07)	V	1.276	42.081	......	44.778	48.772	52.346	55.651	58.775	61.773	64.679
	U	1129.3	2598.6	......	2642.9	2707.6	2764.7	2816.9	2866.1	2913.2	2959.1
	H	1135.3	2796.4	......	2853.3	2936.8	3010.7	3078.5	3142.3	3203.6	3263.1
	S	2.8855	6.0004	......	6.1058	6.2549	6.3811	6.4921	6.5926	6.6853	6.7722
4800 (261.37)	V	1.279	41.161	......	43.604	47.569	51.103	54.364	57.441	60.390	63.247
	U	1135.6	2598.1	......	2639.1	2704.8	2762.5	2815.1	2864.6	2911.9	2958.0
	H	1141.8	2795.7	......	2848.4	2933.1	3007.8	3076.1	3140.3	3201.8	3261.6
	S	2.8974	5.9913	......	6.0887	6.2399	6.3675	6.4794	6.5805	6.6736	6.7608

P (kPa) (Tsat °C)	Prop	Sat. Liq		Sat. Vap							
4900 (262.65)	V	1.282	· · ·	40.278	42.475	46.412	49.909	53.128	56.161	59.064	61.874
	U	1141.9	· · ·	2597.6	2635.2	2701.9	2760.2	2813.3	2863.0	2910.6	2956.9
	H	1148.2	· · ·	2794.2	2843.3	2929.3	3004.8	3073.6	3138.2	3200.0	3260.0
	S	2.9091	· · ·	5.9823	6.0717	6.2252	6.3541	6.4669	6.5685	6.6621	6.7496
5000 (263.91)	V	1.286	· · ·	39.429	41.388	45.301	48.762	51.941	54.932	57.791	60.555
	U	1148.0	· · ·	2597.0	2631.3	2699.0	2758.0	2811.5	2861.5	2909.3	2955.7
	H	1154.5	· · ·	2794.2	2838.2	2925.5	3001.8	3071.2	3136.2	3198.3	3258.5
	S	2.9206	· · ·	5.9735	6.0547	6.2105	6.3408	6.4545	6.5568	6.6508	6.7386
5100 (265.15)	V	1.289	· · ·	38.611	40.340	44.231	47.660	50.801	53.750	56.567	59.288
	U	1154.1	· · ·	2596.5	2627.3	2696.1	2755.7	2809.6	2860.0	2908.0	2954.5
	H	1160.7	· · ·	2793.4	2833.1	2921.7	2998.7	3068.7	3134.1	3196.5	3256.9
	S	2.9319	· · ·	5.9648	6.0378	6.1960	6.3277	6.4423	6.5452	6.6396	6.7278
5200 (266.37)	V	1.292	· · ·	37.824	39.330	43.201	46.599	49.703	52.614	55.390	58.070
	U	1160.1	· · ·	2595.9	2623.3	2693.1	2753.4	2807.8	2858.4	2906.7	2953.4
	H	1166.8	· · ·	2792.6	2827.8	2917.8	2995.7	3066.2	3132.0	3194.7	3255.4
	S	2.9431	· · ·	5.9561	6.0210	6.1815	6.3147	6.4302	6.5338	6.6287	6.7172
5300 (267.58)	V	1.296	· · ·	37.066	38.354	42.209	45.577	48.647	51.520	54.257	56.897
	U	1166.1	· · ·	2595.3	2619.2	2690.1	2751.0	2805.9	2856.9	2905.3	2952.2
	H	1172.9	· · ·	2791.7	2822.5	2913.8	2992.6	3063.7	3129.9	3192.9	3253.8
	S	2.9541	· · ·	5.9476	6.0041	6.1672	6.3018	6.4183	6.5225	6.6179	6.7067
5400 (268.76)	V	1.299	· · ·	36.334	37.411	41.251	44.591	47.628	50.466	53.166	55.768
	U	1171.9	· · ·	2594.6	2615.0	2687.1	2748.7	2804.0	2855.3	2904.0	2951.1
	H	1178.8	· · ·	2790.8	2817.0	2909.8	2989.5	3061.2	3127.8	3191.1	3252.2
	S	2.9650	· · ·	5.9392	5.9873	6.1530	6.2891	6.4066	6.5114	6.6072	6.6963
5500 (269.93)	V	1.302	· · ·	35.628	36.499	40.327	43.641	46.647	49.450	52.115	54.679
	U	1177.7	· · ·	2594.0	2610.8	2684.0	2746.3	2802.1	2853.7	2902.7	2949.9
	H	1184.9	· · ·	2789.9	2811.5	2905.8	2986.4	3058.7	3125.7	3189.3	3250.6
	S	2.9757	· · ·	5.9309	5.9705	6.1388	6.2765	6.3949	6.5004	6.5967	6.6862
5600 (271.09)	V	1.306	· · ·	34.946	35.617	39.434	42.724	45.700	48.470	51.100	53.630
	U	1183.5	· · ·	2593.3	2606.5	2680.9	2744.0	2800.2	2852.1	2901.3	2948.7
	H	1190.8	· · ·	2789.0	2805.9	2901.7	2983.2	3056.1	3123.6	3187.5	3249.0
	S	2.9863	· · ·	5.9227	5.9537	6.1248	6.2640	6.3834	6.4896	6.5863	6.6761
5700 (272.22)	V	1.309	· · ·	34.288	34.761	38.571	41.838	44.785	47.525	50.121	52.617
	U	1189.1	· · ·	2592.6	2602.1	2677.8	2741.6	2798.3	2850.5	2899.9	2947.5
	H	1196.6	· · ·	2788.0	2800.2	2897.6	2980.0	3053.5	3121.4	3185.6	3247.5
	S	2.9968	· · ·	5.9146	5.9369	6.1108	6.2516	6.3720	6.4789	6.5761	6.6663

TABLE E.2 Properties of Superheated Steam (Continued)

TEMPERATURE: t°C
(TEMPERATURE: T kelvins)

P/kPa (t^{sat}/°C)		sat. liq.	sat. vap.	450 (723.15)	475 (748.15)	500 (773.15)	525 (798.15)	550 (823.15)	575 (848.15)	600 (873.15)	650 (923.15)
4100 (251.80)	V	1.256	48.500	77.921	81.062	84.165	87.236	90.281	93.303	96.306	102.26
	U	1089.4	2601.0	3010.4	3054.6	3098.8	3143.0	3187.5	3232.1	3277.1	3367.9
	H	1094.6	2799.9	3329.9	3387.0	3443.9	3500.7	3557.6	3614.7	3671.9	3787.1
	S	2.8099	6.0583	6.9260	7.0037	7.0785	7.1508	7.2210	7.2893	7.3558	7.4842
4200 (253.24)	V	1.259	47.307	75.981	79.056	82.092	85.097	88.075	91.030	93.966	99.787
	U	1096.3	2600.7	3009.4	3053.7	3097.9	3142.3	3186.8	3231.5	3276.5	3367.3
	H	1101.6	2799.4	3328.5	3385.7	3442.7	3499.7	3556.7	3613.8	3671.1	3786.4
	S	2.8231	6.0482	6.9135	6.9913	7.0662	7.1387	7.2090	7.2774	7.3440	7.4724
4300 (254.66)	V	1.262	46.168	74.131	77.143	80.116	83.057	85.971	88.863	91.735	97.428
	U	1103.1	2600.3	3008.4	3052.8	3097.1	3141.5	3186.0	3230.8	3275.8	3366.8
	H	1108.5	2798.9	3327.1	3384.5	3441.6	3498.6	3555.7	3612.9	3670.3	3785.7
	S	2.8360	6.0383	6.9012	6.9792	7.0543	7.1269	7.1973	7.2658	7.3324	7.4610
4400 (256.05)	V	1.266	45.079	72.365	75.317	78.229	81.110	83.963	86.794	89.605	95.177
	U	1109.8	2599.9	3007.4	3051.9	3096.3	3140.7	3185.3	3230.1	3275.2	3366.2
	H	1115.4	2798.3	3325.8	3383.3	3440.5	3497.6	3554.7	3612.0	3669.5	3785.0
	S	2.8487	6.0286	6.8892	6.9674	7.0426	7.1153	7.1858	7.2544	7.3211	7.4498
4500 (257.41)	V	1.269	44.037	70.677	73.572	76.427	79.249	82.044	84.817	87.570	93.025
	U	1116.4	2599.5	3006.3	3050.9	3095.4	3139.9	3184.6	3229.5	3274.6	3365.7
	H	1122.1	2797.7	3324.4	3382.0	3439.3	3496.6	3553.8	3611.1	3668.6	3784.4
	S	2.8612	6.0191	6.8774	6.9558	7.0311	7.1040	7.1746	7.2432	7.3100	7.4388
4600 (258.75)	V	1.272	43.038	69.063	71.903	74.702	77.469	80.209	82.926	85.623	90.967
	U	1122.9	2599.1	3005.3	3050.0	3094.6	3139.2	3183.9	3228.8	3273.9	3365.1
	H	1128.8	2797.0	3323.0	3380.8	3438.2	3495.5	3552.8	3610.2	3667.8	3783.6
	S	2.8735	6.0097	6.8659	6.9444	7.0199	7.0928	7.1636	7.2323	7.2991	7.4281
4700 (260.07)	V	1.276	42.081	67.517	70.304	73.051	75.765	78.452	81.116	83.760	88.997
	U	1129.3	2598.6	3004.3	3049.1	3093.7	3138.4	3183.1	3228.1	3273.3	3364.6
	H	1135.3	2796.4	3321.6	3379.5	3437.1	3494.5	3551.9	3609.3	3667.0	3782.9
	S	2.8855	6.0004	6.8545	6.9332	7.0089	7.0819	7.1527	7.2215	7.2885	7.4176
4800 (261.37)	V	1.279	41.161	66.036	68.773	71.469	74.132	76.768	79.381	81.973	87.109
	U	1135.6	2598.1	3003.3	3048.2	3092.9	3137.6	3182.4	3227.4	3272.7	3364.0
	H	1141.8	2795.7	3320.3	3378.2	3435.9	3493.4	3550.9	3608.5	3666.2	3782.1
	S	2.8974	5.9913	6.8434	6.9223	6.9981	7.0712	7.1422	7.2110	7.2781	7.4072

P kPa (t sat °C)											
4900 (262.65)	V	1.282	40.278	64.615	67.303	69.951	72.565	75.152	77.716	80.260	85.298
	U	1141.9	2597.6	3002.3	3047.2	3092.0	3136.8	3181.7	3226.8	3272.0	3363.5
	H	1148.2	2794.9	3318.9	3377.0	3434.8	3492.4	3549.9	3607.6	3665.3	3781.4
	S	2.9091	5.9823	6.8324	6.9115	6.9874	7.0607	7.1318	7.2007	7.2678	7.3971
5000 (263.91)	V	1.286	39.429	63.250	65.893	68.494	71.061	73.602	76.119	78.616	83.559
	U	1148.0	2597.0	3001.5	3046.3	3091.2	3136.0	3181.0	3226.1	3271.4	3362.9
	H	1154.5	2794.2	3317.5	3375.8	3433.7	3491.3	3549.0	3606.7	3664.5	3780.7
	S	2.9206	5.9735	6.8217	6.9009	6.9770	7.0504	7.1215	7.1906	7.2578	7.3872
5100 (265.15)	V	1.289	38.611	61.940	64.537	67.094	69.616	72.112	74.584	77.035	81.888
	U	1154.1	2596.5	3000.2	3045.4	3090.3	3135.3	3180.2	3225.4	3270.8	3362.4
	H	1160.7	2793.4	3316.1	3374.5	3432.5	3490.3	3548.0	3605.8	3663.7	3780.0
	S	2.9319	5.9648	6.8111	6.8905	6.9668	7.0403	7.1115	7.1807	7.2479	7.3775
5200 (266.37)	V	1.292	37.824	60.679	63.234	65.747	68.227	70.679	73.108	75.516	80.282
	U	1160.1	2595.9	2999.2	3044.5	3089.5	3134.5	3179.5	3224.7	3270.2	3361.8
	H	1166.8	2792.6	3314.7	3373.3	3431.4	3489.3	3547.1	3604.9	3662.8	3779.3
	S	2.9431	5.9561	6.8007	6.8803	6.9567	7.0304	7.1017	7.1709	7.2382	7.3679
5300 (267.58)	V	1.296	37.066	59.466	61.980	64.452	66.890	69.300	71.687	74.054	78.736
	U	1166.1	2595.3	2998.2	3043.5	3088.6	3133.7	3178.8	3224.1	3269.5	3361.3
	H	1172.9	2791.7	3313.3	3372.0	3430.2	3488.2	3546.1	3604.0	3662.0	3778.6
	S	2.9541	5.9476	6.7905	6.8703	6.9468	7.0206	7.0920	7.1613	7.2287	7.3585
5400 (268.93)	V	1.299	36.334	58.297	60.772	63.204	65.603	67.973	70.320	72.646	77.248
	U	1171.9	2594.6	2997.1	3042.6	3087.8	3132.9	3178.1	3223.4	3268.9	3360.7
	H	1178.9	2790.8	3311.9	3370.8	3429.1	3487.2	3545.1	3603.1	3661.2	3777.8
	S	2.9650	5.9392	6.7804	6.8604	6.9371	7.0110	7.0825	7.1519	7.2194	7.3493
5500 (269.93)	V	1.302	35.628	57.171	59.608	62.002	64.362	66.694	69.002	71.289	75.814
	U	1177.7	2594.0	2996.1	3041.7	3086.9	3132.1	3177.3	3222.7	3268.3	3360.2
	H	1184.9	2789.9	3310.5	3369.5	3427.9	3486.1	3544.2	3602.2	3660.4	3777.1
	S	2.9757	5.9309	6.7705	6.8507	6.9275	7.0015	7.0731	7.1426	7.2102	7.3402
5600 (271.09)	V	1.306	34.946	56.085	58.486	60.843	63.165	65.460	67.731	69.981	74.431
	U	1183.5	2593.3	2995.0	3040.7	3086.1	3131.3	3176.6	3222.0	3267.6	3359.6
	H	1190.8	2789.1	3309.1	3368.2	3426.8	3485.1	3543.2	3601.3	3659.5	3776.4
	S	2.9863	5.9227	6.7607	6.8411	6.9181	6.9922	7.0639	7.1335	7.2011	7.3313
5700 (272.22)	V	1.309	34.288	55.038	57.403	59.724	62.011	64.270	66.504	68.719	73.096
	U	1189.1	2592.6	2994.0	3039.8	3085.2	3130.5	3175.9	3221.3	3267.0	3359.1
	H	1196.6	2788.0	3307.7	3367.0	3425.6	3484.0	3542.2	3600.4	3658.7	3775.7
	S	2.9968	5.9146	6.7511	6.8316	6.9088	6.9831	7.0549	7.1245	7.1923	7.3226

TABLE E.2 Properties of Superheated Steam (Continued)

TEMPERATURE: $t°C$
(TEMPERATURE: T kelvins)

P/kPa (t^{sat}/°C)		sat. liq.	sat. vap.	280 (553.15)	290 (563.15)	300 (573.15)	325 (598.15)	350 (623.15)	375 (648.15)	400 (673.15)	425 (698.15)
5800 (273.35)	V	1.312	33.651	34.756	36.301	37.736	40.982	43.902	46.611	49.176	51.638
	U	1194.7	2591.9	2614.4	2645.7	2674.6	2739.1	2796.3	2848.9	2898.6	2946.4
	H	1202.3	2787.0	2816.0	2856.3	2893.5	2976.8	3051.0	3119.3	3183.8	3245.9
	S	3.0071	5.9066	5.9592	6.0314	6.0969	6.2393	6.3608	6.4683	6.5660	6.6565
5900 (274.46)	V	1.315	33.034	33.953	35.497	36.928	40.154	43.048	45.728	48.262	50.693
	U	1200.3	2591.1	2610.2	2642.1	2671.4	2736.7	2794.4	2847.3	2897.2	2945.2
	H	1208.0	2786.0	2810.5	2851.5	2889.3	2973.6	3048.4	3117.1	3182.0	3244.3
	S	3.0172	5.8986	5.9431	6.0166	6.0830	6.2272	6.3496	6.4578	6.5560	6.6469
6000 (275.55)	V	1.319	32.438	33.173	34.718	36.145	39.353	42.222	44.874	47.379	49.779
	U	1205.8	2590.4	2605.9	2638.4	2668.1	2734.2	2792.4	2845.7	2895.8	2944.0
	H	1213.7	2785.0	2804.9	2846.7	2885.0	2970.4	3045.8	3115.0	3180.1	3242.6
	S	3.0273	5.8908	5.9270	6.0017	6.0692	6.2151	6.3386	6.4475	6.5462	6.6374
6100 (276.63)	V	1.322	31.860	32.415	33.962	35.386	38.577	41.422	44.048	46.524	48.895
	U	1211.2	2589.6	2601.5	2634.6	2664.8	2731.7	2790.4	2844.1	2894.5	2942.8
	H	1219.3	2783.9	2799.3	2841.8	2880.7	2967.1	3043.1	3112.8	3178.3	3241.0
	S	3.0372	5.8830	5.9108	5.9869	6.0555	6.2031	6.3277	6.4373	6.5364	6.6280
6200 (277.70)	V	1.325	31.300	31.679	33.227	34.650	37.825	40.648	43.248	45.697	48.039
	U	1216.6	2588.8	2597.1	2630.8	2661.5	2729.2	2788.5	2842.4	2893.1	2941.6
	H	1224.8	2782.9	2793.5	2836.8	2876.3	2963.8	3040.5	3110.6	3176.4	3239.4
	S	3.0471	5.8753	5.8946	5.9721	6.0418	6.1911	6.3168	6.4272	6.5268	6.6188
6300 (278.75)	V	1.328	30.757	30.962	32.514	33.935	37.097	39.898	42.473	44.895	47.210
	U	1221.9	2588.0	2592.6	2626.9	2658.1	2726.7	2786.5	2840.8	2891.7	2940.4
	H	1230.3	2781.8	2787.6	2831.7	2871.9	2960.4	3037.8	3108.4	3174.5	3237.8
	S	3.0568	5.8677	5.8783	5.9573	6.0281	6.1793	6.3061	6.4172	6.5173	6.6096
6400 (279.79)	V	1.332	30.230	30.265	31.821	33.241	36.390	39.170	41.722	44.119	46.407
	U	1227.2	2587.2	2587.9	2623.0	2654.7	2724.2	2784.4	2839.1	2890.3	2939.2
	H	1235.7	2780.6	2781.6	2826.6	2867.5	2957.1	3035.1	3106.2	3172.7	3236.2
	S	3.0664	5.8601	5.8619	5.9425	6.0144	6.1675	6.2955	6.4072	6.5079	6.6006
6500 (280.82)	V	1.335	29.719	. . . . . .	31.146	32.567	35.704	38.465	40.994	43.366	45.629
	U	1232.5	2586.3	. . . . . .	2619.0	2651.2	2721.6	2782.4	2837.5	2888.8	2938.0
	H	1241.1	2779.5	. . . . . .	2821.4	2862.9	2953.7	3032.4	3103.9	3170.8	3234.5
	S	3.0759	5.8527		5.9277	6.0008	6.1558	6.2849	6.3974	6.4986	6.5917

P (T sat)											
6600 (281.84)	V	1.338	29.223	......	30.490	31.911	35.038	37.781	40.287	42.636	44.874
	U	1237.6	2585.5		2614.9	2647.7	2719.0	2780.4	2835.8	2887.5	2936.7
	H	1246.5	2778.3		2816.1	2858.4	2950.2	3029.7	3101.7	3168.0	3232.9
	S	3.0853	5.8452		5.9129	5.9872	6.1442	6.2744	6.3877	6.4894	6.5828
6700 (282.84)	V	1.342	28.741	......	29.850	31.273	34.391	37.116	39.601	41.927	44.141
	U	1242.8	2584.6		2610.8	2644.2	2716.4	2778.3	2834.1	2886.1	2935.5
	H	1251.8	2777.1		2810.8	2853.7	2946.8	3027.0	3099.5	3167.0	3231.3
	S	3.0946	5.8379		5.8980	5.9736	6.1326	6.2640	6.3781	6.4803	6.5741
6800 (283.84)	V	1.345	28.272	......	29.226	30.652	33.762	36.470	38.935	41.239	43.430
	U	1247.9	2583.7		2606.6	2640.6	2713.7	2776.2	2832.4	2884.7	2934.3
	H	1257.0	2775.9		2805.3	2849.0	2943.3	3024.2	3097.2	3165.1	3229.6
	S	3.1038	5.8306		5.8830	5.9599	6.1211	6.2537	6.3686	6.4713	6.5655
7000 (285.79)	V	1.351	27.373	......	28.024	29.457	32.556	35.233	37.660	39.922	42.068
	U	1258.0	2581.8		2597.9	2633.2	2708.4	2772.1	2829.0	2881.8	2931.8
	H	1267.4	2773.5		2794.1	2839.4	2936.3	3018.7	3092.7	3161.2	3226.3
	S	3.1219	5.8162		5.8530	5.9327	6.0982	6.2333	6.3497	6.4536	6.5485
7200 (287.70)	V	1.358	26.522	......	26.878	28.321	31.413	34.063	36.454	38.676	40.781
	U	1267.9	2579.9		2589.0	2625.6	2702.9	2767.8	2825.6	2878.9	2929.4
	H	1277.6	2770.9		2782.5	2829.5	2929.1	3013.1	3088.1	3157.4	3223.0
	S	3.1397	5.8020		5.8226	5.9054	6.0755	6.2132	6.3312	6.4362	6.5319
7400 (289.57)	V	1.364	25.715	......	25.781	27.238	30.328	32.954	35.312	37.497	39.564
	U	1277.6	2578.0		2579.7	2617.8	2697.3	2763.5	2822.1	2876.0	2926.9
	H	1287.7	2768.3		2770.5	2819.3	2921.8	3007.0	3083.4	3153.5	3219.6
	S	3.1571	5.7880		5.7919	5.8779	6.0530	6.1933	6.3130	6.4190	6.5156
7600 (291.41)	V	1.371	24.949	......	......	26.204	29.297	31.901	34.229	36.380	38.409
	U	1287.2	2575.9			2609.7	2691.7	2759.2	2818.6	2873.1	2924.3
	H	1297.6	2765.5			2808.8	2914.3	3001.6	3078.7	3149.6	3216.3
	S	3.1742	5.7742			5.8503	6.0306	6.1737	6.2950	6.4022	6.4996
7800 (293.21)	V	1.378	24.220	......	......	25.214	28.315	30.900	33.200	35.319	37.314
	U	1296.7	2573.8			2601.3	2685.9	2754.8	2815.1	2870.1	2921.8
	H	1307.4	2762.8			2798.0	2906.7	2995.8	3074.0	3145.6	3212.9
	S	3.1911	5.7605			5.8224	6.0082	6.1542	6.2773	6.3857	6.4839
8000 (294.97)	V	1.384	23.525	......	......	24.264	27.378	29.948	32.222	34.310	36.273
	U	1306.0	2571.7			2592.7	2679.9	2750.3	2811.5	2867.1	2919.3
	H	1317.1	2759.9			2786.8	2899.0	2989.9	3069.2	3141.6	3209.5
	S	3.2076	5.7471			5.7942	5.9860	6.1349	6.2599	6.3694	6.4684

TABLE E.2 Properties of Superheated Steam (Continued)

TEMPERATURE: t°C
(TEMPERATURE: T kelvins)

P/kPa (t^{sat}/°C)		sat. liq.	sat. vap.	450 (723.15)	475 (748.15)	500 (773.15)	525 (798.15)	550 (823.15)	575 (848.15)	600 (873.15)	650 (923.15)
5800 (273.35)	V	1.312	33.651	54.026	56.357	58.644	60.896	63.120	65.320	67.500	71.807
	U	1194.7	2591.9	2992.9	3038.8	3084.4	3129.8	3175.2	3220.7	3266.4	3358.5
	H	1202.3	2787.0	3306.2	3365.7	3424.5	3483.0	3541.2	3599.5	3657.9	3775.0
	S	3.0071	5.9066	6.7416	6.8223	6.8996	6.9740	7.0460	7.1157	7.1835	7.3139
5900 (274.46)	V	1.315	33.034	53.048	55.346	57.600	59.819	62.010	64.176	66.322	70.563
	U	1200.3	2591.1	2991.9	3037.9	3083.5	3129.0	3174.4	3220.0	3265.7	3357.9
	H	1208.0	2786.0	3304.9	3364.4	3423.3	3481.9	3540.3	3598.6	3657.0	3774.3
	S	3.0172	5.8986	6.7322	6.8132	6.8906	6.9652	7.0372	7.1070	7.1749	7.3054
6000 (275.55)	V	1.319	32.438	52.103	54.369	56.592	58.778	60.937	63.071	65.184	69.359
	U	1205.8	2590.4	2990.8	3036.9	3082.6	3128.2	3173.7	3219.3	3265.1	3357.4
	H	1213.7	2785.0	3303.5	3363.2	3422.2	3480.8	3539.3	3597.7	3656.2	3773.5
	S	3.0273	5.8908	6.7230	6.8041	6.8818	6.9564	7.0285	7.0985	7.1664	7.2971
6100 (276.63)	V	1.322	31.860	51.189	53.424	55.616	57.771	59.898	62.001	64.083	68.196
	U	1211.2	2589.6	2989.8	3036.0	3081.8	3127.4	3173.0	3218.6	3264.5	3356.8
	H	1219.3	2783.9	3302.0	3361.9	3421.0	3479.8	3538.3	3596.8	3655.4	3772.8
	S	3.0372	5.8830	6.7139	6.7952	6.8730	6.9478	7.0200	7.0900	7.1581	7.2889
6200 (277.70)	V	1.325	31.300	50.304	52.510	54.671	56.797	58.894	60.966	63.018	67.069
	U	1216.6	2588.8	2988.7	3035.0	3080.9	3126.6	3172.2	3218.0	3263.8	3356.3
	H	1224.8	2782.9	3300.6	3360.6	3419.9	3478.7	3537.4	3595.9	3654.5	3772.1
	S	3.0471	5.8753	6.7049	6.7864	6.8644	6.9393	7.0116	7.0817	7.1498	7.2808
6300 (278.75)	V	1.328	30.757	49.447	51.624	53.757	55.853	57.921	59.964	61.986	65.979
	U	1221.9	2588.0	2987.7	3034.1	3080.1	3125.8	3171.5	3217.3	3263.2	3355.7
	H	1230.3	2781.8	3299.2	3359.3	3418.7	3477.7	3536.4	3595.0	3653.7	3771.4
	S	3.0568	5.8677	6.6960	6.7778	6.8559	6.9309	7.0034	7.0735	7.1417	7.2728
6400 (279.79)	V	1.332	30.230	48.617	50.767	52.871	54.939	56.978	58.993	60.987	64.922
	U	1227.2	2587.2	2986.6	3033.1	3079.2	3125.0	3170.8	3216.6	3262.6	3355.2
	H	1235.7	2780.6	3297.7	3358.0	3417.6	3476.6	3535.4	3594.1	3652.9	3770.7
	S	3.0664	5.8601	6.6872	6.7692	6.8475	6.9226	6.9952	7.0655	7.1337	7.2649
6500 (280.82)	V	1.335	29.719	47.812	49.935	52.012	54.053	56.065	58.052	60.018	63.898
	U	1232.5	2586.3	2985.5	3032.2	3078.3	3124.2	3170.0	3215.9	3261.9	3354.6
	H	1241.1	2779.5	3296.3	3356.8	3416.4	3475.6	3534.4	3593.2	3652.1	3770.0
	S	3.0759	5.8527	6.6786	6.7608	6.8392	6.9145	6.9871	7.0575	7.1258	7.2572

P (T)											
6600 (281.84)	V	1.338	29.223	47.031	49.129	51.180	53.194	55.179	57.139	59.079	62.905
	U	1237.6	2585.5	2984.5	3031.2	3077.4	3123.4	3169.3	3215.2	3261.3	3354.1
	H	1246.5	2778.3	3294.9	3355.5	3415.2	3474.5	3533.5	3592.3	3651.2	3769.2
	S	3.0853	5.8452	6.6700	6.7524	6.8310	6.9064	6.9792	7.0497	7.1181	7.2495
6700 (282.84)	V	1.342	28.741	46.274	48.346	50.372	52.361	54.320	56.254	58.168	61.942
	U	1242.8	2584.6	2983.4	3030.3	3076.6	3122.6	3168.6	3214.5	3260.7	3353.5
	H	1251.8	2777.1	3293.4	3354.2	3414.1	3473.4	3532.5	3591.4	3650.4	3768.5
	S	3.0946	5.8379	6.6616	6.7442	6.8229	6.8985	6.9714	7.0419	7.1104	7.2420
6800 (283.84)	V	1.345	28.272	45.539	47.587	49.588	51.552	53.486	55.395	57.283	61.007
	U	1247.9	2583.7	2982.3	3029.3	3075.7	3121.8	3167.8	3213.9	3260.0	3353.0
	H	1257.0	2775.9	3292.0	3352.9	3412.9	3472.4	3531.5	3590.5	3649.6	3767.8
	S	3.1038	5.8306	6.6532	6.7361	6.8150	6.8907	6.9636	7.0343	7.1028	7.2345
7000 (285.79)	V	1.351	27.373	44.131	46.133	48.086	50.003	51.889	53.750	55.590	59.217
	U	1258.0	2581.8	2980.1	3027.4	3074.0	3120.2	3166.3	3212.5	3258.8	3351.9
	H	1267.4	2773.5	3289.1	3350.5	3410.6	3470.2	3529.6	3588.7	3647.9	3766.4
	S	3.1219	5.8162	6.6368	6.7201	6.7993	6.8753	6.9485	7.0193	7.0880	7.2200
7200 (287.70)	V	1.358	26.522	42.802	44.759	46.668	48.540	50.381	52.197	53.991	57.527
	U	1267.9	2579.9	2978.0	3025.4	3072.2	3118.6	3164.9	3211.1	3257.5	3350.7
	H	1277.6	2770.9	3286.1	3347.7	3408.2	3468.1	3527.6	3586.9	3646.2	3764.9
	S	3.1397	5.8020	6.6208	6.7044	6.7840	6.8602	6.9337	7.0047	7.0735	7.2058
7400 (289.57)	V	1.364	25.715	41.544	43.460	45.327	47.156	48.954	50.727	52.478	55.928
	U	1277.6	2578.0	2975.8	3023.5	3070.4	3117.0	3163.4	3209.8	3256.2	3349.6
	H	1287.7	2768.3	3283.2	3345.1	3405.9	3466.0	3525.7	3585.1	3644.5	3763.5
	S	3.1571	5.7880	6.6050	6.6892	6.7691	6.8456	6.9192	6.9904	7.0594	7.1919
7600 (291.41)	V	1.371	24.949	40.351	42.228	44.056	45.845	47.603	49.335	51.045	54.413
	U	1287.2	2575.9	2973.6	3021.5	3068.5	3115.4	3161.9	3208.4	3254.9	3348.5
	H	1297.6	2765.5	3280.3	3342.5	3403.5	3463.8	3523.7	3583.3	3642.9	3762.1
	S	3.1742	5.7742	6.5896	6.6742	6.7545	6.8312	6.9051	6.9765	7.0457	7.1784
7800 (293.21)	V	1.378	24.220	39.220	41.060	42.850	44.601	46.320	48.014	49.686	52.976
	U	1296.7	2573.8	2971.4	3019.6	3066.9	3113.8	3160.4	3207.0	3253.7	3347.4
	H	1307.4	2762.8	3277.3	3339.8	3401.1	3461.7	3521.7	3581.5	3641.2	3760.6
	S	3.1911	5.7605	6.5745	6.6596	6.7402	6.8172	6.8913	6.9629	7.0322	7.1652
8000 (294.97)	V	1.384	23.525	38.145	39.950	41.704	43.419	45.102	46.759	48.394	51.611
	U	1306.0	2571.7	2969.2	3017.6	3065.1	3112.2	3158.9	3205.6	3252.4	3346.3
	H	1317.1	2759.9	3274.3	3337.2	3398.8	3459.5	3519.7	3579.7	3639.5	3759.2
	S	3.2076	5.7471	6.5597	6.6452	6.7262	6.8035	6.8778	6.9496	7.0191	7.1523

TABLE E.2 Properties of Superheated Steam (Continued)

TEMPERATURE: t/°C
(TEMPERATURE: T kelvins)

P/kPa (t^{sat}/°C)		sat. liq.	sat. vap.	300 (573.15)	320 (593.15)	340 (613.15)	360 (633.15)	380 (653.15)	400 (673.15)	425 (698.15)	450 (723.15)
8200 (296.70)	V	1.391	22.863	23.350	25.916	28.064	29.968	31.715	33.350	35.282	37.121
	U	1315.2	2569.5	2583.7	2657.7	2718.5	2771.5	2819.5	2864.1	2916.7	2966.9
	H	1326.6	2757.0	2775.2	2870.2	2948.6	3017.2	3079.5	3137.6	3206.0	3271.3
	S	3.2239	5.7338	5.7656	5.9288	6.0588	6.1689	6.2659	6.3534	6.4532	6.5452
8400 (298.39)	V	1.398	22.231	22.469	25.058	27.203	29.094	30.821	32.435	34.337	36.147
	U	1324.3	2567.2	2574.4	2651.1	2713.4	2767.3	2816.0	2861.1	2914.1	2964.7
	H	1336.1	2754.0	2763.1	2861.6	2941.9	3011.7	3074.8	3133.5	3202.6	3268.3
	S	3.2399	5.7207	5.7366	5.9056	6.0388	6.1509	6.2491	6.3376	6.4383	6.5309
8600 (300.06)	V	1.404	21.627	. . . . . .	24.236	26.380	28.258	29.968	31.561	33.437	35.217
	U	1333.3	2564.9	. . . . . .	2644.3	2708.1	2763.1	2812.4	2858.0	2911.5	2962.4
	H	1345.4	2750.9	. . . . . .	2852.7	2935.0	3006.1	3070.1	3129.4	3199.1	3265.3
	S	3.2557	5.7076	. . . . . .	5.8823	6.0189	6.1330	6.2326	6.3220	6.4236	6.5168
8800 (301.70)	V	1.411	21.049	. . . . . .	23.446	25.592	27.459	29.153	30.727	32.576	34.329
	U	1342.2	2562.6	. . . . . .	2637.3	2702.8	2758.8	2808.8	2854.9	2908.9	2960.1
	H	1354.6	2747.8	. . . . . .	2843.6	2928.0	3000.4	3065.3	3125.3	3195.6	3262.2
	S	3.2713	5.6948	. . . . . .	5.8590	5.9990	6.1152	6.2162	6.3067	6.4092	6.5030
9000 (303.31)	V	1.418	20.495	. . . . . .	22.685	24.836	26.694	28.372	29.929	31.754	33.480
	U	1351.0	2560.1	. . . . . .	2630.1	2697.4	2754.4	2805.2	2851.8	2906.3	2957.8
	H	1363.7	2744.6	. . . . . .	2834.3	2920.9	2994.7	3060.5	3121.2	3192.0	3259.2
	S	3.2867	5.6820	. . . . . .	5.8355	5.9792	6.0976	6.2000	6.2915	6.3949	6.4894
9200 (304.89)	V	1.425	19.964	. . . . . .	21.952	24.110	25.961	27.625	29.165	30.966	32.668
	U	1359.7	2557.7	. . . . . .	2622.7	2691.9	2750.0	2801.5	2848.7	2903.6	2955.5
	H	1372.8	2741.3	. . . . . .	2824.7	2913.7	2988.9	3055.7	3117.0	3188.5	3256.1
	S	3.3018	5.6694	. . . . . .	5.8118	5.9594	6.0801	6.1840	6.2765	6.3808	6.4760
9400 (306.44)	V	1.432	19.455	. . . . . .	21.245	23.412	25.257	26.909	28.433	30.212	31.891
	U	1368.2	2555.2	. . . . . .	2615.1	2686.3	2745.6	2797.8	2845.5	2900.9	2953.2
	H	1381.7	2738.0	. . . . . .	2814.8	2906.3	2983.0	3050.7	3112.8	3184.9	3253.0
	S	3.3168	5.6568	. . . . . .	5.7879	5.9397	6.0627	6.1681	6.2617	6.3669	6.4628
9600 (307.97)	V	1.439	18.965	. . . . . .	20.561	22.740	24.581	26.221	27.731	29.489	31.145
	U	1376.7	2552.6	. . . . . .	2607.3	2680.5	2741.0	2794.1	2842.3	2898.2	2950.9
	H	1390.6	2734.7	. . . . . .	2804.7	2898.8	2977.0	3045.8	3108.5	3181.3	3249.9
	S	3.3315	5.6444	. . . . . .	5.7637	5.9199	6.0454	6.1524	6.2470	6.3532	6.4498

P (T sat)											
9800 (309.48)	V	1.446	18.494	......	19.899	22.093	23.931	25.561	27.056	28.795	30.429
	U	1385.2	2550.0	......	2599.2	2674.7	2736.4	2790.3	2839.1	2895.5	2948.6
	H	1399.3	2731.2	......	2794.3	2891.2	2971.0	3040.8	3104.2	3177.7	3246.8
	S	3.3461	5.6321	......	5.7393	5.9001	6.0282	6.1368	6.2325	6.3397	6.4369
10000 (310.96)	V	1.453	18.041	......	19.256	21.468	23.305	24.926	26.408	28.128	29.742
	U	1393.5	2547.3	......	2590.9	2668.7	2731.8	2786.4	2835.8	2892.8	2946.2
	H	1408.0	2727.7	......	2783.5	2883.4	2964.8	3035.7	3099.9	3174.1	3243.6
	S	3.3605	5.6198	......	5.7145	5.8803	6.0110	6.1213	6.2182	6.3264	6.4243
10200 (312.42)	V	1.460	17.605	......	18.632	20.865	22.702	24.315	25.785	27.487	29.081
	U	1401.8	2544.6	......	2582.3	2662.6	2727.0	2782.6	2832.6	2890.0	2943.9
	H	1416.7	2724.2	......	2772.3	2875.4	2958.6	3030.6	3095.6	3170.4	3240.5
	S	3.3748	5.6076	......	5.6894	5.8604	5.9940	6.1059	6.2040	6.3131	6.4118
10400 (313.86)	V	1.467	17184	......	18.024	20.282	22.121	23.726	25.185	26.870	28.446
	U	1410.0	25418	......	2573.4	2656.3	2722.2	2778.7	2829.3	2887.3	2941.5
	H	1425.2	2720.6	......	2760.8	2867.2	2952.3	3025.1	3091.2	3166.7	3237.3
	S	3.3889	5.5955	......	5.6638	5.8404	5.9769	6.0907	6.1899	6.3001	6.3994
10600 (315.27)	V	1.474	16.778	......	17.432	19.717	21.560	23.159	24.607	26.276	27.834
	U	1418.1	2539.0	......	2564.1	2649.9	2717.4	2774.7	2825.9	2884.5	2939.1
	H	1433.7	2716.9	......	2748.9	2858.9	2945.9	3020.2	3086.8	3163.0	3234.1
	S	3.4029	5.5835	......	5.6376	5.8203	5.9599	6.0755	6.1759	6.2872	6.3872
10800 (316.67)	V	1.481	16.385	......	16.852	19.170	21.018	22.612	24.050	25.703	27.245
	U	1426.2	2536.2	......	2554.5	2643.4	2712.4	2770.7	2822.6	2881.7	2936.7
	H	1442.2	2713.1	......	2736.5	2850.4	2939.4	3014.9	3082.3	3159.3	3230.9
	S	3.4167	5.5715	......	5.6109	5.8000	5.9429	6.0604	6.1621	6.2744	6.3752
11000 (318.05)	V	1.489	16.006	......	16.285	18.639	20.494	22.083	23.512	25.151	26.676
	U	1434.2	2533.2	......	2544.4	2636.7	2707.4	2766.7	2819.2	2878.9	2934.3
	H	1450.6	2709.3	......	2723.5	2841.7	2932.8	3009.6	3077.8	3155.5	3227.7
	S	3.4304	5.5595	......	5.5835	5.7797	5.9259	6.0454	6.1483	6.2617	6.3633
11200 (319.40)	V	1.496	15.639	......	15.726	18.124	19.987	21.573	22.993	24.619	26.128
	U	1442.1	2530.3	......	2533.8	2629.8	2702.2	2762.6	2815.8	2876.0	2931.8
	H	1458.9	2705.4	......	2710.0	2832.8	2926.1	3004.2	3073.3	3151.7	3224.5
	S	3.4440	5.5476	......	5.5553	5.7591	5.9090	6.0305	6.1347	6.2491	6.3515
11400 (320.74)	V	1.504	15.284	......	......	17.622	19.495	21.079	22.492	24.104	25.599
	U	1450.0	2527.4	......	......	2622.7	2697.0	2758.4	2812.3	2873.1	2929.4
	H	1467.2	2701.5	......	......	2823.6	2919.3	2998.7	3068.1	3147.9	3221.2
	S	3.4575	5.5357	......	......	5.7383	5.8920	6.0156	6.1211	6.2367	6.3399

TABLE E.2 Properties of Superheated Steam (Continued)

TEMPERATURE: t°C
(TEMPERATURE: T kelvins)

P/kPa (t^{sat}/°C)		sat. liq.	sat. vap.	475 (748.15)	500 (773.15)	525 (798.15)	550 (823.15)	575 (848.15)	600 (873.15)	625 (898.15)	650 (923.15)
8200 (296.70)	V	1.391	22.863	38.893	40.614	42.295	43.943	45.566	47.166	48.747	50.313
	U	1315.2	2569.5	3015.6	3063.3	3110.5	3157.4	3204.3	3251.1	3298.1	3345.2
	H	1326.6	2757.0	3334.5	3396.4	3457.3	3517.8	3577.9	3637.9	3697.8	3757.7
	S	3.2239	5.7338	6.6311	6.7124	6.7900	6.8646	6.9365	7.0062	7.0739	7.1397
8400 (298.39)	V	1.398	22.231	37.887	39.576	41.224	42.839	44.429	45.996	47.544	49.076
	U	1324.3	2567.2	3013.6	3061.6	3108.9	3155.9	3202.9	3249.8	3296.9	3344.1
	H	1336.1	2754.0	3331.9	3394.0	3455.2	3515.8	3576.1	3636.2	3696.2	3756.3
	S	3.2399	5.7207	6.6173	6.6990	6.7769	6.8516	6.9238	6.9936	7.0614	7.1274
8600 (300.06)	V	1.404	21.627	36.928	38.586	40.202	41.787	43.345	44.880	46.397	47.897
	U	1333.3	2564.9	3011.6	3059.8	3107.3	3154.4	3201.5	3248.5	3295.7	3342.9
	H	1345.4	2750.9	3329.2	3391.6	3453.0	3513.8	3574.3	3634.5	3694.7	3754.9
	S	3.2557	5.7076	6.6037	6.6858	6.7639	6.8390	6.9113	6.9813	7.0492	7.1153
8800 (301.70)	V	1.411	21.049	36.011	37.640	39.228	40.782	42.310	43.815	45.301	46.771
	U	1342.2	2562.6	3009.6	3058.0	3105.6	3152.9	3200.1	3247.2	3294.5	3341.8
	H	1354.6	2747.8	3326.5	3389.2	3450.8	3511.8	3572.4	3632.8	3693.1	3753.4
	S	3.2713	5.6948	6.5904	6.6728	6.7513	6.8265	6.8990	6.9692	7.0373	7.1035
9000 (303.31)	V	1.418	20.495	35.136	36.737	38.296	39.822	41.321	42.798	44.255	45.695
	U	1351.0	2560.1	3007.6	3056.1	3104.0	3151.4	3198.7	3246.0	3293.3	3340.7
	H	1363.7	2744.6	3323.8	3386.8	3448.7	3509.8	3570.6	3631.1	3691.6	3752.0
	S	3.2867	5.6820	6.5773	6.6600	6.7388	6.8143	6.8870	6.9574	7.0256	7.0919
9200 (304.89)	V	1.425	19.964	34.298	35.872	37.405	38.904	40.375	41.824	43.254	44.667
	U	1359.7	2557.7	3005.6	3054.3	3102.3	3149.9	3197.3	3244.7	3292.1	3339.6
	H	1372.8	2741.3	3321.1	3384.4	3446.5	3507.8	3568.8	3629.5	3690.0	3750.5
	S	3.3018	5.6694	6.5644	6.6475	6.7266	6.8023	6.8752	6.9457	7.0141	7.0806
9400 (306.44)	V	1.432	19.455	33.495	35.045	36.552	38.024	39.470	40.892	42.295	43.682
	U	1368.2	2555.2	3003.5	3052.5	3100.7	3148.4	3195.9	3243.4	3290.9	3338.5
	H	1381.7	2738.0	3318.4	3381.9	3444.3	3505.9	3566.9	3627.8	3688.4	3749.1
	S	3.3168	5.6568	6.5517	6.6352	6.7146	6.7906	6.8637	6.9343	7.0029	7.0695
9600 (307.97)	V	1.439	18.965	32.726	34.252	35.734	37.182	38.602	39.999	41.377	42.738
	U	1376.7	2552.6	3001.5	3050.7	3099.0	3146.9	3194.5	3242.1	3289.7	3337.4
	H	1390.6	2734.7	3315.6	3379.5	3442.1	3503.9	3565.1	3626.1	3686.9	3747.6
	S	3.3315	5.6444	6.5392	6.6231	6.7028	6.7790	6.8523	6.9231	6.9918	7.0585

P (T sat)											
9800 (309.48)	V	1.446	18.494	31.988	33.491	34.949	36.373	37.769	39.142	40.496	41.832
	U	1385.2	2550.0	2999.4	3048.8	3097.4	3145.4	3193.1	3240.8	3288.5	3336.2
	H	1399.3	2731.2	3312.9	3377.0	3439.9	3501.9	3563.3	3624.4	3685.3	3746.2
	S	3.3461	5.6321	6.5268	6.6112	6.6912	6.7676	6.8411	6.9121	6.9810	7.0478
10000 (310.96)	V	1.453	18.041	31.280	32.760	34.196	35.597	36.970	38.320	39.650	40.963
	U	1393.5	2547.3	2997.4	3047.0	3095.7	3143.9	3191.7	3239.5	3287.3	3335.1
	H	1408.0	2727.7	3310.1	3374.6	3437.7	3499.8	3561.4	3622.7	3683.8	3744.7
	S	3.3605	5.6198	6.5147	6.5994	6.6797	6.7564	6.8302	6.9013	6.9703	7.0373
10200 (312.42)	V	1.460	17.605	30.599	32.058	33.472	34.851	36.202	37.530	38.837	40.128
	U	1401.8	2544.6	2995.3	3045.2	3094.0	3142.3	3190.3	3238.2	3286.1	3334.0
	H	1416.7	2724.2	3307.4	3372.1	3435.5	3497.8	3559.6	3621.0	3682.2	3743.3
	S	3.3748	5.6076	6.5027	6.5879	6.6685	6.7454	6.8194	6.8907	6.9598	7.0269
10400 (313.86)	V	1.467	17.184	29.943	31.382	32.776	34.134	35.464	36.770	38.056	39.325
	U	1410.0	2541.8	2993.2	3043.3	3092.4	3140.8	3188.9	3236.9	3284.8	3332.9
	H	1425.2	2720.6	3304.6	3369.7	3433.2	3495.8	3557.8	3619.3	3680.6	3741.8
	S	3.3889	5.5955	6.4909	6.5765	6.6574	6.7346	6.8087	6.8803	6.9495	7.0167
10600 (315.27)	V	1.474	16.778	29.313	30.732	32.106	33.444	34.753	36.039	37.304	38.552
	U	1418.1	2539.0	2991.1	3041.4	3090.7	3139.3	3187.5	3235.6	3283.6	3331.7
	H	1433.7	2716.9	3301.8	3367.2	3431.0	3493.8	3555.9	3617.6	3679.1	3740.4
	S	3.4029	5.5835	6.4793	6.5652	6.6465	6.7239	6.7983	6.8700	6.9394	7.0067
10800 (316.67)	V	1.481	16.385	28.706	30.106	31.461	32.779	34.069	35.335	36.580	37.808
	U	1426.2	2536.2	2989.0	3039.6	3089.0	3137.8	3186.1	3234.3	3282.4	3330.6
	H	1442.2	2713.1	3299.0	3364.7	3428.8	3491.8	3554.1	3615.9	3677.5	3738.9
	S	3.4167	5.5715	6.4678	6.5542	6.6357	6.7134	6.7880	6.8599	6.9294	6.9969
11000 (318.05)	V	1.489	16.006	28.120	29.503	30.839	32.139	33.410	34.656	35.882	37.091
	U	1434.2	2533.2	2986.9	3037.7	3087.3	3136.2	3184.7	3233.0	3281.2	3329.5
	H	1450.6	2709.3	3296.2	3362.2	3426.5	3489.7	3552.2	3614.2	3675.9	3737.5
	S	3.4304	5.5595	6.4564	6.5432	6.6251	6.7031	6.7779	6.8499	6.9196	6.9872
11200 (319.40)	V	1.496	15.639	27.555	28.921	30.240	31.521	32.774	34.002	35.210	36.400
	U	1442.1	2530.3	2984.8	3035.8	3085.6	3134.7	3183.3	3231.7	3280.0	3328.4
	H	1458.9	2705.4	3293.4	3359.7	3424.3	3487.7	3550.4	3612.5	3674.4	3736.0
	S	3.4440	5.5476	6.4452	6.5324	6.6147	6.6929	6.7679	6.8401	6.9099	6.9777
11400 (320.74)	V	1.504	15.284	27.010	28.359	29.661	30.925	32.160	33.370	34.560	35.733
	U	1450.0	2527.2	2982.6	3033.9	3083.9	3133.1	3181.9	3230.4	3278.8	3327.2
	H	1467.2	2701.5	3290.5	3357.2	3422.1	3485.7	3548.5	3610.8	3672.8	3734.6
	S	3.4575	5.5357	6.4341	6.5218	6.6043	6.6828	6.7580	6.8304	6.9004	6.9683

Appendix F

Thermodynamic Diagrams

Figure F.1 Methane

Figure F.2 1,1,1,2-tetrafluoroethane (HFC-134a)

Figure F.3 Mollier (*HS*) diagram for steam

Extensive thermodynamic data for these three substances and over 70 additional pure fluids, including permanent gases, refrigerants, and light hydrocarbons, are available in the NIST Chemistry Webbook at http://webbook.nist.gov/chemistry/fluid/. Such data provide a basis for the construction of charts like those shown in this section.

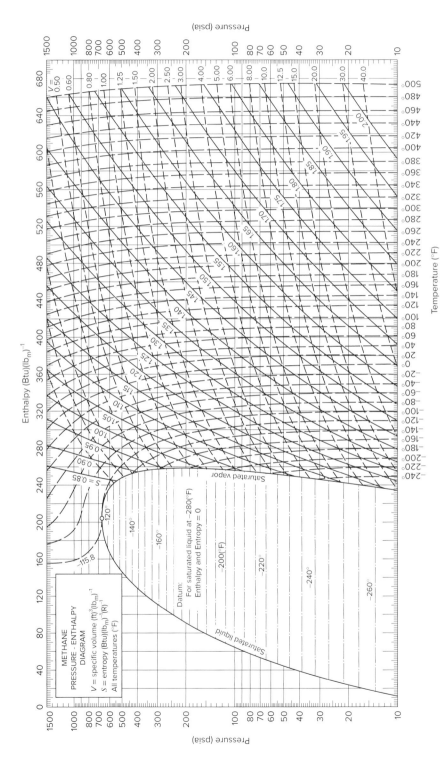

Source: C. S. Matthews and C. O. Hurd, Trans. AIChE, vol. 42, 1946, pp. 55–78.

Figure F.1 *PH* diagram for methane.

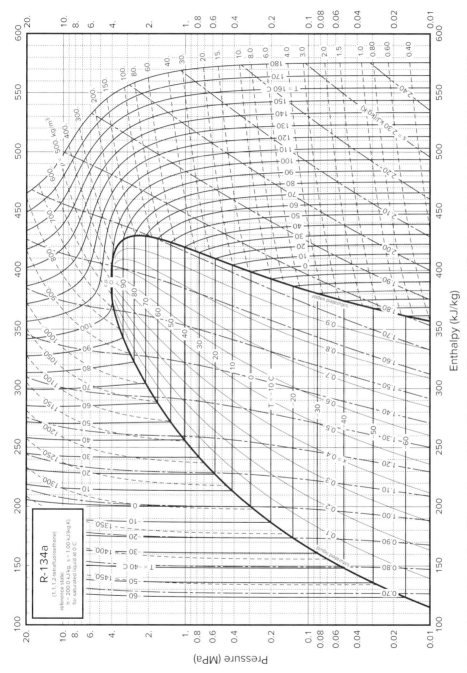

Retrieved from: Green, Don W., and Perry, Robert H. Perry's Chemical Engineers' Handbook (8th Edition). Blacklick, OH, USA: McGraw-Hill Professional Publishing, 2007. Copyright © 2007. McGraw-Hill Professional Publishing. All rights reserved.

Figure F.2 *PH* diagram for tetrafluoroethane (HFC-134a).

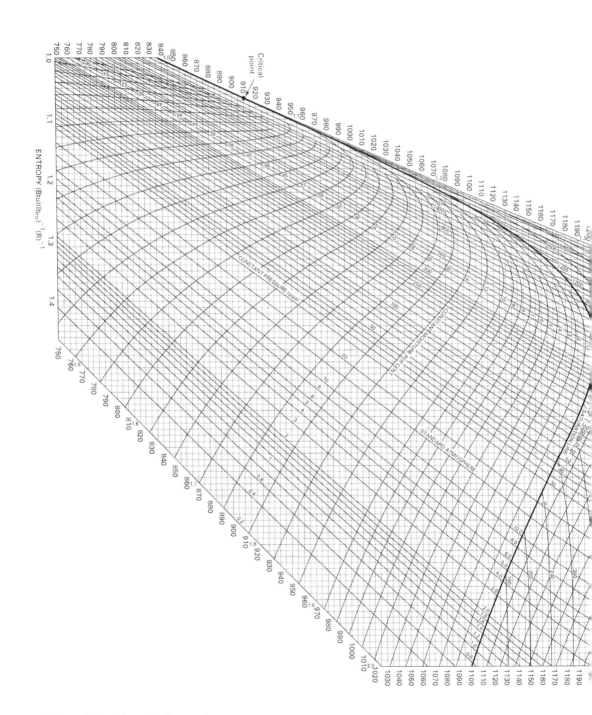

Figure F.3 Mollier (*HS*) diagram for steam.

Mollier diagram for steam.

(Reproduced by permission from "Steam Tables:
Properties of Saturated and Superheated Steam,"
copyright 1940, Combustion Engineering, Inc.

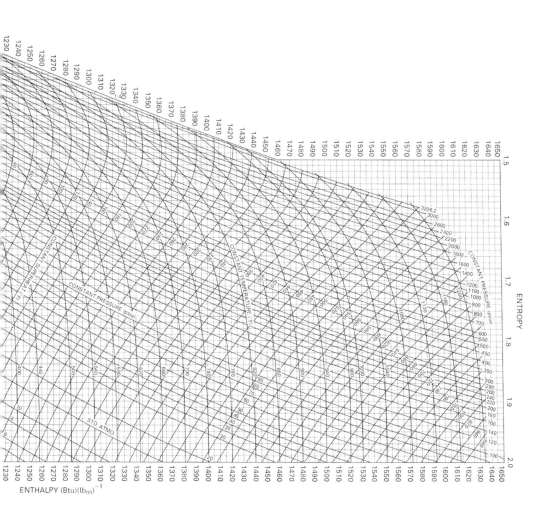

ENTHALPY (Btu)(lb$_m$)$^{-1}$

Appendix G

UNIFAC Method

The UNIQUAC equation[1] treats $g \equiv G^E/RT$ as the sum of two parts, a *combinatorial* term g^C to account for molecular size and shape differences, and a *residual* term g^R (not a residual property as defined in Sec. 6.2) to account for molecular interactions:

$$g \equiv g^C + g^R \tag{G.1}$$

Function g^C contains pure-species parameters only, whereas function g^R incorporates two *binary* parameters for each pair of molecules. For a multicomponent system,

$$g^C = \sum_i x_i \ln \frac{\Phi_i}{x_i} + 5 \sum_i q_i x_i \ln \frac{\theta_i}{\Phi_i} \tag{G.2}$$

$$g^R = -\sum_i q_i x_i \ln \left(\sum_j \theta_j \tau_{ji} \right) \tag{G.3}$$

where

$$\Phi_i \equiv \frac{x_i r_i}{\sum_j x_j r_j} \tag{G.4}$$

$$\theta_i \equiv \frac{x_i q_i}{\sum_j x_j q_j} \tag{G.5}$$

Subscript i identifies species, and j is a dummy index; all summations are over all species. Note that $\tau_{ji} \neq \tau_{ij}$; however, when $i = j$, then $\tau_{ii} = \tau_{jj} = 1$. In these equations r_i (a relative molecular volume) and q_i (a relative molecular surface area) are pure-species parameters. The influence of temperature on g enters through the interaction parameters τ_{ji} of Eq. (G.3), which are temperature dependent:

$$\tau_{ji} = \exp \frac{-\left(u_{ji} - u_{ii} \right)}{RT} \tag{G.6}$$

Parameters for the UNIQUAC equation are therefore values of $(u_{ji} - u_{ii})$.

---

[1] D. S. Abrams and J. M. Prausnitz, *AIChE J.*, vol. 21, pp. 116–128, 1975.

An expression for $\ln \gamma_i$ is found by application of Eq. (13.7) to the UNIQUAC equation for g [Eqs.(G.1) through (G.3)]. The result is given by the following equations:

$$\ln \gamma_i = \ln \gamma_i^C + \ln \gamma_i^R \tag{G.7}$$

$$\ln \gamma_i^C = 1 - J_i + \ln J_i - 5q_i\left(1 - \frac{J_i}{L_i} + \ln\frac{J_i}{L_i}\right) \tag{G.8}$$

$$\ln \gamma_i^R = q_i\left(1 - \ln s_i - \sum_j \theta_j \frac{\tau_{ij}}{s_j}\right) \tag{G.9}$$

where in addition to Eqs. (G.5) and (G.6),

$$J_i = \frac{r_i}{\sum_j r_j x_j} \tag{G.10}$$

$$L_i = \frac{q_i}{\sum_j q_j x_j} \tag{G.11}$$

$$S_i = \tau_{li}\sum_l \theta_l \tag{G.12}$$

Again subscript i identifies species, and j and l are dummy indices. All summations are over all species, and $\tau_{ij} = 1$ for $i = j$. Values for the parameters $(u_{ij} - u_{jj})$ are found by regression of binary VLE data and are given by Gmehling et al.[2]

The UNIFAC method for estimation of activity coefficients[3] depends on the concept that a liquid mixture may be considered a solution of the structural units from which the molecules are formed rather than a solution of the molecules themselves. These structural units are called *subgroups,* and a few of them are listed in the second column of Table G.1. A number, designated k, identifies each subgroup. The relative volume R_k and relative surface area Q_k are properties of the subgroups, and values are listed in columns 4 and 5 of Table G.1. Also shown (columns 6 and 7) are examples of molecular species and their constituent subgroups. When a molecule can be constructed from more than one set of subgroups, the set containing the least number of *different* subgroups is the correct set. The great advantage of the UNIFAC method is that a relatively small number of subgroups combine to form a very large number of molecules.

Activity coefficients depend not only on the subgroup properties R_k and Q_k, but also on interactions between subgroups. Here, similar subgroups are assigned to a main group, as shown in the first two columns of Table G.1. The designations of main groups, such as "CH$_2$," "ACH," etc., are descriptive only. All subgroups belonging to the same main group are

[2]J. Gmehling, U. Onken, and W. Arlt, *Vapor-Liquid Equilibrium Data Collection*, Chemistry Data Series, vol. I, parts 1–8 and supplements, DECHEMA, Frankfurt/Main, 1974–1999.

[3]Aa. Fredenslund, R. L. Jones, and J. M. Prausnitz, *AIChE J.*, vol. 21, pp. 1086–1099, 1975.

Table G.1: UNIFAC-VLE Subgroup Parameters[†]

Main group	Subgroup	k	R_k	Q_k	Examples of molecules and their constituent groups	
1 "CH$_2$"	CH$_3$	1	0.9011	0.848	*n*-Butane:	2CH$_3$, 2CH$_2$
	CH$_2$	2	0.6744	0.540	Isobutane:	3CH$_3$, 1CH
	CH	3	0.4469	0.228	2,2-Dimethyl	
	C	4	0.2195	0.000	propane:	4CH$_3$, 1C
3 "ACH"	ACH	10	0.5313	0.400	Benzene:	6ACH
(AC = aromatic carbon)						
4 "ACCH$_2$"	ACCH$_3$	12	1.2663	0.968	Toluene:	5ACH, 1ACCH$_3$
	ACCH$_2$	13	1.0396	0.660	Ethylbenzene:	1CH$_3$, 5ACH, 1ACCH$_2$
5 "OH"	OH	15	1.0000	1.200	Ethanol:	1CH$_3$, 1CH$_2$, 1OH
7 "H$_2$O"	H$_2$O	17	0.9200	1.400	Water:	1H$_2$O
9 "CH$_2$CO"	CH$_3$CO	19	1.6724	1.488	Acetone:	1CH$_3$CO, 1CH$_3$
	CH$_2$CO	20	1.4457	1.180	3-Pentanone:	2CH$_3$, 1CH$_2$CO, 1CH$_2$
13 "CH$_2$O"	CH$_3$O	25	1.1450	1.088	Dimethyl ether:	1CH$_3$, 1CH$_3$O
	CH$_2$O	26	0.9183	0.780	Diethyl ether:	2CH$_3$, 1CH$_2$, 1CH$_2$O
	CH–O	27	0.6908	0.468	Diisopropyl ether:	4CH$_3$, 1CH, 1CH–O
15 "CNH"	CH$_3$NH	32	1.4337	1.244	Dimethylamine:	1CH$_3$, 1CH$_3$NH
	CH$_2$NH	33	1.2070	0.936	Diethylamine:	2CH$_3$, 1CH$_2$, 1CH$_2$NH
	CHNH	34	0.9795	0.624	Diisopropylamine:	4CH$_3$, 1CH, 1CHNH
19 "CCN"	CH$_3$CN	41	1.8701	1.724	Acetonitrile:	1CH$_3$CN
	CH$_2$CN	42	1.6434	1.416	Propionitrile:	1CH$_3$, 1CH$_2$CN

[†]H. K. Hansen, P. Rasmussen, Aa. Fredenslund, M. Schiller, and J. Gmehling, *IEC Research*, vol. 30, pp. 2352–2355, 1991.

considered identical with respect to group interactions. Therefore parameters characterizing group interactions are identified with pairs of *main* groups. Parameter values a_{mk} for a few such pairs are given in Table G.2.

The UNIFAC method is based on the UNIQUAC equation, for which the activity coefficients are given by Eq. (G.7). When applied to a solution of groups, Eqs. (G.8) and (G.9) are written:

$$\ln \gamma_i^C = 1 - J_i + \ln J_i - 5q_i\left(1 - \frac{J_i}{L_i} + \ln \frac{J_i}{L_i}\right) \tag{G.13}$$

$$\ln \gamma_i^R = q_i\left[1 - \sum_k \left(\theta_k \frac{\beta_{ik}}{s_k} - e_{ki}\ln \frac{\beta_{ik}}{s_k}\right)\right] \tag{G.14}$$

Table G.2: UNIFAC–VLE Interaction Parameters, a_{mk}, in kelvins[†]

		1	3	4	5	7	9	13	15	19
1	CH2	0.00	61.13	76.50	986.50	1318.00	476.40	251.50	255.70	597.00
3	ACH	−11.12	0.00	167.00	636.10	903.80	25.77	32.14	122.80	212.50
4	ACCH2	−69.70	−146.80	0.00	803.20	5695.00	−52.10	213.10	−49.29	6096.00
5	OH	156.40	89.60	25.82	0.00	353.50	84.00	28.06	42.70	6.712
7	H2O	300.00	362.30	377.60	−229.10	0.00	−195.40	540.50	168.00	112.60
9	CH2CO	26.76	140.10	365.80	164.50	472.50	0.00	−103.60	−174.20	481.70
13	CH2O	83.36	52.13	65.69	237.70	−314.70	191.10	0.00	251.50	−18.51
15	CNH	65.33	−22.31	223.00	−150.00	−448.20	394.60	−56.08	0.00	147.10
19	CCN	24.82	−22.97	−138.40	185.40	242.80	−287.50	38.81	−108.50	0.00

[†]H. K. Hansen, P. Rasmussen, Aa. Fredenslund, M. Schiller, and J. Gmehling, *IEC Research*, vol. 30, pp. 2352–2355, 1991.

The quantities J and L are still given by Eqs. (G.10) and (G.11). In addition, the following definitions apply:

$$r_i = \sum_k v_k^{(i)} R_k \tag{G.15}$$

$$q_i = \sum_k v_k^{(i)} Q_k \tag{G.16}$$

$$e_{ki} = \frac{v_k^{(i)} Q_k}{q_i} \tag{G.17}$$

$$\beta_{ik} = \sum_m e_{mi} \tau_{mk} \tag{G.18}$$

$$\theta_k = \frac{\sum_i x_i q_i e_{ki}}{\sum_j x_j q_j} \tag{G.19}$$

$$s_k = \sum_m \theta_m \tau_{mk} \tag{G.20}$$

$$\tau_{mk} = \exp\frac{-a_{mk}}{T} \tag{G.21}$$

Subscript i identifies a species, and j is a dummy index running over all species. Subscript k identifies subgroups, and m is a dummy index running over all subgroups. The quantity $v_k^{(i)}$ is the number of subgroups of type k in a molecule of species i. Values of the subgroup parameters R_k and Q_k and of the group interaction parameters a_{mk} come from tabulations in the literature. Tables G.1 and G.2 show a few parameter values; the number designations of the complete tables are retained.[4]

The equations for the UNIFAC method are presented here in a form convenient for computer programming. In the following example we run through a set of hand calculations to demonstrate their application.

Example G.1

For the binary system diethylamine(1)/*n*-heptane(2) at 308.15 K, find γ_1 and γ_2 when $x_1 = 0.4$ and $x_2 = 0.6$.

Solution G.1

The subgroups involved are indicated by the chemical formulas:

$$CH_3 - CH_2NH - CH_2 - CH_3 \; (1) / CH_3 - (CH_2)_5 - CH_3 \; (2)$$

[4]H. K. Hansen, P. Rasmussen, Aa. Fredenslund, M. Schiller, and J. Gmehling, IEC Research, vol. 30, pp. 2352–2355, 1991.

The following table shows the subgroups, their identification numbers k, values of parameters R_k and Q_k (from Table G.1), and the numbers of each subgroup in each molecule:

	k	R_k	Q_k	$v_k^{(1)}$	$v_k^{(2)}$
CH_3	1	0.9011	0.848	2	2
CH_2	2	0.6744	0.540	1	5
CH_2NH	33	1.2070	0.936	1	0

By Eq. (G.15)

$$r_1 = (2)(0.9011) + (1)(0.6744) + (1)(1.2070) = 3.6836$$

Similarly,

$$r_2 = (2)(0.9011) + (5)(0.6744) = 5.1742$$

In like manner, by Eq. (G.16),

$$q_1 = 3.1720 \qquad \text{and} \qquad q_2 = 4.3960$$

The r_i and q_i values are molecular properties, independent of composition. Substituting known values into Eq. (G.17) generates the following table for e_{ki}:

k	e_{ki} $i = 1$	$i = 2$
1	0.5347	0.3858
2	0.1702	0.6142
33	0.2951	0.0000

The following interaction parameters are found from Table G.2:

$$a_{1,1} = a_{1,2} = a_{2,1} = a_{2,2} = a_{33,33} = 0 \text{ K}$$
$$a_{1,33} = a_{2,33} = 255.7 \text{ K}$$
$$a_{33,1} = a_{33,2} = 65.33 \text{ K}$$

Substitution of these values into Eq. (G.21) with $T = 308.15$ K gives

$$\tau_{1,1} = \tau_{1,2} = \tau_{2,1} = \tau_{2,2} = \tau_{33,33} = 1$$
$$\tau_{1,33} = \tau_{2,33} = 0.4361$$
$$\tau_{33,1} = \tau_{33,2} = 0.8090$$

Application of Eq. (G.18) leads to the values of β_{ik} in the following table:

i	β_{ik} $k = 1$	$k = 2$	$k = 33$
1	0.9436	0.9436	0.6024
2	1.0000	1.0000	0.4360

Substitution of these results into Eq. (G.19) yields:

$$\theta_1 = 0.4342 \qquad \theta_2 = 0.4700 \qquad \theta_{33} = 0.0958$$

and by Eq. (G.20),

$$s_1 = 0.9817 \qquad s_2 = 0.9817 \qquad s_{33} = 0.4901$$

The activity coefficients may now be calculated. By Eq. (G.13),

$$\ln \gamma_1^C = -0.0213 \qquad \text{and} \qquad \ln \gamma_2^C = -0.0076$$

and by Eq. (G.14),

$$\ln \gamma_1^R = 0.1463 \qquad \text{and} \qquad \ln \gamma_2^R = 0.0537$$

Finally, Eq. (G.7) gives:

$$\gamma_1 = 1.133 \qquad \text{and} \qquad \gamma_2 = 1.047$$

Appendix H

Newton's Method

Newton's method is a procedure for the numerical solution of algebraic equations, applicable to any number M of such equations expressed as functions of M variables.

Consider first a single equation $f(X) = 0$, in which $f(X)$ is a function of the single variable X. Our purpose is to find a root of this equation, that is, the value of X for which the function is zero. A simple function is illustrated in Fig. H.1; it exhibits a single root at the point where the curve crosses the X-axis. When it is not possible to solve directly for the root,[1] a numerical procedure, such as Newton's method, is employed.

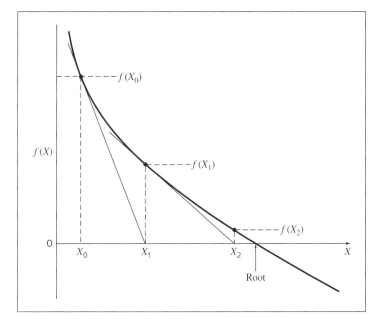

Figure H.1: Newton's method applied to a single function.

---

[1]For example, when $e^X + X^2 + 10 = 0$.

The application of Newton's method is illustrated in Fig. H.1. In the neighborhood of an arbitrary value $X = X_0$ the function $f(X)$ can be approximated by the tangent line drawn at $X = X_0$. The equation of the tangent line is given by the linear relation:

$$g(X) = f(X_0) + \left[\frac{df(X)}{dX}\right]_{X=X_0}(X - X_0)$$

where $g(X)$ is the value of the ordinate at X, as shown in Fig. H.1. The root of this equation is found by setting $g(X) = 0$ and solving for X; as indicated in Fig. H.1, the value is X_1. Because the actual function is not linear, this is not the root of $f(X)$. However, it lies closer to the root than does the starting value X_0. The function $f(X)$ is now approximated by a second line, drawn tangent to the curve at $X = X_1$, and the procedure is repeated, leading to a root for this linear approximation at X_2, a value still closer to the root of $f(X)$. This root can be approached as closely as desired by continued successive linear approximation of the original function. The general formula for iteration is:

$$f(X_n) + \left[\frac{df(X)}{dX}\right]_{X=X_n}\Delta X_n = 0 \tag{H.1}$$

where

$$\Delta X_n \equiv X_{n+1} - X_n \qquad \text{or} \qquad X_{n+1} = X_n + \Delta X_n$$

Equation (H.1), written for successive iterations ($n = 0, 1, 2, \ldots$), produces successive values of ΔX_n and successive values of $f(X_n)$. The process starts with an initial value X_0 and continues until either ΔX_n or $f(X_n)$, or some combination thereof, approaches zero to within a preset tolerance.

Newton's method is readily extended to the solution of simultaneous equations. For the case of two equations in two unknowns, let $f_I \equiv f_I(X_I, X_{II})$ and $f_{II} \equiv f_{II}(X_I, X_{II})$ represent two functions, the values of which depend on the two variables X_I and X_{II}. Our purpose is to find the values of X_I and X_{II} for which both functions are zero. In analogy to Eq. (H.1), we write:

$$f_I + \left(\frac{\partial f_I}{\partial X_I}\right)\Delta X_I + \left(\frac{\partial f_I}{\partial X_{II}}\right)\Delta X_{II} = 0 \tag{H.2a}$$

$$f_{II} + \left(\frac{\partial f_{II}}{\partial X_I}\right)\Delta X_I + \left(\frac{\partial f_{II}}{\partial X_{II}}\right)\Delta X_{II} = 0 \tag{H.2b}$$

These equations differ from Eq. (H.1) in that the single derivative is replaced by two partial derivatives, reflecting the rates of change of each function with each of the two variables. For iteration n the two functions f_I and f_{II} and their derivatives are evaluated at $X = X_n$ from the given expressions, and Eqs. (H.2a) and (H.2b) are solved simultaneously for ΔX_I and ΔX_{II}. These are specific to the particular iteration, and they lead to new values X_I and X_{II}, applicable to the next iteration:

$$X_{I_{n+1}} = X_{I_n} + \Delta X_{I_n} \text{ and } X_{II_{n+1}} = X_{II_n} + \Delta X_{II_n}$$

The iterative procedure based on Eqs. (H.2) is initiated with starting values for X_I and X_{II} and continues until the increments ΔX_{I_n} and ΔX_{II_n} or the computed values of f_I and f_{II} approach zero.

Equations (H.2) can be generalized to apply to a system of M equations in M unknowns; the result for each iteration is:

$$f_k + \sum_{J=1}^{M} \left(\frac{\partial f_k}{\partial X_J} \right) \Delta X_J = 0 \qquad (K = \mathrm{I, II, \ldots, M}) \qquad (H.3)$$

with

$$X_{J_{n+1}} = X_{J_n} + \Delta X_{J_n} \qquad (J = \mathrm{I, II, \ldots, M})$$

Newton's method is well suited to application to multireaction equilibria. As an illustration, we solve Eqs. (A) and (B) of Ex. 14.13 for the case of $T = 1000$ K. From these equations with values given there for K_a and K_b at 1000 K and with $P/P^\circ = 20$, we find the functions

$$f_a = 4.0879 \, \varepsilon_b^2 + \varepsilon_b^2 + 4.0879 \, \varepsilon_a \varepsilon_b + 0.2532 \, \varepsilon_a - 0.0439 \, \varepsilon_b - 0.1486 \qquad (A)$$

and

$$f_b = 1.2805 \, \varepsilon_b^2 + 2.12805 \, \varepsilon_a \varepsilon_b - 0.12805 \, \varepsilon_a + 0.3048 \, \varepsilon_b - 0.4328 \qquad (B)$$

Equations (H.2) are written here as:

$$f_a + \left(\frac{\partial f_a}{\partial \varepsilon_a} \right) \Delta \varepsilon_a + \left(\frac{\partial f_a}{\partial \varepsilon_b} \right) \Delta \varepsilon_b = 0 \qquad (C)$$

$$f_b + \left(\frac{\partial f_b}{\partial \varepsilon_a} \right) \Delta \varepsilon_a + \left(\frac{\partial f_b}{\partial \varepsilon_b} \right) \Delta \varepsilon_b = 0 \qquad (D)$$

The solution procedure is initiated with a choice of starting values for ε_a and ε_b. Numerical values are obtained for f_a and f_b and for their derivatives from Eqs. (A) and (B). Substitution of these values in Eqs. (C) and (D) yields two linear equations which are readily solved for the unknowns $\Delta \varepsilon_a$ and $\Delta \varepsilon_b$. These yield new values of ε_a and ε_b with which to carry out a second iteration. The process continues until $\Delta \varepsilon_a$ and $\Delta \varepsilon_b$ or f_a and f_b approach zero.

Setting $\varepsilon_a = 0.1$ and $\varepsilon_b = 0.7$ as starting values,[2] we find initial values of f_a and f_b and their derivatives from Eqs. (A) and (B):

$$f_a = 0.6630 \qquad \left(\frac{\partial f_a}{\partial \varepsilon_a} \right) = 3.9230 \qquad \left(\frac{\partial f_b}{\partial \varepsilon_b} \right) = 1.7648$$

$$f_b = 0.4695 \qquad \left(\frac{\partial f_b}{\partial \varepsilon_a} \right) = 1.3616 \qquad \left(\frac{\partial f_b}{\partial \varepsilon_b} \right) = 2.0956$$

These values are substituted in Eqs. (C) and (D) to yield:

$$0.6630 + 3.9230 \Delta \varepsilon_a + 1.7648 \Delta \varepsilon_b = 0$$
$$0.4695 + 1.3616 \Delta \varepsilon_a + 2.0956 \Delta \varepsilon_b = 0$$

The values of the increments that satisfy these equations are:

$$\Delta \varepsilon_a = -0.0962 \qquad \text{and} \qquad \Delta \varepsilon_b = -0.1614$$

[2]These are well within the limits, $-0.5 \leq \varepsilon_a \leq 0.5$ and $0 \leq \varepsilon b \leq 1.0$, noted in Ex. 14.13.

from which,

$$\varepsilon_a = 0.1 - 0.0962 = 0.0038 \qquad \text{and} \qquad \varepsilon_b = 0.7 - 0.1614 = 0.5386$$

These values are the basis for a second iteration, and the process continues, yielding results as follows:

n	ε_a	ε_b	$\Delta\varepsilon_a$	$\Delta\varepsilon_b$
0	0.1000	0.7000	−0.0962	−0.1614
1	0.0038	0.5386	−0.0472	−0.0094
2	−0.0434	0.5292	−0.0071	0.0043
3	−0.0505	0.5335	−0.0001	0.0001
4	−0.0506	0.5336	0.0000	0.0000

Convergence is clearly rapid. Moreover, any reasonable starting values lead to convergence on the same answers.

Convergence problems can arise with Newton's method when one or more of the functions exhibit extrema. This is illustrated for the case of a single equation in Fig. H.2. The function has two roots, at points A and B. If Newton's method is applied with a starting value of X smaller than a, a very small range of X values produces convergence on each root, but for most values it does not converge, and neither root is found. With a starting value of X between a and b, it converges on root A only if the value is sufficiently close to A. With a starting value of X to the right of b, it converges on root B. In cases such as this, a proper starting value can be found by trial, or by graphing the function to determine its behavior.

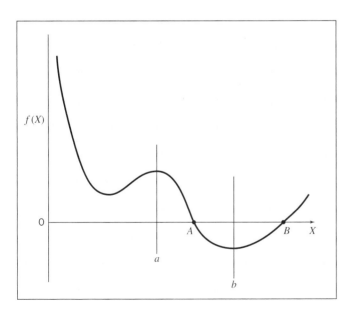

Figure H.2: Finding the roots of a function showing extrema.

Index